Space Physics

This textbook, derived from courses given by three leading researchers, provides advanced undergraduates and graduates with up-to-date coverage of space physics, from the Sun to the interstellar medium. Clear explanations of the underlying physical processes are presented alongside major new discoveries and knowledge gained from space missions, ground-based observations, theory, and modeling to inspire students. Building from the basics to more complex ideas, the book contains enough material for a two-semester course, but the authors also provide suggestions for how the material can be tailored to fit a single semester. End-of-chapter problems reinforce concepts and include computer-based exercises especially developed for this textbook package. Free access to the software is available via the book's website and enables students to model the behavior of magnetospheric and solar plasma. An extensive glossary recaps new terms, and carefully selected further reading sections encourage students to explore advanced topics of interest.

Christopher T. Russell has written over 1600 articles in books and journals on planetary and space physics and has been cited over 45 000 times. He has been awarded the AGU's Macelwane medal, its Fleming medal, and COSPAR's Science Award. He has been a principal investigator on numerous missions including ISEE 1 and 2, Pioneer Venus, the ISTP/Polar mission, and the Magnetospheric Multiscale mission. He is also the Principal Investigator of the ion-propelled Dawn Discovery mission to the asteroid belt.

Janet G. Luhmann has authored or co-authored over 600 publications in areas of space and planetary physics and served as Senior Editor for the *Journal of Geophysical Research: Space Physics*. She has been awarded AGU's Fleming medal and COSPAR's Science Award. She has been an Investigator on numerous NASA and NSF projects involving the Sun's control of the space environments of the Earth and planets, most recently the STEREO mission to observe the three-dimensional effects of solar activity in the inner solar system, and the MAVEN mission to study the Mars atmosphere's escape to space.

Robert J. Strangeway is an author or co-author on over 200 publications covering a variety of space physics topics. He regularly teaches the Introduction to Space Physics course at UCLA, which is the basis for this book. He is currently the Senior Editor for the *Journal of Atmospheric and Solar-Terrestrial Physics*. In addition to serving as an Investigator with the missions AMPTE/CCE, Pioneer Venus, and FAST, he was the Principal Investigator for the magnetometers developed for Space Technology 5.

Space Physics

Christopher T. Russell
University of California, Los Angeles

Janet G. Luhmann
University of California, Berkeley

Robert J. Strangeway
University of California, Los Angeles

CAMBRIDGE
UNIVERSITY PRESS

CAMBRIDGE
UNIVERSITY PRESS

University Printing House, Cambridge CB2 8BS, United Kingdom

One Liberty Plaza, 20th Floor, New York, NY 10006, USA

477 Williamstown Road, Port Melbourne, VIC 3207, Australia

314-321, 3rd Floor, Plot 3, Splendor Forum, Jasola District Centre, New Delhi - 110025, India

79 Anson Road, #06-04/06, Singapore 079906

Cambridge University Press is part of the University of Cambridge.

It furthers the University's mission by disseminating knowledge in the pursuit of education, learning and research at the highest international levels of excellence.

www.cambridge.org
Information on this title: www.cambridge.org/9781107098824

First published 2016
Reprinted 2017

A catalogue record for this publication is available from the British Library

Library of Congress Cataloging in Publication data
Russell, C. T. (Christopher T.), author.
Space physics / Christopher T. Russell (University of California, Los Angeles), Janet G. Luhmann
(University of California, Berkeley), Robert J. Strangeway (University of California, Los Angeles).
pages cm
ISBN 978-1-107-09882-4 – ISBN 1-107-09882-3
1. Astrophysics. 2. Outer space. 3. Solar wind. 4. Sun. 5. Space plasmas. 6. Planets. I. Luhmann, J. G., author. II. Strangeway, R. J. (Robert J.), author. III. Title.
QB461.R87 2016
523.01–dc23

2015026794

ISBN 978-1-107-09882-4 Hardback

Additional resources for this publication at www.cambridge.org/spacephysics

Contents

Preface

We live on a planet in which three states of matter are very familiar. Our feet are firmly planted on the solid ground; not long ago, we poured ourselves a drink of liquid; and right now we are breathing air. Even if we do not understand the mathematical physics behind the behavior of the solids, liquids, and gases that are all around us, we still have an empirical working knowledge that allows us to stand, drink, and breathe. If we but venture a short way off this planet, our intuition begins to fail us. Soon the particles surrounding us have electric charges. Instead of atomic forces holding the molecules together in solids and liquids, and collisions dominating the forces in gases, we find that electric and magnetic forces are dominant. Gravity is still present, but, unlike in the environment in which you read this book, in the plasma universe gravity can be a minor factor in determining motion.

For the first two millennia of the common epoch, we could largely ignore the fourth state of matter, the plasma state, but now that we are living in the third millennium, we have to address how the more distant space around us behaves. In this millennium, "outer space" becomes important, beginning less than 100 km above our heads, a distance that is no longer far out of reach. It is a region in which many of us work every day, tending to our robotic machines. Some of us even live in space, not just for hours or days, but for significant fractions of a year with the hope of extending human presence for even longer periods.

It is important for us to master this fourth state of matter, as it has become important to all of humanity. Technology has advanced rapidly since the beginning of the industrial revolution, so now we depend on technology for communication, weather prediction, and energy transmission. Our planes travel at high altitudes close to the edge of space. With that advance has come new sensitivities that at first caught us by surprise. We did not know how variable was the plasma environment that surrounds the Earth, or how variable was the connectivity of the Sun and the Earth via that plasma. But when electrical systems across the planet began to fail, when spacecraft stopped obeying our commands, and when the radiation hazards in space and within our upper atmosphere were understood, the study of this region became very necessary.

This book is our attempt to codify the basic principles of the plasma state from the Sun to the edge of the solar system. We do not stop at the outermost planets but extend to where the solar system plasma meets the galactic plasma or, as astronomers call it, to the local interstellar medium. Because our daily lives do not prepare us for this journey, it is difficult to know when and how to begin.

There are several possible approaches. We could start with the simplest systems and add complexity. We could order our journey by spatial location beginning at the center of the solar system and moving outward. We could follow the energy flow outward past all the planets. We could move into another dimension, time, and travel in chronological order, following the history of discovery leading to understanding. There is much to justify the spatial/energy flow approach because the Sun is at the center of our solar system and provides much of the energy for the space around it. The solar wind couples to the planets' magnetospheres and ionospheres, and the Sun's photons couple to their atmospheres through heating and photoionization. On the other hand, the chronological approach follows the way that scientists originally came to understand the solar-terrestrial environment. The chronological approach has the advantage of allowing us to start with simple concepts and then build complexity on top of them. The disadvantages of this approach are that some of the early ideas were wrong, and that science does not necessarily progress along the shortest path. So the purely historical path can be inefficient.

In this book, we attempt to combine these approaches, trying to reduce topics to their basics before introducing complications. We start the book with an historical introduction that attempts both to show how the scientific method proceeds in its most general sense and to provide a description of the plasma environment in which the Earth resides. Chapter 1 begins with the ancient observations of the "northern lights" which we now refer to as the aurora borealis (and its sibling, the aurora australis, in the south). We introduce the magnetic state of the Earth, the plasma in the upper atmosphere, and then the advent of the space age and the space exploration program, around our planet and the other planets.

In Chapter 2, we start from the known, the atmosphere of the Earth, and move outwards. The neutral atmosphere has winds and pressure. It varies with altitude. It supports wave phenomena. We encounter these topics later in the book in their plasma forms. Completing this chapter, we examine how the atmosphere becomes ionized, forming the first of the natural plasmas to be discovered.

In Chapter 3, we introduce the physics of plasmas and the mathematical formalism to treat plasmas quantitatively. Since the advent of the space age, the field of space physics has become increasingly more quantitative, and processes are described much more precisely with theory that has rigorous mathematical underpinnings. Some of this is kinetic theory that treats the behavior of the full plasma. Some of this is magnetohydrodynamic theory that averages over the gyro and thermal motions to describe processes in terms of densities, bulk velocities, momenta, and temperatures. Later in the book we describe the numerical simulations that use this mathematical theory to describe the behavior of these complex systems.

In Chapter 4, we head for the center of the solar system where the energy for the plasma processes is produced (and the gravity that binds the planetary system together originates). We have learned much about the Sun over the past few decades, but our ability to determine how and why processes and phenomena occur is limited when we have only remote-sensing data with which to work. Thus, Chapter 4 stresses what happens on the Sun, even when we do not completely understand it.

Missions now in development will soon provide more information about the hows and whys.

In Chapter 5, we move off the Sun and into the heliosphere. In this chapter, we examine how the solar wind is accelerated, how its variability leads to dynamic phenomena, and how it terminates. The solar wind is very important as it couples the Sun's plasma environment to our own, providing a conduit for solar activity to affect terrestrial activity.

Chapter 6 covers the physics of a very non-linear plasma process, the collisionless shock. This irreversible process increases the entropy of the plasma and converts energy from the bulk motion of the plasma into heat. It is this heated and compressed plasma that interacts with the planets and controls the energization of the planetary atmospheres and magnetospheres. These shocks are also responsible for energizing a small fraction of the plasma particles to sometimes very, very high energies, energies that can be damaging to our space electronics and to living organisms, even those just on the edge of space.

Chapter 7 examines how the solar wind interacts with bodies that are strongly magnetized. It also introduces the reader to numerical simulations of four types: gas dynamic, magnetohydrodynamic, hybrid, and fully kinetic. Each of these types of simulations has its advantages, and each has some limitations.

Chapter 8 describes how the solar wind interacts with bodies that are not strongly magnetized. These bodies may, like the Moon, have almost no atmosphere or may, like Venus, have an atmosphere and an ionosphere. This chapter shows how the planetary ionosphere forms a barrier to the solar wind and deflects it around the planet and how comets and asteroids can interact with the solar wind.

In Chapter 9, we examine how energy is extracted from the solar-wind flow and transferred to the Earth's magnetosphere and ultimately to its atmosphere. This chapter treats reconnection at the magnetopause and in the magnetotail. It discusses magnetospheric substorms and storms. These topics have been of interest since the beginning of the space age, and, in the absence of sufficient observations from the sparsely sampled magnetosphere, these topics were initially quite controversial. In writing this chapter, we have not attempted to replay the entire history of these studies, but we simply explain the energy flow and its temporal behavior in terms of physical processes that we now understand well, on a theoretical and operational basis.

In Chapter 10, we examine the inner magnetosphere, the region in which the ionosphere extends upward to form the plasmasphere. This is the most stable of the regions of the magnetosphere, but it is far from being boring. This region is home to the radiation belts and some very interesting and important wave–particle interactions.

In Chapter 11, we step outside the "stable" zone into the auroral and polar ionospheres and the magnetosphere above these regions in which the temporal variation of the magnetosphere is the greatest and where the main coupling of the energy transfer process occurs.

Chapter 12 takes us to the magnetospheres of the magnetized planets other than the Earth: Mercury, Jupiter, Saturn, Uranus, and Neptune, and their moons. We benefit from our knowledge of these systems by understanding more fully the same processes as on Earth, but under very different conditions.

Chapter 13 brings us back to wave processes and gives us a more mathematical treatment of these phenomena that are so important to the evolution, scattering, and thermalization of the plasma in the solar wind and planetary magnetospheres.

Our intention in this book is to provide an understanding of the basic physical principles in space physics for the beginning graduate student. Much of the material is suitable for upper division undergraduates and has been tested on both undergraduate and graduate students in the Department of Earth, Planetary, and Space Sciences at the University of California, Los Angeles.

HOW TO USE THIS BOOK

It is our intention in preparing this text to provide the material necessary to enable a graduate student to embark profitably on a career in space physics. The material has a range of difficulty and has both descriptive and some very mathematical sections. It would be difficult to cover in depth all this material in a single semester or quarter. We would recommend a two-semester approach with the following material in the first semester to give the student a basic working knowledge of the physics of space plasmas. The course would begin with Chapter 1, which covers first the history of the field. All of Chapter 2 would follow, giving an introduction to the physics of neutral gases as a prelude to the analogous physics of plasmas, closing with a discussion of the nearest geophysical plasma, the ionosphere. Following this, we recommend Sections 3.1 to 3.2 together with 3.5, which would cover single-particle theory and fluid theory.

In Chapter 4, Sections 4.1 to 4.3 cover the Sun and solar cycle. Chapter 5's Sections 5.1 and 5.2 bring in the solar wind and coronal sources. Chapter 6's Sections 6.1, 6.2, and 6.4 add a discussion of shock types and shock observations. Chapter 7's Sections 7.1 to 7.6 cover planetary magnetic fields and the formation of a magnetosphere, as well as fluid treatments of the solar-wind flow past planetary obstacles. In Chapter 8, Sections 8.1 to 8.3 treat the basic unmagnetized planet interaction. All of Chapter 9 is recommended in this first course because understanding the energization of the Earth's magnetosphere is perhaps the most important part of space plasma studies. In Chapter 10, Sections 10.1 to 10.2.4 and 10.3 would add cold plasmas in the magnetosphere and the radiation belts. In Chapter 11, we recommend covering Sections 11.1 to 11.3 in the first course, which will allow understanding of auroral emissions. In Chapter 12, Sections 12.1 to 12.4 provide an overview of the size and circulation of planetary magnetospheres. Chapter 13 and the remaining materials in the earlier chapters provide a more intensive look at the physical processes that underlie the phenomenology.

Several features of this book are worthy of note. The book does not simply end with Chapter 13, but continues with four appendices, a glossary, a list of references, and an index. Appendix 1 discusses notation, vector identities, and differential operators of use in treatments of space plasmas. Appendix 2 lists fundamental constants and formulas for the plasma parameters of space physics. Appendix 3 describes many of the coordinate systems used in space plasma physics, and how to transform from one coordinate system to another. Finally, Appendix 4 gives an introduction to power spectral analysis, especially dynamic power spectral analysis including how to determine the direction of propagation of electromagnetic waves in a plasma and their handedness as compared to the motion of charged particles in a plasma, which is the plasma physicist's definition of right-handed and left-handed.

The glossary includes all terms included in bold in the pages of the textbook, usually when the term is first used. The glossary provides a simple definition of the term in bold. The list of references gives those references to classic papers that we felt needed referencing in the text when certain concepts or observations were introduced. Also included are the references in which figures that we did not originate can be found. The index includes those important topics covered in the book that one might like to find without having to read entire sections of the book.

Additional suggested reading and problems are listed at the ends of chapters. We include the reason one might like to consult the additional reading. The problems consist of two types: those that reinforce a concept in the chapter and can be completed by students after having mastered the material in the chapter; and those that are more like laboratory exercises which require use of web-based software developed as an adjunct to the course. Science courses are often accompanied by "labs" in which students do experiments, but in a space plasma the scales are so vast one cannot conceive of building a student laboratory. One must resort to the use of computers. The exercises were developed over a decade of teaching space plasma physics at UCLA to demonstrate the behavior of magnetospheric and solar plasma. They have been recoded and upgraded in conjunction with the publishing of this book and made available over the web. Instructors will soon appreciate that the same software can be used for a variety of different problems and may stray away from the specific problems listed toward problems tailored to the needs of a specific course.

ACKNOWLEDGMENTS

An earlier book, *Introduction to Space Physics*, edited by M. G. Kivelson and C. T. Russell, was published in 1995. This book was a compilation of chapters written by leading researchers in the field. While this was an efficient means of assembling the first comprehensive textbook in the field, it did not result in a uniform treatment of the topics. When it became time to publish a successor volume, three of the original authors volunteered to work together to prepare a new volume with a more even treatment of the topical areas involved.

Chapter 1 contains much of the material in the earlier book, but with additional discussion and tables that summarize the progress made in the two decades since the earlier book. Chapter 2 contains material on the neutral atmosphere and the physics of gases leading to a discussion of the ionosphere which was largely discussed in Chapter 7 of the original edition. Chapter 3 covers the physics of magnetized plasmas, replacing and expanding on the chapter by M. G. Kivelson. Chapter 4 discusses what we know about the Sun and its atmosphere, replacing the earlier chapter by E. R. Priest that emphasized the role of the magnetic field in solar phenomena. Chapter 5 on the heliosphere includes the traditional material found in A. J. Hundhausen's treatment of the solar wind, but expands the discussion to include the increased knowledge we now have on the radial evolution of the solar wind and its interactions with neutrals, dust, cosmic rays, and the interstellar mechanism. Chapter 6 is a complete rewrite of the chapter originally written by D. Burgess with a mathematical derivation of the Rankine–Hugoniot relations that are the key to understanding the behavior of collisionless shocks. Chapter 7 is an update of the material in the chapter by R. J. Walker and C. T. Russell in the earlier book. This chapter has a more complete treatment of the types of numerical simulations used in the study of solar-wind interactions. Chapter 8 is an update of the material on the solar-wind interaction with unmagnetized objects by the original author. Today, we know about interactions that were not even dreamed about two decades ago. Chapter 9 replaces the chapter originally written by W. J. Hughes on solar-wind coupling with the Earth's magnetosphere. This chapter focuses on the dynamics associated with magnetic reconnection between the solar wind and the terrestrial magnetic field. Chapter 10 replaces the chapter originally written by R. A. Wolf. This new chapter preserves the classic treatment of the radiation belts but covers the low-energy plasma and magnetospheric waves in detail not found in the earlier work. Chapter 11 replaces the auroral chapter of H. C. Carlson and A. Egeland. This chapter emphasizes the physics of the auroral processes more than in the earlier chapter. Chapter 12 replaces the chapter originally authored by C. T. Russell and R. J. Walker. This largely preserves the original material but updates it with the results from recent planetary missions. The final chapter in the book covers waves in plasmas and is a complete rewrite of the earlier chapter by C. K. Goertz and R. J. Strangeway, here stressing electromagnetic wave phenomena over electrostatic waves. The book ends with four appendices. Appendices 1, 2, and 3 are updates of the earlier appendices, but Appendix 4 is a completely new appendix that explains fundamental elements in the analysis of time series of magnetic fluctuations in plasmas.

The authors of this book thank the earlier authors for permission to use excerpts from their earlier works where appropriate. We are particularly grateful to the students of the UCLA Department of Earth, Planetary, and Space Sciences who have provided feedback on this project, and to the California Space Grant Consortium for supporting the preparation of some of the graphics for this book and the conversion of the software to a web-based version. We are also grateful to Sharon Uy, Margie Sowmendran, and Richard Sadakane for their help in preparing the volume.

1

Solar-terrestrial physics: the evolution of a discipline

1.1 INTRODUCTION

The physics of the solar-terrestrial environment is governed principally by the interaction of energetic charged particles with electric and magnetic fields in space. Near the Earth, most of these charged particles derive their energy ultimately from the Sun, directly or through the interaction of the solar wind with the Earth's magnetosphere. These interactions are complex and non-linear. The magnetic and electric fields that determine the motion of the charged particles are affected in turn by the motion of these particles. Moreover, the processes that occur on the microscale may affect the behavior of the system on the macroscale. Fortunately, there are approximate methods to treat these systems. We do not have to worry about the motions of the individual particles in most situations.

Some solar-terrestrial research is still carried out on the surface of the Earth with cameras, photometers, spectrometers, magnetometers, and other devices sensitive to the processes occurring high in the upper atmosphere and magnetosphere. However, many processes are invisible to remote sensing and must be studied *in situ*. Thus, the majority of this research is performed using rockets and satellites that enable measurements to be obtained directly in the regions in which the interactions occur. In recent years, these *in situ* data have resulted in explosive growth in our knowledge and understanding of solar-terrestrial processes. Our planetary exploration program has added to this understanding by showing us how such processes operate in quite different settings.

The field of **solar-terrestrial physics** has a long history of investigation, starting well before the advent of satellites and rockets. A convenient way to introduce ourselves to this exotic field, with often non-intuitive, and even counter-intuitive, behavior, is to learn about it in the same sequence as the early pioneers. Thus, we briefly review its historical development to provide context for our later, more physically oriented, presentation of the nature of the processes occurring in the solar-terrestrial environment. We choose to follow a chronological path through solar-terrestrial physics, as that is the way it was revealed to early scientific observers, and this is the way it was originally understood.

1.2 ANCIENT AURORAL SIGHTINGS

The discipline of solar-terrestrial physics began with a growing appreciation of two phenomena: the **auroras** and the variability of the geomagnetic field. Because they can be observed visually, auroras were the first of these phenomena to be recorded. The discovery of the variability of the geomagnetic field had to await the advent of new technology with the invention of the compass.

References to auroras are contained in ancient literature from both the East and West. Chinese literature describes possible auroral sightings, several of which occurred prior to 2000 BC. Several

FIGURE 1.1. Early drawing of an aurora seen on January 12, 1570. (Original print in Crawford Collection, Royal Observatory, Edinburgh.)

passages in the Old Testament appear to have been inspired by auroral sightings. Greek literature includes references to phenomena most likely to have been auroral phenomena. For example, Xenophanes, in the sixth century BC, mentions "moving accumulations of burning clouds." This sequence of discovery may have been influenced by the westward drift of the Earth's tilted **magnetic dipole** that favored auroral displays, first in China, then in Asia Minor, followed by Europe and, today, the eastern United States.

Because the phenomenon was not understood, much fear and superstition surrounded those early sightings of auroras. Figure 1.1, inspired by an auroral display in 1570, illustrates the lack of scientific understanding prevalent at that time. The seventeenth century marked the beginning of scientific theories concerning the origin of the lights in the north. Galileo Galilei, for example, proposed that auroras were caused by air rising out of the Earth's shadow to where it could be illuminated by sunlight. He also appears to have coined the term "aurora borealis," meaning "dawn of the north." At about the same time, Pierre Gassendi, a French mathematician and astronomer, deduced that auroral displays must be occurring at great heights, because they were seen to have the same configuration when observed at places quite remote from one another. His contemporary, René Descartes, is credited with the idea that auroras were caused by reflections from ice crystals in the air at high latitudes. From about 1645 to about 1715, both solar activity and auroral sightings declined, although neither was completely lacking.

Edmund Halley, at the age of 60, after finally observing an auroral display, suggested that auroral phenomena were ordered by the direction of the Earth's magnetic field. In 1731, the French philosopher de Mairan ridiculed the then-popular idea that auroras were a reflection of polar ice and snow, and he also criticized Halley's theory. Instead, he suggested that auroras were connected to the solar atmosphere, and he suspected a connection between the return of sunspots and auroras. Only much later could Halley's ideas and those of de Mairan be reconciled, and studies of geomagnetism and auroras become more firmly linked.

1.3 EARLY MEASUREMENTS OF THE GEOMAGNETIC FIELD

The earliest indication of the existence of the geomagnetic field was the direction-finding capability of the compass. As compasses were improved, more and more was learned about the geomagnetic field. Chinese knowledge that a compass points north or south can be traced to the eleventh century. The encyclopedist Shon-Kau (AD 1030–93) stated that "fortune-tellers rub the point of a needle with the stone of the magnet in order to make it properly indicate the south." In European literature, the earliest mention of the compass and its application to navigation appeared near the end of the twelfth century in *De Untensilibus* and *De Rerum*, two works by Alexander Neekan, a monk of St. Albans, where, coincidentally, much later, one of the authors of this book was born. In the former work, he described the use of the magnetic needle to indicate north and noted that mariners used that means to find their course when the sky was cloudy. In the latter work, he described the needle as being placed on a pivot, a second-generation form of the compass. In neither work did he describe the instrument as a novelty; it was in common use at that time. Official records indicate that, by the fourteenth century, many sailing ships carried compasses. The directions of magnetic north and of geographic or true north differ over most of the globe. The measure of this difference is referred to as the declination. It is not clear when magnetic declination was discovered. However, a letter written in 1544, by Georg Hartmann, vicar of St. Sebald's at Nürnberg, to Duke Albrecht of Prussia showed that he had observed the declination of Rome in 1510 to be 6° east, whereas it was 10° at Nürnberg. Also, it is known that between the years 1538 and 1541, João de Castro made 43 determinations of declination during a voyage along the west coast of India and in the Red Sea.

The geomagnetic field is also inclined to the horizontal. To measure this angle, known as the inclination, one must pivot a needle about a horizontal axis. Georg Hartmann's letter also discussed such an observation, but the angle of inclination was incorrect for his point of observation. William Gilbert ascribed the discovery of the magnetic dip or inclination to an Englishman, Robert Norman, who in 1576 published a work with the title *The newe Attractiue containyng a short discourse of the Magnes or Lodestone, and amongest other his vertues, of a newe discouered secret and subtill propertie, concernyng the Declinyng of the Needle, touched there with onder the plaine of the Horizon. Now first found out by ROBERT NORMAN Hydrographer. Here onto are annexed certaine necessarie rules for the art of Nauigation, by the same R.N. Imprinted at London by John Kyngston, for Richard Ballard, 1581.*

The year 1600 saw the publication of the famous treatise *De Magnete* by William Gilbert, who, in 1601, was appointed chief physician in personal attendance to Queen Elizabeth. This treatise consists of six books containing a total of 115 chapters. The central theme of the book is also the title of Chapter 17, Book 1: "That the globe of the earth is magnetic, a magnet; how in our hands the magnet stone has all the primary forces of the earth, while the earth by the same powers remains constant in a fixed direction in the universe." Figure 1.2 illustrates Gilbert's woodcut, showing the distribution of magnetic inclination or dip over the Earth, and over a small spherical lodestone, which he called a "terrella." Gilbert believed that the terrestrial magnetic field was constant, but it is not. Henry Gellibrand, professor of astronomy at Gresham College, discovered that magnetic declination changed with time, and he published his discovery in a work entitled *A discourse mathematical on the variation of the magneticall needle. Together with its admirable diminuation lately discovered, London, 1635.*

Another early pioneer in the study of geomagnetism was Edmund Halley, who published, in 1683 and 1692, two works on the theory of geomagnetism, but needed to test his theory further. King William III put

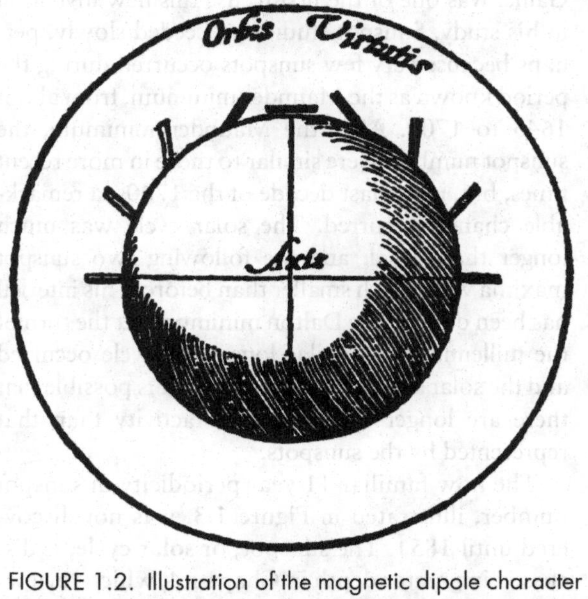

FIGURE 1.2. Illustration of the magnetic dipole character of Earth's main magnetic field, as shown in Gilbert's *De Magnete*.

at his disposal the ship *Paramour Pink*, on which Halley made two voyages: in October 1698 to the North Atlantic, and in September 1700 to the South Atlantic. Those voyages were the first purely scientific expeditions, and they returned measurements of great value, both for practical navigation and for the theory of navigation. Those investigations led to the publication of two geomagnetic charts published in 1701 and 1702, respectively: *New and Correct Chart showing the Variations of the Compass in the Western and Southern Oceans, as observed in year 1700 by his Majesty's Command by Edm. Halley* and *Sea Chart of the whole world, showing the Variations of the Compass.*

1.4 EMERGENCE OF A SCIENTIFIC DISCIPLINE

Despite the fact that the Sun is the most luminous object we can see, the solar half of solar-terrestrial physics awaited technological advance as surely as the study of geomagnetism awaited the development of the compass and its successor, the magnetometer. **Sunspots,** or magnetized cool spots in the solar photosphere, are generally too small to be resolved by the naked eye. Thus, the study of sunspots did not begin until the invention of the telescope. Galileo Galilei was one of the first to use this new invention in his study. Sunspot studies proceeded slowly, perhaps because very few sunspots occurred during the period known as the Maunder minimum, from about 1645 to 1700. After the Maunder minimum, the sunspot numbers were similar to those in more recent times, but, in the last decade of the 1700s, a remarkable change occurred. The solar cycle was much longer than usual, and the following two sunspot maxima were much smaller than before. This interval has been dubbed the Dalton minimum. At the turn of the millennium, a similar long solar cycle occurred and the solar activity plummeted, so it is possible that there are longer cycles of solar activity than that represented by the sunspots.

The now familiar 11 year periodicity in sunspot number, illustrated in Figure 1.3, was not discovered until 1851. The sunspot, or **solar cycle**, is discussed in greater depth in Chapter 4, which reviews our current understanding of the physics of the Sun, in which magnetism plays a significant role that is only gradually being understood. In fact, the deep solar minimum that occurred from 2006 to 2010 as cycle 23 ended was accompanied by weak magnetic fields on the Sun.

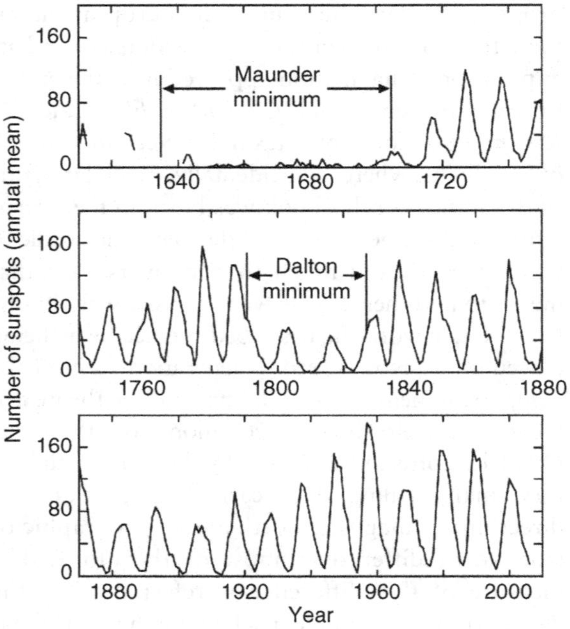

FIGURE 1.3. Sunspot cycle since AD 1610.

Perhaps the first discovery in the emerging discipline we now call solar-terrestrial physics was the observation in 1722 by George Graham, a famous London instrument maker, that the compass is always in motion. Graham's discovery was confirmed in 1740 by Anders Celsius in Uppsala, Sweden. His observations were continued by O. Hiorter to a total of over 20 000 observations made on more than 1000 different days. From those data, Hiorter discovered the diurnal variation of the geomagnetic field. Magnetic perturbations vary systematically with local time, which is determined by the longitudinal separation between the meridian of the observer and that containing the Sun, the noon meridian. These perturbations are due to the rotation of the observer under current systems flowing in the upper atmosphere that are fixed with respect to the Sun.

Even more importantly, on April 5, 1741, Hiorter discovered that geomagnetic and auroral activities were correlated. Simultaneous observations in London by Graham confirmed the occurrence of strong geomagnetic activity on that day. In 1770, J. C. Wilcke noted that auroral rays extend upward along the direction of the magnetic field. In the same year, Captain James Cook first reported the southern counterpart of the aurora borealis, the aurora australis, or "dawn of the south." Twenty years later, the English scientist Henry Cavendish used triangulation to estimate the height of auroras as between 80 and 115 kilometers. Earlier attempts at triangulation by Halley and Mairan had been much less accurate.

The great advance of the early nineteenth century was the development of a network to make frequent simultaneous observations with widely spaced **magnetometers**. C. F. Gauss was one of the leaders of that effort and one of the foremost pioneers in the mathematical analysis of the resulting measurements, which allowed contributions to the geomagnetic field from below the surface of the Earth to be separated from those contributions arising high in the atmosphere.

Meanwhile, Heinrich Schwabe, on the basis of his sunspot measurements taken between 1825 and 1850, deduced that the variation in the number of sunspots was periodic, with a period of about ten years. By 1839, magnetic observatories had spread to the British colonies. Edward Sabine was assigned to supervise four of those observatories (Toronto, St. Helena, Cape of Good Hope, and Hobarton) in 1851. Using these data, he was able to show that the intensity of geomagnetic disturbances varied in concert with the sunspot cycle. Chapter 9 discusses our modern understanding of these disturbances.

The next discovery linking the Sun and geomagnetic activity was Richard Carrington's sighting of a great white-light **flare** on September 1, 1859. Carrington, who was sketching sunspot groups at the time, was startled by the flare, and by the time he was able to summon someone to witness the event a minute later, he was dismayed to find that it had weakened greatly in intensity. Fortunately, it had been simultaneously noted by another observer some miles away. Furthermore, at the moment of the flare, measurements of the magnetic field at the Kew Observatory in London had been disturbed. Today, we realize that that disturbance of the magnetic field was caused by an increase in the electric currents flowing overhead in the Earth's ionosphere. Such currents flow in response to electric fields in the electrically conducting ionosphere. The radiation from extreme-ultraviolet rays and x-rays from the flare increased the ionization high in the atmosphere and hence the electrical conductivity, causing more current to flow in response to the unaltered electric field. Finally, 18 hours later, one of the strongest magnetic storms ever recorded occurred. Auroras were seen as far south as Puerto Rico. To have arrived that quickly, the disturbance would have had to travel from the Sun at over 2300 km s^{-1}. As discussed in Chapter 5, we know today that the Sun and the Earth are linked by the supersonic solar wind, but such a velocity is high even for the disturbed solar wind.

Total solar eclipses occur somewhere on the Earth almost annually, and these solar eclipses reveal the density structure of the solar corona. Now we know that the density structure can change rapidly and significantly in association with the solar events that lead to terrestrial storms, but total solar eclipses last only minutes, far too short to reveal these changes. Thus **coronal mass ejections**, such as the one that certainly led to Carrington's 1859 geomagnetic storm, remained undiscovered until the invention of the coronograph, which could make artificial solar eclipses for automatic cameras. Now the solar corona is regularly measured from high altitude and from space.

In 1861, shortly after Carrington's observations, Balfour Stewart noted the occurrence of pulsations in the Earth's magnetic field, with periods of minutes. Today, we know that the Earth's magnetic field pulsates at a wide variety of frequencies. These pulsations provide diagnostics of the state of the magnetosphere and the processes occurring therein. They are described in further detail in Chapter 10. The energy for these waves in the plasma can be exogenic, from the solar wind and its interaction with the magnetosphere, or endogenic, powered by the free energy in unstable plasmas produced in the dynamic magnetosphere. The exchange of energy between the electrons and ions in a plasma and waves that can transport the exchanged energy is a very active and important area of research in space plasma physics. This is described in more detail in Chapter 10.

The nineteenth century also brought another simple but important observation of auroras. Captain John Franklin, the ill-fated English Arctic explorer whose party perished in 1845 attempting to discover the Northwest Passage (now open as a result of recent global warming), noted that auroral frequency did not increase all the way to the pole; this was according to observations made during his 1819–22 journeys. In 1860, Elias Loomis of Yale was one of the first to plot the zone of maximum auroral occurrence, which roughly corresponds to what we today refer to as the **auroral zone**. This zone is an oval band around the magnetic pole, roughly 20–25° from the pole. Precursors to our modern understanding of the auroras began to appear in the late nineteenth century. About 1878, H. Becquerel suggested that particles were shot off from the Sun and were guided by the Earth's magnetic field to the auroral zone. He believed that sunspots ejected protons. A similar theory was espoused by E. Goldstein.

In 1897, the great Norwegian physicist Kristian Birkeland made his first auroral expedition to northern Norway. However, it was not until after his third expedition in 1902–03, during which he obtained extensive data on the magnetic perturbations associated with auroras, that he concluded that large electric currents flowed along magnetic field lines during auroras. The invention of the vacuum tube led to the understanding that auroras were in some way similar to the cathode rays in those devices. Soon, Sir William Crookes demonstrated that cathode rays were bent by magnetic fields, and shortly thereafter J. J. Thomson showed that cathode rays consisted of the tiny, negatively charged particles we now call electrons. Birkeland adopted those ideas for his auroral theories and attempted to verify them with both field observations and laboratory experiments. Specifically, he conducted experiments with a magnetic dipole inside a model Earth, which he called a terrella. Figure 1.4 shows Birkeland in his laboratory beside his terrella experiment. Those experiments demonstrated that electrons incident on the terrella would produce patterns quite reminiscent of the auroral zone. He believed, as we do today, that those particles came from the Sun.

K. Birkeland's work inspired the Norwegian mathematician Carl Størmer, whose subsequent calculations of the motion of charged particles in a dipole magnetic field in turn supported Birkeland. Figure 1.5 shows Størmer and his assistant Bernt Johannes (not Kristian) Birkeland. As is evident from this

FIGURE 1.4. Kristian Birkeland (left) in his laboratory with his terrella and with his assistant, O. Devik (right), about 1909.

photograph, the advent of the camera was an important advance in the study of auroras. It was through measurements such as these that Størmer accurately determined the height of the auroras.

Figure 1.6 illustrates one of Størmer's charged-particle orbit calculations in a forbidden region to which charged particles from the Sun did not have direct access. In such a region, charged particles would spiral around the magnetic field and bounce back and forth along the field, reflected by the converging magnetic field geometry. Størmer's contributions became much more relevant and appreciated after the discovery of the Earth's radiation belts, whose particle motions resemble those of Figure 1.6. Kristian Birkeland's work was not appreciated until later, when spacecraft found field-aligned currents linked to the auroras. A more detailed discussion of the trajectories of charged particles in the Earth's magnetic field can be found in Chapter 10, and more about auroras can be found in Chapter 11.

This work on the solar-terrestrial connection proceeded, despite Lord Kelvin's 1882 argument that he had provided absolutely conclusive evidence against the supposition that terrestrial magnetic storms were due to magnetic action in the Sun or to any kind of dynamic action taking place within the Sun. Lord Kelvin also claimed "that the supposed connection between magnetic storms and sunspots is unreal, and the seeming agreement between periods has been a mere coincidence." More telling was the criticism of

FIGURE 1.5. Auroral physicists C. F. Størmer, standing, and Bernt Johannes Birkeland, seated, in northern Norway, c. 1910.

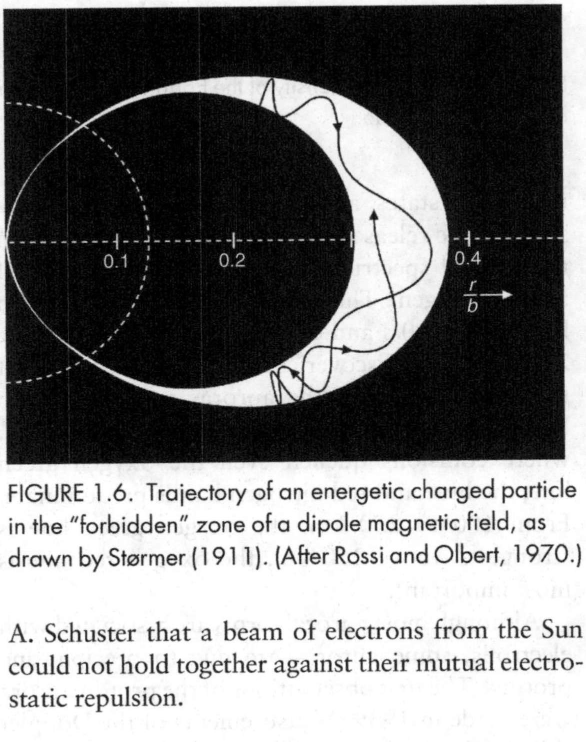

FIGURE 1.6. Trajectory of an energetic charged particle in the "forbidden" zone of a dipole magnetic field, as drawn by Størmer (1911). (After Rossi and Olbert, 1970.)

A. Schuster that a beam of electrons from the Sun could not hold together against their mutual electrostatic repulsion.

1.5 THE IONOSPHERE AND MAGNETOSPHERE REVEALED

The electrically conducting region above roughly 100 km altitude that we now call the **ionosphere** may rightly be claimed to have been discovered by Balfour Stewart. In his 1882 *Encyclopedia Britannica* article, entitled "Terrestrial magnetism," he concluded that the upper atmosphere was the most probable location of the electric currents that produce the solar-controlled variation in the magnetic field measured at the surface of the Earth. He noted that "we know from our study of aurora that there are such currents in these regions – continuous near the poles and occasional in lower latitudes."

He proposed that the primary causes of the daily variations in the intensity of the surface magnetic field were "convective currents established by the Sun's heating influence in the upper regions of the atmosphere." These currents "are to be regarded as conductors moving across lines of magnetic force and are thus the vehicle of electric currents which act upon the magnetic field." Those statements are very close to modern atmospheric-dynamo theory. However, it was left to A. Schuster to put the dynamo theory into quantitative form.

The turn of the twentieth century brought another new invention that was used to probe the solar-terrestrial environment: the radio transmitter and receiver. In 1902, A. E. Kennelly and O. Heaviside independently postulated the existence of a highly electrically conducting ionosphere to explain G. Marconi's transatlantic radio transmissions. Verification of the existence of the ionosphere did not come until much later, in 1925, when E. V. Appleton and M. A. F. Barnett in the United Kingdom, and shortly thereafter G. Breit and M. A. Tuve in the United States, established the existence and altitude of the Kennelly–Heaviside layer, as it was known then. The only contemporary usage of the term "Heaviside layer" occurs in the musical *Cats*.

The original method of Breit and Tuve, using short pulses of radio energy at vertical incidence and timing the arrival of a reflected signal in order to infer the altitude of the electrically reflecting layer, is still used today for sounding the ionosphere. In drawing diagrams of the electromagnetic waves reflected by the ionosphere, Appleton used the letter E for the electric vector of the downcoming wave. When he found reflections from a higher layer, he used the letter F for the electric vector of those reflected waves, and when he occasionally got reflections from a lower layer, he naturally used the letter D. When it came time to name these layers, he chose the same letters, leaving the letters A, B, and C for possible later discoveries that never came. So, today, the ionospheric layers are referred to as the D, E, and F layers, as illustrated in Figure 1.7. We now know that all planets with atmospheres have electrically conducting ionospheres. Chapter 2 discusses how these are formed.

At about that same time, progress was being made in understanding the auroral glows. Spectroscopy, together with photography, permitted first the determination of the wavelength and then the identity of the excited molecules that were radiating. There were initial successes, beginning with Lars Vegard's work in Norway, relating auroral emissions to emission bands from known atmospheric gases such as nitrogen. However, identification of the yellow-green line at 557.7 nm was elusive. Finally, H. Babcock's precise measurements in 1923 allowed John McLennan to identify this line as a metastable transition of atomic oxygen. At atmospheric pressures close to that at the surface of the Earth, collisions between molecules de-excite the molecules before they have a chance to radiate if they happen to become excited into one of the metastable states. However, at the altitude of the auroras, collisions are so rare that the time between collisions is longer than the lifetimes of the

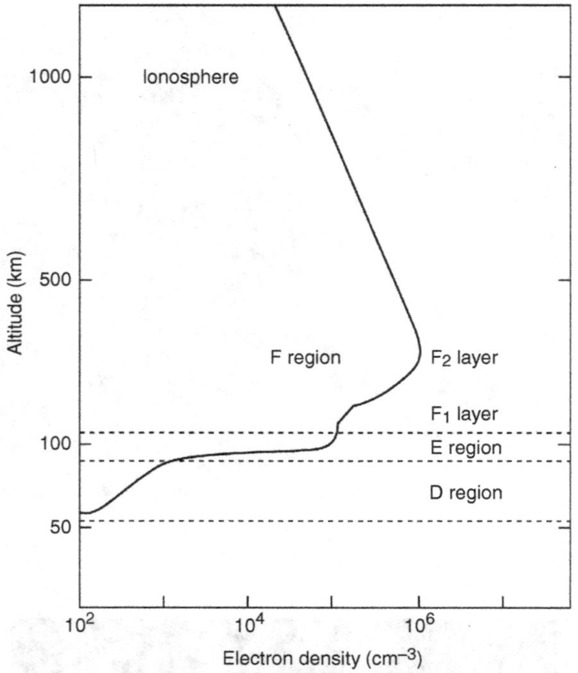

FIGURE 1.7. Electron density of the Earth's ionosphere as a function of altitude.

metastable states, and the excitation energy of the state can be released by radiation. A similar line in the auroral spectrum is the 630.0 nm red line of atomic oxygen. This metastable transition has a lifetime of 110 s and can radiate only above some 250 km. Those discoveries led to the realization that the varied colors of the auroras were related to height. In low-altitude auroras, below 100 km, where collisions quench even the oxygen green line, the blue and red nitrogen bands predominate. From 100 to 250 km, the oxygen green line is strongest. Above 250 km, the oxygen red line is most important.

Although most auroral forms are associated with electrons, some auroras are due to precipitating protons. The first observations of the proton aurora were made in 1939. Measurements of the Doppler shifts of the proton emissions permitted estimates of the energy of the precipitating particles from the ground. Chapter 11 contains a more detailed discussion of auroras and the auroral ionosphere.

With the concept of the ionosphere firmly established, scientists began to wonder about the upper extension of the ionosphere that is linked magnetically to the Earth; today, this is called the magnetosphere. In 1918, Sydney Chapman postulated a singly charged beam from the Sun as the cause of worldwide magnetic disturbances, a revival of an old idea that had previously been criticized by Schuster. Chapman was soon challenged by Frederick Lindemann, who pointed out that mutual electrostatic repulsion would destroy such a stream. Lindemann instead suggested that the stream of charged particles contained particles of both signs in equal numbers. We now call the material in such a stream a **"plasma."** That proposal was a breakthrough, and it permitted Chapman and his co-workers, in a series of papers beginning in 1930, to lay the foundations for our modern understanding of the interaction of the solar wind with the magnetosphere. Today we understand plasmas sufficiently to use plasma beams to both orient and propel our spacecraft, but it was not until the 1960s that the acceleration of the solar plasma was understood.

In the rarefied conditions of outer space, where collisions between particles are infrequent, the ion–electron gas, or plasma, is highly electrically conducting. Thus, Chapman and Ferraro proposed that, as the plasma from the Sun approached, the Earth would effectively see a mirror magnetic dipole moment advancing, as illustrated in Figure 1.8. The net result of that advancing mirror field would be to compress the terrestrial field. Eventually, as sketched in Figure 1.9, the plasma would surround the Earth on all sides, and a cavity would be carved out of the solar plasma by the terrestrial magnetic field. That is very similar to our modern concept of the geomagnetic cavity, the formation of which is discussed in greater detail in Chapter 7.

After the compression of the **magnetosphere**, which is detected by ground-based magnetometers as a sharp increase in the magnetic field, the magnetosphere becomes inflated. Chapman and Ferraro correctly interpreted this subsequent decrease in the magnetic field at the surface of the Earth as the appearance of energetic plasma deep inside the magnetosphere, forming a ring of current in the near-equatorial regions. The development of this ring current in what we now call a geomagnetic storm is discussed at greater length in Chapter 9.

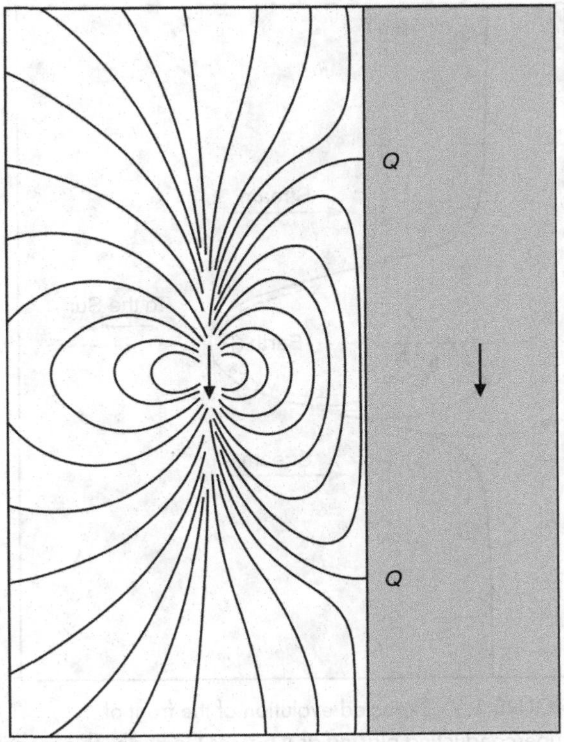

FIGURE 1.8. Compression of a dipole field by an advancing infinite, superconducting slab. The magnetic field is due to the original dipole plus an image dipole an equal distance behind the front, as shown by the right-hand arrow. (After Chapman and Bartels, 1940.)

At the same time as the ionosphere was being discovered by virtue of its effects on man-made radio signals, natural radio emissions were also being explored, and the magneto-ionic theory, developed for the man-made signals, was being applied to those natural emissions. The first report of electromagnetic signals in the audio-frequency range was an observation of what have become known as **"whistlers,"** detected using a 22 km telephone line in Austria in 1886. Whistlers are short bursts of audio-frequency radio noise of continuously decreasing pitch. In 1894, British telephone operators heard "tweeks," possibly whistlers generated by lightning, and a "dawn chorus" generated deep in the magnetosphere during a display of the aurora borealis. Little work was done on those observations because of the lack of suitable analysis equipment at the time. During World War I,

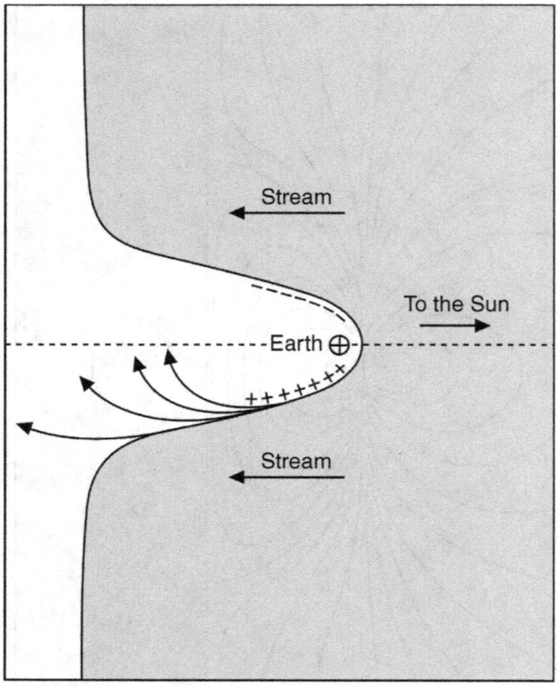

FIGURE 1.9. Expected evolution of the front of superconducting plasma as it passes the Earth. This model was proposed by Chapman and Ferraro in the 1930s to explain the phenomena of the geomagnetic storm. (After Chapman and Bartels, 1940.)

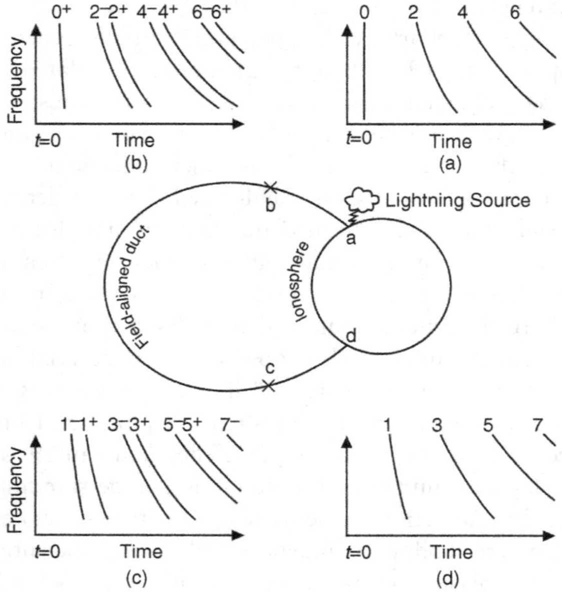

FIGURE 1.10. Dispersion of whistler-mode waves generated by lightning, as seen at four different locations. The different velocities of propagation as functions of wave frequency (dispersion) cause the wave arrival to be delayed by a different amount at each frequency. The delay depends on the distance traveled and the properties of the plasma traversed by the wave. (From Russell, 1972.)

equipment installed to eavesdrop on enemy telephone conversations picked up whistling sounds. Soldiers on the front would say, "You could hear the grenades fly." H. Barkhausen reported on those observations in 1919, and suggested that they were correlated with meteorological influences. However, he could not duplicate the phenomenon in laboratory experiments.

In 1925, T. L. Eckersley also described that phenomenon but incorrectly ascribed it to the dispersion of an electrical impulse in a medium loaded with free ions, causing signals at different frequencies to travel at different speeds. Eventually, in 1935, after much work and several incorrect explanations, Eckersley concluded that the distinctive swooping sound of whistlers was due to electron-caused dispersion of a burst of electromagnetic noise traveling through the ionosphere.

Very little work was done on whistlers until the early 1950s, at which time L. R. O. Storey, with a homemade spectrum analyzer, conducted a thorough study of whistlers. He found that whistlers are caused by lightning flashes, whose electromagnetic energy then echoes back and forth along field lines in the upper ionosphere, as illustrated in Figure 1.10. A major implication of those findings was that the electron density in the outer ionosphere, which is now called the plasmasphere, was unexpectedly high. Storey also found other types of audio-frequency, or very low frequency (VLF), emissions that are not associated with lightning and are now known to be generated within the magnetospheric plasma. Chapters 10 and 13 discuss the generation and propagation of these waves.

Because of atmospheric drag, it is difficult to study the upper atmosphere and ionosphere from space. Table 1.1 lists several missions that have attempted to explore part of this low-altitude environment. The Alouette 1 and 2 spacecraft used radio sounding to study the topside ionosphere. ISIS 1, 2

Table 1.1. Aeronomy missions

Mission name	Orbit	Type of mission	Launch	End of mission
Alouette 1	low-altitude polar	ionosonde	Sept. 29, 1962	1972
Alouette 2	low-altitude polar	ionosonde	Nov. 29, 1965	Aug. 1, 1975
ISIS I	low-altitude polar	ionospheric *in situ*	Jan. 30, 1969	Jan. 24, 1990
ISIS II	low-altitude polar	ionospheric *in situ*	Apr. 1, 1971	1984
DMSP[a]	low-altitude polar	auroral ionosphere	Aug. 23, 1962	
DE 1	elliptical polar	magnetosphere *in situ*	Aug. 3, 1981	Feb. 19, 1983
DE 2	low-altitude polar	ionosphere *in situ*	Aug. 3, 1981	Feb. 28, 1991
SME	low-altitude polar	upper atmosphere	Oct. 6, 1981	Mar. 5, 1991
UARS	low-altitude polar	upper atmosphere	Sept. 12, 1991	Sept. 24, 2011
FAST	low-altitude polar	auroral physics	Aug. 21, 1996	May 1, 2009

ISIS, International Satellite for Ionospheric Sounding; DE, Dynamics Explorer; SME, Solar Mesosphere Explorer; UARS, Upper Atmosphere Research Satellite; FAST, Fast Auroral SnapshoT.

[a] Continued under different names.

were *in situ* measuring missions. DMSP, SME, and UARS were low-altitude polar orbiting missions, and DE-1 and 2 were a pair of satellites innovatively sampling at high and low altitudes. FAST was an auroral particles and fields mission.

1.6 THE SOLAR WIND DISCOVERED

The importance of the Earth's magnetic field in controlling phenomena in the upper atmosphere is well known, but the importance of the solar magnetic field awaited the 1908 invention of the solar magnetograph by George Ellery Hale, who revolutionized our understanding of the solar system's central body. These magnetic fields are the key to understanding how the Sun can affect the Earth's magnetosphere so far away. It provides the heating processes ultimately responsible for the acceleration of the solar wind that in turn is ultimately responsible for creating the Earth's auroras.

Auroras are caused by energized electrons. If those electrons came from the Sun, as was commonly believed among solar-terrestrial researchers in the first half of the twentieth century, then those electrons would have to travel in the company of an equal number of ions, otherwise the beam would disrupt. This idea can be considered the first model of the streaming ionized plasma of interplanetary space that we know as solar wind. It was an essential element of the geomagnetic-storm model of Chapman and Ferraro, but in their model the solar wind was intermittent: it flowed only at active times. However, in 1943, C. Hoffmeister noted that a comet tail was not strictly radial, but lagged behind the comet's radial direction by about 5°. In 1951, L. Biermann correctly interpreted that lag in terms of an interaction between the comet tail and a solar wind. That wind was deduced to flow at about 450 km s^{-1} at all times and in all directions from the Sun, although Biermann assumed that the electron density was about 600 cm^{-3}, two orders of magnitude too high. Several years later, in 1957, Hannes Alfvén postulated that the solar wind was magnetized and that the solar-wind flow draped that magnetic field over the comet, forming a long magnetic tail downstream in the antisolar direction, as illustrated in Figure 1.11. The cometary ions were confined by that tail in a narrow ribbon between the two tail "lobes."

The gradual acceptance of the existence of a solar wind led to the explanation of a long studied phenomenon, the sudden impulse, in which several days after a solar eruption the magnetic field strength on the surface of the Earth is rapidly (in minutes) enhanced. This explanation in turn led to the postulate of the existence of collisionless shocks. In 1955, T. Gold, who also gave us the term "magnetosphere," postulated that the sudden impulse must be caused by a **collisionless shock** because the kinetic pressure front was so thin. This postulate, taken as predicting the existence of collisionless shocks in magnetized plasmas, is certainly correct. If the bulk velocity on either side of a wave front changes less than the velocity of

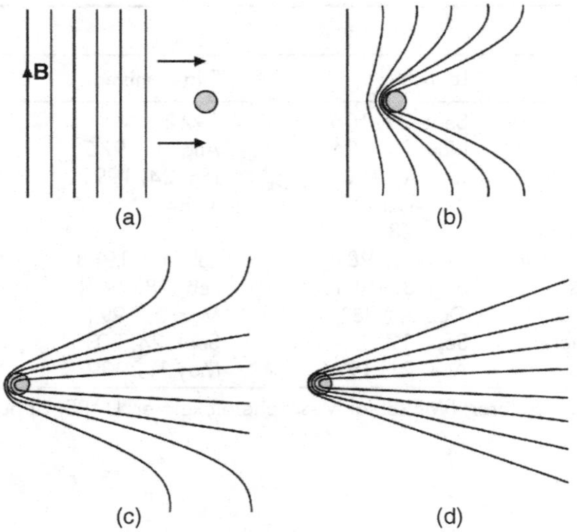

FIGURE 1.11. Original model of the formation of a type 1 (plasma) cometary tail, according to H. Alfvén. In this model the solar-wind magnetic field is draped over the comet by the motion of the plasma from left to right. (After Alfvén, 1957.)

the pressure wave, the wave is subsonic. If a pressure wave in the solar wind were subsonic, it would be much broader than the wave causing the sudden impulse. Taken as a postulate on the cause of sudden impulses, it is not completely correct, for not all sudden impulses are created by propagating pressure waves. Pressure-balanced tangential discontinuities are also quite thin. These structures do not propagate but are carried past the magnetosphere rapidly. It is the dynamic pressure or momentum flux measured in the Earth's frame that is important in determining the compressional state of the magnetosphere. It is the motion of that front of increased dynamic pressure past the magnetosphere that causes the sudden impulse. A detailed discussion of collisionless shocks can be found in Chapter 6.

In 1958, E. W. Parker provided the theoretical underpinning for a supersonic flow of magnetized plasma, and in 1962 he showed that, in order to be consistent with the geomagnetic records, the electron density of the solar wind should seldom exceed 30 cm^{-3}. Confirmation was not long in coming. The 1960s marked the dawn of the space age, and soon observations were being returned by both Soviet and American space probes that clearly confirmed the existence of the solar wind and its entrained magnetic field, measured its properties, and demonstrated its pivotal role in controlling geomagnetic activity and the auroras. Instrumentation developed by Konstantin I. Gringauz for the Soviet Luna probes provided the first measurements of the solar wind, while instrumentation developed by Conway Snyder and Marcia Neugebauer for Mariner 2, on its voyage to Venus in 1962, provided the first long-term, nearly continuous observation of the interplanetary medium. Chapter 5 discusses the solar wind and interplanetary magnetic fields in greater detail.

1.7 INTERACTION OF THE SOLAR WIND WITH THE MAGNETOSPHERE

Rockets provided the initial opportunity to explore the magnetosphere. In the early and middle 1950s, James Van Allen and his colleagues launched a series of rocket flights into the Arctic and Antarctic ionospheres, reaching heights up to 110 km. Those flights detected either energetic electrons or the bremsstrahlung radiation caused by the deceleration of such electrons. The year 1957 marked the beginning of an International Geophysical Year (IGY), an 18 month period of worldwide geophysical studies. It also marked the launch of Sputnik 1. The attendant space race began a period of explosive growth in our knowledge of the terrestrial magnetosphere and its interaction with the solar wind. In 1958, Explorer 1 carried a Geiger counter that enabled Van Allen to discover the trapped radiation belts.

Battery-powered Explorer 10, launched in 1961, was the first spacecraft to provide measurements across the magnetopause, the boundary between the flowing solar wind and the Earth's magnetic field, but the first detailed examination of that boundary awaited measurements with a solar-powered spacecraft, Explorer 12, which provided four months of data and coverage from the noon meridian to the dawn meridian. Figure 1.12 shows magnetic measurements obtained by Explorer 12 during a traversal of the outer magnetosphere, through the **magnetopause**, and out into the **magnetosheath**. It was clear from the data provided by

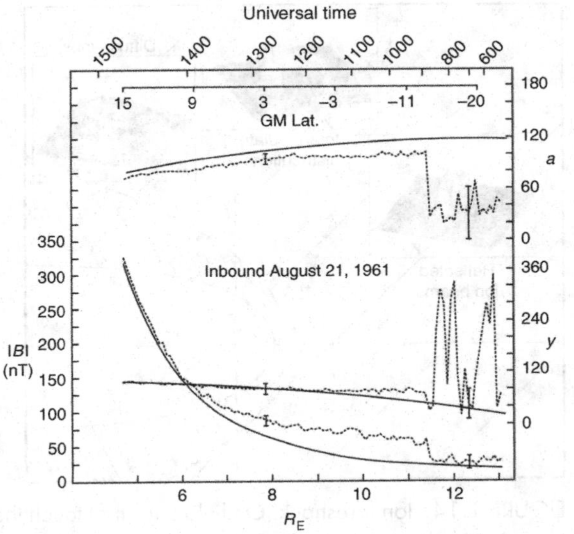

FIGURE 1.12. Measurements of magnetic field strength and direction in the outer magnetosphere and through the magnetopause by the Explorer 12 spacecraft. Smooth lines are dipole values. (After Cahill and Patel, 1967.)

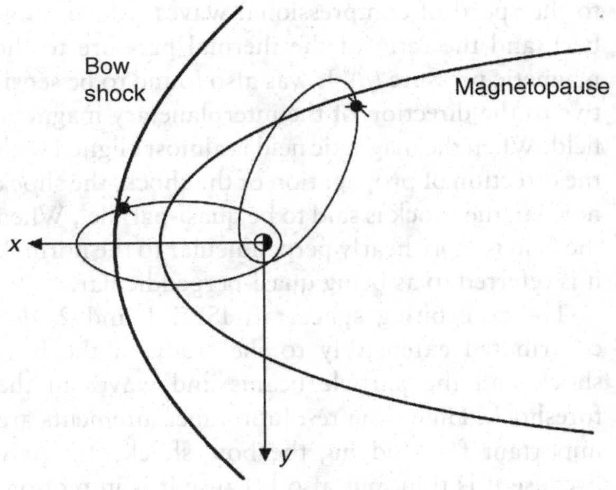

FIGURE 1.13. Mapping of the magnetopause and bow shock as the Earth goes around the Sun. During the course of a year, the magnetopause and bow-shock surfaces maintain their orientation about the Earth–Sun line, while the spacecraft orbit is fixed in inertial space. Thus the orbits appear to sweep through these boundaries in the course of a year.

the many spacecraft that were launched into the solar wind during those early years that the solar wind undergoes an abrupt transition prior to reaching the magnetopause. This transition is caused by a standing bow shock produced as the high-speed solar wind encounters the Earth as an obstacle to its flow. This standing bow shock is akin to the shock wave that forms in air in front of a supersonic aircraft.

The concept of a standing shock in front of the magnetosphere was not proposed until seven years after Gold's proposal of the existence of collisionless interplanetary shocks, but when it was it surprisingly appeared simultaneously in two articles in the same issue of the *Journal of Geophysical Research* (Axford, 1962; Kellogg, 1962). By this time, interplanetary spacecraft were being built, and, soon, many high-altitude measurements were returned. The measurements needed to resolve the structure of the bow-shock discontinuity had to await the launch of the first Orbiting Geophysical Observatory (Holzer, McLeod, and Smith, 1966), but the mapping of the location of the bow shock could be done with the measurements of lower cadence on the Interplanetary Monitoring Platforms, and the permanence of this feature of the interaction was quickly confirmed (e.g., Fairfield, 1971).

In the intervening years, it has become clear that the electric and magnetic fields in the plasma can alter the motion of the particles in a manner similar to ordinary collisions. These changes provide the dissipation needed to form a shock. The physics of this process is discussed in Chapter 6. The shock allows the solar wind, which flows faster than the speed of compressional waves in a plasma, to be slowed, heated, and deflected around the planet. It was not until the launch of the first Orbiting Geophysical Observatory (OGO) in 1964 that scientists obtained measurements of sufficient time resolution to study the bow shock. The OGO 1, 3, and 5 spacecraft in highly eccentric orbits mapped the locations of both boundaries as the Earth orbited the Sun, and the orbits precessed relative to the magnetosphere, as shown in Figure 1.13. Those measurements and data from other spacecraft launched in the 1960s, such as the Interplanetary Monitoring Platform (IMP) spacecraft and the VELA spacecraft, revealed that the structure of the bow shock was very sensitive to the conditions in the plasma, the ratio of the speed of the solar wind to the speed of compressional waves (Mach number), and the ratio of the thermal pressure to the magnetic pressure (β). It was also found to be sensitive to the direction of the interplanetary magnetic field. When the magnetic field is almost aligned with the direction of propagation of the shock, the shock normal, the shock is said to be quasi-parallel. When the field is more nearly perpendicular to the normal, it is referred to as being quasi-perpendicular.

The co-orbiting spacecraft ISEE 1 and 2 also contributed extensively to the study of the bow shock and the particle beams and waves in the **foreshock**. High time-resolution measurements are important for studying the bow shock, not only because it is thin, but also because it is in motion. Determining this motion was an objective of ISEE 1 and 2. The two ISEE spacecraft had an adjustable separation that allowed the shock velocity relative to the spacecraft to be deduced from the timing of the appearance of the shocks at the two locations (Russell and Greenstadt, 1979). Ion beams are found upstream of the quasi-parallel shock, as illustrated in Figure 1.14, and these ion beams interact with the incoming solar-wind plasma to produce copious large-amplitude waves, called upstream waves, shown in Figure 1.15. Astrophysical shocks are believed to be responsible for accelerating cosmic rays to ultra-relativistic energies. These particles in turn produce elementary particles when they collide with the Earth's atmosphere. Once, the only means to study these elementary particles was with high-altitude balloons.

The shock is important because it modifies the properties of the solar-wind plasma before the flow interacts with the Earth's magnetic field. These properties control the processes acting at the magnetopause that are finally responsible for determining how much energy the magnetosphere receives from the solar-wind flow. One can imagine a very inviscid interaction in which the solar wind is completely diverted by the magnetosphere and there is very little drag and, hence, little momentum transfer across the boundary. In fact, this situation does occur when the interplanetary magnetic field is northward, but when the interplanetary field is southward the momentum transfer from the solar wind increases markedly.

Laboratory plasma measurements have been useful in understanding the interaction of the solar

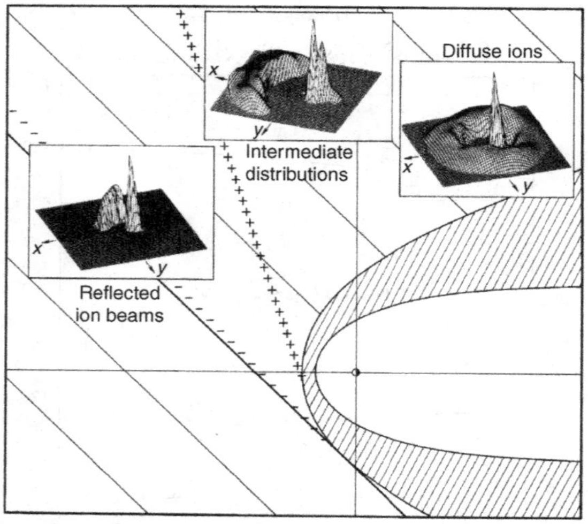

FIGURE 1.14. Ion foreshock. On field lines that touch the bow shock, charged particles can spiral upstream along the magnetic field. The solar-wind electric field causes these particles to drift antisunward. The fastest particles (electrons) are affected the least, and the slowest particles the most, by this drift. Representative distribution functions of ions are shown for various locations observed with plasma instruments on ISEE 1 and 2. The X-direction is to the Sun. The narrow peak represents the unperturbed solar-wind beam. The broader distributions represent the back-streaming ions. (After Russell and Hoppe, 1983.)

wind with the magnetosphere. Figure 1.16 shows a wire model of the magnetic field lines in the outer magnetosphere developed by Igor Podgorny and his colleagues at the Space Research Institute in Moscow, based on laboratory experiments undertaken in the 1960s, illustrating the development of cusp-shaped openings in the field pattern on the dayside and a long tail at night. Figure 1.17 shows a three-dimensional sketch of the magnetosphere, representing the structure that has been inferred from spacecraft observations. It is deep within this magnetosphere that the radiation-belt particles bounce and drift, as illustrated in Figure 1.18. As noted earlier, the converging magnetic field lines of the dipole magnetic field stop the forward motion of the particles and accelerate them back toward the equatorial regions. While gyrating and bouncing, these particles also drift, because their gyro-radii are greater when they gyrate into weaker fields than when they are in the stronger fields on the

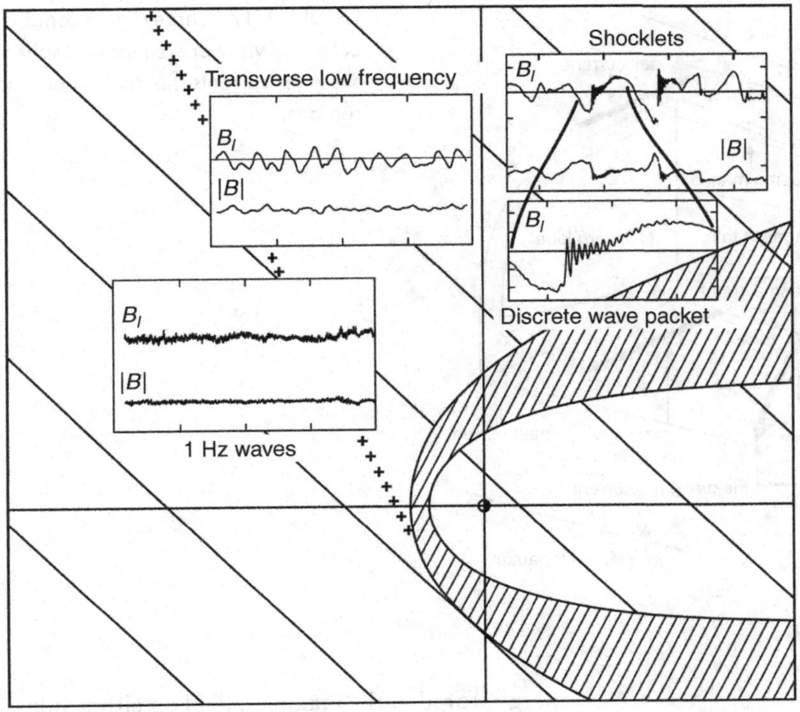

FIGURE 1.15. Back-streaming ions and electrons can stimulate a variety of low-frequency waves (with periods of seconds to many tens of seconds). Waves representative of different regions of the near-Earth solar wind are illustrated using data from the ISEE 1 and 2 magnetometers. B_l is the component of largest wave amplitude; $|B|$ is the magnetic field magnitude. (After Russell and Hoppe, 1983.)

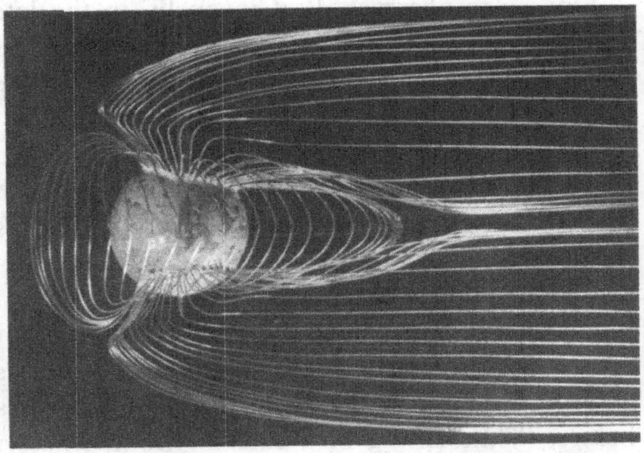

FIGURE 1.16. Three-dimensional wire model of the magnetosphere based on the laboratory experimental data of Podgorny and colleagues. (From Podgorny, 1976.)

inner part of their trajectories. The dipole magnetic field of the Earth can confine particles over a wide range of energies, the more energetic of which Van Allen encountered when he and his colleagues discovered the **radiation belts.**

Figure 1.19 shows the intensities of very energetic protons and electrons in the inner magnetosphere. These particles enter the radiation belts through a variety of means, including radial diffusion from more distant regions, with accompanying acceleration, and the decay of neutrons from the sputtering of the atmosphere by cosmic rays. Whereas energetic protons form a single belt, as illustrated in the top panel of Figure 1.19, electrons, as illustrated in the bottom panel, form two belts separated by a region called the slot. This slot in the flux of electrons of fixed energy is formed when naturally occurring electromagnetic waves interact with the gyro-motion of the electrons, causing them to spiral into the atmosphere, where they are lost through collisions. The inner electron belt is quite stable and the outer belt quite variable. The radiation belts and the motions of charged particles are discussed further in Chapters 3 and 10.

1.8 ENERGY STORAGE AND RELEASE

Both the Sun and the Earth exhibit rapid energy release in their magnetized plasma envelopes.

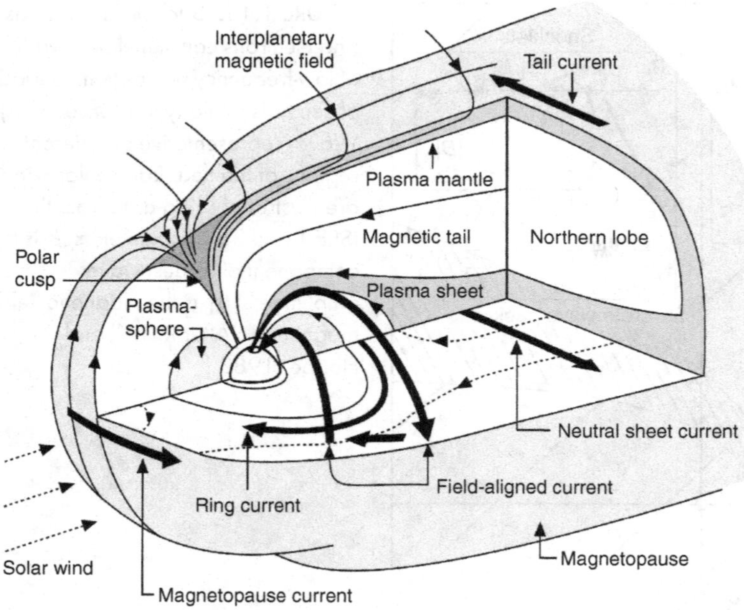

FIGURE 1.17. Three-dimensional cut-away view of the magnetosphere showing currents, fields, and plasma regions.

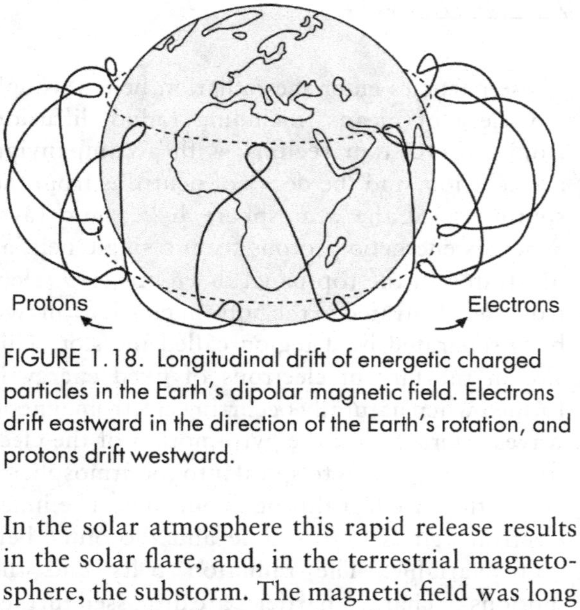

FIGURE 1.18. Longitudinal drift of energetic charged particles in the Earth's dipolar magnetic field. Electrons drift eastward in the direction of the Earth's rotation, and protons drift westward.

In the solar atmosphere this rapid release results in the solar flare, and, in the terrestrial magnetosphere, the substorm. The magnetic field was long believed to be the agent for the storage of the energy, but how to release it rapidly was a mystery. Many workers, such as V. Ferraro, P. Sweet, and E. Parker, tried unsuccessfully in the 1950s and 1960s to find ways in which the current layers that separated regions of oppositely directed magnetic flux could become resistive and liberate the energy stored in the magnetic field on either side of the current sheet.

One solar observer, R. Giovanelli, in the late 1940s, was struck by the presence of neutral points in the magnetic field where the energy in solar flares was released. He was joined in 1953 by postdoctoral student J. W. Dungey who had been assigned earlier the task of applying this mechanism, which had become known as **reconnection**, to the Earth's magnetosphere. Dungey focused on the role of neutral points and put the physics of magnetic neutral points on a rigorous and quantitative basis, but not until 1961 did he solve the problem that his thesis advisor, F. Hoyle, had assigned him a dozen years before. While sitting in a cafe in Montparnasse, prior to a seminar, Dungey finally realized how reconnection could lead to momentum transfer from the solar wind and the stirring of the magnetospheric plasma.

Dungey's solution is shown in Figure 1.20(b), in which the solar wind and terrestrial field become linked at the subsolar region on the magnetopause and the solar wind drags magnetospheric plasma over the polar cap and into the tail region. When the magnetic field pointed northward, a reversed convection pattern arose, but this flow was much

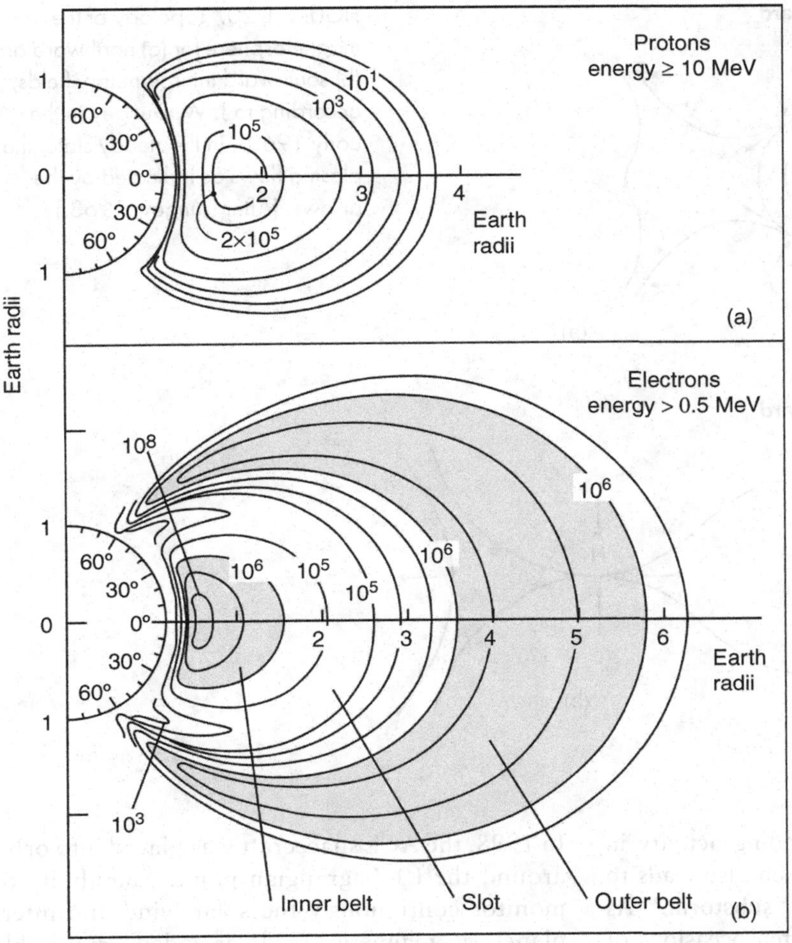

FIGURE 1.19. Earth's radiation belts. (a) Contours of the omni-directional flux (particles per square centimeter per second) of protons with energies greater than 10 MeV. (b) Contours of the omni-directional flux of electrons with energies greater than 0.5 MeV.

weaker. Time variations of the direction of the interplanetary magnetic field lead to storage and release. The neutral point, or x-point, geometry allowed the release to be rapid. While this work was in progress, advocates of the current sheet were also making progress, fortunately converging with the ideas of the advocates of the neutral point. H. Petschek in 1964 showed how magnetohydrodynamic (MHD) waves coupled with a neutral-point geometry would enable the rapid acceleration of plasma at a current sheet.

The testing of these ideas did not take long, although their acceptance did. The most important clue that the nature of the processes at the magnetopause changes with variations in the properties of the solar wind is the observation that geomagnetic activity is controlled by the north–south component of the interplanetary magnetic field. The availability of extended measurements in the solar wind, especially from Explorer 33 and 35, allowed researchers such as Roger Arnoldy and Joan Hirshberg (nee Feynman) to study this control. As can be deduced from Figure 1.20, if interplanetary and planetary magnetic fields become linked, magnetic flux will be transported from the dayside of the magnetosphere to the nightside. This magnetic flux builds up in the tail until reconnection occurs there too and returns the magnetic flux to the magnetosphere proper. Spacecraft such as OGO 5, launched in 1968, showed the erosion of the dayside

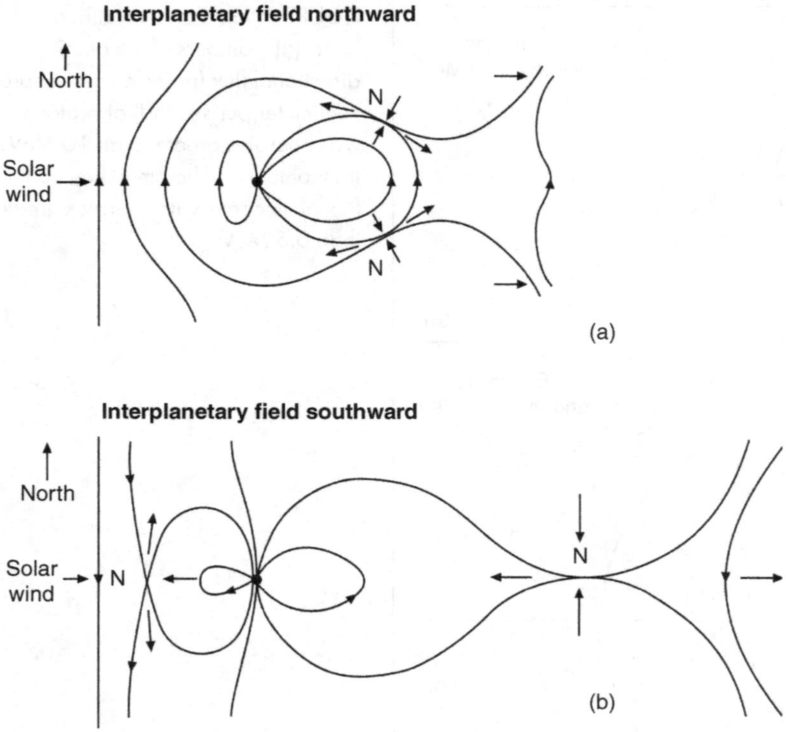

FIGURE 1.20. Topology of the magnetosphere for (a) northward and (b) southward interplanetary fields, according to J. W. Dungey in the early 1960s. In the steady state, the plasma flows as indicated by the arrows. (After Dungey, 1963.)

magnetosphere and the corresponding activity in the **magnetotail**. This process, which also leads to activation of auroras, is called a "**substorm.**" As implied by the name, there are often occasions of more major activity covering the entire magnetosphere, known as geomagnetic storms, but these phenomena are not linked in the way the coiners of the terms had originally envisioned. In the mid 1960s, a series of Explorer spacecraft called the Interplanetary Monitoring Platforms were launched to study the outer boundaries of the magnetosphere and the magnetotail.

In the late 1970s, after the launch of the dual co-orbiting satellites ISEE 1 and 2, the reconnection mechanism finally gained general acceptance in the magnetosphere community. These satellites returned plasma data of sufficiently high temporal resolution to show the accelerated flows at the magnetopause and in the magnetotail. These were followed in the 1990s by the Geospace mission consisting of Geotail, Polar, and Wind that studied the solar-wind interaction with the Earth. In 1998, the ACE spacecraft was placed into orbit around the L1 Lagrangian point. The ability to monitor continuously the solar wind and interplanetary magnetic field provided irrefutable proof that reconnection must be responsible for the energy transfer from the solar wind to the magnetosphere. Chapter 9 provides further details about the processes on the magnetopause and in the magnetotail. However, even today, there is debate as to where reconnection is initiated and how important it is relative to other processes. Furthermore, it has been found that three-dimensional structures are present, and their exploration requires not just two spacecraft, but four or more, as flown on ESA's recent Cluster mission and the five of NASA's THEMIS mission. Furthermore, the time scales and spatial scales of relevance are quite short, so new multi-spacecraft missions with close spacing and high sampling rates are being planned. Table 1.2 lists some of the earlier missions that contributed to magnetospheric exploration.

Table 1.2. Earth-orbiting space physics missions

Mission name	Target	Trajectory	Launch	End of mission
Sputnik 1	Earth	low-alt. circular	Oct. 4, 1957	Jan. 3, 1958
Explorer 1	Earth	low-alt. circular	Feb. 1, 1958	Mar. 31, 1970
Explorer 10	Earth m'sphere	low-incl. elliptical	Mar. 25, 1961	June 1, 1968
Explorer 12	Earth m'sphere	low-incl. elliptical	Aug. 16, 1961	Dec. 6, 1961
VELA (12 s/c)	nuclear detection	high-alt. circular	Oct. 17, 1963	1984
IMP 1	Earth m'sphere	low-incl. elliptical	Dec. 26, 1963	May 10, 1965
IMP 2	Earth m'sphere	low-incl. elliptical	Oct. 4, 1964	Oct. 13, 1965
OGO 1	Earth m'sphere	low-incl. elliptical	Sept. 5, 1964	Nov. 1, 1971
IMP 3	Earth m'sphere	low-incl. elliptical	May 29, 1965	May 12, 1967
OGO 3	Earth m'sphere	low-incl. elliptical	June 7, 1966	Sept. 14, 1981
Explorer 33	solar wind	Earth orbit	July 1, 1966	Sept. 21, 1971
Explorer 35	Moon orbiter	low-incl. elliptical	July 19, 1967	June 24, 1973
OGO 5	Earth m'sphere	low-incl. elliptical	Mar. 4, 1968	July 2, 2011
ISEE 1/ISEE 2	Earth m'sphere	low-incl. elliptical	Oct. 22, 1977	Sept. 1987
ISEE 3[a]	solar wind	L1 libration point	Aug. 12, 1978	May 1997
Geotail	Earth m'sphere	low-incl. elliptical	July 24, 1992	in progress
Wind	solar wind, m'tail	ecliptic plane	Nov. 1, 1994	in progress
Polar	polar m'sphere	low-incl. elliptical	Feb. 24, 1996	Apr. 2008
ACE	solar wind	L1 Lagrangian point	Aug. 25, 1997	in progress
Cluster (4 s/c)	Earth m'sphere	high-incl. elliptical	July 16, 2000	in progress
THEMIS (5 s/c)	Earth m'sphere	low-incl. elliptical	Feb. 17, 2007	in progress
Van Allen probes (2)	radiation belts	equatorial circular	Aug. 30, 2012	in progress
Magnetospheric multi-scale (4 s/c)	Earth m'sphere	low-incl. elliptical	Mar. 13, 2015	in progress

[a] Renamed International Cometary Explorer and directed to fly by comet Giacobini–Zinner on December 11, 1985.

1.9 PLANETARY AND INTERPLANETARY EXPLORATION

The Earth is but one testbed for the physical processes occurring in space plasmas. These same processes under different conditions occur at the other planets and in the interplanetary plasma. The solar-wind properties evolve with heliocentric distance, and on the way through the solar system the solar wind encounters a variety of obstacles to its flow, including magnetized bodies, unmagnetized bodies, some with atmospheres and others without, and eventually the heliopause, where the solar and the interstellar winds meet. Inside the various planetary magnetospheres, we find other boundary conditions that alter the way the plasmas behave. Volcanoes and plumes on moons in outer planetary magnetospheres may add neutral gas particles to these magnetospheres. The neutrals, in turn, become ionized. A rapidly rotating giant magnetosphere like Jupiter's produces significant centrifugal force in this magnetospheric plasma that stretches the magnetic-field lines into a magnetodisk. Processes that are thought to play minor roles at Earth, such as charge exchange, may play major roles in these settings. Conservation of angular momentum, of minor interest at Earth, may become a dominant factor as plasmas rush inward or outward after reconnection. Often these effects make us reconsider our understanding of terrestrial processes.

Planetary exploration began with missions to the Moon. This atmosphereless body covered with basaltic flows was studied with orbiters and landers culminating with the Apollo manned missions from 1969–73. These missions detected the magnetized crust and showed that the Moon had an electrically conducting core. More recently, the Moon has been studied with the Discovery Lunar Prospector, the Japanese Kaguya Mission, the Indian Chandrayan, NASA's Lunar Reconnaissance Orbiter and its companion LCROSS, an impactor, and the Lunar Atmosphere and Dust Environment Explorer, LADEE.

The earliest deep-space probes were the Mariner 2, 4, and 5 spacecraft that in the 1960s went to Venus, Mars, and Venus again. Those missions showed that both Venus and Mars were quite different from the Earth in their interactions with the solar wind, because neither planet had a significant magnetic moment. However, it was not until after the Venera 9 and 10 orbiters in 1975, the Pioneer Venus orbiter in 1978, and the short-lived but significant return from the orbiters Mars 3, Mars 5, and the Phobos mission that the details of the solar-wind interactions with these planets became better understood. Mars Global Surveyor and Mars Express missions in the first decade of the twenty-first century have made important single-instrument contributions. Finally, in 2006, a well-instrumented ESA probe, Venus Express, was put into orbit around Venus, and, most recently, the MAVEN mission was inserted into Mars orbit with a full complement of aeronomy instruments. Details of these missions are given in Table 1.3.

At Venus and Mars, the extreme-ultraviolet radiation from the Sun ionizes the upper atmosphere and creates a hot neutral atmosphere, or exosphere, that extends into the solar wind. As shown in Figure 1.21, the ionospheric pressure, consisting of both thermal and magnetic components, balances the dynamic pressure of the solar-wind flow. The neutral atmosphere that extends into the solar wind becomes ionized and adds to the solar-wind flow, further decelerating it. The slowing down of the magnetized flow around the planet leads to the draping of magnetic field lines over the obstacle and the formation of a long tail. In this respect, the solar-wind interaction with Venus and Mars resembles that with a comet. The cometary interaction was probed by the International Cometary Explorer (ICE) spacecraft at comet Giacobini–Zinner in 1985 and by the VEGA 1 and 2,

Table 1.3. Inner solar system missions contributing to space physics

Mission name	Target	Type of mission	Launch	End of mission
Mariner 2	Venus flyby	flyby	Aug. 27, 1962	Jan. 3, 1963
Mariner 4	Mars flyby	flyby	Nov. 28, 1964	Dec. 21, 1967
Mariner 5	Venus flyby	flyby	June 14, 1967	Nov. 1967
Mars 3	Mars orbit	low-incl. elliptical	May 28, 1971	Aug. 22, 1972
Mars 5	Mars orbit	low-incl. elliptical	July 24, 1973	Feb. 28, 1974
Mariner 10	Venus, Mercury	flyby (1V, 3M)	Nov. 3, 1973	Mar. 24, 1975
Venera 9	orbit/surface	low-incl. elliptical	June 8, 1975	~Dec. 25, 1975
Venera 10	orbit/surface	low-incl. elliptical	June 14, 1975	~Nov. 1, 1975
Pioneer Venus	Venus orbit	polar, elliptical	May 20, 1978	Aug. 1992
Sakigake	comet Halley	flyby	Jan. 7, 1985	Nov. 15, 1995
VEGA 1	Venus, Halley	balloon flyby	Dec. 15, 1984	Jan. 30, 1987
VEGA 2	Venus, Halley	balloon flyby	Dec. 21, 1984	Mar. 24, 1987
Giotto	comet Halley	flyby	July 2, 1985	July 23, 1992
Suisei	comet Halley	flyby	Aug. 18, 1985	Aug. 20, 1992
Phobos	Mars orbit	low-incl. elliptical	July 12, 1988	Mar. 27, 1989
Mars Global Sur.	Mars orbit	elliptical, polar	Nov. 7, 1996	Nov. 2, 2006
MESSENGER	Mercury orbit	elliptical, high ind.	Aug. 3, 2004	Apr. 30, 2015
Mars Express	Mars orbit	elliptical	June 2, 2003	in progress
Venus Express	Venus orbit	elliptical, polar	Nov. 9, 2005	Dec. 16, 2014
MAVEN	Mars orbit	elliptical, polar	Nov. 18, 2013	in progress

Flyby dates: Mariner 2, Venus, Dec. 14, 1962; Mariner 4, Mars, July 14, 1965; Mariner 5, Venus, Oct. 19, 1967; Mariner 10, V – Feb. 5, 1974; Me – Mar. 29, 1974; Me – Sept. 21, 1974; Me – Mar. 16, 1975; Sakigake, Mar. 11, 1986; Suisei, Mar. 8, 1986; VEGA 1, V – June 11, 1985; H – Mar. 6, 1986; VEGA 2, V – June 15, 1985; H – Mar. 9, 1985; Giotto, Mar. 13, 1986.

Orbital insertion dates: Mars 3, Dec. 2, 1971; Mars 5, Feb. 12, 1974; Venera 9, Oct. 20, 1975; Venera 10, Oct. 23, 1975; Pioneer Venus, Dec. 4, 1978; Phobos, Jan. 29, 1989; MGS, Sept. 12, 1997; MESSENGER, Mar. 18, 2011; MEX, Dec. 25, 2003; VEX, May 7, 2006; MAVEN, Sept. 22, 2014.

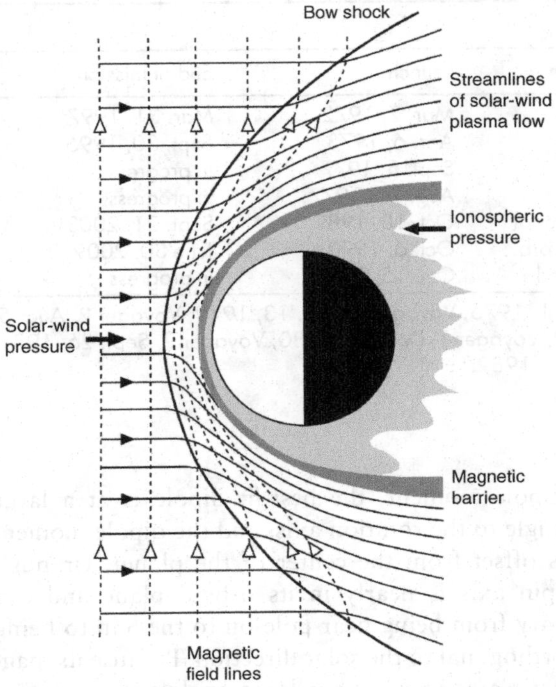

FIGURE 1.21. Solar-wind interaction with an unmagnetized planet. The ionospheric pressure stands off the solar-wind flow, so that the streamlines going from left to right flow around the planet. The magnetic field, which is shown here perpendicular to the flow in the solar wind, is bent around the obstacle by this interaction. (After Luhmann, 1986.)

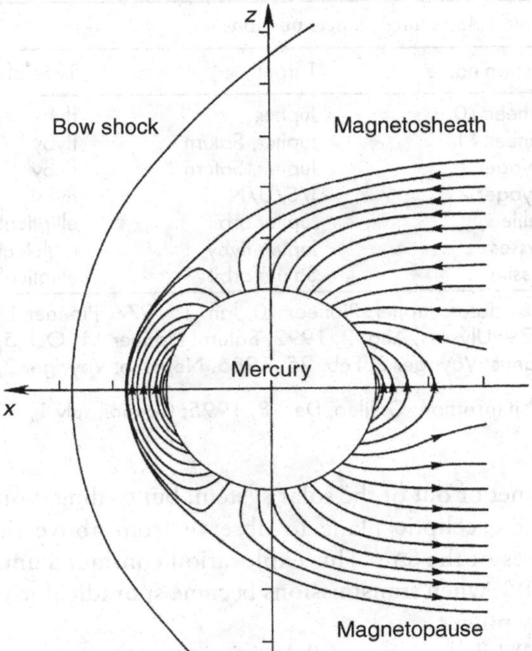

FIGURE 1.22. Magnetic field configuration and noon–midnight cross section of Mercury's magnetosphere. (From Russell, Baker, and Slavin, 1988.)

Giotto, Suisei, and Sakigake spacecraft at the comet Halley in 1986. Chapter 8 describes in greater detail the solar-wind interaction with such unmagnetized bodies.

Mariner 10, with a gravitational assist from Venus, made three passes by Mercury in 1974 and 1975. As illustrated in Figure 1.22, Mariner 10 found a mini-magnetosphere very much like that of the Earth. Mercury, however, has almost no atmosphere, so that the ionospheric current systems that we believe to be so important for the terrestrial magnetosphere must be absent from Mercury. This small magnetosphere also helps show where physical size matters. Mercury's magnetosphere should be different in many respects from that of the Earth. Presently, the MESSENGER mission has successfully completed its Mercury observations, and the two-spacecraft BepiColombo mission is in its development phase.

In 1972 and 1973, the first spacecraft to the outer solar system, Pioneer 10 and 11, were launched, reaching Jupiter in December 1973 and 1974, with Pioneer 11 going on to Saturn in 1979. Now, Pioneer 10 and 11 are heading out of the solar system, with Pioneer 10 going downwind relative to the interstellar medium, and Pioneer 11 upwind. Voyager 1 and 2 were launched in 1977, and reached Jupiter in 1979 and Saturn in 1980 and 1981. Voyager 2 then went on to successful encounters with Uranus in 1986 and Neptune in 1989. Both Voyager 1 and 2 are now heading upwind toward the heliopause with Voyager 1 close to and Voyager 2 through the termination shock. In 1990, Ulysses was also launched to fly by Jupiter, but in this case, it used Jupiter not to sling the spacecraft toward another

Table 1.4. Outer planets missions

Mission name	Target	Type of mission	Launch	End of mission
Pioneer 10	Jupiter	flyby	Mar. 2, 1972	Mar. 31, 1997
Pioneer 11	Jupiter, Saturn	flyby	Apr. 6, 1973	Sept. 30, 1995
Voyager 1	Jupiter, Saturn	flyby	Sept. 5, 1977	in progress
Voyager 2	J/S/U/N	flyby	Aug. 20, 1977	in progress
Galileo	Jupiter orbit	elliptical equatorial	Oct. 18, 1989	Sept. 21, 2003
Ulysses	Jupiter flyby	high-incl. solar orb.	Oct. 6, 1990	June 30, 2009
Cassini	Saturn orbit	elliptical inclined	Oct. 15, 1997	in progress

Flyby dates: Jupiter: Pioneer 10, Jan. 1, 1974; Pioneer 11, Jan. 1, 1975; Voyager 1, Apr. 13, 1979; Voyager 2, Aug. 5, 1979; Ulysses, Aug. 2, 1992. Saturn: Pioneer 11, Oct. 5, 1979; Voyager 1, Dec. 14, 1980; Voyager 2, Sept. 25, 1981. Uranus: Voyager 2, Feb. 25, 1986. Neptune: Voyager 2, Oct. 2, 1989.

Orbit insertion: Galileo, Dec. 8, 1995; Cassini, July 1, 2004.

planet or out of the solar system, but to fling it out of the ecliptic plane to observe from above the poles of the Sun. This exploration continued until 2008, when transmissions became sporadic due to low power.

While exploratory flyby missions such as Pioneer 10 and 11, Voyager 1 and 2, and Ulysses are very useful, orbiters provide the long-term measurements over a significant volume of the plasma environment of a planet necessary for greater understanding. The arrival at Jupiter in 1995 of the Galileo orbiter, even with its crippled communication system, marked a major increase in our understanding of the jovian magnetosphere. The arrival of Cassini in 2004 has done the same for Saturn.

Those missions revealed well-developed magnetospheres at the four gas giants, each with a bow shock, magnetopause, and magnetotail. The rapid rotation of the magnetosphere of Jupiter, coupled with a strong plasma source at the moon Io, causes the magnetosphere to be distorted into a disk-like geometry. Jupiter is also a source of intense radio waves. A source of mass at the icy moon Enceladus enables the rapid rotation of Saturn's magnetospheric plasma also to produce a magnetodisk. Cassini's high telemetry rate enables study of the saturnian system in far greater detail than was possible in the jovian magnetosphere. These missions are detailed in Table 1.4.

Uranus and Neptune have unusual orientations of their planetary magnetic fields. Both fields are very complex, and, when each is fitted with a dipole moment, the best-fit dipole is at a large angle to the rotation axis, and the dipole moment is offset from the center of the planet. Uranus's spin axis is nearly in its orbital plane and can vary from being near pole-on to the Sun to being orthogonal to the solar direction. Because its magnetic axis is at such a large angle to its rotation axis, its magnetosphere undergoes large oscillations during the course of a day. Neptune has a more customary rotation axis, roughly perpendicular to its orbit around the Sun, but its planetary magnetic field is even more complex than that of Uranus. In both magnetospheres, the radiation belts are much more benign than those of the Earth. Chapter 12 describes the phenomena occurring in the magnetospheres of these outer planets. Finally, the New Horizons spacecraft flew by Pluto on July 4, 2015, with a complement of plasma and energetic particle instruments, finding no evidence for a magnetosphere and little evidence for a solar-wind interaction.

1.10 SOLAR PHYSICS

Another frontier that is as difficult to reach as the outer planets is the Sun. Most solar work in space is done by remote sensing, as listed in Table 1.5, but some missions, such as Helios 1, 2 have attempted to get close to the Sun, as close as 0.3 AU. Future missions will get even closer. See Table 1.5 for a list of solar missions.

Table 1.5. Solar missions

Mission name	Trajectory	Type of mission	Launch	End of mission
Skylab	Earth orbit	remote sensing	May 14, 1973	July 11, 1979
Solwind	Earth orbit	remote sensing	Feb. 24, 1979	Sept. 13, 1985
Helios A	solar 0.3–1.0 AU	*in situ* measurements	Dec. 10, 1974	Feb. 18, 1985
Helios B	solar 0.3–1.0 AU	*in situ* measurements	Jan. 15, 1976	Dec. 23, 1979
GOES 1-N	geosynchronous	solar x-rays	Oct. 16, 1975	in progress
Yohkoh	Earth orbit	solar x-rays imager	Aug. 30, 1991	Sept. 12, 2005
SOHO	L1 point	remote sensing	Dec. 2, 1995	in progress
RHESSI	Earth orbit	solar flares	Feb. 5, 2002	in progress
TRACE	Earth orbit	transition region obs.	Apr. 2, 1998	June 21, 2010
SORCE	Earth orbit	remote sensing	Jan. 25, 2003	in progress
STEREO A, B	1 AU orbit	*in situ* and remote	Oct. 26, 2006	in progress
Hinode	Earth orbit	remote sensing	Sept. 22, 2006	in progress
SDO	Earth orbit	remote sensing	Feb. 11, 2010	in progress

1.11 SUMMARY

The discipline of solar-terrestrial physics has come far in the five centuries since its inception, and its importance to the inhabitants of this planet has grown as society has advanced in the sophistication of its technology, as illustrated in Figure 1.23. Had mankind not left the surface of the planet, engaging in both robotic and human exploration, it still would have experienced the effects of space weather, as it is now called. Long conductors across the surface of the Earth are subject to very large voltage drops and high currents when the Earth's magnetic field changes suddenly in response to changes in the solar wind. Energetic particles affect humans, spacecraft, and the ionosphere and communications through it. Thus it is imperative to understand the solar-terrestrial environment better. We now have physical models for most of the observed phenomena.

The discipline has evolved from one of remote sensing to *in situ* observations, theory, and computer modeling. In fact, it is now more appropriate to refer to the field as "space physics," as one of the major journals in the field does, rather than solar-terrestrial physics, and we have done so in choosing the title for this book. In the chapters that follow, we provide a discussion of the physical principles that underlie the observed phenomena as well as detailed descriptions of the phenomena.

Additional reading

BREKKE, A. and A. EGELAND (1983). *The Northern Light: From Mythology to Space Research*. Berlin: Springer-Verlag. This popular book covers the history of myth, understanding, and exploration of the auroral zone and auroras.

CHAPMAN, S. and J. BARTELS (1940). *Geomagnetism*. Oxford: Oxford University Press. This scientific compendium covers the history of the early studies of the solar-terrestrial environment and was drawn on heavily for Chapter 1.

EATHER, R. H. (1980). *Majestic Lights*. Washington, D.C.: American Geophysical Union. A popular book on the history and observations of the auroras.

EGELAND, A. and W. J. BURKE (2012a). The ring current: a short biography. *Hist. Geo. Space Sci.*, 3, 131–142. This paper outlines how the understanding of the ring current evolved from Carl Størmer's original postulates until observations were available in the magnetosphere in the 1960s and 1970s.

EGELAND, A. and W. J. BURKE (2012b). Carl Størmer's auroral discoveries. *Can. J. Phys.*, 90, 785–793. This paper describes Størmer's studies of auroral processes and charged-particle trajectories, for which he had to develop new numerical methods.

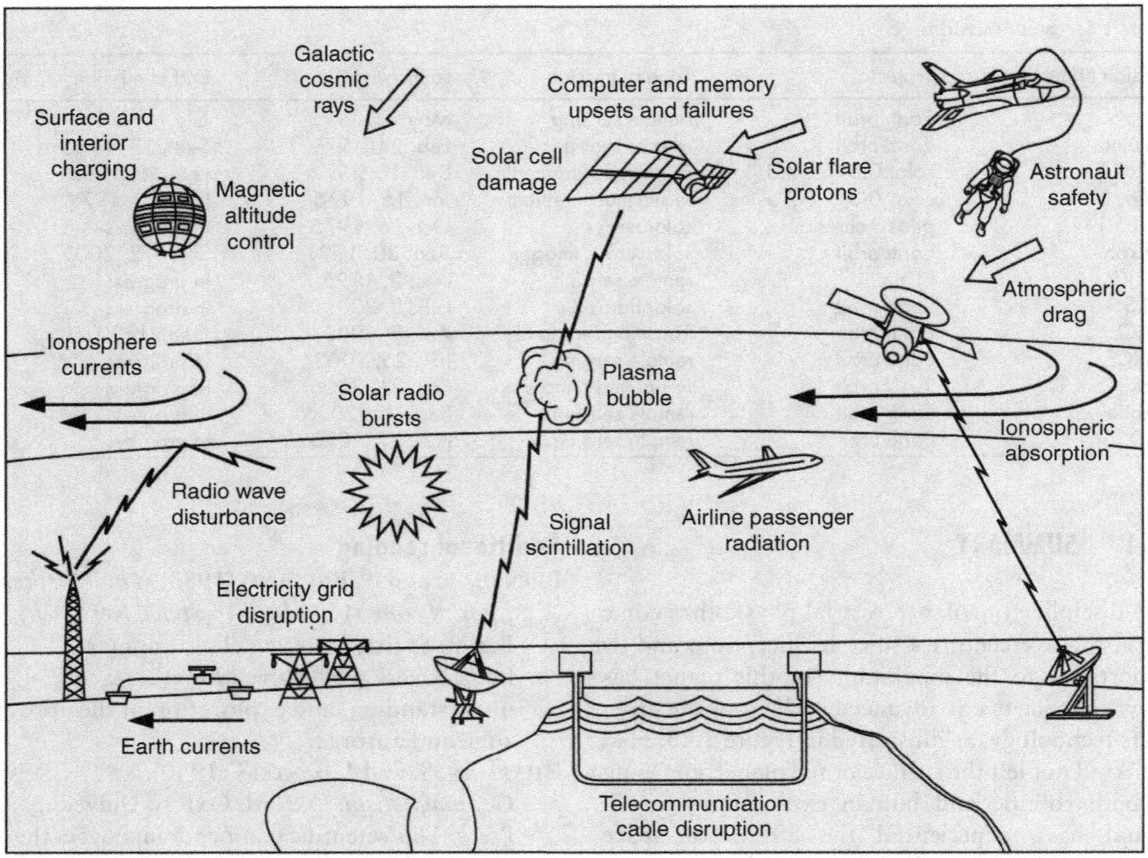

FIGURE 1.23. Some of the ways in which we and our technology are sensitive to the space environment. (Source L. J. Lanzerotti, Lucent Technologies, Bell Labs, with permission.)

GILBERT, W. (1893). *De Magnete*. Trans. P. FLEURY MOTTELAY. New York: Dover. Reprinted 1958. William Gilbert's book, originally published in 1600, in which he proposed that the Earth is a magnet, is still available in the 1893 translation, reprinted in 1958.

HELLIWELL, R. A. (1965). *Whistlers and Related Ionospheric Phenomena*. Stanford, CA: Stanford University Press. This early important magnetospheric treatise covers wave phenomena in the magnetosphere.

Space Science Reviews has published a series of special issues on space missions. Special issues of particular interest to readers of the chapter include:

The Galileo mission, *Space Sci. Rev.*, 1992.

The Global Geospace mission, *Space Sci. Rev.*, 1995.

The Cluster and Phoenix missions, *Space Sci. Rev.*, 1997.

The Advanced Composition Explorer mission, *Space Sci. Rev.*, 1998.

The Cassini/Huygens mission, *Space Sci. Rev.*, **104**, 2002; **114**, 2004; **115**, 2004.

Rosetta: Mission to Comet 67P/Churyumov–Gerasimenko, *Space Sci. Rev.*, **128**, 2007.

The MESSENGER mission to Mercury, *Space Sci. Rev.*, **131**, 2007.

The STEREO mission, *Space Sci. Rev.*, **136**, 2008.

The THEMIS mission, *Space Sci. Rev.*, 141, 2008.

The Acceleration, Reconnection with Turbulence and Electrodynamics of the Moon's Interaction with the Sun (ARTEMIS) mission, *Space Sci. Rev.*, 165, 2011.

Problems

1.1 At its typical velocity of 440 km s^{-1}, how long does it take the solar wind to arrive at Mercury, Earth, Jupiter, and Pluto? How long does it take radio waves to go the same distance?

1.2 What arguments might have Edmund Halley made to King William III of England to finance the voyage of the *Paramour Pink* in 1698 and 1700? Were there scientific, commercial, or military returns on investment? Navies are still interested in magnetism today and are thought to install magnetometers on the sea floor. Why?

1.3 Discuss the role of new technology in pacing the development of solar-terrestrial physics prior to the International Geophysical year in 1957. Describe specifically what scientific discoveries were led by new developments. Describe a more recent technological development and how it has led to new scientific discovery. Note, not every scientific discovery can be traced to a specific new technical development.

1.4 In the mid 1950s, what arguments could Van Allen have made to justify putting a Geiger counter on Explorer 1? What arguments might have been applicable to the flight of Pioneer 10 and 11 to Jupiter? How could you justify a mission to sit at the L1 Lagrangian point on the Earth–Sun line where the combined force of gravity on a spacecraft from these two bodies allows a spacecraft to orbit the Sun in one Earth year even though it does not orbit around the Earth, thereby staying on the Earth–Sun line? This point is about 0.01 AU from Earth.

2 The upper atmosphere and ionosphere

2.1 INTRODUCTION

As inhabitants of Earth, our lives are affected by each of the four states of matter: solid, liquid, gas, and plasma. Most of the mass of the Earth is in the solid state. It provides a place for us to stand and it provides the gravitational field that keeps us as well as everything else bound to the planet, including the oceans (liquid), the atmosphere (gas), and the ionosphere (plasma). Our bodies need liquids and gases, especially water and oxygen. The gases of the atmosphere protect us from the most energetic of the photons from the Sun, those in the ultraviolet, the extreme-ultraviolet (EUV), and the x-ray regions of the solar spectrum. This protection arises from photochemical processes in which molecules in the upper atmosphere absorb energetic photons at the cost of their molecular makeup. Ozone (O_3) forms a protective gaseous envelope that absorbs solar ultraviolet radiation. Extreme-ultraviolet and x-ray photons are also absorbed by the atmosphere, but these photons are so energetic that they can strip molecules and atoms of one or more of their electrons, turning them into ions and producing a plasma. These plasmas persist at high altitudes where the atmosphere has become "thin" and the collision rates between molecules, atoms, ions, and electrons has become quite low. The plasmas are generally benign, and can be quite useful to humanity by allowing over-the-horizon communication without the expense of launching satellites into space or installing communication lines across continents or on ocean bottoms. The gravitational field affects these plasmas just as it does the terrestrial solids, liquids, and gases, but the plasmas are affected additionally by the Earth's magnetic and electric fields.

While the Earth's gravitational force is strong, its pull is not strong enough to prevent the escape of all the gas in the atmosphere, and a small amount of the atmosphere is continuously escaping into space. In order to calculate how fast the atmosphere is escaping, we must first learn about the gravitational field, the thermodynamics of gases, and the structure of the atmosphere and the ionosphere. This brief examination of the atmosphere will introduce concepts that we shall use again in later chapters. An example is the propagating wave, which can both provide remote diagnostics of a process occurring within the gas and also transport energy into and out of the gas. We also examine how the ionosphere is formed and lost, and how it conducts the electrical currents that lead to the geomagnetic variations that have been linked to solar activity.

2.1.1 The gravitational field

The first breakthrough in understanding gravity arose from Johannes **Kepler's laws** of planetary motion, which were derived from Tycho Brahe's studies of planetary positions.

(1) All planets move in elliptical paths with the Sun at one focus.
(2) The line connecting the Sun to the planet sweeps out equal areas in equal times.

(3) The square of the orbital period of a planet about the Sun is proportional to the cube of the semi-major axis of its orbit.

These laws are consequences of a set of more fundamental laws later formulated by Isaac Newton. Newton's universal **law of gravity** states that two bodies exert attractive forces on each other proportional to the product of the masses of the two bodies and inversely proportional to the square of the distance between them. The universal law of gravity may be expressed as

$$\mathbf{F} = -(Gm_1m_2/r^2)\hat{\mathbf{r}}\ , \tag{2.1}$$

where F is the force (downward on the surface of the Earth, consistent with the negative sign in this equation) on a mass m_2 due to the mass m_1 (the Earth in the case of the inhabitants of this planet). The universal constant of gravitation, G, is 6.672×10^{-11} N m^2 kg^{-2}, and, when combined with the mass of the Earth and its radius, we find that the acceleration of gravity, g, on the surface of the Earth is 9.80 m s^{-2}.

Newton's laws of motion combined with the universal law of gravitation in turn explain Kepler's laws of planetary motion. These laws in their simplest forms are:

(1) A body either remains at rest or moves at a constant velocity unless acted upon by an outside force.
(2) The acceleration of a body is directly proportional to the force applied and inversely proportional to its mass.
(3) For every action, there is an equal but opposite reaction.

While inspired by the motion of the planets that are macroscopically observable, these same laws govern the microscopic world of gases. Useful quantities for studying the motion of both planets and the molecules in gases is energy, both **kinetic energy** and **potential energy**. The kinetic energy of a particle of mass m and velocity v is

$$\text{KE} = mv^2/2. \tag{2.2}$$

The gravitational potential energy in a uniform gravitational field with acceleration g is

$$\text{PE} = mgh, \tag{2.3}$$

where h is the distance moved in the gravitational field.

If gravity is the only force acting, then the sum of the gravitational potential energy, which is a negative number, and the kinetic energy is constant. Thus, when we drop an object of mass m in a gravitational field of acceleration g, its terminal velocity after falling distance h is $(2gh)^{1/2}$ when its initial velocity was zero. When a body orbits about a central mass, it will exchange potential and kinetic energy if it is in an elliptical orbit. If a particle is to move to a new elliptical orbit farther away from the central body, work must be done on the particle.

A particular application of Eq. (2.1) that is important to all of us is the balance between the Sun's gravitational field and the centrifugal force associated with the orbit of the Earth around the Sun. The **centrifugal force** is $m\omega^2 r$, where ω is the orbital frequency in radians per second and r is the distance in meters (1.496×10^{11} m from the Sun to the Earth). Knowing that the Sun has a mass of 1.98×10^{30} kg, we find that the orbital period of the Earth around the Sun is one year. The gravitational field of the Earth is critical to preserving our atmosphere over the age of the solar system. This atmosphere in turn supports life on the planet and protects it.

2.1.2 Our atmospheric shield

The Earth's upper atmosphere and ionosphere provide the boundary and the boundary conditions for the transition into space. The upper atmosphere is our shield against the harmful extreme-ultraviolet radiation and x-rays from the Sun. The ionization produced when these energetic photons are absorbed creates the **ionosphere**, which in turn guides radio waves over the horizon so that we can communicate around the planet even without cables or satellites. As we proceed upwards into the upper atmosphere and ionosphere, they change dramatically. At low altitudes, we have a cold collisional gas, fully mixed and chemically uniform. At higher altitudes, the density falls until the gas becomes collisionless, the species become decoupled, and the atmosphere is diffusive, not convective. The properties of the ionosphere are similarly affected by the decreasing collision frequency. At lowest altitudes in the ionosphere, the collisions produce classical resistivity and currents flow according to the direction and strength of the electric field. When the ionosphere is only weakly collisional, application of an electric field drives a current at an angle to the magnetic field direction, and at greater heights there is no current at all, only plasma drifts.

In order to understand this behavior even in simple atmospheres, we need to understand the physics of gases. In this chapter we briefly review that physics and its application to the upper atmosphere and ionosphere of the Earth. We find that in order to understand this gaseous envelope, we need to examine both its fluid behavior parameterized by its density, bulk velocity, pressure, and temperature, and also kinetic behavior in processes such as diffusion, viscosity, and heat conduction.

2.2 CHARACTERIZING A GAS

A solid object has a specific shape and therefore a volume, and that volume has a mass. By dividing the total mass by the volume, we obtain the average density. A gas does not have a specific shape and its density can easily change as it moves, and so its density is not so much a number as a function, a function of space and time, $n(\mathbf{r}, t)$. The mean separation of particles in this volume element is $d = n^{-1/3}$. There are three different velocities that characterize a gas: its bulk velocity, $\mathbf{u}$; the velocity of individual particles, $\mathbf{v}_i$; and the random thermal velocity, $\mathbf{w}_i$, of the particles about the mean or bulk velocity, where $\mathbf{w}_i = \mathbf{v}_i - \mathbf{u}$. The gases with which we come into daily contact are collisional. One can calculate the **collision frequency** by taking a pseudo-particle with a radius of the sum of the radii of the colliding particles, moving at a mean, relative velocity to the other stationary point particles. As the pseudo-particle travels, it sweeps out a cylinder of cross section $\pi(r_1 + r_2)^2$ of length $\mathbf{w}_{1,2}\Delta t$, encountering n_2 particles per Δt. Here n_2 is the number density of the particles with which the collisions are taking place and $\mathbf{w}_{1,2}$ is the average relative velocity. We define the **collisional cross section** to be $\sigma_{1,2} = \pi(r_1 + r_2)^2$. To obtain the average relative velocity, we assume that equal numbers of particles move in each of the six coordinate directions. In this scenario, one-sixth of the particles will collide head-on, one-sixth of the particles will collide in overtaking collisions, and two-thirds of the particles will collide with a speed equal to the square root of the squared sum of the velocities. The resulting average relative velocity is

$$\overline{w}_{1,2} = \overline{w}_1[1 + (w_2/w_1)^2]^{1/2}. \tag{2.4}$$

For a gas in thermodynamic equilibrium, the mean random velocity is $\overline{w} = (8KT/\pi m)^{1/2}$, where K is the Boltzmann constant, T is the temperature, and m is the particle mass. If we define the reduced temperature to be

$$T_{1,2} = (m_1 T_1 + m_2 T_2)/(m_1 + m_2) \tag{2.5}$$

and the reduced mass to be

$$m_{1,2} = m_1 m_2/(m_1 + m_2), \tag{2.6}$$

we find that the collision frequency becomes

$$\nu_{1,2} = \sigma_{1,2} n_2 (8KT_{1,2})/(\pi m_{1,2})^{1/2}. \tag{2.7}$$

The mean free path is then

$$l_{1,2} = [\sigma_{1,2} n_2 (1 + m_1 T_2/m_2 T_1)]^{-1}. \tag{2.8}$$

If the gas particles are colliding with their own species, then

$$\sigma_{1,1} = 4\pi r^2, \tag{2.9}$$

$$\nu_{1,1} = 4\pi^{-1/2} n \sigma_{1,1} (kT/m)^{1/2}, \tag{2.10}$$

$$l_{1,1} = 2^{-1/2}(n\sigma_{1,1})^{-1}. \tag{2.11}$$

In a gas, scalar quantities, such as mass or energy, may be carried across a reference surface normal to that flow. The amount of the quantity carried is called the flux, Φ. For example, if the flux of particles is needed and the number density is n and the bulk velocity is u_x, the particle flux $\Phi_x^n = nu_x$. In three dimensions, $\mathbf{\Phi}_n = n\mathbf{u}$. The momentum flux along x is $\Phi_x^{\rho u} = \rho u_x^2$, where $\rho = mn$, the mass density. If there is a gradient in a flux, then the quantity increases within a volume. The one-dimensional equation describing this for density is

$$\frac{\partial n}{\partial t} = \frac{-\partial(nu_x)}{\partial x}. \tag{2.12a}$$

In general,

$$\frac{\partial n}{\partial t} = -\nabla \cdot (n\mathbf{u}). \tag{2.12b}$$

This is called the **continuity equation.**

The random motions of a gas exert a pressure on the walls of any enclosure surrounding the gas. To understand this pressure, we examine the momentum transport provided by these thermal motions. This pressure is exerted only by motions along the normal to the surface. If we define $\Phi_i^{\rho u}$ as the flux of momentum in the i direction delivered by the random motion of the particles, the pressure is given by

$$P = (\Phi_x^{\rho u} + \Phi_y^{\rho u} + \Phi_z^{\rho u})/3. \tag{2.13}$$

If we use our simplistic model above where all particles have the same speed and let one-sixth of the particles move in each of the six directions $\pm x, \pm y, \pm z$, we can derive the transport of momentum across any of those six directions. If we divide a box of gas with a central plane as shown in Figure 2.1, the momentum flux from the left side to the right side is $nm\overline{w}^2/6$ and from the right side to the left side is $-nm\overline{w}^2/6$, where the minus sign denotes the direction (right to left) of the momentum flow. The net momentum flux is the difference of these two fluxes, or $nm\overline{w}^2/3$. While this illustration uses a simplistic model of thermal motion, the same result is obtained for a more realistic maxwellian distribution. We note that pressure is a measure of the energy density in the random motions of the gas. We also note that had we examined the change in momentum flux at a reflecting wall, we would obtain the same answer.

A gas also has a dynamic pressure associated with the bulk flow. For a flow with number density n of particles with mass m and flow velocity u, the dynamic pressure is mnu^2 or ρu^2. We note that the internal or thermal pressure of a gas is defined in the reference frame of the gas where the bulk velocity is zero. In contrast, dynamic pressure requires a bulk velocity to be established by either the gas moving versus a stationary obstacle or an object moving through the gas.

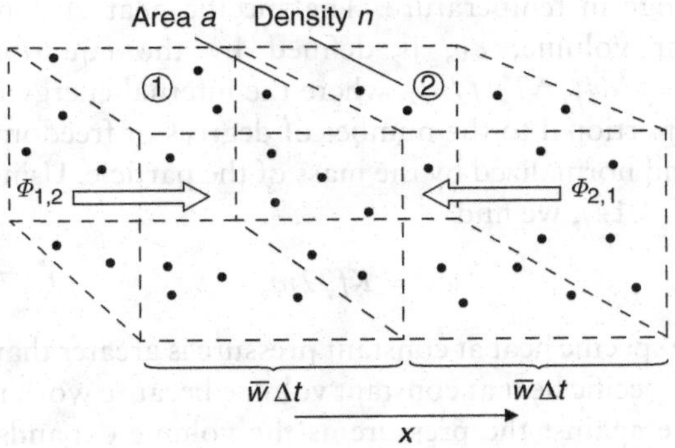

FIGURE 2.1. Pressure exerted by a gas. The momentum flux across the central plane of area A in direction x is $nm\overline{w}^2/6$ from the particles to the left and $-nm\overline{w}^2/6$ from the right. The pressure at the central plane is $nm\overline{w}^2/3$.

The temperature of a gas is defined as the amount of heat stored per degree of freedom, where a degree of freedom can be, in the case of an atom, any one of the three orthogonal directions. In the case of a molecule, there can be rotational and vibrational degrees of freedom as well. The collisions in the gas spread the energy equally in each of these degrees of freedom. We define temperature via the following equation:

$$KT/2 = \overline{U}_f, \tag{2.14}$$

where $\overline{U}_f$ is the average heat or internal energy per degree of freedom and K is the Boltzmann constant. In an atomic gas with three degrees of freedom, the temperature will be

$$T = m\overline{w}^2/3K, \tag{2.15}$$

where $m\overline{w}^2/2$ is the average random translational energy of each particle. We note that in some situations, such as in a magnetized plasma, to be discussed later, the temperatures in two orthogonal directions may be quite different. Finally, we note the relationship between pressure, number density, and temperature:

$$P = nKT. \tag{2.16}$$

This equation is called the ideal gas law or the equation of state of an ideal gas.

The heat capacity of a gas at constant volume is the change in internal energy accompanying a change in temperature. The specific heat at constant volume, c_V, is defined by the equation $c_V = (\Delta \overline{U}_f / \Delta T)(f/m)$, where the internal energy is proportional to the number of degrees of freedom, f, and normalized by the mass of the particle. Using Eq. (2.14), we find

$$c_V = Kf/2m. \tag{2.17}$$

The specific heat at constant pressure is greater than the specific heat at constant volume because work is done against the pressure as the volume expands. Using the ideal gas law, we can show that the work per unit mass is

$$\Delta W = -K/m\Delta T \tag{2.18}$$

and

$$c_P \Delta T = c_V \Delta T + K/m\Delta T. \tag{2.19}$$

Thus

$$c_P = (K/m)(f/2 + 1), \tag{2.20}$$

and the ratio of specific heats is

$$\gamma = c_P/c_V = (f + 2)/f. \tag{2.21}$$

The index γ is called the adiabatic exponent and also the polytropic index. It is useful for determining the change in a gas when that change occurs adiabatically at the expense of its own internal energy. This expansion thus requires that the work done dW (= $-P\,dV$) equals the change in internal energy dU (= $NfK\,dT/2$).

Since $N = nV$ and $P = nKT$, we may write

$$dT/T = (-2/f)(dV/V). \tag{2.22}$$

Integration of this equation yields the **adiabatic relation**

$$T = T_0\,(V/V_0)^{-2/f} \tag{2.23}$$

or

$$TV^{2/f} = \text{constant}. \tag{2.24a}$$

Using the ideal gas law, we may rewrite this as

$$(PV^{\gamma})/NK = \text{constant}, \tag{2.24b}$$

where $\gamma = (f + 2)/f$.

Since the number of particles in the expansion is constant, we may then write, using Eq. (2.21),

$$PV^{\gamma} = \text{constant}. \tag{2.24c}$$

And since $\rho \propto V^{-1}$,

$$P/\rho^{\gamma} = \text{constant}. \tag{2.24d}$$

These are all forms of the adiabatic relation.

2.3 THE DISTRIBUTION FUNCTION

Thus far, we have been treating gases with parameters such as temperature, pressure, number density, and bulk velocity with sometimes overly simplifying assumptions. In most situations, we cannot assume that all particles have an identical thermal velocity, for example. It is important that they have a continuous distribution of velocities with some low velocities and some high. It is also important to allow particles to have properties that vary with position. For example, density and temperature vary in the atmosphere with altitude. We need a formulation that allows this. This tool is the **six-dimensional phase space density,** $f(\mathbf{r}, \mathbf{v}, t)$, sketched in Figure 2.2. The six coordinates consist of x, y, z locations in configuration space

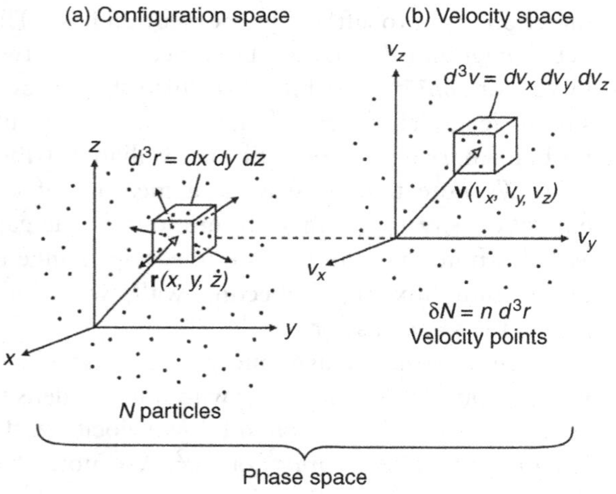

FIGURE 2.2. Six-dimensional phase space that allows the density of particles to be described as a function of where they are (a) and what velocity they have (b). (After Prölss, 2003.)

and v_x, v_y, v_z locations in velocity space. A differential volume in phase space is expressed by $d\mathbf{r}d\mathbf{v} = dx\,dy\,dz\,dv_x\,dv_y\,dv_z$ at some location $(\mathbf{r}, \mathbf{v})$ at which the number of particles in the differential phase space volume is $f(\mathbf{r}, \mathbf{v}, \mathbf{t})d\mathbf{r}\,d\mathbf{v}$. If there are several species present, then it may be necessary to follow the distribution function of each one separately, i.e., $f_s(\mathbf{r}, \mathbf{v}, t)$.

To determine the number density at any point, we integrate the phase space density over all velocities:

$$n_s(\mathbf{r}, t) = \int d\mathbf{v} f_s(\mathbf{r}, \mathbf{v}, t). \qquad (2.25)$$

Here, $n_s(\mathbf{r},t)$ is the number density of species s, and it equals the mass density, $\rho_s(\mathbf{r}, \mathbf{v}, t)$, divided by the particle mass. The number density is the zero-order moment of the distribution function. Other parameters that characterize the average properties of the gas are also moments of the distribution function. The bulk velocity, for example, is a first-order moment of the distribution function

$$\mathbf{u}_s(\mathbf{r}, t) = \int d\mathbf{v}\mathbf{v} f_s(\mathbf{r}, \mathbf{v}, t) / \int d\mathbf{v} f_s(\mathbf{r}, \mathbf{v}, t), \qquad (2.26)$$

where we normalize by the gas density. To find the average thermal or kinetic energy for species s due to the random motions about the average or bulk velocity, we find the moment

$$\begin{aligned} \frac{1}{2} m_s w^2 &= \langle \frac{1}{2} m_s (\mathbf{v} - \mathbf{u}_s)^2 \rangle \\ &= \int d\mathbf{v} \frac{1}{2} \mathbf{m}_s\,(\mathbf{v} - \mathbf{u}_s)^2 f_s(\mathbf{r}, \mathbf{v}, t) / d\mathbf{v} f_s(\mathbf{r}, \mathbf{v}, t). \end{aligned} \qquad (2.27)$$

We recall that the partial pressure of species s is related to its average random energy by

$$P_s/n_s = (2/f)\langle \tfrac{1}{2} m_s w^2 \rangle. \qquad (2.28)$$

In SI units, pressure is measured in pascals (Pa), with one pascal equal to one newton per square meter (N m^{-2}).

For systems in equilibrium, the phase space distribution is the maxwellian distribution. It is given by

$$f_s(\mathbf{r}, \mathbf{v}) = n_s (m_s/2\pi KT)^{3/2} \exp[-m_s(\mathbf{v} - \mathbf{u}_s)^2/2KT_s]. \qquad (2.29)$$

We note that this equilibrium distribution function is independent of time. If we are interested solely in the speed of the particles independent of direction, when the bulk velocity is zero, we can integrate over all directions:

$$\begin{aligned} g_s(\mathbf{r}, \mathbf{v})dv &= \int f_s(\mathbf{r}, \mathbf{v}) d\Omega_v v^2 dv \\ &= 4\pi f_s(\mathbf{r}, v) v^2 dv. \end{aligned} \qquad (2.30)$$

The distribution function $g(\mathbf{r}, \mathbf{v})dv$ is the number of particles per unit speed between v and $v + dv$. This quantity varies quadratically with v for *small* v and falls off exponentially with v for *large* v. This function maximizes at zero velocity and falls off to larger negative and positive values. Figure 2.3(b) shows the speed distribution that maximizes at the thermal speed $(2KT_s/m_s)^{1/2}$, where we recall that the average energy of the random motion is related to the temperature by $\langle \frac{1}{2} m_s (\mathbf{v} - \mathbf{u})^2 \rangle = fKT_s/2$. Figure 2.3(c) shows a maxwellian distribution along the direction x with a finite bulk velocity in this same direction.

2.4 VERTICAL DISTRIBUTION OF THE ATMOSPHERE

The Earth and planets are warmed by the Sun largely through visible radiation that is absorbed by the planet and then reradiated in the infrared portion of the spectrum. The radiating layer may be the surface or a higher layer in the atmosphere. Radiation from a higher layer of atmosphere occurs when the lower atmospheric layers are opaque to the IR radiation. The "greenhouse gases" contribute to closing windows to space that the surface of the Earth (or other planet) might have. The equilibrium radiating temperature at which the planet balances the incoming and outgoing radiation (visible coming in and IR going out) will rise to some high level in the atmosphere. This may seem to be a benign result, but, since the equation of state, $P\rho^{-\gamma}$ = constant, links T and n, the dense lower atmospheric temperature rises. The Earth's lower

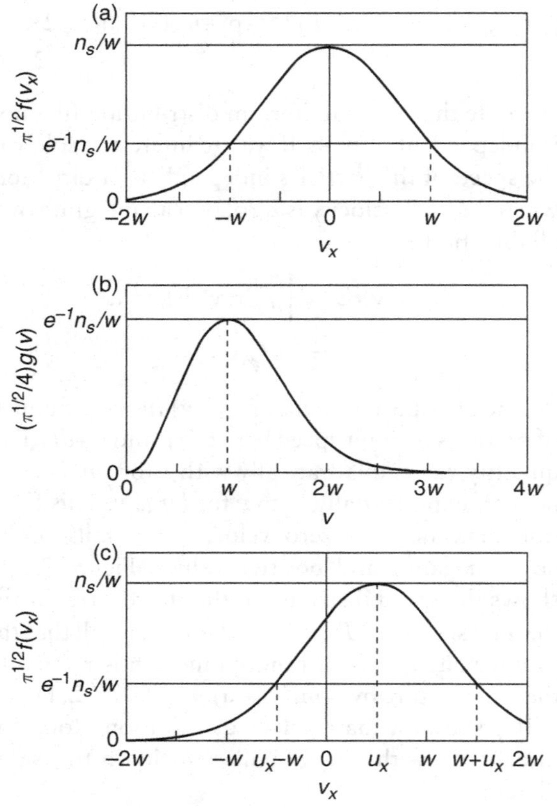

FIGURE 2.3. (a) The height of a maxwellian distribution as a function of a particular velocity component, v_x, where the bulk velocity is zero. The dependence upon the v_y- and v_z-components has been removed by integrating over v_y and v_z. (b) The height of the speed distribution function defined in Eq. (2.30) as a function of the speed v normalized by the thermal speed. (c) The height of the distribution function versus the v_x-component when there is a bulk velocity u_x to the right.

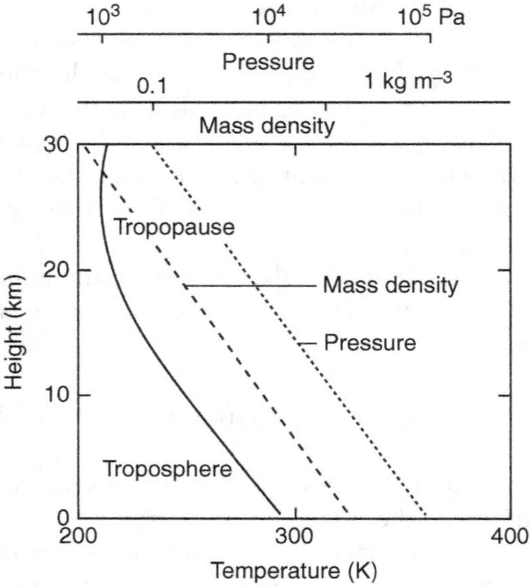

FIGURE 2.4. Altitude profiles of temperature, pressure, and mass density in the troposphere. Note that pressure and density are on a logarithmic scale.

atmosphere has not yet become entirely opaque to IR radiation, and the Earth's surface is still largely cooled by radiation into space. Thus, as a first approximation, it is safe to assume that the conversion from visible input to IR output occurs at the surface of the Earth. We shall see that the upper atmosphere is a complex region with absorption and radiation at different levels. Moreover, it is not just the thermal balance of the surface of the Earth that concerns us, but also the lower flux and more energetic photons at shorter wavelengths in the ultraviolet, far-ultraviolet, extreme-ultraviolet, and x-ray regions that are sufficiently energetic to damage living organisms. Thus we must be careful to continue to block certain emissions and not to block others. Ozone protects us from the solar UV radiation, and any gas that would destroy the ozone in the stratosphere where UV is absorbed is undesirable.

The atmospheric envelope is surprisingly thin on Earth. Its highest mountains extend to altitudes at which it is difficult for humans to breathe. We illustrate this in Figure 2.4, which shows how the density, pressure, and temperature decline in the first 30 km above the surface of the Earth. The pressure and density scales are logarithmic, while the temperature scale is linear. The temperature falls only to about 215 K with altitude before it begins to increase, but the pressure and density decrease over an order of magnitude in the first 30 km. The reason for the fall-off is that the force of gravity on the atmosphere must be balanced by the pressure gradient force in the atmosphere below acting upwards to support its weight. In mathematical terms,

$$\Delta P/\Delta h = -\rho(h)g(h), \qquad (2.31)$$

where ΔP is the change of pressure in an interval of height Δh, and $\rho(h)$ and $g(h)$ are the mass density of the gas and the acceleration of gravity as functions of height, h.

We can express the pressure differential in terms of the mass of an atom and the acceleration of gravity if both are constant,

$$dP = -mgn(h)dh, \qquad (2.32a)$$

where $n(h)$ is the number density as a function of height.

If T is constant, we may further write

$$dn/n = -(mg/KT)dh, \qquad (2.32b)$$

where K is the Boltzmann constant.

We can integrate this expression to obtain

$$n = n_0\exp(-h/H), \qquad (2.33)$$

where $H = KT/mg$ and is termed the scale height.

Since our equation of state ($P\rho^{-\gamma}$ = constant) links T and n through $T = \text{constant} \times n^{\gamma-1}$, the temperature falls with altitude as n decreases. Ultimately, above about 25 km altitude, as shown in Figure 2.4, there is heating due to the absorption of UV radiation and the temperature stops dropping.

We note that the atmosphere transports heat through convection as well as radiation and also somewhat through conduction. The most obvious manifestation of convection is seen in cumulus clouds, especially storm clouds, which tower above the landscape. A parcel of gas will rise until it reaches the level at which it has the same density as the surrounding gas, i.e., it has neutral buoyancy. It will also expand (adiabatically) until it is in pressure equilibrium with its surroundings. The decrease in temperature with altitude (the temperature lapse rate) can be faster than the adiabatic lapse rate of the rising air parcel. The rising air parcel stays less dense (and therefore more buoyant) than the surrounding air and the convection will continue upward. This situation is termed unstable to convection. Stable air would suppress convection.

If we examine the temperature above the tropopause, we obtain the variation seen in Figure 2.5.

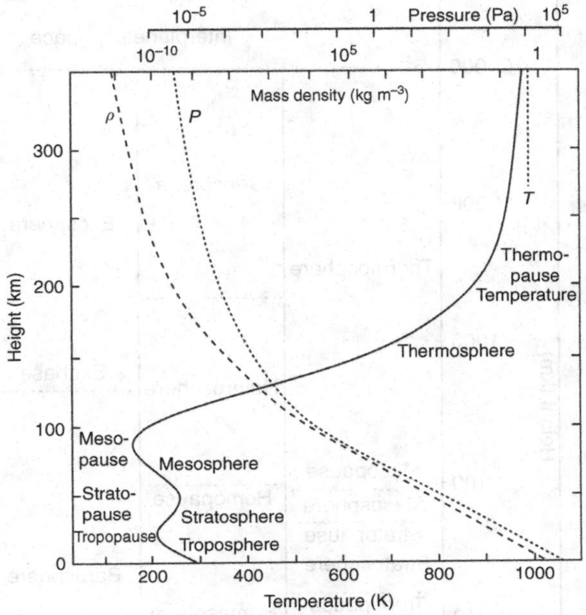

FIGURE 2.5. Altitude profiles of temperature, pressure, and mass density from the surface of the Earth to the thermosphere. Note that the pressure and density scales are logarithmic.

First, we reach the **stratosphere**, bounded by the **tropopause** below and the **stratopause** above. The stratopause, a local temperature maximum, is a local heating maximum. The temperature drops in the mesosphere until about 90 km altitude where the temperature minimizes and then climbs rapidly to fairly high values at or above 1000 K. The region is called the thermosphere. It exists on planets that have magnetospheres because the processes in the magnetosphere can accelerate the ions and electrons in the ionosphere and they in turn exchange heat with the neutrals. We emphasize again that the temperature and density are displayed on a logarithmic scale and have decreased by ten orders of magnitude by 300 km altitude, while the temperature is shown on a linear scale and has increased by only about a factor of three.

It is most common to call regions in the upper atmosphere by their temperature-associated names: troposphere, stratosphere, mesosphere,

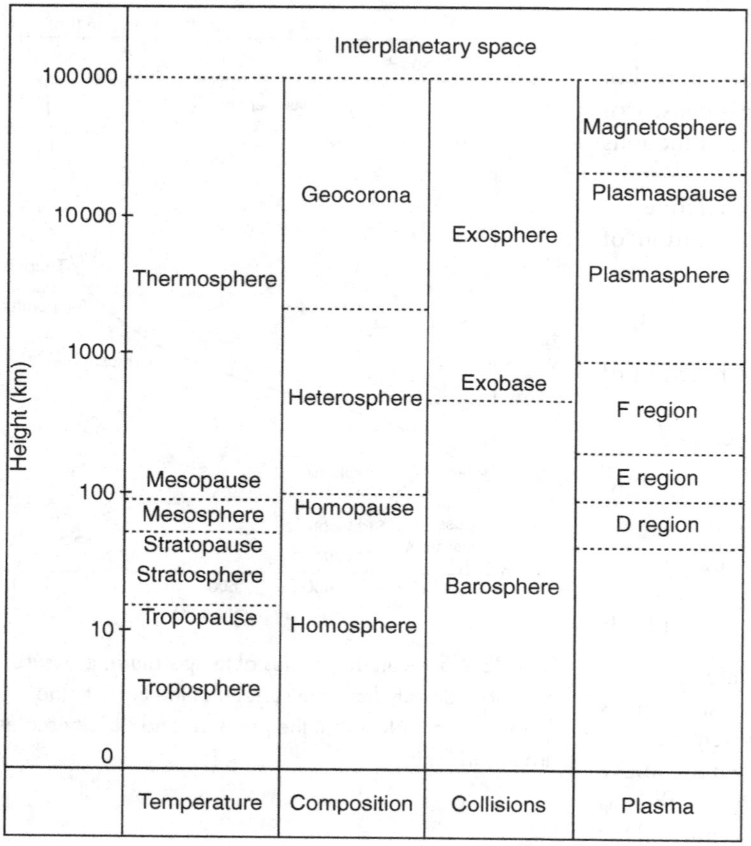

FIGURE 2.6. Atmospheric nomenclature based on four different properties.

and thermosphere, and their extrema, the tropopause, stratopause, and mesopause. However, there are other characteristics of the upper atmosphere by which we may classify these regions and they do not always have the same boundaries. These terms are given schematically in Figure 2.6. For example, if we choose to label the composition of the atmosphere, the convection of the atmosphere and eddy diffusion keep the composition constant up to about 100 km (the **homopause**). This boundary is also called the **turbopause**. Above this level, mixing ceases and gravitational forces establish scale heights for each species mass, causing the composition to change with altitude. This is illustrated in Figure 2.7. Eventually the heavy atoms and molecules have dropped in density sufficiently that a minor constituent in the lower atmosphere, H, becomes dominant. This is the **geocorona** or hydrogen-sphere and, as shown in Figure 2.8, it begins at about 2000 km altitude.

We can also classify the atmospheric regions by their collision frequency. If the atoms and molecules collide frequently enough to transmit pressure to the other constituents, then the term barosphere is appropriate. However, above about 500 km, particles seldom collide and follow trajectories that rise and fall back to the Earth without encountering other particles. We call this the **exosphere**; the bottom of the exosphere is the **exobase**.

The ionic content of the lower atmosphere is insignificant, but, above about 85 km altitude, the number of ions and electrons, even though always smaller than the number of neutrals, becomes electrically significant. For historical reasons, the lowest layer of the ionosphere has been called the **D region**. This is deep inside the barosphere where collisions are frequent, but energetic photons and energetic charged particles can penetrate this region and create an electron–ion plasma that can interact with electromagnetic waves. The

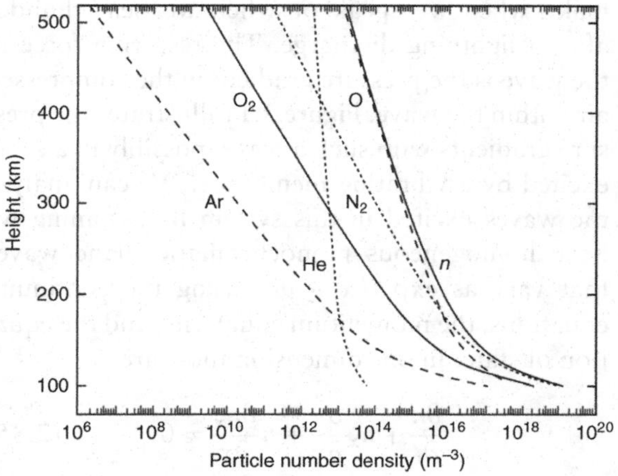

FIGURE 2.7. Variation with altitude of the number density of the five most abundant atmospheric species.

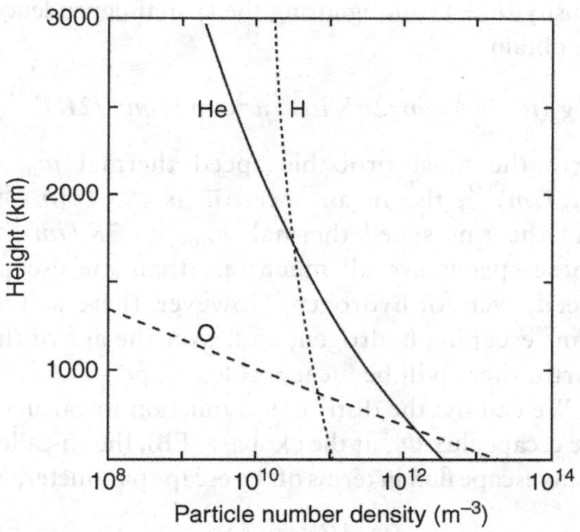

FIGURE 2.8. Altitude variation of the composition of the thermosphere.

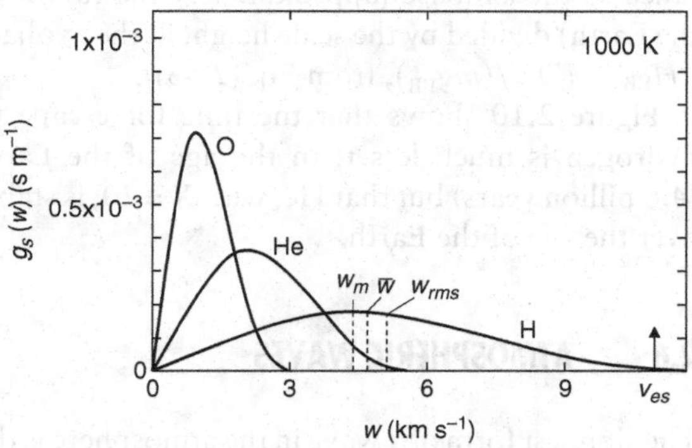

FIGURE 2.9. Thermal distributions of O, He, and H at 1000 K versus speed; w_m is the most probable thermal speed, $\overline{w}$ is the average thermal speed, w_{rms} is the rms thermal speed (the square root of the average squared thermal speed), and v_{es} is the terrestrial escape speed at the top of the exosphere. (After Prölss, 2003.)

waves lose energy to the electrons which in turn collide with neutrals, losing that energy, and thus dissipating the energy in the electromagnetic wave. Hence a strong D region is bad for electromagnetic communication. At higher altitudes, we reach the E **layer**, where the electron density becomes much greater and the collision frequency much less. These electrons can reflect the electromagnetic waves with little loss, enabling worldwide (over-the-horizon) communication. Here, high densities are beneficial to communications. Above the E layer is the F **layer**, a region of comparable density but even lower collision rate. Above that is the **plasmasphere**, which is basically the top of the ionosphere and is a big reservoir of plasma that fills during the day and drains at night. At these altitudes for charged particles, the magnetic field controls their motion. Its dipolar nature (discussed in Chapter 7) confines the plasmasphere to a region that extends in latitude almost to the auroral zone and in altitude above the equator to about four Earth radii. Because of the magnetic control exerted by the dipole field, the plasmapause is more a latitudinal boundary than an altitude boundary. As one moves upward along a magnetic-field line at low latitudes (~45°), one would not reach the plasmapause. As one moves upward along a magnetic-field line in the auroral zone, one would be always beyond the plasmapause.

2.5 ATMOSPHERIC ESCAPE

The **escape velocity** from the surface of the Earth is 11 km s^{-1}. The thermal speed of the atoms and molecules in the Earth's atmosphere is generally much less than this speed, as shown in Figure 2.9. The equation of this distribution of speed is given in Eq. (2.30). Rewriting the distribution function for a normalized density (n = 1) and ignoring the spatial dependence, we obtain

$$g_s(w) = 4\pi(m/2\pi KT)^{3/2}w^2\exp(-mv^2/2KT).$$

Here the most probable speed thermal w_m is $(2KT/m)^{1/2}$, the mean speed $\overline{w}$ is $(8KT/\pi m)^{1/2}$, and the rms speed thermal w_{rms} is $(3KT/m)^{1/2}$. These speeds are all much less than the escape speed, even for hydrogen. However, there will be some escaping hydrogen, and, over the age of the Earth, there will be measurable escape.

We can use the distribution function to calculate the escape flux Φ_{es} at the exobase (EB), the so-called **Jeans escape flux** in terms of the escape parameter, X,

$$\Phi_{es} = n_{EB}\ w_m/(2\pi^{1/2})\ (1+X)\exp(-X), \quad (2.34)$$

where w_m is the most probable thermal speed and $X = (w_{es}/w_m)^2_{EB}$, which equals the geocentric distance to the exobase (approximately the radius of the Earth) divided by the scale height at the exobase ($H_{EB} = KT_\infty/(mg_{EB})$, from Eq. (2.33)).

Figure 2.10 shows that the time for escape of hydrogen is much less than the age of the Earth (4.6 billion years) but that He, with X = 20, is stable over the age of the Earth.

2.6 ATMOSPHERIC WAVES

The simplest form of a wave in the atmosphere is the **acoustic wave.** This wave carries the sounds we make when we speak or when we hear thunder after a lightning discharge. The restoring force in the wave is the pressure gradient in the compressed air within the wave. Figure 2.11 illustrates the pressure gradients with such a wave, possibly in a tube, excited by a vibrating membrane. We can analyze the waves excited in this system by assuming we have homogeneous monochromatic plane waves that vary as $\exp i(kx - \omega t)$ using the continuity equations, the momentum equations, and the equation of state. In one dimension these are

$$\frac{\partial n}{\partial t} + u_x\frac{\partial n}{\partial x} + n\frac{\partial u_x}{\partial x} = 0, \quad (2.35)$$

$$\rho\frac{\partial u_x}{\partial t} + \rho u_x\frac{\partial u_x}{\partial x} + \frac{\partial P}{\partial x} = 0, \quad (2.36)$$

and

$$n \propto P^{1/\gamma}. \quad (2.37)$$

If we substitute Eq. (2.37) into Eq. (2.35), we obtain

$$\frac{\partial P}{\partial t} + u_x\frac{\partial P}{\partial x} + \gamma P\frac{\partial u_x}{\partial x} = 0. \quad (2.35a)$$

If we assume small-amplitude waves, we may write

$$P(x,t) = P_0 + P_1(x_1 t),$$
$$\rho(x_i t) = \rho_0 + \rho_1(x,t),$$
$$u_x(x,t) = u_{1x}(x,t).$$

We assume that we are working in the reference frame of the bulk flow of the gas, so u_0 = 0.

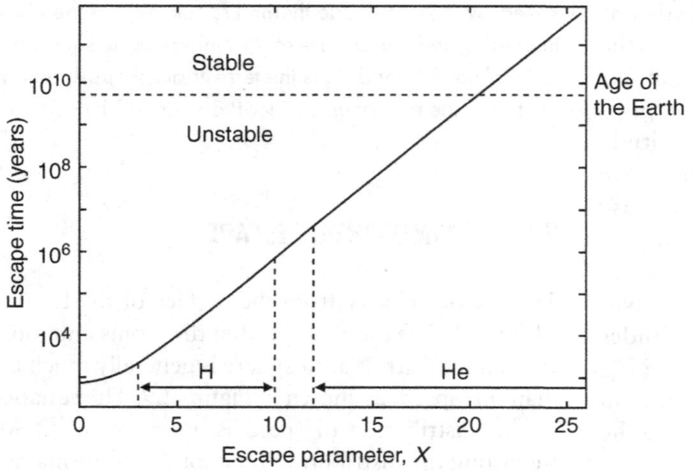

FIGURE 2.10. Escape times for hydrogen and helium as a function of the escape parameter, which in turn is a function of the thermopause temperature. (After Prölss, 2003.)

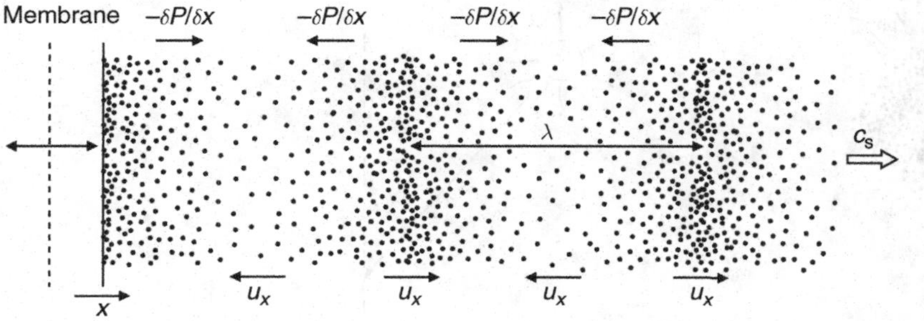

FIGURE 2.11. Distribution of density and pressure maxima in an acoustic wave excited by a membrane. Phase velocity is to the right. Gas molecules oscillate to the right and the left as the pressure gradients move through the system. The wavelength is given by λ. The speed of sound is c_s. (After Prölss, 2003.)

Substituting these small-amplitude perturbation forms for P, ρ, and u in Eqs. (2.35a) and (2.36) and discarding the smallest terms consisting of products of two small terms, we obtain

$$\frac{\partial u_{1x}}{\partial t} + \frac{1}{\rho_0}\frac{\partial P_1}{\partial x} = 0, \tag{2.36a}$$

$$\frac{\partial P_1}{\partial t} + \gamma P_0 \frac{\partial u_{1x}}{\partial x} = 0. \tag{2.35b}$$

Assuming plane monochromatic waves, we may substitute for derivatives

$$\frac{\partial}{\partial t} \rightarrow -i\omega \text{ and } \frac{\partial}{\partial x} \rightarrow ik,$$

and Eqs. (2.36a) and (2.35b) become

$$-i\omega u_{1x} + \frac{i}{\rho_0} kP_1 = 0, \tag{2.38}$$

$$i\omega P_1 + i\gamma P_0 k u_{1x} = 0. \tag{2.39}$$

For a solution of this equation, the determinant must equal zero. Hence,

$$(i\omega)^2 - (ik)^2 \frac{P_0}{\rho_0}\gamma = 0 \tag{2.40}$$

and

$$v_{\text{phase}} = \left(\frac{\omega}{k}\right) = \left(\frac{\gamma P_0}{\rho_0}\right)^{1/2}. \tag{2.41}$$

Thus the acoustic wave has a phase velocity of $(\gamma \cdot P_0/\rho_0)^{1/2}$, or $(\gamma KT/m)$, where K is the Boltzmann constant. The velocity is called the sound speed and is usually abbreviated as c_s.

A **gravity wave** is one in which the restoring force is the action of gravity on the displaced body of a fluid such as an ocean wave. Ocean waves exist because of the abrupt change in density at the ocean's surface. In the atmosphere, the change in density is not as abrupt, but there still is a vertical density gradient that allows gravity waves to exist. If we assume as above that the density, pressure, and velocity variations are small, that the variation is adiabatic, and that there is a two-dimensional solution of the form $\exp i(\omega t - k_x x - k_z z)$, we obtain the dispersion relation

$$\omega^4 - \omega^2 c_s^2 (k_x^2 + k_z^2) + (\gamma - 1) g^2 k_x^2 + \omega^2 \gamma^2 g^2 / 4c_s^2 = 0, \tag{2.42}$$

where g is the force of gravity and c_s is the speed of sound. If $g = 0$, this is the acoustic wave.

If we substitute $\omega_a = \gamma g/2c_s$ and $\omega_B = (\gamma - 1)^{1/2} g/c_s$, we obtain

$$k_z^2 = (1 + \omega_a^2/\omega^2)\omega^2/c_s^2 - k_x^2(1 - \omega_B^2/\omega^2). \tag{2.43}$$

If $\omega^2/k_z^2 \ll c_s^2$,

$$k_z^2 = k_x^2(\omega_B^2/\omega^2 - 1). \tag{2.44}$$

This is a gravity wave that propagates only if k_x, k_z are real and positive so that $\omega < \omega_B$. It propagates at an angle to the horizontal

$$\theta = \tan^{-1}(\omega_B^2/\omega^2 - 1). \tag{2.45}$$

For an upward phase velocity, as shown in Figure 2.12, the group velocity and energy flow must be downward. The atmospheric motions, indicated by arrows, are transverse to the phase velocity. The frequency ω_B is called the **Brunt–Väisälä frequency** and the region below this frequency is called the gravity range. The region above ω_a, the acoustic frequency, is called the acoustic range.

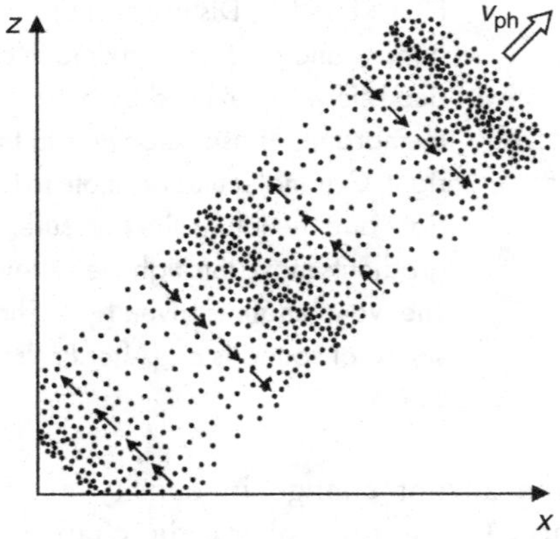

FIGURE 2.12. Density and pressure maximum in a gravity wave. Phase velocity is upward and to the right. Group velocity is downward to the right. The bulk velocity of the gas is alternatively in the direction given by the arrows as the phase fronts move upward. (After Prölss, 2003.)

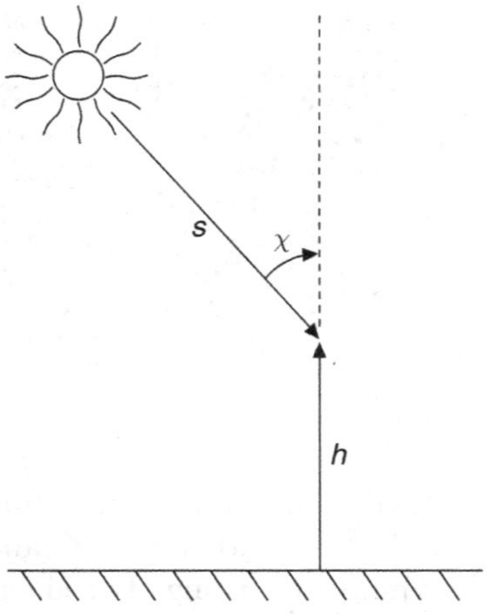

FIGURE 2.13. Illustration showing the line-of-sight path length s, the solar zenith angle χ, and the altitude h.

2.7 ION PRODUCTION

So far, we have been concerned only about the behavior of neutral gases, but, as we continue to higher altitudes, we encounter the ionosphere, where the atmosphere has become partially ionized. We refer to this region as the ionosphere, even though, in fact, the neutral density is greater than the ion density.

The ingredients for a planetary ionosphere are simple: the only requirements are a neutral atmosphere and a source of ionization for the gases in that atmosphere. Sources of ionization include photons and energetic particle "precipitation." The process involving the former is referred to as **photoionization**, and the latter is often labeled **impact ionization.** The photons come primarily from the Sun. Ionizing particles can come from the galaxy (cosmic rays), the Sun, the magnetosphere, or from the ionosphere itself if a process for local ion or electron acceleration is operative. Precipitating energetic electrons produce additional ionizing photons within the atmosphere by a process known as **bremsstrahlung**, or braking radiation. The only requirement on the ionizing photons and particles is that their energies ($h\nu$ in the case of photons, and kinetic energy in the case of particles) exceed the ionization potential or binding energy of a neutral-atmosphere atomic or molecular electron. In nature, atmospheric ionization usually is attributable to a mixture of these various sources, but one often dominates. Solar photons in the "extreme-ultraviolet" (EUV) and ultraviolet (UV) wavelength range of approximately 10–100 nm typically produce at least the dayside ionospheres of most planets.

Ions are produced from the constituents of the upper atmosphere that usually obey the hydrostatic equation we developed earlier:

$$n_{\rm n} m_{\rm n} g = -\frac{dP}{dh} = -\frac{d}{dh}(n_{\rm n} K T_{\rm n}). \qquad (2.46)$$

When T is independent of temperature, we can express the altitude dependence as an exponential:

$$n_n = n_0 \exp[-(h - h_0)/H_{\rm n}], \qquad (2.47)$$

where h_0 is a reference level. However, $T_{\rm n}$ many depend on h, so that a simple exponential distribution does not always provide an accurate description.

2.7.1 Photoionization

To "model" an ionosphere produced in a given neutral atmosphere, one must first calculate the altitude profile of the rate of ion production Q. For photoionization, this entails a consideration of the **radiative transfer** of photons through the neutral gas. When handled rigorously, this is a very complex problem, because it requires detailed knowledge of all of the photon-absorption cross sections of the atmospheric constituents and some method of keeping track of absorption events that cause excitation of bound electrons as well as those that cause removal of **photoelectrons**. Fortunately, some simplifying assumptions can be made that together will allow an analytical approach to ionosphere modeling known as **Chapman theory**. Before delving into Chapman theory, it is important to appreciate that the altitude profile of ion production will have a peak at some altitude, because the rate of ionization depends on both the neutral density (which decreases with height) and the incoming solar-radiation intensity (which increases with height). In Chapman theory, the goal is to describe ion production as a function of height for the simple case in which the details of photon absorption are hidden in a radiation-absorption cross section a, and in which ion production is assumed to depend only on the amount of radiative energy absorbed. We define the following variables:

n_n = density of neutrals (per cubic meter),
h = height,
I = intensity of radiation (energy flux, electronvolts per square meter per second),
σ = photon-absorption cross section (square meters),
Q = rate of ion production (photoionization rate, electrons per cubic meter per second),
s = line-of-sight path length,
χ = zenith angle,
C = number of electrons produced in the absorber per unit energy absorbed (electrons per electronvolt).

The path length s and zenith angle χ are illustrated in Figure 2.12. The atmosphere is presumed to be exponential, planar, and horizontally stratified (assumptions that are idealizations of the real world, where atmospheres are only approximately exponential and curved, and the scale height H_n depends on χ because of the effects of global circulation and chemistry). As radiation is absorbed, its intensity decreases as

$$-\frac{dI}{ds} = \sigma n_n I. \tag{2.48}$$

Because the rate of ion production should be proportional to the rate at which radiation is absorbed, we can write

$$Q = -C\frac{dI}{ds} = C\sigma n_n I, \tag{2.49}$$

where C is the constant of proportionality (≈1 ion pair per 35 eV in air). The production rate Q reaches a peak (along s) when

$$C\sigma\left(I\frac{dn_n}{ds} + n_n\frac{dI}{ds}\right) = 0 \tag{2.50}$$

or

$$\frac{dQ}{ds} = 0. \tag{2.51}$$

But s is related to h by $ds = -dh \sec\chi$ (Figure 2.13), and because

$$\frac{1}{n_n}\frac{dn_n}{ds} = -\frac{1}{n_n}\frac{dn_n}{dh}\cos\chi = \frac{\cos\chi}{H_n} \tag{2.52}$$

for the peak or maximum of production (subscript m), the foregoing equations give

$$\sigma H_n n_m \sec\chi = 1 \tag{2.53}$$

or

$$\sigma N_{nm} = 1, \tag{2.54}$$

where N_{nm} is the integrated density $N_{nm} = \int_{\infty}^{s_m} n_n ds$ along the line of sight up to the position of the peak s_m. A useful term is the **optical depth**, which describes the attenuation of the ionizing radiation. The optical depth arises naturally in the expression for intensity at position s along the line of sight relative to the intensity at infinity. From Eq. (2.49),

$$\frac{dI}{I} = d\ln I = -\sigma n_n ds. \tag{2.55}$$

Integrating,

$$\ln\left(\frac{I(s)}{I(\infty)}\right) = -\sigma\int_0^s n_n ds = -\sigma N_{ns}$$

or

$$I(s) = I(\infty)\exp(-\sigma N_{ns}) = I(\infty)\exp(-\tau),$$

where N_{ns} is the integrated density along the line of sight.

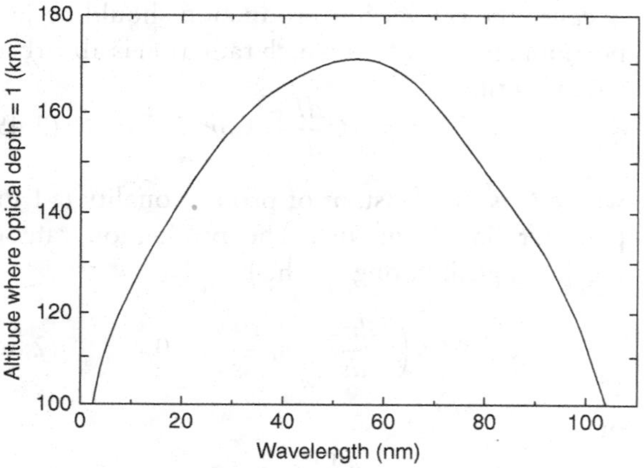

FIGURE 2.14. Photon wavelength versus the altitude in the Earth's atmosphere where the optical depth is equal to 1 (effectively, the depth of penetration of a photon). (Adapted from Rishbeth and Garriott, 1969.)

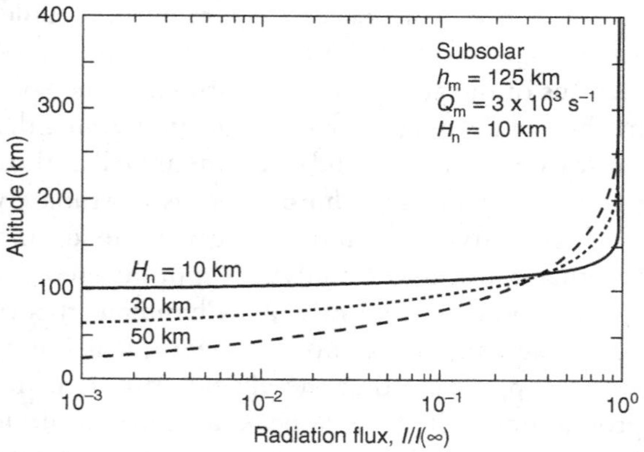

FIGURE 2.15. Radiation flux $I/I(\infty)$ as a function of altitude in a neutral atmosphere with a fixed scale height and a single frequency of illumination (monoenergetic photons). The maximum production function is 3×10^3 ions s^{-1} at 125 km for a neutral scale height of 10 km.

At the peak, $\sigma N_{ns} = \sigma N_{nm} = 1$, so the peak is the altitude where the optical depth is unity. Both the radiation intensity at the top of the atmosphere (which we call $I(\infty)$) and σ vary for different wavelengths, but it is σ that determines τ. The altitudes at which photons of various wavelengths reach unit optical depth in the Earth's atmosphere are illustrated in Figure 2.14. The photoionization process is clearly complex, and calculating the production rate as a function of altitude requires a knowledge of the ionizing flux as a function of wavelength. This function varies with the solar cycle and with location on the Sun, so, as the Sun rotates, the ionization rate in the Earth's atmosphere changes. To simplify our discussion, we assume that the ionization is accomplished at a single wavelength and that the atmosphere has a constant scale height H_n from the surface of the Earth upward. Figure 2.15 shows how the incoming ionization is absorbed as a function of height directly under the Sun ($\chi = 0$). The altitude of peak production is 125 km, which is where the intensity reaches e^{-1} of its initial value. This altitude gives the optical depth of the radiation. The largest scale height has the least steep extinction rate at this point. The smallest scale height gives the most rapid extinction rate at this point.

Now we return to the analysis of the production rate Q. The peak production rate is

$$Q_m = C\sigma n_m I_m = C\sigma(\sigma H_n \sec\chi)^{-1}(I(\infty)\exp(-1))$$
$$= CI(\infty)\cos\chi/(H_n \exp(1)). \tag{2.56}$$

Given the neutral-gas altitude profile $n_n = n_0 \exp(-h - h_0)/\, H_n$, we can determine the height of the production peak h_m by writing

$$\sigma H_n n_m \sec\chi = 1 = \sigma H_n n_0 \exp[-(h_m - h_0)/H_n]\sec\chi \tag{2.57}$$

and solving for h_m. Similarly, we can determine the dependence of the radiation intensity I on h by using the earlier expression for $I(s)$ and noting that N_{ns} is the integrated density along the line of sight:

$$I(h) = I(\infty)\exp[-\sigma n_0 H_n \sec\chi \exp\{-(h - h_0)/H_n\}]. \tag{2.58}$$

The dependence of the production rate Q on h is then given by

$$Q = C\sigma n_n I = C\sigma n_0 I(\infty)\exp[-(h - h_0)/H_n - n_0\sigma H_n \sec\chi \exp\{-(h - h_0)/H_n\}].$$

With $CI(\infty) = Q_m \exp(1)H/\cos\chi$, as before, we finally obtain

$$Q = Q_m \exp[1 + (h_m - h)/H_n - \exp\{(h_m - h)/H_n\}]. \quad (2.59)$$

This can be simplified by defining $y = (h - h_m)/H_n$, whence

$$Q = Q_m \exp[1 - y - \exp(-y)], \quad (2.60)$$

which is the Chapman production function. Note that far above the peak ($y \geq 2$), a good approximation is

$$Q \propto \exp(-y), \quad (2.61)$$

which says that because the radiation intensity is practically constant at high altitudes (not much absorption occurs), Q is proportional to neutral density. The rate Q is the rate of production of both ions and photoelectrons, because these are usually produced in pairs (most ions are singly ionized in this process). It is notable that in the expression for Q, the photon-absorption cross section does not explicitly appear. The properties of the absorber are contained in the constant C. The subsolar ($\chi = 0$) Chapman production function or ion production profile in atmospheres with three different scale heights is plotted in Figure 2.16.

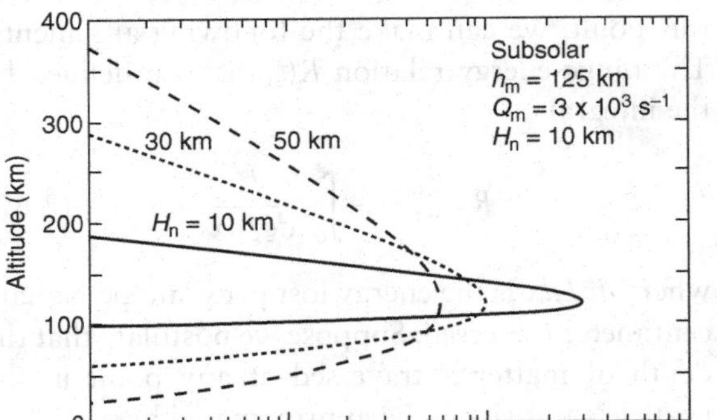

FIGURE 2.16. Chapman production function for h_m = 125 km, $\chi = 0$, and $Q_m = 3 \times 10^3$ s^{-1} for a neutral scale height 10 km. Variation in Q also given for H_n = 30 and 50 km. Note that the peak production rate Q_m varies with the scale height since the number of photons $I(\infty)$ is constant.

Our derivation of the Chapman production function has assumed a flat Earth in which the solar zenith angle has a fixed value. In practice, the solar zenith angle varies with both latitude and solar longitude. We can reference the local production rate not only to the local production maximum, but also to the production maximum at the subsolar point Q_{m0} at a height h. Doing this, we find that

$$Q = Q_{m0} \exp(1 - z - \sec x \exp(-z)),$$
$$h_m = h_{m0} + \ln(\sec\chi),$$
$$Q_m = Q_{m0} \cos\chi, \quad (2.62)$$

where

$$z = (h - h_{m0})/H. \quad (2.63)$$

Thus, in an isothermal atmosphere, the altitude of the peak production rate increases with solar zenith angle, and the rate of ionization decreases.

2.7.2 Particle-impact ionization

In many situations of interest, solar photons can be assumed to compose the dominant source of atmosphere ionization. However, on occasion, energetic (energies $\geq$ 1 keV) precipitating charged particles are more important. This can occur, for example, during the night at high magnetic latitudes in the case of planets with dipolar magnetic fields, or on satellites with atmospheres submerged in planetary magnetospheres, such as Io and Titan. The altitude profiles of ion production from particle impacts are derived from considerations different from those for photoionization, except that the reasoning related to the production of a **peak** is the same. In this case, it is the particle energy flux, rather than the photon flux, that is attenuated with decreasing altitude as the atmosphere density increases. The transport of charged particles and the loss of their energy to the atmosphere are unlike those processes for photons, because a particle gradually loses its energy in the excitation or removal of bound electrons from many particles by Coulomb collisions as it travels, whereas a photon is absorbed in a single event. To complicate matters further, a **primary particle** can produce **secondary electrons** that are energetic enough themselves to cause impact ionization, and both primary and secondary electrons can lose additional energy by radiating bremsstrahlung photons when their paths are deflected by the Coulomb collisions. In the case of ions, charge exchange may occur, in which case the energetic ion becomes an energetic neutral that distributes its energy by collisional processes other than Coulomb collisions. To deal with this problem, investigators may resort to such rigorous alternatives as Monte Carlo calculations. However, as for photoionization, there are some simpler ways to estimate the effects of energetic particle absorption.

One simplified approach involves the use of an empirically determined function $R(\xi_0)$ called the **range-energy relation.** The range-energy relation gives the depth of penetration in a particular medium as a function of the incident-particle energy ξ_0. Because the important quantity is the amount of matter traversed by the particle, rather than the particle path length (which depends on the density distribution of the matter), the range is typically expressed in units of grams per square centimeter, instead of centimeters or meters. The relationship between this equivalent distance x and the matter density n_n along the path of the particle s is given by

$$x = \int_0^{\eta} n_n(s)ds,$$

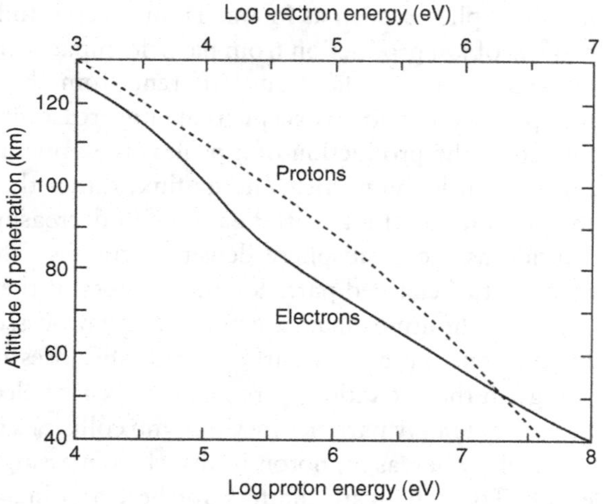

FIGURE 2.17. Penetration altitudes for electrons and protons in the Earth's atmosphere versus their incident energies.

where η is the point of interest along s. Similarly, for vertical incidence, x is related to altitude h by

$$x = \int_{\eta}^{\infty} n_n(h)dh, \tag{2.64}$$

where η is, in this case, the altitude of interest. For protons in air, for example, the range-energy relation given by Rees (1989),

$$R(\xi_0) = 5.05 \times 10^{-6}\xi_0^{0.75}\text{g cm}^{-2}, \tag{2.65}$$

is a good approximation for the incident-proton energy range between 1 keV and 100 keV. For electrons, another expression (Rees, 1989),

$$R(\xi_0) = 4.30 \times 10^{-7} + 5.36 \times 10^{-6}\xi_0^{1.67}\text{g cm}^{-2}, \tag{2.66}$$

applies for 0.2 keV $< \xi_0 <$ 50 keV. Once the range (in g cm^{-2}) is known, all that we need to do to calculate the corresponding stopping altitude is to perform the foregoing integration over the atmospheric-density distribution, that is, to find η in Eq. (2.64) for which $x = R(\xi_0)$. For example, some stopping altitudes for protons and electrons incident on the Earth's atmosphere are shown in Figure 2.17. The range-energy relation tells us about the depth to which the particle penetrates, but it does not give us the altitude distribution of its energy loss, which is what we require for the ionization-rate profile. At this point, we can make the following arguments. The range-energy relation $R(\xi_0)$ is also defined by the integral

$$R(\xi_0) = -\int_0^{\xi_0} \frac{d\xi}{d\xi/dx}, \tag{2.67}$$

where $d\xi/dx$ is the energy lost per gram per square centimeter traversed. Suppose we postulate that the depth of matter x traversed at any point in the particle's transit can be approximated by

$$x = -\int_{\xi_{loc}}^{\xi_0} \frac{d\xi}{d\xi/dx} = R(\xi_0) - R(\xi_{loc}), \tag{2.68}$$

where ξ_{loc} is the energy at x. Because $R(\xi_0)$ has the general functional form

$$R(\xi_0) = A_1 + A\xi_0^{\gamma}, \tag{2.69}$$

where A_1 and A are constants, this equation for x then gives

$$\xi_{loc} = [(A\xi_0^{\gamma} - x)/A]^{1/\gamma}. \tag{2.70}$$

It follows that the energy deposited at a given x is

$$\frac{d\xi_{loc}}{dx} = -\frac{\xi_{loc}^{1-\gamma}}{A\gamma}. \tag{2.71}$$

The altitude profile of $d\xi_{loc}/dx$ times the local-atmosphere mass density $\rho(h)$ gives the energy deposition in electronvolt per meter for the particle at x. The curve of $\rho(h)\, d\xi_{loc}/dx$ versus h is then the required profile. Some examples obtained with the foregoing expressions for $R(\xi_0)$in air, assuming the density profile for the Earth's atmosphere, are shown in Figure 2.18. Because the incident particles have a distribution of energies described by a flux spectrum $J(\xi_0)$ (in particles cm^{-2} s), a total energy-deposition profile can be built up by weighting the profiles for the single particles by $J(\xi_0)$, giving the energy deposition in electronvolt seconds per cubic centimeter.

Finally, for a particular gas mixture, one uses an empirically determined value for the energy required to produce an ion pair. As mentioned earlier, for air, approximately 35 eV will produce an ion pair. Division of the energy-deposition profile by this constant gives the ion production profile (in ions produced per cubic centimeter per second versus altitude) for the incident-particle flux described by $J(\xi_0)$. The first-generation secondary-electron production profile is the same (an electron avalanche can result if the secondary electrons are themselves capable of additional ionizing impacts). One part of the energy deposition due to particle precipitation that has been neglected in the preceding discussion is peculiar to electrons, be they primary or secondary. Because of the small mass of electrons relative to that of the target particles, electrons traveling through a gas "scatter" much more than ions. Their direction of motion can be significantly altered by a Coulomb collision. This causes radiation, because accelerating electric charges produce electromagnetic waves.

In the case of the precipitating electron energies of interest here, this bremsstrahlung or braking radiation tends to be in the x-ray range, at energies capable of additional photoionization of the gas. Thus, strictly speaking, one must deal with the complicated problem of keeping track of both the radiation losses of the precipitating electrons and the energy lost to the bound electrons, and then carry out a radiative transport calculation for the bremsstrahlung photons. The latter is not as straightforward as that for solar photons, because the bremsstrahlung photons are produced at different points within the same medium that is absorbing them. The bremsstrahlung transport calculation provides an additional ion production profile that must be added to that for the ionization loss. Fortunately, this non-solar source of photoionization is usually

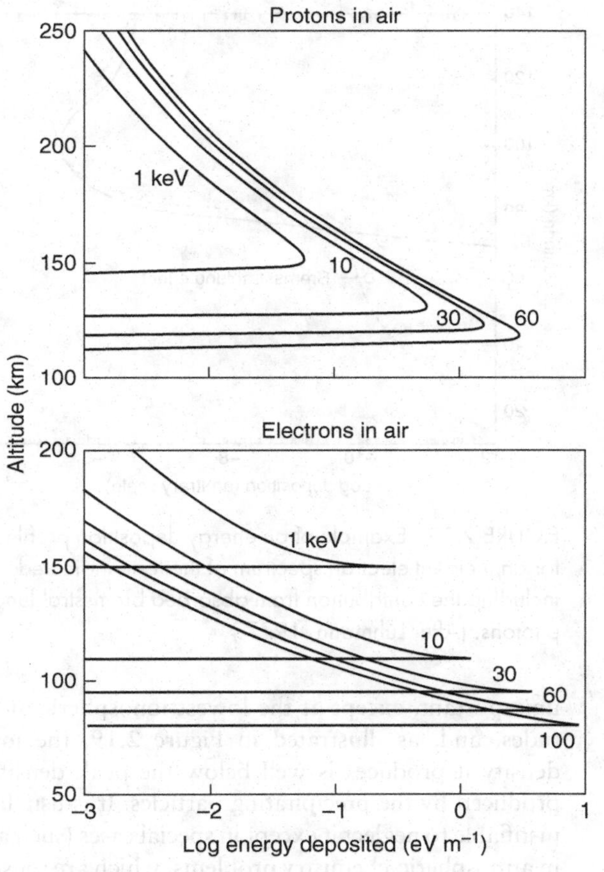

FIGURE 2.18. Energy-deposition profiles for protons (a) and electrons (b) of various energies incident on the Earth's atmosphere, as calculated from the method described in the text.

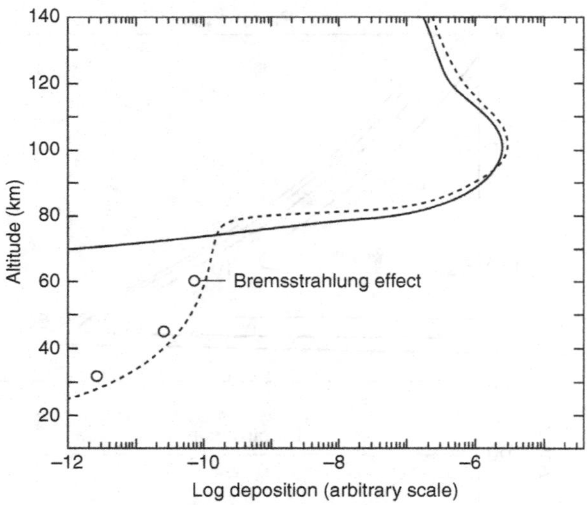

FIGURE 2.19. Example of an energy-deposition profile for an incident electron spectrum of the form indicated, including the contribution from absorbed bremsstrahlung photons. (After Luhmann, 1977.)

unimportant, except at the lowest ionospheric altitudes, and, as illustrated in Figure 2.19, the ion density it produces is well below the peak density produced by the precipitating particles. It is usually justifiable to neglect it except in special cases (such as in atmospheric-chemistry problems, which are sensitive to the local rates of ion production in the Earth's middle atmosphere).

2.8 ION LOSS

Once the ion production rate is known, the next critical quantity that must be specified for an ionospheric model is the ion or electron loss rate L. Ionospheric electrons disappear by virtue of three types of recombination:

(1) radiative recombination $e + X^+ \rightarrow X + h\nu$
(2) dissociative recombination $e + XY^+ \rightarrow X + Y$
(3) attachment $e + Z \rightarrow Z^-$

The first two of these are most important throughout the bulk of the ionosphere. (Radiative recombination is responsible for many types of observed airglows. In contrast, most of the emission that one sees in an aurora occurs when atomic and molecular electrons are excited to higher energy levels by Coulomb collisions with the passing precipitating particles and then undergo radiative de-excitation.) Recombination occurs at rates that depend on the local concentrations of the ions and electrons:

$$L = \alpha n_e n_i,$$

where n_e and n_i are the electron and ion densities, and α is a recombination coefficient. The recombination coefficient is determined by empirical and theoretical methods. For the more important atmospheric dissociative recombination reactions, such as

$$O_2^+ + e \rightarrow O + O$$

and

$$N_2^+ + e \rightarrow N + N,$$

the recombination coefficients (in units of $m^{-3}\ s^{-1}$) are $1.6 \times 10^{-1}(300/T_e)^{0.55}$ and $1.8 \times 10^{-1}(300/T_e)^{0.39}$, respectively, where T_e is the electron temperature in the ionosphere. These quantities can generally be found in tables in the aeronomy literature (e.g., Banks and Kockarts, 1973; Schunk and Nagy, 1980). We see that the altitude profiles of the loss rates depend on altitude through the electron and ion densities and T_e. If there is one major ion constituent, such that $n_i \approx n_e$, the loss rate at a particular altitude is proportional to n_e^2. It is notable that α is often assumed constant, although T_e generally depends on altitude h.

2.9 DETERMINING IONOSPHERIC DENSITY FROM PRODUCTION AND LOSS RATES

Once the ion production rates and loss rates are established, we can consider the problem of finding the altitude distribution of the ionospheric electron density n_e. If the electrons and ions do not move very far from where they are produced (e.g., by virtue of a strong horizontal ambient magnetic field), we can say that n_e (and thus n_i) obeys the equilibrium-continuity or particle-conservation equation

$$\frac{\partial n_e}{\partial t} = Q - L = 0, \qquad (2.72)$$

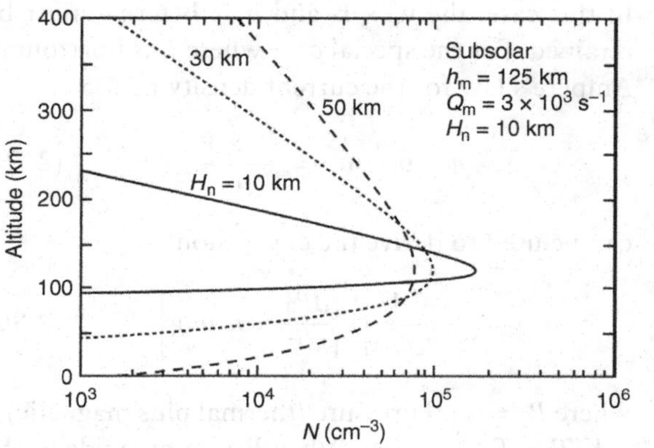

FIGURE 2.20. The number density as a function of altitude in a neutral atmosphere with a fixed scale height and a single wavelength of incident illumination assuming photochemical equilibrium so that production matches loss at each altitude. Thus there is no vertical transport in this one-dimensional model. Here the maximum production function is 3×10^3 ions s^{-1} at 125 km, at a neutral scale height of 10 km. In a realistic ionosphere, there would be a broad band of incident radiation absorbed at different altitudes and horizontal and vertical transport.

and, if the loss rate is due to electron–ion collisions,

$$Q = L = \alpha n_e^2,$$

and hence

$$n_e = (Q/\alpha)^{1/2} \tag{2.73}$$

describes the spatial distribution of the electrons or ions. This particular distribution is called a photochemical equilibrium distribution because it involves only local photochemistry. We call models that invoke the Chapman production function and the assumption of photochemical equilibrium, α, Chapman-layer models, and the layer, an α-Chapman layer. Figure 2.20 illustrates such an α-Chapman-layer model for three neutral scale heights.

If the electrons are removed by attachment to neutrals of density N, then the steady-state production and loss rates are

$$Q = L - \beta N.$$

Such a layer is called a β-Chapman layer.

In a realistic ionosphere, the electrons and ions move a significant distance from their points of creation before they recombine, and so we must consider a transport term in a more general continuity equation. Because vertical transport is usually of most interest, given the relative horizontal and vertical scales of atmospheres (atmospheres often can be approximated by thin slabs), we can restrict our attention to the vertical transport attributable to the vertical velocity u_h. The equilibrium electron density distribution must then satisfy the vertical-continuity equation

$$\frac{\partial n_e}{\partial t} = Q - L \frac{\partial (n_e u_{eh})}{\partial h}, \tag{2.74}$$

where the new influence on the electron density is the vertical-flux gradient, which describes the difference between the flux of electrons entering and leaving a given altitude. The subscript h is used to denote the vertical component of a vector. The flux gradient $\partial(n_e u_{eh})\partial h$ can represent a source of electrons, if more arrive than depart, or a loss if the reverse is true. To determine the velocity u_{eh}, we need to invoke another equation, the momentum or force-balance equation. Ionospheric electrons can be said to obey the following steady-state vertical-momentum equation:

$$-\frac{dP_e}{dh} - n_e m_e g - e n_e [E_h + (u_e \times B)_h]$$
$$= n_e m_e \nu_{en} (u_{eh} - u_{nh}) + n_e m_e \nu_{ei} (u_{eh} - u_{ih}), \tag{2.75}$$

where

P_e = electron thermal pressure $n_e K T_e$,
g = gravitational acceleration,
E = electric field,
B = magnetic field,
$\mathbf{u}_e$ = electron velocity,
$\mathbf{u}_i$ = ion velocity,
ν_{en} = electron–neutral collision frequency,
ν_{ei} = electron–ion (Coulomb) collision frequency,
$\mathbf{u}_n$ = neutral velocity.

From left to right, the terms represent the pressure-gradient force, the gravitational force, the force due to both externally applied and convection electric fields, and friction from collisions with other types of particles. This equation can be solved for u in terms of all of the other variables, but it is particularly useful to obtain an expression that is independent of seldom-measured quantities like E. By adding the ion-momentum equation,

$$-\frac{dP_i}{dh} - n_e m_i g + q n_e [E_h + (\mathbf{u}_i \times \mathbf{B})_h] = n_e m_i \nu_{in}(u_{ih} - u_{nh}) + n_e m_i \nu_{ie}(u_{ih} - u_{eh}), \tag{2.76}$$

where P_i is ion pressure $n_e K T_i$, m_i is ion mass, q is ion charge, and ν_{in} is the ion–neutral collision frequency, to the electron-momentum equation, one can eliminate E. It can further be assumed that the ions are singly charged ($q = e$) and that the ions and electrons vertically drift together at velocity u_{ph} (to maintain charge neutrality in the plasma), assumptions that eliminate the $\mathbf{u}_i \times \mathbf{B}$ and $\mathbf{u}_e \times \mathbf{B}$ terms. Finally, with the additional knowledge that $m_i \gg m_e$ and $m_i \nu_{in} \gg m_e \nu_{en}$, so that certain other terms can be neglected, we obtain

$$u_{ph} - u_{nh} \approx -\frac{1}{n_e m_i \nu_{in}}\left[\frac{d}{dh}(P_i + P_e) + n_e m_i g\right]. \tag{2.77}$$

2.10 DETERMINING IONOSPHERIC DENSITY

Furthermore, if the temperatures are all assumed to be independent of h, and the vertical neutral velocity u_{nh} is zero, Eq. (2.77) can be written in the form of a diffusion equation for n_e:

$$n_e u_{ph} = D\left[\frac{dn_e}{dh} + \frac{n_e}{H_p}\right], \tag{2.78}$$

where $D = K(T_i + T_e)/m_i \nu_{in}$, is called the ambipolar diffusion coefficient, and H_p is the "plasma scale height" $K(T_i + T_e)/m_i g$. The "ambipolar diffusion" nomenclature derives from the fact that, in the absence of externally imposed E fields, the vertical drift given by u_p, is caused by the charge-separation (or polarization) electric field because of gravity acting on the different masses of the ions and electrons, which must maintain equal scale heights to conserve local charge neutrality. In general, u_i and u_e have horizontal as well as vertical components. In this case, the $\mathbf{u}_e \times \mathbf{B}$ and $\mathbf{u}_i \times \mathbf{B}$ terms must be retained. For the special case where B is horizontal, Ampère's law for the current density j,

$$\mathbf{j} = n_e e(\mathbf{u}_i - \mathbf{u}_e) = \frac{\nabla \times \mathbf{B}}{\mu_0}, \tag{2.79}$$

can be used to derive the expression

$$u_{ph} = -\frac{1}{n_e m_i \nu_{in}}\left[\frac{dP_T}{dh} + n_e m_i g\right], \tag{2.80}$$

where P_T = total pressure (thermal plus magnetic) = $n_e K(T_i + T_e) + B^2/2\mu_0$. When B is at an angle to the horizontal, other magnetic terms besides the vertical magnetic-pressure gradient must be considered. It should be noted here that, in some planetary ionospheres, the magnetic and thermal pressures are comparable, whereas in others either thermal or magnetic pressure will dominate. At the Earth, where the intrinsic magnetic field is strong, **B** can be assumed to be the planetary dipole field, but at weakly magnetized Venus (where, as we shall see in Chapter 8, the ionospheric field is of interplanetary origin), **B** must be calculated from Maxwell's equations. This greatly complicates matters. Temperatures can also be derived from another equation for the heat balance, but the most basic calculations generally assume constant temperatures, or empirical values, or other simple temperature models. With the height-dependent vertical velocity $u_{ph}(h)$ put into the continuity equation, we can proceed to solve for $n_e(h)$. It should be appreciated that u_{ph} can be either upward or downward, depending on the sign of the total pressure gradient and the latter's size in comparison with the gravitational force. Large collision frequencies tend to keep u_p small. Ratcliffe (1972) considered the special case of zero vertical drift ($u_{ph} = 0$) to make the point that, under such circumstances, for zero B, and for equal electron and ion temperatures, one can solve for the polarization electric field,

$$E = \frac{1}{2} gM/e. \tag{2.81}$$

This upward electric field makes both electrons and ions behave as if they had a mass of $M/2$ in the gravitational field. Because $T_i = T_e$, the plasma scale height is twice that of a neutral gas made up of atoms of mass M. This is because the polarization electric field buoys the heavier ions up and "weights" the light electrons down. If different ion species are present, one can write individual ion-momentum equations with $u_{ph} = 0$ to see that, for equal temperatures, the M in the plasma scale height becomes the mean ion mass $\langle M \rangle$, and the vertical electric field is given by $E = \frac{1}{2} g \langle M \rangle / e$. Under such circumstances, individual ion species can behave as if their mass were negative if it is less than $\frac{1}{2}\langle M \rangle$. Thus, height profiles of individual species densities in a multi-component ionosphere can have intervals in which the gradient is not what one would expect if that ion were the only constituent. Strictly speaking, one must simultaneously solve the continuity and momentum equations for all ion species in order to find the correct height profiles of their densities. In general, an ionosphere will not have the same composition as the neutral atmosphere, and the altitude of the peak electron density will not coincide with the peak in production. The ion composition and the peak altitude depend on production and loss rates and transport.

2.11 AN EXAMPLE: THE EARTH'S IONOSPHERE

In situ measurements with rocket- and satellite-borne instruments, combined with the remote-sensing data from topside and bottomside ionospheric sounders (wherein time delays of reflected radio signals, transmitted from a satellite or the ground, give the altitude profile of the plasma frequency (and thus n_e) on either side of the peak), have given the picture of the Earth's dayside ionosphere shown in Figure 2.21. The density and composition of the neutral atmosphere are also shown, both to emphasize the weakness of atmospheric ionization in comparison to the neutral density at the Earth, and to illustrate the contrast in composition and vertical structure between the ions and the neutrals. It is seen that, although there is one main peak in electron density at approximately 250 km altitude, there is considerable substructure. The discovery of this substructure led early observers to designate three major ionospheric layers or regions: the D region (below 90 km), the E region (between 90 and 130 km), and the F region (above 130 km). The F region is usually further divided into F_1 and F_2 layers, because a second ledge sometimes appears in its profile below the main (F_2) peak.

The concepts introduced earlier can be applied to an understanding of these layers as follows. One can think of the layers as being independently produced by the absorption of solar radiation by specific constituents of the neutral atmosphere, which respond differently to different parts of the incident solar photon spectrum. Two of these layers, the E and F_1 layers, are often thought to be fair approximations to Chapman layers. On the other hand, the

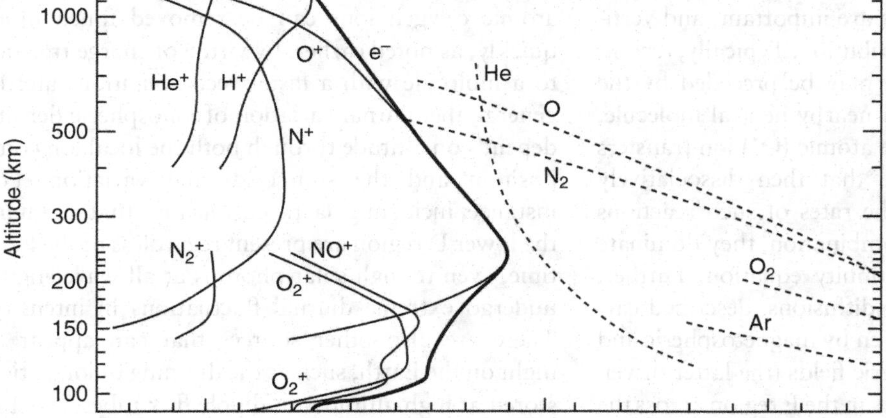

FIGURE 2.21. International quiet solar year (IQSY) daytime ionospheric and atmospheric composition based on mass-spectrometer measurements. (After Johnson, 1969.)

F_2 layer, which is the highest, seems to require a more exotic explanation of its profile in terms of photochemistry or vertical movements driven by neutral drag or magnetospheric effects. The D layer, the deepest, is connected with the most energetic radiations (x-ray photons and cosmic-ray particles) and poorly understood loss processes.

The E layer is usually clearly noticeable as a change in slope in daytime electron density profiles near 110 km. The ions in this layer are mainly O_2^+ and NO^+ (Figure 2.21) that have been produced by ultraviolet radiation in the 100–150 nm range and solar x-rays in the 1–10 nm range. The peak density of this layer is close to the peak of the production rate Q for these ions. An effective recombination rate α can be deduced by dividing Q by the observed n_e^2 (see Section 2.9). Vertical transport of ions is considered to play a minor role in the formation of this layer. The F_1 layer is composed primarily of O^+. The maximum electron density in this layer occurs near approximately 170 km, which is close to the level of the maximum ion production by photons in the spectral range from about 17 to 91 nm. This layer, which is more of a ledge, is not obvious in Figure 2.21, because it almost merges into the F_2 layer, which contains the main peak in the ionosphere density.

It is somewhat unfortunate (albeit not without interest) that the major peak in the terrestrial ionosphere cannot be described by simple Chapman-layer theory. This F_2 (or overall F layer) peak density is also in a region dominated by O^+. However, at the altitudes where it occurs, other chemical processes besides simple direct recombination between the O^+ and surrounding electrons are important, and vertical drifts affect the ion distribution. Typically, recombination in the F_2 region may be preceded by the reaction of the ion with a nearby neutral molecule, with the net effect that the atomic (O^+) ion transfers its charge to a molecule that then dissociatively recombines. As long as the rates of such reactions exceed that of simple recombination, they dominate the loss term in the continuity equation. Further, collisional and ambipolar diffusions, described earlier, and vertical drifts driven by magnetospheric and atmospheric-dynamo electric fields (the latter driven by neutral drag forcing ions in the E region across the geomagnetic field), significantly affect the ion motion near the F-region peak. Thus, the reader is advised to consult more specialized references to appreciate fully all the features of the F region.

The lowest ionosphere, the D region, is of practical interest because of its role in commercial radio communication. The high ion–neutral collision frequencies make radio-wave absorption there important, and so the electron density is of primary concern. Only the most energetic ionization sources can penetrate to D-region altitudes. Between about 80 and 90 km, 0. 1–1 nm x-rays from the Sun are the primary sources; the very intense Lyman-α (121.6 nm) radiation from the Sun has its peak ion production rate at about 70–80 km, and ionization of cosmic-ray particles dominates below. The predominant ions, NO^+ and O_2^+, can recombine with electrons, but at these low altitudes the electrons can also attach themselves to neutrals to form negative ions. Thus, a treatment of the D-layer "equilibrium" profile is not straightforward. Moreover, the aforementioned sources each undergo variations, depending on the prevailing solar activity and interplanetary conditions. These considerations leave the D region, like the F_2 region, a topic of ongoing research.

Finally, it is worth mentioning what happens at night when at least the solar photon sources are turned off (except for some scattered radiation that may make the disappearance of some wavelengths, such as Lyman-α, more gradual after sunset). At heights near 250 km, the effective time constant for recombination, which depends on the ion species, can be as short as 10 s for molecular nitrogen or as long as 300 h for atomic oxygen. However, the atomic oxygen ions can be removed much more quickly, as noted earlier, by virtue of charge transfer to a molecule with a faster recombination rate. In general, the diurnal variation of ionospheric density depends on altitude through both the local ion composition and the source diurnal variation. For instance, incident galactic cosmic rays that maintain the lower D region are present regardless of the local time, even though solar photons at all wavelengths undergo extreme diurnal fluctuations in intensity. There are also other sources that can appear at night on the Earth, such as the draining of ionization stored at high altitudes in dipole flux tubes, and the highly spatially and temporally variable source of ionization from auroral-particle precipitation.

2.12 OTHER CONSIDERATIONS RELATING TO IONOSPHERES

The escape of atmospheric neutral gas was described in Section 2.5, but ion escape also plays an important role in space physics. As we shall see in Chapter 8, the processes that are important for ion escape depend on whether a planetary magnetic field is an organizing influence. For planets like Earth with an important internal field, the picture of ionospheric outflows is still undergoing major conceptual changes as observational results and theoretical modeling address the problem. Nevertheless, there are some basic ideas that merit mention here. In particular, the concepts of the ionosphere as both a source of magnetospheric plasma, as well as *polar* ion outflows (sometimes referred to as *polar wind*) at high latitudes arise because there are both observations indicating upward-moving fluxes of primarily H^+, He^+, and O^+ from the ionosphere, as well as physical reasons to expect such outflows.

2.12.1 Ionospheric outflow

A curiosity of particle conservation and transport reveals itself in the light elements at high altitudes in situations where production and loss can be ignored and the magnetic field is essentially vertical. The continuity equation for an ion of species α is the law of flux conservation,

$$\partial(n_\alpha u_{h\alpha} A)/\partial h = 0, \tag{2.82}$$

where A is the flux tube area, and the momentum equation can be written

$$n_\alpha m_\alpha u_{h\alpha}\frac{\partial u_{h\alpha}}{\partial h} + n_\alpha q_\alpha \frac{K(T_i + T_e)}{n_e}\frac{\partial n_e}{\partial h} + n_\alpha m_\alpha g = -n_\alpha m_\alpha \nu_{\alpha n} u_{h\alpha}, \tag{2.83}$$

where the *inertial term* on the left has been introduced in anticipation of its importance. It is usually negligible in lower ionospheres, where flows are generally subsonic. The polarization electric field set up by the major species is here expressed as

$$E = -\frac{1}{en_e}\frac{\partial p_e}{\partial h} = -\frac{KT_e}{en_e}\frac{\partial n_e}{\partial h} \tag{2.84}$$

under the assumption that all other forces but the pressure-gradient force are negligible for the electrons. The foregoing continuity equation gives

$$\frac{\partial n_\alpha}{\partial h} = -\frac{n_\alpha}{u_\alpha}\frac{\partial u_\alpha}{\partial h} - \frac{n_\alpha}{A}\frac{\partial A}{\partial h}. \tag{2.85}$$

For singly charged ions ($q_\alpha = 1$), this can be used together with the thermal velocity definition $w = (K(T_i + T_e)/m_\alpha)^{1/2}$ to cast the ion-momentum equation in the form

$$(u_\alpha^2 - w^2)\frac{1}{u_\alpha}\frac{\partial u_\alpha}{\partial h} + g = -\frac{w^2}{A}\frac{\partial A}{\partial h} - \nu_{\alpha n} u_\alpha \tag{2.86}$$

or

$$\frac{1}{M}\frac{\partial M}{\partial h} = \left[\frac{w^2}{A}\frac{\partial A}{\partial h} - g - \nu_{\alpha n} w M\right] / [w^2(M^2 - 1)], \tag{2.87}$$

where $M = |u_\alpha|/w$ is the species **Mach number**. This latter equation resembles the solar-wind equation to be derived in Chapter 5. In that chapter, the idea that a flow can theoretically undergo a subsonic–supersonic transition when the numerator and denominator in the flow equation become zero together is applied to the solar corona. The density, obtainable from the preceding flux-conservation form of the continuity equation, decreases with altitude in accord with the velocity increase. Indeed, a supersonic outflow of light ions is seen emanating from regions like the Earth's polar cap. Of course, a detailed treatment of these outflows is much more involved, requiring simultaneous solution of the equations for other species of ions and special treatments of collision frequencies for high-speed ions (Schunk and Nagy, 2009). Also of interest here is the role of collisions with neutrals and the effects of the magnetic field.

Between collisions, charged particles gyrate around the magnetic field at frequencies $\Omega_i = qB/m_i$ for ions and $\Omega_e = qB/m_e$ for electrons. If $\nu_{in} \gg \Omega_i$, the magnetic field exerts little direct influence on the motion of the ions. However, if the gyromotion is not negligible, other forces such as magnetic mirroring and drift motions across the field can come into play. A comparison of some collision frequencies and gyro-frequencies in the ionospheres of Earth and Venus is shown in Figure 2.22. In calculating plasma velocities, the best policy is to evaluate all of the terms in the momentum equation before deciding which is negligible. Even a term that seems relatively small can be important if the other larger terms practically cancel each other out. In addition, the relative velocities that enter into the

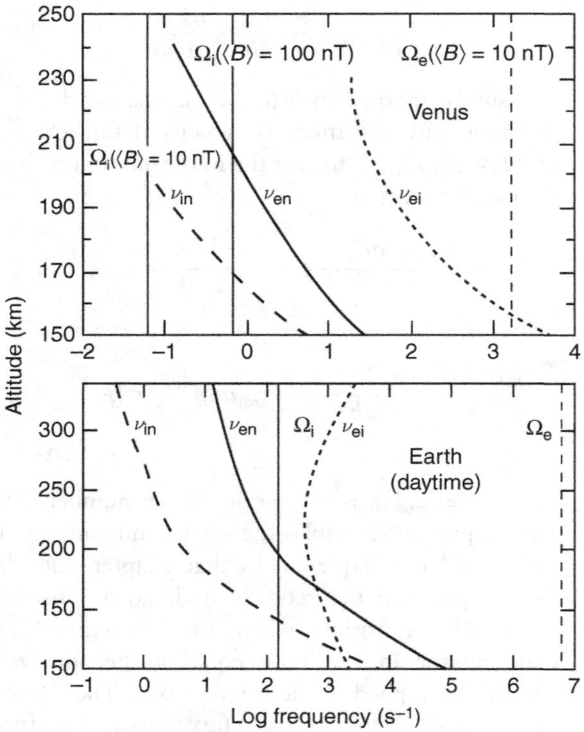

FIGURE 2.22. Comparison of collision frequencies and gyro-frequencies in the ionospheres of Venus (top) and Earth (bottom). (After Luhmann and Elphic, 1985.)

collision term can determine whether or not collisions are important in the overall force balance at a particular altitude. In general, the problem of ion outflows is extremely complicated. Observations at the Earth show a variety of time- and space-dependent behaviors, suggesting that a number of processes contribute under different circumstances. For example, the outflow compositions, energy spectra, and fluxes have distinctive characteristics in the magnetospheric cusp, polar cap, and auroral zone. Recent treatments include multifluid and multispecies kinetic simulations, including neutrals, assuming various approximations and physical settings. Interested readers are encouraged to investigate the details of this still active area of research (e.g., Schunk and Nagy, 2009).

2.12.2 Conductivity

The competition between the magnetic field and collisions for control of the ion motion is particularly important in problems requiring an evaluation of the electrical conductivity of the ionosphere. If only the electric-field and collision terms participated in the force balance, the steady-state momentum equations for ions and electrons would be

$$qE = m_i \nu_{in} u_i \tag{2.88}$$

and

$$-eE = m_e \nu_{in} u_e. \tag{2.89}$$

Thus, for this simple medium, j is related to the electric field by

$$\mathbf{j} = \sigma_0 \mathbf{E}, \tag{2.90}$$

where the conductivity σ_0 is a scalar that depends only on the collision frequencies. If a magnetic field is present, magnetic terms are present in the momentum equations, and the velocities $\mathbf{u}_i$ and $\mathbf{u}_e$ are no longer expressed so simply in terms of E. However, if we solve for $\mathbf{u}_i$ and $\mathbf{u}_e$ for the case where the magnetic field is in the z-direction (and $q = e$), we find that j can be written in the concise form

$$\mathbf{j} = \begin{pmatrix} \sigma_1 & \sigma_2 & 0 \\ -\sigma_2 & \sigma_1 & 0 \\ 0 & 0 & \sigma_0 \end{pmatrix} \begin{pmatrix} E_x \\ E_y \\ E_z \end{pmatrix}, \tag{2.91}$$

where

$$\sigma_1 = \left[\frac{1}{m_e \nu_{en}} \left(\frac{\nu_{en}^2}{\nu_{en}^2 + \Omega_e^2} \right) + \frac{1}{m_i \nu_{in}} \left(\frac{\nu_{in}^2}{\nu_{in}^2 + \Omega_i^2} \right) \right] n_e e^2, \tag{2.92}$$

$$\sigma_2 = \left[\frac{1}{m_e \nu_{en}} \left(\frac{\Omega_e \nu_{en}}{\nu_{en}^2 + \Omega_e^2} \right) - \frac{1}{m_i \nu_{in}} \left(\frac{\Omega_i \nu_{in}}{\nu_{in}^2 + \Omega_i^2} \right) \right] n_e e^2, \tag{2.93}$$

$$\sigma_0 = \left[\frac{1}{m_e \nu_{en}} + \frac{1}{m_i \nu_{in}} \right] n_e e^2. \tag{2.94}$$

The conductivity is thus a tensor quantity. It is a tensor because the magnetic field makes the medium anisotropic in its response to an applied electric field. It can be seen from the form of the tensor that if the electric field is applied perpendicular to the

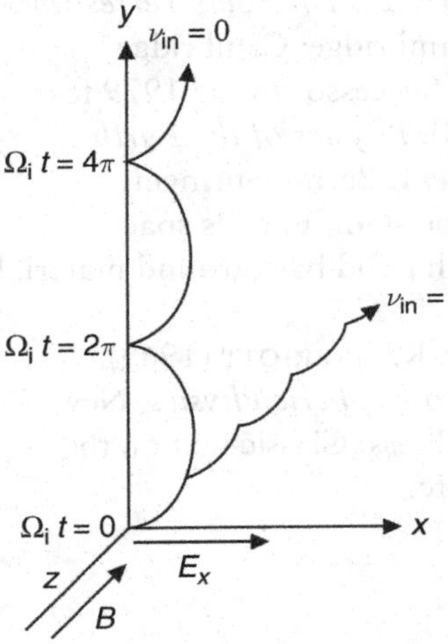

FIGURE 2.23. Trajectory of a picked-up ion in uniform magnetic and electric fields when the ion–neutral collision frequency is zero and when the gyro and collision frequencies are equal.

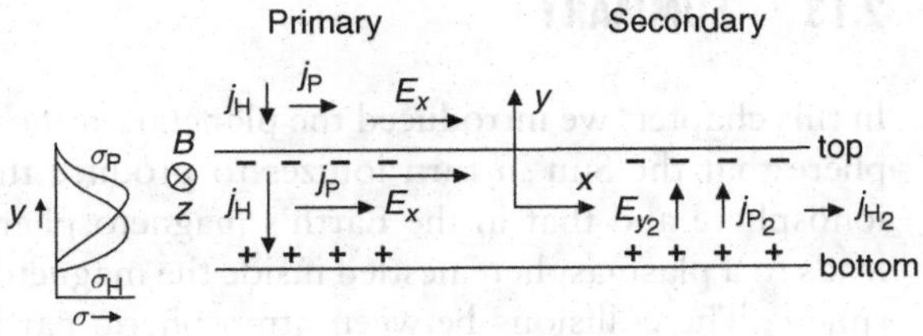

FIGURE 2.24. The formation of a Cowling channel of "enhanced" electrical conductivity. At the magnetic equator, the low-altitude Hall conductivity and the high-altitude Pedersen conductivity provide a bottom and the top of a channel called the equatorial electrojet. The primary conductivity produces a charge separation that leads to a secondary electric field that enhances the horizontal current.

magnetic field ($\mathbf{E} = E_x\,\mathbf{i} + E_y\,\mathbf{j}$), σ_1 is the conductivity in the direction of the applied electric field. It is called the Pedersen conductivity. The component σ_2 is the conductivity perpendicular to the direction of the applied field. It is called the Hall conductivity. If an electric field is applied parallel to the magnetic field, the conductivity is the same as that derived earlier for zero field. It is called the direct or longitudinal conductivity and depends only on the collision frequencies. Hence, we can deduce the values of the conductivities in an atmosphere once n_e is derived from the foregoing considerations, as long as the behavior of the magnetic field is known.

Figure 2.23 illustrates what is happening in a collisional and a collisionless ionosphere. The magnetic field is horizontal (along z) and the electrical field is to the right. A neutral atom is ionized at the origin and drifts in a cycloid with a net displacement in the y-direction if the ion–neutral collision frequency is zero. In this situation the ions and electrons drift together and there is no net electric current. In the presence of collisions, the drift path is altered and the ions and electrons move along different paths with their different collision rates. Here we have chosen to make the ion collision frequency equal to the ion gyro-frequency.

At the equator, where the magnetic field is horizontal, the application of an eastward electric field E_x will generate a Pedersen current along x. If there is a top and a bottom to the ionosphere, the Hall current will lead to charge buildup on the top and bottom, as shown in Figure 2.24. The vertical electric field associated with this charge then causes an electric field in the y-direction E_{y_2} that leads to a secondary Hall current j_{H_2}. The secondary Pedersen current cancels the primary Hall current, but the primary Pedersen current and the secondary Hall current reinforce each other. This situation occurs in the Earth's ionosphere at the magnetic equator and is called the equatorial electrojet. The effective conductivity along the electrojet (x here) is called the Cowling conductivity.

2.13 SUMMARY

In this chapter, we introduced the planetary atmosphere that the Sun in turn ionizes to produce the ionosphere and that in the Earth's magnetosphere leads to a plasmasphere nestled inside the magnetosphere. The collisions between atmospheric particles lead to pressure in the gas. Pressure gradients support the gas at higher altitudes. We find both sound and gravity waves in this gas. These concepts prepare us for examining the plasma state later in this book. Plasmas have pressure too, and gradients in that pressure exert forces. Waves have important roles in plasmas as well.

The ionosphere is a partially ionized gas. Collisions are important here. In later chapters, we move to regions which are collisionless, but the concepts of pressure, temperature, density, and waves serve us well even in the collisionless regime. Thus it is important to master the understanding of concepts presented here. They are universal in this field.

We should also caution the reader that, in order to stress the physical principles, we have presented simplified scenarios and treatments. We have not dwelt on details of atmospheric chemistry, time dependence, or horizontal structure, which may affect the observed phenomena.

Additional reading

Bauer, S. J. and H. Lammer (2004). *Planetary Aeronomy*. Springer: Berlin. Comprehensive study of planetary atmospheres.

Hargreaves, J. K. (1979). *The Upper Atmosphere and Solar Terrestrial Relations*. New York: Van Nostrand Reinhold. Classic textbook on solar-terrestrial physics.

Hargreaves, J. K. (1992). *The Solar Terrestrial Environment*. Cambridge: Cambridge University Press. Successor to the 1979 text.

Prölss, G. W. (2003). *Physics of the Earth's Space Environment*. Berlin: Springer. Elementary textbook on Earth's space environment with good background material for this chapter.

Rishbeth, H. and O. K. Garriott (1969). *Introduction to Ionospheric Physics*. New York: Academic Press. Classic text on the Earth's ionosphere.

Problems

2.1 The electric current j and the direction and magnitude of electron and ion drift ($\mathbf{v}_e$ and $\mathbf{v}_i$) relative to an ionosphere electric field $\mathbf{E}$ parallel to the x-axis (while $\mathbf{B}_E$ is along the z-axis) are plotted in Figure 2.25 for three different heights. The angle a between electrons and ions relative to $\mathbf{E}$ is given, for a charged particle k, by $a_k = \tan^{-1}(\Omega_k/\nu_{kn})$ (Ω and ν are the gyro-frequency and collision frequency, respectively). For a height of about 180 km, $\nu_{kn} \ll \Omega_k$; electrons and ions move in the same direction, with a velocity $\mathbf{E} \times \mathbf{B}/B^2$. This drift produces no net current. Discuss the importance of collisions for two cases. In the first case, assume that the electron–neutral collision frequency greatly exceeds the electron gyro-frequency (as, for example, near 70 km). In the second case, assume that the electron gyro-frequency greatly exceeds the electron–neutral collision frequency (as, for example, near 180 km).

2.2 Find the expression for h_m, the height of the production peak for the Chapman production function in terms of neutral atmosphere quantities

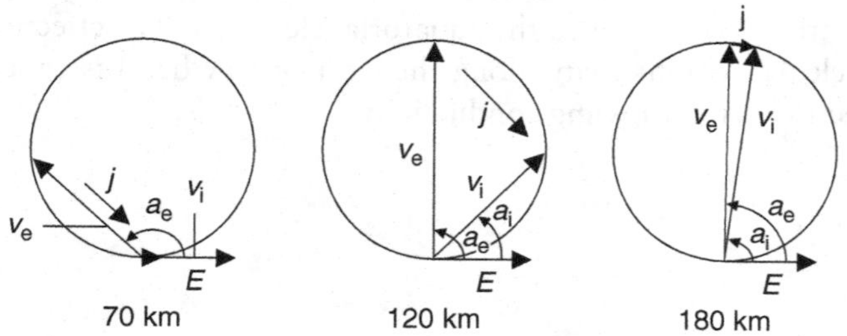

FIGURE 2.25. Electric current j and the direction and magnitude of electron and ion drift ($\mathbf{v}_e$ and $\mathbf{v}_i$) relative to an ionospheric electric field $\mathbf{E}$ parallel to the x-axis (while $\mathbf{B}_E$ is along the z-axis) for three different heights.

only. [Hint: Use Eq. (2.57).] Does the production peak height double if some event such as a dust storm increases the neutral gas density so that n_0 occurs at twice the original reference height h_0 (with no change to the neutral scale height)?

2.3 Figure 2.18 shows some altitude profiles of energy deposition for protons in Earth's atmosphere (air). If you had a monoenergetic 10 keV proton flux of 10^5 cm^{-2} s^{-1} entering the atmosphere from the magnetosphere or a solar source, what would the approximate peak ion production rate be (ions cm^{-3} s^{-1})? How does this compare with the peak subsolar ion production rate from solar EUV photons (Figure 2.16)? If, instead of protons, 10 keV electrons precipitated, how would the altitude and magnitude of the peak production change? Describe how you would calculate the peak altitude if these were alpha particles (He^{2+}). [Note: dE/dx is independent of the incident ion mass, but depends on the square of its charge.] [Hint: Remember the range of the particles is determined in x with units of g cm^{-2}, which must be translated to altitude.]

2.4 Compare the ranges of protons and electrons in air (see Figure 2.17). At a particular energy, which are most penetrating? Using Figure 2.14, how do the energies of photons that penetrate to the same altitudes compare?

2.5 Using the assumption that B is along z, derive the expression for the Pedersen and Hall conductivities from the electron- and ion-momentum equations.

2.6 Use the Space Physics Exercises*, selecting the Chapman Layer Altitude Profile option with the right-hand mouse button.

(a) Turn on the overlay option and use the default values for scale height and the plotting range. Using three different colors, plot the ionospheric profiles for zenith angles of the sun of 0°, 80°, and 89°. Print out your results. Describe how the changing height of the Sun affects the ionospheric density profile.

(b) Turn on the overlay option after erasing the old plots and reset the solar zenith angle to 0°. Set the scale height to 20, 40, and 60 km using different colors. Print out your results. How does the increased scale height affect the absorption of radiation and the density of electrons with altitude?

(c) Enter the Solar Zenith Angle Plot module using the right-hand mouse button. Turn on the overlay option. For scale heights of 20, 40, and 60 km, plot the peak electron density and altitude of the maximum altitude versus solar zenith angle. Print out your results. How does the peak density change? How does the altitude of the maximum density change?

* Space Physics Exercises are at http://spacephysics.ucla.edu.

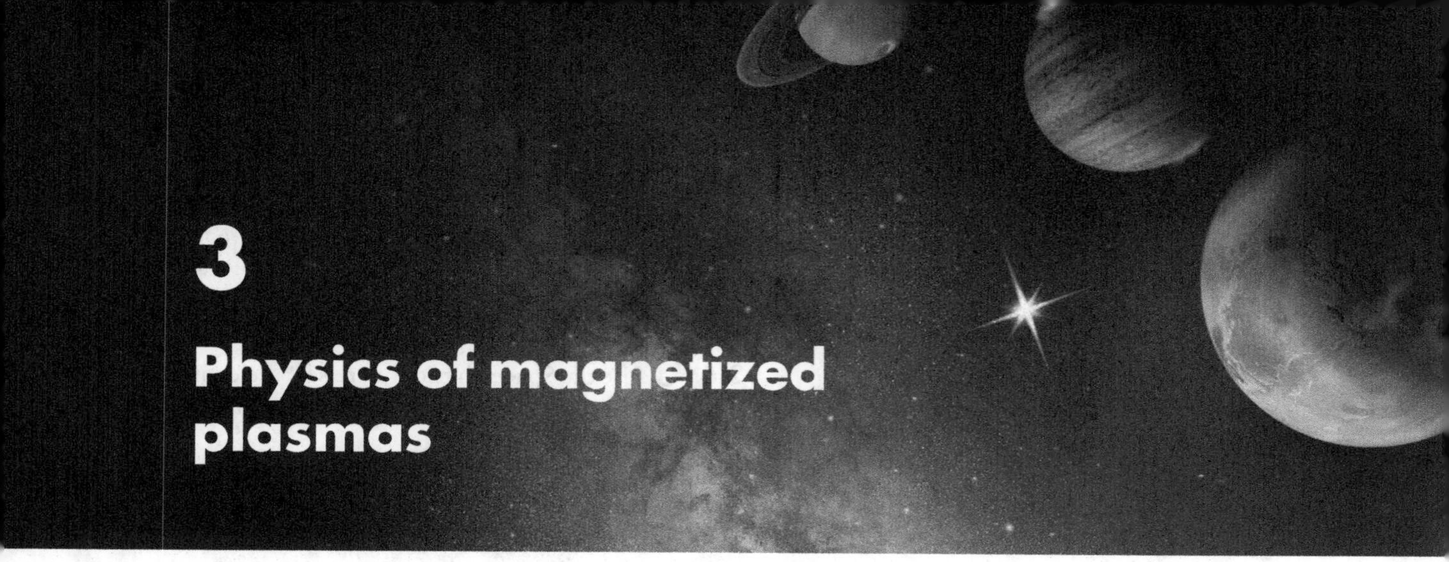

3 Physics of magnetized plasmas

3.1 INTRODUCTION

In this chapter we present the theoretical foundation for the physics of magnetized plasmas. By necessity we must take some results as given, and we also restrict the scope to two principal areas: the development of **particle orbit theory** and **magnetohydrodynamics** (MHD). A starting point for the discussion will be **Maxwell's equations** and the **Lorentz force law**, which are discussed in Section 3.2. From this, in Section 3.3, we shall investigate **single-particle motion**, and introduce the concept of particle **gyration**, with its characteristic frequency, the **gyro-frequency**, and **guiding center** motion. As a result we derive equations describing the guiding center drift, in particular the **electric field drift**, and **gradient** and **curvature drifts**. We also develop a generalized drift equation that we subsequently compare with the results of MHD.

Before developing the governing equations for MHD, in Section 3.4 we first discuss the **distribution function**, which specifies the distribution of particles in the six-dimensional **phase space**, defined by position and velocity, and the **Boltzmann equation**, which governs the variation of the distribution function. From this we discuss the properties that make a mixture of ions and electrons behave as a plasma, introducing concepts such as the **plasma frequency** and the **Debye length**.

Next we use the Boltzmann equation to derive the conservation equations that result in MHD in Section 3.5. To do this we take velocity-moment integrals of the Boltzmann equation. We show that this results in a hierarchy of equations, where the time rate of change of one moment integral, e.g., the species number density, depends on the divergence of the next higher moment, which would be the number flux in this case. In principle, this would result in an infinite set of conservation equations, but in practice the set is terminated at the energy equation with the simplifying assumption that the divergence of the heat flux vanishes.

The conservation equations are combined with Maxwell's equations to form a set of equations that are almost complete. The **generalized Ohm's law**, derived in Section 3.6, is required to give a closed set of equations. The **idealized Ohm's law**, which is derived from the generalized Ohm's law, leads to a fundamental theorem for plasmas – the **frozen-in theorem**. This theorem states that the magnetic field can be treated as if it were frozen to the fluid, i.e., the magnetic field lines "move" with the fluid (we use quotation marks here as the concept of moving field lines is a construct to aid in interpreting how the magnetic field evolves).

In Section 3.7 we give three applications of the MHD equations: an introduction to MHD or **magnetohydrodynamic waves**, a discussion of the equivalence between MHD and guiding center theory when specifying the perpendicular current in a plasma, and a discussion of how the divergence of the perpendicular current results in field-aligned currents. Section 3.8 provides a brief summary.

3.2 MAXWELL'S EQUATIONS AND THE LORENTZ FORCE

The fundamental feature of a plasma is that it consists of dissociated matter, usually consisting of a gas of electrons and positive ions. As such, then, the particles that constitute the plasma will both create and be acted upon by electric and magnetic fields. The equations that govern the fields are, of course, Maxwell's equations:

Faraday's law,

$$\nabla \times \mathbf{E} = -\frac{\partial \mathbf{B}}{\partial t}; \tag{3.1}$$

Ampère's law,

$$\nabla \times \mathbf{B} = \mu_0 \mathbf{j} + \frac{1}{c^2}\frac{\partial \mathbf{E}}{\partial t}; \tag{3.2}$$

Gauss's law,

$$\varepsilon_0 \nabla \cdot \mathbf{E} = \rho_q; \tag{3.3}$$

and the absence of magnetic monopoles,

$$\nabla \cdot \mathbf{B} = 0. \tag{3.4}$$

The last Maxwell equation, Eq. (3.4), is implicit in Faraday's law, as can be seen by taking the divergence of Eq. (3.1).

We have used SI units in Eqs. (3.1)–(3.4), and we use SI units throughout the chapter. In SI units $\mathbf{E}$ is the electric field in volts per meter and $\mathbf{B}$ is the magnetic field in nanotesla; $\mathbf{B}$ is also referred to as the magnetic induction, and $\mathbf{H}$ is the magnetic intensity, with $\mathbf{B} = \mu_0 \mathbf{H}$ in vacuum, where μ_0 is the permeability of free space. In Eq. (3.3), ρ_q is the charge density, and ε_0 is the permittivity of free space. In Eq. (3.2) $\mathbf{j}$ is the current density, and we have made use of the identity $c^2 = 1/\mu_0\varepsilon_0$.

In space plasmas we usually treat all charge and current carriers as free, as opposed to bound, and assume that the **relative permittivity** is unity, i.e., the displacement $\mathbf{D} = \varepsilon_0 \mathbf{E}$. The main exception to this approach is when considering high-frequency waves, where it is mathematically convenient to include the response of the plasma as part of a dielectric tensor. By the same token, treating all currents as free is another way of stating that the magnetization $\mathbf{M} = 0$.

Ampère's law also yields **charge conservation** by taking the divergence of Eq. (3.2) and combining with Gauss's law, Eq. (3.3),

$$\frac{\partial \rho_q}{\partial t} + \nabla \cdot \mathbf{j} = 0. \tag{3.5}$$

With the field behavior given by Maxwell's equations, we now specify how the fields affect the particles. This is given by the momentum equation

$$m\frac{d\mathbf{v}}{dt} = q(\mathbf{E} + \mathbf{v} \times \mathbf{B}) + \mathbf{F}_g + m\left.\frac{d\mathbf{v}}{dt}\right|_c. \tag{3.6}$$

In Eq. (3.6) the symbols have their usual meaning, with $\mathbf{F}_g$ representing non-electromagnetic forces, such as gravity. The term $m d\mathbf{v}/dt|_c$ represents the momentum change through collisions. We have kept the gravity and collisional terms for completeness in Eq. (3.6). In particular they can be important in an ionosphere, but they can often be neglected. In that case Eq. (3.6) becomes

$$m\frac{d\mathbf{v}}{dt} = q(\mathbf{E} + \mathbf{v} \times \mathbf{B}). \tag{3.7}$$

The force on the right-hand side of Eq. (3.7) is known as the Lorentz force. We use this equation for much of the discussion in the rest of this chapter.

Equation (3.7) is written in non-relativistic form, and is invariant under **galilean (non-relativistic) transformations**. This means that $\mathbf{E} + \mathbf{v} \times \mathbf{B}$ is frame invariant. The magnetic field $\mathbf{B}$ is frame invariant under galilean transformations, but clearly $\mathbf{v}$ is frame dependent. As a consequence, the electric field $\mathbf{E}$ is also frame dependent. It is therefore important to know the frame being used when discussing the electric field.

As noted at the beginning of this section, Maxwell's equations (3.1)–(3.4) govern the fields, and the Lorentz force law (3.7) describes how the fields affect charged-particle motion. A plasma is a collection of charged particles, and understanding how a plasma behaves, and how the plasma affects the fields, means that we have to understand the behavior of the collection of particles. But first we can develop an understanding of how individual particles respond to specified fields. This is known as single-particle motion, and exploring this requires the development of particle orbit theory. Particle orbit theory is an important first step in understanding space plasmas, and we shall develop the theory in the following section.

3.3 SINGLE-PARTICLE MOTION – PARTICLE ORBIT THEORY

The motion of charged particles in prescribed electric and magnetic fields is often referred to as particle orbit theory, and the implicit assumption is that the particle motion does not itself affect the fields. As such the particles are considered to be test particles, whose motion is given by Eq. (3.7) (i.e., as already noted, we are ignoring non-electromagnetic forces and collisions).

A perhaps obvious point from Eq. (3.7) is the difference between motion perpendicular and parallel to the magnetic field. The magnetic field enters through the cross product in Eq. (3.7), and therefore only affects motion perpendicular to the magnetic field. Thus we naturally consider motion perpendicular to the magnetic field separately from parallel motion, and we use the subscripts "$\perp$" and "$\parallel$" to denote perpendicular and parallel motion with respect to the ambient magnetic field.

Before considering the details of particle trajectories, it is also worth noting that magnetic fields do not change particle energy. This can be seen by taking the dot product of Eq. (3.7) with **v**.

3.3.1 Particle gyration

The first approximation we shall make is to consider the case of a uniform magnetic field, with no electric field or any other force. If we define $\mathbf{v}_\perp$ as the velocity perpendicular to the magnetic field then, from Eq. (3.7),

$$\frac{d\mathbf{v}_\perp}{dt} = \frac{q}{m}\mathbf{v}_\perp \times \mathbf{B}. \tag{3.8}$$

Thus, on taking the time derivative of Eq. (3.8),

$$\frac{d^2\mathbf{v}_\perp}{dt^2} = \frac{q}{m}\frac{d\mathbf{v}_\perp}{dt} \times \mathbf{B} = -\frac{q^2B^2}{m^2}\mathbf{v}_\perp. \tag{3.9}$$

This is the equation for a simple harmonic oscillator, with the oscillation frequency given by $\Omega = |qB/m|$. We refer to Ω as the gyro-frequency, and the period of the oscillation as the gyro-period; Ω is an angular frequency and has units of radian per second. If we use the symbol f_g to denote the gyro-frequency in hertz, then $\Omega = 2\pi f_g$. Here we have chosen to define Ω as a positive quantity. But it is important to remember that the force in Eq. (3.8) depends on the sign of the charge. In some applications Ω may be defined as a signed quantity.

The frequency is the gyro-frequency because this corresponds to circular motion, or gyration, about the magnetic field. If we assume the magnetic field defines the z-axis of a right-handed cartesian coordinate system, then the x-component of Eq. (3.8) gives

$$dv_x/dt = \pm\Omega v_y, \tag{3.10}$$

where the upper sign corresponds to positively charged particles (usually ions) and the lower sign corresponds to negatively charged particles (usually electrons). For reasons that will become clear shortly, we assume

$$v_x = -v_\perp \sin(\Omega t). \tag{3.11a}$$

Then,

$$v_y = \mp v_\perp \cos(\Omega t). \tag{3.11b}$$

On integrating Eqs. (3.11a) and (3.11b) we find

$$x = x_0 + \frac{v_\perp}{\Omega}\cos(\Omega t) \tag{3.12a}$$

and

$$y = y_0 \mp \frac{v_\perp}{\Omega}\sin(\Omega t), \tag{3.12b}$$

where x_0 and y_0 are constants of integration.

A positively charged particle (upper sign in Eqs. (3.11b) and (3.12b)) will be at the position $(x_0 + v_\perp/\Omega, y_0)$ at $t = 0$, and, one quarter gyro-period later, the particle will be at the position $(x_0, y_0 - v_\perp/\Omega)$ (we chose Eq. (3.11a) to make this sense of rotation clear). Thus a positive particle performs a left-handed rotation, or gyration, about the direction defined by the magnetic field. A negatively charged particle will gyrate in a right-handed sense about the magnetic field. In a typical plasma, then, the positive ions gyrate in a left-handed sense, while the electrons gyrate in right-handed sense. The radius of the gyration is known as the **Larmor radius** (typically denoted by r_L), or the gyro-radius (ρ_g), and $r_L = \rho_g = v_\perp/\Omega$.

3.3.2 Electric field (E × B) drift

The next approximation we shall make is to assume that a uniform electric field is present in addition to a uniform magnetic field. In that case,

$$\frac{d\mathbf{v}}{dt} = \frac{q}{m}(\mathbf{E} + \mathbf{v} \times \mathbf{B}). \tag{3.13}$$

For motion parallel to the magnetic field, we have the solution

$$\frac{dv_{\|}}{dt} = \frac{q}{m}E_{\|}. \tag{3.14}$$

This equation is often used to argue that a plasma cannot have a large-scale parallel electric field as particles will be accelerated to relativistic energies. There are exceptions to this statement, but in that case the fields are either localized or time varying. Typical spatial and temporal scales for which parallel electric fields are important are the **plasma skin depth** c/ω_{pe} and the plasma-wave period $2\pi/\omega_{pe}$, where ω_{pe} is the electron plasma frequency, as derived in Section 3.4. If we assume $E_{\|} = 0$, then $v_{\|}$ is constant.

For perpendicular motion we take the time derivative of Eq. (3.13), as we did for Eq. (3.9), and find

$$\frac{d^2\mathbf{v}_{\perp}}{dt^2} = \frac{q^2}{m^2}\left(\mathbf{E} \times \mathbf{B} - \mathbf{v}_{\perp}B^2\right). \tag{3.15}$$

We further assume that $\mathbf{v}_{\perp} = \tilde{\mathbf{v}}_{\perp} + \mathbf{v}_{\mathrm{E}}$, where $\tilde{\mathbf{v}}_{\perp}$ is time varying and $\mathbf{v}_{\mathrm{E}}$ is constant. In that case, Eq. (3.15) becomes

$$\frac{d^2\tilde{\mathbf{v}}_{\perp}}{dt^2} = -\frac{q^2B^2}{m^2}\tilde{\mathbf{v}}_{\perp}, \tag{3.16}$$

which is the same as Eq. (3.9), and

$$\mathbf{v}_{\mathrm{E}} = \frac{\mathbf{E} \times \mathbf{B}}{B^2}. \tag{3.17}$$

Thus, in the presence of a uniform perpendicular electric field, the particle motion separates into gyration about the magnetic field and a uniform drift motion. The drift given by Eq. (3.17) is perpendicular to **E** and **B** and is known as the **E** × **B** drift. The **E** × **B** drift is independent of charge and mass, so all particles drift with this velocity. Figure 3.1 shows the **E** × **B** drift for positive ions and electrons, with Figure 3.2 showing drifts for different gyrational velocities.

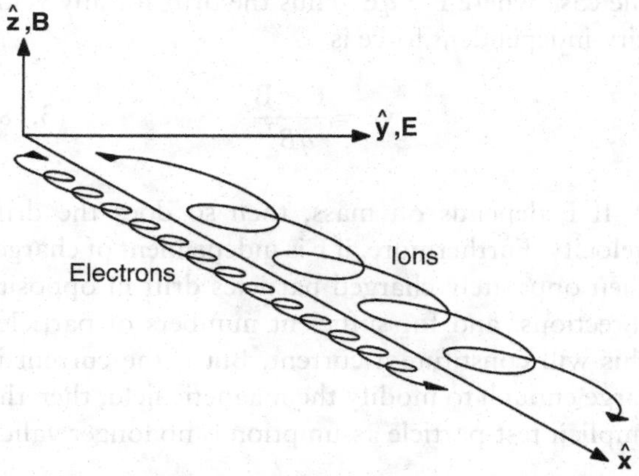

FIGURE 3.1. **E** × **B** drifts for positive ions and electrons. The smaller mass electrons have much smaller gyro-orbits (in order to make the electron orbits visible an artificially low ion to electron mass ratio has been used).

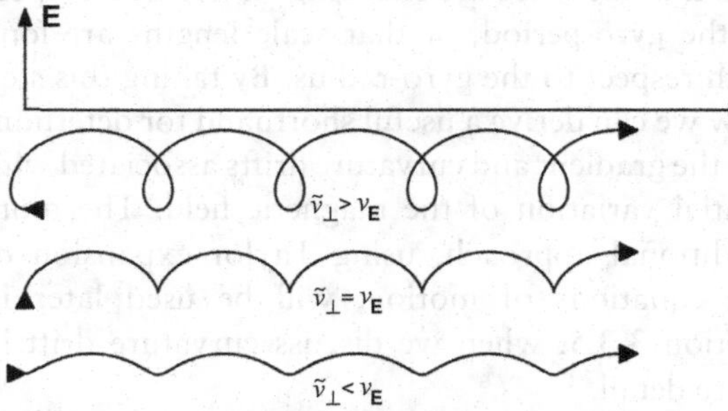

FIGURE 3.2. **E** × **B** drifts for positive ions with different gyrational velocities. Particles with $\tilde{v}_{\perp} << v_E$ have nearly straight-line trajectories, while those with $\tilde{v}_{\perp} > v_E$ have prolate cycloidal trajectories.

For strictly time-stationary magnetic fields there is no induction electric field, and we can replace the electric field with $\mathbf{E} = -\nabla\phi$, where ϕ is the electric potential. In that case the **E** × **B** drift velocity follows an equipotential. But again it must be remembered that this only applies for time-stationary magnetic fields.

The formalism used to derive the **E** × **B** drift can be generalized to any velocity-independent force, **F**, by recognizing that the electric field corresponds to the case where $\mathbf{F} = q\mathbf{E}$. Thus the drift for any velocity-independent force is

$$\mathbf{v}_\mathrm{F} = \frac{\mathbf{F} \times \mathbf{B}}{qB^2}. \tag{3.18}$$

If F depends on mass, then so does the drift velocity. Furthermore, if F is independent of charge, then oppositely charged particles drift in opposite directions, and for sufficient numbers of particles this will constitute a current. But if the current is large enough to modify the magnetic field, then the implicit test-particle assumption is no longer valid.

3.3.3 Guiding center motion and magnetic moment

At this stage, rather than derive the drift motion associated with a non-uniform magnetic field, which is the usual next step, we instead establish the concepts of guiding center motion and the magnetic moment. Both are dependent on the assumption that the fields are changing slowly with respect to the gyro-period, or that scale lengths are long with respect to the gyro-radius. By taking this step now we can derive a useful shorthand for determining the gradient and curvature drifts associated with spatial variation of the magnetic field. The more traditional approach, using Taylor expansion of the equations of motion, will be used later in Section 3.3.5, when we discuss curvature drift in more detail.

In the preceding section we derived the $\mathbf{E} \times \mathbf{B}$ drift velocity. If we assume that the magnetic field defines the z-axis, and the electric field is along the y-axis of a cartesian coordinate system, then the particle motion consists of a perpendicular and parallel drift given by $\mathbf{v}_\mathrm{d} = (v_\mathrm{E}, 0, v_{\|})$, and a gyration about this drift velocity. This leads to the concept of the guiding center motion. Although we formally derived the $\mathbf{E} \times \mathbf{B}$ drift assuming uniform fields, provided the fields vary only slightly over both the gyro-period and the gyro-radius, we can assume that the fields are approximately constant for the purposes of determining the drift velocity. The particle motion can then be thought of as a slowly varying guiding center motion given by the trajectory defined by $\mathbf{v}_\mathrm{d}(\mathbf{r}, t)$, where this velocity is now a function of time and position, and a faster gyro-period rotation about the guiding center.

Having separated the motion into gyrational and guiding center motion, we can now define what is known as the **first adiabatic invariant**. This is the **magnetic moment** of the particle, and it is an invariant for changes that are much slower than the gyro-period, or equivalently have scale lengths much larger than the gyro-radius.

Formally, adiabatic invariants are derived from action integrals, but in this case we demonstrate that the magnetic moment, denoted by μ, is an adiabatic invariant from the momentum equation (3.7). We show this first for time-varying magnetic fields.

From the integral form of Faraday's law, the azimuthal electric field is given by

$$2\pi\rho_\mathrm{g} E_\phi = -\pi\rho_\mathrm{g}^2 \frac{\partial B}{\partial t}, \tag{3.19}$$

where the line and surface integrals are defined by the particle gyration. The adiabatic approximation applies because we are assuming that the gyro-radius ρ_g is constant over a gyration of the particle.

Remembering that positively charged particles gyrate in a left-handed sense with respect to the ambient field, and in a right-handed sense for negatively charged particles, from Eq. (3.8) the azimuthal speed varies as

$$m\frac{\partial \tilde{v}_\perp}{\partial t} = \frac{|q|}{2}\rho_\mathrm{g}\frac{\partial B}{\partial t} = \frac{m\tilde{v}_\perp}{2B}\frac{\partial B}{\partial t}, \tag{3.20}$$

where, again, $\tilde{v}_\perp$ corresponds to the gyrational speed of the particles.

Therefore,

$$\frac{\partial W_\perp}{\partial t} = \frac{W_\perp}{B}\frac{\partial B}{\partial t} \quad \text{or} \quad \frac{\partial}{\partial t}\left(\frac{W_\perp}{B}\right) = \frac{\partial \mu}{\partial t} = 0, \tag{3.21}$$

where $W_\perp = \frac{1}{2}mv_\perp^2$.

To see why μ is the magnetic moment of a particle, we note that it takes one gyro-period for the particle to complete a gyro-orbit, so the current in the loop defined by the gyration is

$$\mathbf{I} = -\frac{|q|\Omega}{2\pi}\hat{\mathbf{q}}, \tag{3.22}$$

where $\hat{\mathbf{q}}$ is the azimuthal unit vector of a right-handed cylindrical coordinate system, with the

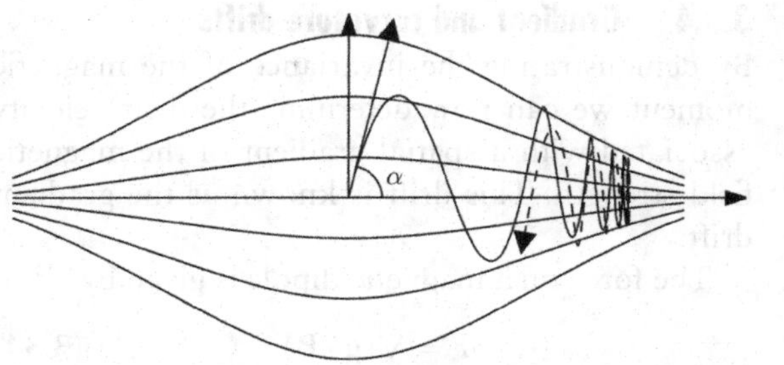

FIGURE 3.3. Magnetic-field lines and a representative particle trajectory in a mirror configuration. In order to visualize the trajectory better, which has been projected into the plane of the figure, the vertical scale has been increased by a factor of ten. The dashed line corresponds to the mirrored particle.

z-axis defined by the magnetic field. The direction of the current is independent of the sign of the charge. Given that the magnitude of the dipole moment for a current loop is IA, where A is the area of the current loop, then

$$\boldsymbol{\mu} = -\frac{|q|\Omega}{2\pi}\frac{\pi \widetilde{v}_\perp^2}{\Omega^2}\hat{\mathbf{b}} = -\frac{W_\perp}{B}\hat{\mathbf{b}}. \tag{3.23}$$

The magnetic moment is given as a vector in Eq. (3.23) to emphasize that the magnetic moment is opposite to the externally applied field. Thus the particles in a plasma are diamagnetic – the magnetic field associated with the gyration of the particles opposes the applied field.

So far we have demonstrated that μ is an adiabatic invariant for time-varying magnetic fields. We next consider the case of a magnetic mirror, where the magnetic field is spatially varying. Examples of such field geometries include the Earth's dipole and mirror devices in laboratory plasmas. Figure 3.3 shows such a magnetic field geometry. A representative particle trajectory is also shown. The particle initially starts with both a parallel and a perpendicular velocity. The angle α that the velocity vector makes with respect to the magnetic field is known as the pitch angle. The figure shows that as the particle moves into a region of increasing magnetic field, the pitch angle increases, until eventually the angle is 90° and the particle is reflected.

This process can be inferred from the previous discussion concerning the conservation of the magnetic moment for time-varying magnetic fields. As the particle moves into a region of increasing magnetic field strength, the particle senses a time-varying magnetic field. If the particle is moving with velocity $\mathbf{v}_\parallel$ along the magnetic field, then the time variation in the frame moving with the particle is given by

$$\frac{\partial B}{\partial t} = \mathbf{v}_\parallel \cdot \nabla B, \tag{3.24}$$

where we have denoted the axial or z-component of the magnetic field as B. Thus the particle perpendicular energy would increase, but, since the magnetic field is in fact time stationary, there is no induction field in the stationary frame. The total energy of the particle is constant, and hence the parallel velocity must decrease.

To show this formally we note that, since $\nabla \cdot \mathbf{B} = 0$,

$$\frac{1}{\rho}\frac{\partial}{\partial \rho}(\rho B_\rho) = -\frac{\partial B}{\partial z}, \tag{3.25}$$

where B_ρ is the radial component in the cylindrical coordinate system shown in Figure 3.3.

Thus

$$\delta B_\rho \approx -\frac{\rho_g}{2}\frac{\partial B}{\partial z} \approx -\frac{\widetilde{v}_\perp}{2\Omega}\frac{\partial B}{\partial z}, \tag{3.26}$$

and the z-component of Eq. (3.7) gives

$$m\frac{dv_\parallel}{dt} = -|q|\widetilde{v}_\perp \delta B_\rho = -\frac{|q|\widetilde{v}_\perp^2}{2\Omega}\frac{\partial B}{\partial z} = -\mu\frac{\partial B}{\partial z}, \tag{3.27}$$

while the azimuthal component gives

$$m\frac{d\widetilde{v}_\perp}{dt} = |q|v_\parallel \delta B_\rho = \frac{|q|\widetilde{v}_\perp v_\parallel}{2\Omega}\frac{\partial B}{\partial z} = \mu\frac{v_\parallel}{\widetilde{v}_\perp}\frac{\partial B}{\partial z}. \tag{3.28}$$

These two equations are independent of the sign of the charge because positively or negatively charged particles gyrate in left- or right-handed sense about the ambient field, respectively.

Combining Eqs. (3.27) and (3.28), not surprisingly gives $d(W_\parallel + W_\perp)/dt = 0$, but, in addition, rearranging terms in Eq. (3.28),

$$\frac{dW_\perp}{dt} = \mu v_\parallel \frac{\partial B}{\partial z} = \mu \frac{dB}{dt}. \quad (3.29)$$

Therefore

$$\frac{d(\mu B)}{dt} = \mu \frac{dB}{dt}, \text{ hence } \frac{d\mu}{dt} = 0. \quad (3.30)$$

Thus the magnetic moment is an adiabatic invariant for both temporally and spatially varying magnetic fields. The invariance of μ leads to a phenomenon known as **betatron acceleration.** If the magnetic field slowly increases with time, then the particle perpendicular energy will increase.

There is another adiabatic invariant associated with slow changes to the field in a magnetic mirror. This is the **second adiabatic invariant**, given by the action integral

$$J = \oint p_\parallel ds, \quad (3.31)$$

where $p_\parallel$ is the parallel momentum, ds is along the guiding center path, and the integral is between the mirror points.

Since

$$\frac{p_\parallel^2}{2m} = W - \mu B = \mu(B_m - B), \quad (3.32)$$

where B_m is the magnetic field at the mirror point,

$$J = \sqrt{2m\mu} \oint (B_m - B)^{1/2} ds. \quad (3.33)$$

Provided the magnetic field only changes slowly over the bounce period, J is an invariant.

One consequence of J as an invariant is **Fermi acceleration,** much like the first adiabatic invariant is associated with betatron acceleration. Fermi acceleration occurs when the magnetic mirror points move closer together, such as occurs for particles trapped between a planetary bow shock and the bow shock of an approaching coronal mass ejection (CME). The former is fixed relative to the planet, while the CME is moving at roughly the solar-wind speed. Thus the trapped particle mirror points move closer together, and the particle is accelerated.

3.3.4 Gradient and curvature drifts

By demonstrating the invariance of the magnetic moment we can now determine the drift velocity associated with a spatial gradient of the magnetic field strength. This drift is known as the gradient drift.

The force on a magnetic dipole is given by

$$\mathbf{F} = \nabla(\boldsymbol{\mu} \cdot \mathbf{B}). \quad (3.34)$$

From Eq. (3.23), since $\boldsymbol{\mu}$ is a constant and antiparallel to the external field,

$$\mathbf{F} = -\frac{W_\perp \nabla B}{B}. \quad (3.35)$$

For slowly varying fields this force is constant, and we can use the generalized drift velocity Eq. (3.18) to give the gradient drift velocity associated with a perpendicular gradient in the magnetic field strength:

$$\mathbf{v}_g = \frac{W_\perp \mathbf{B} \times \nabla B}{qB^3}. \quad (3.36)$$

Figure 3.4 shows a positive ion trajectory in the presence of a perpendicular gradient in the magnetic field strength. The drift can be understood in

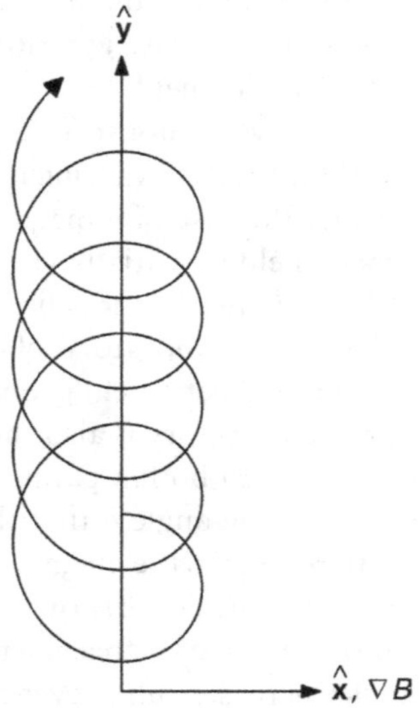

FIGURE 3.4. Positive ion trajectory in the presence of a gradient in the magnetic field strength.

terms of the change in the particle gyro-radius, which is larger for smaller magnetic field strengths. The gradient drift depends on the charge of the particles. In the Earth's magnetosphere this means that ions drift in the opposite sense to the planetary rotation, while the electrons drift in the same sense as the Earth's rotation. This is often used to suggest that the ions and electron drifts therefore correspond to the ring current, but, as noted earlier, particle orbit theory is strictly a test-particle approach. Any macrophysical consequences of the guiding center drifts need to be assessed carefully, taking into account collective effects not present in particle orbit theory alone.

We can also use Eq. (3.18) to determine the curvature drift. In this case, however, some care is required in deciding what force to use. Because particles can move easily along the field, but gyrate around the field, intuition suggests that a particle will tend to follow a curved field line. In that case the particle will sense a centrifugal force in the frame rotating with the particle. That centrifugal force is given by

$$\mathbf{F}_{\rm cf} = \hat{\mathbf{r}}_{\rm c} m v_{\|}^2 / r_{\rm c} = 2W_{\|}\hat{\mathbf{r}}_{\rm c}/r_{\rm c}, \qquad (3.37)$$

where $\mathbf{r}_{\rm c}$ is the radius of curvature vector, defined as positive outward.

It can further be shown that

$$\frac{\hat{\mathbf{r}}_{\rm c}}{r_{\rm c}} = \frac{\mathbf{B}(\mathbf{B}\cdot\nabla B)}{B^3} - \frac{(\mathbf{B}\cdot\nabla)\mathbf{B}}{B^2}. \qquad (3.38)$$

As a consequence, on substituting Eqs. (3.37) and (3.38) into Eq. (3.18), the curvature drift is given by

$$\mathbf{v}_{\rm c} = \frac{2W_{\|}\mathbf{B}\times(\mathbf{B}\cdot\nabla)\mathbf{B}}{qB^4}. \qquad (3.39)$$

3.3.5 Curvature drift – Taylor series approach

The derivation of the curvature drift given in the preceding section may be counter-intuitive, given that it is couched in terms of a rotating frame of reference and the centrifugal force. We therefore also derive this drift using the more standard Taylor series expansion of the momentum equation. The geometry is shown in Figure 3.5, where

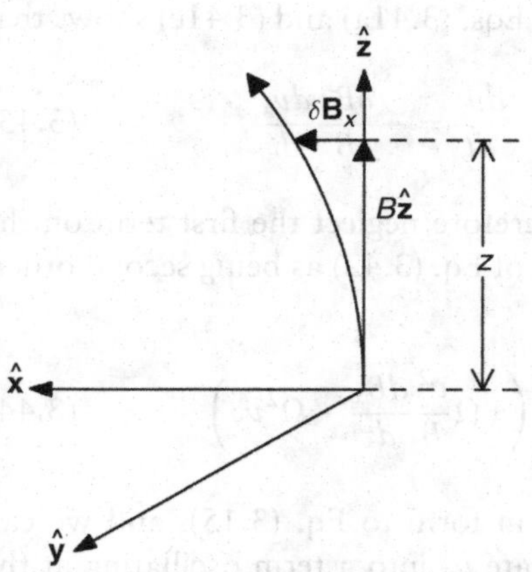

FIGURE 3.5. Field-line geometry for calculating curvature drift.

the magnetic field is initially in the z-direction, but bends toward the x-direction for increasing z.

We assume that the x-component of the magnetic field is small, increasing from zero at $z = 0$, and is given by

$$\delta B_x = z\frac{dB_x}{dz}. \qquad (3.40)$$

At $t = 0$ we assume the particle has a parallel velocity $v_{\|}$. Hence $z = v_{\|}t$. We further assume that the z-component of the magnetic field, denoted by B, is constant, and use the standard definition of the gyro-frequency, $\Omega = |qB/m|$.

The momentum equation (3.8) for this geometry becomes

$$\frac{dv_x}{dt} = \pm\Omega v_y, \qquad (3.41a)$$

$$\frac{dv_y}{dt} = \pm\Omega\left(v_z\frac{\delta B_x}{B} - v_x\right), \qquad (3.41b)$$

$$\frac{dv_z}{dt} = \pm\Omega v_y\frac{\delta B_x}{B}, \qquad (3.41c)$$

where the upper sign corresponds to a positively charged particle.

Taking the total time derivative of Eq. (3.41b) we find

$$\frac{d^2 v_y}{dt^2} = \pm\Omega\left(\frac{dv_z}{dt}\frac{\delta B_x}{B} + \frac{v_z^2}{B}\frac{dB_x}{dz} - \frac{dv_x}{dt}\right). \qquad (3.42)$$

Comparison of Eqs. (3.41a) and (3.41c) shows that

$$\frac{dv_z}{dt} = -\frac{\delta B_x}{B}\frac{dv_x}{dt}, \tag{3.43}$$

and we can therefore neglect the first term on the right-hand side of Eq. (3.42) as being second order. Consequently

$$\frac{d^2 v_y}{dt^2} = \left(\pm\Omega\frac{v_z^2}{B}\frac{dB_x}{dz} - \Omega^2 v_y\right). \tag{3.44}$$

This is similar in form to Eq. (3.15), and we can therefore separate v_y into a term oscillating at the gyro-frequency and a drift term as follows:

$$v_y = \widetilde{v}_\perp \cos\,(\varphi \mp \Omega t) + \frac{2W_\parallel}{qB^2}\frac{dB_x}{dz}, \tag{3.45}$$

where φ is an arbitrary phase angle, and it should be remembered that the z-direction is parallel to the magnetic field at the origin of the coordinate system shown in Figure 3.5. Inspection will show that, for the assumed geometry, the second term on the right-hand side of Eq. (3.45) is identical to the y-component of Eq. (3.39). This term is again the curvature drift.

Given v_y, we can integrate Eq. (3.41a):

$$v_x = -\widetilde{v}_\perp \sin\,(\varphi \mp \Omega t) + \frac{v_\parallel^2 t}{B}\frac{dB_x}{dz}. \tag{3.46}$$

Since $z = v_\parallel t$,

$$v_x = v_\parallel \frac{\delta B_x}{B} - \widetilde{v}_\perp \sin\,(\varphi \mp \Omega t). \tag{3.47}$$

Similarly, on integrating Eq. (3.41c), ignoring second-order terms, and noting that $v_z = v_\parallel$ at $t = 0$,

$$v_z = v_\parallel + \frac{\delta B_x}{B}\widetilde{v}_\perp \sin\,(\varphi \mp \Omega t). \tag{3.48}$$

Thus both v_x and v_z have a constant or slowly varying term (the first term on the right-hand side of Eqs. (3.47) and (3.48)) and a gyrational term (the second term on the right-hand side of the same equations). Moreover, the ratio of these terms is such that

$$\frac{\overline{v}_x}{\overline{v}_z} = \frac{\delta B_x}{B}, \tag{3.49}$$

where $\overline{v}_x$ and $\overline{v}_z$ are the slowly varying terms, and

$$\frac{\widetilde{v}_x}{\widetilde{v}_z} = -\frac{B}{\delta B_x}, \tag{3.50}$$

with $\widetilde{v}_x$ and $\widetilde{v}_z$ being the gyrational terms.

These ratios are those required such that the guiding center follows the field line, whereas the gyrational motion is always perpendicular to the field line, as assumed when deriving the curvature drift using the centrifugal force.

In a dipole magnetic field both the gradient and curvature drifts will be azimuthal with respect to a coordinate system defined by the dipole axis. Thus, in this coordinate system, the particle drifts lie on a shell defined by the magnetic field line. In this axisymmetric field geometry, the equatorial radial distance of the field line is constant for all drift paths, regardless of pitch angle. This leads to the **third adiabatic invariant** – the drift shell. Figure 3.6 shows the different types of particle motion in a dipole magnetic field, along with the associated invariants.

3.3.6 Generalized drift motion

Most texts end their discussion of particle orbit theory once they have established the concepts of $\mathbf{E} \times \mathbf{B}$ drift and gradient and curvature drifts. And, indeed, that is sufficient for most purposes. In this chapter, however, we are presenting both single-particle and fluid approaches to space plasmas, and we want to emphasize that the approaches have elements in common. To this end we develop a generalized equation for the drift motion of the guiding center, which we return to in Section 3.7.2, where we show the equivalence between the current density as derived from the plasma momentum equation and the current density derived by adding up the individual drift currents and the effect of magnetization. To derive the generalized drift motion we use a similar approach to that used in the preceding section, where we assume any gradients in the electric and magnetic fields are first-order quantities in the particle momentum equation.

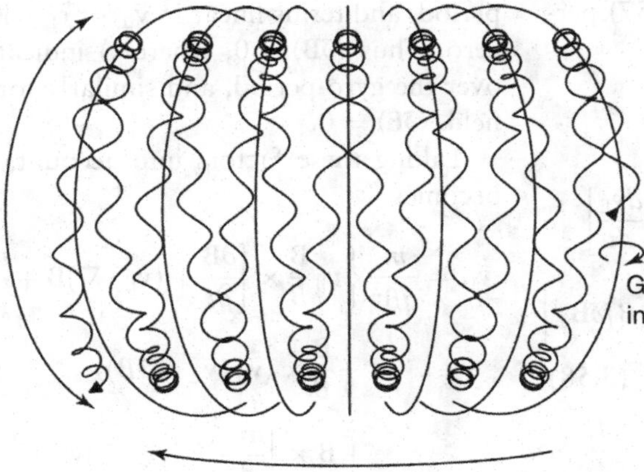

FIGURE 3.6. Particle motion and the associated adiabatic invariants.

From Eq. (3.7),

$$\frac{d}{dt}\mathbf{v}(\mathbf{r},t) = \frac{q}{m}\left[\mathbf{E}(\mathbf{r},t) + \mathbf{v}(\mathbf{r},t) \times \mathbf{B}(\mathbf{r},t)\right]. \quad (3.51)$$

All the terms in Eq. (3.51) are evaluated at particle position $\mathbf{r}$ at time t.

We now make use of the guiding center approximation and define the particle velocity as a sum of the slowly varying guiding center velocity and a rapidly varying term, which will correspond to particle gyration:

$$\mathbf{v}(\mathbf{r},t) = \mathbf{v}_{\mathrm{gc}}(\mathbf{r}_{\mathrm{gc}},t) + \widetilde{\mathbf{v}}_{\perp}(\mathbf{r},t), \quad (3.52)$$

where $\mathbf{v}_{\mathrm{gc}}$ is the guiding velocity, $\widetilde{\mathbf{v}}_{\perp}$ is the gyrational velocity, $\mathbf{r}_{\mathrm{gc}}$ is the guiding center position, and

$$\mathbf{r} = \mathbf{r}_{\mathrm{gc}}(t) + \widetilde{\mathbf{r}}_{\mathrm{L}}(t), \quad (3.53)$$

where $\widetilde{\mathbf{r}}_{\mathrm{L}}$ corresponds to the gyration of the particle about the guiding center. We can then Taylor expand the magnetic field:

$$\mathbf{B}(\mathbf{r},t) = \mathbf{B}(\mathbf{r}_{\mathrm{gc}}) + \widetilde{\mathbf{r}}_{\mathrm{L}} \cdot \nabla\mathbf{B}|_{\mathbf{r}=\mathbf{r}_{\mathrm{gc}}} = \mathbf{B} + \delta\mathbf{B}, \quad (3.54)$$

where $\delta\mathbf{B}$ is a first-order quantity since we are assuming gradient scale lengths much longer than the particle gyro-radius. Both terms on the right-hand side of Eq. (3.54) are implicitly functions of time, and are evaluated at $\mathbf{r}_{\mathrm{gc}}$. Looking forward, we assume the variables are functions of $\mathbf{r}_{\mathrm{gc}}$ and t unless otherwise stated.

Similarly, the electric field can be expressed as

$$\mathbf{E}(\mathbf{r},t) = \delta\mathbf{E} - \mathbf{v}_{\mathrm{E}} \times \mathbf{B}, \quad (3.55)$$

with $\mathbf{v}_{\mathrm{E}}$ being the $\mathbf{E} \times \mathbf{B}$ drift velocity evaluated at the guiding center.

On taking the time derivative of Eq. (3.51) we find

$$\begin{aligned}\frac{d^2\widetilde{\mathbf{v}}_{\perp}}{dt^2} &= \frac{q}{m}\left[\left(\mathbf{v}_{\mathrm{gc}} + \widetilde{\mathbf{v}}_{\perp} - \mathbf{v}_{\mathrm{E}}\right) \times \frac{d\mathbf{B}}{dt} + \mathbf{B} \times \frac{d\mathbf{v}_{\mathrm{E}}}{dt}\right] \\ &\quad - \frac{q^2}{m^2}(\mathbf{B} + \delta\mathbf{B}) \\ &\quad \times \left[\delta\mathbf{E} + \left(\mathbf{v}_{\mathrm{gc}} + \widetilde{\mathbf{v}}_{\perp} - \mathbf{v}_{\mathrm{E}}\right) \times (\mathbf{B} + \delta\mathbf{B})\right].\end{aligned} \quad (3.56)$$

In Eq. (3.56) we have already made the first-order approximation that $d^2\mathbf{v}_{\mathrm{gc}}/dt^2$ can be ignored as being second order. Also, since the time derivatives on the top row of the right-hand side of Eq. (3.56) are already first-order terms, the first-order $\delta\mathbf{E}$ and $\delta\mathbf{B}$ terms have been ignored in this row. For the time being they have been included to second order in the second and third rows.

We shall now further separate the guiding center velocity into three components,

$$\mathbf{v}_{\mathrm{gc}} = \mathbf{v}_{\mathrm{D}} + \mathbf{v}_{\mathrm{E}} + v_{\parallel}\frac{\mathbf{B}}{B}, \quad (3.57)$$

where $\mathbf{v}_{\mathrm{D}}$ is perpendicular to $\mathbf{B}$.

Taking only the components of Eq. (3.57) perpendicular to **B**, we find

$$\frac{d^2\tilde{\mathbf{v}}_\perp}{dt^2} = \frac{q}{m}\left[v_\parallel \frac{\mathbf{B}}{B}\times\frac{d\mathbf{B}}{dt} + (\mathbf{v}_\mathrm{D}+\tilde{\mathbf{v}}_\perp)\times\frac{\mathbf{B}}{B}\frac{dB}{dt} + \mathbf{B}\times\frac{d\mathbf{v}_\mathrm{E}}{dt}\right] - \frac{q^2}{m^2}\left[(B^2+2\mathbf{B}\cdot\delta\mathbf{B})(\mathbf{v}_\mathrm{D}+\tilde{\mathbf{v}}_\perp) + \mathbf{B}\times\delta\mathbf{E} - (v_\parallel B)\delta\mathbf{B}_\perp\right] \tag{3.58}$$

to first order in $\delta\mathbf{E}$ and $\delta\mathbf{B}$.

We can now separate Eq. (3.58) into two equations:

$$\frac{d^2\tilde{\mathbf{v}}_\perp}{dt^2} = -\frac{q^2B^2}{m^2}\tilde{\mathbf{v}}_\perp \tag{3.59a}$$

and

$$\frac{q^2B^2}{m^2}\mathbf{v}_\mathrm{D} = \frac{q}{m}\left[v_\parallel \frac{\mathbf{B}}{B}\times\frac{d\mathbf{B}}{dt} + (\mathbf{v}_\mathrm{D}+\tilde{\mathbf{v}}_\perp)\times\frac{\mathbf{B}}{B}\frac{dB}{dt} + \mathbf{B}\times\frac{d\mathbf{v}_\mathrm{E}}{dt}\right] - \frac{q^2}{m^2}\left[2\mathbf{B}\cdot\delta\mathbf{B}(\mathbf{v}_\mathrm{D}+\tilde{\mathbf{v}}_\perp) + \mathbf{B}\times\delta\mathbf{E} - (v_\parallel B)\delta\mathbf{B}_\perp\right]. \tag{3.59b}$$

Clearly, the solution of Eq. (3.59a) corresponds to the particle gyration, with gyro-frequency $\Omega = |qB/m|$.

The first step in evaluating Eq. (3.59b) is to note that all the terms on the right-hand side of Eq. (3.59b) involving $\mathbf{v}_\mathrm{D}$ are at least one order smaller than those on the left-hand side and so can be ignored. Second,

$$\frac{d}{dt} = \frac{\partial}{\partial t} + \mathbf{v}_\mathrm{E}\cdot\nabla + \tilde{\mathbf{v}}_\perp\cdot\nabla + \frac{v_\parallel \mathbf{B}\cdot\nabla}{B} + \mathbf{v}_\mathrm{D}\cdot\nabla. \tag{3.60}$$

But the last term on the right-hand side of Eq. (3.60), $\mathbf{v}_\mathrm{D}\cdot\nabla$, can also be ignored because it results in low-order terms in comparison to the left-hand side of Eq. (3.59b). Finally, since all the terms except the gyration-dependent terms are slowly varying, we can average Eq. (3.59b) over a gyro-period, and terms linear in $\tilde{\mathbf{v}}_\perp$ or $\tilde{\mathbf{r}}_\mathrm{L}$ will average to zero. Thus $\langle\delta\mathbf{B}\rangle = 0$, where $\langle\rangle$ indicate averaging over the gyro-period, and similarly for the electric field, $\langle\delta\mathbf{E}\rangle = 0$.

Taking these factors into account, Eq. (3.59b) becomes

$$\mathbf{v}_\mathrm{D} = \frac{m}{qB^2}\left\{v_\parallel\frac{\mathbf{B}}{B}\times\left[\frac{\partial\mathbf{B}}{\partial t} + (\mathbf{v}_\mathrm{E}\cdot\nabla)\mathbf{B} + \frac{v_\parallel(\mathbf{B}\cdot\nabla)\mathbf{B}}{B}\right] - \frac{\mathbf{B}}{B}\times\langle\tilde{\mathbf{v}}_\perp(\tilde{\mathbf{v}}_\perp\cdot\nabla B)\rangle + \mathbf{B}\times\left[\frac{\partial\mathbf{v}_\mathrm{E}}{\partial t} + (\mathbf{v}_\mathrm{E}\cdot\nabla)\mathbf{v}_\mathrm{E} + \frac{v_\parallel(\mathbf{B}\cdot\nabla)\mathbf{v}_\mathrm{E}}{B}\right] - \frac{2q\langle\tilde{\mathbf{v}}_\perp(\mathbf{B}\cdot\delta\mathbf{B})\rangle}{m}\right\}. \tag{3.61}$$

The last term in the square brackets on the top line of Eq. (3.61) is the curvature drift. The gradient drift arises from the $\tilde{\mathbf{v}}_\perp$-dependent terms at the end of lines 2 and 4. To determine the gradient drift we need to evaluate the gyro-period averages of the gyro-velocity-dependent terms. If we assume, as usual, that the magnetic field defines the z-axis of a local coordinate system, then without loss of generality we can define the x-axis and y-axis so that

$$\tilde{\mathbf{v}}_\perp = (\tilde{v}_x, \tilde{v}_y, 0) = \tilde{v}_\perp(\cos\Omega t, \mp\sin\Omega t, 0), \tag{3.62}$$

where the upper sign corresponds to a positively charged particle.

On averaging over a gyro-period,

$$\langle\tilde{v}_x^2\rangle = \langle\tilde{v}_y^2\rangle = \tilde{v}_\perp^2/2 \text{ and } \langle\tilde{v}_x\tilde{v}_y\rangle = 0. \tag{3.63}$$

Consequently,

$$\langle\tilde{\mathbf{v}}_\perp(\tilde{\mathbf{v}}_\perp\cdot\nabla B)\rangle = \frac{\tilde{v}_\perp^2}{2}\nabla_\perp B = \frac{W_\perp}{m}\nabla_\perp B, \tag{3.64}$$

where again we have defined $W_\perp$ as the perpendicular energy associated with the gyration of the particles, i.e., $W_\perp = \frac{1}{2}m\tilde{v}_\perp^2$, and $\nabla_\perp$ is the perpendicular component of the divergence operator. Since $\mathbf{B}\times\nabla_\perp B = \mathbf{B}\times\nabla B$, on substituting Eq. (3.64) into Eq. (3.61) we get a term that has the same form as the gradient drift, but of the opposite sign.

We must now evaluate the $\delta\mathbf{B}$-dependent term in Eq. (3.61), which at first sight does not appear to correspond to a gradient-drift term. If, however, we integrate Eq. (3.62), then

$$\widetilde{\mathbf{r}}_{\rm L} = \frac{\widetilde{v}_\perp}{\Omega}(\sin \Omega t, \pm \cos \Omega t, 0) = \frac{m}{qB^2}\mathbf{B} \times \widetilde{\mathbf{v}}_\perp, \qquad (3.65)$$

and the last term in Eq. (3.61) becomes

$$\frac{2q\langle\{\mathbf{B}\cdot\delta\mathbf{B}\}\widetilde{\mathbf{v}}_\perp\rangle}{m} = \frac{2\langle\widetilde{\mathbf{v}}_\perp\{\mathbf{B}\cdot[(\mathbf{B}\times\widetilde{\mathbf{v}}_\perp)\cdot\nabla]\mathbf{B}\}\rangle}{B^2} = -\frac{2\langle\widetilde{\mathbf{v}}_\perp[\widetilde{\mathbf{v}}_\perp\cdot(\mathbf{B}\times\nabla B)]\rangle}{B}. \qquad (3.66)$$

The right-hand side of Eq. (3.66) is functionally similar to Eq. (3.64). Consequently, using the definition of $W_\perp$ and again averaging over the gyro-period, we find

$$\frac{2q\langle\{\mathbf{B}\cdot\delta\mathbf{B}\}\widetilde{\mathbf{v}}_\perp\rangle}{m} = -\frac{2W_\perp\mathbf{B}\times\nabla B}{mB}. \qquad (3.67)$$

We can now give the final form of the generalized drift equation:

$$\begin{aligned}\overline{\mathbf{v}}_\perp = \mathbf{v}_{\rm E} + \frac{m}{qB^2}\mathbf{B}\times\Bigg\{&\frac{W_\perp\nabla B}{mB} + \frac{v^2{}_{\parallel}(\mathbf{B}\cdot\nabla)\mathbf{B}}{B^2}\\ &+\frac{v_\parallel}{B}\left[\frac{\partial\mathbf{B}}{\partial t} + (\mathbf{v}_{\rm E}\cdot\nabla)\mathbf{B} + (\mathbf{B}\cdot\nabla)\mathbf{v}_{\rm E}\right]\\ &+(\mathbf{v}_{\rm E}\cdot\nabla)\mathbf{v}_{\rm E} + \frac{\partial\mathbf{v}_{\rm E}}{\partial t}\Bigg\},\end{aligned} \qquad (3.68)$$

and the guiding center velocity is given by $\mathbf{v}_{\rm gc} = \overline{\mathbf{v}}_\perp + v_\parallel\mathbf{B}/B$ (in Eq. (3.57) we explicitly included $\mathbf{v}_{\rm E}$ as a separate term). The terms given on the first row of Eq. (3.68) correspond to the already defined drifts: the E × B, gradient, and curvature drifts. The remaining terms are associated with additional changes in the electric and magnetic fields. Usually they are not named, apart from the last term, which is known as the polarization drift. This drift is mass dependent, and is larger for more massive particles.

Interestingly, the generalized drift does not include any advective terms dependent on $\mathbf{v}_{\rm D}\cdot\nabla$. This does not mean that $|\mathbf{v}_{\rm D}|$ is small in comparison to $|\mathbf{v}_{\rm E}|$. The $\mathbf{v}_{\rm D}\cdot\nabla$-dependent terms vanished because they were small in comparison to the leading term in $\mathbf{v}_{\rm D}$, regardless of the size of $\mathbf{v}_{\rm E}$. There is no constraint on the relative size of these two drifts, except that both drifts assume the fields are varying slowly over a gyro-period and have scale lengths much larger than the gyro-radius.

3.3.7 Summary of particle orbit theory

Equation (3.68) contains all the different drifts experienced by charged particles in the presence of electric and magnetic fields, but in general the three primary drifts are the electric field drift,

$$\mathbf{v}_{\rm E} = \frac{\mathbf{E}\times\mathbf{B}}{B^2}, \qquad (3.17)$$

the gradient drift,

$$\mathbf{v}_{\rm g} = \frac{W_\perp\mathbf{B}\times\nabla B}{qB^3}, \qquad (3.36)$$

and the curvature drift,

$$\mathbf{v}_{\rm c} = \frac{2W_\parallel\mathbf{B}\times(\mathbf{B}\cdot\nabla)\mathbf{B}}{qB^4}. \qquad (3.39)$$

In the special case where there are no local currents within the plasma, $(\mathbf{B}\cdot\nabla)\mathbf{B} = B\nabla B$, and the gradient and curvature drift can be combined as

$$\mathbf{v}_{\rm g+c} = \left(W_\perp + 2W_\parallel\right)\frac{\mathbf{B}\times\nabla B}{qB^3}. \qquad (3.69)$$

Each of these drifts require that the spatial and temporal scales be large in comparison to the particle gyro-radius and gyro-period, respectively.

Under this same assumption we find that the magnetic moment of a particle is constant over a gyro-period. This is the first adiabatic invariant. The three adiabatic invariants are the magnetic moment,

$$\mu = \frac{W_\perp}{B}, \qquad (3.70)$$

the bounce invariant,

$$J = \sqrt{2m\mu}\oint(B_{\rm m} - B)^{1/2}ds, \qquad (3.33)$$

and the L-shell. The L-shell invariant is also known as the flux invariant, as it is equivalent to the statement that the magnetic flux enclosed by a drifting particle is invariant; J and the L-shell are invariant provided changes are slow with respect to the particle bounce period and drift period, respectively.

3.4 KINETIC THEORY – THE VLASOV EQUATION

The preceding section described particle motion in prescribed electric and magnetic fields. But the particles themselves are moving charges and so carry current and charge, resulting in their own magnetic and electric fields. For sufficient numbers of particles these fields can become significant, and the particles themselves modify the fields. These fields also result in interactions between the particles. This strong coupling between particles and fields means that a plasma is not just a collection of non-interacting particles, but instead has more fluid-like properties. As such, the collective effects of the particles within the plasma need to be addressed. This is done through **kinetic theory**. Kinetic theory is the basis for many theoretical aspects of plasma physics, and we introduce some of the concepts of kinetic theory here. The primary objective in this chapter is to use kinetic theory to derive magnetohydrodynamics (MHD). But first we discuss one phenomenon that requires the plasma to be treated as a collection of particles, i.e., plasma oscillations.

3.4.1 Collective phenomena – plasma oscillations

In Section 3.3.1 we showed that there is a characteristic frequency associated with particle motion in a magnetized plasma. This is the cyclotron or gyrofrequency. The other characteristic frequency depends on the collective properties of the plasma. This is the plasma frequency. In deriving this frequency we use a technique that we also use later, specifically in the context of MHD waves at the end of this chapter, and also in Chapter 13, where we discuss plasma waves in more detail. Specifically, we use Eq. (3.7) to determine the effect of an electric field on a plasma, under the assumption that the electric field, and hence the velocity, are small and are therefore only retained to first order. This is known as the **linearization** of equations. We shall assume that n_e is the electron density and that there is only one ion species, with density n_i. The respective masses are m_e and m_i, e is the magnitude of the electron charge, and the ions are singly charged positive ions. For an unmagnetized plasma we can add Eq. (3.7) for both ions and electrons, giving the time rate of change of the current density, $\mathbf{j} = (n_i\mathbf{u}_i - n_e\mathbf{u}_e)e$. In doing this we have implicitly averaged over all the particles that constitute the electron or ion species, in that Eq. (3.7) was written in terms of an individual particle velocity $\mathbf{v}$. For a cold plasma, i.e., one without any thermal velocity, all particles have the same velocity and $\mathbf{u} = \mathbf{v}$. Hence,

$$\frac{\partial \mathbf{j}}{\partial t} = \left(\frac{n_e e^2}{m_e} + \frac{n_i e^2}{m_i}\right)\mathbf{E}. \tag{3.71}$$

The linearization process removes the advective $(\mathbf{u}\cdot\nabla)$ term in the total time derivative, and allows us to ignore the effect of the change in density when calculating the current density. If we now take the divergence of Eq. (3.71) and use Gauss's law (3.3) and charge conservation (3.5), we find, for the charge density $\rho_q = (n_i - n_e)e$,

$$\frac{\partial^2}{\partial t^2}\rho_q = -\left(\frac{n_e e^2}{\varepsilon_0 m_e} + \frac{n_i e^2}{\varepsilon_0 m_i}\right)\rho_q = -\left(\omega_{pe}^2 + \omega_{pi}^2\right)\rho_q. \tag{3.72}$$

In other words, the plasma oscillates at a characteristic frequency known as the plasma frequency $\omega_p = \left(\omega_{pe}^2 + \omega_{pi}^2\right)^{1/2}$, with each species having its own plasma frequency,

$$\omega_{ps}^2 = n_s q^2/\varepsilon_0 m_s \tag{3.73}$$

for the species s. Usually the mass of the ions is much larger than the mass of the electrons and $\omega_p \approx \omega_{pe}$. The period of these oscillations is referred to as the plasma period.

The difference in mass between the ions and electrons is a major factor in understanding the behavior of a plasma. The number densities, on the other hand, are roughly equal as a consequence of quasineutrality, which we discuss in Section 3.4.3. Hence the charge density is usually small, and the current density is mainly given by the difference in the velocities of the two species, but the momentum in the plasma is mainly carried by the ions. Furthermore, the characteristic temporal and spatial scales, which depend on the species mass, are much larger for the ions. In Section 3.2 we presented the concept of guiding center motion. This guiding center motion tends to break down more easily for the ions, in which case the particle trajectories are referred to as non-adiabatic, since the adiabatic invariants are no longer constant.

In deriving Eq. (3.71) we assumed that all the particles in each species were identical, not only in terms of mass and charge, but also in terms of the velocity of the particles (they were assumed to be stationary in the absence of the wave electric field). In general, however, each species consists of a collection of particles, each with its own position and velocity. We need to specify this distribution of particles, and how this distribution varies in response to the forces on the plasma.

3.4.2 The distribution function and the Boltzmann equation

The concept of the distribution function was introduced in Chapter 2, and we repeat some of that discussion here. The distribution function is defined as the density of particles in a phase space given by position and velocity. The distribution function $f(\mathbf{r}, \mathbf{v}, t)$ is a function of position ($\mathbf{r}$), velocity ($\mathbf{v}$), and time (t). By definition, $f(\mathbf{r}, \mathbf{v}, t)\, d^3r\, d^3v$ is the number of particles occupying a volume $d^3r\, d^3v$ in the six-dimensional **r**-**v** phase space at time t. (A useful analogy might be to consider the density: $n(\mathbf{r}, t)\, d^3r$ is the number of particles occupying a volume d^3r in three-dimensional configuration space ($\mathbf{r}$).) For this reason, $f(\mathbf{r}, \mathbf{v}, t)$ is often referred to, interchangeably, as either the distribution function or the **phase space density**. By analogy to density, $f(\mathbf{r}, \mathbf{v}, t)\mathbf{v}$ is the flux in configuration space, and $f(\mathbf{r}, \mathbf{v}, t)\mathbf{a}$ is the flux in velocity space, where $\mathbf{a}$ is the acceleration of the particles, $\mathbf{a} \equiv d\mathbf{v}/dt$. We can write a conservation equation for phase space density:

$$\frac{\partial f}{\partial t} + \frac{\partial}{\partial \mathbf{r}} \cdot (f\mathbf{v}) + \frac{\partial}{\partial \mathbf{v}} \cdot (f\mathbf{a}) = \left.\frac{\partial f}{\partial t}\right|_c \quad (3.74)$$

In Eq. (3.74) and subsequent equations $f(\mathbf{r}, \mathbf{v}, t)$ has been replaced by f for convenience. The term on the right-hand side of Eq. (3.74) represents the change in f due to collisions. We do not attempt to specify this term. The left-hand side of Eq. (3.74) can be understood in terms of conservation of particles in the six-dimensional phase space.

In Eq. (3.74) $\mathbf{v}$ is a coordinate within the phase space, and as such is independent of position ($\mathbf{r}$). Usually a particle's acceleration $\mathbf{a}$ is independent of velocity (e.g., $q\mathbf{E}/m$ for electric fields, or $\mathbf{g}$ for gravity), with the exceptions of the $\mathbf{v} \times \mathbf{B}$ force and friction. Friction is generally due to collisions, and so could be included in the collisional term. Because $\mathbf{v} \times \mathbf{B}$ is perpendicular to $\mathbf{v}$, $\nabla_\mathbf{v} \cdot \mathbf{v} \times \mathbf{B} = 0$, where, similarly to the divergence operator in configuration space, $\nabla_\mathbf{v} \equiv \partial/\partial\mathbf{v}$. We can consequently rewrite Eq. (3.74) as

$$\frac{\partial f}{\partial t} + \mathbf{v} \cdot \nabla f + \mathbf{a} \cdot \nabla_\mathbf{v} f = \left.\frac{\partial f}{\partial t}\right|_c. \quad (3.75a)$$

This is the Boltzmann equation.

If collisions can be neglected we can set the right-hand side of Eq. (3.75a) to zero:

$$\frac{\partial f}{\partial t} + \mathbf{v} \cdot \nabla f + \mathbf{a} \cdot \nabla_\mathbf{v} f = 0, \quad (3.75b)$$

which is the collisionless Boltzmann or **Vlasov equation**.

We gave Eq. (3.74) without proof, using the albeit reasonable assumption that phase space density is a conserved quantity in the absence of collisions. The Boltzmann equation can be derived formally using statistical mechanics. Most classical textbooks on plasma physics, however, use arguments similar to those presented here to introduce the concept of phase space density and the Boltzmann and Vlasov equations, and defer the detailed statistical mechanics argument until later. While a fully rigorous derivation of Eqs. (3.75a) and (3.75b) is desirable, we do not present that here, but rather refer the reader to other texts, as given in the Additional reading section at the end of this chapter.

We shall use Eqs. (3.75a) and (3.75b) to derive the MHD equations. These equations are also of fundamental importance for plasma physics, beyond the realm of MHD. For example, we use kinetic theory to take into account phenomena such as wave–particle interactions, and Eq. (3.75b) allows us to develop wave dispersion analyses for warm plasmas that include such effects as **Landau damping.** One theorem that is the basis of much plasma physics is the **Liouville theorem.** This theorem states that the phase space density is constant along a particle trajectory. Thus if we assume that, at any instant in time t, a particle at the position $\mathbf{r}, \mathbf{v}$ in the phase space subsequently moves to $\mathbf{r}', \mathbf{v}'$ at time t', and if the phase density at $\mathbf{r}, \mathbf{v}$ is $f(\mathbf{r}, \mathbf{v}, t)$, and the phase density at $\mathbf{r}', \mathbf{v}'$ is $f(\mathbf{r}', \mathbf{v}', t')$, then the Liouville theorem states that $f(\mathbf{r}, \mathbf{v}, t) = f(\mathbf{r}', \mathbf{v}', t')$. This can be seen from Eq. (3.75b) once it is recognized that the operator $\partial/\partial t + \mathbf{v} \cdot \nabla + \mathbf{a} \cdot \nabla_{\mathrm{v}} \equiv \mathscr{L}$ is the gradient along a particle trajectory, since $\mathbf{a} \equiv d\mathbf{v}/dt$ and $\mathbf{v} \equiv d\mathbf{r}/dt$. We refer to $\mathscr{L}$ as the Liouville operator.

3.4.3 Debye shielding and the plasma state

Having introduced the concept of the distribution function, we now use the Liouville theorem to derive the Debye length. This is the characteristic length over which an isolated charge in a plasma is shielded from the plasma. The electric field of the isolated charge vanishes for distances large compared to the Debye length. First, we assume a maxwellian velocity distribution

$$f = \frac{n}{(2\pi KT)^{1/2}} \exp\left(\frac{-v^2}{2mKT}\right), \quad (3.76)$$

where n is the density, m is the species mass, T is the temperature, and K is the Boltzmann constant.

We now assume a potential ϕ that is associated with a charge q, where the electric field associated with the charge is given by $\mathbf{E} = -\nabla\phi$, where ϕ is the **electric potential.** We choose to set $\phi = 0$ at infinity. If q is positive, then the ions in the plasma will be slowed down as they approach the charge, with some fraction of the distribution being reflected. The Liouville theorem then states that

$$f_{\mathrm{i}} = \frac{n}{(2\pi KT_{\mathrm{i}})^{1/2}} \exp\left(-\frac{e\phi}{KT_{\mathrm{i}}} - \frac{v^2}{2mKT_{\mathrm{i}}}\right), \quad (3.77\mathrm{a})$$

where the subscript "i" denotes ions, and we have further assumed these are singly charged.

For the electrons, on the other hand, we find

$$f_{\mathrm{e}} = \frac{n}{(2\pi KT_{\mathrm{e}})^{1/2}} \exp\left(\frac{e\phi}{KT_{\mathrm{e}}} - \frac{v^2}{2mKT_{\mathrm{e}}}\right). \quad (3.77\mathrm{b})$$

In applying the Liouville theorem for the electrons we are now taking account of those electrons that are initially moving away from the charge but are then reflected before they reach infinity. At infinity the electrons and ions have the same density. If q is negative, then ϕ is also negative.

We now make the simplifying assumption that $|e\phi| \ll KT$ for both species, in which case the charge density is given by

$$\rho_{\mathrm{q}} = -\frac{ne^2}{K}\left(\frac{1}{T_{\mathrm{e}}} + \frac{1}{T_{\mathrm{i}}}\right)\phi. \quad (3.78)$$

From Gauss's law

$$\nabla^2\phi = \frac{ne^2}{\varepsilon_0 K}\left(\frac{1}{T_{\mathrm{e}}} + \frac{1}{T_{\mathrm{i}}}\right)\phi. \quad (3.79)$$

This has the solution

$$\phi = \frac{q}{4\pi\varepsilon_0 r}\exp(-r/\lambda) \quad (3.80)$$

for an isolated charge q, with

$$\lambda^2 = \frac{\varepsilon_0 K}{ne^2}\frac{T_{\mathrm{e}}T_{\mathrm{i}}}{(T_{\mathrm{e}} + T_{\mathrm{i}})}. \quad (3.81)$$

If we further assume that the electrons and ions have the same temperature, we get the classical result that $\lambda = \lambda_{\mathrm{D}}/\sqrt{2}$, with the Debye length

$$\lambda_{\mathrm{D}} = \left(\frac{\varepsilon_0 KT}{ne^2}\right)^{1/2}. \quad (3.82)$$

Taking the thermal velocity as $\frac{1}{2}mv_{\mathrm{T}}^2 = KT$ (the definition of the thermal velocity is somewhat arbitrary; often the factor 1/2 is not included in the definition), we can also give the Debye length as

$$\lambda_{\mathrm{D}} = \frac{1}{\sqrt{2}}\frac{v_{\mathrm{T}}}{\omega_{\mathrm{p}}} \quad (3.83)$$

for equal ion and electron temperatures.

If the ions and electrons have different temperatures, we can rewrite Eq. (3.81) as

$$\frac{1}{\lambda^2} = \frac{1}{\lambda_{\mathrm{De}}^2} + \frac{1}{\lambda_{\mathrm{Di}}^2}, \quad (3.84)$$

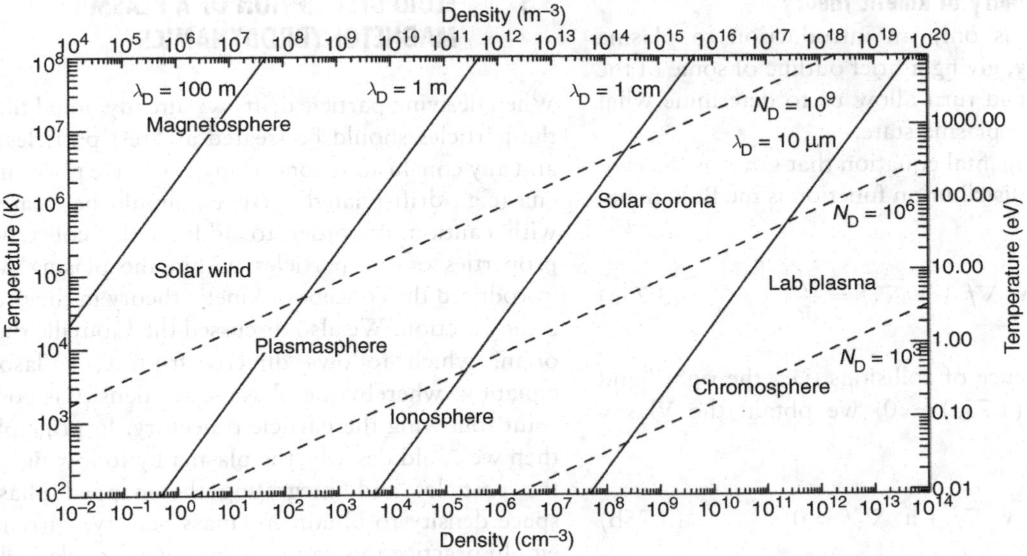

FIGURE 3.7. Characteristic densities and temperatures for a variety of plasmas and the associated Debye lengths (λ_D) and number of particles per Debye sphere (N_D).

where the Debye lengths are determined using Eq. (3.82) (or Eq. (3.83)) separately for each species.

Equation (3.80) shows that the potential, and hence the electric field, from an isolated charge is shielded from the plasma over a distance scale length given by the Debye length. We have implicitly assumed, however, that there are sufficient particles available for the potential to be shielded in that length. This leads to a fundamental definition of a plasma: that the number of particles in a sphere with a Debye length radius is $\gg 1$. That is,

$$N_D = \frac{4}{3}\pi n\lambda_D^3 = \frac{4}{3}\pi n\left(\frac{\varepsilon_0 KT}{ne^2}\right)^{3/2} \gg 1, \qquad (3.85)$$

where N_D is the number of particles in a Debye sphere. Figure 3.7 shows characteristic densities and temperatures for a variety of plasmas and the associated λ_D and N_D.

Closely related to Debye shielding is the concept of quasi-neutrality, in the sense that any large charge imbalances with the plasma cannot exist over long time or spatial scales. The characteristic time for the plasma to adjust and remove the charge imbalance is the plasma period. We give an example that emphasizes the robustness of the assumption of quasi-neutrality. At the Earth the inner magnetospheric plasma, especially the plasmasphere, is strongly coupled to the ionosphere and, through collisions, the neutral atmosphere, and as such tends to corotate with the Earth. At the equator the corotation velocity is $460L$ m s^{-1}, where L is the radial distance in Earth radii (i.e., $L = 1$ corresponds to the surface of the Earth, and $L = 6.6$ is the radial distance of geosynchronous orbit). The magnetic field is roughly dipolar, and, assuming a surface magnetic field of 30 000 nT, from Eq. (3.17) we get an electric field of $14/L^2$ mV m^{-1}. If we then assume that the divergence of the electric field varies as $1/L$, from Gauss's law (3.3) we find that the particle density required to maintain the electric field (δn) is ~$10^{-7}/L^3$ cm^{-3}, assuming each particle carries a charge e. Within the ionosphere the plasma density is ~10^5 cm^{-3}, and $\delta n/n$ ~ 10^{-12}. At $L = 3$, within the plasmasphere, n ~ 10^3 cm^{-3}, and $\delta n/n$ ~ 4×10^{-12}. In the outer magnetosphere, electric fields are typically of the order 1 mV m^{-1}, and the magnetic field strength is around 10 nT. Again, somewhat arbitrarily assuming a 1 Earth radius scale length, we get δn ~ 10^{-8} cm^{-3}. Even in the magnetospheric lobes we expect $\delta n/n \ll 1$.

3.4.4 Summary of kinetic theory

This section is only an introduction to plasma kinetic theory, giving a brief outline of some of the concepts that in turn allow us to determine what constitutes the plasma state.

The fundamental equation that governs the evolution of the distribution function is the Boltzmann equation,

$$\frac{\partial f}{\partial t} + \mathbf{v} \cdot \nabla f + \mathbf{a} \cdot \nabla_{\mathrm{v}} f = \left. \frac{\partial f}{\partial t} \right|_{\mathrm{c}}. \tag{3.75a}$$

In the absence of collisions (i.e., the right-hand side of Eq. (3.75a) = 0) we obtain the Vlasov equation,

$$\frac{\partial f}{\partial t} + \mathbf{v} \cdot \nabla f + \mathbf{a} \cdot \nabla_{\mathrm{v}} f = 0. \tag{3.75b}$$

This can be rewritten as

$$\mathscr{L} f = 0, \tag{3.86}$$

where $\mathscr{L}$ is the Liouville operator given by

$$\mathscr{L} \equiv \partial/\partial t + \mathbf{v} \cdot \nabla + \mathbf{a} \cdot \nabla_{\mathbf{v}}. \tag{3.87}$$

Equation (3.86) also leads to the Liouville theorem, which states that the phase space density is constant along a particle trajectory.

Finally, the three parameters that characterize the plasma state are the electron plasma frequency,

$$\omega_{\mathrm{pe}}^2 = n_{\mathrm{e}} e^2 / \varepsilon_0 m_{\mathrm{e}}, \tag{3.88}$$

the Debye length

$$\lambda_{\mathrm{D}} = \left(\frac{\varepsilon_0 KT}{ne^2} \right)^{1/2}, \tag{3.82}$$

and the number of particles in a Debye sphere,

$$N_{\mathrm{D}} = \frac{4}{3} \pi n \lambda_{\mathrm{D}}^3, \tag{3.89}$$

where we have assumed that the ion and electron species have the same temperature T. A mixture of ions and electrons constitutes a plasma provided $N_{\mathrm{D}} \gg 1$.

3.5 FLUID DESCRIPTION OF A PLASMA – MAGNETOHYDRODYNAMICS

When deriving particle drifts we already noted that the particles should be treated as "test particles," and any conclusions concerning collective phenomena (e.g., drift-related currents) should be treated with caution. In order to address the collective properties of the particles within the plasma we introduced the concept of kinetic theory in the preceding section. We also discussed the Liouville theorem, which follows directly from the Vlasov equation, whereby the phase space density is constant following the particle trajectory. In principle then we could describe the plasma by following all the particles and integrating the mapped phase space density to obtain the mass density, current, etc. In practice this can only be done for relatively well-defined geometries, or small volumes, such that the number of particles being traced is manageable. Instead we use the Boltzmann equation (3.75a) to derive a set of equations that describe the evolution of the bulk plasma parameters themselves. These equations are analogous to the fluid equations of hydrodynamics, and are therefore called magnetohydrodynamics or MHD equations.

The most significant assumption in deriving the MHD equations is that the plasma is "localized." This means that rates of change of plasma properties depend only on properties of the plasma in the immediate vicinity of the point where the rates of change are determined. In a normal fluid this is done through collisions. For example, even though the atoms or molecules within a gas can have very high individual speeds, we do not need to follow the trajectories of all the particles in the gas to determine bulk properties of the gas. In space, plasma collisions can be infrequent, and as such the plasma may not appear to be localized. The magnetic field can act to localize the plasma partially, since individual particles gyrate around the magnetic field, but even that localization is not ideal because there are no collisions to mix the plasma fully. As far as motion along the magnetic field is concerned, clearly the magnetic field cannot act to localize the plasma. It is possible that wave–particle interactions could do this by removing sharp gradients in the phase space density on relatively short time scales. While the MHD equations seem to require an assumption for validity that is difficult to justify for collisionless plasmas, they provide a very good means for understanding collective plasma phenomena.

The Liouville theorem indicates why localization is an issue. If we consider a region in phase space at time t that is occupied by particles that have significantly different histories, i.e., they have come from significantly different regions within the phase space, then the corresponding phase space density at time t includes contributions from particles that themselves may have significantly different phase space densities. For example, let us assume that one of the local particles came from a region of hot plasma, while another particle came from a region of cold plasma. In that case the local phase space density will include portions of the phase space density from both the hot and the cold plasma. Collisions will act to average these different phase densities in a collision-dominated gas. Without collisions, other processes, such as plasma-wave generation, are required to mix the phase space density. This is therefore what localization means. At every position local processes act to average out the structure implied by the Liouville theorem. The history of the individual particle trajectories is lost.

3.5.1 Generalized moment equation

Now we proceed with the derivation of the fluid equations for MHD. We first derive a general form, and then derive the continuity, momentum, and energy equations. To do this we take moment integrals of the Boltzmann equation (3.75a), that is, we perform integrals over velocity space of the form $\int \varphi(\mathbf{v})\mathscr{L} f d^3v$, where again $\mathscr{L}$ is the Liouville operator and $\varphi(\mathbf{v})$ is some yet to be specified function of $\mathbf{v}$. The moment integral of Eq. (3.75a) is

$$\int \varphi(\mathbf{v})\frac{\partial f}{\partial t}d^3v + \int \varphi(\mathbf{v})\mathbf{v}\cdot\nabla f d^3v + \int \varphi(\mathbf{v})\mathbf{a}\cdot\nabla_v f d^3v = \int \varphi(\mathbf{v})\left.\frac{\partial f}{\partial t}\right|_c d^3v. \tag{3.90}$$

As noted earlier, since $\mathbf{v}$ is a coordinate of the phase space, $\partial\varphi(\mathbf{v})/\partial t = 0$ and $\nabla\varphi(\mathbf{v}) = 0$. We can therefore rewrite Eq. (3.90) as

$$\frac{\partial}{\partial t}\int \varphi(\mathbf{v}) f d^3v + \nabla\cdot\int \mathbf{v}\varphi(\mathbf{v}) f d^3v + \int \varphi(\mathbf{v})\mathbf{a}\cdot\nabla_v f d^3v = \int \varphi(\mathbf{v})\left.\frac{\partial f}{\partial t}\right|_c d^3v. \tag{3.91}$$

We can now rewrite the first two integrals in Eq. (3.91) in terms of expectation or average values. For example,

$$\frac{\int \varphi(\mathbf{v}) f d^3v}{\int f d^3v} = \langle\varphi(\mathbf{v})\rangle \tag{3.92a}$$

or

$$\int \varphi(\mathbf{v}) f d^3v = n\langle\varphi(\mathbf{v})\rangle, \tag{3.92b}$$

where n is the particle density, $\int f d^3v = n$, and n is implicitly a function of $\mathbf{r}$, t, as is $\langle\varphi(\mathbf{v})\rangle$. Thus

$$\frac{\partial}{\partial t}\left(n\langle\varphi(\mathbf{v})\rangle\right) + \nabla\cdot\left(n\langle\mathbf{v}\varphi(\mathbf{v})\rangle\right) + \int \varphi(\mathbf{v})\mathbf{a}\cdot\nabla_v f d^3v = \left.\frac{\partial}{\partial t}\left(n\langle\varphi(\mathbf{v})\rangle\right)\right|_c. \tag{3.93}$$

The right-hand side of Eq. (3.93) is not well defined. As noted earlier we have not yet considered what the collisional processes are. For now we use the collisional term as a place-holder for a term that needs to be specified later.

The third term on the left-hand side of Eq. (3.93) is evaluated through integration by parts. By definition, the phase density f is assumed to approach zero sufficiently rapidly as $v\to\infty$ that $\varphi(\mathbf{v})f\to 0$ for any function $\varphi(\mathbf{v})$. Hence

$$\int \varphi(\mathbf{v})\mathbf{a}\cdot\nabla_v f d^3v = -\int f\nabla_v\cdot\left(\mathbf{a}\varphi(\mathbf{v})\right)d^3v. \tag{3.94}$$

Earlier, when deriving the Boltzmann equation, we noted that $\mathbf{a}$ is independent of velocity or only includes $\mathbf{v}\times\mathbf{B}$ terms. Equation (3.93) therefore becomes

$$\frac{\partial}{\partial t}\left(n\langle\varphi(\mathbf{v})\rangle\right) + \nabla\cdot\left(n\langle\mathbf{v}\varphi(\mathbf{v})\rangle\right) - n\langle(\mathbf{a}\cdot\nabla_v)\varphi(\mathbf{v})\rangle = \left.\frac{\partial}{\partial t}\left(n\langle\varphi(\mathbf{v})\rangle\right)\right|_c. \tag{3.95}$$

Equation (3.95) is a generalized form of the moment equations that together with Maxwell's equations constitute MHD. The next stage is to choose the functions $\varphi(\mathbf{v})$, but before we do that inspection of Eq. (3.95) shows one of the fundamental issues in MHD. In particular, for any function $\varphi(\mathbf{v})$, the time rate of change $\partial\langle\varphi(\mathbf{v})\rangle/\partial t$ depends on the divergence $\nabla\cdot\langle\mathbf{v}\varphi(\mathbf{v})\rangle$. That is, closure of the equations for any order moment requires knowledge of a higher-order moment. As we shall see later, the issue for MHD is what assumptions can be made that close the equations as a finite set.

3.5.2 Continuity equations

The first equation we shall derive assumes $\varphi(\mathbf{v}) = 1$. We have already noted that $\int f d^3v = n$ (clearly $\langle 1\rangle = 1$). We shall denote $\langle\mathbf{v}\rangle = \bar{\mathbf{v}}$, which corresponds to the average or bulk flow velocity of the particle species. Sometimes this is denoted by $\mathbf{u}$, but to avoid potential later confusion we reserve $\mathbf{u}$ without any subscripts to denote the flow velocity of the plasma, not the flow velocity of the constituent species within the plasma. The last term on the left-hand side of Eq. (3.95) will vanish since the velocity gradient vanishes. Thus for the zero-order moment

$$\frac{\partial n}{\partial t} + \nabla\cdot(n\bar{\mathbf{v}}) = \left.\frac{\partial n}{\partial t}\right|_{\mathrm{c}}. \tag{3.96}$$

Equation (3.96) is typically referred to as the continuity equation (the higher-order moments of Eq. (3.95) are also continuity equations in the general sense, but they are often distinguished from Eq. (3.96), being referred to as the momentum equation, etc.). Equation (3.96) applies for each species in the plasma. As yet we have still not specified the nature of the collisional term on the right-hand side of Eq. (3.96), but we note that if the collisions do not involve any dissociation or recombination processes, then no particles are created in the plasma, and the right-hand side vanishes.

We also note that Eq. (3.96) shows that $\bar{\mathbf{v}}$ must be specified in order to determine $\partial n/\partial t$. Thus we need a first-order equation, and we assume $\varphi(\mathbf{v}) = m\mathbf{v}$, where m is the particle species mass. In that case Eq. (3.95) becomes

$$\frac{\partial}{\partial t}(mn\bar{\mathbf{v}}) + \nabla\cdot(mn\langle\mathbf{v}\mathbf{v}\rangle) - nm\langle\mathbf{a}\rangle = \left.\frac{\partial}{\partial t}(mn\bar{\mathbf{v}})\right|_{\mathrm{c}}. \tag{3.97}$$

This is the momentum equation, but before we proceed we need to evaluate $\langle\mathbf{v}\mathbf{v}\rangle$ and $\langle\mathbf{a}\rangle$. First, in order to clarify the second term in Eq. (3.97), we make use of the **summation convention.** To do this we denote the i-component of a vector $\mathbf{v}$ as v_i. By convention, i is a number (1, 2, or 3) corresponding to the first, second, or third coordinate (e.g., 1 corresponds to x, 2 to y, and 3 to z for cartesian coordinates). In addition we assume that if two indices repeat then it is implied that there is a summation over the components, e.g., the scalar dot product $\mathbf{a}\cdot\mathbf{b}$ can be represented as a_ib_i. With this notation the i-component of the second term on the left-hand side of Eq. (3.97) can be written as

$$\frac{\partial}{\partial x_j}(nm\langle v_jv_i\rangle),$$

where the repetition of the index j indicates the scalar dot product.

To proceed further we return to the integral representation,

$$n\langle v_jv_i\rangle = \int fv_jv_id^3v. \tag{3.98}$$

We now define $\mathbf{w} = \mathbf{v} - \bar{\mathbf{u}}$, where $\mathbf{w}$ is the velocity with respect to some bulk velocity $\bar{\mathbf{u}}$. For the time being, we do not define $\bar{\mathbf{u}}$, but, depending on the final form of the MHD equations, $\bar{\mathbf{u}}$ will be the plasma flow velocity for single-fluid MHD, or the flow velocity of each species in two-fluid or multi-fluid MHD. An aside may be in order here. So far we have not indicated species in our derivation and the equations as given apply separately for each species in the plasma. We indicate species separately when we combine the equations later. The velocity $\mathbf{w}$ is sometimes referred to as the random or **peculiar velocity**. Previously we noted that $\langle\mathbf{v}\rangle = \bar{\mathbf{v}}$, and therefore $\langle\mathbf{w}\rangle = \bar{\mathbf{v}} - \bar{\mathbf{u}}$. We can now expand Eq. (3.98):

$$\int f v_j v_i d^3 v$$
$$= \int f(w_j + \bar{u}_j)(w_i + \bar{u}_i) d^3 v$$
$$= n[\langle w_j w_i \rangle + (\bar{v}_j - \bar{u}_j)\bar{u}_i + \bar{u}_j(\bar{v}_i - \bar{u}_i) + \bar{u}_j \bar{u}_i]$$
$$= n(\langle w_j w_i \rangle + \bar{v}_j \bar{u}_i + \bar{u}_j \bar{v}_i - \bar{u}_j \bar{u}_i). \quad (3.99)$$

We note that in a gas, pressure is defined as $mn\langle w_j w_i \rangle = P_{ji}$, and we shall use this definition here. Using pressure for $mn\langle w_j w_i \rangle$ is why we leave $\bar{\mathbf{u}}$ unspecified. In single-fluid MHD, where the mass, momentum, and energy equations are combined to form a single set of equations for the plasma, $\bar{\mathbf{u}} = \mathbf{u}$, the bulk flow of the plasma. In that case the pressures for each species are defined with respect to $\mathbf{u}$. In multi-fluid MHD, the mass, momentum, and energy equations are specified for each species separately. In that case $\bar{\mathbf{u}} = \bar{\mathbf{v}}$ for each species, and the pressures for each species are defined with respect to the individual species flow velocities. These flow velocities need not be equal to $\mathbf{u}$, the average velocity for the plasma, and so the pressures will be defined differently in single-fluid and multi-fluid MHD. The differences are usually small, but formally we keep the distinction between $\bar{\mathbf{v}}$, $\mathbf{u}$, and $\bar{\mathbf{u}}$.

The last term in Eq. (3.97) to be evaluated is $\langle \mathbf{a} \rangle$ (again we defer our discussion of the collisional terms until later). In general, $m\mathbf{a} = q(\mathbf{E} + \mathbf{v} \times \mathbf{B}) + \mathbf{F}_g$, where $\mathbf{F}_g$ can include the force due to gravity ($m\mathbf{g}$) and any other non-electromagnetic velocity-independent forces. Since the integration over velocity only includes $\mathbf{v}$ to first order, $m\langle \mathbf{a} \rangle = q(\mathbf{E} + \bar{\mathbf{v}} \times \mathbf{B}) + \mathbf{F}_g$. The momentum equation for each species therefore becomes

$$\frac{\partial}{\partial t}(mn\bar{v}_i) + \frac{\partial}{\partial x_j}[P_{ji} + nm(\bar{v}_j \bar{u}_i + \bar{u}_j \bar{v}_i - \bar{u}_j \bar{u}_i)]$$
$$- n\left[q\left(E_i + (\bar{\mathbf{v}} \times \mathbf{B})_i\right) + F_{gi}\right] = \frac{\partial}{\partial t}(mn\bar{v}_i)\bigg|_c. \quad (3.100)$$

For clarity, we have used subscript notation to show the form of the gradient term more clearly.

As with the continuity equation, Eq. (3.100) requires that a higher-order term (P_{ji}) be specified. Clearly one method to close the equations is to assume a cold plasma, and the pressure term vanishes. If the plasma is warm, however, we must investigate the next higher-order moment equation. For this purpose we take $\varphi(\mathbf{v}) = \frac{1}{2}mv^2$, i.e., particle energy. In subscript notation $v^2 = v_i v_i$, and so Eq. (3.95) becomes

$$\frac{\partial}{\partial t}\left(n\left\langle \frac{1}{2}mv_i v_i \right\rangle\right) + \frac{\partial}{\partial x_j}\left(n\left\langle \frac{1}{2}mv_j v_i v_i \right\rangle\right)$$
$$- n\left\langle a_j \frac{\partial}{\partial v_j}\frac{1}{2}mv_i v_i \right\rangle = \frac{\partial}{\partial t} n\left\langle \frac{1}{2}mv_i v_i \right\rangle\bigg|_c. \quad (3.101)$$

From Eq. (3.99) and the definition of pressure,

$$n\left\langle \frac{1}{2}mv_i v_i \right\rangle = \frac{1}{2}P_{ii} + nm\left(\bar{v}_i \bar{u}_i - \frac{1}{2}\bar{u}_i \bar{u}_i\right). \quad (3.102)$$

Similarly, for the second term on the left-hand side of Eq. (3.101), we can show that

$$n\left\langle \frac{1}{2}mv_j v_i v_i \right\rangle = n\left\langle \frac{1}{2}m(w_j + \bar{u}_j)(w_i + \bar{u}_i)(w_i + \bar{u}_i) \right\rangle$$
$$= \frac{1}{2}nm\langle w_j w_i w_i \rangle + P_{ji}\bar{u}_i + \frac{1}{2}P_{ii}\bar{u}_j$$
$$+ nm\left(\frac{1}{2}\bar{v}_j \bar{u}_i \bar{u}_i + \bar{u}_j \bar{v}_i \bar{u}_i - \bar{u}_j \bar{u}_i \bar{u}_i\right). \quad (3.103)$$

We shall define the **heat flux** per species as

$$q_j = \frac{1}{2}nm\langle w_j w_i w_i \rangle. \quad (3.104)$$

Note that we also use the symbol q for particle charge, but there should be no ambiguity as the heat flux is a vector. Also, as with the pressure term, $\mathbf{q}$ will be slightly different in the single-fluid and multi-fluid forms of MHD because of the different species flow velocities.

Again setting aside the collision terms for later consideration, the remaining term to be evaluated is the last term on the left-hand side of Eq. (3.101):

$$n\left\langle a_j \frac{\partial}{\partial v_j}\frac{1}{2}mv_i v_i \right\rangle = n\langle a_j m \delta_{ji} v_i \rangle = nm\langle \mathbf{a} \cdot \mathbf{v} \rangle. \quad (3.105)$$

Once again, we note that the only velocity-dependent terms in $\mathbf{a}$ are of the form $\mathbf{v} \times \mathbf{B}$, and so $nm\langle \mathbf{a} \cdot \mathbf{v} \rangle = n(q\mathbf{E} + \mathbf{F}_g) \cdot \bar{\mathbf{v}}$. We therefore come to the energy equation for each species:

$$\frac{\partial}{\partial t}\left[\frac{1}{2}P_{ii}+nm\left(\overline{v}_i\overline{u}_i-\frac{1}{2}\overline{u}_i\overline{u}_i\right)\right]+\frac{\partial}{\partial x_j}\left[q_j+P_{ji}\overline{u}_i+\frac{1}{2}P_{ii}\overline{u}_j\right.$$

$$\left.+\,nm\left(\frac{1}{2}\overline{v}_j\overline{u}_i\overline{u}_i+\overline{u}_j\overline{v}_i\overline{u}_i-\overline{u}_j\overline{u}_i\overline{u}_i\right)\right]-n(qE_i+F_{gi})\overline{v}_i$$

$$=\frac{\partial}{\partial t}n\langle\frac{1}{2}mv_iv_i\rangle\Big|_{\mathrm{c}}. \tag{3.106}$$

We could continue the hierarchy of equations, deriving an equation that relates the time rate of change of the heat flux to a higher-order moment. Conventionally, however, we usually stop at the energy equation (3.106). At first sight this equation appears complicated, but we shall see later that the energy equation reduces to a relatively simple form.

One other point to be made with regard to Eq. (3.106) is that heat flux $\mathbf{q}$ has a very specific meaning. This term is derived from the third-order moment of the phase space density integral and is related to the skew of the distribution with respect to the reference velocity $\overline{\mathbf{u}}$. There are other energy flux terms in Eq. (3.106), but these are associated with the advection of thermal and dynamic pressure. These energy fluxes are specified through multiples of lower-order moments. Only $\mathbf{q}$ as defined here requires the evaluation of a third-order moment.

3.5.3 Single-fluid magnetohydrodynamics

To proceed further, we now make a choice as to whether we should derive single-fluid or multi-fluid equations for MHD. Here we derive single-fluid MHD by assuming $\overline{\mathbf{u}} = \mathbf{u}$ and defining the following quantities:

$$\text{mass density } \rho=\sum_s n_s m_s, \tag{3.107}$$

$$\text{charge density } \rho_{\mathrm{q}}=\sum_s n_s q_s, \tag{3.108}$$

$$\text{flow velocity } \mathbf{u}=\sum_s n_s m_s \mathbf{u}_s/\rho,\ \text{with } \mathbf{u}_s\equiv\overline{\mathbf{v}}_s, \tag{3.109}$$

$$\text{current density } \mathbf{j}=\sum_s n_s q_s \mathbf{u}_s, \tag{3.110}$$

$$\text{total pressure tensor } P_{ij}=\sum_s P_{sij}, \tag{3.111}$$

$$\text{heat flux } \mathbf{q}=\sum_s \mathbf{q}_s, \tag{3.112}$$

where we have summed over all species in the plasma, as indicated by the subscript "s." We now also use $\mathbf{u}_s$ as the species flow velocity, and again q_s corresponds to species charge, and $\mathbf{q}_s$ to species heat flux.

The collisional terms also simplify somewhat when deriving single-fluid MHD. In particular, for collisions between species within the plasma, there is no change in total mass, momentum, or energy. Thus, on combining the continuity, momentum, and energy equations, only changes due to collisions with external scattering centers remain (e.g., neutral atoms in the ionosphere, or charge exchange in a gas torus).

On multiplying Eq. (3.96) by mass and summing over species

$$\frac{\partial\rho}{\partial t}+\nabla\cdot(\rho\mathbf{u})=\frac{\partial\rho}{\partial t}\Big|_{\mathrm{n}\to\mathrm{p}}, \tag{3.113}$$

where on the right-hand side we have replaced the subscript c with n→p to make it clear that any mass addition or loss involves processes external to the plasma, such as dissociation or recombination. These processes are external in the sense that, while recombination, for example, does not destroy mass, when an ion and electron recombine to form a neutral particle the mass is lost from the plasma. The sign convention implied by the arrow in Eq. (3.113) is that mass is added to the plasma from the neutrals.

If instead we multiply the continuity equation (3.96) by charge and sum over species, we obtain the charge conservation equation

$$\frac{\partial\rho_{\mathrm{q}}}{\partial t}+\nabla\cdot\mathbf{j}=0. \tag{3.114}$$

In Eq. (3.114) there are no external sources of charge, and the right-hand side equals zero. This is because charge cannot be created, regardless of the collisional process. For example, recombination may remove mass from the plasma, but the net charge is the same. Equation (3.114) is therefore identical to Eq. (3.5), which derived charge conservation from Maxwell's equations, while Eq. (3.114) used moment integrals of the Boltzmann equation.

Summing Eq. (3.100) over species gives

$$\frac{\partial}{\partial t}(\rho \mathbf{u})+\nabla\cdot(\mathbf{P}+\rho\mathbf{uu})-\rho_q\mathbf{E}-\mathbf{j}\times\mathbf{B}-\rho\mathbf{g}=\left.\frac{\partial \mathbf{p}}{\partial t}\right|_{n\to p}, \tag{3.115}$$

which is the single-fluid momentum equation. For simplicity we have assumed that $\mathbf{F}_g = m\mathbf{g}$, otherwise the last term on the left-hand side of Eq. (3.114) is slightly more complicated. In Eq. (3.115) $\mathbf{P}$ is a tensor, and, using the summation convention, the divergence term is given by

$$\nabla\cdot(\mathbf{P}+\rho\mathbf{uu})=\frac{\partial}{\partial x_j}(P_{ji}+\rho u_j u_i). \tag{3.116}$$

In Eq. (3.115) the term on the right-hand side indicates changes in momentum ($\mathbf{p}$) through external processes. One process that can change the plasma momentum is charge exchange, where an ion gains an electron to become a neutral particle, and a neutral particle loses an electron to become an ion. Since the initial ion and neutral particle will in general have different velocities, the effect of charge exchange is to add or remove momentum from the plasma without changing the total number of ions or electrons within the plasma. Charge exchange may also change the mass density if the ion and neutral species are different, e.g., when a proton exchanges charge with an oxygen atom.

Finally, summing (3.106) over species, we get the energy equation

$$\frac{\partial}{\partial t}\left(\frac{1}{2}P_{ii}+\frac{1}{2}\rho u^2\right)+\nabla\cdot\left(\mathbf{q}+\mathbf{P}\cdot\mathbf{u}+\frac{1}{2}P_{ii}\mathbf{u}+\frac{1}{2}\rho u^2\mathbf{u}\right)$$
$$-\mathbf{j}\cdot\mathbf{E}-\rho\mathbf{u}\cdot\mathbf{g}=\left.\frac{\partial W}{\partial t}\right|_{n\to p}. \tag{3.117}$$

Similarly to Eqs. (3.113) and (3.115), the term on the right-hand side of Eq. (3.117) is the rate of change of the energy density due to interactions with particles not included in the plasma. Also, as in Eq. (3.115), $\mathbf{P}$ is a tensor and

$$\nabla\cdot(\mathbf{P}\cdot\mathbf{u})=\frac{\partial}{\partial x_i}P_{ij}u_j. \tag{3.118}$$

The set of equations (3.113), (3.115), and (3.117) is essentially complete and as written allows for the inclusions of collisions in a relatively straightforward manner. It is also clear that the equations will close if we assume $\nabla\cdot\mathbf{q}=0$. This is the usual assumption made to close the MHD equations. Sometimes it is stated that the closure is achieved by assuming $\mathbf{q} = 0$. Certainly this assumption will close the set of equations, but the condition $\nabla\cdot\mathbf{q}=0$ allows for the plasma to have a heat flux. The only time the heat flux is of concern, however, is when $\nabla\cdot\mathbf{q}\neq 0$.

In most applications of MHD it is assumed that there are no "external" processes, and the right-hand sides of Eqs. (3.113), (3.115), and (3.117) can be set to zero. In that case, the equations reduce to the form usually considered as constituting MHD:

$$\frac{\partial\rho}{\partial t}+\nabla\cdot(\rho\mathbf{u})=0, \tag{3.119}$$

$$\rho\frac{D\mathbf{u}}{Dt}+\nabla\cdot\mathbf{P}-\mathbf{j}\times\mathbf{B}-\rho\mathbf{g}-\rho_q\mathbf{E}=0, \tag{3.120}$$

$$\frac{D}{Dt}\left(\frac{1}{2}P_{ii}\right)+\frac{1}{2}P_{ii}\nabla\cdot\mathbf{u}+\mathbf{P}:\nabla\mathbf{u}+\nabla\cdot\mathbf{q}$$
$$+\left(\rho_q\mathbf{u}-\mathbf{j}\right)\cdot(\mathbf{E}+\mathbf{u}\times\mathbf{B})=0, \tag{3.121}$$

where

$$\frac{D}{Dt}\equiv\frac{\partial}{\partial t}+\mathbf{u}\cdot\nabla, \tag{3.122}$$

":" indicates a double dot product,

$$\mathbf{P}:\nabla\mathbf{u}\equiv P_{ij}\frac{\partial}{\partial x_j}u_i, \tag{3.123}$$

and we have used $P_{ij} = P_{ji}$. In deriving Eq. (3.121) we have used the momentum equation (3.120) to remove the gravitational term from Eq. (3.117), and the continuity equation (3.119) to remove the remaining terms that are dependent on the dynamic pressure.

Before proceeding further we shall discuss the relative importance of the last terms on the left-hand side of Eqs. (3.120) and (3.121). As noted earlier when discussing the plasma state, it was pointed out that the plasma is quasi-neutral, i.e., $n_e \approx n_i \approx n$, where n is the plasma number density, and consequently $\rho_q \ll nq$ (here the e and i subscripts refer to electrons and ions, respectively). Because of quasi-neutrality we ignore the last term on the left-hand side of Eq. (3.120). In addition, the current density $j \approx ne(u_i - u_e)$, where e is the magnitude of the electron charge. Usually the difference in velocities is small, and so $(\rho_q \mathbf{u} - \mathbf{j}) \sim 0$. Moreover, as we show later, $(\mathbf{E} + \mathbf{u} \times \mathbf{B}) \sim 0$. This is the "frozen-in" condition. The last term on the left-hand side of Eq. (3.121), which has the form of Joule dissipation, can therefore be ignored.

At this stage therefore we drop the last term on the left-hand side of Eqs. (3.121) and (3.120). This has the effect of removing any explicit dependence on the electric field in both the momentum and energy equations. We make the further assumption that $\nabla \cdot \mathbf{q} = 0$, and finally assume that the pressure is isotropic, i.e., $P_{ij} = P\delta_{ij}$. With these approximations, Eq. (3.121) becomes

$$\frac{D}{Dt}\left(\frac{3}{2}P\right) + \frac{5}{2}P\nabla \cdot \mathbf{u} = 0. \qquad (3.124)$$

Rewriting the mass continuity equation as

$$\frac{D\rho}{Dt} + \rho\nabla \cdot \mathbf{u} = 0, \qquad (3.125)$$

and noting that in thermodynamics the ratio of specific heats $\gamma = (N+2)/N$, where N is the number of degrees of freedom, with $\gamma = 5/3$ for $N = 3$, Eq. (3.124) becomes

$$\frac{DP}{Dt} - \frac{\gamma P}{\rho}\frac{D\rho}{Dt} = 0 \qquad (3.126)$$

or

$$\frac{D}{Dt}(P\rho^{-\gamma}) = 0, \qquad (3.127)$$

which is the relationship between pressure and density for adiabatic processes. This is not surprising since we assumed $\nabla \cdot \mathbf{q} = 0$, and no heat is gained or lost from the plasma.

So far we have not discussed the relationship between pressure, internal energy, and temperature for the plasma. In most texts, however, MHD is derived with explicit reference to the ideal gas law, $P = \sum_s n_s KT$, and the relationship between internal energy and temperature, $U = N\sum_s n_s KT/2$, where N is the number of degrees of freedom, K is the Boltzmann constant, and T is the plasma temperature. It should be noted that the ideal gas law implicity assumes that all the constituents within the plasma have the same temperature, which is the case for a highly collisional plasma. In addition, the number density in the ideal gas law is the total number density, thus $P = 2nKT$ for a two-species plasma with both species having the same temperature T. Given that we typically consider n to be the plasma number density, not $2n$, there is a potential ambiguity in how the ideal gas law applies to a plasma. Provided we apply the ideal gas law separately to each species, i.e., $P_s = n_s KT_s$, and define $U_s = N_s n_s KT_s/2$, then this potential ambiguity is removed. In particular, we do not require that each species have the same temperature. For a two-species plasma T_e need not equal T_i. Indeed, to make the connection with the ideal gas law clear, we can define an average temperature, $T = \sum_s n_s T_s / \sum_s n_s$, and use the average temperature to define the pressure and internal energy, without requiring each species to have the same temperature.

3.5.4 Summary of fluid theory

In this section we have given the steps that allow us to derive fluid equations for a plasma. We took the Boltzmann equation as our starting point. From the Boltzmann equation we derived a generalized moment equation that forms the basis of any fluid derivation:

$$\frac{\partial}{\partial t}\Big(n\langle\varphi(\mathbf{v})\rangle\Big) + \nabla \cdot \Big(n\langle\mathbf{v}\varphi(\mathbf{v})\rangle\Big) - n\langle(\mathbf{a}\cdot\nabla_{\mathbf{v}})\varphi(\mathbf{v})\rangle = \frac{\partial}{\partial t}\Big(n\langle\varphi(\mathbf{v})\rangle\Big)\Big|_c. \qquad (3.95)$$

The right-hand side is a place-holder for collisional processes that have not been defined. In addition, the acceleration term, $\mathbf{a}$, is assumed to be independent of velocity, except for the $\mathbf{v} \times \mathbf{B}$ term.

Equation (3.95) can be used to derive the hierarchy of moment equations for each species separately, but here we have presented the derivation of single-fluid MHD. In this derivation plasma pressures are defined relative to the plasma bulk flow velocity $\mathbf{u}$.

For single-fluid MHD, ignoring collisions, we have the following moment equations: mass density,

$$\frac{\partial \rho}{\partial t} + \nabla \cdot (\rho \mathbf{u}) = 0, \tag{3.119}$$

charge density,

$$\frac{\partial \rho_q}{\partial t} + \nabla \cdot \mathbf{j} = 0, \tag{3.114}$$

momentum,

$$\rho \frac{D\mathbf{u}}{Dt} + \nabla \cdot \mathbf{P} - \mathbf{j} \times \mathbf{B} = 0, \tag{3.128}$$

and energy,

$$\frac{D}{Dt}(P\rho^{-\gamma}) = 0. \tag{3.127}$$

In Eq. (3.128) we have ignored gravity and invoked quasi-neutrality (cf. Eq. (3.120)), while in Eq. (3.127) we have assumed that the terms associated with Joule dissipation and heat flux can be ignored. This is often referred to as ideal MHD.

3.6 THE GENERALIZED OHM'S LAW AND THE FROZEN-IN CONDITION

Equations (3.119), (3.128), and (3.127) describe the conservation of mass, momentum, and energy in a plasma under the ideal MHD assumptions. One of these assumptions is quasi-neutrality, under which there is no explicit reference to the electric field in the conservation equations. The electric field, however, does enter Maxwell's equations (3.1), (3.2), and (3.3). As written, Ampère's law (3.2) includes the displacement current. While this can formally be retained in the set of equations, within the MHD context it can often be ignored.

3.6.1 Displacement current and MHD

To understand under what conditions the displacement current can be ignored, we divide the electric field into "longitudinal" and "transverse" components, i.e., $\mathbf{E} = \mathbf{E}_L + \mathbf{E}_T$, and $\nabla \cdot \mathbf{E} = \nabla \cdot \mathbf{E}_L$, $\nabla \times \mathbf{E} = \nabla \times \mathbf{E}_T$. An explicit formalism that shows this separation is to consider the electric field expressed in terms of scalar and vector potentials, with $\mathbf{B} = \nabla \times \mathbf{A}$ and $\mathbf{E} = -\nabla\phi - \partial\mathbf{A}/\partial t$, and ϕ and $\mathbf{A}$ are the scalar and vector potentials, respectively. Assuming the Coulomb gauge, $\nabla \cdot \mathbf{A} = 0$, which itself implies that $\mathbf{A}$ can be expressed as curl of yet another vector potential, then $\mathbf{E}_L = -\nabla\phi$ and $\mathbf{E}_T = -\partial\mathbf{A}/\partial t$.

The longitudinal portion of the displacement current ensures conservation of charge, through the time variation of Gauss's law (3.3)

$$\varepsilon_0 \nabla \cdot \frac{\partial \mathbf{E}_L}{\partial t} = \frac{\partial \rho_q}{\partial t} = -\nabla \cdot \mathbf{j}. \tag{3.129a}$$

On the other hand, the transverse component enters through Faraday's law, which on substitution in the curl of Ampère's law (3.2) gives

$$-\mu_0 \nabla \times \mathbf{j} = -\nabla \times (\nabla \times \mathbf{B}) + \frac{1}{c^2}\nabla \times \frac{\partial \mathbf{E}_T}{\partial t}$$

$$= \nabla^2 \mathbf{B} - \frac{1}{c^2}\frac{\partial^2 \mathbf{B}}{\partial t^2}. \tag{3.129b}$$

This reduces to the equation for light waves if $\nabla \times \mathbf{j} = 0$. The transverse component of the displacement current ensures that signals do not travel faster than the speed of light.

From Eqs. (3.129a) and (3.129b), the displacement current can be ignored if the characteristic velocities within the plasma are much less than the speed of light, and if quasi-neutrality applies. In that case the conservation of charge condition becomes

$$\nabla \cdot \mathbf{j} = 0 \tag{3.130}$$

and Ampère's law becomes

$$\nabla \times \mathbf{B} = \mu_0 \mathbf{j}. \tag{3.131}$$

Technically, the condition for neglecting the longitudinal component of the displacement current is not $\rho_q = 0$, but rather $\partial\rho_q/\partial t = 0$. Thus, as we shall see later, $\nabla \cdot \mathbf{E}$ need not be zero, but as long as the time scales are sufficiently long we can assume Eq. (3.130) applies. Similarly, $\nabla \times \mathbf{E}$ need not be zero, otherwise Faraday's law would require the time-stationary magnetic fields, but ignoring the displacement current assumes that $\varepsilon_0 \partial \mathbf{E}/\partial t$ is small in Ampère's law.

3.6.2 Generalized Ohm's law

Because we can neglect the displacement current, Faraday's law (3.131) is the only equation of the set derived so far that includes the electric field. We now need a second equation that relates the electric field to the other quantities. This is the generalized Ohm's law, which is obtained by multiplying the momentum equation for each species (3.100) by q_s/m_s, and summing over species:

$$\frac{\partial \mathbf{j}}{\partial t} + \nabla \cdot \left(\sum_s \frac{q_s \mathbf{P}_s}{m_s} + \mathbf{ju} + \mathbf{uj} - \rho_q \mathbf{uu} \right) - \sum_s \frac{n_s q_s^2}{m_s} (\mathbf{E} + \mathbf{u}_s \times \mathbf{B}) - \rho_q \mathbf{g} = -\sum_{s,k} n_s q_s \nu_{sk} (\mathbf{u}_s - \mathbf{u}_k). \tag{3.132}$$

Following our approach with the mass, momentum, and energy equations, we have formally retained all the terms in Eq. (3.132), but as we shall see there are several simplifying assumptions. The term on the right-hand side of Eq. (3.132) is the collisional term, where ν_{sk} is the collision frequency for collisions between species s and k. We have also assumed that the collisions do not include electron–neutral or ion–neutral collisions.

Our first simplifying assumption is that $\rho_q \approx 0$. In that case all the terms explicitly involving ρ_q vanish, as do the $\nabla \cdot \mathbf{j}$ terms in the divergence term. Thus, to this degree of approximation,

$$\frac{D\mathbf{j}}{Dt} + \mathbf{j} \cdot \nabla \mathbf{u} + \mathbf{j} \nabla \cdot \mathbf{u} + \nabla \cdot \left(\sum_s \frac{q_s \mathbf{P}_s}{m_s} \right) - \sum_s \frac{n_s q_s^2}{m_s} (\mathbf{E} + \mathbf{u}_s \times \mathbf{B}) = -\sum_{s,k} n_s q_s \nu_{sk} (\mathbf{u}_s - \mathbf{u}_k). \tag{3.133}$$

Next we assume that the electron mass is much less than the mass of the ions, $m_e \ll m_i$. In that case the two summations on the left-hand side only include the electron mass, and

$$\frac{D\mathbf{j}}{Dt} + \mathbf{j} \cdot \nabla \mathbf{u} + \mathbf{j} \nabla \cdot \mathbf{u} - \nabla \cdot \left(\frac{e \mathbf{P}_e}{m_e} \right) - \frac{n_e e^2}{m_e} (\mathbf{E} + \mathbf{u}_e \times \mathbf{B}) = \sum_i n_e e \nu_{ei} (\mathbf{u}_e - \mathbf{u}_i). \tag{3.134}$$

The right-hand side also simplifies under this assumption since the collisions conserve momentum, and $\nu_{ie} = m_e \nu_{ei}/m_i$.

Finally, although we have formally allowed for multiple ion species, we assume that all the ion species move with the same bulk velocity, and again, since $m_e \ll m_i$, $\mathbf{u}_i = \mathbf{u}$. In that case, $\mathbf{j} = n_e e (\mathbf{u}_i - \mathbf{u}_e)$.

Using these approximations and rearranging terms gives

$$\frac{m_e}{n_e e^2} \left(\frac{D\mathbf{j}}{Dt} + \mathbf{j} \cdot \nabla \mathbf{u} + \mathbf{j} \nabla \cdot \mathbf{u} \right) = \mathbf{E} + \mathbf{u} \times \mathbf{B} - \frac{\mathbf{j} \times \mathbf{B}}{n_e e} + \frac{1}{n_e e} \nabla \cdot \mathbf{P}_e - \frac{m_e}{n_e e^2} \bar{\nu}_{ei} \mathbf{j}, \tag{3.135}$$

where $\bar{\nu}_{ei}$ is the sum of the electron–ion collision frequencies. The collision term is often expressed in terms of a conductivity,

$$\sigma = \frac{n_e e^2}{m_e \bar{\nu}_{ei}}. \tag{3.136}$$

We also further assume that terms quadratic in $\mathbf{u}$ and $\mathbf{j}$ can be dropped, and Eq. (3.135) reduces to the classical form of the generalized Ohm's law

$$\mathbf{j} = \sigma \left(\mathbf{E} + \mathbf{u} \times \mathbf{B} - \frac{\mathbf{j} \times \mathbf{B}}{n_e e} + \frac{1}{n_e e} \nabla \cdot \mathbf{P}_e - \frac{m_e}{n_e e^2} \frac{\partial \mathbf{j}}{\partial t} \right). \tag{3.137}$$

For vanishingly small collision frequencies (or infinite conductivity) the term in parentheses must ≈ 0. If we further assume that the $\mathbf{j} \times \mathbf{B}$ (or Hall) term, the electron pressure term, and the $\partial \mathbf{j}/\partial t$ term all vanish, we then get

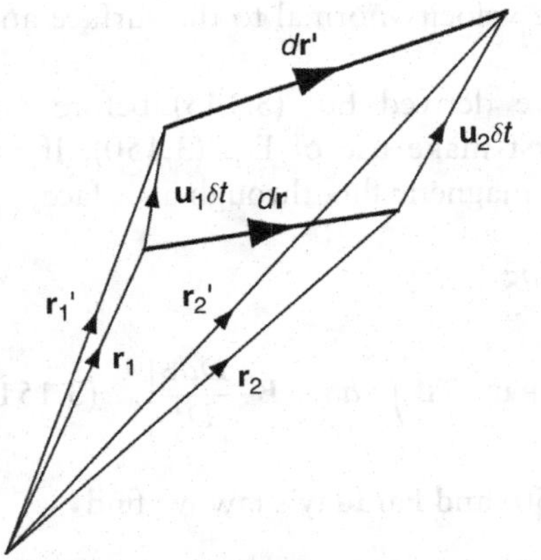

FIGURE 3.8. Change in a line element $d\mathbf{r}$ that is transported in the flow field defined by **u**.

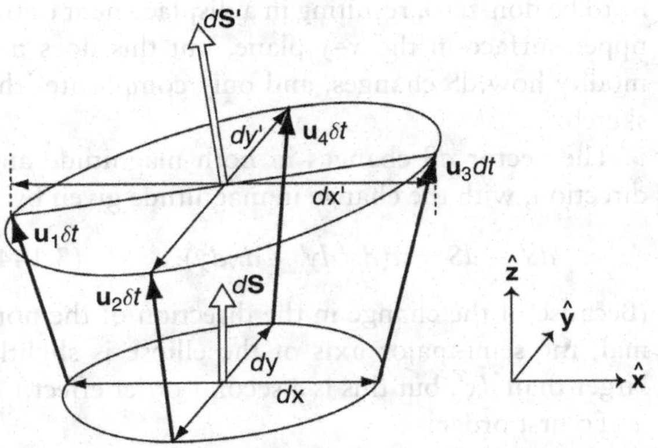

FIGURE 3.9. Change in an area element $d\mathbf{S}$ that is transported in the flow field defined by **u**.

$$\mathbf{E} + \mathbf{u} \times \mathbf{B} = 0, \tag{3.138}$$

which is the Ohm's law for ideal MHD. This is why the last term on the left-hand side of Eq. (3.121) could be neglected. Equation (3.138) also justifies the assumption that $\rho_q \approx 0$ in Eq. (3.120). If we assume that the plasma flow velocity has a shear over scale length L, comparing the electromagnetic force terms in Eq. (3.120) yields

$$\rho_q E : jB \approx \frac{\varepsilon_0 E^2}{L} : \frac{B^2}{\mu_0 L} \approx \frac{u^2}{c^2} : 1, \tag{3.139}$$

where we have assumed that the scale length L also applies for the current density, i.e., $B/L \approx \mu_0 j$.

3.6.3 The frozen-in theorem

The idealized Ohm's law (3.138) has an important consequence, known as the frozen-in theorem; the idealized Ohm's law (3.138) is therefore also referred to as the frozen-in condition. The frozen-in theorem states that the magnetic field can be treated as if it were frozen to the fluid. The theorem consists of two parts: first, the magnetic flux through any enclosed area remains constant as the area is transported with the fluid; second, any magnetic field line that connects two fluid elements remains connected with the fluid elements.

To prove these two statements we first need to derive two identities that relate changes in line and area elements to gradients in the flow. We consider a line element first, to demonstrate the approach used, and Figure 3.8 shows how a line element $d\mathbf{r}$ changes to $d\mathbf{r}'$ on being advected by the flow given by **u**.

Based on the geometry given in Figure 3.8,

$$\begin{aligned} d\mathbf{r}' &= \mathbf{r}_2{}' - \mathbf{r}_1{}' = \mathbf{r}_2 - \mathbf{r}_1 + (\mathbf{u}_2 - \mathbf{u}_1)\delta t \\ &= d\mathbf{r} + (\mathbf{u}_2 - \mathbf{u}_1)\delta t. \end{aligned} \tag{3.140}$$

From Taylor series expansion,

$$\mathbf{u}_2 = \mathbf{u}_1 + (d\mathbf{r} \cdot \nabla)\mathbf{u}_1 \tag{3.141}$$

and

$$d\mathbf{r}' - d\mathbf{r} = [(d\mathbf{r} \cdot \nabla)\mathbf{u}_1]\delta t. \tag{3.142}$$

As $\delta t \to 0$,

$$\frac{Dd\mathbf{r}}{Dt} = (d\mathbf{r} \cdot \nabla)\mathbf{u}. \tag{3.143}$$

For changes in elemental areas, we show a simplified geometry in Figure 3.9. We have assumed an ellipsoidal area element $d\mathbf{S}$ with the normal to the surface defining the direction of the vector, which in turn defines the z-axis of a local cartesian coordinate system. The flow velocity is again denoted by **u**, and defining the divergence operator as $\nabla = \nabla_\perp + \partial/\partial z$, we assume that the direction given by $\nabla_\perp u_z$ defines the x-axis of the coordinate system. Also, for simplicity in drawing Figure 3.9, we have assumed the average value of u_x and u_y is zero. We could allow for the average value of u_x and

u_y to be non-zero, resulting in a displacement of the upper surface in the x–y plane, but this does not modify how dS changes, and only complicates the sketch.

The vector $d\mathbf{S}$ changes in both magnitude and direction, with the change in magnitude given by

$$dS' - dS = \pi(dx'dy' - dxdy). \quad (3.144)$$

(Because of the change in the direction of the normal, the semi-major axis of the ellipse is slightly larger than dx', but this is a second-order effect.)

To first order

$$dx' - dx = (u_{3x} - u_{1x})\delta t = dx(\partial u_x/\partial x)\delta t \quad (3.145a)$$

and

$$dy' - dy = (u_{3y} - u_{1y})\delta t = dy(\partial u_y/\partial y)\delta t. \quad (3.145b)$$

Consequently, as $\delta t \to 0$ we find that

$$\hat{\mathbf{n}}\frac{DdS}{Dt} = d\mathbf{S}\left(\frac{\partial u_x}{\partial x} + \frac{\partial u_y}{\partial y}\right) = d\mathbf{S}\left(\nabla\cdot\mathbf{u} - \frac{\partial u_z}{\partial z}\right), \quad (3.146)$$

given that $\hat{\mathbf{n}}dS = d\mathbf{S}$ and the normal $\hat{\mathbf{n}}$ is assumed to be parallel to the z-axis.

From Figure 3.9, the change in direction of the normal is given by

$$\delta\hat{\mathbf{n}} = -\hat{\mathbf{x}}(u_{3z} - u_{1z})\delta t/dx. \quad (3.147)$$

Therefore, again by Taylor series expansion, as $\delta t \to 0$

$$\frac{D\hat{\mathbf{n}}}{Dt} = -\hat{\mathbf{x}}\frac{\partial u_z}{\partial x}. \quad (3.148)$$

Since we have chosen the x-axis as the direction of the gradient of u_z in the plane of the surface dS, we can rewrite Eq. (3.148) more generally as

$$dS\frac{D\hat{\mathbf{n}}}{Dt} = -\left(dS\nabla - d\mathbf{S}\frac{\partial}{\partial z}\right)u_z. \quad (3.149)$$

Combining Eqs. (3.146) and (3.149),

$$\frac{Dd\mathbf{S}}{Dt} = d\mathbf{S}\nabla\cdot\mathbf{u} - dS\nabla u_{\mathrm{n}}, \quad (3.150)$$

where u_{n} is the velocity normal to the surface and $\mathbf{u}\cdot d\mathbf{S} = u_{\mathrm{n}}dS$.

Although we derived Eq. (3.143) before Eq. (3.150), we first make use of Eq. (3.150). If we define Φ as the magnetic flux through a surface,

$$\frac{D\Phi}{Dt} = \frac{D}{Dt}\int \mathbf{B}\cdot d\mathbf{S}$$
$$= \int\left[\left(\frac{\partial\mathbf{B}}{\partial t} + \mathbf{u}\cdot\nabla\mathbf{B}\right)\cdot d\mathbf{S} + \mathbf{B}\cdot\frac{Dd\mathbf{S}}{Dt}\right]. \quad (3.151)$$

From Eq. (3.150) and Faraday's law we find

$$\frac{D\Phi}{Dt} = \int d\mathbf{S}\cdot[-\nabla\times\mathbf{E} + \mathbf{u}\cdot\nabla\mathbf{B} + \mathbf{B}\nabla\cdot\mathbf{u} - (\mathbf{B}\cdot\nabla)\mathbf{u}]. \quad (3.152)$$

Given the standard vector identity for $\nabla\times(\mathbf{u}\times\mathbf{B})$ and $\nabla\cdot\mathbf{B} = 0$, Eq. (3.152) becomes

$$\frac{D\Phi}{Dt} = -\int\nabla\times(\mathbf{E} + \mathbf{u}\times\mathbf{B})\cdot d\mathbf{S}. \quad (3.153)$$

Clearly, if the idealized Ohm's law (3.138) applies, then

$$D\Phi/Dt = 0, \quad (3.154)$$

which is the first part of the frozen-in theorem.

For the second part of the frozen-in theorem we shall make use of the same identities that we used to derive Eq. (3.153) from Eq. (3.152). First we assume that the idealized Ohm's law applies, in which case

$$\nabla\times(\mathbf{E} + \mathbf{u}\times\mathbf{B}) = 0$$
$$= -\partial B/\partial t - \mathbf{u}\cdot\nabla\mathbf{B} - \mathbf{B}\nabla\cdot\mathbf{u} + (\mathbf{B}\cdot\nabla)\mathbf{u}. \quad (3.155)$$

Hence,

$$\frac{D\mathbf{B}}{Dt} + \mathbf{B}\nabla\cdot\mathbf{u} = (\mathbf{B}\cdot\nabla)\mathbf{u}. \quad (3.156)$$

Making use of the continuity equation (3.119), we can rewrite Eq. (3.156) as

$$\frac{D\mathbf{B}}{Dt} - \frac{\mathbf{B}}{\rho}\frac{D\rho}{Dt} = (\mathbf{B}\cdot\nabla)\mathbf{u} \quad (3.157)$$

or

$$\frac{D}{Dt}\left(\frac{\mathbf{B}}{\rho}\right) = \left(\frac{\mathbf{B}}{\rho}\cdot\nabla\right)\mathbf{u}. \quad (3.158)$$

This should be compared with Eq. (3.143), which gives the total time derivative for a fluid element $d\mathbf{r}$. Equations (3.143)and (3.158) have the same functional form. As a consequence, if $d\mathbf{r}$ and $\mathbf{B}/\rho$ are initially parallel, they remain parallel. In other words, if the magnetic field connects two fluid elements, then the field continues to connect these fluid elements as they are transported by the fluid.

3.6.4 Comments on the frozen-in theorem

The frozen-in theorem is very powerful, as it allows us to determine how the magnetic field in a plasma evolves based solely on the motion of the fluid. As an example, it is the frozen-in theorem that shows why structures in the interplanetary magnetic field (IMF) are convected by the solar wind. The changing magnetic field changes how the solar wind interacts with the magnetosphere. At the same time, it is the violation of the frozen-in condition, manifested as magnetic field reconnection, that allows the solar wind to drive magnetospheric convection.

Nevertheless, the frozen-in theorem should be used with caution. The theorem states that the magnetic field can be treated *as if* it were frozen to the fluid. Sometimes this is misstated as the fluid being frozen to the field. These two statements may appear to be equivalent, but the latter mixes cause and effect. It is the motion of the fluid that gives the change in the field. Furthermore, while the frozen-in theorem leads to the concept of moving field lines, this is in fact a construct. Plasma motion can be measured directly, but there is no method by which field-line motion can itself be measured. Clearly changes in field-line magnitude or direction can be measured, as these result in an induction electric field, but there is no induction electric field directly given by a "moving" uniform magnetic field. Nor does a "moving" magnetic field directly affect the Lorentz force (3.7) acting on a particle. Again, to be clear, there is an electric field associated with a moving plasma, and the frozen-in theorem allows us to assume that the electric field corresponds to moving field lines, but it is the plasma motion that results in an electric field, not the field-line motion.

One example that may clarify this point is to consider how corotation is imposed on a magnetospheric plasma. In particular, while there is an induction electric field associated with the change in the magnetic field of a spinning tilted dipole, this induction electric field does not impose corotation. To make this clear, consider a dipole that is aligned along the planetary spin axis, such as Saturn. In that case there is no induction electric field associated with the rotation, yet the magnetospheric plasma still tends to corotate. One often reads the following as an explanation of corotation: because the ionosphere is fixed to the planet and corotates, and the magnetic field lines are equipotentials, then the ionospheric corotation electric field maps to the equator and the magnetospheric plasma corotates. But this is not a statement of cause and effect. Indeed, the gas giant magnetospheres show how corotation is imposed by a rotating planet. The moons at these planets are sources of plasma that initially rotates with the keplerian velocity. Currents flow between the ionosphere and newly created plasma so the resultant $\mathbf{j} \times \mathbf{B}$ force accelerates the plasma to corotation velocities. The generator of the currents is the differential motion of the planetary atmosphere and ionosphere, with the ionosphere coupled to the atmosphere through collisions. The same principle applies at the Earth, but with the additional complication that the magnetosphere is strongly coupled to the solar wind. If, for example, the flow at the magnetopause changes because of changes in the solar-wind driver, then currents flow through the system to provide the forces required to change the motion of the plasma within the magnetosphere.

3.6.5 The "B, U" paradigm

One other consequence of the frozen-in theorem is that we can remove the electric field from Faraday's law, as in Eq. (3.156). Neglecting the displacement current also means that we can replace the current in the momentum equation (3.120) using Ampère's law as given by Eq. (3.131). In particular the $\mathbf{j} \times \mathbf{B}$ force can be rewritten as

$$\mathbf{j} \times \mathbf{B} = \frac{(\mathbf{B} \cdot \nabla)\mathbf{B}}{\mu_0} - \nabla \frac{B^2}{2\mu_0}. \tag{3.159}$$

The first term on the right-hand side of Eq. (3.159) corresponds to field-line curvature, and can be thought of as a force resulting from the tendency for field lines to straighten. This is often referred to as the field-line tension force, but it must be remembered that in Eq. (3.159) there is no net force along the magnetic field, and the parallel component of the tension force is balanced by the parallel component of the second term on the right-hand side of Eq. (3.159), which is the magnetic-pressure force.

The next step would be to use Eq. (3.159) to replace the $\mathbf{j} \times \mathbf{B}$ term in the momentum equation (3.158), but before that we consider the special case where $\mathbf{j} \times \mathbf{B} = 0$. This is often referred to as the force-free condition as there is no net force associated with the magnetic field. An archetypal example of a force-free structure is a flux rope. Flux ropes have been observed in the solar wind (Chapter 5), in the Venus ionosphere (Chapter 8), and at the Earth's magnetopause (Chapter 9), and are characterized as having a strong axial field along the major axis of the rope at the center of the rope and an increasingly azimuthal field on moving away from the center of the rope. This is a case, then, where the pressure force associated with the axial field is balanced by the curvature force of the azimuthal field.

For a force-free rope the current is given by $\mathbf{j} = \alpha \mathbf{B}$. If α is a constant, the flux rope is said to be in the Taylor state, and the magnetic field can be described using Bessel functions.

We shall now use Eq. (3.159) to rewrite the ideal MHD limit of the momentum equation (3.120) as

$$\rho \frac{D\mathbf{u}}{Dt} = \frac{(\mathbf{B} \cdot \nabla)\mathbf{B}}{\mu_0} - \nabla \left(\frac{B^2}{2\mu_0} + P \right), \qquad (3.160)$$

where we have again assumed isotropic plasma pressure for simplicity, together with quasi-neutrality and the neglect of gravitational forces. The ratio of the two pressure terms in Eq. (3.160) is known as the plasma beta, and is defined as

$$\beta = \frac{2\mu_0 P}{B^2}. \qquad (3.161)$$

Having replaced $\mathbf{j}$ in the momentum equation, we now have a closed set of equations that does not involve $\mathbf{j}$ or $\mathbf{E}$: the mass continuity equation (3.119), the momentum equation as given by Eq. (3.160), the energy equation (3.127), and Faraday's law written as a magnetic field transport equation, Eq. (3.156). In the MHD regime, then, the fundamental parameters are $\mathbf{B}, \mathbf{u}, \rho$, and P; and $\mathbf{E}$ and $\mathbf{j}$ are secondary quantities that can be derived from these more fundamental parameters (either the frozen-in condition for $\mathbf{E}$, or Ampère's law for $\mathbf{j}$). These derived quantities may be important at boundaries. For example, in the coupled magnetosphere–ionosphere system, the ionosphere is collisional and the resultant **Pedersen** and **Hall conductivities** provide a linkage between the current and electric field in the ionosphere. The divergence of these horizontal currents leads to vertical currents, while the electric field corresponds to a flow pattern that is imposed on the magnetosphere at the ionospheric boundary. In this case, the use of $\mathbf{E}$ and $\mathbf{j}$ makes the problem of characterizing the boundary condition more tractable. But some care must be taken if either $\mathbf{E}$ or $\mathbf{j}$ is taken as a given, as that assumes a particular state of the ionosphere, say, without knowing how the system evolved to that state.

We shall also see that $\mathbf{E}$ and $\mathbf{j}$ can be useful in describing magnetohydrodynamic waves, for example, as we do in Section 3.7.1. We also discuss why field-aligned currents flow in a plasma. As noted above, these currents provide a pathway for magnetosphere–ionosphere coupling. But again, from an MHD perspective, $\mathbf{E}$ and $\mathbf{j}$ are secondary parameters.

3.6.6 Summary of the frozen-in theorem

The fluid momentum equations can be recast into a generalized Ohm's law,

$$\mathbf{j} = \sigma \left(\mathbf{E} + \mathbf{u} \times \mathbf{B} - \frac{\mathbf{j} \times \mathbf{B}}{n_e e} + \frac{1}{n_e e} \nabla \cdot \mathbf{P}_e - \frac{m_e}{n_e e^2} \frac{\partial \mathbf{j}}{\partial t} \right), \qquad (3.137)$$

under the assumption that terms quadratic in $\mathbf{u}$ and $\mathbf{j}$ can be neglected, and recasting the electron–ion collision frequency as a conductivity, Eq. (3.136).

In space plasmas the conductivity σ is assumed to be large, and the first two terms in the parentheses on the right-hand side of Eq. (3.137) are usually the leading terms. In that case we obtain the ideal MHD Ohm's law

$$\mathbf{E} + \mathbf{u} \times \mathbf{B} = 0. \qquad (3.138)$$

Equation (3.138) in turn results in the frozen-in theorem, where the magnetic field can be treated as if it were frozen to the plasma. This is a very powerful theorem that is frequently invoked when trying to understand the interaction between plasmas and electromagnetic fields.

3.7 APPLICATIONS OF MHD

We end this chapter by giving three applications of the MHD equations, as these will provide a useful framework for future discussions. First we introduce the concept of MHD waves. Second, we demonstrate the equivalence of the momentum equation in MHD and guiding center drift theory. Third, we derive a classical result that relates field-aligned currents to the plasma inertia and thermal pressure.

Much of the discussion in this section relies on the MHD momentum equation (3.120). In using this equation we shall neglect gravity and assume quasi-neutrality:

$$\rho \frac{D\mathbf{u}}{Dt} = \mathbf{j} \times \mathbf{B} - \nabla \cdot \mathbf{P}. \qquad (3.162)$$

As written, Eq. (3.162) allows for a full pressure tensor, but, except for the discussion in Section 3.7.2, we assume that the pressure is isotropic and that $\nabla \cdot \mathbf{P}$ is replaced by ∇P.

3.7.1 Introduction to MHD waves

One of the most important consequences of MHD is that waves involving motion of the mass of the plasma have a characteristic speed of propagation governed in part by the **Alfvén speed.** We address this in more detail in Chapter 13, but we want to introduce the concept of MHD waves here.

In deriving the dispersion relation (that is, the relationship between frequency and wave number) for MHD waves, we make the usual assumption that the waves are small in amplitude, and can therefore be treated as first-order perturbations to the background plasma. We also assume that the background plasma is stationary, and the only zero-order quantities are the magnetic field ($\mathbf{B}_0$), the plasma density (ρ_0), and the plasma pressure (P_0). We further assume that the perturbations are harmonic, i.e., they vary as $\exp[-i(\omega t - \mathbf{k} \cdot \mathbf{r})]$. We can therefore replace the differential operators in Maxwell's equations and the MHD equations with $\partial/\partial t \equiv -i\omega$ and $\nabla \equiv i\mathbf{k}$. Last, because we are linearizing the equations, we ignore the $\mathbf{u} \cdot \nabla$ terms in the operator D/Dt as being second order. Under these assumptions Maxwell's equations become

$$\mathbf{k} \times \mathbf{E} = \omega \mathbf{b}, \qquad (3.163)$$

$$\mathbf{k} \times \mathbf{b} = -i\mu_0 \mathbf{j}, \qquad (3.164)$$

where $\mathbf{E}$, $\mathbf{b}$, and $\mathbf{j}$ are the wave electric field, magnetic field, and current density, respectively. We have again neglected the displacement current for simplicity, and we have also ignored Gauss's law because of quasi-neutrality. From Eq. (3.163), $\mathbf{k} \cdot \mathbf{b} = 0$.

On linearization, the MHD conservation equations (3.125), (3.162), and (3.127) become

$$\omega\rho - \rho_0 \mathbf{k} \cdot \mathbf{u} = 0, \qquad (3.165)$$

$$\omega\rho_0 \mathbf{u} - \mathbf{k}P - i\mathbf{j} \times \mathbf{B}_0 = 0, \qquad (3.166)$$

$$P = \frac{\gamma P_0}{\rho_0} \rho = c_s^2 \rho, \qquad (3.167)$$

with $\mathbf{u}$, ρ, and P being the first-order plasma velocity, mass density, and pressure, respectively. We have also assumed the pressure is isotropic, and we have defined c_s as the **speed of sound** for the plasma.

The last equation needed to close the equations is the linearized frozen-in condition:

$$\mathbf{E} + \mathbf{u} \times \mathbf{B}_0 = 0. \qquad (3.168)$$

From Eqs. (3.165) and (3.167) we can rewrite Eq. (3.166) as

$$\omega\rho_0 \left[\mathbf{u} - c_s^2 \frac{\mathbf{k}(\mathbf{k} \cdot \mathbf{u})}{\omega^2} \right] = i\mathbf{j} \times \mathbf{B}_0. \qquad (3.169)$$

If we consider the component of Eq. (3.169) parallel to the ambient magnetic field, $\mathbf{B}_0$, then

$$u_\parallel = \frac{k_\parallel c_s^2}{\omega^2} (\mathbf{k} \cdot \mathbf{u}). \qquad (3.170)$$

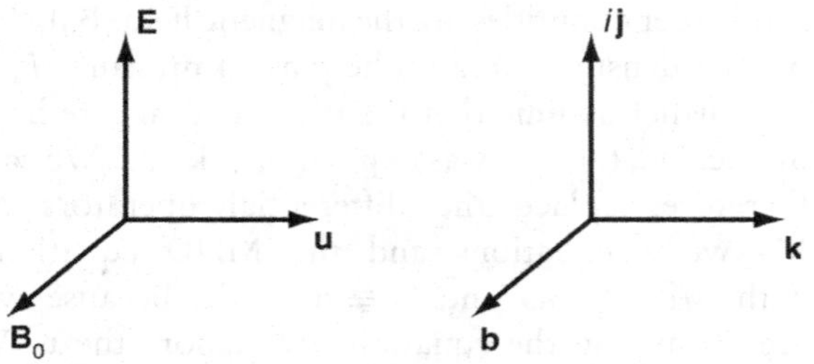

FIGURE 3.10. The two MHD-wave triads, $\mathbf{B}_0$, **u, E** at left, and **b, k,** $i\mathbf{j}$ at right.

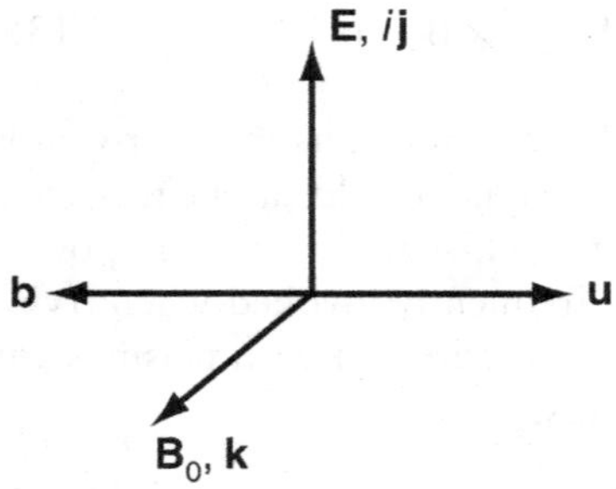

FIGURE 3.11. The two MHD-wave triads combined so that $\mathbf{k}||\mathbf{B}_0$.

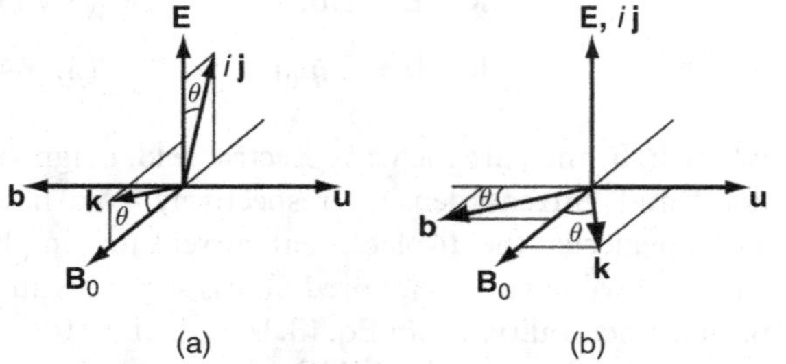

FIGURE 3.12. Rotation of the **b, k,** $i\mathbf{j}$ triad about the *b*-axis (a) and the *ij*-axis (b). The angle θ is the propagation angle of the wave.

For parallel propagation, Eq. (3.170) gives the dispersion relation $\omega^2 = k_{||}^2 c_s^2$. This corresponds to longitudinal sound waves in a classical gas.

Although we consider wave dispersion for a warm plasma later in this section, we now make the cold plasma approximation, and $c_s = 0$. In that case $u_{||} = 0$, and Eq. (3.169) becomes

$$\omega\rho_0 \mathbf{u} = i\mathbf{j} \times \mathbf{B}_0. \tag{3.171}$$

Interestingly, Eqs. (3.163), (3.164), (3.168), and (3.171) all have a similar form, where a vector is related to the cross product of two other vectors. There are six vectors in these equations (the four wave quantities, **E, b, u,** and **j**, the wave vector **k**, and the ambient magnetic field $\mathbf{B}_0$). This suggests that the vectors can be combined into two right-angled triads. Specifically, Eq. (3.167) requires that $\mathbf{E}\perp\mathbf{u}$ and $\mathbf{B}_0$ while Eq. (3.171) requires $\mathbf{u}\perp\mathbf{B}_0$, and $\mathbf{B}_0$, **u**, and **E** therefore form a right-angled triad. Equation (3.169) also requires that $\mathbf{u}\perp i\mathbf{j}$. This constraint will be considered when we combine the two triads.

From Eq. (3.164), $i\mathbf{j}\perp\mathbf{k}$ and **b** while from Eq. (3.163), $\mathbf{b}\perp\mathbf{k}$. Thus the second triad is **b, k,** and $i\mathbf{j}$. Equation (3.163) provides an additional constraint that $\mathbf{b}\perp\mathbf{E}$, which also controls how the triads can be combined. When combining the triads we should also take into account the handedness of the cross products in Eqs. (3.163) and (3.171).

The two triads are shown in Figure 3.10. In Figure 3.11, the **b, k,** $i\mathbf{j}$ triad is rotated about the $-ij$-axis so $\mathbf{k}||\mathbf{B}_0$, corresponding to parallel propagation of the waves. This configuration also satisfies the two additional constraints, $\mathbf{u}\perp i\mathbf{j}$ and $\mathbf{b}\perp\mathbf{E}$, and maintains the handedness of the cross products given in Eqs. (3.163) and (3.170). Indeed, these additional constraints mean that we cannot rotate the **b, k,** $i\mathbf{j}$ triad about the *k*-axis, and the relative orientation shown in Figure 3.11 is the only one allowed for parallel propagation.

In order to consider propagation of waves at an angle to the magnetic field we must rotate one triad with respect to the other. But to do this and satisfy the two additional constraints ($\mathbf{u}\perp i\mathbf{j}$ and $\mathbf{b}\perp\mathbf{E}$), we must keep $i\mathbf{j}$ in the $\mathbf{E}$–$\mathbf{B}_0$ plane and **b** in the $\mathbf{u}$–$\mathbf{B}_0$ plane. To do this we can rotate the **b, k,** $i\mathbf{j}$ triad about either the *b*-axis or the *ij*-axis. These two options are shown in Figure 3.12.

Figure 3.12 also leads very quickly to the wave dispersion relation. For the mode on the left-hand side, Eqs. (3.163) and (3.164) become

$$kE\cos\theta = \omega b \tag{3.172}$$

and

$$kb = -i\mu_0 j. \tag{3.173}$$

Table 3.1. MHD wave properties as deduced from Figure 3.12; the phase velocity dependence on propagation angle is included for completeness

Property	Alfvén (shear) mode	Fast mode
Phase velocity	max. $\parallel \mathbf{B}_0$, Eq. (3.179)	isotropic, Eq. (3.180) (cold plasma)
Magnetic compression	no, $\mathbf{B}_0 \cdot \mathbf{b} = 0$	yes, $\mathbf{B}_0 \cdot \mathbf{b} \neq 0$, for $\theta \neq 0$
Plasma compression	no, $\mathbf{k} \cdot \mathbf{u} = 0$	yes, $\mathbf{k} \cdot \mathbf{u} \neq 0$, for $\theta \neq 0$
b transverse to $\mathbf{B}_0$	yes, all θ	only when $\theta = 0$
Field-aligned current	yes, for $\theta \neq 0$	no, for all θ
Poynting vector	field aligned	parallel to **k** (cold plasma)

Combining these two equations,

$$k^2 E\cos\theta = -i\omega\mu_0 j, \tag{3.174}$$

and Eqs. (3.168) and (3.170) become

$$E = -uB_0 \tag{3.175}$$

and

$$\omega\rho_0 u = ijB_0\cos\theta. \tag{3.176}$$

Defining the **Alfvén speed** as

$$v_{\mathrm{A}} = \frac{B_0}{\sqrt{\mu_0\rho_0}}, \tag{3.177}$$

we can combine Eqs. (3.175) and (3.176) to give

$$E = -i\mu_0 j\frac{v_{\mathrm{A}}^2}{\omega}\cos\theta. \tag{3.178}$$

Finally, combining Eqs. (3.173) and (3.178),

$$\omega^2 = k^2 v_{\mathrm{A}}^2\cos^2\theta. \tag{3.179}$$

This is the dispersion relation for the **Alfvén or shear mode.**

For the triads on the right-hand side of Figure 3.12, the only difference is that **k** and **E** are always perpendicular to each other, as are *i***j** and $\mathbf{B}_0$. The cos θ term is therefore absent from the equations equivalent to Eqs. (3.171) and (3.175) for this mode, and the dispersion relation is

$$\omega^2 = k^2 v_{\mathrm{A}}^2. \tag{3.180}$$

This is the **fast mode** dispersion relation for a cold plasma.

The vector relationships also allow us to determine several properties of the waves simply by inspection. These are summarized in Table 3.1.

There is no compression of the magnetic field or plasma for the shear mode. For this reason the shear mode properties are unaltered even if the plasma is warm (i.e., $P_0 \neq 0$). The fast mode will be modified by finite plasma pressure.

For the fast mode the phase relationship between plasma and magnetic field pressure can also be deduced from Figure 3.12. When $\mathbf{B}_0 \cdot \mathbf{b} > 0$, $\mathbf{k} \cdot \mathbf{u} = \omega p/\rho_0 > 0$. Thus, the plasma pressure and magnetic-pressure variations are in phase for the fast mode.

While it is implicit in the statement that the shear mode is non-compressive, the wave magnetic field is perpendicular to the background field for all propagation angles. The wave magnetic field for this mode corresponds to bending, or shearing, of the magnetic field, hence the name for this mode.

The second to last row in Table 3.1 shows that only the shear mode carries a field-aligned current. This is important for issues such as magnetosphere–ionosphere coupling, where the shear mode provides field-aligned currents that are a consequence of dynamical processes.

The last row in Table 3.1 indicates that the wave Poynting vector is along the magnetic field for the shear mode, and along the wave vector for the fast mode. The wave group velocity is parallel to the Poynting vector, so the shear-mode group velocity is parallel to $\mathbf{B}_0$ for all wave normal angles. The fast mode group velocity is parallel to the wave vector in the cold plasma limit.

A minor point from Figure 3.12 is that, for the shear mode, $\mathbf{k} \cdot \mathbf{E} \neq 0$. Thus $\rho_{\mathrm{q}} \neq 0$, but again, as discussed in Section 3.6.1, provided $\partial\rho_q/\partial t \approx 0$, the conservation of charge reduces to Eq. (3.130).

For completeness we now derive the dispersion relation for MHD waves in a warm plasma when $c_s \neq 0$. We provide a more detailed discussion of the warm plasma MHD wave modes in Chapter 13, but these waves will also be considered in Chapter 6, where we discuss shock phenomena, and Chapter 10 on the Earth's magnetosphere, for example.

From Ampère's law (3.164),

$$\mathbf{j} \times \mathbf{B}_0 = \frac{-iB_0}{\mu_0}\left(\mathbf{k}b_{\parallel} - \mathbf{b}k_{\parallel}\right). \qquad (3.181)$$

Thus, making use of the definition of the **Alfvén speed** (3.176), we can rewrite Eq. (3.169) as

$$\mathbf{u} - \mathbf{k}\frac{c_s^2}{\omega^2}(\mathbf{k}\cdot\mathbf{u}) = \frac{v_A^2}{\omega^2}\frac{\omega}{B_0}\left(\mathbf{k}b_{\parallel} - \mathbf{b}k_{\parallel}\right). \qquad (3.182)$$

Taking the dot product of Eq. (3.182) with **k**,

$$\left(1 - \frac{k^2c_s^2}{\omega^2}\right)(\mathbf{k}\cdot\mathbf{u}) = \frac{k^2v_A^2}{\omega^2}\frac{\omega b_{\parallel}}{B_0} \qquad (3.183a)$$

since $\mathbf{k}\cdot\mathbf{b} = 0$.

From the parallel component of the cross product of Eq. (3.182) with **k**,

$$\mathbf{B}_0\cdot(\mathbf{k}\times\mathbf{u}) = \frac{-k_{\parallel}v_A^2}{\omega^2}\frac{\omega}{B_0}\mathbf{B}_0\cdot(\mathbf{k}\times\mathbf{b}). \qquad (3.183b)$$

The reason for deriving these two equations may become clear on considering Figure 3.12. On the left-hand side of the figure, the two triads are such that the velocity **u** is perpendicular to the $\mathbf{k}$–$\mathbf{B}_0$ plane, and $\mathbf{k}\cdot\mathbf{u} = 0$. For the right-hand triads **u** is in the $\mathbf{k}$–$\mathbf{B}_0$ plane, and this is still the case even if we allow $u_{\parallel} \neq 0$. For the right-hand triads $\mathbf{B}_0\cdot(\mathbf{k}\times\mathbf{u}) = 0$. Thus $\mathbf{k}\cdot\mathbf{u}$ only includes components of **u** that are in the $\mathbf{k}$–$\mathbf{B}_0$ plane, while $\mathbf{B}_0\cdot(\mathbf{k}\times\mathbf{u})$ only contains the component of **u** that is perpendicular to the $\mathbf{k}$–$\mathbf{B}_0$ plane.

Making use of Faraday's law (3.163) and the frozen-in condition (3.168),

$$\omega\mathbf{b} = \mathbf{k}\cdot(\mathbf{B}_0\times\mathbf{u}) = \mathbf{B}_0(\mathbf{k}\cdot\mathbf{u}) - \mathbf{u}k_{\parallel}B_0. \qquad (3.184)$$

Hence

$$\frac{\omega b_{\parallel}}{B_0} = \mathbf{k}\cdot\mathbf{u} - k_{\parallel}u_{\parallel} = \left(1 - \frac{k_{\parallel}^2c_s^2}{\omega^2}\right)(\mathbf{k}\cdot\mathbf{u}), \qquad (3.185a)$$

where we have made use of Eq. (3.170), and

$$\frac{\omega}{B_0}\mathbf{B}_0\cdot(\mathbf{k}\times\mathbf{b}) = -k_{\parallel}\mathbf{B}_0\cdot(\mathbf{k}\times\mathbf{u}) \qquad (3.185b)$$

since the first term in Eq. (3.184) is parallel to $\mathbf{B}_0$ and vanishes in the scalar triple product.

Combining Eqs. (3.183a) and (3.183b) with Eqs. (3.185a) and (3.185b) we get

$$\left(1 - \frac{k^2\left(c_s^2 + v_A^2\right)}{\omega^2} + \frac{k^2v_A^2}{\omega^2}\frac{k_{\parallel}^2c_s^2}{\omega^2}\right)(\mathbf{k}\cdot\mathbf{u}) = 0 \qquad (3.186a)$$

and

$$\left(1 - \frac{k_{\parallel}^2v_A^2}{\omega^2}\right)\mathbf{B}_0\cdot(\mathbf{k}\times\mathbf{u}). \qquad (3.186b)$$

These two equations must be satisfied simultaneously. We have already shown that $\mathbf{k}\cdot\mathbf{u}$ and $\mathbf{B}_0\cdot(\mathbf{k}\times\mathbf{u})$ are independent of each other, and one solution has $\mathbf{k}\cdot\mathbf{u} = 0$. In that case Eq. (3.186b) becomes the **Alfvén mode** dispersion relation (3.178).

The other solution has $\mathbf{B}_0\cdot(\mathbf{k}\times\mathbf{u}) = 0$ and

$$\frac{\omega^2}{k^2} = \frac{c_s^2 + v_A^2 \pm \sqrt{\left(c_s^2 + v_A^2\right)^2 - 4c_s^2v_A^2cos^2\theta}}{2}, \qquad (3.187)$$

where we have made use of the standard relation for the roots of a quadratic equation and replaced $k_{\parallel} = k\cos\theta$.

Equation (3.187) gives the dispersion for two compressional modes in a warm plasma. For parallel propagation these modes propagate at the **Alfvén speed** (v_A) and the sound speed (c_s). In a low-β plasma $c_s < v_A$, and these modes are consequently referred to as the fast and slow modes, respectively. For a high-β plasma, the fast and slow modes are still present, but here the fast mode corresponds to the sound speed for parallel propagation.

For perpendicular propagation the slow mode vanishes, while the fast mode propagates at the phase speed given by $\omega^2/k^2 = c_s^2 + v_A^2$.

Inspection of Eq. (3.183a) shows that

$$\left(1 - \frac{k^2 c_s^2}{\omega^2}\right)\frac{P}{\gamma P_0} = \frac{k^2 v_A^2}{\omega^2}\frac{\mathbf{B}_0 \cdot \mathbf{b}}{B_0^2}, \qquad (3.188)$$

where we have made use of Eqs. (3.165) and (3.167). This shows that for the fast mode the plasma pressure and magnetic pressure are in phase, as we noted when discussing Figure 3.12, whereas for the slow mode the plasma and magnetic pressure are in antiphase. This phase relation is also the case when $c_s > v_A$, since the phase velocity $> c_s$ for the fast mode.

Finally, when we include the shear mode Eq. (3.178), this mode propagates at a phase speed between the fast and slow modes. Consequently, the shear mode is also referred to as the **intermediate mode.**

3.7.2 Equivalence between MHD and guiding center theory

The second example we shall consider here is to show that the current that can be derived from the MHD momentum equation (3.162) and the current derived from summing the generalized drift equation (3.68) over species give the same results provided the latter includes the **magnetization current** associated with the particle gyration.

Because we are not considering the full set of MHD equations (in particular the energy equation) for this discussion, we allow the plasma pressure to be different in the parallel and perpendicular directions. In that case we can write the pressure tensor as

$$\mathbf{P} = P_\perp \mathbf{I} + \mathbf{B}\mathbf{B}(P_\parallel - P_\perp)/B^2, \qquad (3.189)$$

where $\mathbf{I}$ is the unit diagonal tensor (δ_{ij} in the index notation used in Section 3.5).

On taking the cross product of Eq. (3.162) with $\mathbf{B}$ we find

$$\mathbf{j}_\perp = \frac{\mathbf{B}}{B^2} \times \left\{\rho\frac{D\mathbf{u}}{Dt} + \nabla P_\perp + \frac{(P_\parallel - P_\perp)}{B^2}(\mathbf{B}\cdot\nabla)\mathbf{B}\right\}. \qquad (3.190)$$

In order to calculate the current density from Eq. (3.68) we need to integrate this equation over all the particles. To do these we need to make some use of several identities.

First, given the definition of pressure,

$$n\langle W_\perp\rangle = P_\perp, \qquad (3.191)$$

where n is the species density. In this case the pressure is defined with respect to the guiding center velocity, and there are two degrees of freedom in the perpendicular direction. As in Section 3.5, $\langle\rangle$ denotes the average or expectation value.

Second, for the parallel velocity terms, we shall define

$$\langle v_\parallel\rangle = \overline{v}_\parallel. \qquad (3.192)$$

Hence, on averaging over all the particles,

$$\langle v_\parallel^2\rangle = \overline{v}_\parallel{}^2 + P_\parallel/nm, \qquad (3.193)$$

where $P_\parallel$ is defined with respect to $\overline{v}_\parallel$.

Last, we note that

$$\frac{v_\parallel\mathbf{B}}{B} \times \frac{\partial\mathbf{B}}{\partial t} = \mathbf{B} \times \frac{\partial}{\partial t}\left(\frac{v_\parallel\mathbf{B}}{B}\right), \qquad (3.194)$$

with a similar identity for the other terms that involve a differential operation on $\mathbf{B}$.

We can now average Eq. (3.68) over all the particles in a species to get the average drift velocity for that species:

$$\overline{\mathbf{v}}_\perp = \mathbf{v}_E + \frac{m}{qB^2}\mathbf{B} \times \left\{\frac{P_\perp\nabla B}{nmB} + \frac{P_\parallel(\mathbf{B}\cdot\nabla)\mathbf{B}}{nmB^2} + \frac{d}{dt}[\overline{\mathbf{v}}_\parallel + \mathbf{v}_E]\right\}, \qquad (3.195)$$

where $\overline{\mathbf{v}}_\parallel = \overline{v}_\parallel\mathbf{B}/B$ and $d/dt = \partial/\partial t + \overline{\mathbf{v}}_\parallel\cdot\nabla + \mathbf{v}_E\cdot\nabla$.

To derive the current density, we multiply Eq. (3.195) by nq for each species and sum over species. In addition we have to include the magnetization current, which is present because the particles have a magnetic moment. Unlike $\overline{\mathbf{v}}_\perp$, this current is not related to any motion of the guiding center.

The perpendicular component of the magnetization current is given by

$$\mathbf{j}_{m\perp} = -\left[\nabla \times \left(\frac{P_\perp \mathbf{B}}{B^2}\right)\right]_\perp$$
$$= \frac{\mathbf{B}}{B} \times \nabla\left(\frac{P_\perp}{B}\right) - \frac{P_\perp}{B^4}\mathbf{B} \times (\mathbf{B}\cdot\nabla)\mathbf{B}. \quad (3.196)$$

Consequently, the perpendicular current from guiding center drift and magnetization is

$$\mathbf{j}_\perp = \frac{\mathbf{B}}{B^2} \times \left\{\frac{P_\perp \nabla B}{B} + B\nabla\left(\frac{P_\perp}{B}\right) + \frac{(P_\parallel - P_\perp)(\mathbf{B}\cdot\nabla)\mathbf{B}}{B^2} + \rho\frac{d}{dt}[\bar{\mathbf{v}}_\parallel + \mathbf{v}_E]\right\}. \quad (3.197)$$

We note that in deriving Eq. (3.68) we neglected terms on the right-hand side of the order $\mathbf{v}_D$, where this velocity corresponds to all the perpendicular drift terms except $\mathbf{v}_E$, which was treated separately. To the same precision, then, we can replace the $d(\bar{\mathbf{v}}_\parallel + \mathbf{v}_E)/dt$ term in Eq. (3.197) with $D\mathbf{u}/Dt$, since these only differ by $\mathbf{v}_D$-dependent terms. Additionally, the pressures are the sum of the ion and electron pressures, while $\bar{\mathbf{v}}_\parallel$ is only the parallel velocity of the ions since $m_e \ll m_i$. The other important aspect of Eq. (3.197) is that the current from gradient drift, which is proportional to ∇B, is exactly canceled by the ∇B-dependent term from the magnetization current. Thus the current given by guiding center drift plus magnetization, Eq. (3.196), is the same as the current derived from MHD, Eq. (3.190).

3.7.3 Field-aligned currents and MHD

Having established the equivalence of the currents obtained from MHD and guiding center theory, provided the latter includes the magnetization current, we now derive an important result that is frequently used for the inner magnetosphere. This is the relationship between field-aligned currents and plasma and magnetic field gradients. As noted previously, field-aligned currents play a significant role in magnetosphere–ionosphere coupling.

In this case, however, we use the isotropic pressure approximation. With this approximation, the usual assumption of quasi-neutrality, and neglecting gravity, the curl of Eq. (3.162) gives

$$\nabla \times \left(\rho\frac{D\mathbf{u}}{Dt}\right) = (\mathbf{B}\cdot\nabla)\mathbf{j} - (\mathbf{j}\cdot\nabla)\mathbf{B}. \quad (3.198)$$

Because of the isotropic pressure assumption, the pressure term vanishes on taking the curl, and we have also made use of $\nabla\cdot\mathbf{B} = 0$ and $\nabla\cdot\mathbf{j} = 0$.

One approximation frequently made at this stage is the "slow-flow" approximation, where the ion inertia term on the left-hand side of Eq. (3.198) is dropped. For completeness, we keep this term.

The next step is to take the dot product of Eq. (3.198) with $\mathbf{B}$ and rearrange terms:

$$\mathbf{B}\cdot(\mathbf{B}\cdot\nabla)\mathbf{j} = B\mathbf{j}\cdot\nabla B + \mathbf{B}\cdot\nabla\times\left(\rho\frac{D\mathbf{u}}{Dt}\right). \quad (3.199)$$

From the dot product of Eq. (3.159) with $\mathbf{j}$,

$$\mathbf{j}\cdot(\mathbf{B}\cdot\nabla)\mathbf{B} = B\mathbf{j}\cdot\nabla B. \quad (3.200)$$

Hence,

$$(\mathbf{B}\cdot\nabla)(\mathbf{j}\cdot\mathbf{B}) = 2B\mathbf{j}\cdot\nabla B + \mathbf{B}\cdot\nabla\times\left(\rho\frac{D\mathbf{u}}{Dt}\right). \quad (3.201)$$

We now can determine the field-aligned gradient of the current density per unit magnetic flux as

$$(\mathbf{B}\cdot\nabla)\left(\frac{\mathbf{j}\cdot\mathbf{B}}{B^2}\right) = 2\mathbf{j}\cdot\left[\frac{\nabla B}{B} - \frac{\mathbf{B}}{B^3}(\mathbf{B}\cdot\nabla B)\right] + \frac{\mathbf{B}}{B^2}\cdot\nabla\times\left(\rho\frac{D\mathbf{u}}{Dt}\right). \quad (3.202)$$

Recognizing that the term in square brackets is $\mathbf{B}\times(\nabla B\times\mathbf{B})/B^3$, and using the permutation rule for the scalar triple product, we have

$$(\mathbf{B}\cdot\nabla)\left(\frac{\mathbf{j}\cdot\mathbf{B}}{B^2}\right) = \frac{2}{B^3}(\mathbf{j}\times\mathbf{B})\cdot(\nabla B\times\mathbf{B}) + \frac{\mathbf{B}}{B^2}\cdot\nabla\times\left(\rho\frac{D\mathbf{u}}{Dt}\right). \quad (3.203)$$

We can now use the momentum equation itself to replace the $\mathbf{j}\times\mathbf{B}$ term, and again, using the permutation rule, we get a compact expression for the field-aligned current:

$$(\mathbf{B} \cdot \nabla)\left(\frac{\mathbf{j} \cdot \mathbf{B}}{B^2}\right) = \frac{\mathbf{B}}{B^2} \cdot \left[2\left(\nabla P + \rho \frac{D\mathbf{u}}{Dt}\right) \times \frac{\nabla B}{B} + \nabla \times \left(\rho \frac{D\mathbf{u}}{Dt}\right)\right]. \tag{3.204}$$

The "slow-flow" approximation to this equation is obtained by neglecting the terms that are dependent on the mass density. In general, however, these terms are associated with plasma inertia and should be kept. With suitable rearrangement of the terms, they can be related to flow braking (mainly through the first inertia term), and **vorticity** (through the $\nabla \times D\mathbf{u}/Dt$ term). In particular, the vorticity term on the right-hand side of Eq. (3.204) corresponds to the shear mode. Inspection of Figure 3.12 shows that $\mathbf{B}_0 \cdot (\mathbf{k} \times \mathbf{u}) \neq 0$ for the mode on the left-hand side (the shear mode), but $\mathbf{B}_0 \cdot (\mathbf{k} \times \mathbf{u}) = 0$ for the fast mode. As already noted, but confirmed by Eq. (3.204), it is the shear mode that carries field-aligned current.

3.7.4 Summary of MHD applications

We have given three applications of MHD here, with a view to what follows in subsequent chapters. The MHD waves described here are the principal component by which the mass and momentum of the plasma interact with the electromagnetic fields. The bow shock, for example, is a standing fast mode wave (see Chapter 6), while the shear mode is an intrinsic part of magnetosphere–ionosphere coupling (Chapters 10 and 11). In discussing MHD waves it should be noted that the shear mode is often referred to as the **Alfvén mode**. We showed the equivalence of the perpendicular current as derived from MHD with that derived using guiding center drifts. At one level, then, it seems that the results should be the same regardless of which theory is used. But comparison between models that use MHD and models that use guiding center drifts give different results. Reconciling the two approaches is a major effort in the modeling community. Finally, we showed how the divergence of the perpendicular current results in field-aligned current. This is important for magnetosphere–ionosphere coupling (Chapters 10 and 11).

3.8 SUMMARY

In this chapter we have presented the theoretical foundation for the physics of magnetized plasmas. Of necessity we took as given some fundamental equations, such as Maxwell's equations, the Lorentz force law, and the Boltzmann equation. From that we derived the two main representations of plasma physics: particle drift theory and magnetohydrodynamics (MHD). These are often thought as being disparate approaches, but this chapter has shown that the two approaches have elements in common, most notably the idea that the current derived from drift motion is also contained within MHD.

Based on Maxwell's equation and the Lorentz force, we determined the motion of charged particles in prescribed electric and magnetic fields, leading to the concept of the guiding center. We found the primary drifts of the guiding center to be the electric field ($\mathbf{E} \times \mathbf{B}$) drift, and the gradient and curvature drifts. We also presented the generalized drift, which included additional drifts such as the polarization drift. Guiding center drifts are often used to understand the motion of charged particles in the inner magnetosphere (see Chapter 10), where the background magnetic field is not significantly distorted by currents. But guiding center theory is essentially a test-particle theory. Fluid theory, on the other hand, can be used to determine how the plasma affects the fields.

In order to derive MHD theory, which is the basic fluid theory for space plasmas, we discussed the concept of collective phenomena, such as plasma oscillations. We then introduced the Boltzmann equation, which in turn was used to discuss Debye shielding, where the electric field of an individual particle is shielded from the rest of the particles in the plasma. The Boltzmann equation is also the starting point for the branch of plasma-wave theory that considers wave–particle resonance (see Chapter 13), and for the derivation of fluid theory. Fluid theory is based on moment integrals of the Boltzmann equation, which interestingly removes any effects of wave–particle resonances. The moment integrals form the basis of MHD.

Single-fluid MHD is the usual starting point for understanding fluid theories, and the equations that govern single-fluid MHD are Maxwell's equations, most notably Ampère's law and Faraday's law, together with the moment equations that result in mass, momentum, and energy conservation, and the generalized Ohm's law.

The ideal version of the generalized Ohm's law is given by $\mathbf{E} + \mathbf{u} \times \mathbf{B} = 0$, which can be interpreted as stating that the electric field vanishes in the frame moving with the fluid. This condition also results in the frozen-in theorem, which in turn states that the magnetic field can be assumed to be frozen to the plasma as the plasma moves.

Finally we discussed three applications of MHD. The first was an introduction to MHD waves. These are low-frequency waves that are responsible for moving the material that constitutes the plasma. Higher-frequency waves are mainly associated with differential motion between the electron and ion species (i.e., currents or charge-separation electric fields), with the exception of waves associated with ion gyration. We also discussed the equivalence between particle drift theory and MHD in terms of perpendicular currents. We closed with a derivation of the general form for the source of field-aligned currents in the MHD regime, based on the divergence of perpendicular current. This will be useful in Chapters 10 and 11, where we discuss magnetosphere–ionosphere coupling.

Additional reading

Much of the derivations given in this chapter are based on two classical textbooks:

Boyd, T. J. M. and J. J. Sanderson (1969). *Plasma Dynamics*. London: Nelson, which provides a concise introduction to plasma physics.

Clemmow, P. C. and J. P. Dougherty (1969). *Electrodynamics of Particles and Plasmas*. Reading, MA: Addison-Wesley Publ. Co., which gives a strong theoretical foundation for plasma physics, including a discussion on the derivation of kinetic theory.

The following two references both demonstrate that the perpendicular current from fluid theory is equivalent to that derived on the basis of guiding center drift, with the inclusion of the magnetization current:

NORTHROP, T. G. (1963). Adiabatic charged-particle motion. *Rev. Geophys.*, 1(3), 283–304.

PARKER, E. N. (1957). Newtonian development of the dynamical properties of ionized gases of low density. *Phys. Rev.*, 107(4), 924–933.

Other reading material includes:

SISCOE, G. L. (1983). Solar system magnetohydrodynamics, in *Solar-Terrestrial Physics*. Eds. R. L. CAROVILLANO and J. M. FORBES. Hingham, MA: D. Reidel, pp. 11–100, which provides a detailed derivation of magnetohydrodynamics.

VASYLIUNAS, V. M. (1970). Mathematical models of magnetospheric convection and its coupling to the ionosphere, in *Particles and Fields in the Magnetosphere*. Ed. B. MCCORMAC. Hingham, MA: D. Reidel, pp. 60–71, where the "slow-flow" approximation to the plasma momentum equation is used to determine field-aligned currents.

Problems

3.1 Use the Particle Motion option of the Space Physics Exercises* to study particle motion in a uniform magnetic field.

(a) Set the magnetic field strength to 250 nT. Start the particles at $x = 0$, $y = 30$, $z = -30$ km, with $v_x = 50$, $v_y = 0$, $v_z = 10$ km s^{-1}. Follow the motion of a proton, a He^+ ion, and an O^+ ion for about 500 ms, as given in the time box. Measure the gyro-radii by clicking on the screen or by using the maximum and minimum range boxes. Show how the calculation is made. Plot the log of gyro-radius versus the log of mass. What is the dependence? The pseudo-electron used has a mass of $m_p/43$ not $m_p/1836$. Calculate its gyro-radius based on your formula for the ions. Use only the number of significant figures appropriate to your measurement. Show a sample screen capture.

(b) Vary the magnetic field strength from 25 to 250 nT, keeping $v_x = 50$ km s^{-1} and $v_y = v_z = 0$, and determine for protons the variation of gyro-radius and gyro-frequency with field strength. Plot both on log-log scales versus field strength. Show how your measurements were made and how you derived the gyro-radius and gyro-frequency. What is the dependence?

(c) Starting particles at $x = 0$, $y = 30$, $z = -30$ km, vary the perpendicular velocity by factors of 2 from 50 to 400 km s^{-1} using a magnetic field strength of 200 nT to determine the variation of proton gyro-radius and gyro-frequency with velocity. Explain how you derived measurements and calculated quantities. Plot both on a log-log scale versus velocity. What is the dependence of each?

(d) Based on parts (a), (b), and (c), write down a formula for the dependence of gyro-radius and gyro-frequency on mass, field strength, and perpendicular velocity in MKS units with correct normalization. Use the experimentally derived slopes of the plots given in (a), (b), and (c).

3.2 Use the Particle Motion option of the Space Physics Exercises* to study particle motion in crossed electric and magnetic fields (**E** × **B**).

(a) Set the magnetic field strength to 100 nT. With a 1 mV m^{-1} electric field at 90° to the magnetic field, choose a variety of proton velocities perpendicular to the magnetic field (e.g., 25, 50, 100) and show graphically how the velocity of the guiding center of the proton gyro-orbit depends on the starting velocity. Indicate how you determined the starting and ending gyro-center.

(b) Vary the magnetic field strength at fixed electric field (1 mV m^{-1}) and show graphically how the proton guiding center drift velocity depends on the magnetic field strength. Use $v_x = 50$ km s^{-1}; $v_y = v_z = 0$.

(c) Vary the electric field strength at fixed magnetic field (100 nT) for $v_x = 50$ km s^{-1} ($v_y = 0$) protons and show graphically how the guiding center drift depends on the electric field.

* Space Physics Exercises are at http://spacephysics.ucla.edu.

(d) Use the module to determine whether mass or charge affects the drift velocity. Show how electrons drift in the same electric field. From (a)–(d) determine the correct formula for the drift velocity including proper normalization for constants. Show how you obtain the formula and normalization from your results. Do not force your result to agree with theory. State units as appropriate.

(e) Set the angle of the electric field to the magnetic field to 87°. For $B = 100$ nT, $E = 1$ mV m^{-1}, calculate the functional dependence of acceleration on mass, charge, and electric field. Use your results and normalization to get this formula. Do not force your formula to agree with theory. State units as appropriate. Use protons, singly charged helium, and oxygen for your varying mass. Use singly and doubly charged (alpha particle) helium to check charge dependence.

3.3 Use the Particle Motion option of the Space Physics Exercises* to examine the motion of protons in magnetic gradients. The magnetic field strength setting gives the field strength on the left-hand side of the left-hand panel. The value of the gradient gives the change in magnetic field every 10 km from left to right across the panel that is 100 km wide.

(a) Determine how the motion of the guiding center depends on an ion's velocity perpendicular to the magnetic field. Illustrate the dependence with a log-log plot and least-square fit a straight line to the logarithms. Suggestion: start the ion at (0, –30, 0) km with $v_y = v_z = 0$ and $v_x = 30, 45, 60, 75$ km s^{-1}. Use 10 nT field strength, and a gradient of 10 nT per 10 km. The field strength at the center line of the left-hand panel should be 60 nT.

(b) Determine how the motion of the guiding center depends on the gradient in the magnetic field strength. Illustrate the dependence graphically on a log-log scale and least-square fit a straight line to the data. Note that it is very important to keep the magnetic field at the center of the orbit constant in this exercise. Suggestion: start the ion at (0, –30, 0) km with $v = (60, 0, 0)$ km s^{-1} with magnetic fields and gradients of (1 nT, 20 nT/10 km); (26 nT, 15 nT/10 km); (51 nT, 10 nT/10 km), and (76 nT, 5 nT/10 km). What is the magnetic field at the center of the left-hand panel if these values are used?

(c) Determine how the motion of the guiding center depends on the magnetic field strength at the gyro-center. Illustrate the dependence with a log-log plot and least-square fit a straight line to the logarithms. Suggestion: start the ion at (0, –30, 0) km with a velocity of (60, 0, 0) km s^{-1}. Use a gradient of 5 nT/10 km and field strengths of 10, 20, 30, and 40 nT. What are the resulting field strengths at the gyro-center of the particle given these parameters?

(d) How do electrons drift in gradients? Do gradient drifts in a plasma lead to currents? You may use a negative ion to substitute for an electron.

3.4 Use the Particle Motion option of the Space Physics Exercises* to examine the motion of protons in curved magnetic fields. Determine how the motion of the guiding center of the motion depends on the perpendicular and parallel energy of the particle, the magnetic field, and the curvature of the field line. When testing the effects of perpendicular and parallel velocity be careful to vary only the parameter being studied. This can be done by launching the particle with velocities only along either the perpendicular or parallel direction. Illustrate with graphs. Use the program to determine whether curvature drifts are charge dependent and lead to currents. We suggest initial parameters of (–40, 0, 0) km; (25, 0, 0) km s^{-1} for perpendicular motion and (0, 25, 0)km s^{-1} for parallel motion.

3.5 Use the Magnetic Mirror option of the Space Physics Exercises* to study the magnetic mirroring of charged particles. With the particle in the center of the magnetic bottle ($x = y = z = 0$) and $v_x = 0$, trace the motion of protons and alphas with equal v_y and v_z velocities ($v_y = v_z = 1, 2, 5$, etc.).

* Space Physics Exercises are at http://spacephysics.ucla.edu.

Does the mirror distance depend on the absolute value of velocity or on the mass of the particle? Change the ratio of the mirror field from 100 to 50. What happens to the trajectories of the mirroring particles used above? Repeat with different ratios of v_y and v_z, both greater than and less than unity. Explain your results.

3.6 Use the Dipole Magnetic Field option of the Particle Motion exercise of the Space Physics Exercises.* This module allows you to follow particle motion in a realistic model of a planetary magnetosphere in which there is mirror trapping along the magnetic field and gradient and curvature drift across the magnetic field. Curvature drift arises only when there is a velocity of the particle parallel to the magnetic field.

(a) Launch protons at right angles to the magnetic field with $v_y = 30$ km s^{-1}, $v_x = v_z = 0$ at three distances $x = 25$, 28, and 32 km, $y = z = 0$. Measure their drift velocity, i.e., the motion of the center of gyration. The magnetic field falls off as the inverse third power of the distance. How does this affect the drift velocity? Use the formula for drift velocity to show what dependence is expected theoretically in a dipole field.

(b) Launch protons at right angles to the magnetic field with $v_y = 30$, 40, and 50 km s^{-1}, and $v_x = v_z = 0$ at $x = 30$ km and $y = z = 0$. Measure their drift velocity. How does the drift velocity depend on the perpendicular energy of the particle, i.e., v_y^2?

(c) Launch protons with $v_y = 30$ km s^{-1}, $v_x = 0$, and $v_z = 15$, 30, 45, and 60 km s^{-1}, and $x = -30$, $y = 0$, and $z = 0$ km. Measure the drift velocity around the dipole and then the mirror latitude. The mirror latitude can be measured at the first reflection point in the x–z plane. How does the drift velocity depend on the total energy, i.e., $v_y^2 + v_z^2$? How does the mirror latitude depend on the ratio

$$v_z/\left(v_y^2 + v_z^2\right)^{1/2}?$$

* Space Physics Exercises are at http://spacephysics.ucla.edu.

3.7 Determine the equation of a magnetic field line and sketch the magnetic field lines corresponding to $B_x = y$, $B_y = x$, making sure that their relative spacing indicates the field strength. Calculate the magnetic curvature and pressure forces for this field. Include on the diagram several isocontours of magnetic field strength and several vectors indicating the direction and strength of the curvature and pressure forces. [Hint: Magnetic-pressure force is $-\nabla(B^2/2\mu_0)$ and curvature force is $(\mathbf{B}\cdot\nabla)\mathbf{B}/\mu_0$.]

3.8 The single-fluid moment equations that form the basis of magnetohydrodynamics (MHD) are as follows (neglecting collisions):

$$\text{mass}: \frac{\partial \rho}{\partial t} + \nabla\cdot(\rho\mathbf{u}) = 0;$$

$$\text{momentum}: \frac{\partial}{\partial t}(\rho\mathbf{u}) + \nabla\cdot(\mathbf{P} + \rho\mathbf{u}\mathbf{u}) - \rho_q\mathbf{E} - \mathbf{j}\times\mathbf{B} - \rho\mathbf{g} = 0;$$

$$\text{energy}: \frac{\partial}{\partial t}\left(\frac{1}{2}P_{ii} + \frac{1}{2}\rho u^2\right) + \nabla\cdot\left(\mathbf{q} + \mathbf{P}\cdot\mathbf{u} + \frac{1}{2}P_{ii}\mathbf{u} + \frac{1}{2}\rho u^2\mathbf{u}\right) - \mathbf{j}\cdot\mathbf{E} - \rho\mathbf{u}\cdot\mathbf{g} = 0.$$

(See Eqs. (3.113), (3.115), and (3.117).)

From these equations, derive the momentum and energy equations of MHD in their more familiar form, as given below, under the assumption that the pressure is isotropic (i.e., the pressure in tensor notation is given by $P\delta_{ij}$):

$$\text{momentum}: \rho\frac{D\mathbf{u}}{Dt} = \mathbf{J}\times\mathbf{B} - \nabla P;$$

$$\text{energy}: \frac{D}{Dt}(P\rho^{-\gamma}) = 0.$$

Note the approximations that are required to give the final form of the momentum and energy equations.

4 The Sun and its atmosphere

4.1 INTRODUCTION

The Sun plays a central role in virtually all of space physics, and is itself a rich plasma physics laboratory for everything from nuclear fusion to shocks, plasma waves, and particle acceleration. In this chapter we focus on general attributes of the Sun that control planetary upper atmospheres and their ionospheres, and the interplanetary medium with all its phenomena. In particular, the many processes in the interplanetary medium that control the dynamics of planetary magnetospheres have their origins at the Sun.

It is worth keeping the perspective that the Sun is one particular star in a galaxy of myriad stars, in a universe full of countless galaxies. It is a rather typical middle aged G-type star by astronomical standards, formed about 4.6 billion years ago from the collapsing, rotating cloud of magnetized interstellar gas and dust that also created our solar system. The Sun lies 1.496×10^8 km from Earth, or 215 solar radii, away. Thus the solar diameter subtends an angle of about one-half degree in the sky, coincidentally at this time close to the angle that the Moon subtends. We are comfortably in the Sun's habitable zone. At the same time, we are aware that the Sun is variable. It has been different in the past and will change in the future. It has semi-regular cycles that include outbursts of solar activity.

As in many other areas of research, our knowledge is limited both by available measurements and the human capacity to understand unfamiliar physical conditions and highly complex systems in often inaccessible places. While we have benefitted from accumulating measurements on the ground and in space, many areas have not advanced much beyond scientific debate and cartoon stages. Still, the progress made since the beginning of the space age in the 1960s in this area, as in others, has been remarkable.

In this chapter we describe some key observations of the Sun and its atmosphere obtained using three basic techniques of remote sensing: imaging, spectroscopy, and radio signals. The obtained information on solar phenomenology at many wavelengths with increasingly higher spatial and temporal resolution both provides insights and challenges our understanding. Research on the theoretical front has also kept pace with observational progress, though by necessity we provide only limited detail here. The reader will recognize many basic concepts from other chapters, and especially Chapter 3 on the basic physics of space plasmas, which should be read before proceeding. We include brief descriptions of selected models that lend physical insight and serve as valuable aids to the data interpretations. The aim is to give the reader a concise overview of some of the main observational results, concepts, terminology, and methods used in solar physics toward a working background for space physics research, or for independently pursuing further details. Solar physics and solar astronomy are highly active fields in their own right, with many dedicated books and journals that can be consulted for in-depth descriptions of any of the subtopics mentioned and more.

Table 4.1. Solar percent composition: top eight by mass

Element	H	He	O	C	Fe	Ne	N	Si
Photosphere	73.5	24.9	0.8	0.3	0.2	0.1	0.1	0.1
Corona	81.0	18.0	0.04	0.08	0.01	0.006	0.01	0.01
Ratio	1.3	0.76	0.05	0.27	0.05	0.06	0.1	0.1

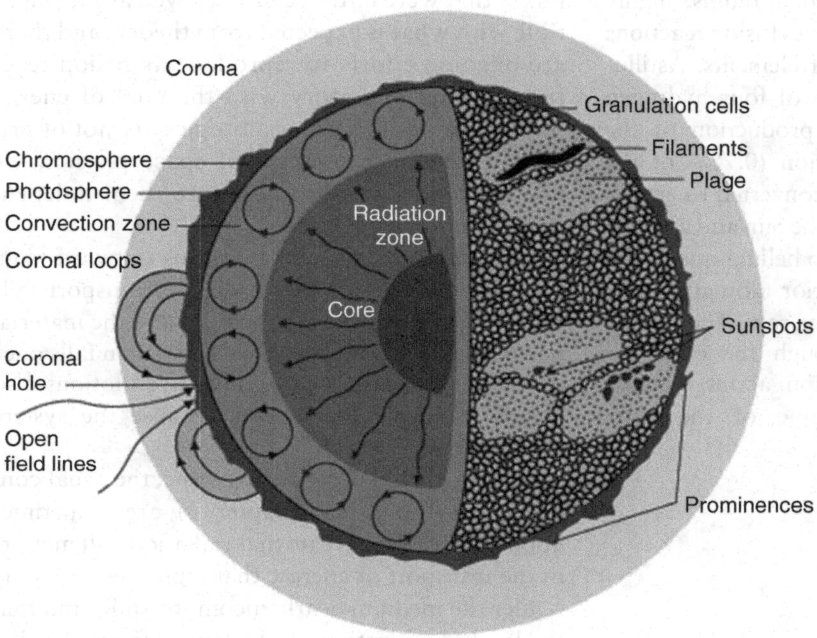

FIGURE 4.1. Basic structure of our star, the Sun. The center of the figure shows the core, the radiative zone, and the convective zone. The right side illustrates features found in the photosphere and chromosphere, and on the far left are features found above these in the corona.

4.2 THE STRUCTURE OF THE SUN AND ITS ATMOSPHERE

The radius of the Sun is slightly over 109 times that of the Earth with a mass of 1.99×10^{30} kg, resulting in a surface gravity of 274 m s^{-2}, which drops to about 2.4×10^{-3} m s^{-2} at the orbit of Earth. The **escape velocity** from the Sun is 618 km s^{-1} compared with a velocity of 42 km s^{-1} to escape from the orbit of the Earth or 11 km s^{-1} to escape from the surface of the Earth to nearby space. The mean density of the Sun is 1400 kg m^{-3}, consistent with its composition of 90% H and 10% He under pressure. The Sun's rotation axis is tilted at 7.25° to the ecliptic plane, and the Sun rotates about this axis relative to the stars every 25 days. The photons we receive on Earth have the spectrum of a black body radiating at a temperature of 5785 K. These photons release 3.86×10^{33} erg s^{-1} into space.

The basic structure of a star like the Sun, sketched in Figure 4.1, consists of concentric nearly spherical shells of gravitationally stratified hot gas. The relative composition of main elements found in the Sun (Table 4.1) is inferred from a combination of spectral evidence, including absorption and emission lines and continuum signatures. The bulk of the Sun's mass and its highest gas densities exist in the core where fusion of hydrogen to make helium is occurring. Surrounding this is the radiative zone where the energy slowly makes its way toward the surface. About two-thirds of the way to the surface of the Sun is the **convective zone,** where heat transport is made more efficient by overturning of the material, similar to the motion of the fluid in a pot of water heated from the bottom. Most of the light reaching us at 1 AU comes from its effective surface, the **photosphere.** Above the photosphere stretches the extended solar atmosphere that also affects us in its own distinctive ways (as described in Chapter 5).

The center of Figure 4.1 indicates the conditions and processes at work in the different layers of the solar interior. In the deepest interior, or *core*, inside about $0.2R_S$, where R_S is the solar radius, highly compressed hydrogen gas undergoes fusion reactions that form deuterium and heavier elements. As illustrated in Figure 4.2, the fusion of four hydrogen nuclei is enabled through the production of the helium-3 isotope. A small portion (0.7%) of the mass of the original particles is converted to energy in the fusion reaction, powering the Sun and all of its associated outputs. In addition to helium nuclei, the main reactions in the solar interior ultimately produce neutrinos, x-rays, and gamma rays. The weakly interacting neutrinos pass through the overlying solar strata, carrying information about what is going on in the nuclear engine of the core.

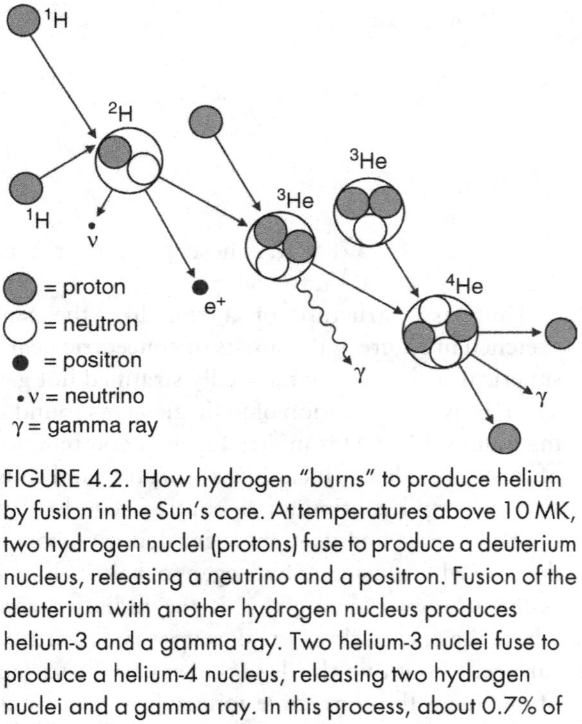

FIGURE 4.2. How hydrogen "burns" to produce helium by fusion in the Sun's core. At temperatures above 10 MK, two hydrogen nuclei (protons) fuse to produce a deuterium nucleus, releasing a neutrino and a positron. Fusion of the deuterium with another hydrogen nucleus produces helium-3 and a gamma ray. Two helium-3 nuclei fuse to produce a helium-4 nucleus, releasing two hydrogen nuclei and a gamma ray. In this process, about 0.7% of the mass of the four hydrogen nuclei that formed the helium-4 nucleus is lost in the form of mainly radiative energy.

However, the energy released in the form of radiation (the x-rays and gamma rays) diffuses outward and heats the overlying **radiative zone,** which absorbs much of what is generated within about $0.7R_S$. There are some interesting problems with the understanding of the deep solar interior, including the measurements of solar neutrino fluxes that were until recently viewed as incompatible with what is expected from theory, and there are ongoing efforts to reproduce its fusion reactions in the laboratory with the goal of energy production. But as these subtopics are not of primary importance to most of space physics, we leave it to the reader to explore them independently as interest dictates.

Eventually, as the energy moves outward from the core, the efficiency of radiative transport falls below that of convective transport and the material begins to rise toward the surface and then falls once it cools. These large-scale convective motions also generate turbulence, in part because the system rotates.

Within the Solar convective zone, the usual concepts of MHD (see Chapter 3) are sometimes applied, with the caveats that radiation still matters in the transport of energy, that high pressures may render the medium nearly incompressible, and that the level of ionization can be low. Estimated radial profiles of some of the physical properties of the gas in the solar interior are shown in Figure 4.3.

In both the radiative and convective zones, the gases are nearly opaque or "optically thick" due to photon scattering and absorption/re-emission. But at the **photosphere** where the radiation escapes, it is dominated by visible-light wavelengths, as shown by the sketch in Figure 4.4(a). Thermal emissions are characteristic of an equilibrium state of a gas with Maxwell–Boltzmann velocity or energy distributions. Other emissions come from different parts of the solar atmosphere and other, often "non-thermal," processes described below. The main part of this spectrum is what largely determines the **solar constant,** a parameter that is a measure of the total energy output of the Sun. It has remained nearly the same (to within a few tenths of a percent) for as long as it has been observed. In fact, the sources of the small variations of the solar constant are a target of ongoing investigation and are the main solar

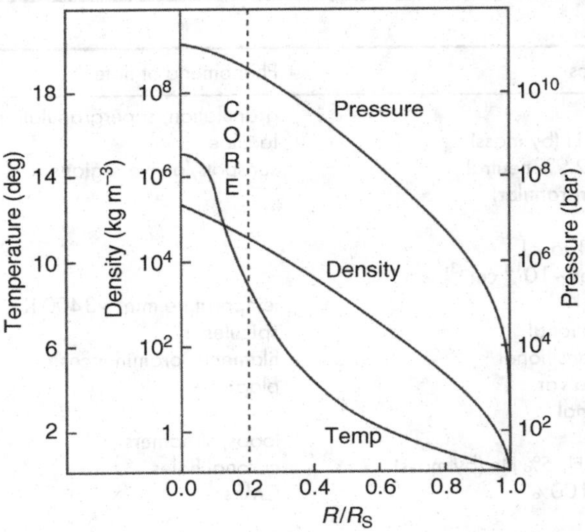

FIGURE 4.3. Radial profiles of solar interior's basic properties, identifying the main layers, including core, radiative zone, and convective zone. Most of the solar mass resides in the core. The main energy transport mechanism evolves from radiative (photon carried) to mechanical (convection carried) at the inner boundary of the convective zone at about $0.67R_S$. (After Sexl *et al.*, 1980.)

property used in long-term climate modeling. Its measurement has been improved over the space age by obtaining a more complete and accurate spectrum of the **total solar irradiance** (referred to as TSI) above the Earth's selectively absorbing atmosphere, and by observing the specific spectral contributions of different solar features in more detail.

To the unaided eye, the quiet photosphere appears nearly featureless; however, on closer examination with high spatial resolution solar telescopes and filters, the photosphere exhibits non-uniformities including networks of small-scale convection cells that are constantly moving and evolving. These cells, shown in the solar image in Figure 4.5, are called granules on the smallest scales (thousands of kilometers or megameters, Mm), and supergranules when they exhibit larger-scale coherent behavior (scales of tens of thousands of kilometers or tens of megameters) that appears to be organized from below. Granules have average lifetimes of tens of minutes, while supergranules last longer, perhaps days. They are considered to occupy a relatively thin layer at the top of the convection zone with a depth of about the same scale as their diameter. Different filters reveal details of this visible skin of the Sun. Doppler

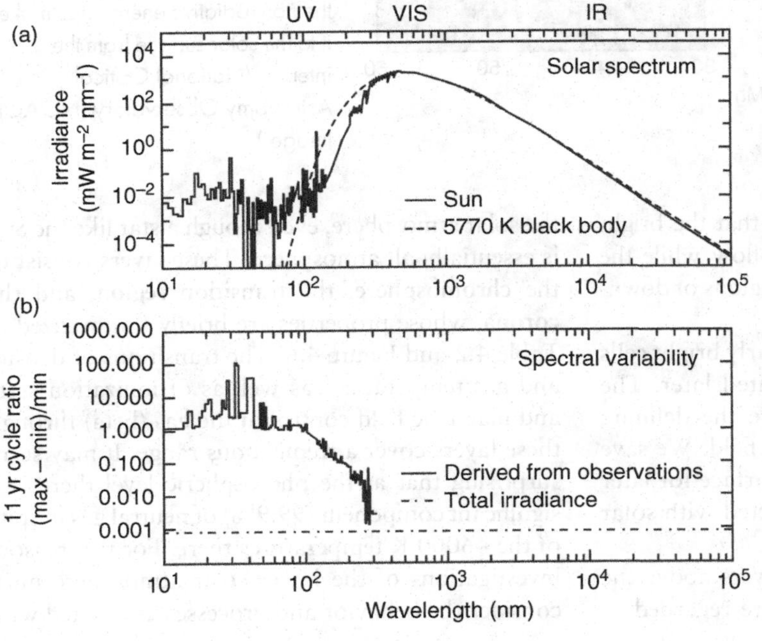

FIGURE 4.4. Solar spectrum (a) and variation with wavelength (b). The main portion of the solar spectrum, dominated by visible wavelengths, is a black body distribution characteristic of an emitting body at ~6000 K. The extremes of the spectrum, especially at wavelengths shortward of ~100 nm, exhibit the largest variations. The integration of this spectrum over all wavelengths is a measure of the total solar irradiance (TSI) or "solar constant," so-named because it varies so little over time in the current epoch of the Sun's life. (After Lean, 1991.)

Table 4.2. Key properties of solar atmospheric regions

Region	Main characteristics	Phenomena of note
Photosphere	$T \sim 6000$ K composition: 74% H (by mass) ionization state: 99.9% neutral $B \sim 0$–100 G (intergranular) (1 G = 10^{-4}T) beta transitional >1 to <1 collisional (densities $\sim 10^{14}$ cm^{-3})	granulation, supergranulation faculae sunspots/active regions
Chromosphere and transition region	$T \sim 6000$–30 000 K composition: transitional ionization state: transitional $B \sim$ several μT, beta var. collisions: transitional	temperature min. ~3400 K spicules filaments, prominences plage
Corona	$T \sim 1$–2 MK composition: 95% H, 5% He (by mass) ionization state: ~100% $B \sim 1$–100 μT beta < 1 collisionless (densities $\sim 10^{8}$ cm^{-3})	loops, streamers coronal holes CMEs

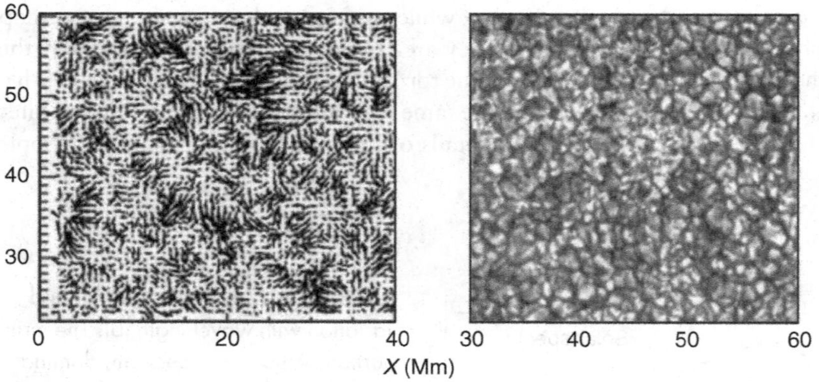

FIGURE 4.5. Images of solar granulation seen on the photosphere, the effective solar surface. Projected unit velocity vectors are superposed on the left image. A typical granule size is about the area of Texas. The convective motion contains most of the non-radiative energy that makes it to the solar surface from the interior. (National Optical Astronomy Observatory (NOAO) Image.)

shifts of strong spectral lines suggest that the bright centers of granules are regions of upflow while the dark areas between the granules are regions of downflow (see Figure 4.5).

Some distributed areas of particularly bright cells are also present; these will be revisited later. The granular and supergranular cells are the defining structures of the quiet Sun magnetic field. We save other features found on the visible surface for later discussions because these are connected with solar magnetism and solar activity.

The sequentially more tenuous layers above the photospheric "surface" of the Sun are regarded as the solar atmosphere, even though a star like the Sun is essentially all atmosphere. These layers consist of the **chromosphere**, the **transition region**, and the **corona**, whose properties are briefly summarized in Table 4.2 and Figure 4.6. The transitions of density and gas temperature, as well as of ionization state and magnetic field control of the gas (beta) through these layers cover a tremendous range. It may seem surprising that at the photospheric level there is a significant component (99.9%) of neutral gas in spite of the ~6000 K temperatures there. For this reason, investigations of the lower solar atmosphere must consider the behavior and processes associated with

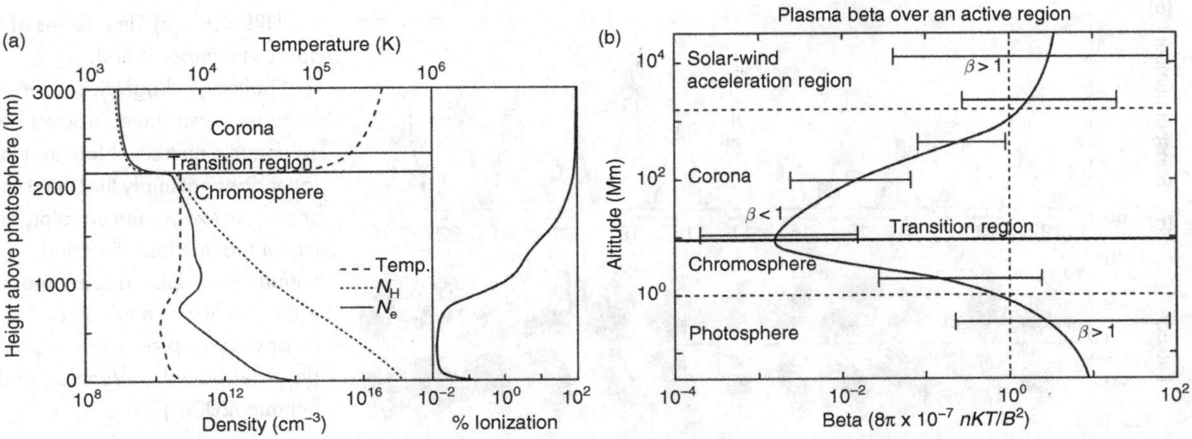

FIGURE 4.6. Radial or altitude profiles of the solar atmosphere basic properties, showing the major divisions into chromosphere, transition region, and corona. (a) Density and percent ionization (after Avrett and Loeser, 2008); (b) ratio of the thermal energy of the plasma to the magnetic energy density or the plasma beta (after Gary, 2001).

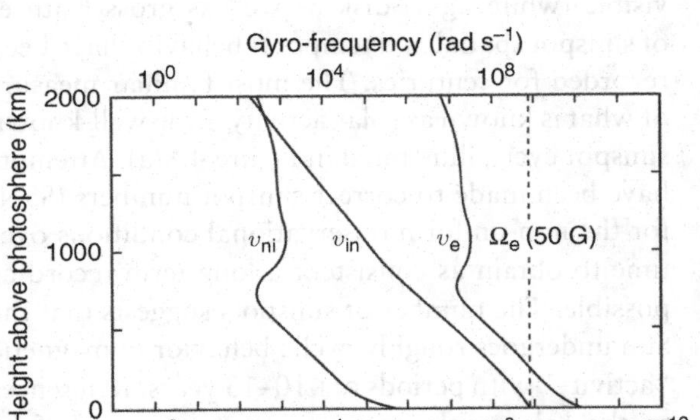

FIGURE 4.7. Altitude profiles of various collision frequencies compared to the proton gyro-frequency in the solar atmosphere. The vertical dashed line is the electron gyro-frequency for a 50 gauss magnetic field. The implication is that collisions dominate particle motions in the photosphere and chromosphere and give way to magnetic field control in the corona. (After Song and Vasyliunas, 2011.)

neutrals as well as with ions. This includes the nonradiative ionization and collision processes that also affect planetary upper atmospheres and ionospheres (see Chapter 2), such as impact ionization and charge exchange. Figure 4.7 shows altitude profiles of some of the different collision frequencies of interest in the solar atmosphere.

The local minimum in the gas temperature in the chromosphere (see Figure 4.6) reflects the loss of energy by radiation into space. But somewhere between the upper chromosphere and the transition region a small (~10^{-4}) fraction of the mechanical energy seen at the surface goes into local heating, turning the temperature gradient into a contrasting rapid rise that terminates in a ~1 MK corona. At the same time, the local gases become highly ionized due to increased collisional ionization by the heated electrons. The higher temperatures increase the local scale heights and reduce the densities, until there is a great reduction in the importance of collisions of all types (see Figure 4.7). The ambient magnetic field then increases in importance, transitioning from its relatively passive convected behavior in the photospheric granulation layer to dominating the now largely ionized gases in the corona. In the following it will become apparent why the question of how this transition happens continues to occupy researchers after decades of investigation. It is important to remember that none of these layers of the solar atmosphere is either well defined or in a steady state. Like Earth's upper atmosphere (see Chapter 2) they are also greatly influenced by changes in the energy or momentum inputs at the complex dynamical boundary below.

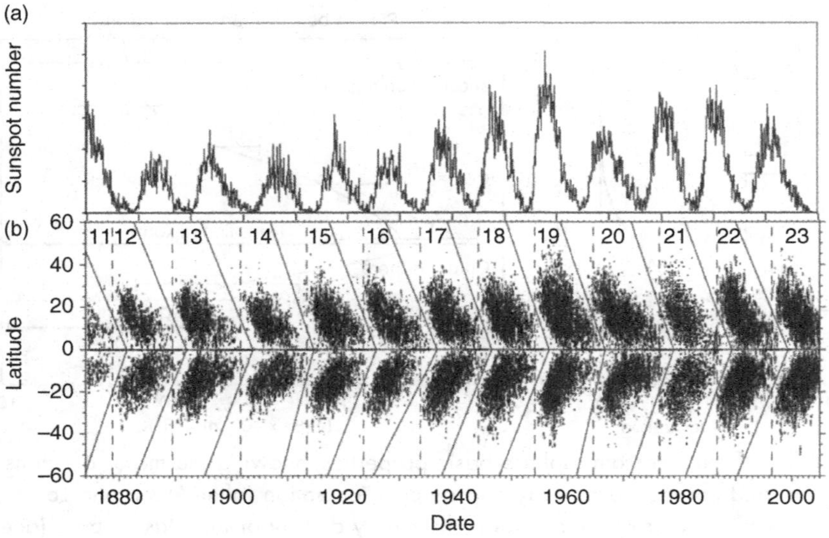

FIGURE 4.8. (a) Time series of sunspot number (SSN); (b) "butterfly" diagram of sunspot latitudes versus time, labeled by solar cycle number. Note that the latter does not imply that the same spots drift toward the equator, but rather that the latitude band within which spots appear moves equatorward as a whole as the sunspot cycle progresses. ((b) After Solanki, Wenzler, and Schmitt, 2008.)

The main solar outputs that are of interest to space physics are the Sun's photon emissions in the EUV (extreme-ultraviolet) to x-ray wavelengths (combined, referred to as XUV), and the small fraction of the solar atmosphere that escapes from the Sun as the solar wind. The solar wind is so important in space physics that it has its own chapter (coming up next) although its coronal roots are described here. The EUV or extreme UV emissions adjacent to x-ray wavelengths (wavelengths shortward of 100 nm) are most responsible for ionizing both local interstellar neutrals and planetary atmospheres (as discussed in Chapter 2). As indicated in Figure 4.4, this part of the solar spectrum, together with radio wavelengths, is much more variable than the rest. This variability is primarily due to the effects of the solar magnetic field, the presence of which controls much of the phenomena described in this chapter and the next, making it a good place to start a more in-depth discussion.

4.3 THE SOLAR MAGNETIC FIELD

4.3.1 Basic observational properties

The most obvious and earliest evidence of a magnetic field of the Sun was the presence of dark spots on the visible solar disk, the **sunspots**. As mentioned in the historical perspective of Chapter 1, the simple parameter of sunspot numbers counted on the visible (white light) disk as well as gross features of sunspot spatial and temporal behavior have been recorded for centuries. The most familiar measure of what is known as solar activity is the well-known **sunspot cycle**, illustrated in Figure 4.8(a). Attempts have been made to correct sunspot numbers (SSN) for the non-uniform observational conditions over time to obtain as consistent a long-term record as possible. The number of sunspots suggests that the Sun undergoes roughly cyclic behavior in magnetic "activity" with periods of ~10–13 years. References to the solar cycle concept often distinguish four phases: the solar minimum when sunspot number is at its lowest value of the cycle; the rising phase, during which it is increasing; the maximum, defined by sunspot numbers reaching their highest values; and the declining phase. Closer examination of Figure 4.8(a) reveals that, in addition to different magnitudes (maximum SSN) and slightly different durations, each cycle has a different shape, although rising phases tend to be shorter than declining phases. The reasons for the variations in the maximum SSN are the subject of ongoing studies. Although sunspots are only one piece of evidence of a magnetic field of the Sun, they are so deeply ingrained in our definitions and perceptions of the solar cycle, and so often used as a proxy, that knowing their basic attributes and behavior is an advantage.

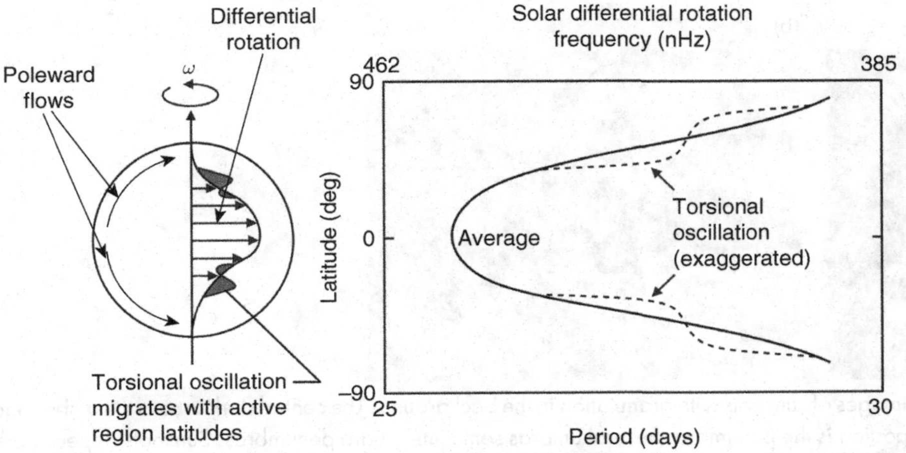

FIGURE 4.9. Illustration of solar differential rotation determined from tracking features like sunspots. The implication is that the polar regions take longer (~30 days) to rotate than the equatorial regions (~25 days). Much slower poleward (meridional) flows are also seen in each hemisphere.

In addition to counting their numbers, observers have generally recorded information about the spatial locations of the sunspots on the visible disk, monitoring their trends over time. Sunspots occur in restricted latitude bands that migrate equatorward from the start to the end of their cycle. When the latitudes at which they are observed are plotted as a function of time, one obtains the aptly named **butterfly diagram** reproduced in Figure 4.8(b). In the rising phase of the sunspot cycle, the spots first appear around 35° from the solar equator, usually in both northern and southern hemispheres within a year or two of each other. The comparison of Figures 4.8(a) and (b) shows that the centers of the butterfly "wings" occur roughly at the times when the maximum number of sunspots is present for each cycle, while the ends of the butterfly wings occur at the low points of sunspot number between cycles. The butterfly wings of adjacent cycles often overlap by a year or two. Remember that the butterfly diagram does not represent a snapshot of a spatial distribution, but shows that the latitude belts in which sunspots occur move equatorward as the sunspot number cycle progresses.

Individual sunspots may appear and disappear in only a day or two, but others can be recognized and tracked for weeks to months. Apparent sunspot movement across the visible solar disk provided early solar observers with the first indications that the solar surface is rotating in a right-handed sense around an axis tilted ~7° with respect to the ecliptic plane, with a ~27.3 day period as seen from Earth. This led to the counting of **Carrington rotations**, mentioned in Chapter 1, that began in the mid 1800s. (As of 2016 we have reached Carrington rotation numbers in the 2170s.) Moreover, it was found that the speed with which sunspots transit the disk depends on their solar latitude. This inferred **differential rotation** is illustrated in Figure 4.9. While 27.3 days is an often cited value for the solar rotation period, the solar equatorial region more closely follows a ~25 day cycle, while the polar regions follow a roughly 30 day cycle. Thus latitudinal or zonal bands of the Sun's visible surface appear to slip with respect to one another, suggesting the existence of mid-latitude surface velocity gradients or shears. A relatively recent finding is that a perturbation exists in the average differential rotation trend around the latitude of the sunspot band, where the rotation is slightly faster than the general trend toward the poles, and slightly slower toward the equator. This **torsional oscillation** (also illustrated in Figure 4.9) migrates equatorward with the sunspot belts as the cycle progresses. Its cause and importance, as well as the origin of the solar differential rotation itself, are still topics of active research.

Some details of sunspot appearance can be seen in the high-resolution white-light image in Figure 4.10. The presence of the dark center, or umbra, enveloped

(a) (b)

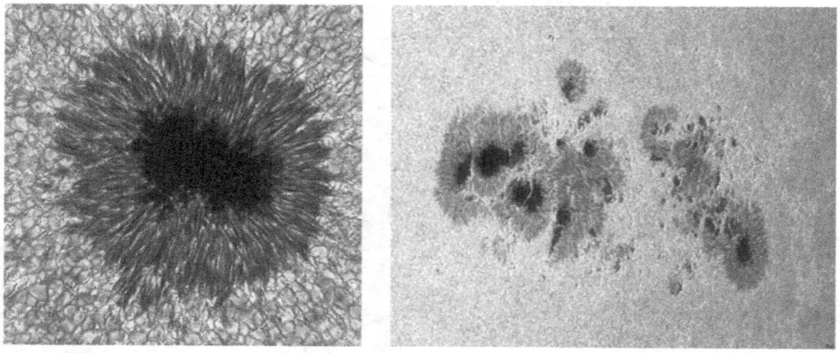

FIGURE 4.10. Visible-light images of sunspots with granulation in the background. The central dark area is the umbra and the lighter-appearing outer portion is the penumbra. Several umbras sometimes share penumbras. Sunspots appear dark on the visible disk because at ~4000 K they are cooler than the surrounding ~6000 K background photosphere. ((a) F. Woeger (Kiepenheuer Inst. Sonnenphysik), NSO archive; (b) E. Roel, spaceweather.com (11/5/2011 archive).)

by the less dark area (called the penumbra) indicates that their temperatures are lower than their surroundings. (Thus their dark appearance is only relative.) But their first identification as magnetic features required an innovation in observational techniques. Sunspots were first established as the sites of the strongest magnetic fields on the solar surface in the early 1900s, from spatially targeted spectral measurements by George Ellery Hale and then Horace Babcock (see Chapter 1), who recognized that **Zeeman splitting** of spectral lines could be used to infer magnetic field strength from solar spectra and filtered images. Their invention, the magnetograph, revealed a variety of physical features and phenomena. An example of a full-disk magnetogram, obtained from images taken in a Zeeman-split spectral line, is shown in Figure 4.11 together with its white-light counterpart. Here the line-of-sight field direction or polarity and magnitude are indicated on a scale from white (strongest field toward the observer) to black (strongest field away), with gray indicating areas of weak or mixed, spatially unresolved polarities. These observations established that the sunspots are sites of roughly radial or vertical magnetic fields with strengths of at least hundreds of gauss. As magnetographs became more sensitive, it was observed that many of the areas with higher field concentrations seen in magnetograms, referred to as **active regions**, do not have associated visible- or white-light sunspots – supporting the idea that there is a field threshold for spot formation. These observations have important implications for interpreting sunspot cycle anomalies such as the **Maunder minimum**, mentioned in the introductory chapter, when spotless active regions may have been present. A fuller understanding of the solar magnetic cycle can be obtained by observing active regions instead of sunspots, but for long-term historical studies, sunspot numbers are all we have.

Sunspots may occur in pairs, with members having opposite polarities. When they are not paired with an opposite polarity spot there is generally a diffuse region of moderately strong field with opposite polarity nearby, often in the trailing position in the sense of right-handed solar rotation. Roughly east–west axis alignment between bipolar partners is seen, with the leading and trailing members having opposite polarities in the northern and southern hemispheres. Moreover, this polarity ordering reverses at each solar minimum, defining the ~22 year **Hale cycle** that incorporates two adjacent 11 year sunspot number cycles. When a simple bipolar active region is seen it usually has an axial tilt of ~5° from constant latitude (where its axis is the line connecting the centers of the opposite polarity spots or patches) with the leading partner of the pair equatorward of the trailing partner (leading and tailing are defined by the sense of solar rotation as before). This property, illustrated by Figure 4.12

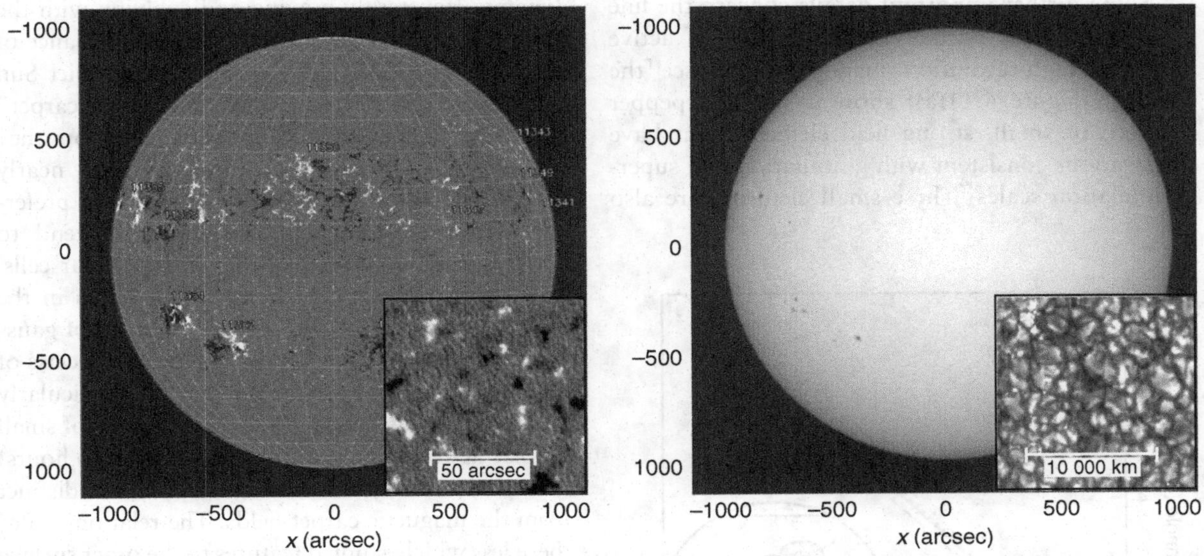

FIGURE 4.11. (a) Example of a full-disk magnetogram; the inset shows a close-up of the weak (quiet Sun) fields. (b) Visible (white) light counterpart; the inset shows the granulation pattern. In the magnetogram, the black areas identify the line-of-sight field pointing toward the Sun and white regions designate the field pointing away, also called negative and positive polarities, respectively. The dividing line between these polarities is referred to as the magnetic neutral line. Gray areas are unresolved "salt and pepper" fields on the granulation scale that give on-average weaker and more sign-neutral field contributions. Note the difference between the areas of the surface covered by sunspots in the visible image and the much larger areas covered by active regions (black and white concentrations in the magnetogram). The number and total area of active regions follows the sunspot cycle trends. (SOHO MDI images (NASA/ESA/Stanford archive).)

and known as **Joy's law**, will turn out to be important later. Some active regions – especially those with large area – may include multiple sunspots in a more complex distribution than a bipolar patch pair, with several spots often sharing the same penumbra. Figure 4.10(b) shows an example. Active regions are designated by various Greek letters depending on their complexity. The main types are alpha (having one spot, with a diffuse opposite magnetic polarity patch), beta (having a pair of spots with opposite polarities), gamma (having a complex, irregular polarity distribution that may include spots or groups of spots), or delta (spots of opposite polarity within a single umbra). Later we shall see that complex active regions are often the sites in which major solar activity originates.

Magnetograms also established the presence of weaker (≲10 G) field (1 G = 10^{-4} T) distributed over the surface of the Sun. The gray areas in Figure 4.11 are "quiet Sun" fields that are present everywhere on the solar disk not occupied by an active region. These weaker fields have a cycle that is not everywhere in phase with the sunspot magnetic field (Hale) cycle. In particular, the solar high-latitude or polar regions, far from the active region latitudes, are where the global solar dipole field is defined by the quiet Sun field polarity biases there. Note that these polar field biases are barely visible in the magnetograms shown. The polar field also reverses with the sunspot cycle period. However, as shown in Figure 4.13, where its magnitude and sign are plotted together with the sunspot number, there is a ~90° phase difference between the two. The polar surface magnetic fields decline in strength and then reverse their sign or polarity around the sunspot cycle maximum when active regions make their strongest contributions. These cycles and their relative phases provide critical observational constraints for theories and models of the solar magnetic field generation discussed in Section 4.3.2.

Some further important details concern the fine structure of the weaker fields. Outside of active regions, high-resolution magnetograms (see the inset in Figure 4.11(a)) show a salt and pepper pattern of small, strong field elements that have separations consistent with granulation and supergranulation scales. These small elements are also dynamic, constantly moving and evolving with the granulation pattern (see Figure 4.5). The balance of positive and negative polarities in these quiet Sun fields, also referred to as the "magnetic carpet" because of the network of low-lying field loops they must create, is variable, ranging from nearly balanced to showing a significant polarity preference. Their ~10 G field concentrations tend to occur at the boundaries of the supergranular cells. This compares to up to hundreds of gauss in the larger active regions, and up to thousands of gauss in some sunspots. In addition, there is a subclass of magnetic carpet field elements that are particularly large and strong, called pores, and a family of small bipolar active regions with short (minutes to hours) lifetimes, called ephemeral regions, that are distinct from the magnetic carpet fields. The relationship of these less widely studied features to the other surface fields, and their overall importance, remains unclear.

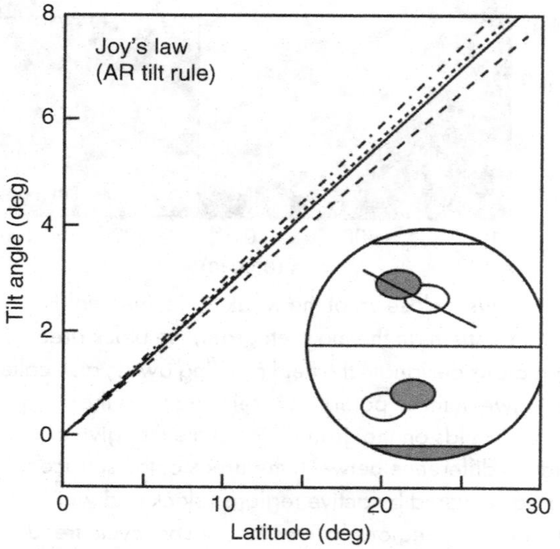

FIGURE 4.12. Illustration of the tilt angle of observed bipolar active regions (ARs) on the disk that defines "Joy's law." The tilts are such that the leading spot is usually equatorward of the trailing spot. This property turns out to be very important for some solar dynamo theories.

Before continuing, it is useful to introduce more of the nomenclature associated with solar images with the help of Figures 4.14 and 4.15. First, although the Sun rotates in the same sense as Earth, the right-hand half of the visible solar disk is referred to as the western side and the left-hand half as the eastern side (see Figure 4.14).

These are separated by the **central meridian**, through which the line connecting the centers of the Sun and the Earth passes. The edge of the disk is referred to as the limb. A much-used solar image product is a ~27.3 day synthesis constructed from full-disk images into a **synoptic map**. Several

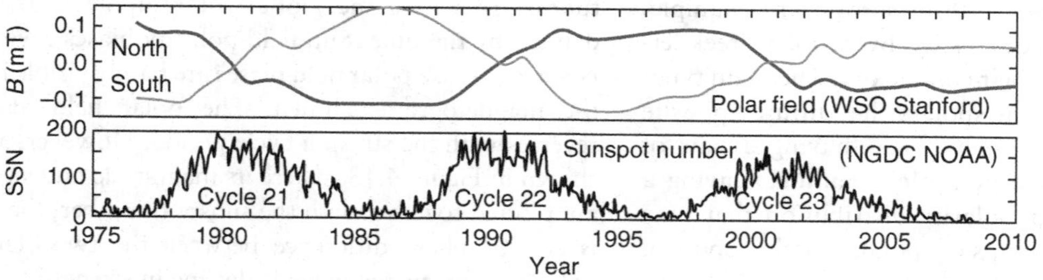

FIGURE 4.13. The polar field cycle and sunspot number cycle compared. Areas of the quiet Sun fields often contain some polarity (radial field sign) bias. In the polar regions the biases are usually opposite in the north and south. These fields, though at ~10 G are much weaker than active region fields, define the large-scale axial dipole field of the Sun. Because it takes two sunspot cycles to return to the same polar field biases, the solar polar field cycle is ~22 years in length. Its relative phase is such that, when sunspot activity is at its minimum, the polar fields are at their maximum strength. (After Li *et al.*, 2011.)

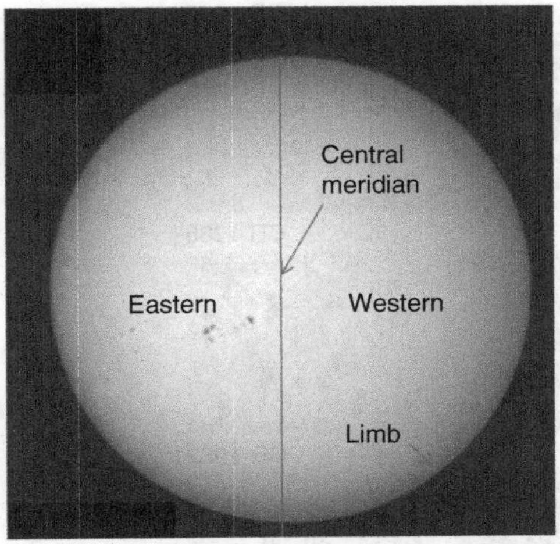

FIGURE 4.14. Definitions of solar disk terminology that will be used in this chapter and in Chapter 5. The central meridian, the line bisecting the Sun from north to south, is usually defined with respect to the Earth's perspective. Because the Sun has no permanent features, it serves as the "prime meridian," or line of zero longitude, at the beginning of Carrington rotations. The western half of the disk is the receding half and the eastern half is approaching. The limb is the visible edge of the disk.

examples of synoptic photospheric magnetic field maps from different times in the solar cycle are shown in Figure 4.15. To make these maps, slices of each full-disk image taken around the central meridian are merged together to represent a snapshot of the global solar field for a nominal solar rotation. While this does not give a strictly correct picture for any single time, it provides a reasonable approximation to the global photospheric field for periods when the fields are changing relatively slowly. Techniques of synoptic map making involve a host of necessary corrections, especially in the highest-latitude portions of the map. The polar regions are always viewed obliquely compared to the central disk and low latitudes, making them particularly subject to measurement errors. In addition, a slightly changing perspective, due to the ~7° tilt of the solar rotation axis relative to the ecliptic pole, makes one solar pole alternately less visible than the other through the year. Nevertheless, the synoptic maps, used as proxies for the whole Sun's surface field, are an important research tool for global representations.

4.3.2 Dynamo origins of the observed magnetic fields

The solar magnetic field, like other stellar magnetic fields, is believed to be caused by an interior dynamo. One of the main problems in applying the dynamo concept (see Chapter 1) to the Sun prior to the 1980s was the lack of knowledge of both rotational and other flows, and diffusion processes, in the solar interior. A critical breakthrough came in the 1970s with the development of **helioseismology.** This technique, which is reminiscent of solid planet seismology, utilizes the Doppler shifts in the solar surface spectrum caused by acoustic waves to probe the Sun's interior structure and motions. High-spatial-resolution, visible-light pictures of the Sun, such as the one in Figure 4.16(a), taken in the wings of a spectral line such as sodium 589.6 nm, reveal the constant motions of the photosphere on the granulation and supergranulation scales. In this **dopplergram** the gray to white spots indicate where the solar surface is moving toward and away from the observer, respectively. These mottled patterns are the surface manifestation of sound waves that permeate the convection zone, similar to what might be imaged if we could take a snapshot of Earth's surface motions following a major earthquake. The photosphere is literally always quaking with typical vertical oscillations of up to hundreds of meters per second and periodicities of ~2–17 minutes (~1–7 mHz). Helioseismology requires observations obtained continuously over long periods to obtain useful results. Thus early experiments were conducted in the summer in Antarctica, and more recent observations are from space and worldwide ground-based observatory networks with 24/7 observations of the Sun.

The interpretation of helioseismology data is a major subfield of solar physics and merits further independent exploration by interested readers. Figure 4.17(a) illustrates a standard data display associated with the technique, an acoustic wave spectrum that can be interpreted to derive the interior structure and dynamics. The observed acoustic waves are mainly "p modes," in which pressure

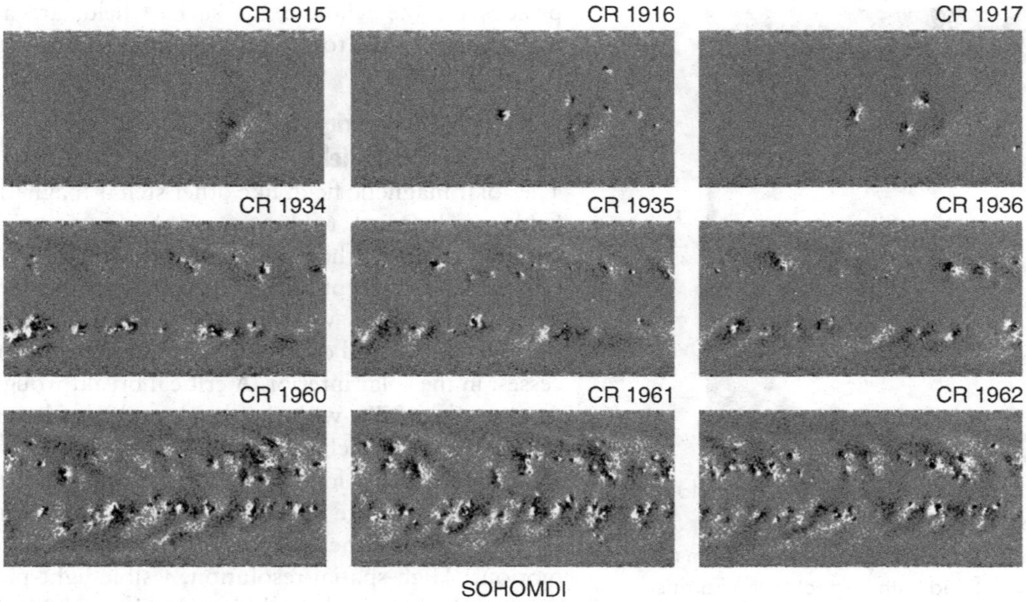

FIGURE 4.15. Examples of synoptic magnetic maps of the radial fields on the solar surface constructed from full-disk magnetograms taken at different times of the solar cycle, as indicated by the Carrington rotation (CR) number. These are often constructed from the central portions of at least 27 daily images of the field, with special corrections in the polar regions where radial fields are poorly observed at large line-of-sight angles. They are an important observational product for those seeking to model the global characteristics of the solar atmosphere. (SOHO MDI images (NASA/ESA/Stanford archive).)

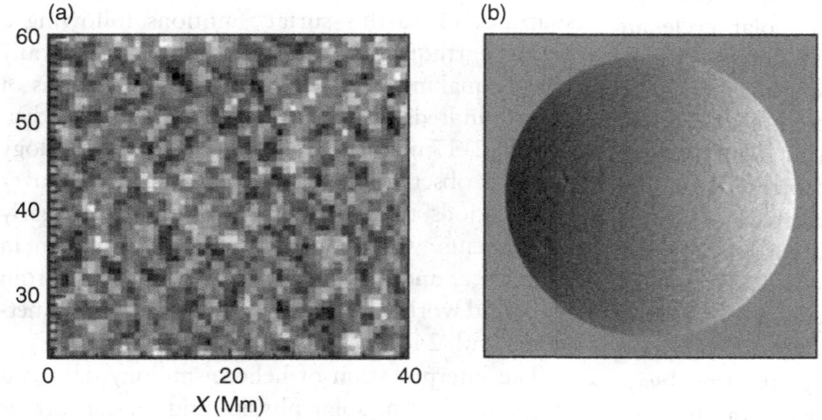

FIGURE 4.16. Examples of dopplergrams, where white and black indicate motions toward and away from the observer, respectively. The full-disk image (b) shows the effects of solar rotation. The close-in image (a) was corrected for the effects of solar rotation, so that only the small-scale motions on supergranular scales are visible. The light and dark spots oscillate between colors with periods of ~5–10 minutes. Images like these are the observational basis for the field of helioseismology. ((a) NASA SDO/HMI/Stanford archive; (b) after Duvall and Birch, 2010.)

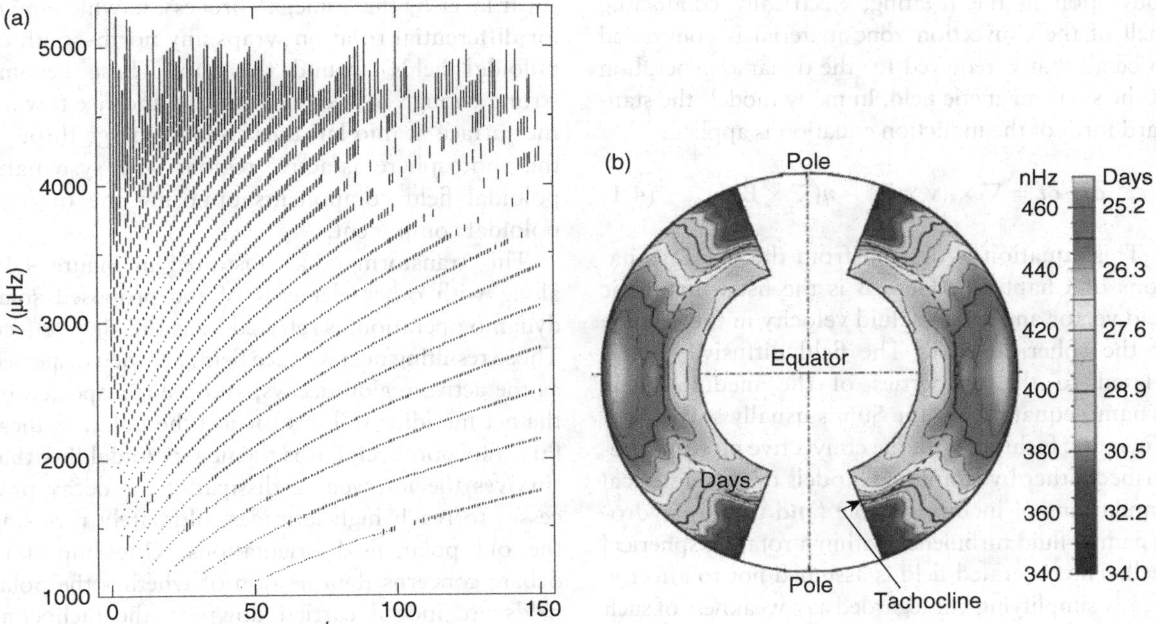

FIGURE 4.17. (a) Example of an acoustic wave spectrum obtained from helioseismological analysis of the surface oscillations of the Sun and (b) the interior rotation of the Sun inferred from such measurements. The research breakthrough represented by the result in (b) revolutionized studies of the solar interior and solar dynamo. They showed that, while the central portions of the Sun rotate rigidly, the differential rotation seen on the surface (see Figure 4.9) occurs throughout the convection zone. The relatively narrow boundary between these two rotational zones is a shear layer called the tachocline. There is also another shear layer near the solar surface, probably associated with the granulation layer. Iota, plotted along the abscissa in (a), is the harmonic order. ((a) After Hill *et al.*, 1996; (b) MSFC Solar Science website.)

gradients provide the restoring force for a perturbed gas element, although there are also "g modes," in which gravity plays that role. The different frequencies probe different depths, reaching down to the solar core. Although the diagnosis of small-scale motions is limited to the shallower depths, helioseismology techniques have been used to infer the larger-scale more orderly motions in the solar convection zone, including differential rotation, as shown in Figure 4.17(b).

At the beginning of this chapter we sketched a picture of the solar interior that was largely unknown before helioseismology. Helioseismology tells us that the solar convection zone occupies roughly the outer third of the solar radius (Figure 4.1), and that the solar differential rotation seen on the surface (Figure 4.9) exists throughout the convection zone while the radiative zone interior to it rigidly rotates. The rotation rate of the radiative zone is roughly the same as that of the convection zone at around 35–40° latitude, near the locations where active regions emerge at the start of a new sunspot cycle. These observations also imply the existence of a relatively narrow layer of high radial and latitudinal shear between the rigidly rotating interior and the convection zone at its base. Both the interior solar rotation and this layer, called the **tachocline**, play key roles in solar dynamo theories and models.

Modern **solar dynamo** theories generally assume that a thick spherical shell of convecting fluid, about one-third of the solar radius thick, surrounds the energy-generating interior. As mentioned earlier, the heat from this interior results in both larger-scale overturning and turbulent motions within this convection zone that redistribute the interior's heat input and reduce the temperature gradient. This turbulent convection in the rotating, electrically conducting shell of the convection zone material is considered to be all that is required for the dynamo generation of the solar magnetic field. In many models the standard form of the induction equation is applied:

$$\partial B/\partial t = \nabla \times (\mathbf{v} \times \mathbf{B}) - \eta(\nabla \times \mathbf{B}). \qquad (4.1)$$

This equation is derived from the MHD equations of Chapter 3. Here **B** is the usual magnetic field vector and **v** is the fluid velocity in the volume of the spherical shell. The field diffusivity, η, is related to the properties of the medium. The dynamo equation for the Sun is usually solved in a kinematic fashion, with the convective motions prescribed either by simplified models or by numerical simulations of incompressible (and usually hydrodynamic) fluid turbulence within a rotating spherical shell. The generated field is assumed not to affect **v**, greatly simplifying but regarded as a weakness of such calculations. The diffusivity is used as a parameter or estimated from turbulence theory or observations. A seed field is then evolved by solution of the equation in two or three dimensions (often using spherical harmonic approaches) to obtain the magnetic field in the interior of the spherical shell. The solutions are sensitive to the competition between the advection (described by **v**) and diffusion (described by η) terms. Some of these models produce northern and southern hemisphere toroidal wreaths of twisted, filamentary magnetic flux that could be considered the subphotospheric source of the active region belts. Some solutions even exhibit cyclic behavior with field polarity reversals. The challenge is to find combinations that produce results consistent with the emerging fields seen on the Sun's surface, as well as the time scales of their polarity cycles.

In addition to the differential rotation, Figure 4.9 also illustrates another type of motion inferred to exist in the convection zone. Surface feature tracking for some time suggested the existence of large-scale, slow (a few meters per second), on average poleward "meridional" flows superposed on the differential rotation. Some dynamo theories assume this flow is connected with a deep return path near the tachocline that makes a flux "conveyor belt." In this picture, strong toroidal belts of twisted fields encircling the Sun are produced in the tachocline shear layer by the "omega" process, in which interior differential rotation wraps any north–south or poloidal fields around the Sun. These become strong enough to become buoyant and rise toward the surface at mid latitudes. Some emerge through the photosphere as active regions with systematic poloidal field components opposing the original poloidal component.

This transformation, illustrated in Figure 4.18 along with other elements of the proposed solar dynamo operation, is referred to as the alpha effect. Those resulting near-surface fields are then dispersed as the active region decays, and are transported by the net meridional flows toward the poles. In these flux transport cycles, it is the new poloidal flux that survives the intervening dissipation or decay processes to reach high latitudes, ultimately reversing the old polar field orientations. One important debate concerns the question of whether the polar fields are indeed carried down to the tachocline where they provide seed fields for the active regions of a future cycle. Such a deep convection zone circulation process would be expected to take around two to three solar cycles. But helioseismic research suggests that the single hemispheric cell circulation picture may need revision. Multiple, vertically stacked and latitudinally adjacent cells with shorter circulation paths (Figure 4.18) may be common and vary over each solar cycle. The generated field may also have important effects on the presumed flows, causing feedback. Nevertheless, poleward near-surface flows, combined with pervasive differential rotation, are the two robustly established, observed velocity components of the convection zone dynamics that we now know must be included in any viable dynamo model.

The small-scale magnetic fields on the surface are often neglected in global dynamo calculations for reasons of both spatial resolution and uncertainty about their influences. Yet observations suggest that an additional shallow ($<0.01R_S$) shear layer near the surface, associated with the observed granulation and supergranulation cells (Figure 4.5), likely contributes a separate "surface convective dynamo" – locally making the small-scale fields of the **magnetic carpet**. A sketch of this complicated system of fields is attempted in Figure 4.19. The magnetic carpet field may in fact influence the

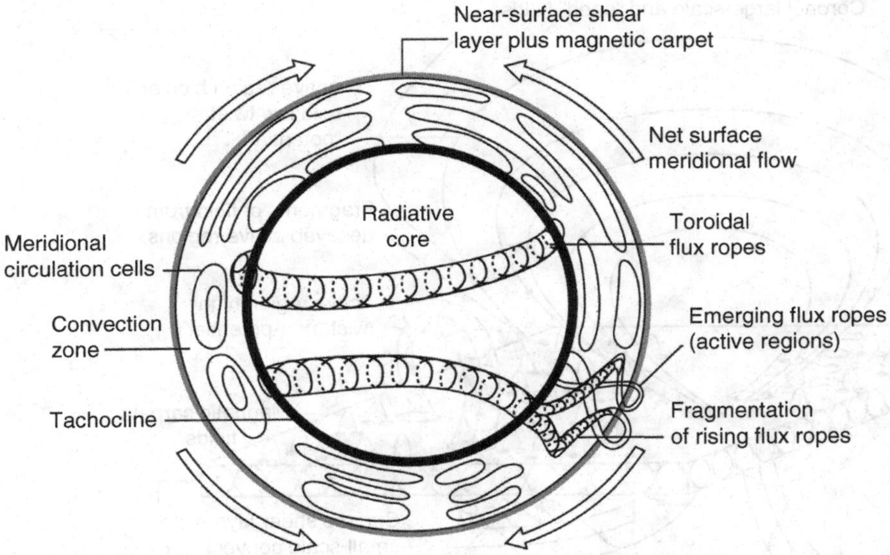

FIGURE 4.18. Illustration of the magnetic fields emerging from the solar interior, producing poloidal or north–south field components at the surface from strong toroidal fields generated at the base of the convection zone around the tachocline (Figure 4.17). These fields are generated by the solar dynamo. Ideas on how they evolve and impact the subsequent dynamo cycles are discussed in the text.

near-surface evolution of the larger-scale active region fields generated in the tachocline by enabling (or discouraging) their emergence and/or dispersal. This raises the question of the extent to which activity cycle differences are a matter of differences in flux delivery to the surface rather than deep dynamo changes in flux generation. Numerical modeling suggests the original flux ropes generated in the tachocline should shred on their way through the convection zone, and that sunspots are formed from reconstituted concentrations of these fields by thermally enabled convective focusing near the surface. But if this refocusing process becomes ineffective for some reason, significant surface field concentrations and all their associated effects may not appear. The Sun's near-surface convection zone "weather" may thus play an important role in the solar dynamo, and our sunspot proxy for its cycles, separate from what goes on in the deeper interior.

4.3.3 The solar (sunspot) cycle

Almost universally the solar cycle is simply defined by the time series of sunspot numbers (Figure 4.8(a)). As mentioned in Chapter 1, over the long term, cycles have had significantly different sunspot number maxima, rise times and decay times, and lengths. Usually the rise time is faster than the declining phase even though the cycle shape may vary. The duration of the sunspot cycle can be a few years more or less than the nominal 11 years, and, as mentioned earlier, it takes two adjacent sunspot cycles to make one magnetic cycle. Decades of scientific investigations have been invested in analyzing the sunspot number record for patterns and hints of predictive behavior. This activity is spawned in part by the practical impacts of very active cycles on human technologies (as noted in Section 1.11 and discussed in detail in Section 9.8.4 on "space weather") and in part by the interesting coincidence of the very small Maunder minimum sunspot numbers (see Chapter 1) with a period of apparent climate change. However, while sunspot number continues to be the canonical measure of activity, it only scratches the surface of the changes occurring on the Sun and in its atmosphere through the different solar cycle phases.

The fact that not all emerged magnetic flux shows up as sunspots makes a butterfly diagram derived from magnetograms, such as that in Figure 4.20, a

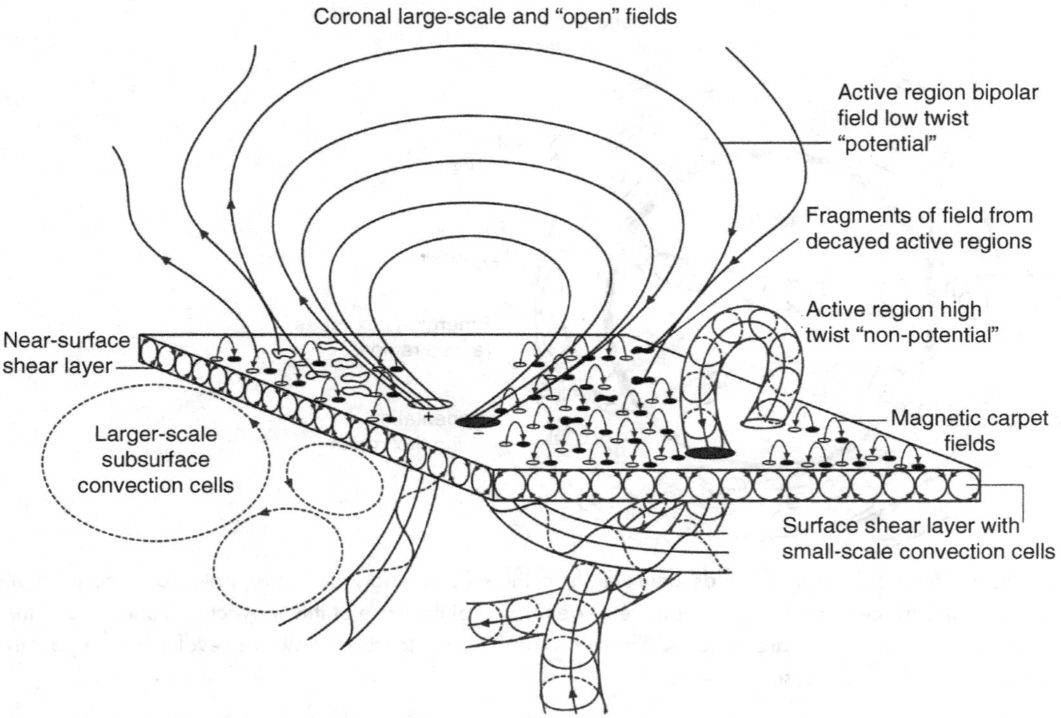

FIGURE 4.19. Illustration of the solar fields near the surface showing the distinction between the small-scale near-surface granular convection scale fields and those emerging from the solar interior, which make the active region fields (see also Figure 4.17). The existence of these two sources of magnetic field may be key to understanding how the solar cycle works.

better representation of the solar activity cycle. This display was constructed from longitude-averaged synoptic maps (see Figure 4.15) and illustrates how dominant the active region fields are on the solar surface. As the cycle progresses new regions emerge at ever lower latitudes while the previously emerged regions smear out and eventually disappear or merge into the magnetic carpet, leaving some net polarity in the quiet Sun fields. The tilt of the active regions causes the systematic displacement of their positive and negative (outward and inward) contributions. The differently phased polar field cycle is also clearly visible at the top and the bottom of Figure 4.20. How does this surface magnetic field evolution process work, and how does it fit into the overall solar dynamo picture described in this chapter?

Interestingly, as observed active regions spread out and lose their initial field strength and coherence, their fields appear to be dispersed as if they are not anchored by deep roots. One can effectively model the evolution of the photospheric field observed in synoptic maps by numerically solving a two-dimensional version of the dynamo equation for the surface radial field, assuming it is governed purely by surface motions and dissipation:

$$\frac{\partial B_r}{\partial t} + \frac{1}{R_S}\frac{1}{\sin\theta}\frac{\partial}{\partial\theta}(v_\varphi(\theta)B_r \sin\theta) + v_\theta(\theta)\frac{\partial B_r}{\partial\varphi}$$
$$+\frac{\eta}{R_S^2}\frac{1}{\sin\theta}\frac{\partial}{\partial\theta}\left(\sin\theta\frac{\partial B_r}{\partial\theta}\right)+\frac{\eta}{R_S^2}\frac{1}{\sin^2\theta}\frac{\partial^2 B_r}{\partial\varphi^2}$$
$$= S(\theta,\varphi,t). \tag{4.2}$$

Here, B_r is the radial field component, v_φ and v_θ are the horizontal surface velocity components in longitude and colatitude, respectively, η is the diffusivity, and S, absent in the previous version of the dynamo equation, is a source term representing the emergence of new bipolar active regions. The description of S is typically constrained by sunspot

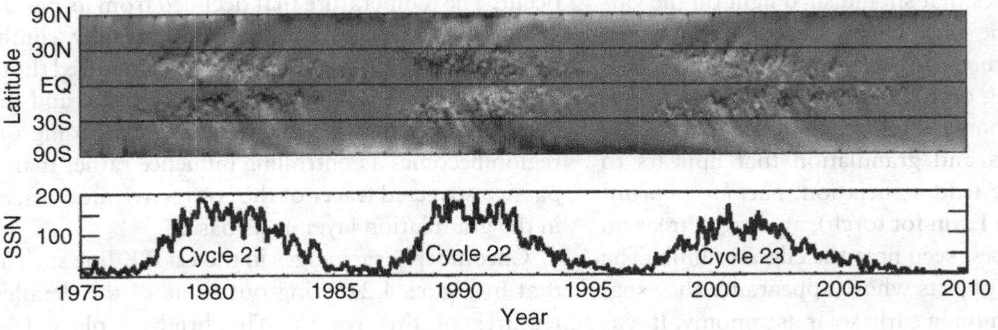

FIGURE 4.20. Butterfly diagram made from the magnetic synoptic maps constructed for many consecutive Carrington rotations by longitudinally averaging their radial fields. This type of display captures the evolution of the magnetic field polarities in both the active region belts (where sunspots are located) and the polar regions, and shows their phase differences. Note the active region belt polarities show opposite north–south biases in each hemisphere that are due to the Joy's law tilts of bipolar active region axes (see Figure 4.12). (The magnetic butterfly diagram (upper panel) is after Hathaway/MSFC/NASA; sunspot number (lower panel) is provided by NOAA.)

number, expected active region polarity, and the assumed latitude and tilt angle of each active region emergence. The large-scale surface convection velocity (v_φ and v_θ), including differential rotation and poleward meridional flows, shares control of the field evolution on the plane of the map, with the diffusion (η terms) representing the effects of the small-scale (e.g. granules) random motions. Magnetic flux is assumed lost when opposite polarities of equal strength meet within a computational grid cell, as if a process such as flux submergence beneath the solar surface or cancelation by merging/reconnection of the opposing fields occurs there. It is noteworthy that the granular motions that produce the magnetic carpet, themselves produced by a secondary near-surface dynamo, influence the overall surface field evolution.

This superficial treatment of synoptic map evolution, called the surface flux transport approach, does a remarkably good job of replicating the details of the observed solar surface field cycle. It can mimic the cyclical formation of the polar fields, and hence the large-scale solar dipole and its reversals, by choosing reasonable v_θ, η, and S. Various manipulations of these values can replicate the surface fields for long periods over many sunspot cycles, although regular magnetograms for validation have only been available since the mid 1970s. It is interesting to note that the polar magnetic fluxes that result from the diffusive–convective transport described by the above dynamo equation are only about 0.1% of the flux that emerges in the active regions. Yet these relatively weak polar fields both define the solar cycle polarity in many researchers' eyes and exert major influences on the large-scale coronal structure.

The reason why a treatment of photospheric magnetic field evolution that neglects the roots of the field below the surface works so well, treating the interior as a mere source and sink of flux, is a matter of research and debate central to solving the dynamo problem. Whether the generation of the active region magnetic fluxes of the next cycle in the interior is affected by what is left over in the poles from the previous cycle remains an open question. However, an empirical correlation is observed between the solar minimum polar fields of the three previous cycles and the following cycle maximum sunspot numbers that is so accurate it is used in sunspot maximum forecasting. The newest helioseismology observations are more sensitive to smaller scales and deeper depths than previous measurements, allowing more exploration of these issues. In particular, the question of whether or not a return (equatorward) flow at depth exists would distinguish between a dynamo that produces solar cycles via a magnetic flux conveyor belt versus one that is controlled by other processes. Numerical dynamo models are gradually achieving more realistic spatial resolutions and parameter regimes that should shed light on the various roles of the tachocline, the convection zone properties, and the near-surface processes.

Before moving on to higher atmospheric levels, it is worth mentioning another photospheric feature besides sunspots and granulation that appears to have a magnetic field association. *Faculae* (a word derived from the Latin for torch), are bright areas on the visible disk best seen near the edges or limb. The use of names of objects whose appearance they suggested was common in early solar astronomy. It was realized decades ago that the faculae are associated with active region magnetic fields outside of sunspots. Recent observations show that these stronger magnetic fields cause density reductions that produce greater transparency into hotter, deeper layers. Faculae are of particular interest to those studying the solar constant because they can more than make up for the loss of disk emission from the dark areas of sunspots. Now that we have a good picture of the magnetic jungle that is present on the solar surface, we can consider the gases that mingle with it.

4.4 CHROMOSPHERE AND TRANSITION REGION

The next two relatively thin layers (around hundreds to thousands of kilometers thick) of the solar atmosphere above the photosphere, mentioned in the introduction to this chapter and characterized by their main physical properties in Figure 4.1, are the **chromosphere** and the **transition region**. Historically, the chromosphere was identified by features observed on the limb and through filters on the disk in spectral lines such as Ca II K and H-alpha. It was particularly associated with the minimum temperature in the solar atmosphere. The transition region was a relatively recent addition invoked to distinguish the particularly sharp changes in physical parameters between the top of the chromosphere and the base of the corona. Figure 4.6 summarized the overall parameter ranges involved. The solar atmosphere at photospheric levels (0 km in Figure 4.6(a)) has a significant neutral gas component and is still controlled in part by the effects of the radiation passing through it from the interior. As one moves to higher altitudes, fundamental physical transitions occur. The temperature that declined from its photospheric value in the chromosphere rapidly climbs from its ~3800 K minimum to several hundred thousand kelvin, the gas is increasingly ionized, and the beta indicates that the magnetic field threading the region becomes a controlling influence rather than a passive advected tracer of the convective motions, as in the granulation layer at its base.

Chromospheric images in the Ca II K line such as that in Figure 4.21 bring out some of the detailed features of this region. The brighter **plage** (the French word for beach) areas, like the photospheric faculae, roughly coincide with the spotless regions of stronger fields in the solar magnetograms. There is a bright network-like pattern that emulates the supergranular cell boundaries on the photosphere where quiet Sun/magnetic carpet fields are concentrated. Thus emissions in this line are indicators of the presence of stronger magnetic fields. For this reason, they are also useful for remotely inferring magnetic fields and their activity cycles on other stars.

Filaments are dark, often long and narrow, meandering features seen in H-alpha (121.6 nm) disk images (see Figure 4.21). They are dark because they are relatively cool material suspended like clouds above the surface that absorb the photospheric emissions in the spectral line. **Prominences** are these same features seen on the limb as arches or curtains of bright material, sometimes extending out to tenths of a solar radius, and sometimes exhibiting time-dependent behavior that suggests either collapse or eruption of the structure. More will be said about the latter behaviors later. The suspension of the material in filaments and prominences has been suggested by both observations and models to be magnetic in nature, although the field geometry that enables this suspension is still a matter of debate. Figure 4.22 illustrates two often-shown concepts. The suspended material lies roughly along the polarity reversal or neutral line of an active region or decayed active region or polarity-biased quiet sun magnetic fields (see the earlier discussion of the photospheric fields). Whether the filament material is in a hammock-like field structure or inside a helical field conduit as suggested in Figure 4.22, it must involve magnetic forces such as the magnetic-pressure gradient and/or curvature

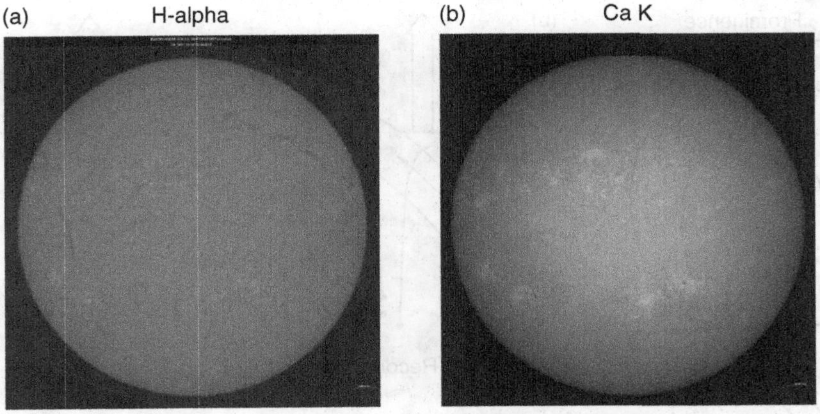

FIGURE 4.21. Examples of chromospheric images for the same time as the magnetogram and visible images in Figure 4.11. (a) Obtained in the H-alpha line of neutral hydrogen and shows the filaments and prominences of colder material suspended over some magnetic neutral lines on the surface. (b) Obtained in the ionized calcium (Ca II) emission line called the K line and shows the bright regions of plages in the strong-field region and the more widespread chromospheric network on a smaller scale. (Images from Big Bear Solar Observatory.)

force(s) to resist gravity. These features demonstrate how important magnetic field control of the solar atmosphere becomes within a few thousand kilometers above the photosphere.

H-alpha images taken at and near the limb (Figure 4.23(a)) also reveal the existence of **spicules**, jets of hot gas a few hundred kilometers in cross section and several thousand kilometers in vertical extent. The shapes and locations of spicules within the chromospheric network suggest they are at least partly channeled by the local magnetic fields in the supergranular boundaries, as sketched in Figure 4.23(b). These highly dynamic features appear and disappear within minutes and show spectral evidence (i.e., Doppler shifts) indicating rapid motion of their material at tens of kilometers per second. It is expected that much of this material is gravitationally bound and falls back toward the Sun, but some of it may escape into the atmosphere above as hot, accelerated plasma jets. Numerical simulations suggest that spicules have underlying causes involving a complicated combination of Lorentz forces, currents and related resistive heating, and deflection by magnetic fields that are mainly vertical and connected to higher levels of the atmosphere. Spicules may thus provide an important energy and material link to the corona.

The transition region is even less clearly associated with an altitude range than the chromosphere. It nominally coincides with the location and width of the sharpest gradients in temperature, density, ionization, and beta (the ratio of thermal plasma pressure to magnetic pressure) at the top of the chromosphere (see Figure 4.6). Across it, the gas temperature rises from $\sim 10^4$ K, which existed in the chromosphere, toward 10^6 K in the corona at its upper boundary; the density decreases precipitously to maintain pressure balance; the percent ionization changes from <1% to ~50% (see Table 4.2); and the magnetic field influence indicated by plasma beta goes from passive to dominant. The details of these transitions, moreover, seem to vary from place to place, depending on the setting. Important spectral lines for transition region observations include emission lines of partially ionized heavy ions that exist under its temperature and density conditions but are faint in the photospheric emission spectrum: 17.1 nm (Fe IX-X), 19.5 nm (Fe XII), and 28.4 nm (Fe XV) (see Figure 4.24(a)). In addition, a strong collisionally excited emission line at 30.4 nm provides a particularly interesting set of transition region images that suggest its magnetic connections to both the photospheric field below and coronal features above. Both active region connected bright loops and dark coronal holes,

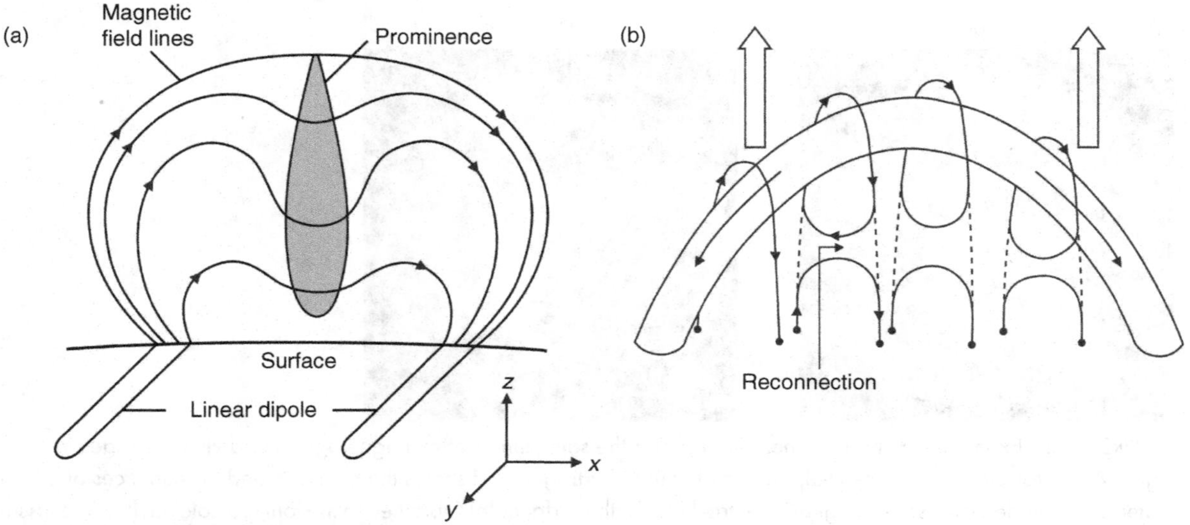

FIGURE 4.22. Models of filaments (disk) and prominences (limb filaments) showing various concepts of magnetic field enabled suspension above the photosphere. (After Kivelson and Russell, 1995.)

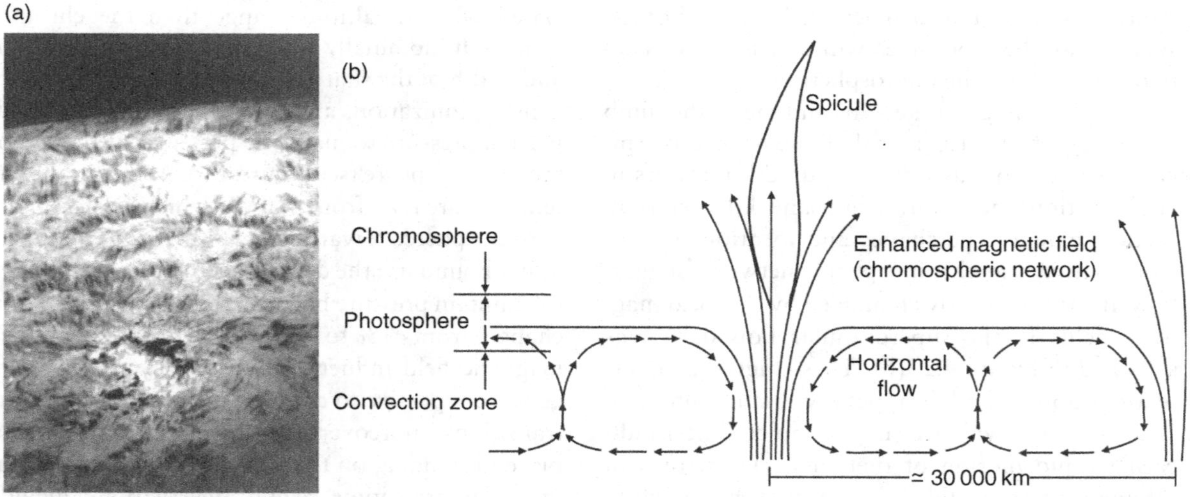

FIGURE 4.23. Image of spicules (a) with a sketch (b) of the likely related convection patterns and magnetic fields. These chromospheric features are thought to be controlled by the complicated interplay of dynamics and energetics that occurs in this boundary region between the photosphere and the corona. They may be an important source of accelerated coronal material. (Solar image from NSO/NOAO/NSF.)

where the field opens to the corona above, are seen in many of the images (Figure 4.24(b)). The conditions in the transition region also lead to an optically thin state of the solar atmosphere for most wavelengths in the solar spectrum. This means that the role of radiation as a means of transporting energy is greatly reduced from what it is in the photosphere and chromosphere. As we shall see,

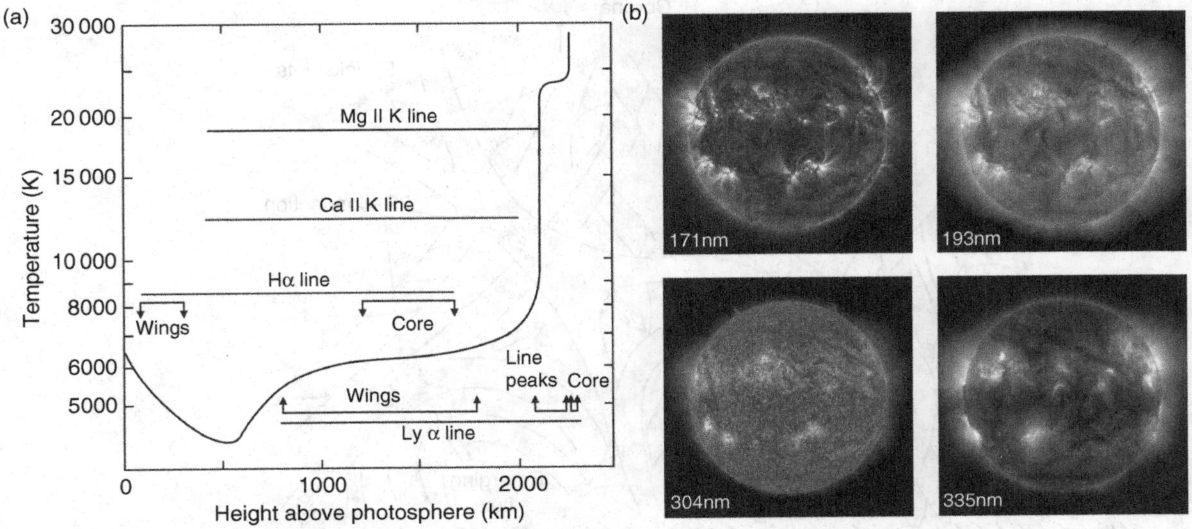

FIGURE 4.24. (a) Altitudes where different ionization states occur in the solar atmosphere due to the local temperatures increasing with altitude. These determine what features are observed in the images obtained at various wavelengths. (b) Examples of solar images obtained in the EUV (e.g., Fe IX at 171 nm, Fe XII at 195 nm, Fe XV at 284 nm) corresponding to the times of the other images in Figures 4.11 and 4.21.) ((a) After Vernazza, Avrett, and Loeser, 1973; (b) SOHO EIT (ESA/NASA).)

magnetic field controlled plasma pressures, flows, and processes take over.

The level of uncertainty regarding details of the physics of the chromosphere and transition region limits our discussion here. When encountering the topic or related research, it is worth keeping in mind the artist's conception in Figure 4.25 that communicates some essentials about this complicated zone. This representation reinforces the idea that no single altitude range defines these regions of the atmosphere. The evolution from what is considered photosphere to what is considered corona depends on local conditions and is different in a place where magnetic fields are concentrated in the boundaries of supergranule cells, in an active region, and in between such features. Overall, the combination of chromosphere and transition region constitutes a complex, structured, and dynamic boundary layer between the photosphere and the corona, and this is daunting from a physics and research perspective. Yet what goes on in the transition region is largely responsible for most non-radiative effects of the Sun that extend throughout the corona and solar wind and affect the planets. Frontier areas for making progress include numerical simulations that follow the fate of the dynamo-generated magnetic flux that, together with mechanical and radiative energy fluxes, emerges from beneath the photosphere with the corona at the upper boundary. These simulations generally solve resistive MHD equations including radiative transport, solar gravity, ionization and recombination processes, and heating and cooling terms, together with an equation of state that allows for all the physics involved. Although many simplifying assumptions are made, these capture some of the features, including granulation-scale convection, magnetic carpet-like fields, and even spicule-like structures, at least on local scales. Together with the continually more detailed observations, they are shedding light on the physical phenomena and interconnections that exist in this important transitional part of the solar atmosphere.

4.5 THE CORONA

The **corona** is the outermost solar atmosphere that most concerns space physicists because of its direct

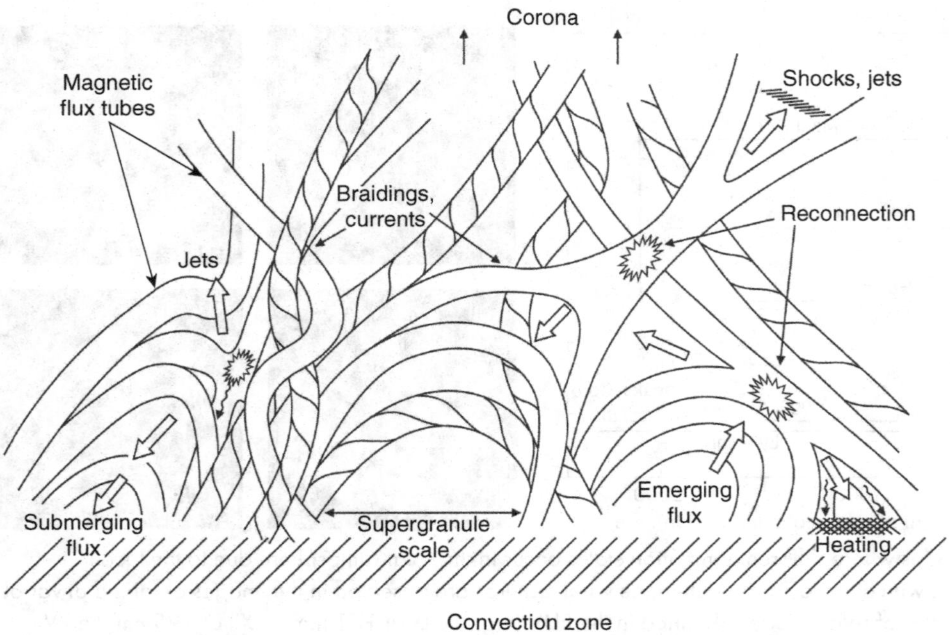

FIGURE 4.25. Sketch showing a number of the chromosphere–transition region phenomena that have been suggested to occur in this complicated boundary layer between the photosphere below and the corona above.

linkages to the solar wind and interplanetary field. Lyot's invention of the coronagraph, and its eventual space deployment, made it possible to view regularly its off-limb features with a perspective resembling total eclipse observations. Coronagraph images show two distinctive coronal emissions, one at distances $>2R_S$ produced by sunlight scattered from circumsolar dust particles associated with zodiacal light, called the **F corona**, and the other, closer-in component from "Thompson scattering" of the sunlight by the coronal electrons. It is this latter component, the **K corona**, that is of primary interest in space physics research. The K corona can be separated from the F corona by observing in polarized light because the dust-scattered component is unpolarized. In the literature such images are sometimes referred to as pB (polarized brightness) images. The examples of eclipse images in Figure 4.26 give an idea of the variety of coronal appearances, the reasons for which are the subject of this section.

In the corona, the solar atmosphere becomes so rarefied due to the combination of gravitational stratification and high temperature that it is practically collisionless. Moreover, because its main constituent, hydrogen, is fully ionized, the magnetic field largely dictates its behavior – especially of the electrons whose gyro-radii in the coronal fields are small. This is what makes the electrons so useful for Thompson-scattered light observations. However, the ion motions are also highly constrained by the local fields, as indicated by the beta values and collision frequencies in Figures 4.6 and 4.7. The coronal images in Figure 4.26 can thus be thought of as pictures of the three-dimensional coronal magnetic field structure as well as its density, projected on the sky plane outside the bright solar disk as the solar field goes from a high to low sunspot number state. In contrast to the white-light loops seen by the coronagraph, the EUV emissions from the bright loops in the transition region and low corona shown earlier (Figure 4.24) exceed the photospheric emission at the same wavelengths, allowing on-disk features to be imaged. Also as mentioned earlier, the on-disk images show the dark regions called coronal holes in between the large-scale bright structures that show up in profile in the off-limb white-light

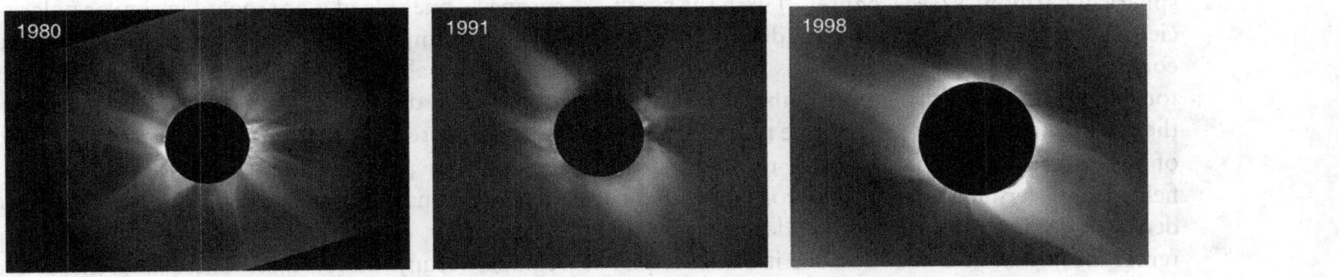

FIGURE 4.26. Examples of coronal white-light images from solar eclipses, which can show more detail than coronagraph images due in part to the occulting geometry of the Sun–Moon–Earth system. These examples illustrate the solar cycle variation seen in coronal structure, which is dominated by its magnetic field. Because these images are projections of a semi-transparent three-dimensional structure, they are interpreted using three-dimensional models. (From the HAO Online Eclipse Image Set, with permission.)

projections between the coronal streamers. Corresponding images of the larger loops that include the on-disk features can be obtained in soft x-rays caused by scattering of the hot coronal electrons (the thermal bremsstrahlung or free-free emission process). This emission provides less spatial detail than the EUV but allows the on-disk imaging to higher altitudes. The physics underlying these observed coronal structures involves two related problems: the coronal magnetic field structure that controls its basic organization, and coronal heating, which fills the coronal loops with plasma and elsewhere enables its escape. The second topic in particular overlaps with Chapter 5, so it is worth keeping in mind while reading that you are also learning about the roots of the solar wind.

4.5.1 The corona's magnetic skeleton

The fact that the corona's high beta (Figure 4.6) allows us to think of its magnetic field as its skeleton makes it a natural topic from which to start a deeper discussion of this atmospheric region. It also turns out that, to a first approximation, one can neglect the complicated physics of the chromosphere and transition region and assume that the magnetic field observed on the photosphere is a good representation of the coronal field boundary. Then a relatively simple approach to visualizing the real coronal field is the **potential field source surface** (PFSS) model. The PFSS model first appears in the literature in the early 1960s, inspired in part by the increasing availability of coronagraph and soft x-ray images. It assumes the coronal field is time independent, current free (potential), and neglects any effects of the coronal plasma on the field other than at its boundaries. Later on we discuss departures from these assumptions, but a good way to think about the model is in the same way that a dipole added to a uniform external field is used to describe a magnetosphere to first order – an approximation whose value for providing insights far outweighs its limitations.

The PFSS model coronal field is calculated within a spherical shell via solutions of Laplace's equation for the scalar magnetic potential:

$$\nabla^2\phi = 0, \ \mathbf{B} = -\nabla\phi,$$

$$B_r(r,\theta,\varphi) = -\frac{\partial\phi}{\partial r},$$

$$B_\theta(r,\theta,\varphi) = -\frac{1}{r}\frac{\partial\phi}{\partial\theta},$$

$$B_\varphi(r,\theta,\varphi) = -\frac{1}{r\sin\theta}\frac{\partial\phi}{\partial\varphi}, \tag{4.3}$$

where the inner boundary field is a global magnetic field description (often the line-of-sight or radial component magnetogram-derived synoptic maps described earlier) and it is assumed that at an outer spherical boundary, typically located at ~1.50–3.25R_S, the field is everywhere radial. This model was developed around the time the existence of a solar wind was established by observations, making the latter a reasonable expectation. The solution is usually expressed in terms of spherical harmonics (see Chapter 7) with the coefficients of the various components determining the coronal field's appearance. Various solar observatories make these coefficients available, but users of these results must be aware that the normalizations of the various harmonics used to reconstruct the fields can differ depending on the formulas used to derive them. Overall, the PFSS model has proved a remarkably powerful tool because it makes it possible to understand how the large-scale coronal structure is related to the magnetic field seen on the photosphere. It is worth mentioning that nothing in the photospheric images, including the magnetograms, shows where the footpoints of the coronal open and closed fields are located. The coronal magnetic field topology can only be inferred from EUV, soft x-ray, and coronagraph images and/or such models.

Figure 4.27 displays a photospheric field synoptic map and the large-scale PFSS model coronal field topologies that go with it. The choice of Carrington rotation was made to show the field at an intermediate phase of solar activity. In the model illustrations, shading is used to represent the photospheric footprints of coronal fields that are open (e.g., connected to the source surface), while the larger-scale closed fields are shown as field lines. The coronal field is simplest at solar minimum when it becomes roughly dipolar, with open field regions mainly at the poles. But for most of the solar cycle it is much more complex. The warps and folds in the field line arcades, and irregular shapes of the open coronal hole regions seen here, are typical of the rising and declining phases of the solar cycle.

Figure 4.28 identifies the main coronal field topology features seen in the models. The largest-scale continuous field arcade that encircles the Sun is called the helmet streamer belt, and the locus of its outermost point, or cusp, produces a magnetic "neutral line" on the PFSS source surface. This feature is what produces the main coronal streamers or rays in coronagraph images. The associated neutral line represents the base of a current sheet that extends into the heliosphere, as will resurface in the following chapter. It separates the radial open fields of opposite sign or polarity on the source surface and is typically a single line except around solar maximum when it becomes highly contorted. Thus all open fields in the corona to one side of it are rooted in similar polarity fields. It is important to remember that the coronal field is generally not well represented by a dipole, or even a tilted dipole. In addition to the helmet streamer belt there are other more localized closed field regions that produce additional white-light streamers in coronagraph images called pseudostreamers. These occur more and more frequently as active regions occupy the surface. These and the active region fields add the tilts and warps to the source surface neutral line (see Figure 4.27(c)) as the solar cycle goes through its more active phases. A relevant comparison here is the amount of flux in the solar surface polar areas versus what emerges in the active regions. About 1000 times greater magnetic flux is contained in the active regions relative to the solar polar regions around solar maximum, which explains why the source surface neutral line gets so complicated. In general, the active regions win the competition for control of the coronal structure when they are present.

While PFSS models show how the large-scale coronal fields are divided into what we have called open and closed field regions, it is important to remember that there are still "quiet Sun," magnetic carpet fields within the coronal holes. Moreover, decayed active region fields (see earlier discussion of the photospheric field evolution) give these quiet Sun fields some net polarity that determines whether the open field of the coronal hole is directed outward or inward. In addition, magnetic arcades of smaller scales are nested inside the large-scale closed fields. Thus, at every radius inside the source surface there are neutral lines within the locally closed field regions that become increasingly complicated as one moves closer to the surface. This network of closed fields includes magnetic null points, "spines" and "separators" that define topologically distinct volumes or cells within the closed fields and are particularly susceptible to reconfiguration. It is important to keep in mind that the photospheric boundary field is always evolving due to differential rotation, convection, and diffusion, as well as new flux emergence, and so the coronal field must always be adjusting to it.

On closer inspection of the coronal field geometries in Figures 4.27(c), (e), and (f), it is apparent that the relationship between the active regions and

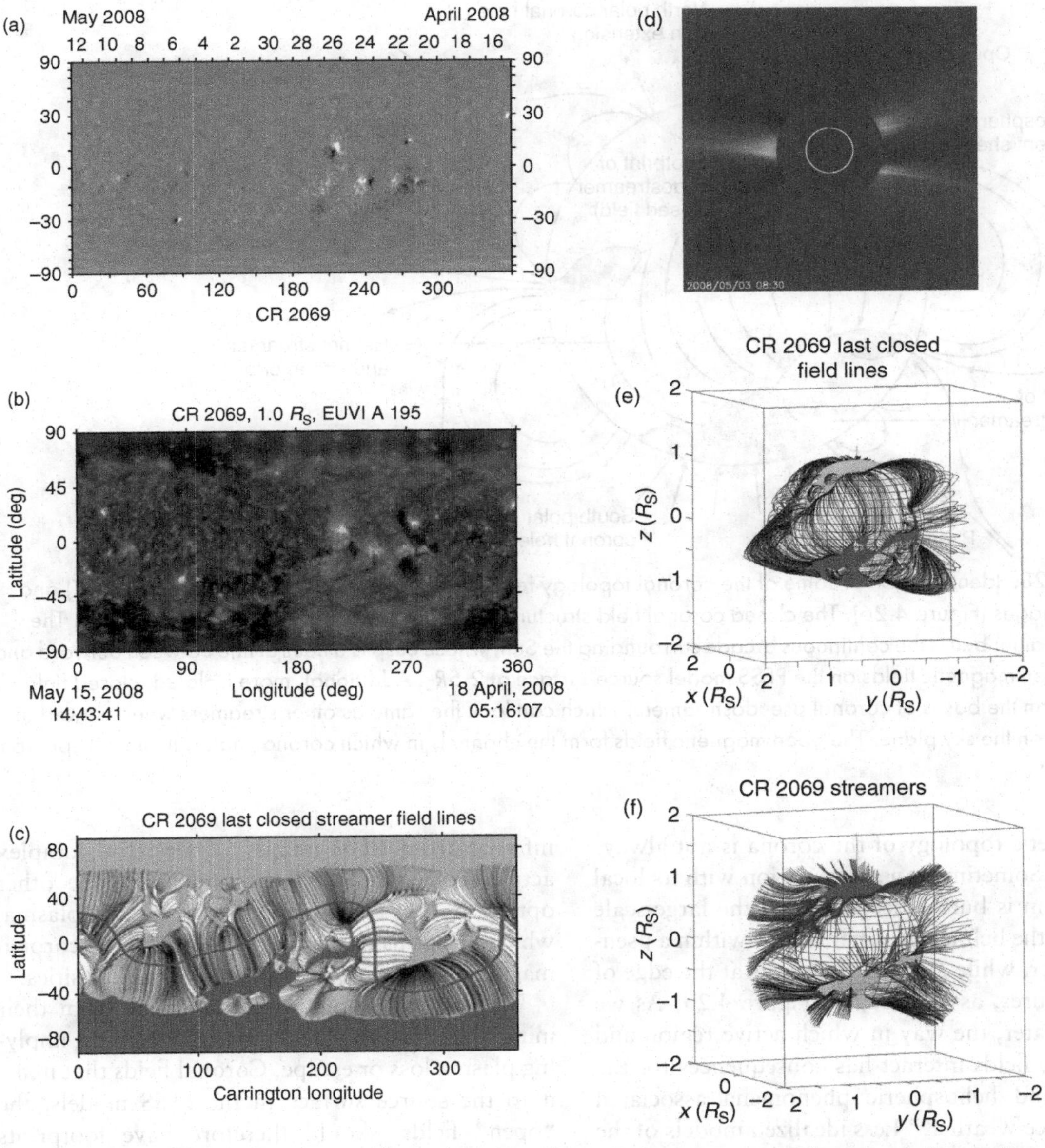

FIGURE 4.27. Illustration of the use of potential field source surface (PFSS) models in the understanding of coronal structure. (a) Solar magnetogram, on which the field model is based; (b) EUV map showing coronal holes (dark areas); (c) map of closed magnetic field lines; (d) coronagraph image; (e) three-dimensional rendering of closed magnetic field lines; (f) three-dimensional rendering of open magnetic flux or streamers. The main features of interest are the large-scale coronal streamer arcades and the footpoints of the open field. These illustrations make clear the need for models like the PFSS because these cannot be inferred simply by looking at magnetograms. ((a) NASA; (b) STEREO/SECCHI-EUVI; (d) ESA/NASA SOHO LASCO.)

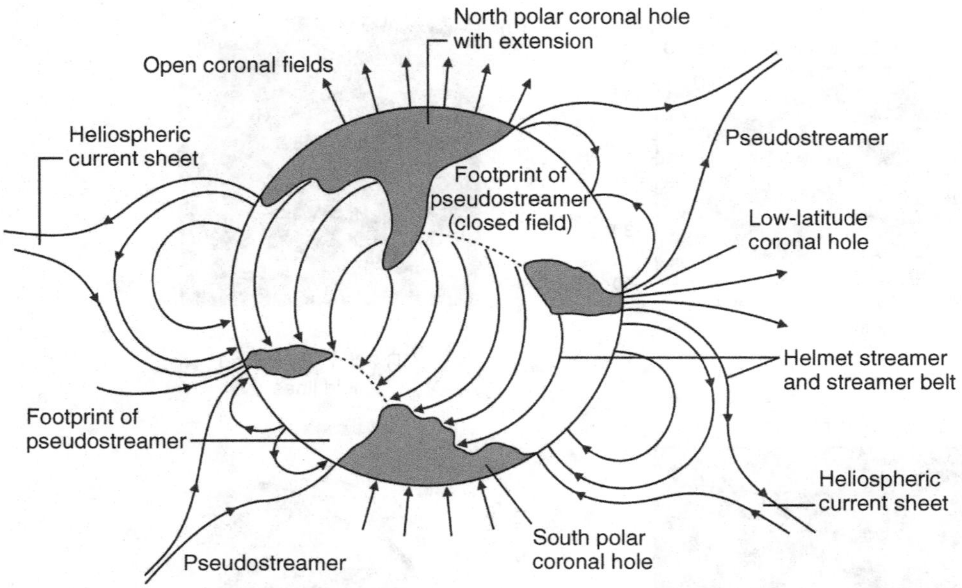

FIGURE 4.28. Identification of some of the coronal topology features seen in the PFSS models (Figure 4.27) and coronal images (Figure 4.26). The closed coronal field structures are at the bases of the coronal streamers. The helmet streamer belt is the continuous arcade surrounding the Sun, whose cusp is a neutral line between outward and inward open magnetic fields on the PFSS model source surface at $2.5R_S$. Additional, more isolated, closed field regions form the bases of coronal pseudostreamers, which can look the same as other streamers when viewed in projection on the sky plane. The open magnetic fields form the channels in which coronal material can escape from the Sun.

the magnetic topology of the corona is not always the same. Sometimes an active region with its local field system is buried deep beneath the large-scale arcade of the helmet streamer belt or within a pseudostreamer, while at other times it is at the edge of these features, as sketched in Figure 4.29. As we shall see later, the way in which active region and large-scale fields interact has consequences for the coronal and heliospheric phenomena associated with "space weather." Less idealized models of the coronal field are sometimes used when there is cause to expect the fields are not current free, where plasma pressures and dynamics are an issue, or where the goals are to represent more accurately the outer corona than the imposition of a spherical source surface approximation allows. These include the force-free field approximation, where the assumption is made that any currents are field aligned, or that $\mathrm{j} \times \mathbf{B} = 0$ (also see the discussion in Chapter 3). This approximation allows for current-related twists in the fields, which are often inferred from EUV images of fresh or complex active regions. Full MHD models are the other option, but these include both current and plasma, which brings us to the problem of how the coronal magnetic skeleton gets fleshed out with densities.

The coronal holes were recognized from their initial observation as regions of low density, implying plasma loss or escape. Coronal fields that make it to the source surface in the PFSS models, the "open" fields, should therefore have footprints resembling the dark coronal holes in EUV and x-ray images. Similarly, the "closed" fields that form loops and confine plasma within the source surface should resemble the bright EUV and soft x-ray features as well as the bases of the bright coronal rays or streamers in coronagraph and eclipse images. Figure 4.27(c), with Figure 4.27(b), includes an example comparing a synoptic map of PFSS model features with a synoptic map constructed from EUV images for the same Carrington rotation.

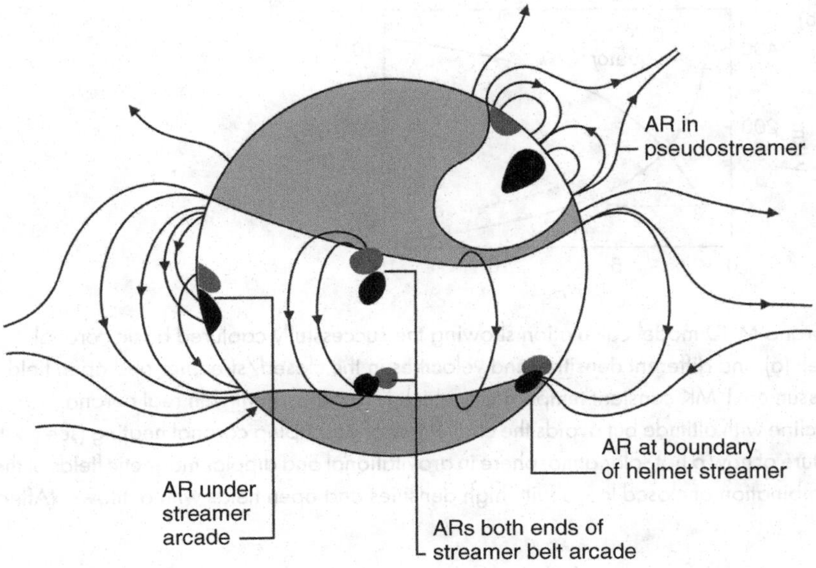

FIGURE 4.29. Different relationships that the large-scale coronal field structure can have with the active regions (ARs). Active regions can be at one or both ends of a closed field loop, or may be buried deep inside streamer arcades or loops. The coronal field can be thought of as a global network of interconnecting field loops, where local connections are often determined by the fields within a very large area and sometimes over the globe.

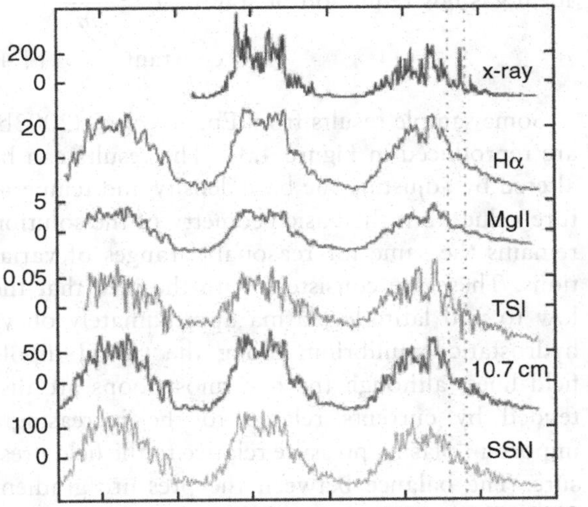

FIGURE 4.30. Solar cycle variation of selected EUV flux and flux proxies, showing they follow the sunspot number trends. These variations in part reflect the emissions that come from the more magnetized regions of the chromosphere including the plage and chromospheric network. The x-ray emissions in contrast are controlled by the closed coronal loops. These emissions as a whole are important to the photochemistry and related time variations of planetary atmospheres.

The overall brightness of the corona at all wavelengths varies with the solar cycle because the coronal atmosphere is different in the loops than it is in the open field regions where the plasma escapes. The loops have temperatures ~2 MK and densities ~10^9–10^{11} cm^{-3}, while the open field regions have temperatures of ~1 MK and densities ~10^8 cm^{-3}. Approximations to observed EUV and soft x-ray images using the PFSS model closed fields outlined with emitting plasma can capture the general appearance of the coronal features at these wavelengths. These reconstructions also hint at the coronal emission mechanisms because the parameterization of brightness needed is found to depend on both the magnetic field strength and loop length. The magnetically controlled coronal structure also causes a solar rotational (~27 day) modulation in virtually all solar emissions from the corona, especially in the EUV and soft x-ray emissions. It also has a major influence on the solar cycle dependence of these emissions, as shown in Figure 4.30. What is needed to make a more physically consistent picture is a model that includes both the realistic magnetic structure and coronal plasma including density and temperature – a challenge still facing space physics researchers.

The most complete models of the global corona use MHD fluid treatments that may include the

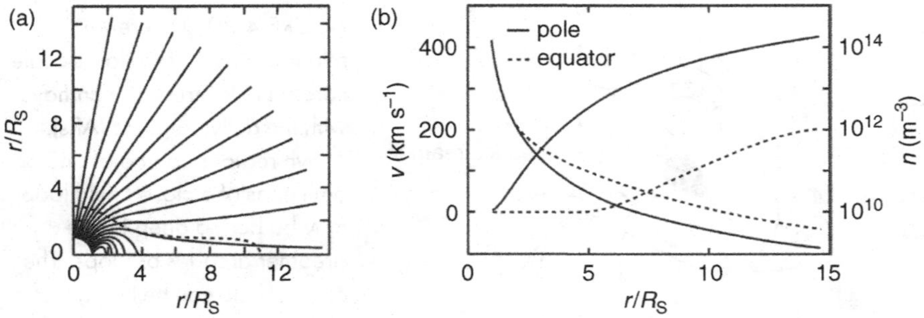

FIGURE 4.31. Isothermal dipolar corona MHD model calculation showing the successfully captured basic coronal features of an equatorial streamer belt (a) and different densities and velocities in the closed/streamer and open field regions (b). This model was for an assumed 1 MK constant temperature which is not consistent with real corona temperatures (see Figure 4.6) that decline with altitude but avoids the challenges of describing coronal heating (see text). Nevertheless, it gives a first-order picture of how a hot solar atmosphere in gravitational and dipolar magnetic fields of the right strength make a corona-like combination of closed loops with high densities and open fields with outflows. (After Endeve *et al.*, 2003b.)

resistive terms depending on the approximations used to solve the equations (see Chapter 7). The simplest solutions are for axially symmetric dipolar field topologies that assume an ideal gas law with isothermal conditions. These can be considered as working around the coronal heating issues while still containing some essential elements such as geometry and the competition between outward pressure gradient forces and confining magnetic forces. For coronal temperatures, even these simple models self-consistently produce a description that includes regions of magnetically trapped plasma on closed fields at low to mid latitudes, and lower densities on high-latitude open fields along which plasma escapes to the outer boundary. One example from Endeve, Leer, and Holzer (2003a) uses this system of equations:

$$\text{continuity equation}: \frac{\partial \rho}{\partial t} + \nabla \cdot (\rho \mathbf{u}) = 0,$$

$$\text{momentum equation}: \frac{\partial (\rho \mathbf{u})}{\partial t} + \nabla \cdot (\rho u \mathbf{u})$$

$$= -\nabla P - \frac{GM_S}{r^2}\rho \hat{\mathbf{e}}_r + \mathbf{j} \times \mathbf{B},$$

$$\text{Ampére's law}: \mathbf{j} = \frac{1}{\mu_0}\nabla \times \mathbf{B},$$

$$\text{induction equation}: \frac{\partial B}{\partial t} = \nabla \times (\mathbf{u} \times \mathbf{B}),$$

$$\text{ideal gas law (equation of state)}: P = \frac{\rho KT}{m},$$

$$T = \text{constant}. \qquad (4.4)$$

Some sample results from Endeve *et al.* (2003b) are reproduced in Figure 4.31. The results can be altered by adjusting the base density and temperature values, but the basic geometry of the solution remains the same for reasonable ranges of variations. These are consistent with the idea that the low to mid latitude plasma approximately obeys hydrostatic equilibrium along the closed dipole field lines, although the outermost loops are distended by currents related to the increasingly important plasma pressure relative to the field pressure. The balance between the pressure gradient force of the plasma and the confining $\mathrm{j} \times \mathbf{B}$ force of the dipole field lines is no longer possible at higher latitudes, thus determining the "last closed field line" with the cusp-like shape. One can use this simple model to investigate where the plasma flow along the field becomes supersonic and superalfvenic. The outline of the rotationally symmetric surface defining these singular points for the chosen test case is indicated by the dashed line in Figure 4.31(a). While these results should be regarded as an extremely rough approximation given the assumed constant temperature, they give an idea of how some first-order solar coronal properties arise and illustrate some important concepts. The field lines look like the PFSS model for a dipolar field, but the outermost field lines and plasma properties including outflows are now a more physically self-consistent part of the picture. The reader can find descriptions of much more sophisticated three-dimensional MHD models in the literature that address some of the challenges facing coronal simulations, including realistic, and generally more complex, magnetic field geometries. Simulated images from these models rendered in white-light, EUV, and soft x-ray emissions can look quite realistic, but the heating is still parameterized.

What heats the corona is a question that has existed since its temperature was first inferred from its spectral lines. At its most basic level, the Sun must get rid of all of the energy regularly generated in its interior if it is to remain a stable and viable star. Photon radiation carries most of it away, but the mechanical energy of the convection zone transports much of the rest to the surface where it must be dissipated in the solar atmosphere. But how is it dissipated and thus transformed to coronal heating by a few thousand kilometers above the surface? For perspective, it is useful to keep in mind that the fraction of solar energy that goes into its atmosphere rather than being radiated away in its photon spectrum is an almost negligible fraction of that generated in its interior. Nevertheless, how and where it is deposited is critical for both the production of the Sun's excess radiated energy in the EUV and soft x-ray wavelengths (important to planetary atmospheres) and for what escapes to become the solar wind.

In our earlier discussions we reviewed some of the observations that tell us that mechanical energy from the solar convection zone is making it into the chromosphere and transition region. It is generally agreed that much of this mechanical energy gives itself up to thermal energy in the chromosphere–corona transition. The problem is how to turn the mechanical energy into thermal energies of the particles. One problem is that the resonant wave–particle interactions discussed in Chapter 13 do not work here because the Alfvén wave frequencies associated with the inferred footpoint motions of field lines – moving with the granular convection patterns – are simply too low compared to the particle gyro-frequencies. The question is whether the power in these waves can somehow transfer their energy to a frequency range more compatible with ion cyclotron resonance in particular. Two proposals with possibilities have been made. In one, some of the outgoing waves are reflected in the chromosphere-transition region due to the large gradients in the conditions there, and then collide with the outgoing waves to produce a turbulent cascade to higher frequencies. The other is that some of the waves steepen to form localized shocks, which then give up their bulk energy to thermal energy in the shock dissipation process.

The complexity of the solar magnetic field may also play a key role here. In particular, the mixed polarity of the photospheric field implies that magnetic flux bundles with fields of opposite sign and different strengths are always being forced together by the complicated surface footpoint motions. One result may be the magnetic reconnection of oppositely directed adjacent fields, with its associated acceleration of local plasma flows (see Chapter 9) and related shocks. Moreover, small-scale current sheets exist at the boundaries between these flux tubes that can resistively heat the local plasma, especially at locations where some neutrals are still present.

There is actually ample evidence of both mechanisms at work on the Sun. In particular, small transient bright points in x-rays and other emissions sometimes referred to as nanoflares may be localized heating events associated with the reconnections. Polar plumes, bright elongated structures that are seen mainly in polar coronal holes off the limb in various emissions, may result from the interaction of the small bipolar ephemeral regions in the quiet Sun fields with surrounding open fields. The spicules, mentioned earlier, have appearances and dynamics consistent with small current sheets produced on granular motion scales. In fact, the Alfvén waves with periods of ~5–10 min and amplitudes of a few tens of kilometers per second that are detected in the motions of the spicules may be the cause of their braiding. Coronal heating is in fact likely to be carried out by a combination of processes, including these two. It is also likely that the dominant mechanism depends on the local context, e.g., whether one is considering regions rooted in quiet Sun magnetic fields, or above the stronger, more polarity-biased areas of active region fields. Similarly, it may depend on whether one is considering a magnetically closed or open field region.

So far we have mainly discussed the subject of coronal heating in a broad, single-fluid sense. But the electrons are likely to be responsible for most of the heat (thermal energy) transport in the corona. Their low mass makes them both more tightly tied to the magnetic field and also more easily energized. Some coronal models attach special importance to the electrons, applying the exospheric concepts mentioned in the discussion of upper atmospheres in Chapter 2. In these models the lighter electrons set up a polarization electric field directed outward, drawing the heavier ions out of the solar gravitational well in essentially electron-driven coronal plasma outflows. Such theories also provide useful insights into the transitions from collisional to collisionless behavior that must occur at the base of the corona. The electrons can carry energy in any direction relative to where they obtain it, including downward into the chromosphere. With their small gyro-radii, the electrons can also stay within narrow regions (such as current sheets) where strong electric fields may be present, or can become trapped in small volumes such as reconnection sites where they are subject to statistical energization processes. Magnetic mirroring on closed coronal loops produces counterstreaming distributions, with possible consequences for plasma-wave instabilities (Chapter 13) and related electron heating at the loop tops. The important point is that the electrons likely reach suprathermal to high energies in the chromosphere–corona transition layer by interacting with the local fields. Whether subsequent energy transfer to ions is via electric fields or Coulomb collisions depends on the prevailing conditions and locale. The ions are also affected by the more easily generated lower-frequency waves and can more easily cross flux tube and other boundaries via gradient and curvature drifts because of their larger gyro-radii. Where the electrons' influences on coronal heating give way to ion-dominated processes such as ion cyclotron wave heating is another consideration. A quantity known as FIP for first ionization potential is sometimes used to estimate the location at which at least the ionizing collisions with electrons stop. The FIP is the energy required to remove the outermost, least-bound electron from an atom. Inferences of the ion ionization states from solar spectral lines are used to infer the "freezing in" temperature and thus the location where coronal ionization states are set, which turns out to depend on the coronal setting. The bottom line is that the coronal heating problem ultimately requires one to incorporate species-dependent processes in the many different contexts that make up the corona. In addition, as we shall see next, there are coronal plasma energizations and outflows that occur for reasons other than this regular heating.

4.6 SOLAR CORONAL ACTIVITY

As discussed in Chapter 1, the importance of space physics has increased as our technological society has become more dependent on infrastructure that depends on the space environment. The weather in space: the state of the solar wind and the interplanetary field, and the intensities of energetic particle fluxes, are all dependent on the activity of processes in the solar corona.

It is easy to appreciate from the widely available movies of coronal images that the solar atmosphere is essentially never steady, especially when the sunspot number is high. Two types of transient activity are **flares** and **coronal mass ejections**. Not long ago, a distinction between these was not made or at best remained ambiguous. Localized, short-lived bursts of emissions on the disk may briefly intensify the solar spectrum at shorter wavelengths, sometimes by orders of magnitude (Figure 4.4(b). These "flares" were first recognized as explosive energy releases on the Sun that were sometimes followed, in the ~8 min speed of light transit time, by measurable **sudden ionospheric disturbances** (SIDs). They were also sometimes associated with geomagnetic storms and auroras that followed the outburst by a few days. While anecdotal evidence of coronal transients had been reported in some early coronal images, their connection to the apparent delayed terrestrial effects of some flares awaited the space age to be revealed. It was only in the 1970s and 1980s that several spacecraft carrying coronagraphs firmly established the existence of large eruptions of coronal material related to but distinct from flares. These became known as coronal mass ejections, or CMEs. Both flares and CMEs were found to occur more frequently when active regions and sunspots are present on the solar surface. In the 1990s, there was a general shift in perspective on whether flares or CMEs were the main cause of significant terrestrial responses to solar activity. Since then, both flares and CMEs have been investigated in detail so that today we have a clearer picture of each phenomenon, their connections, and their different consequences.

4.6.1 Solar flares

Flares, defined by the occasionally visible, impulsive brightenings in the vicinity of sunspots, occur over a large range of magnitudes. Regular solar x-ray monitoring, together with extended periods of x-ray and EUV imaging, provided the ability to examine flare intensities, locations, and coronal contexts. Major flares have a clear connection to active regions with strong (relative to quiet Sun, typically >50 mT) fields that often have complex magnetic polarity patterns and large areas. Gamma and delta spots (described earlier in this chapter) are the sites of some of the largest, brightest flares, making their mere appearance on the surface a predictive tool. Figure 4.32(c) gives several examples of the full-disk appearance of major flares captured in the EUV images. The accompanying magnetograms (Figure 4.32(a)) and visible-light images (Figure 4.32(b)) show their context with respect to active regions and sunspots, respectively. A few more spatially resolved images of large flares in various emissions, are shown in Figure 4.33. At the other end of the flare size spectrum are nanoflares, small brightenings in the magnetic carpet that may be associated with coronal heating via similar processes, but they will not concern us here.

The major flares are mainly characterized and classified by the intensities of their outbursts in radio through x-ray wavelengths. A summary of a widely used classification scheme, based on their H-alpha and soft x-ray emissions, is provided in Table 4.3. A useful rule of thumb to remember is that flares at and above C1 class are of primary interest for their effects on Earth's ionosphere and because they are more likely than lesser flares to be accompanied by other kinds of events including CMEs. The largest flares for which we have modern observations are estimated as X20+ class events. However, the frequency of occurrence of a particular flare severity roughly follows a power law where the number of flares declines with severity as (class designation)$^{-2}$; so, for example, a C2 flare is about four times less likely than a C1 flare. Only a few X-class flares occur per ~11 year activity cycle, while several small flares occur per day at active times. The most intense flares can also be seen in white light and/or they produce gamma ray line emissions from the excitation and decay of the states of local nuclei energized at the Sun by flare processes. They may also produce observable "sunquakes" or Moreton waves in chromospheric emissions, which show a disturbance moving through the atmosphere away from the flare site like ripples from a stone thrown in a pond. An especially important observation for interpreting the physics of flares is that their highly resolved images show distinctive "two-ribbon" emission patterns in the H-alpha and EUV emissions.

As mentioned earlier, the occurrence of pre-flare filaments within an active region can be regarded as a sign of sheared, stressed, or twisted magnetic fields, indicating the presence of currents and related available energy buildup in the field (recall the configurations in Figure 4.22). These filaments generally show motion ("activate") and often at least partly disappear ("disrupt") at the time of a flare. In fact, some flaring regions repeat similar filament reformation and flare sequences several times as they transit the visible disk, or may flare in different subareas of the filament over a period (e.g., the associated filament disrupts in sections). Of course not all filament-containing active regions flare, and a filament may collapse or quietly

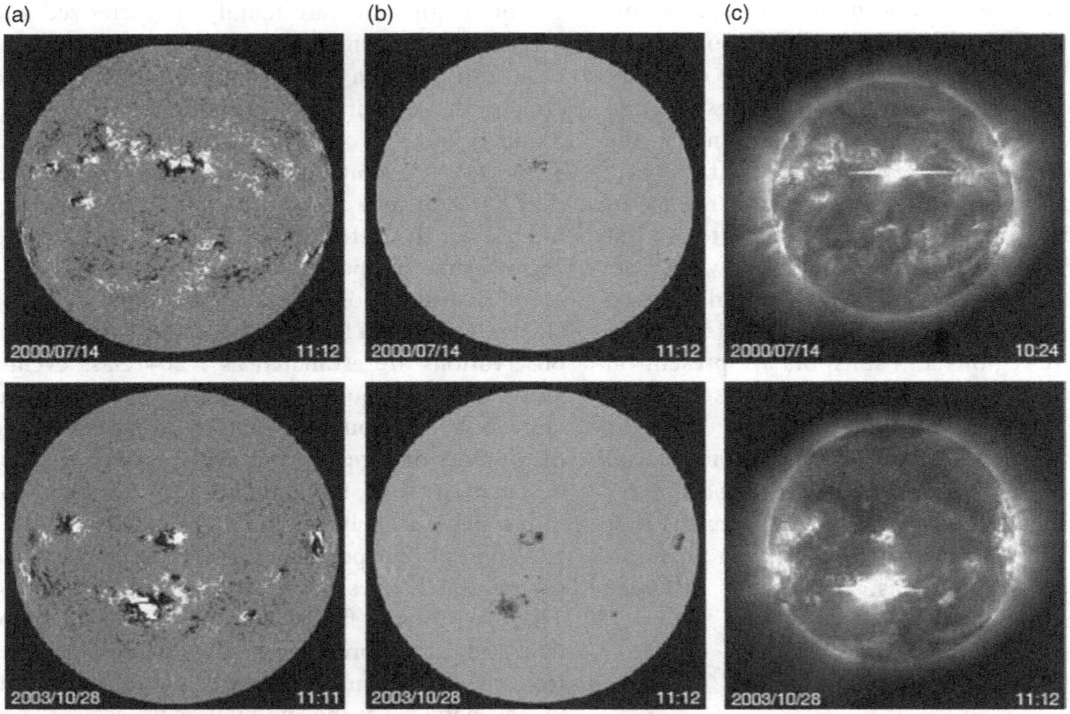

FIGURE 4.32. Examples of full-disk images for two major flares showing, from right to left, the EUV image of the flare, a visible-light image, and the magnetogram at the time. (NASA/ESA images from SOHO MDI and EIT.) The top images are of a flare that occurred on July 14, 2000. The lower images are from October 28, 2003. Major flares usually occur at the sites of large-area active regions with complex magnetic polarity.

disappear as an active region evolves. Similarly, flares can occur in active regions without a visible filament. The flare process sometimes results in an observable post-flare simplification of a complex active region, including transformation to a more potential (current free) state of the overlying field loops. However, detecting these magnetic changes usually requires vector field observations and high-resolution EUV images. Fortunately for the research, both of these are now a routine part of space-based solar observing. In general, as active regions age and their fields become weaker and less complex, their likelihood of producing flares diminishes. While highly decayed active regions can still host filament formation and eruptions that produce extra emissions, as we shall see this setting better fits the other category of solar event, the CME. Thus when reading the literature one must keep in mind that there is a range of flare-like phenomena that results in a sometimes ambiguous boundary between the definitions of flares and CMEs.

Figure 4.34 is a sketch of a widely used **standard flare model** geometry for a flare site, showing coronal field lines with a magnetic "x-point" or "x-line" at the loop top depending on whether this geometry occurs in three or two dimensions (e.g., is extended in the dimension out of the page or axially symmetric). The scaling of this field structure for flares can be viewed as having its footpoints within an active region on the photosphere. Although the details of what is shown in Figure 4.35 are debated, the weight of observational evidence for the basic properties is strong. In particular, the double "ribbons" of H-alpha emission can be explained as the signature of chromospheric energy deposition by the impact of electrons accelerated near the cusp of the field arcade. These are expected to stream

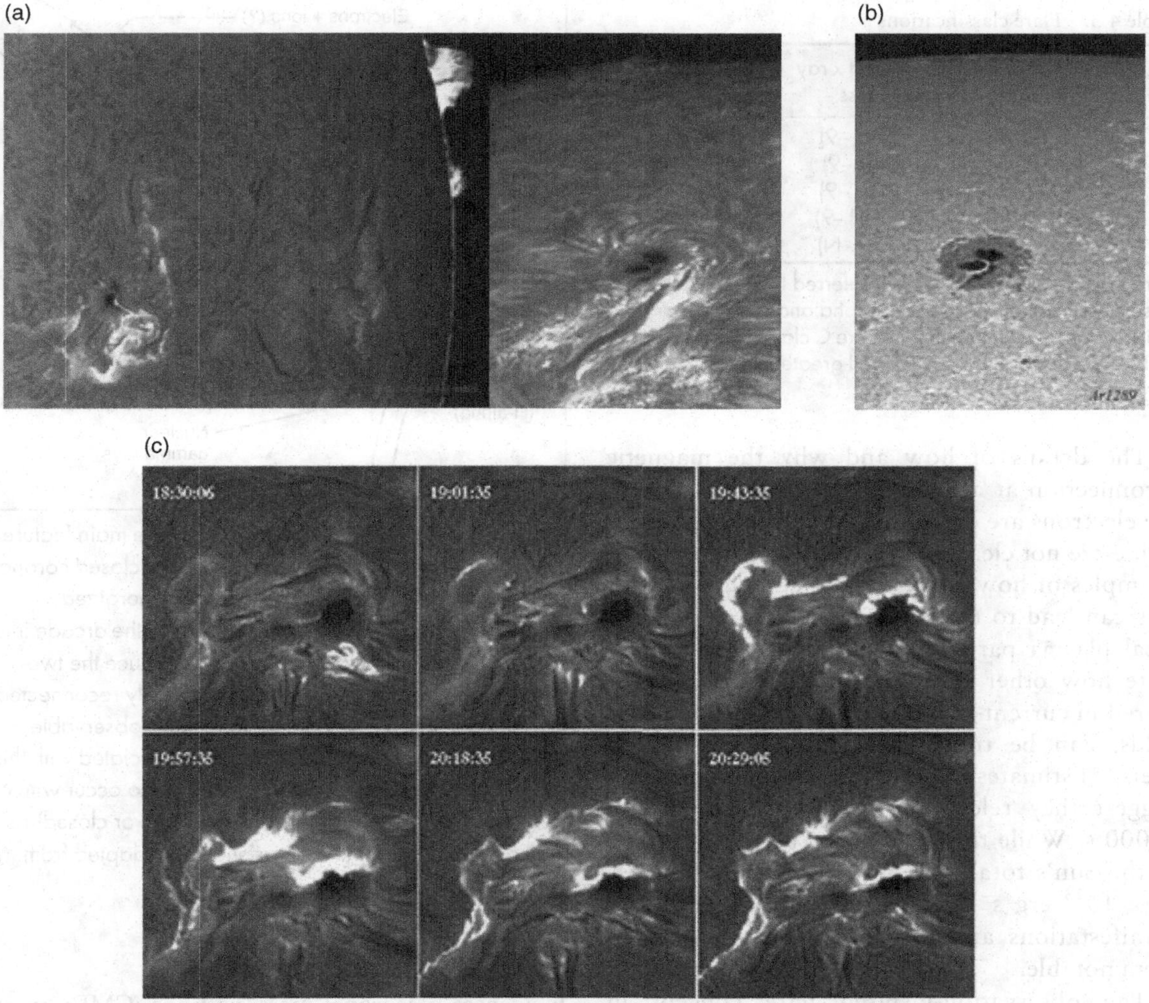

FIGURE 4.33. Examples of flare images in (a) H-alpha, (b) visible, and (c) EUV wavelengths illustrating the different appearances flares take, including the "two-ribbon" emission pattern. In (c), the two ribbons roughly bracket the underlying active region's magnetic neutral line. These are transient emissions sometimes lasting only minutes, and constantly evolving in appearance. They represent solar emission intensifications at wavelengths generally shortward of 100 nm of sometimes orders of magnitude. The effects of these intensifications can be seen in planetary atmospheres as increased ionization and photodissociation. (NASA images.)

down the field lines to the footpoints of the reconnecting fields at the top of the loop(s) straddling the active region magnetic neutral line. X-ray emissions also arise from both those footpoints as well as from the x-point or line at the arcade apex. This distinctive triple x-ray source pattern is best observed at the limb, where different x-ray spectra from the footpoints and loop top indicate different source details at the different sites. The same basic geometry in fact occurs over a broad range of scales in the corona, from magnetic carpet scales (perhaps explaining nanoflares) to the largest coronal loops. However, for the latter, the activity takes a somewhat different character, more appropriately associated with the CMEs, to be discussed in the following section.

Table 4.3. Flare classifications

H-alpha class	Area (deg^2)	Soft x-ray class	Log flux at 0.1–0.8 nm
S	2.0	A(1–9)	–8 to –7
1	2.0–5.1	B(1–9)	–7 to –6
2	5.2–12.4	C(1–9)	–6 to –5
3	12.5–24.7	M(1–9)	–5 to –4
4	>24.7	X(1–N)	> –4

Flare intensity and size are often referred to in terms of these classifications based on H-alpha and soft x-ray emissions. Most significant flares are C class. Large flares are ~X15–17, although a few X28 and greater events have been reported.

The details of how and why the magnetic reconnection at a flare site is initiated and how the electrons are energized around the x-point or x-line are not clear. Overall, flares are important examples of how coronal magnetic field topologies can lead to the energization of some of the local plasma particles. In particular, they illustrate how other forms of energy, such as that stored in current-carrying active region magnetic fields, can be transformed to particle kinetic energy. Estimates of the energy produced in flares suggest they release up to 10^{32} erg in times ~1000 s. While this is small on the overall scale of the Sun's total rate of normal energy release (4×10^{33} erg s^{-1}), their sometimes spectacular manifestations and spatial concentration make them notable.

The observation-inspired flare concept in Figure 4.34 has led to numerous theories and models for producing the different observed flare phenomena. Among these, more complex field geometries with the same essential components have been explored, with the roles of null points and magnetic connectivities within the active region and with the surrounding coronal fields in mind. The standard flare model is but one simple representation of the ingredients of a potential flare site. Flares have a long history as Earth ionosphere perturbers, yet they are not the type of solar activity that is of most concern to space physicists. That distinction is reserved for a family of coronal transients called coronal mass ejections.

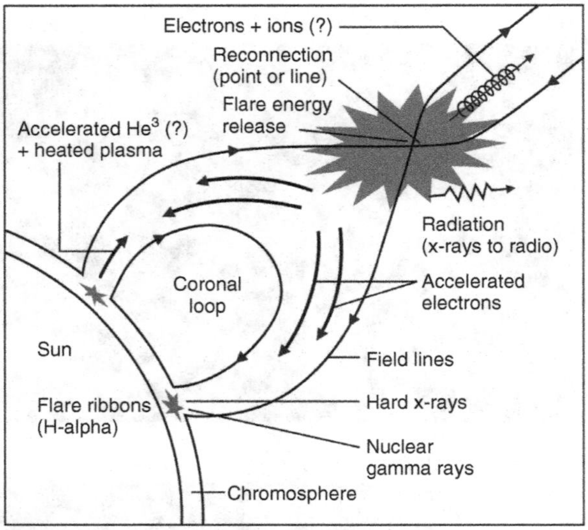

FIGURE 4.34. "Standard" flare model. The main features of the model are magnetic reconnection of closed coronal fields near the loop top, which results in energized electrons that stream both down the legs of the arcade into the chromosphere below (where they produce the two-ribbon emissions) and outward along newly reconnected field lines. Triple x-ray sources, sometimes observable when a flare occurs near the limb, are associated with this model. Note that the reconnection may also occur with a neighboring magnetic field structure (open or closed) as well as within the same loop as shown. (Adapted from Lang, 2010.)

4.6.2 Coronal mass ejections

Incidences of coronal mass ejections, CMEs, were first recorded in significant numbers in Skylab soft x-ray images obtained in the 1970s, but their detailed study gained momentum when white-light coronagraph images became available from spacecraft. As mentioned earlier, coronal electrons confined by the magnetic loops of the corona make it possible to view the field topology and its changes via Thompson scattering of sunlight. In images such as those reproduced in Figure 4.35, CMEs typically appear as expanding loops or bubbles, or sometimes as narrow jets. They may evolve slowly, taking a good fraction of a day to leave the coronagraph field of view of up to tens of solar radii, or explosively in tens of minutes. Observations suggest there is probably more than one mechanism for producing

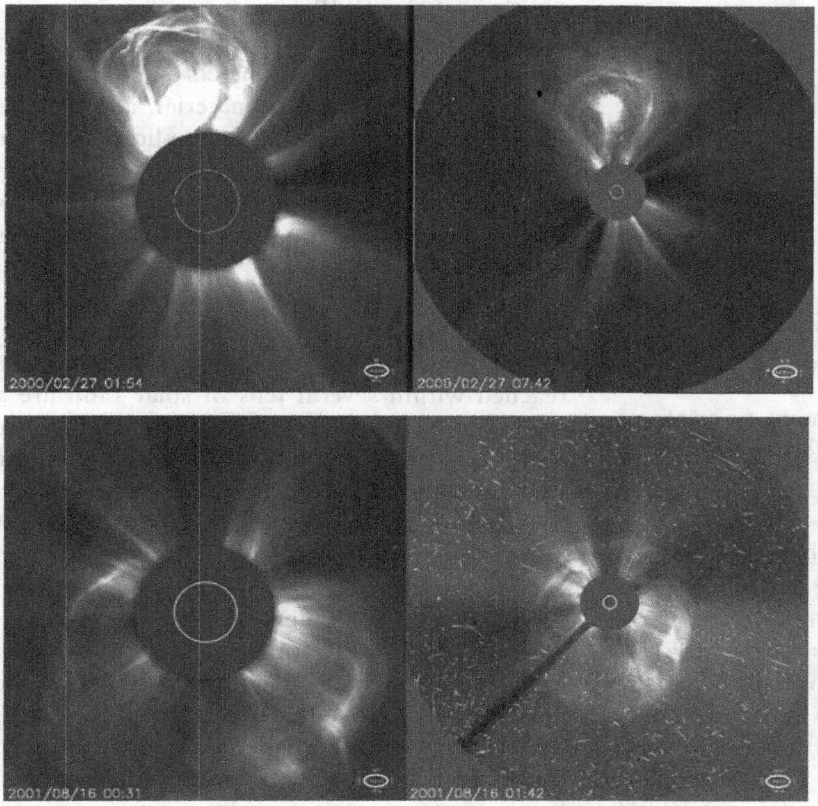

FIGURE 4.35. Coronagraph images of coronal mass ejections (CMEs) in visible light. The canonical CME has the appearance of a bright loop or bubble seen in projection on the sky plane. Sometimes there is a particularly bright, dense additional thread of material involved that may be associated with a filament that has erupted with the CME. CMEs may appear to come from the vicinity of active regions or from coronal streamers without clear active region connections. The latter are sometimes referred to as "streamer blowout" CMEs. Flares are often associated with the active region site CMEs. (SOHO LASCO images (ESA/NASA).)

CMEs. Here we consider why the corona may reasonably be expected to have several kinds of eruptive behaviors as part of its normal solar cycle evolution.

At very low solar activity times, slow coronal changes observed by coronagraphs can be simply attributed to rotation of the non-axisymmetric coronal structure (e.g., see Figure 4.27). However, the coronal field is also constantly adjusting to the evolving photospheric magnetic fields at its base, described earlier in this chapter. The related evolution of the corona's density is also noticeable in the coronagraph images as material outflows or inflows, often located in the vicinity of coronal streamers – suggesting that adjustments of the open and closed field boundaries are regularly occurring. The natural follow-on question might then be what happens when the necessary adjustments become more extreme, e.g., in terms of volume or flux content of the required coronal field changes. As shown by the coronal images in Figure 4.35, CMEs often first appear in or in the vicinity of coronal streamers. Slow changes are always occurring in the coronal field topology. However, sometimes the streamers drastically change their position or shape, and/or new streamers may appear while old ones disappear. The latter are sometimes referred to as **streamer blowout** CMEs. They may look like a gradual expansion of the streamer whose front moves slowly outward at speeds between a few tens and a few hundreds of kilometers per second. Sometimes their images give the impression of a large expanding flux rope anchored to the Sun at both ends, seen in projection on the plane of the sky. Figure 4.36 shows a version of this concept. Occasionally a bright central core is seen at the base of the bubble or loop cavity giving the impression of a light bulb with a glowing filament inside (see Figure 4.35). These can be associated with the involved chromospheric filaments mentioned above. As the erupting

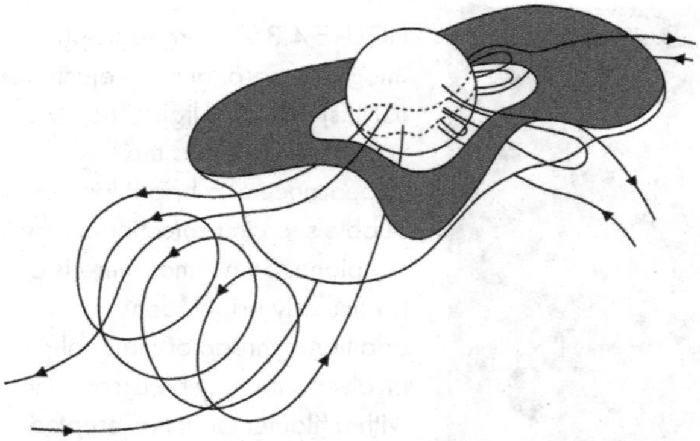

FIGURE 4.36. Illustration of how a CME may be related to the coronal streamer belt. In this concept, the erupting structure is embedded within the heliospheric current sheet. (After Crooker *et al.*, 1993.)

coronal structure moves outward it may entrain the filament material suspended over the local magnetic neutral line (see Figure 4.22), and also visibly interact with its surroundings, compressing and deflecting adjacent coronal loops and coronal hole boundaries as it passes. It may also become distorted by these interactions. Classic examples of this type of CME are the so-called polar crown filament eruption events, which arise from mid- to high-latitude decayed active region field sites with longitudinally extended magnetic neutral lines.

A second class of CME can have similar appearance or show much more complexity including twisted and/or multiple loops. These occur in association with an active region that is in the process of emerging or noticeably evolving, although sometimes vector magnetograms are needed to observe the changes. To distinguish these from streamer blowout CMEs, we refer to them here as active region CMEs. Distinguishing between this CME type and the other is sometimes difficult because active region involvement can come in a number of field structure settings.

Figure 4.29 illustrates some of the many possible connection geometries of active region fields with the larger coronal fields. Any of these may spawn an eruption of an active region CME. A few specific examples where PFSS models have been used to visualize the CME coronal field settings are shown in Figure 4.37. The occurrence of a flare within the associated active region around the time of the CME (sometimes before, sometimes after it) is common in these cases, although the flare can be modest. The coronal ejecta may again contain filamentary material, although its spatial distribution can be complicated. These events have a much broader range of speeds than the streamer blowout CMEs, on rare occasions leaving the Sun at up to a few thousand kilometres per second.

The general distribution of CME speeds from coronagraph image sequences sketched in Figure 4.38(a) shows that the most typical values reached within several tens of solar radii are a few hundred kilometers per second. Only those few events in a small tail of the distribution are very fast, up to several thousands of kilometers per second. As might be expected, the fastest events generally involve exceptionally strong, complex active regions, which are typically, though not always, found during periods of moderate to large sunspot number. They are also often linked to major flares. Figure 4.38(b) shows the apparent angular width distribution of the CMEs projected on the sky plane above the limb. The fastest events tend to have large angular widths. Some events appear to involve a significant portion of the magnetically connected global corona. However, there are also large angular extent events whose appearance is determined by the viewer's perspective.

A subclass of all CMEs, the "halo" event, is of special interest to Earth observers. Halo events are those that involve active regions or streamers near the central solar disk. As they move toward the observer, these CMEs expand beyond the disk to create a widening circular or elliptical white-light halo around the Sun. This family of events produces the small peak at 360 degrees in the angular width statistics in Figure 4.38(b). Because most coronagraphs are at Earth, halo events can be used for forecasting Earth responses. However, their velocities, widths, and precise directions are more challenging to determine than limb events because they are seen projected on the plane of the sky, and may even be headed away from the observer (e.g., "back-sided" halos). Thus velocities in particular must be estimated by assuming a CME geometry. Techniques have been developed, such as fitting

(a)

(b)

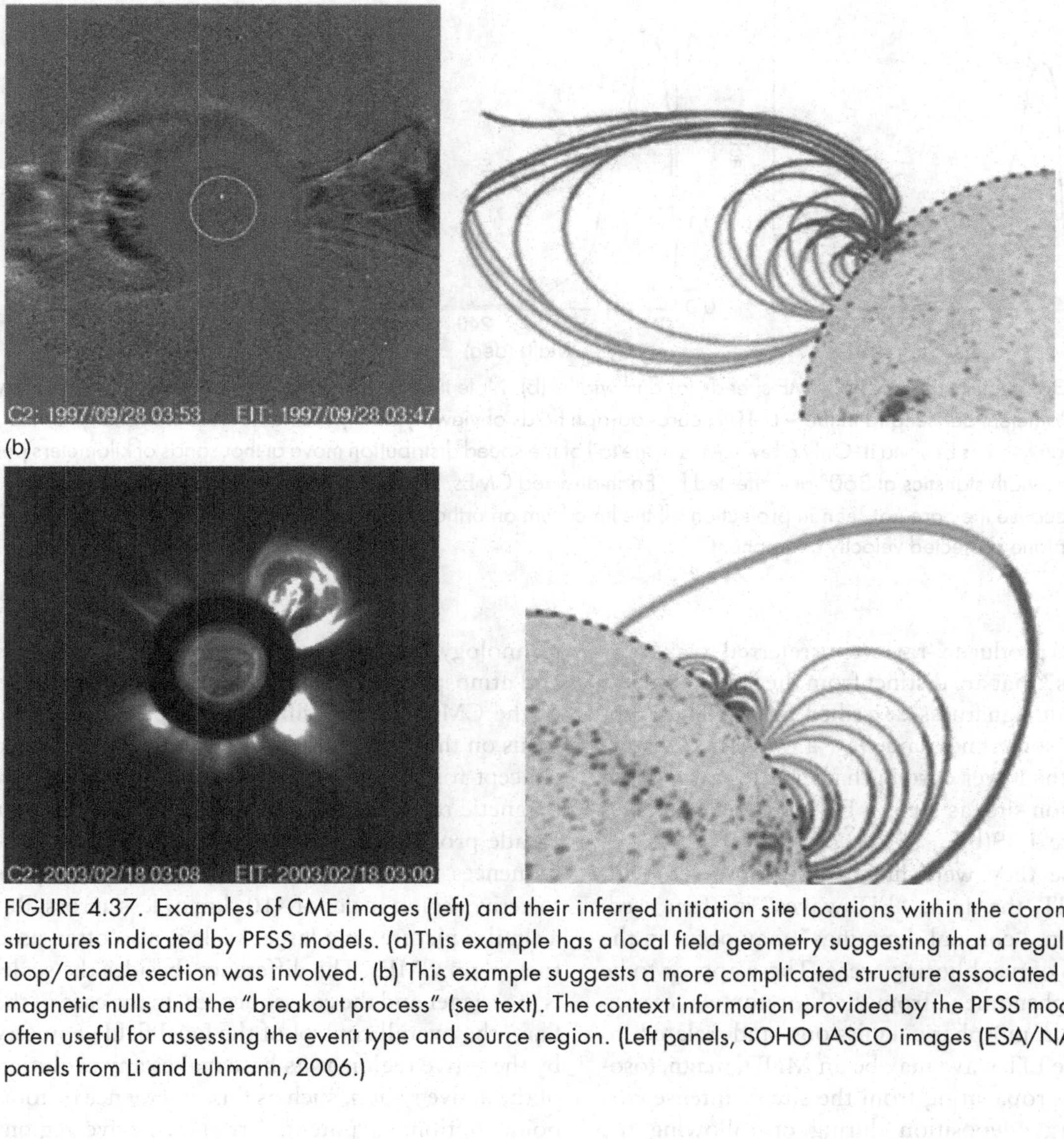

FIGURE 4.37. Examples of CME images (left) and their inferred initiation site locations within the coronal field structures indicated by PFSS models. (a)This example has a local field geometry suggesting that a regular coronal loop/arcade section was involved. (b) This example suggests a more complicated structure associated with magnetic nulls and the "breakout process"(see text). The context information provided by the PFSS models is often useful for assessing the event type and source region. (Left panels, SOHO LASCO images (ESA/NASA). Right panels from Li and Luhmann, 2006.)

cone shapes to coronagraph image sequences, to provide an idea of both the actual size and direction, and hence the true CME velocity. While these "cone models" of CMEs have already seen numerous applications, they contain no details of the coronal ejecta (more discussion of cone models can be found in Chapter 5).

CMEs also have other characteristic observational signatures seen in both soft x-ray and EUV images of coronal loops before, during, and after the eruption. Before eruptions, active region coronal loops may brighten. The illuminated field lines sometimes take on a distinctively non-potential, twisted "sigmoid" appearance (see Figure 4.39(a) for an example). These participate in the eruption, which can leave behind a more potential-looking (e.g., less twisted) field arcade whose afterglow gradually fades over the next hours. The latter

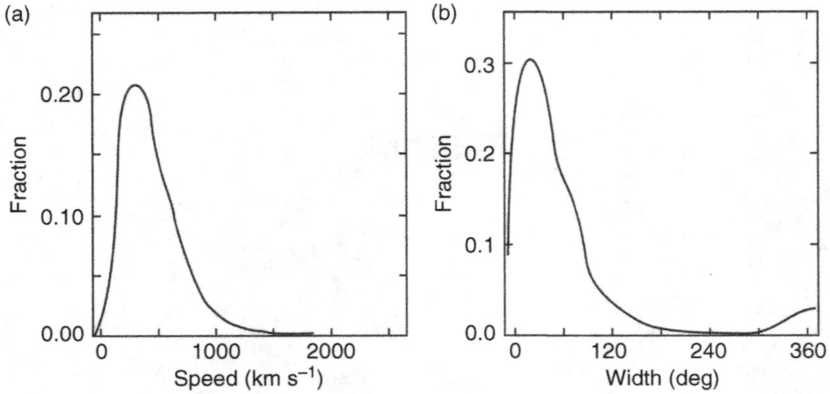

FIGURE 4.38. Statistical trends of CME speeds (a) and widths (b). Note that the large majority of CMEs move at a few hundred kilometers per second in the ~1–10R_S coronagraph fields of view, typically taking hours to half a day for the leading edge to cross beyond it. Only a few CMEs in the tail of the speed distribution move at thousands of kilometers per second. The width statistics at 360° are affected by Earth-directed CMEs. The speed of these "halo" CMEs is hard to measure because they are not seen in projection off the limb from an orthogonal viewpoint. Instead their images show only a sky-plane projected velocity component.

sometimes produce x-ray events referred to as "gradual flares" that are distinct from the flare impulsive x-ray event signatures described above. The eruption also sometimes launches a wave-like disturbance in the lower corona that radiates away from the eruption site, as seen in EUV difference images (see Figure 4.39(b)).

Because they were first clearly observed with SOHO EIT, the name "EIT waves" is often used. In addition, "coronal dimmings" may occur in the emissions of neighboring areas. These appear to be different phenomena from the Moreton waves seen in the chromospheric emissions and related to flares. The EIT wave may be an MHD, magnetosonic wave propagating from the site of intense coronal energy deposition during or following the CME eruption process. However, dimmings may also be the magnetic field footprints of a reconnection front following the progression of newly opening and/or closing coronal fields. The interpretation for these features and their relationships is a key part of the investigation of the physics of CMEs.

Understanding the origin and initiation of CMEs is one of the major research goals in solar and space physics because of their importance to understanding "space weather" storms that affect life and technology, as well as planetary magnetospheres and atmospheres. The interplanetary consequences of the CMEs are described in Chapter 5; here we focus on the solar end. Earlier we looked at a basic concept for solar flares, in particular the idea that magnetic reconnection involving a coronal loop or arcade produces particle energization and its consequences. The flare scenario (Figure 4.34) applies to active region scales of ~10^5 km or less across. But a similar picture on a larger scale is also often used to explain CMEs. The difference, besides the scale, is the trigger and the consequences. In the case of the flare, the overall control of the local field geometry by the active region at its base makes any evolution of the active region, such as flux emergence or footpoint motions, a potential trigger. Active regions may also host multiple magnetic null points, or complex neutral lines that enable flux cancellation by diffusion or submergence of opposing fields. In the case of the CME-scale version of the eruptive setup, triggers can also take many forms, but in the CMEs there is a substantial amount of large-scale coronal volume and mass involved.

Figure 4.40 illustrates some CME initiation mechanisms that have been suggested by observations. The common thread here is the introduction

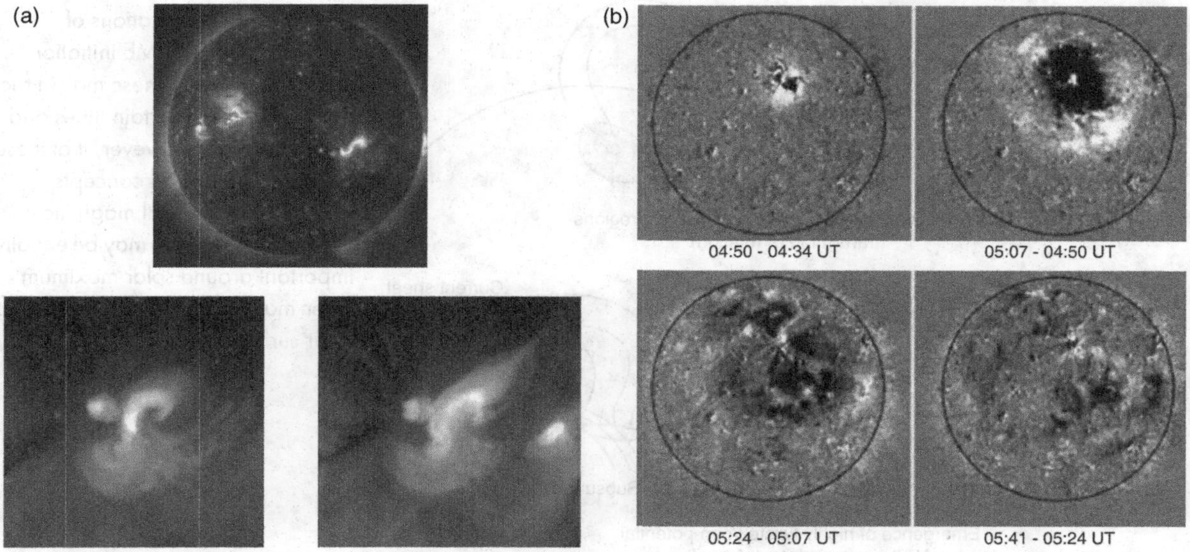

FIGURE 4.39. (a) Image of a soft x-ray "sigmoid," an evidently twisted, emitting coronal loop seen over an active region that often ultimately gives rise to a CME, and the candle flame-like evolution of the sigmoid after an eruption. This appearance suggests involvement of twisted fields that have either emerged or evolved during the event. (b) Example of an "EIT wave" seen in EUV difference images that suggests a disturbance front propagating from a CME site. This is considered to be a physically different phenomenon from similar appearing Moreton waves that occur after large flares in chromospheric emission images. It has been explained as an MHD wave or shock but is considered distinct from other coronal emission "dimmings" that sometimes follow a CME (see text). Both sigmoids and EIT waves are used to predict Earth-directed CME effects. ((a) Yohkoh SXT Images (JAXA/NASA archive); (b) SOHO EIT EUV difference images (ESA/NASA).)

of non-potential or twisted magnetic fields, implying that magnetic stresses and currents are present. Another term that is used to describe these is "helicity," which has a formal definition that essentially describes field twistedness in analytical form, and includes details such as the handedness of the twist.

The so-called streamer blowout CMEs, which do not require an active region association, may derive their stressed/twisted fields simply from the shearing action of differential rotation (Figure 4.40(a)). The slow buildup of stress in these cases may require particularly steady coronal conditions. Thus they may dominate the solar minimum CMEs. Another initiation scenario involves streamers with a multipolar field geometry. For example, if the local coronal field is quadrupolar, a null point or line exists above a central magnetic loop or arcade, as sketched in Figure 4.40(c). In this case, if the central arcade is sheared along its axis or neutral line, it expands and forces reconnection to occur at the overlying null point. The side lobes of the streamer then peel away and the sheared central structure is able to expand or "break out" into the space above, forming the CME ejecta. The breakout mechanism has been shown to explain the observed characteristics of more than a few real events. Even a weak spot or seam in the coronal field overlying a stressed magnetic structure may be enough to enable those fields to erupt, regardless of the details of the field geometry. However, though both footpoint shearing and the presence of magnetic nulls are important considerations in the CME initiation scenarios, newly emerging magnetic flux is generally considered the most common source of highly twisted or stressed fields in the corona (see Figure

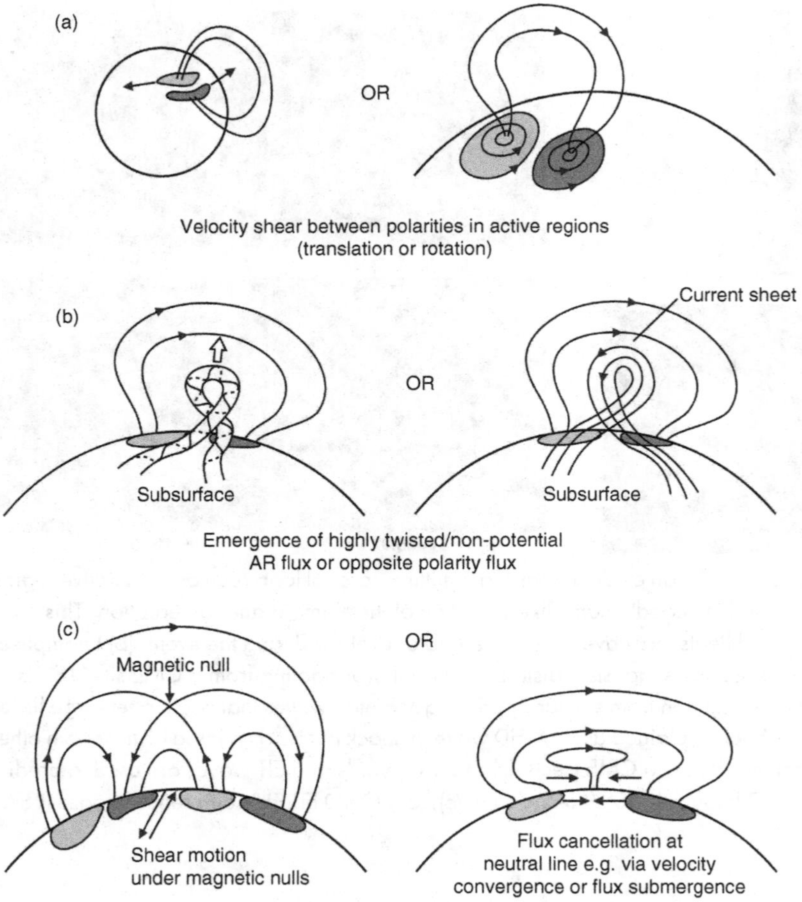

FIGURE 4.40. Illustrations of various proposed CME initiation mechanisms. All of these may in fact be candidates at certain times and locations. Note, however, that these are mainly localized concepts, whereas more global magnetic connection concepts may be equally important around solar maximum when many active regions are on the solar surface at once.

4.40). Fresh twisted flux tubes are imagined to exist above active regions (Figure 4.19), and these usually emerge into surroundings ripe with possibilities for null points and currents at their interfaces. Thus many real CMEs are likely not pure examples of any single concept but a combination of several.

In spite of the increased understanding of both phenomena, the relationship between CMEs and flares remains a subject of many different viewpoints. Timing studies indicate that a flare may either precede or follow a CME, or occur at about the same time. Then there is confusing terminology where post-CME loops and arcades seen glowing in x-rays and EUV are sometimes said to produce gradual flares. Here we have used the term "flare" to refer only to the occurrence of short-lived photon flux enhancements from the chromosphere and low-lying coronal loops within active regions. However, it is good to be aware when reading the general literature that not everyone makes this distinction. Larger, more complex active regions have the necessary ingredients for both flaring and generating CMEs. Another point of interest is that CMEs and flaring active regions can be remotely connected, so that a flare may occur as a CME takes off some distance away or vice versa. Sometimes it appears that solar hemisphere-scale or even global complexes of magnetically interconnected regions exhibit combinations of these two phenomena, suggesting coupled triggering or sympathetic activity. As might be expected, this is more likely when there are a significant number of active regions on the disk. In general, and especially at solar maximum, the

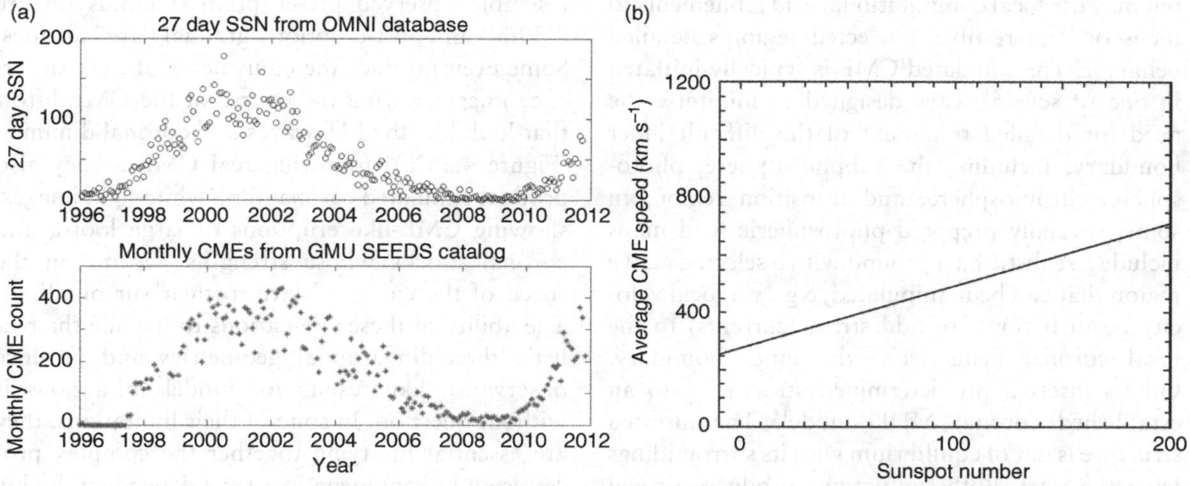

FIGURE 4.41. (a) The 27 day (solar rotation) sunspot number (top) from the OMNI data base (OMNIWeb Plus, 2015) compared to a monthly CME rate based on events observed on the SOHO spacecraft by the LASCO coronagraph (George Mason University website SEEDS catalog at http://spaceweather.gmu.edu/seeds/). (b) A rough sketch of the trend of average CME speeds with increasing sunspot number. The greater number of faster CMEs occurring when sunspot number is high may reflect the transition from mainly streamer blowout type events around solar minimum to mainly active region related events around solar maximum. However, this trend may result from a few very fast events more than an overall increase.

corona may be envisioned as a complex network of stressed fields with shared topological connections sensitive to perturbations anywhere. The result is a CME rate that has the same basic trend as the sunspot cycle (see Figure 4.41(a)). But it is not surprising that most case studies of real CMEs focus on isolated events involving single active regions, as even these are extremely challenging.

At its most basic level, the CME problem, like anything else in the corona, is about force balance and processes than can work to restore them. In principle, anything that creates a coronal structure out of force balance with its surroundings may lead to a CME as part of the readjustment process. One can also think of the CME process as the corona's way of returning to a more potential, lower-energy state. A related question of interest in connection with the CME initiation problem is what determines their size, shape, speed, and magnetic field geometry. Any theoretical treatment or model must be able to explain the range of appearances and behavior, some of which are seen in Figure 4.35. Several additional observational results that have physical implications are: (1) The great majority of CMEs start slow, <300 km s^{-1} at the Sun, although they may accelerate as they move out. (2) The average apparent CME angular width near the Sun is ~50°. (Remember the "360°" wide halo CMEs are an artifact of the viewing geometry rather than a true measure of their width.) (3) CME speed and width are positively correlated. (4) The largest, fastest, most explosive CMEs are seen to reach their speeds of several thousand kilometres per second very close to the Sun. In addition, average CME speeds tend to increase with sunspot number (a rough sketch of this dependence is shown in Figure 4.41(b)), although this may reflect the effect of occurrence of a few very fast events rather than an overall increase. The reason why a few CMEs reach exceptionally high speeds of up to several thousand kilometres per second is still a major topic for investigations.

State-of-the-art models of CMEs use numerical MHD approaches that include the global corona but may use local computational grid refinements to focus on and resolve a selected region's detailed behavior. The simulated CME is typically initiated in one of several ways designed to minimize the need for detailed treatment of the difficult inner boundary, including the subphotosphere, photosphere, chromosphere, and transition region. In some, specially prepared photospheric field maps include a realistic background with a selected active region that can be manipulated, e.g., via local velocity perturbations, to add stress (currents) to the local coronal fields from the inner boundary. Others insert a pre-determined structure into an established coronal MHD model. The inserted structure is out of equilibrium with its surroundings from the start. Both simulations produce coronal transients with various sizes, shapes, speeds, and styles of evolution, including interactions with their surroundings. The fact that even these simplified treatments produce a remarkable variety of CME-like phenomena provide valuable lessons whether or not they reproduce a specific observed event. In particular, once they are launched, few retain the idealized croissant-shaped magnetic flux ropes often invoked to explain the coronagraph images. The simulated ejecta usually contain a flux-rope-like field structure, but twists and kinks can develop as they are released, and then interactions with their surroundings further modify what was ejected. Thus one should be cautious about fixating on a single simplified picture of CMEs to describe all events, with the possible exception of the relatively modest streamer blowout events. Those, at least, look roughly compatible with what is produced in simulations by shearing the footpoints of an extended streamer arcade.

Do numerical simulations of CMEs shed any light on the CME–flare relationship? While they do not (yet) include the physics and spatial resolution to reproduce details of flare emissions, such as the flare ribbons, they do involve reconnection processes on many scales. One pervasive concern is that the simulation of reconnection physics is parameterized or numerically determined, and thus different for every model and not necessarily realistic. Nevertheless, some do produce features in the process of the CME initiation and eruption, whose simulated emissions in EUV and soft x-rays resemble observed pre-eruption sigmoids (Figure 4.39(a)) and post-eruption "gradual flare" arcades. Some even produce the equivalents of EUV difference images during the course of the CME liftoff that look like the EIT waves and coronal dimming (Figure 4.39(b)) following real CMEs. They also produce simulated coronagraph white-light images, showing CME-like eruptions of large loops, and coronal shocks whose strengths depend on the speed of the ejecta relative to their surroundings. The ability of these simulations to include the realistic three-dimensional geometries and produce observation-like results for model validations is without question. In spite of their limitations, they are essential for tying together the complex processes and phenomena in a time-dependent, highly structured corona.

4.6.3 Shocks

In the introduction to collisionless shocks in plasmas in Chapter 6, we learn that the steepening of MHD waves into these discontinuities depends on the combined presence of a wave source plus ambient conditions in which the speed of the wave front exceeds the magnetosonic velocity. In the case of the corona, where Alfvén speed dominates the magnetosonic speed, the strongest shocks occur in connection with flares and/or CMEs. There are several proposals for the shock sources, including the explosive pressure enhancements from localized flare heating ("blast waves"), and the bow waves formed ahead of CME ejecta (which act as drivers or pistons). Some consider that both may exist when an associated flare and CME eruption take place, creating a two-shock picture.

Evidence of the occurrence of coronal shocks is not always present in the observations, e.g., as a Moreton wave or EIT wave (see earlier discussions). Values of the Alfvén and magnetosonic speeds near the compression source and along its propagation path can take a wide range of changing values in the real corona. Figure 4.42 shows a sketch of the magnetosonic speed versus altitude for a particular trajectory through a simple density and magnetic field model of the corona. Typical values for both closed field/active regions and open field, coronal hole conditions are estimated here. The local minimum in the trend at ~$1.5R_S$ reflects the changeover, with the

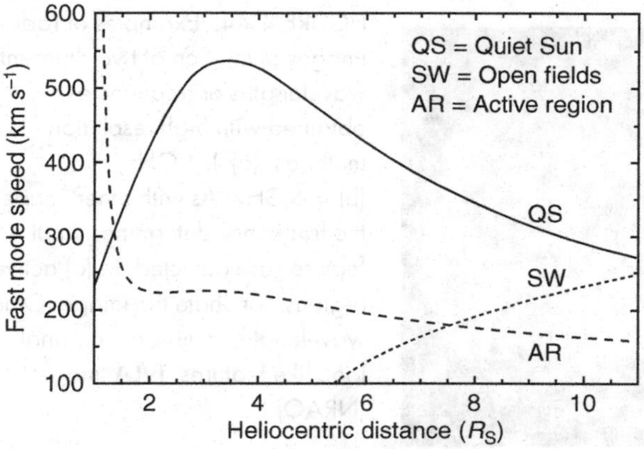

FIGURE 4.42. Altitude dependence of magnetosonic (fast mode) speed in the corona, showing the different radial trends for closed field/active region sites and coronal hole regions. This contrast makes a radial minimum in the critical speed that may cause a coronal shock to fade out and then return as its CME driver propagates outward. This behavior is an important consideration for solar energetic particle acceleration by coronal shocks. ((a) After Gopalswamy and Kaiser, 2002; (b) adapted from the CDAW SOHO LASCO CME catalog (S. Yashiro, NASA).)

implication that a steepening magnetosonic wave taking this path may form a shock at low coronal heights which weakens and then reforms or strengthens again at larger distances. This may explain the occasional inference of a "two-shock" picture suggested by some observations, without invoking two separate sources like flare heating and a CME.

One important way to detect and characterize coronal shocks remotely is via radio emissions. **Solar radio emissions** have been observed essentially as long as radio waves have been detected. The interested reader can find entire books, including texts, devoted to solar radio astronomy; however, here we cover only the types of emissions that are often mentioned in discussions of coronal diagnostics and solar activity. Solar radio noise, seen at the low-frequency end of the extended solar photon spectrum in Figure 4.43, takes over from the end of the infrared (~10^{12} Hz) and includes millimeter (~100 GHz), centimeter (10 GHz), and meter (~100 MHz) wavelengths (frequencies). A variety

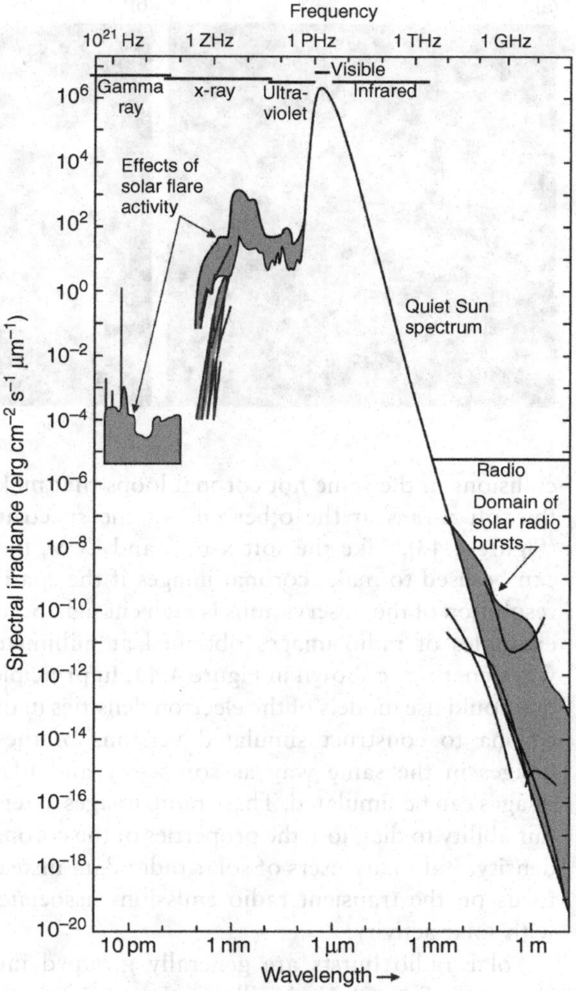

FIGURE 4.43. A more expansive view of the overall solar emission spectrum, with greater emphasis on its more variable extremes. This gives a perspective on the intensities in the x-ray/EUV and radio wavelength ranges compared to the visible. Radio bursts are seen to be relatively weak signals among the solar emissions. (Adapted from Golub and Pasachoff, 1997.)

of antenna types and detectors are thus required to span this range, although even a simple commercial receiver can sometimes detect the effects of intense solar activity.

The quiet corona regularly emits radio waves in the millimeter and centimeter wavelengths due to the thermal bremsstrahlung process. These emissions come from electrons undergoing glancing

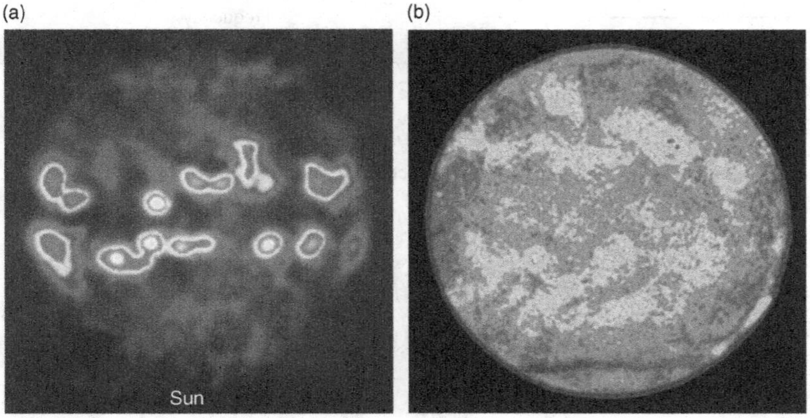

FIGURE 4.44. Examples of radio images of the Sun at two different wavelengths or frequencies obtained with high-resolution methods: (a) 1.4 GHz; (b) 4.6 GHz. As with other images, the frequency determines what features are detected; in (a) active regions dominate the image. Other wavelengths may show coronal loop-like features. (VLA images (NRAO).)

collisions in the same hot coronal loops that make the soft x-rays at the other end of the spectrum (Figure 4.43). Like the soft x-rays and EUV, they can be used to make coronal images if the spatial resolution of the observations is high enough. Some examples of radio images obtained at millimeter wavelengths are shown in Figure 4.44. In principle, one could use models of the electron densities in the corona to construct simulated versions of these images in the same way as soft x-ray and EUV images can be simulated. These radio images extend our ability to diagnose the properties of the coronal density. But many users of solar radio data instead focus on the transient radio emissions associated with solar activity.

Solar radio bursts are generally grouped into classes as Types I–V. An illustration showing the characteristics of the most common emissions and their timing relationships relative to x-ray events is shown in Figure 4.45 in a frequency versus time display (a "dynamic spectrum," or spectrogram). Type I "storms," occurring at a few hundreds of megahertz frequencies, are not tied to individual flare or CME occurrences but are attributed to emission from energetic electrons trapped in coronal loops at times when the Sun is generally active. Of particular interest here are the Type II and Type III events, which we discuss in reverse order because of their relative timing (Figure 4.45). The Type III events at tens of megahertz occur at the time of the initial impulsive flare x-ray bursts, sweeping rapidly from high to low frequency at the local plasma frequency and its harmonics, suggesting a source traveling outward through the diminishing coronal density. They are attributed to gyro-synchrotron emission (similar to the process responsible for jovian radio emissions) by electron beams with energies in the tens of kiloelectronvolts. These are considered evidence of electron acceleration well beyond thermal levels in or near the reconnection site (see Figure 4.34), and are associated with impulsive onset suprathermal electron events detected near Earth following a flare (more on this topic in Chapter 5). A possible emission mechanism is the generation of plasma waves via an instability associated with the streaming electron subpopulation (Chapter 13), followed by a mode conversion process that changes some of the plasma waves to radio waves. This makes them a useful remote-sensing diagnostic of the acceleration processes in the flare site itself, and moreover suggests that there are open coronal field lines at or near the flare site on which the flare-accelerated electrons stream outward. Type V emission is associated with Type III bursts (Figure 4.45), following at lower frequencies, but is not generally discussed as a separate diagnostic.

Type II bursts in contrast have a relatively slow (~hours) high- to low-frequency dispersion in radio dynamic spectra at frequencies below ~100 MHz. These are sometimes seen to coincide with the relatively long-lived ("gradual") soft x-ray flare that follows a CME release. As suggested by Figure 4.45, these generally follow any Type III burst that occurs at the event onset. Their behavior indicates that the sources of the Type IIs,

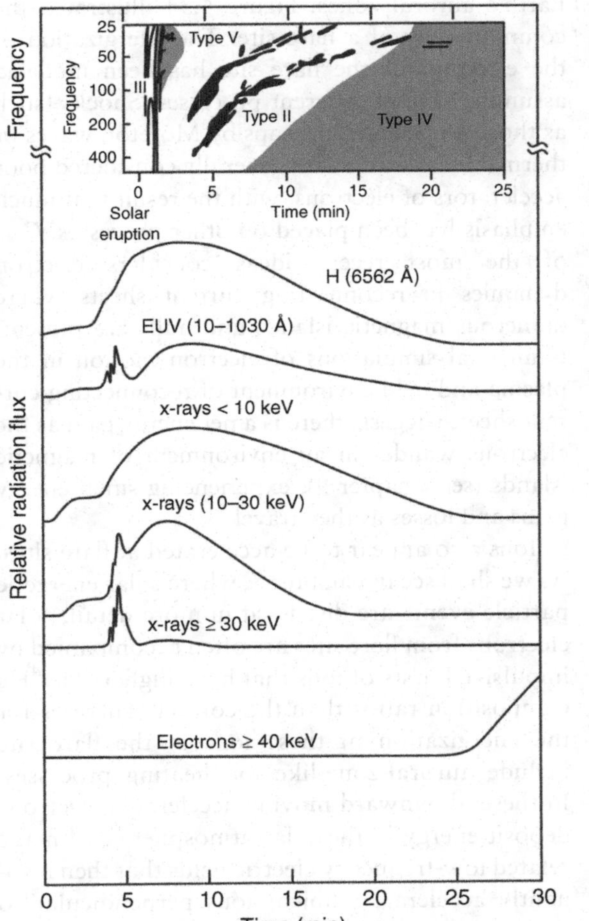

FIGURE 4.45. Illustration of the common features of a solar radio dynamic or frequency–time spectrum and their timing relationships to other solar activity phenomena and emissions. The proposed physical sources of the various radio bursts with their characteristic frequency bands and sweeps are discussed in the text. The fast sweep Type III emissions are often-used indicators of escaping flare-accelerated electrons, while the slower sweep Type II emissions are often associated with traveling CME-driven shocks. (Time lines of various emissions after Lang, 2010; radio burst dynamic spectrum inset after Palmer, Davies, and Large, 1962.)

which also track the coronal plasma frequency profile, are longer lived than the flare itself. Coronal shock fronts are one candidate. Type IIs are thus often used as an indication that a CME has occurred, and their timing conjunctions with the gradual soft x-ray flares further support this interpretation. Electrons accelerated at the shock front are the probable cause of the radio emissions through the same plasma wave to radio mode conversion process as for Type IIIs. The Type II drift rates are also consistent with the shock exciter of the plasma emissions traveling at the order of, or faster than, the local magnetosonic speeds. Type IV emissions appear as more of a continuum and are thought to arise from flare-accelerated electrons that remain trapped in the flare site coronal fields following the main event(s) of the solar activity. It is worth noting that when several radio receivers are available to observe the Sun from widely different perspectives (e.g., at separated spacecraft locations), spatially resolved radio signals can be used in triangulation mode to track moving, emitting features such as the shocks and flare-generated electron streams on open field lines.

4.6.4 Energetic particle acceleration

In reading the broader literature, the reader will see that there is sometimes a fine line between what is considered "heating" and what is considered "acceleration." Both imply the addition of energy to particles, although in the first case it is usually to the general or bulk population while in the latter it is usually to a small fraction of the particles present. "Heating" also usually implies that the resulting particle velocities or energies have a Maxwell–Boltzmann distribution that is broadened by some process. On the other hand, acceleration mechanisms generally produce changes in the bulk velocity or "non-thermal tails" on the particle distributions that may have a quite different energy spectrum than the main population, such as a power law. The thermal population is also often viewed as isotropic or perhaps having different temperatures parallel and perpendicular to the background magnetic field, while the non-thermal tail of the distribution may be highly anisotropic or non-gyrotropic, or both. Yet a number of the coronal "heating" mechanisms discussed earlier closely resemble acceleration mechanisms in their physical descriptions. These include two basic types: stochastic processes, where energy gain (and loss) occurs in a statistical fashion, and electric field acceleration. The former may involve resonant and/or non-resonant particle interactions with waves (see Chapter 13) and/or reflections due to magnetic mirroring and interaction(s) with shocks (see Chapter 7). The latter may involve localized double layers (see Chapter 11) or other special topological locations, such as magnetic x-points, nulls, or islands, where the different ion and electron gyro-radius scales come into play (see Chapter 9). Whether these processes end up producing particle energization that is redistributed and appears as bulk "heating" or an energized minority population that remains distinct depends on the ambient conditions.

As we have seen, the coronal environment is complicated, so it is fortunate that in many cases energetic particle acceleration can be considered from a test-particle perspective, rather than a collective plasma one. Recall that in Chapter 3 we discussed plasma collective behavior related to the Coulomb forces that couple the motions of ions and electrons to maintain overall charge neutrality. The smaller populations of suprathermal and energetic electrons and ions that concern us here instead can be envisioned as independently moving in the force fields provided by the background corona. In Chapter 3 the equation of motion of individual charged particles was shown to be relatively simple, involving mainly the Lorentz force $\mathbf{E} + \mathbf{v} \times \mathbf{B}$ in cases where gravity can be neglected. The complicated part is solving it when the description of the electric and magnetic fields is both spatially complex and time dependent, and even before that, describing the fields in the first place. In addition, relativistic modifications to the particle equation of motion are sometimes necessary. In spite of these challenges, the charged particle equations of motion for both electrons and ions have been solved for a variety of coronal plasma and field scenarios toward better understanding how a few can obtain such high energies.

The brightenings in EUV and x-ray emissions that are the hallmark of flares provide remote diagnostics of flare site particle acceleration. In particular, as pointed out above, energized electrons are likely responsible for the H-alpha flare ribbons, Type III radio bursts, and x-ray emissions from flare loops, with analogies sometimes drawn to Earth's auroral zones. Figure 4.34 illustrated the common view of a flare site. The energization of the electrons at the flare site has been modeled assuming several different processes. Shocks (such as those represented perhaps by Moreton waves or thermal blast waves) are generally considered poor accelerators of electrons, with the result that much emphasis has been placed on other processes. One of the most recent ideas considers electron dynamics in reconnecting current sheets where numerous magnetic islands and nulls are present. Numerical simulations of electron motion in the plasma and field environment of reconnecting current sheets suggests there is a net energy gain as the electrons wander in an environment of magnetic islands (see Chapter 9), experiencing small energy gains and losses as they travel.

Ions also appear to be accelerated at flare sites. As we shall see in Chapter 5, where solar energetic particle events are discussed in more detail, solar electrons from flare sites are often accompanied by impulsive bursts of ions that have higher ^{3}He/^{4}He composition ratios than the corona. Theories for the energization of these ions at the flare site include auroral-zone-like ion heating processes. In these, downward moving accelerated electrons deposit energy in the solar atmosphere and make related low-frequency electric fields that then resonantly accelerate ambient ions perpendicular to the magnetic field (see Chapter 11). The heated ions are then driven outward by the mirror force associated with the rapidly diverging open coronal magnetic fields in the vicinity. In this scenario, ^{3}He is favored because the spectral content of the related electric fields has higher power at its gyrofrequency. While there are also other ideas on this topic, the resonant wave-heating hypothesis continues to be invoked because it provides a relatively well understood mechanism for ion mass-selective acceleration.

Another remote diagnostic of ion acceleration at flare sites in extreme cases is the detection of nuclear gamma ray lines in the electromagnetic spectra of some particularly intense events. The gamma rays are presumably excited by collisional excitation of the nuclei at the flare site, which then subsequently de-excite by photon emission. As it is

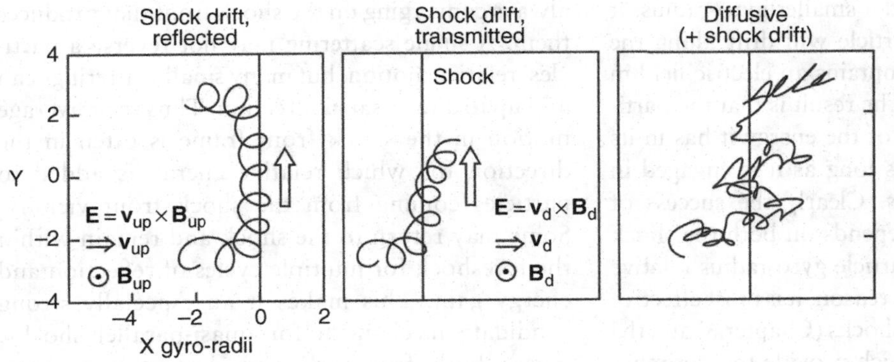

FIGURE 4.46. Illustration of shock acceleration processes involving the shock gradients in plasma parameters. These include the more statistical diffusive and Fermi accelerations and the shock drift mechanisms. (After Decker, 1988.)

much more difficult to excite nuclear energy levels, these events are relatively rare. These same types of events may also give rise to the most energetic solar energetic particle (ion) events called GLEs for ground level events. The GLEs have such a hard (large high-energy flux) proton spectrum that their effects on Earth's atmosphere can be detected at the surface in the form of showers of secondary particles extending to mid latitudes. A major debate in this area of research concerns the question of whether the GLE particles are accelerated in the flare site by an extreme version of the flare ion heating processes, or whether they are accelerated at shocks that form low in the corona in association with the major flare or a CME accompanying the flare. GLE events will come up again in the following chapter in the context of the discussion of the larger picture of solar energetic particle events.

The energization of a fraction of the solar plasma particles in the vicinity of collisionless shock waves has been a subject of space physics research for decades, and especially following the ISEE mission measurements in the late 1970s. The ISEE mission probed the details of the Earth's bow shock and foreshock (see Chapters 1 and 6), providing ion and electron distribution functions and electric and magnetic field measurements with sufficient detail to reveal some of the processes by which shocks produce suprathermal particles. These suprathermals can be regarded as a by-product of the shock formation because they require either the presence of the waves (for statistical acceleration) or the presence of the shock ramp gradients in the plasma parameters and magnetic field (electric field or shock drift acceleration). Before we look at some of the proposed mechanisms for shock-related particle acceleration, we caution the reader that these processes are described differently by different authors. Some of the differences involve the adopted reference frame, while others involve a historical viewpoint. Here we try to provide descriptions of the most basic concepts only, providing additional references at the end for the interested reader.

The sketches in Figure 4.46 illustrate two specific forms of shock-related acceleration that are considered viable candidates for explaining most observations. These are drawn from the perspective where the shock, shown as a plane in cross section, is stationary. The so-called **shock drift mechanism** involves interaction of a particle with the shock itself, and is most easily described for a purely perpendicular shock (see Chapter 6 for a discussion of shock types and their characteristics). Note that the shock is not a wall and that the real scenario involves the particle's motion in the plasma and field gradients associated with the shock ramp. However, here we can presume that upstream conditions exist on one side of the shock (left here) and downstream conditions on the other side (right). As a particular particle gyrates in the ambient fields, it oscillates in velocity around the inflow bulk speed V_{up} until it encounters the shock front in its path toward the right at a particular phase in its gyration. On the other side the ($\mathbf{E} = -\mathbf{v} \times \mathbf{B}$) electric and magnetic fields are different, and effectively reinitialize the particle motion at the shock but with non-zero velocity and a smaller gyro-radius. It is easy to see that the particle will drift along the shock plane, which also contains an electric field in the direction of the drift. The result is that the particle gains energy in excess of the energy it has in its gyrations on either side as long as it is engaged in shock-crossing trajectories. Clearly the success of shock drift acceleration depends on both the shock front geometry and the particle gyro-radius relative to the shock ramp. For this reason, it is most effective for narrow perpendicular shocks (Chapter 6) and the larger gyroradius ions, which provide the optimum scaling.

So-called **diffusive shock acceleration**, a form of Fermi acceleration, involves multiple particle reflections with many related small energy changes, and as such is considered a statistical acceleration process. The basic idea is the same as the more general converging reflectors picture that can be found in introductory physics texts. A ball elastically colliding with a moving wall will increase its velocity by the wall velocity if its reflected motion is in the direction of the wall motion, and lose that amount if the wall is moving away. If the moving wall can go both ways but has a statistically greater chance of moving with the reflected ball, energy can be gained at the expense of the moving wall energy. The reflectors in the case of diffusive shock acceleration include the shock itself, but also the field fluctuations in and around it. For example, foreshock field fluctuations represent particle scatterers that are always converging on the shock front that produces them. A single scattering may not reverse a particles' relative motion, but many small scatterings can add up to a reversal in direction. Thus their average motion in the shock front frame is often in the direction for which relative energy is added to particles coming from the shock front vicinity. Some may return to the shock and remain within the foreshock for multiple cycles of reflection and energy gain. This makes it an especially strong candidate mechanism for quasi-parallel shocks, where both foreshock and downstream (post-shock) waves and field fluctuations are often ubiquitous and large in amplitude. This process also depends on the particle species because the ability of the shock and/or field fluctuations to alter their trajectories depends on the gyro-radius relative to the scales of these scattering/reflecting structures. Ion scales are often more amenable to this mechanism.

So it seems as if current ideas of particle acceleration in the corona favor electron acceleration in flare reconnection site scenarios and ion acceleration by shocks or, under special circumstances, by wave–particle interactions. Further discussion of solar particle acceleration, including information about energy spectra, pitch angle distributions, and ion composition, requires that we move into areas including particle transport and *in situ* measurements at 1 AU that are best placed in the context of Chapter 5.

4.7 SUMMARY

In a few words, in this chapter on the Sun we started with a picture of the solar interior structure and rapidly proceeded into the solar atmosphere, which provides the most direct connections between solar and space physics interests. We learned that the solar magnetic field, generated by some dynamo-type action in the conducting, convective solar interior, produces the cyclic behaviors that are observed on the solar surface and in the solar atmosphere above it. The solar atmosphere undergoes a remarkably fast transition within the chromosphere and transition region from an initially cooling, heavily neutral and collision-dominated state in the photosphere where the dynamo-generated field is subject to its dynamics, to one that is fully ionized, collisionless, and magnetically controlled. The effective magnetic skeleton of the solar corona, the Sun's outermost ionized, mostly hydrogen atmosphere observed by coronagraphs through Thompson scattering of visible light by the electrons, determines much of what affects us on Earth outside of sunlight. We reviewed some recent ideas about how the hot coronal plasma that fleshes out the magnetic skeleton is heated and how that relates to imaged features in the EUV and x-ray wavelengths. The existence of coronal holes in these images that result from outflow of coronal plasma at locations where its pressure gradient force exceeds any confining force of the magnetic field is central to the subject of the following chapter. We moved on from a relatively steady, quiet corona picture to one in which transient, dynamic events occur, considering what occurs in the solar surface magnetic field that leads to the eruptive state. We noted that flares and coronal mass ejections (CMEs), the two most energetic classes of coronal transients, can be thought of as solar cycle-related evolutionary behaviors. We considered other outcomes of this activity, including the emissions of radio bursts and the production of shock waves and energetic particles. We finally reached the point where a natural transition to the phenomenology and physics of the even higher-altitude solar atmosphere, the solar wind, is in order. However, it has been clear throughout this chapter that significant gaps remain in the overall physical picture developed here. From the workings of the solar dynamo to coronal heating to the reason(s) for extremely fast CMEs, there is plenty of important fodder for future research. Readers who found the topics of this chapter especially interesting are encouraged to follow up on their own using the many resources now available in both print and digital media, including a few that are listed below.

ADDITIONAL SOURCES ON THE WEB

The websites below have proven to provide informative solar archival data:

solarscience.msfc.nasa.gov/images/internal_rotation_mjt.jpg. Images provided by M. J. Thompson.

mlso.hao.ucar.edu/hao-eclipses.php. High Altitude Observatory eclipse data.

ase.tufts.edu/cosmos/. Images by K. R. Lang.

cdaw.gsfc.nasa.gov/CME_list/. Images collected by S. Yashiro.

soi.stanford.edu (SOHO MDI data archive access). Archive of SOHO MDI images.

Additional reading

ASHWANDEN, M. (2006). *Physics of the Solar Corona.* Springer. This book contains a comprehensive survey of research results up to the time of writing, with many details for both interested students and researchers.

GOLUB, L. and J. PASACHOFF (2010). *The Solar Corona.* Cambridge: Cambridge University Press. An updated introduction to the solar corona that takes into account the research carried out to date by dedicated solar missions and ground-based investigations. Ideal for exploring more about subjects only briefly touched upon here.

PRIEST, E (2014). *Solar Magnetohydrodynamics of the Sun.* Cambridge: Cambridge University Press. Provides access to the theoretical and mathematical aspects of the physics underlying the study of the Sun.

SCHRIJVER, C. and G. SISCOE (2010). *Heliophysics, Vol. 1.* Cambridge: Cambridge Univeristy

Press. A compendium of lectures based on a summer school held in Boulder, Colorado, over a number of years, with contributions from experts in the field. Useful perspectives on the state of knowledge in subareas.

Problems

4.1 The solar radius is 6.960×10^5 km; the Sun has mass 1.989×10^{30} kg; it emits 3.9×10^{26} J s^{-1}, and rotates with a 25.34 day period. Calculate the acceleration due to gravity at the surface of the Sun and the escape velocity from the solar surface. (Note that a neutral particle of mass m that leaves the surface of the Sun in a radial direction moving at the *escape velocity* will reach infinity with zero kinetic energy.) If all the energy emitted comes from fusion in the core, how much mass is burned (converted into energy) per second on the Sun? What is the energy flux from the Sun at 1 AU (149.6×10^6 km)? What is the apparent rotational period of the Sun as viewed from the Earth?

4.2 Some investigations have suggested that solar energetic particles (SEPs) can become temporarily trapped in closed coronal magnetic field loops, where they can radiate by exciting emissions of various kinds, and leak out slowly, altering their arrival times in interplanetary space.

(a) Assume the solar corona has a dipolar magnetic field structure with an equatorial field strength of 10 Gauss (10^{-3} Tesla). Calculate the gyro-radius of a 100 keV proton, and also of a 10 MeV proton, at the equator at 1.1, 1.5, and 2.0 solar radii. Next compare the 100 keV proton gyro-radius with that of a 100 keV electron and a 100 keV alpha particle (He^{++} ion). Could any of these particles drift around tht Sun due to gradient-curvature drift (see chapter 10)? [Hint: For each equatorial distance compare the gyro-radius-scale excursions inward and outward from the guiding centers of these particles.]

(b) The coronal magnetic field contains fluctuations from many sources that can scatter and further accelerate SEPs via wave–particle interactions. Gyro-resonance often plays a role in both processes. What fluctuation (or wave) frequencies would resonate with the particles in part (a) at $2R_{\text{Sun}}$? Collisions can interfere with a gyro-resonant interaction (as well as charged particle trapping and drifts). Given what you have read in this chapter about the collision frequencies in the corona as a function of height, would gyro-resonant processes be effective for protons at the chosen solar altitudes above the surface? [Hint: See Figures 4.6 and 4.7.]

4.3 Show that equation $\frac{\partial B}{\partial t} = \eta \frac{\partial^2 B}{\partial x^2}$ possesses solutions of the form $B(x)$, with $x = t^n$, only when $n = -1/2$. Solve the equation for $B(x)$, with $n = -1/2$, subject to the conditions that $B \rightarrow B_0$ when $X \rightarrow -\infty$ and $B \rightarrow B_1$ when $x \rightarrow \infty$.

4.4 A classic model for the support of a prominence assumes a uniform temperature T, a horizontal field B_x, a vertical field $B_z(x)$, and a pressure $P(x)$ that vary with horizontal position, x. Horizontal and vertical force balance is then expressed by $P + B^2/2\mu_0 = \text{constant}$ and $\rho g = \frac{dB_z}{dx}\frac{B_x}{\mu_0}$. Show, given $\rho = p/RT$, that $B_z = B_0 \tanh(x/l)$ and $P = (B_0^2/2\mu_0)\,\text{sech}^2(X/l)$. What is the width of the prominence, l? Find the solution for the magnetic field and pressure in a prominence for which the temperature is a given function of x.

5 The solar wind and heliosphere

5.1 INTRODUCTION

In the preceding chapter we learned that solar plasma outflows arise as a natural result of the corona's physical state. Simply put, the corona is the hot, mainly hydrogen atmosphere of the Sun, heated at its base and surrounded by relatively empty space. In spite of the Sun's deep gravitational well, the thermal pressure gradient forces and ionizing photon flux together produce an extended, highly ionized corona with a fraction of its thermal speed distribution above the Sun's 618 km s^{-1} escape velocity. But the nearly complete ionization of the corona and the solar magnetic field together ensure that this atmosphere and its escaping component are not spherically symmetric. The coronal magnetic field is strong enough to confine the hot gas in magnetic loops and arcades rooted in the photosphere, but it cannot contain it globally. Coronal field loops that reach altitudes of at least a few solar radii are forced to expand outward by the plasma pressure gradients that essentially blow the field open. Additional physics beyond normal coronal heating imparts even higher bulk velocities to the plasma on these open coronal fields than the temperature gradients can alone produce. The reduced coronal densities from this non-uniform escape are seen in coronagraph and eclipse images as dark channels between the bright coronal rays rooted in the still closed fields, and as dark coronal hole footprints among the bright loops in soft x-ray and EUV images (e.g., see Figure 4.27). But coronal hole outflows are only part of what makes up the extended solar atmosphere. The patterns of closed fields and coronal holes in the corona are in a constant state of evolution as the photospheric field evolves through the solar activity cycle. As we shall see, these changes and their consequences are an important part of the modern picture of the "solar wind."

A major part of the field of heliophysics is focused on understanding the solar wind. Phenomena such as auroras, geomagnetic storms, and the conditions in the radiation belts, our local "space weather," all depend on the interaction of the solar wind with Earth's intrinsic magnetic field and atmosphere (see Chapter 1). Similarly, the solar-wind interactions with planetary magnetospheres and ionospheres, the Moon, comets, and asteroids, also described elsewhere in this book (Chapters 7 and 8), represent other natural experiments that give us perspective on our own planet in the larger context of the solar system. An expression sometimes used to describe these areas of study is "living with a star." Stellar winds like the solar wind, as well as more extreme versions, are inferred to exist during the evolution of most stars. In fact, the solar wind is a rather weak stellar wind compared to those of some young stars, with the consequence that solar-wind-like stellar winds are extremely hard to detect remotely. Nevertheless, they are gaining new interest in the growing field of extrasolar planet studies, in part because they may affect the remote sensing of other worlds. These studies may also teach us more about our own evolving Sun's effects on our solar system through time.

In this chapter, basic solar-wind observational results and concepts are introduced, including some simplified models of its coronal connections. Our current understanding of these topics is based on a several decades' history of theory and space- and ground-based observations, including both solar imaging and *in situ* measurements. We consider what is known about the temporal and spatial evolution of the solar wind with solar activity and heliocentric distance, respectively. Many of these details of the heliospheric plasma and field behavior come from *in situ* measurements on a long line of near-Earth spacecraft, as well as others obtained from planetary missions en route to their destinations. Special perspectives on conditions into ~0.3 AU came from the twin spacecraft Helios mission in the 1970s, on the high heliospheric latitude solar wind from Ulysses in its near solar-polar orbit, on the global perspective at 1 AU from STEREO, and on the farthest reaches of the heliosphere from the originally planetary Voyager mission (and the other missions listed in Tables 1.3 and 1.4).

The reader may find it useful to adopt the viewpoint that solar cycle changes in the solar wind are part of a continuum of time-dependent behavior, ultimately tied to the solar dynamo. We learn that the coronal source makes a solar wind that is often unsteady, and always spatially non-uniform, sometimes resulting in patterns of behavior that are predictable in some respects. This is important because what we see in the ecliptic at the heliocentric distances of solar system bodies is not always representative of the global conditions. While we do not attempt to address the long-term history of the solar wind, at the end of the chapter we briefly consider the larger heliosphere that includes the volume the solar wind carves out in interstellar space. In fact, the solar wind may be considered to define the outer limits of our solar system. Beyond the heliopause is the interstellar wind. The location of the heliopause is largely determined by the balance between interior solar-wind dynamic pressure and the external local interstellar medium pressure. However, it is permeable to certain constituents, including the flowing interstellar neutral gases and dust in the Sun's neighborhood. This has especially significant effects on the solar wind, beginning roughly outside the orbit of Jupiter, although its influences are evident in some 1 AU observations as well. As part of the big picture we also consider the main features of both solar energetic particle populations and cosmic rays, the high-energy tail of the heliospheric particle spectrum. As in Chapter 4, the aim is to provide the reader with a broad background on the subject with some perspective on the still outstanding questions.

5.2 THE SOLAR WIND: THE EXPANDING OUTERMOST SOLAR ATMOSPHERE

The introduction to this book briefly reviewed the historical development of ideas on what exists in the space around the Earth. Early thinking on the reasons why occurrences of auroras and geomagnetic storms followed the solar cycle was influenced by spectral line observations suggesting the corona was hot enough for its ionized gases to escape into space. Initially, debate focused on whether this escape was steady or sporadic in nature, but the observation that comet tails essentially always had a nearly radially directed feature not attributable to light pressure alone reinforced the idea of a constantly present particle outflow. E. N. Parker first recognized that the conditions should exist for a fluid-like pressure-driven **solar wind**, in which the plasma thermal pressure gradient forces are sufficient to overcome solar gravity and thus drive a robust, supersonic outflow – a magnetically channeled version of hydrodynamic escape on open coronal field lines (see Chapter 4). In contrast, J. W. Chamberlain envisioned an exosphere-like escape of the ionized coronal gas similar to what occurs in the uppermost atmospheres of planets where collisions become unimportant (see Chapter 2). In this case, the role of a polarization electric field due to the differing masses of the ions and electrons is key in enabling the outflow (see Chapter 11). But the fluid picture predictions matched the first measurements in space outside of the Earth's sphere of influence so well, especially the observed high bulk velocities of the plasma and the behavior of the interplanetary magnetic field, that the decades of research that followed have primarily used MHD fluid descriptions of the solar wind. Nevertheless, as in much of space plasma physics, the particle kinetic aspects are not completely out of the picture, as we shall see later.

A simplified MHD treatment of a solar dipole field controlled corona was described in the preceding chapter. Details of the coronal heating processes were neglected and it was simply assumed the corona was uniformly heated to ~1 MK consistent with coronal spectral line observations.

The configuration of the resulting corona included a coronal helmet streamer structure confining low- to mid-latitude coronal plasma inside a few solar radii inside an equatorial arcade or belt of closed field loops, and outflows on the higher-latitude fields opening into the surrounding space. At larger radial distances, where the channeling effects of the coronal magnetic field give way to control by the fluid forces, much can be learned from Parker's simplified isothermal solar-wind model.

A form of the single-fluid equations for a spherically symmetric isothermal (T = constant) atmosphere is found in virtually every introduction to the physics of the solar wind because it includes the essentials of the problem: the spherical geometry with an assumed "source surface" of radial outflow at an inner boundary, and the Sun's gravitational field. Even though we know the real coronal source produces a rarefied ionized gas with complicated particle distribution functions and not a collisional, isotropic maxwellian (as is assumed in most fluid approximations), and that coronal holes represent a highly structured solar-wind source geometry (we return to these matters later), and that the outflows may not be entirely radial, this simple picture provides important insights into how the solar wind works. Note that here the magnetic field role is implicit, in this case hidden in the radial outflow assumption. So these equations are essentially hydrodynamic or gas dynamic. In particular, the family of solutions for the velocity of the gas-dynamic (fluid) outflow includes an important transition at a "critical point" R_c in radial distance, from subsonic to supersonic speed. The basic equations are

$$\text{continuity/mass conservation}: \ \nabla \cdot \rho u = 0, \quad (5.1)$$

$$\text{momentum}: \rho u \cdot \nabla u = -\nabla p + \mathbf{j} \times \mathbf{B} + \rho F_g, \quad (5.2)$$

$$\text{radial flow assumption}: \ u = u(r)\hat{e}_r, \quad (5.3)$$

$$\text{gravitational force}: \ F_g = -\frac{GM_S}{r^2}\hat{e}_r, \quad (5.4)$$

$$\text{thermal pressure gradient}: \ \nabla P = \frac{dP}{dr}\hat{e}_r. \quad (5.5)$$

Usually, $\mathbf{j} \times \mathbf{B}$ is dropped, giving the

$$\text{revised momentum equation}: \quad -\frac{dP}{dr} - \rho\frac{GM_S}{r^2} = 0. \quad (5.6)$$

One solution, where R_S is the radius of the Sun, is a static atmosphere:

$$P(r) = P_0 \exp\left\{\frac{GM_S m}{2KT}\left(\frac{1}{r} - \frac{1}{R_S}\right)\right\}. \quad (5.7)$$

Another solution with non-zero outflow at infinity assuming that $\rho u r^2$ = constant and that $P = nKT$ has the velocity described by

$$\left(u^2 - \frac{2KT}{m}\right)\frac{1}{u}\frac{du}{dr} = \frac{4KT}{mr} - \frac{GM_S}{r^2}. \quad (5.8)$$

This solution has a "critical point" $r = R_c = \dfrac{GM_S m}{4KT}$, when

$$u = (2KT/m)^{1/2} (\text{sound speed}). \quad (5.9)$$

The solution of these equations for u passes smoothly from sub- to supersonic, describing a wind physically consistent with both the coronal and outer boundary requirements (see Figure 5.1).

This problem is sometimes compared to that of the de Laval nozzle in fluid mechanics where a constricted cross section of a flow channel can enable a similar transition in regular gas or fluid flows. (The interested reader is encouraged to look up this analogy.) However, the critical point in this solar version where the magnetic field and flow are strictly radial is determined by the combination of coronal temperature, which, with density, controls the radial thermal pressure gradient and the Sun's gravitational force (Eq. (5.4)). Some sample solutions for selected temperatures and a common base density are shown in Figure 5.1. Parker's prediction that the solar plasma flows at Earth would resemble the solution with supersonic speeds was confirmed by the first *in situ* plasma detections, ushering in the modern concept of the solar wind. Now, some 50 plus years and numerous observations later, research continues on the one hand on why the outermost extension of the solar atmosphere behaves this way, and on the other hand on the many ways the real solar wind departs from this idealized concept.

The configuration of the **interplanetary magnetic field** associated with this simple fluid picture of the solar wind is another part of its success story. In the discussion of the dipolar coronal model counterpart

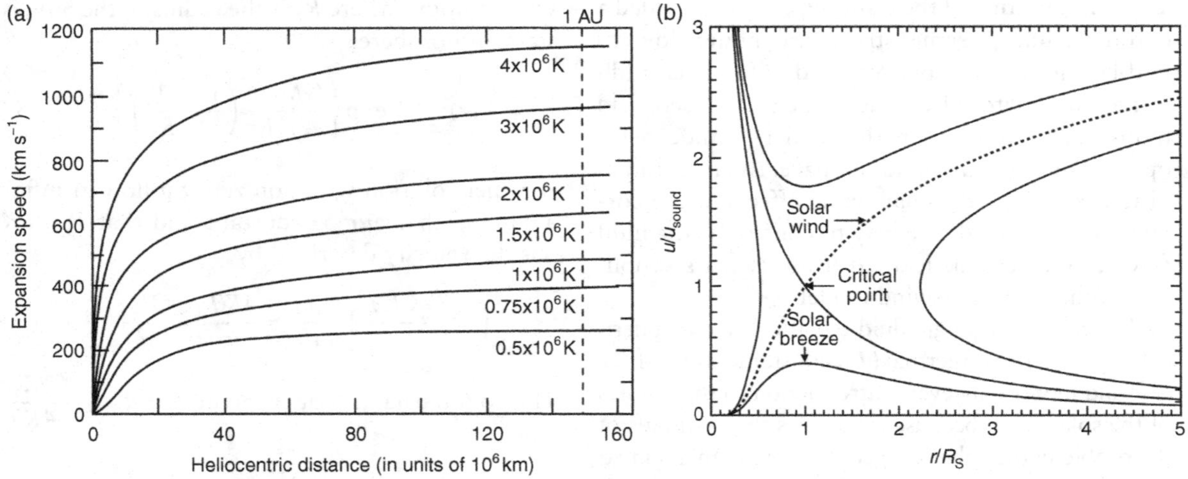

FIGURE 5.1. Illustration of a family of solutions to Parker's fluid equation description of the solar-wind expansion. (a) Solutions out to 1 AU showing how the asymptotic velocity of the flow depends on the assumed (constant) temperature. (b) Zoomed in view of a more complete set of solutions for one particular temperature for the region close to the Sun, where the solar-wind solution passes through a "critical point" to continue its supersonic ($u > u_{sound}$) outward acceleration.

in Chapter 4, we were able to neglect the rotation of the Sun because the low-beta nature of the corona forces near-corotation of the solar atmosphere. In this case, making rotating or non-rotating views is simply a matter of the preferred frame of reference. But the corona to solar-wind transition is an example of an MHD setting where the conditions with radial distance change from field-dominated to flow-dominated plasma behavior (see Figure 4.6). The real critical points include magnetosonic and Alfvén speed transitions. Within a few tens of solar radii (inside the Alfvén critical point at which the coronal plasma and magnetic pressures essentially balance), the solar coronal plasma and field nearly corotate with the Sun at its ~27 day period. However, outside of this critical point the plasma flow moves on a roughly radial outward path, carrying the open coronal field effectively embedded ("frozen-in" fields of plasmas are described in Chapter 3) within it. Parker recognized the "garden sprinkler" analogy of this situation, likening the solar-wind gas behavior to the dispersal of the droplets emitted from a rotating lawn sprinkler head with several spouts. The sources in this analogy rotate under the released water that continues moving outward. The result is a spiral pattern of outflow from each source. Plasma elements in the solar wind follow a like path, tracing the time history of fluid elements launched outward from a particular inner boundary location. The basic interplanetary field geometry is thus an archimedean spiral called the Parker spiral. A mathematical description of the field in spherical coordinates, based on the kinematic picture illustrated in Figure 5.2, is given below. Here the subscript 0 refers to the inner boundary of the field model (the solar-wind source surface, say), and ω and V are solar rotation rate (assumed to hold at the inner boundary) and solar-wind radial velocity. From flux conservation,

$$B_R = \pm B_0 \left(\frac{R_0}{R}\right)^2, \tag{5.10}$$

and

$$B_\theta = 0. \tag{5.11}$$

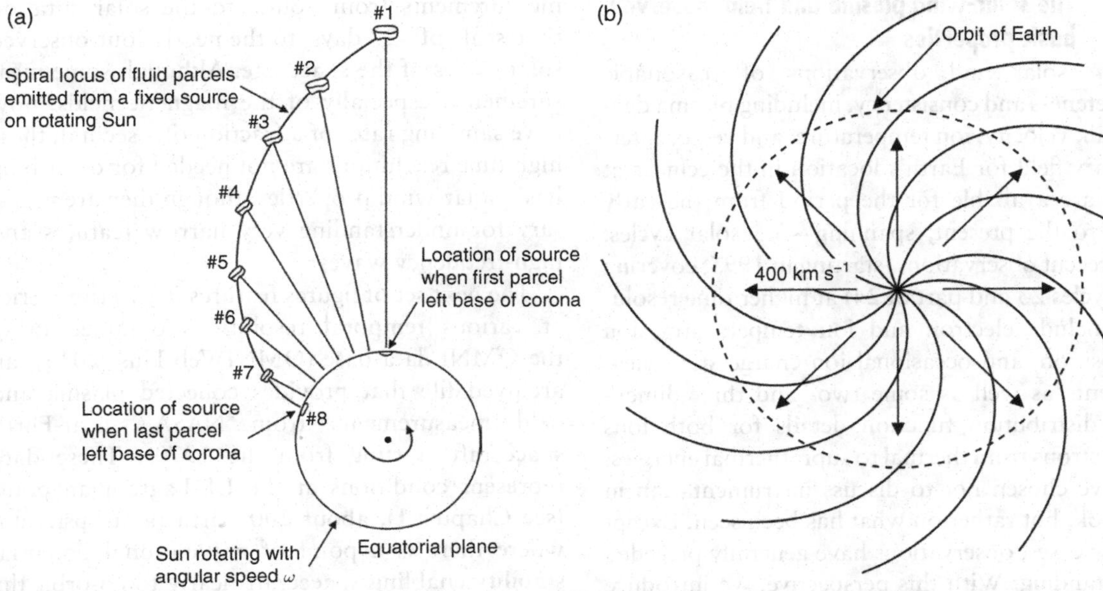

FIGURE 5.2. The interplanetary field in Parker's model takes on a spiral geometry because of solar rotation. (a) Motion of a single parcel of solar-wind fluid carrying the Sun's "open" (in this case, radial) magnetic field outward, as the Sun rotates beneath it. (b) "Parker spiral" field in the equatorial plane for a particular solar-wind speed. The tightness of the spiral depends on the solar-wind velocity, which is radially directed. Only the field is in a spiral shape, because it traces the trajectories of sequential parcels from the same site on the Sun (see (a)).

The effect of solar rotation adds

$$B_\varphi = -\beta B_R (R/R_0) \sin\theta, \qquad (5.12)$$

where $\beta = R_0\omega/V$. The form of the magnetic field lines are then archimedean spirals:

$$r - R = -\frac{u}{(\omega \sin\theta)}(\varphi - \varphi_0). \qquad (5.13)$$

Because the details of this spiral field depend on radial distance, latitude, and the assumed solar-wind velocity, one must take into account the location of interest in considering the field geometry. For example, the field is nearly parallel to the ecliptic at low heliolatitudes, is a more tightly wound spiral for slower solar wind, is fairly radial at the orbit of Mercury (~0.3 AU), and is nearly perpendicular to the radial solar-wind flow at Saturn's orbit (at ~10 AU).

An interesting consequence of the magnetization of the solar wind is the process of "magnetic braking." Even though the interplanetary magnetic field is essentially convected with the solar-wind plasma, it produces a torque that acts to brake the rotating source of the outflow at its base. Like an ice skater controlling her spin rate by extending or retracting her folded arms, the Sun's rotation is modified by the outflow of the solar wind that is still magnetically connected to it. Over the long term this is thought to have established the Sun's current rotation rate, and to be responsible for the observed anticorrelation between stellar ages and rotation rates. We do not delve into this topic here, but the interested reader is encouraged to investigate further. The current solar coronal braking by the solar wind is barely measurable on the time scales of our ability to determine it, although it has been attempted. As there is a connection between stellar rotation rate and dynamo activity, this magnetic braking has likely played a critical role in the long-term history of solar activity and related solar system evolution.

5.2.1 The solar-wind plasma and field: observed basic properties

In situ solar-wind observations of reasonable completeness and consistency, including plasma density, bulk velocity, ion temperature, and vector interplanetary field for Earth's location in the ecliptic at 1 AU, are available for the period from the early 1970s to the present, spanning ~3.5 solar cycles. More recent observations (starting in 1995, covering solar cycles 23 and part of 24) at higher time resolutions include electron and ion temperatures, ion composition, and occasional ion charge state measurements as well as some two- and three-dimensional distribution function details for both ions and electrons from thermal to suprathermal energies. We have chosen not to discuss instrumentation in this book, but rather on what has been seen. Except in a few cases, observations have generally preceded understanding. With this perspective, we introduce the "real" solar wind as many original investigators learned about it, without prior interpretational expectations or bias, and with the particular perspective of near-Earth measurements.

We begin with examples of the most basic solar-wind measurements – plasma density, velocity and temperature moments, and magnetic field. Many of these long-term observations of the solar wind are not composition specific, but the knowledge that the solar corona is primarily hydrogen and a small fraction of helium (see Table 4.2) means the detected ions are mostly protons. The ions dominate the solar-wind momentum, and, as we shall see later, have thermal velocity distributions that are narrow compared to their bulk speeds, making their moments easy to determine with ion spectrometers. The further expectation that the "bulk" solar-wind plasma is electrically neutral means that any reference to density also applies to electrons, although the ions are typically used for the plasma moments for reasons that will become clear. The magnetic fields are determined with magnetometers that generally measure all the field vector components. The solar wind varies on time scales from less than the current typical plasma measurement cadence of a minute up to a solar cycle, and so it is important to look at different temporal resolutions and durations of time series of measurements from hours, to the solar rotation time scale of ~27 days, to the nearly four observed solar cycles of the space age. Although some measurements, especially of the magnetic fields, now have sampling rates of a fraction of a second, these high time resolutions are not needed for describing basic solar-wind properties, though they are necessary for understanding very narrow features and high-frequency waves.

The next set of figures features *in situ* time series at various temporal resolutions obtained using the OMNI data base (OMNIWeb Plus, 2015), an archived file that provides collected plasma and field measurements from NASA's near-Earth spacecraft starting from the 1970s. These data represent conditions at the L1 Lagrangian point (see Chapter 1), about 200 Earth radii upstream, where there is a point of gravitational potential stability enabling spacecraft nearly to co-orbit the Sun with the Earth, and where the solar-wind plasma and field set to impact the Earth is relatively undisturbed by the Earth's presence (beyond the foreshock; see Chapter 7). Most of the long-term *in situ* measurements have either been obtained at L1 or in Earth orbits where the spacecraft spent considerable time outside of the Earth's bow shock in this undisturbed region. Measured vector quantities, including bulk plasma velocity and magnetic field, are usually given in a coordinate system that includes the predominantly radial velocity and spiral interplanetary magnetic field in "RTN," "GSE," or "GSM" (see Appendix 3 for a discussion of frequently used coordinate systems in space physics and their interrelationships). The user of these data needs to be aware of the system in which they are given because some are Earth centered and others are Sun centered. Here we display the standard OMNI GSE choice, which gives vector components in locally radial, north–south, and east–west directions with respect to the Earth. In the GSE system, the x-axis points toward the Sun; the y-axis points toward the east as seen from the Sun, opposite the direction of Earth's orbital motion; and the z-axis points roughly northward, normal to the ecliptic. Elsewhere in this chapter, RTN coordinates may be used instead because they are a standard often used in heliospheric studies. RTN

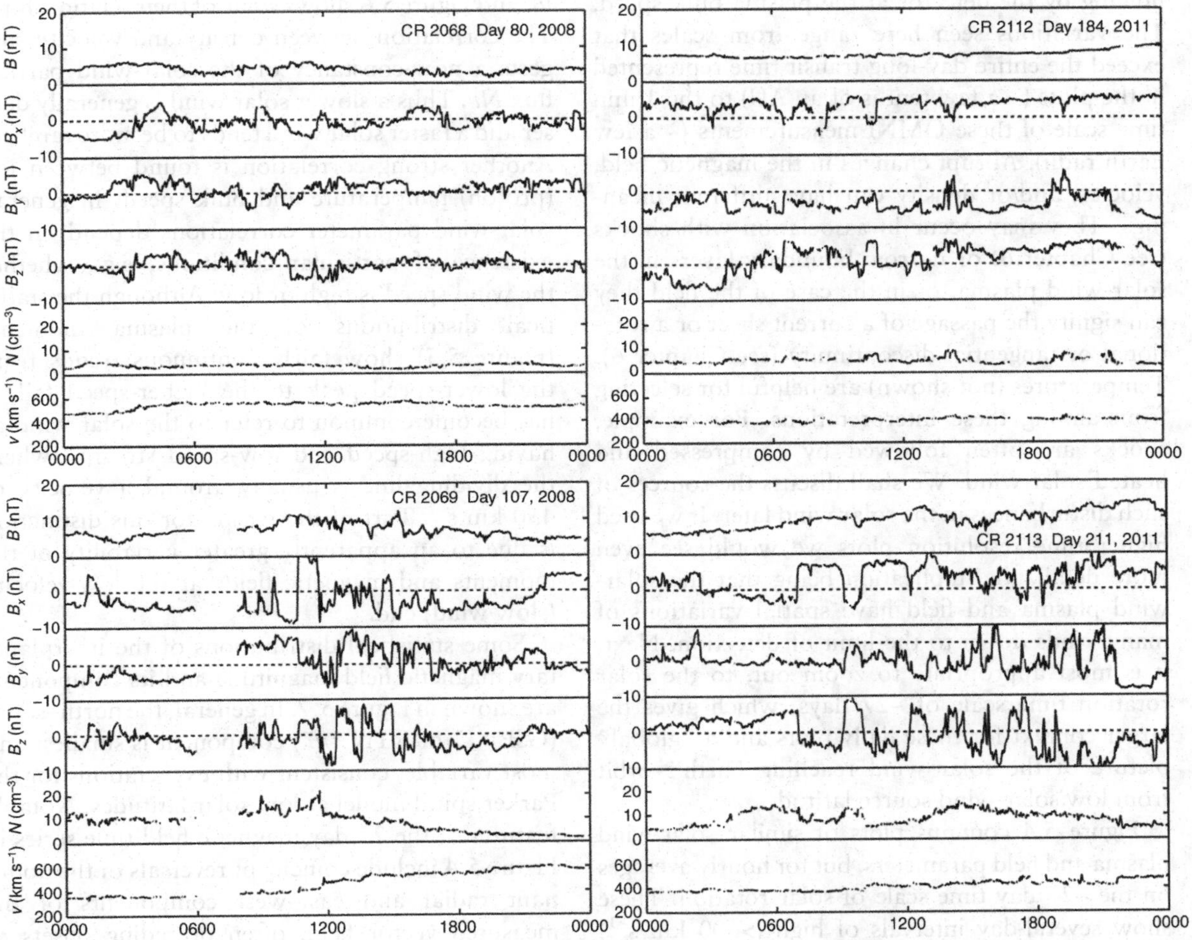

FIGURE 5.3. Daily plots of 1 min solar-wind data obtained at 1 AU. These illustrate the ranges of key solar-wind parameters for several arbitrarily selected days and the variations to which the Earth's magnetosphere must respond on relatively short time scales. (From OMNIWeb Plus, 2015.)

differs from GSE in that it is centered at the measuring spacecraft (at Earth or not), with the "r" component similar to GSE "x" (sunward), the "t" component roughly like GSE "y" ('eastward'), and the "n" component analogous to GSE "z" ('northward'). The main orientation difference is that the GSE system is referenced to Earth's equatorial plane whereas the RTN system is referenced to the ecliptic. When details of field strengths in certain orientations are important, coordinate transformations are necessary, but often the similar directions provided by the three components in either system are sufficient for more qualitative uses.

The daily plots in Figure 5.3 illustrate the typical ranges of solar-wind plasma bulk speeds, which are usually between about 300 and 800 km s^{-1}, and densities, from a few to several tens of ions per cubic centimeter. The magnetic field magnitudes usually vary from a few to a few tens of nanotesla. The time cadence of the measurements shown here is 1 min. There are many irregularities and variations in these parameters, including oscillatory or wave-like behavior as well as step-like changes in magnitude and/or direction. The velocities give an idea of the scale sizes of these because the plasma with its frozen-in magnetic field (see Chapter 3) is flowing by the detector at the plasma bulk speed. The variations seen here range from scales that exceed the entire day-long transit time represented in the plots (~ a few tenths of an AU) to the 1 min time scale of these OMNI measurements (~ a few Earth radii). Abrupt changes in the magnetic field, velocity, and/or density can have different meanings. They may occur in association with shocks (see Chapter 6) or narrow boundary layers in the solar-wind plasma, or in the case of the field they can signify the passage of a current sheet or a rotational or tangential discontinuity (see Chapter 6). Temperatures (not shown) are helpful for selecting from among these interpretations. For example, shocks are often followed by compressed and heated solar wind. We shall discuss the sources of such disturbances in the solar wind later. If we used higher time resolution plots we would see even more detail, the implication being that the solar-wind plasma and field have spatial variations of many kinds down to the limit of detection. Next, it is most appropriate to zoom out, to the solar rotation time scale of ~27 days, which gives the larger context for these daily plots and a "global" picture of the solar wind reaching Earth's orbit from low solar-wind source latitudes.

Figure 5.4 contains plots of similar solar-wind plasma and field parameters, but for hourly averages on the ~27 day time scale of solar rotation. These show several-day intervals of high (>600 km s^{-1}) bulk speed and low to moderate density (~10 cm^{-3}) solar wind interspersed with intervals of low speed (<350 km s^{-1}), and on average higher density (few tens of cubic centimeters) solar wind. Moreover, they show relatively narrow (fractions of a day) density and magnetic field magnitude enhancements at the leading edges of the higher-speed streams. Some of the plots from consecutive 27 day intervals show similar features. This apparent corotation of the solar-wind stream structure occurs most often when solar activity is low, and thus presumably when coronal sources are only slowly evolving. The time scale shown also brings out some other well-documented features, including parameter interdependences, which provide clues to the solar-wind generation and propagation physics.

Figure 5.5 shows statistical distributions of the main plasma moments (density N and bulk velocity v), and Figure 5.6 shows some of their relationships. The correlation between density and velocity suggests a near-constancy of the solar-wind particle flux Nv. Thus a slower solar wind is generally denser and a faster solar wind tends to be more rarefied. Another strong correlation is found between ion (proton) temperature and bulk speed. In general, solar-wind parameter correlations depend on the existence of particular conditions, e.g., whether the wind speed is high or low. Although the statistical distributions of the plasma moments (Figure 5.5) show fairly continuous trends from the lower-speed peak to the higher-speed tail, it has become common to refer to the solar wind as having high-speed and low-speed streams, where the dividing line is usually around a velocity of 450 km s^{-1}. Part of the reason for this distinction is due to an apparently greater variability of the moments and magnetic fields at the low-velocity (slow wind) end.

Some statistical distributions of the interplanetary magnetic field magnitude and its components are shown in Figure 5.7. In general, the north–south (GSE "z" or RTN "n") component is smallest and most variable, consistent with expectations for the Parker spiral model at low solar latitudes. Notable features of the 27 day magnetic field time series in Figure 5.4 include coincident reversals of the dominant radial and east–west components of the measured vector field, often preceding onsets of high-speed wind streams. While not universal, this association is quite common. Note that the field is often most radial in the declining speed portions of high-speed streams, in contrast to Parker spiral expectations that slower wind should show a tighter spiral. The reasons for this departure and other complications in the observed field, considered later on in this chapter, remind us that the simplified Parker picture of a uniformly outflowing solar wind has its limitations.

We shall also see that consecutive 27 day plots like those shown in Figure 5.4 are particularly useful for investigating the coronal source connections of various features seen near the Earth. For times around solar maximum this recurrent behavior due to solar rotation becomes less distinguishable, suggesting significant coronal source evolution is occurring over the period of the plot.

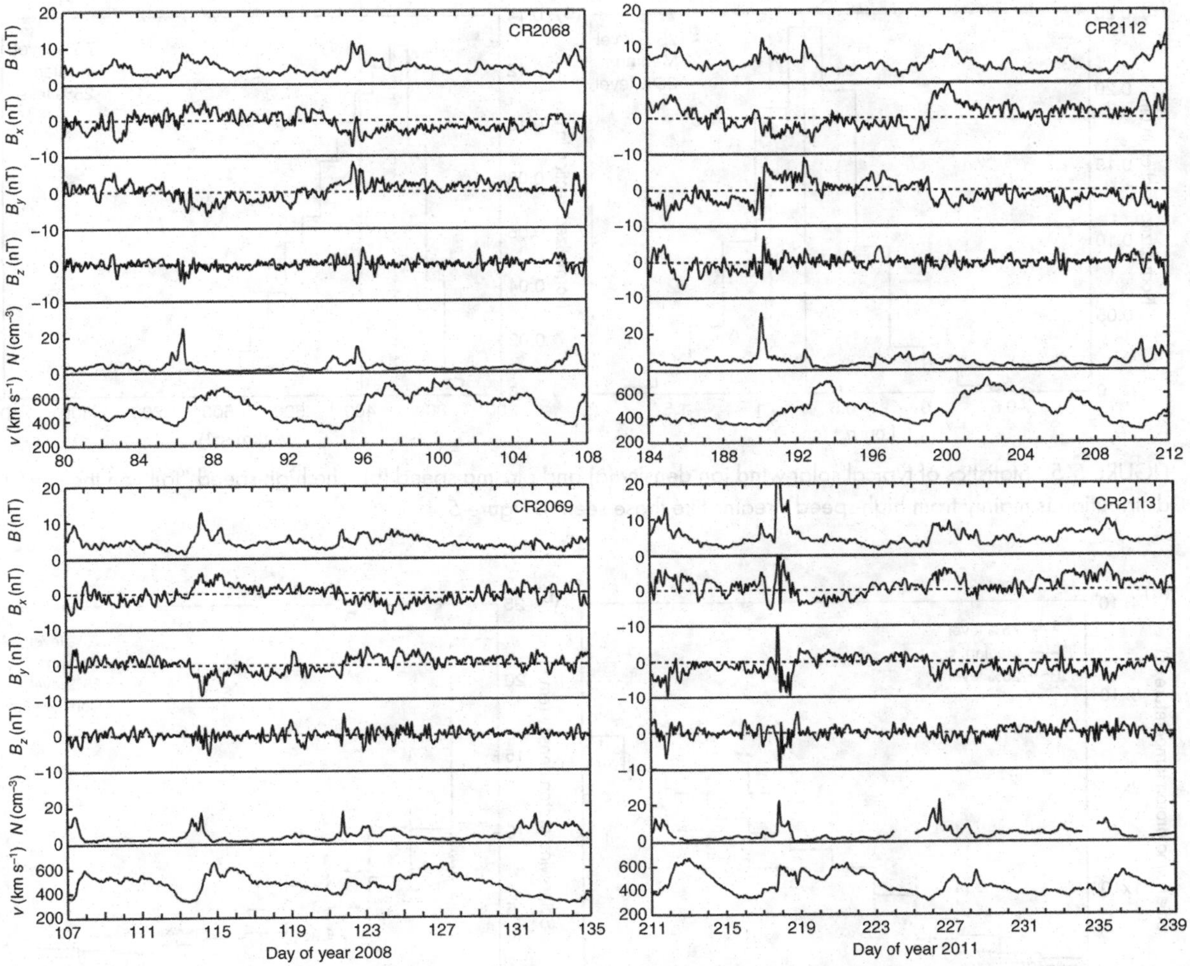

FIGURE 5.4. Examples of 1 h time resolution solar-wind data like that in Figure 5.3, but for several 27 day intervals. The 27 day periods of these plots correspond to a solar rotation time scale. The start and stop times are chosen to correspond to Carrington rotations (see Chapter 4) and are labeled accordingly. On this time scale the high- and low-speed stream structure of the solar wind, and the related sign changes in the interplanetary field components, are evident. For the examples that occur during a fairly quiet period of the solar cycle, one can observe repeating patterns in the parameters. This is referred to as "corotating" solar-wind structure because it appears to rotate with the Sun.

The solar-wind plasma and field features are also more generally variable, sometimes exhibiting several-day-long intervals of exceptional enhancements in a number of parameters. These enhancements may appear as a sudden onset that has the plasma and field characteristics of a shock (see Chapter 6). They were recognized quite early as the likely results of transient fast outflows following flares, but only later as the interplanetary counterparts of the coronal mass ejections in coronagraph images (see Chapter 4). Some details of these are best examined together with the solar-wind stream-related structures, since together they produce the main departures in solar-wind parameters from the simple Parker picture. We return to these dynamical and transient aspects after a look at the *in situ* measurements on solar cycle time scales.

Solar cycle dependences of basic solar-wind parameters in the ecliptic are not as notable as one

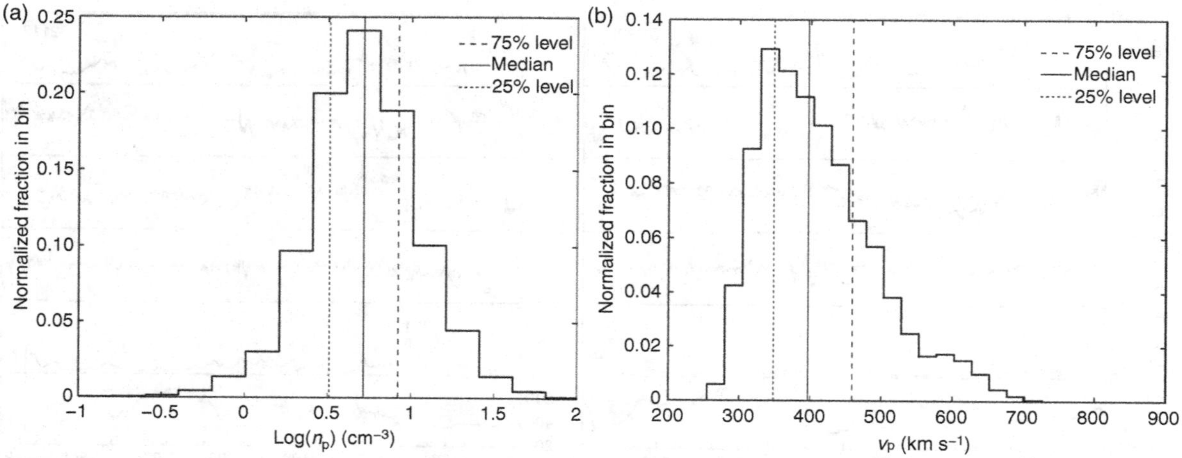

FIGURE 5.5. Statistics of typical solar-wind ion density (a) and plasma speed (b). The high-speed "tail" on the speed distribution is mainly from high-speed streams like those seen in Figure 5.4.

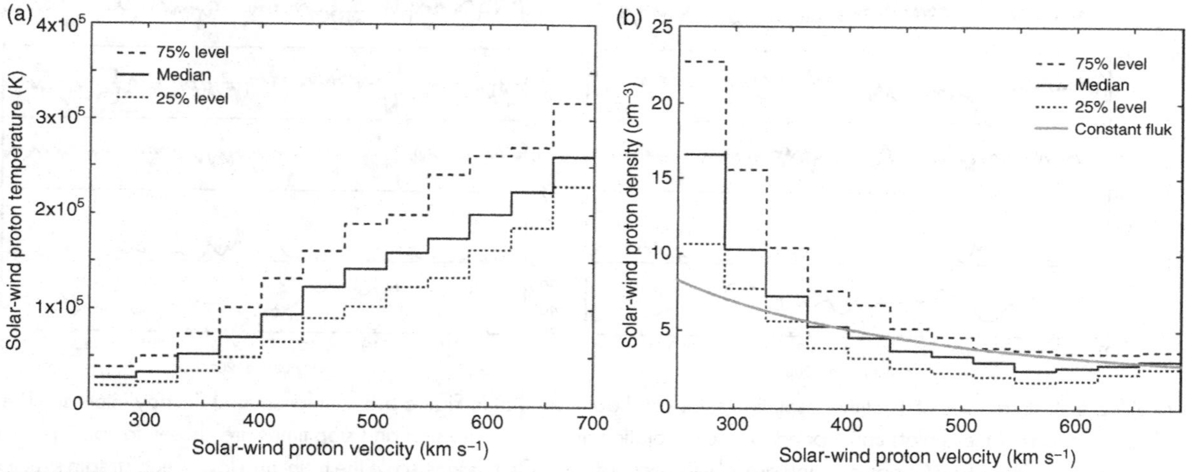

FIGURE 5.6. Illustration of some of the relationships that are observed to exist in measured solar-wind parameters. (a) The close relationship between the speed and the ion (proton) temperature. (b) The density–speed relationship, which approaches, but does not exactly match, a constant flux behavior.

might expect from the striking changes in the solar corona described in Chapter 4. Recall that the corona is most nearly dipolar and slowly varying at solar minimum, with polar coronal holes a characteristic feature, while the solar maximum corona shows streamer structures and coronal holes over a wide range of latitudes that exhibit sometimes explosive evolution. Figure 5.8 shows selected *in situ* measurements on a highly compressed time scale with 27 day (solar rotation) averages so that stream structure details are smoothed out. Sunspot number is added at the bottom to allow visual correlations with these low-time-resolution OMNI data, which span part of cycle 21, all of cycles 22 and 23, and the beginning of cycle 24. The only parameters shown where there are clear sunspot-number trends are the magnetic field magnitude and the alpha to proton ratio. The reasons why the other solar-wind parameters do not show such a clear solar cycle signature is best considered by looking

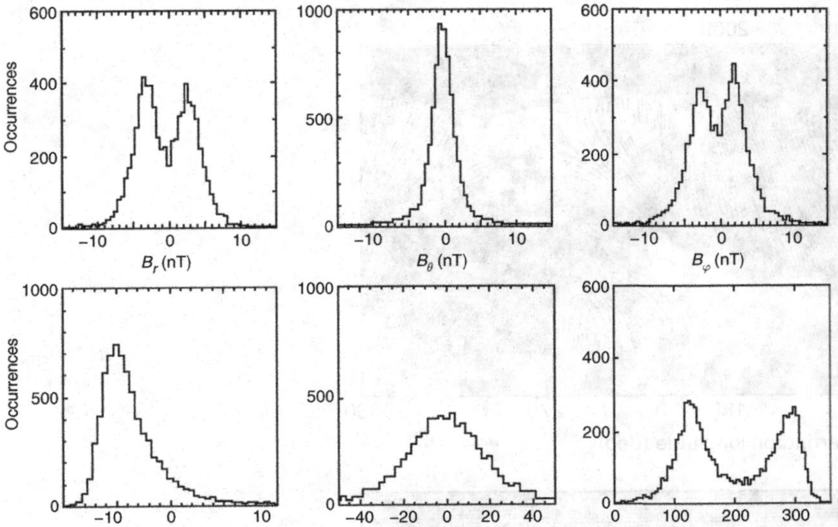

FIGURE 5.7. Statistics of some interplanetary magnetic field measurements at 1 AU, showing the behavior of the different components of the field vector. These show that the field generally lies close to the ecliptic plane and has nearly equal radial and azimuthal (φ) components – as expected at 1 AU for the Parker spiral geometry for a typical plasma speed of ~400 km s^{-1}. They also show that these two components have alternating signs, as was seen in the 27 day time series shown earlier (Figure 5.4). (After Lee *et al.*, 2009.)

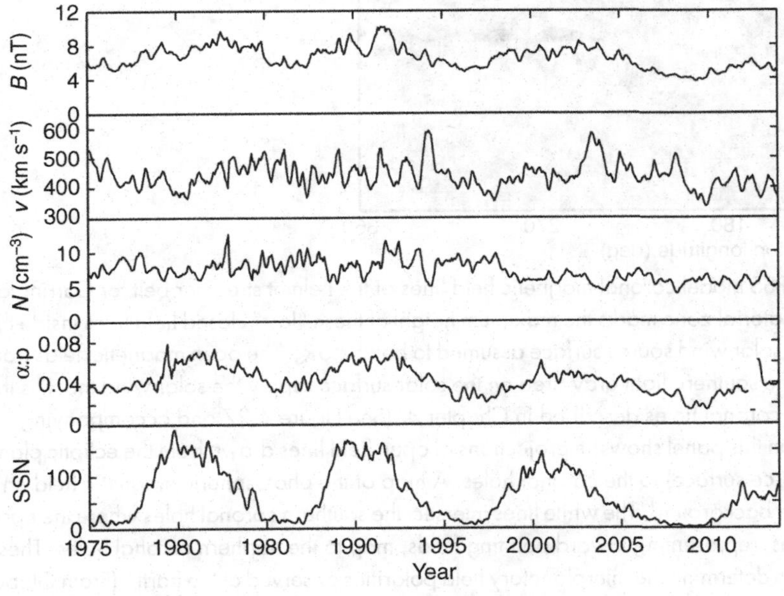

FIGURE 5.8. Smoothed 27 day (solar rotation) averages of solar-wind parameters at 1 AU compared to the sunspot number, SSN. This illustrates the extent to which the longer-term behavior of the solar wind follows the solar activity cycle. The magnetic field, magnitude B, and alpha to proton ratios, α:p, show the clearest solar cycle signatures, with higher values occurring in both at more active times. This plot also shows the decreasing trend in solar activity over this period. (From OMNIWeb Plus, 2015.)

into how both source(s) and propagation influence their characteristics.

5.2.2 Influences of the coronal source(s)

The evolving solar surface magnetic field and its control of the coronal field structure was described in the preceding chapter. The changing distribution of closed and open field areas as the solar cycle progresses between minimum and maximum depends on the strength of the polar region fields versus the strengths, areas, and locations of active region fields that emerge at middle to low latitudes. At solar minimum the dipolar corona and polar coronal hole source picture discussed in Section 4.5.1 may apply for a time, but for the rest of the cycle the coronal open field distributions, and

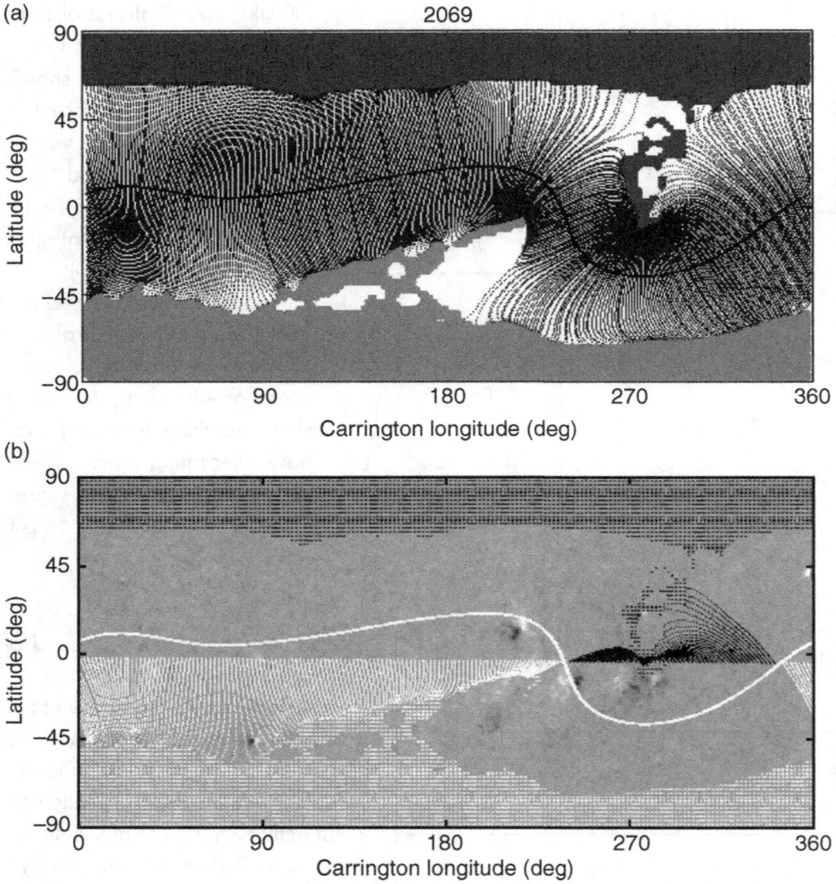

FIGURE 5.9. (a) Example illustrating PFSS model coronal magnetic field lines of the helmet streamer belt for Carrington rotation 2069. The black line in the equatorial zone marks the maximum height of these "last" closed field lines inside the model's spherical outer boundary – the solar-wind source surface assumed to be at $2.5R_S$. The open magnetic fields from the northern dark gray area and from the southern light gray area on the solar surface supply the solar wind that fills the entire heliosphere. These represent the coronal holes described in Chapter 4. (See Figure 4.27 and accompanying material.) (b) The black and white lines in this panel show the projections of open field lines drawn from the ecliptic plane (roughly latitude = 0° on the model source surface) to the coronal holes. A map of the photospheric magnetic field on which the model is based is shown in the background. The white lines intersect the southern coronal holes where the open field direction is outward. The dark lines, representing inward pointing fields, map to the northern coronal holes. These open field directions at the ecliptic plane determine the interplanetary field polarities observed at the Earth. (From Global Oscillation Network Group, 2015.)

thus the related solar-wind sources, are more complicated. The end result is that different solar-wind streams from different coronal hole source regions diverge with the coronal field to fill the space above. The PFSS model of the coronal magnetic field described in Chapter 4, an example of which is seen in Figure 5.9 has been particularly useful for analyzing solar-wind stream sources. Figure 5.9(b) illustrates how open coronal field lines, defining the solar-wind outflow channels, can be traced from the coronal hole footprints on the surface (stippled) to the low solar latitudes on the source surface supplying the ecliptic. These mappings suggest that the solar wind all planets experience comes mainly from the edges of the polar open field regions, from low- and mid-latitude polar coronal hole extensions, and from isolated low-latitude coronal holes. The different line shading indicates how the magnetic field polarity (outward or inward at the Sun's surface) maps to the PFSS model source surface at $2.5R_S$. Field polarity may or may not change as the mappings jump from one source region to the next, depending on the solar magnetic field at the base of adjacent stream sources. Thus some but not all stream boundaries are also sites of field reversals in the solar wind. The important result to remember here is that polar coronal hole descriptions of solar-wind sources do not generally apply for most of the solar cycle, especially at low heliospheric latitudes. In evaluating the solar wind observed upstream of the Earth, one needs to consult PFSS or similar models to obtain a picture of the prevailing solar source connections.

PFSS model mappings of open coronal fields like those described provide a basis for constructing realistic models of the solar wind. But because the model describes only the magnetic field and provides no plasma information, comparisons with *in situ* measurements at 1 AU are used to obtain empirical relationships between attributes such as speed and the inferred source location. Such studies suggest that proximity to a coronal hole boundary, or the coronal field divergence at the solar site, determine what is observed. For example, an important result from these modeling efforts is the inference that lower-speed wind (<400 km s^{-1}) often comes from coronal hole boundaries, which are also coronal streamer boundaries. In contrast, the fastest wind (>600 km s^{-1}) tends to come from the central regions of large-area coronal holes where the open field diverges less rapidly than radial. More physically consistent coupled MHD coronal and solar-wind models have also been developed. Coronal hole and ecliptic source mappings with PFSS models are now widely available, and a few global MHD models can be found in community model libraries. However, as mentioned above, the different coronal hole sources are not the only things affecting comparisons to heliospheric *in situ* observations, e.g., at Earth at 1 AU. Dynamical effects both in the corona and during solar-wind transit must also be taken into account. We postpone discussion of further details of the coronal contributions until after we have considered the effects of this complication.

5.2.3 Dynamical alterations in transit

The solar wind flowing out of the corona on its generally diverging open fields almost completely surrounds the Sun by a few solar radii. However, the radius where it is no longer controlled by solar gravity and corotation enforcement by the solar magnetic fields is not reached until the critical point for its magnetosonic velocity at ~10–$20R_S$ – well outside of the last closed coronal field loops and/or the PFSS source surface. Beyond this radius, all information carried in the plasma (including by waves) is flowing outward only, making this the most physically meaningful **source surface** or inner boundary of the solar wind. At this radius, which is not necessarily the same at all locations, the plasma takes control from the magnetic field. As we have seen in the discussion of the Parker model, the field thereafter behaves as a tracer frozen in the plasma outflow rather than actively channeling it. The imprint of the source regions in the corona, both plasma and (open) magnetic field characteristics, are convected into interplanetary space along roughly spiral paths. But as described above, rather than the uniform radial outflow with Parker spiral magnetic field of early solar-wind visualizations, the outflow is a mixture of roughly spiral-shaped streams connected by their fields to their different source regions, each bounded by low-speed wind (Figure 5.10).

The details of the solar-wind structure thus always reflect the state of the corona at the time. Only at solar minimum does an approximately dipolar corona provide roughly symmetrical polar coronal hole sources, with Parker spiral fields of opposite magnetic polarity occupying the northern and southern hemispheres of the heliosphere. But even when such near-ideal conditions are realized, the ~5° angle offset between the solar rotation axis and ecliptic north ensures varying near-Earth solar-wind conditions.

Above we examined solar-wind plasma and field observations at 1 AU and noted the occasionally repetitive ~27 day patterns in its properties

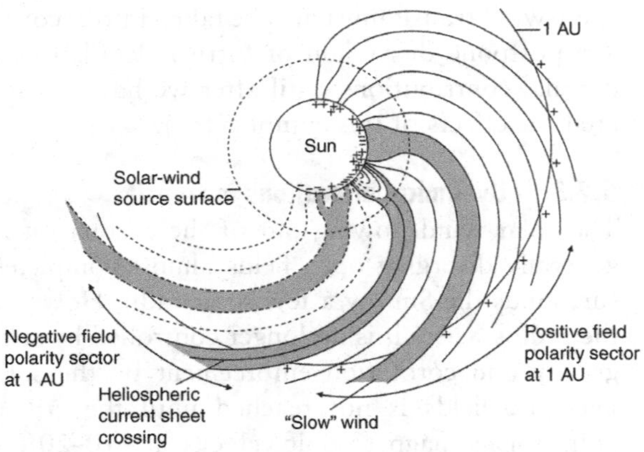

FIGURE 5.10. Extension of a complex solar-wind source picture like that in Figure 5.9 into interplanetary space (not to scale; ecliptic view). The + symbols indicate positive outward magnetic field sources while the – symbols and dashed lines indicate negative (inward) fields. The observations indicate that lower-speed solar wind (shaded) comes from the boundaries of the open field regions, thus separating the higher-speed streams from different sources in the measurements at 1 AU. The existence of evolving or transient structures at some of those boundaries is also suggested. (After Schatten, Wilcox and Ness, 1969.)

indicating corotation (Figure 5.4). Of course, this local "corotation" occurs with a ~4 day solar-wind transit time lag relative to the time the solar source(s) pass the central meridian. Also clearly seen in these time series are apparent compressions in density and magnetic field at the onsets of the high-speed streams. These stream interaction regions (SIRs), called corotating interaction regions (CIRs) if they recur with a ~27 day periodicity, can be interpreted as the passage of spiral density ridges between streams from the different coronal source regions (recall Figure 5.9). Such compression regions, illustrated by Figure 5.11(b), may occur because of the presence of the low speeds coming from coronal hole boundaries, or contrasts in radial outflow speeds from adjacent stream source regions. They occur anywhere that the projected Parker spirals of adjacent streams cross one another because collisionless magnetized plasma interpenetration is not generally allowed. In the reverse case, spiral-shaped rarefactions occur.

A particularly clear example in Figure 5.11(a) illustrates some of the typical characteristics of a compressive SIR/CIR in *in situ* data: a fairly symmetric density and magnetic field magnitude enhancement typically lasting less than a day as velocity ramps up; pairs of opposing deflections in the bulk plasma velocity away from radial, and the vector magnetic field out of the ecliptic; and a total plasma pressure (thermal plus magnetic plus dynamic) time profile that is fairly symmetric and cusp-like in shape. They are also sometimes bracketed by a weak forward-reverse shock pair with magnetosonic Mach number $M_{ms} \sim 2$. These stream interactions do more than just local compressions in between lower-speed and higher-speed solar wind. They also act to redistribute momentum between the different outward-flowing solar-wind parcels, slowing down the fast wind and speeding up the slow wind, so that velocity differences between adjacent streams are eventually smeared out as radial distance increases. But the streams from the larger sources are often still quite identifiable at 1 AU, sometimes beginning with highly peaked densities (and thus dynamic pressures), and followed by deep rarefactions. As the "angle of attack" of adjacent streams (essentially the Parker spiral angle for their speeds) increases with radial distance from the Sun due to solar rotation, SIR compressions generally become stronger and bracketing shocks become more common (most are found beyond 1 AU). MHD models have been used with some success to propagate magnetized plasma outflows from the complex coronal source regions described by the coronal models to and beyond 1 AU. There are also some approximations that mimic the plasma propagation effects, but the bottom line is that they must be included to obtain reasonable descriptions of the solar wind at a distance from the source surface.

When we considered the PFSS models above, the magnetic neutral line on the source surface, dividing the regions of outward and inward open magnetic fields, was another feature whose appearance was determined by the coronal source mappings (Figure 5.9). In the ecliptic one observes the related interplanetary field reversals, called **sector boundaries**, as (sometimes multiple) abrupt changes between field orientations toward and

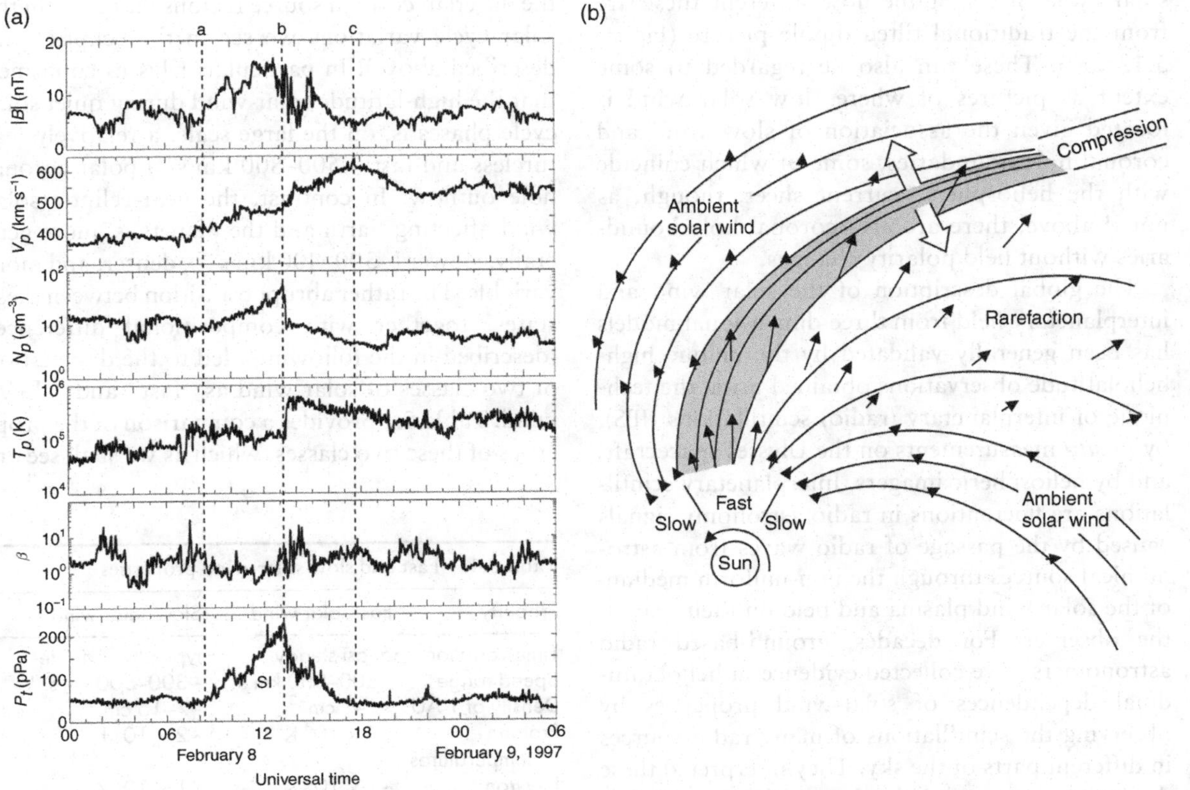

FIGURE 5.11. (a) Example of solar-wind observations from the WIND spacecraft obtained at a solar-wind stream interface, showing the existence of a compression where high-speed solar-wind speed follows low-speed wind. Dashed lines a, b, and c mark the leading edge of the stream interaction, the stream interface (SI), and the trailing edge, respectively. P_t is the total static pressure, magnetic and thermal. (b) Formation of a stream interaction region (SIR) or corotating interaction region (CIR) when a radially directed high-speed flow runs into preceding lower-speed flows. Note that a rarefaction must also occur behind the high-speed "front" if another lower-speed stream follows it. (After Pino and Gosling, 1999.)

away from the Sun along the Parker spiral. These tell us when the heliospheric current sheet extension of the source surface neutral line passes by. They are particularly clear in the 27 day *in situ* data plots in Figure 5.4. The three-dimensional appearance of the heliospheric current sheet is also of interest because it carries information about the solar magnetic field structure, is possibly subject to compression-related magnetic reconnection in stream interaction regions, and has significant implications for cosmic ray propagation (discussed in Section 5.4.4) and solar-wind interactions with planets and comets. For some time, various investigations were conducted looking for evidence of magnetic reconnection at the heliospheric current sheet near the orbit of Earth, but the presence of this evidence was appreciated only recently. In fact, reconnection can occur at almost any boundary between adjacent solar-wind streams or structures that have different magnetic fields. But such heliospheric reconnection does not appear to alter our basic view of the heliosphere and its relationship to the coronal field and sources of the solar wind. Some three-dimensional renderings of the heliospheric current sheet, based on MHD model extrapolations of the coronal source-determined magnetic neutral line, are shown in Figure 5.12A for a few times during the solar cycle. It is notable how different these are from the traditional tilted dipole picture (Figure 5.12A(a)). These can also be regarded to some extent as pictures of where slow solar wind is located given the association of slow wind and coronal hole boundaries, some of which coincide with the heliospheric current sheet, though, as noted above, there are also coronal hole boundaries without field polarity changes.

The global description of the solar wind and interplanetary field from three-dimensional models has been generally validated by the unique high-heliolatitude observations obtained using the technique of interplanetary (radio) scintillations (IPS), by *in situ* measurements on the Ulysses spacecraft, and by heliospheric imagers. Interplanetary scintillations are fluctuations in radio astronomy signals caused by the passage of radio waves from astronomical sources through the non-uniform medium of the solar-wind plasma and field on their way to the observer. For decades, ground-based radio astronomers have collected evidence of heliolatitudinal dependences of solar-wind properties by observing the scintillations of many radio sources in different parts of the sky. They interpreted these data to suggest a global solar-wind structure that was quieter and more uniform above the solar poles than at low heliolatitudes during low solar activity periods. The Ulysses spacecraft confirmed this general picture with *in situ* plasma and field measurements in a high inclination heliocentric orbit. Its observations provided proof that the heliosphere is filled with solar wind with different properties from the different coronal source regions that explain the solar cycle variations we see in the ecliptic (and described above). In particular, Ulysses confirmed that the high-latitude solar wind during quiet solar cycle phases is, on the large scale, a relatively featureless and fast (~500–800 km s^{-1}) polar coronal hole outflow. In contrast, the near-ecliptic solar wind affecting Earth and the planets is, more generally, slower (~300–400 km s^{-1}), denser, and more variable. The rather abrupt transition between these states, together with compositional differences (described in the following), led to the designation of two classes of solar wind as "fast" and "slow" wind. Table 5.1 provides a comparison of the properties of these two classes, which as we shall see are

Table 5.1. Fast and slow solar-wind properties

Property	Fast solar wind	Slow solar wind
Time variation	quasi-steady	typically variable
Speed range	~600–800 km s^{-1}	~300–500 km s^{-1}
Density at 1 AU	~1–7 cm^{-3}	~7–15 cm^{-3}
Proton temperatures	~4 × 10^4 K	~2 × 10^5 K
Electron temperatures	~1 × 10^5 K	~1 × 10^5 K
Composition	higher He^{++} (~4%)	higher O^{+7}/O^{+6}, Fe/O
Field structure	Alfvén waves	current sheet(s), rotational discontinuity
Sources	coronal hole centers	coronal hole streamers, boundaries

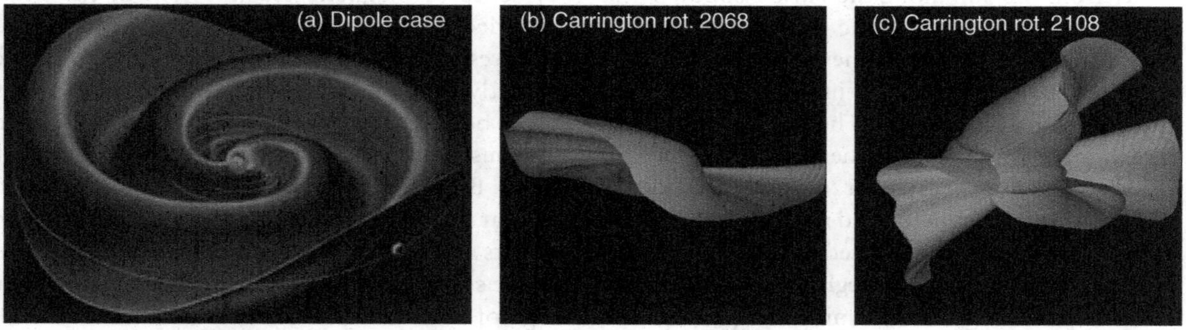

FIGURE 5.12A. Sample three-dimensional renderings of the heliospheric current sheet from an MHD model of the solar wind with realistic source distributions, compared to what would be expected for a tilted dipole corona solar-wind source (P. Riley, personal communication, 2014.)

Table 5.2. Typical solar-wind composition (relative to hydrogen) and charge state for major elements

Element	Hydrogen	Helium	Carbon	Nitrogen	Oxygen	Iron	Silicon
Relative composition	1.0	4.0×10^{-2}	2.3×10^{-4}	7.9×10^{-5}	5.3×10^{-4}	1.3×10^{-4}	1.1×10^{-4}
Charge (+)	1	2	4–6	5–7	5–7	7–13	7–11

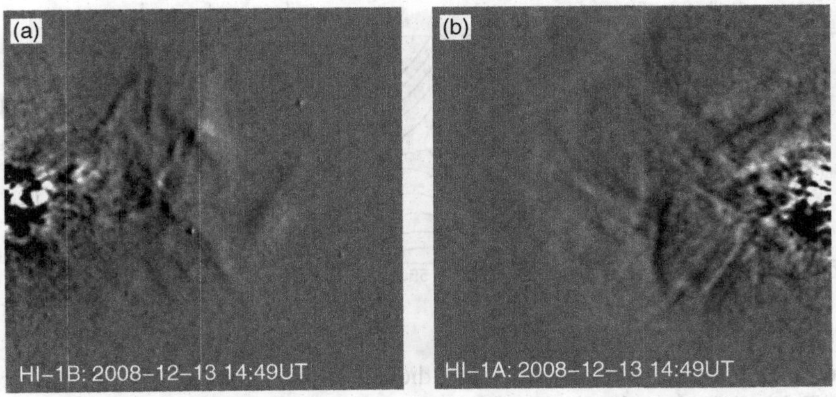

FIGURE 5.12B. STEREO Heliospheric Imager images showing the highly structured appearance of the near-Sun solar wind off the western (a) and eastern (b) limbs as viewed by STEREO B and STEREO A, (see Figure 5.20(b)). These are "difference images," where changes occurring between two times are brought out by subtracting the intensity of light in each pixel. This type of structure is common and has the appearance of arising from the vicinity of the streamer boundaries where solar rotation and coronal evolution adjust the geometry of the solar-wind source regions. (STEREO SECCHI images (NASA).)

influenced by different physical processes at their sources.

The most recent observations relevant to global solar-wind structure come from heliospheric imagers that observe extremely faint white light from the same Thompson scattering seen by coronagraphs, but well beyond their ~$30R_S$ limits. Images from several perspectives out to and beyond 1 AU, such as those shown in Figure 5.12B from the STEREO twin spacecraft, provide sky plane projections of line-of-sight integrated density enhancements. In addition to capturing stream-interaction-related density compressions, these images suggest the presence of ubiquitous structured outflows arising from the vicinity of the coronal streamers, even at quiet times of the solar cycle. The implication of both the IPS measurements and heliospheric images is that the solar wind, most generally the slow wind, always has a significant time-dependent component, the nature of which is more appropriately discussed in Section 5.3.1. At more active times, when the polar coronal holes shrink and the coronal plasma outflows are complicated by the increased frequency of transients, this two-state picture of the interplanetary plasma is no longer applicable.

5.2.4 Insights from ion composition, kinetic/microscopic properties, and waves

In situ measurements of the solar wind available to researchers today often include ion composition, charge state, and/or three-dimensional distribution functions of ions and electrons. Table 5.2 summarizes the major elements and change states of solor-wind ions. For the most part, the charge states of solar-wind ions are set in the corona. Minor contributions from singly ionized ions, including He^+ from the chromospheres and heliospheric pickup ions, will be described later in this chapter. As noted in Chapter 4, the coronal abundances

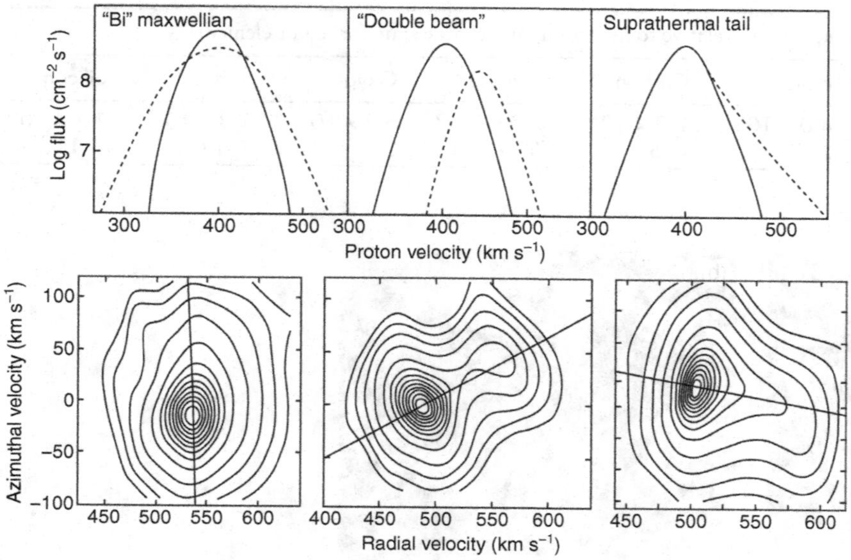

FIGURE 5.13. Upper panels: Proton distribution functions in a 400 km s^{-1} solar-wind flow illustrating three types of departures from a simple maxwellian. Lower panels: Contour plots in the spacecraft frame of actual proton distribution density contours, where with magnetic field directions shown by the black lines. These correspond to the distribution function types in the top row. (Lower panels: After Marsch *et al.*, 1982.)

show a preference for the lighter elements over the photospheric abundances (see Table 4.1), but the solar-wind composition closely matches the coronal composition. While the clear sunspot cycle trends seen in the alpha (He^{++}) to proton ratio (Figure 5.8) suggest a role of active regions in modulating the helium abundance in the solar-wind source regions, the same explanation does not seem to apply to – or at least dominate – the behavior of the other heavy solar-wind ions. Instead, composition and ionization state analyses have found that abundances of elements with low and high first ionization potential (FIP), such as magnesium and oxygen, respectively, and their average charge states, are organized by the same categories of fast and slow wind described earlier (see also Table 5.1). In particular, Mg/O and O^{+7}/O^{+6} ratios are generally enhanced in slow wind. Thus the alpha to proton ratio may reflect overall coronal heating trends associated with the presence of active regions, while the oxygen charge state ratio and low FIP ion relative abundance are associated with the source(s) of the slow wind (more on this in the following).

Increasingly detailed measurements of plasma distribution functions have also been obtained over a broad range of thermal and suprathermal energies for both ions and electrons. Some examples, organized by the particle velocities parallel and perpendicular to the local magnetic field, are shown in Figures 5.13 and 5.14. These complement the studies of the larger-scale properties as well as add to the debates concerning the coronal source processes and regions. For example, in the preceding chapter it was mentioned that remote sensing, via spectral line widths, of selected heavy ion temperatures in the corona indicate that some ions have significantly higher perpendicular than parallel (to the field) coronal temperatures. *In situ* solar-wind plasma distribution functions measured in fast solar-wind streams show that even the dominant protons and helium ions sometimes have perpendicular (to the field) local temperatures measurably higher than their parallel temperatures. These "perpendicular" anisotropies exist in spite of the continual action of the mirror force acting to focus the ion pitch angle distributions along the diverging open heliospheric fields. Waves such as ion cyclotron waves can resonantly accelerate (see Chapter 12) the ions anywhere. However, if the focusing then exceeds the threshold for some streaming/beam plasma instability, waves may be generated locally that isotropize the distribution but then perpendicularly accelerate other ions in an ongoing cycle of perpendicular "heating," mirror force focusing, beam disruption/wave generation, and so on. The proton-generated waves can also result in heavier ion heating – including of the second

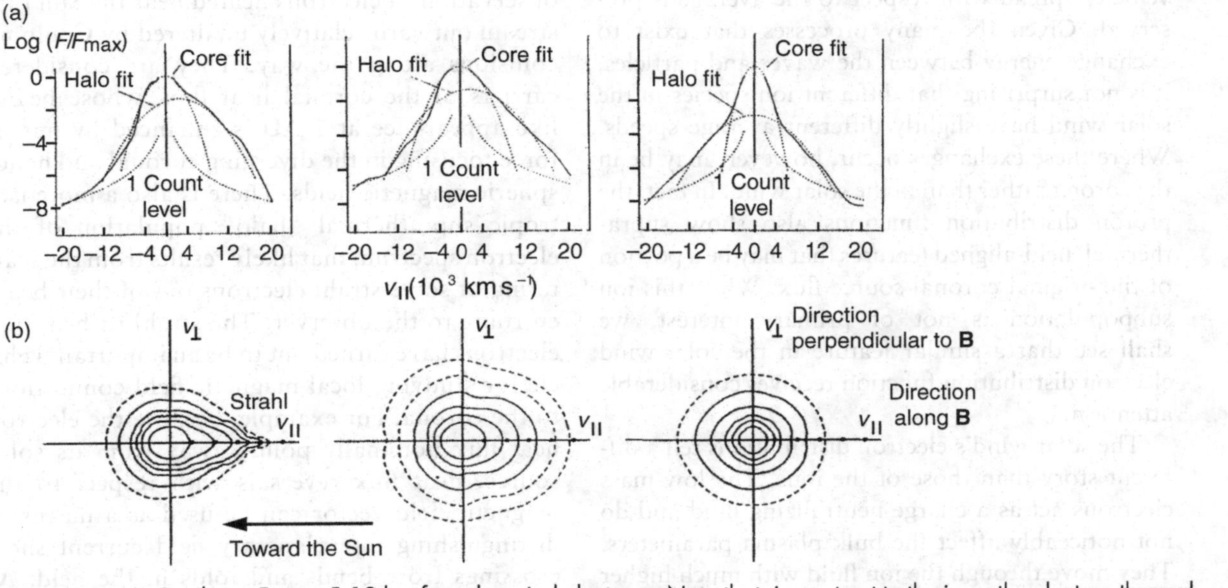

FIGURE 5.14. Similar to Figure 5.13 but for solar-wind electrons. Note that, in contrast to the ions, the electron thermal velocities are high compared to the solar-wind bulk speed (~400 km s^{-1}). The electrons have the special feature of a regular suprathermal component ("halo") that also has a highly field-aligned anisotropic part (the "strahl"). Flux F has been normalized by the maximum flux, F_{max}. (After Marsch, 2006.)

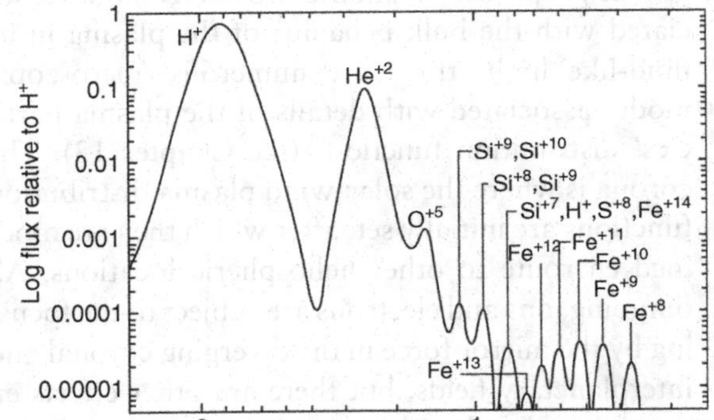

FIGURE 5.15. An example of a typical solar-wind ion spectrum illustrating both the ion composition and the shapes of the different ion peaks (here in energy per charge rather than versus velocity). The shapes and locations of the peaks are approximately co-moving maxwellians, but detailed analysis reveals some anisotropies and relative speed differences whose origins are still studied. (After Bame *et al.*, 1975.)

most abundant element helium, whose gyro-frequency is two or four times lower than the proton gyro-frequency depending on whether it is doubly or singly ionized.

Figure 5.15 shows that most solar-wind ions have total velocity distributions whose main components are shifted maxwellians characterized by a width (temperature) that is narrow (cold) compared to the peak (bulk plasma) speed. This means most of the energy (and momentum) in these species is in their bulk flow, but the heating, focusing, and beam instability/isotropization cycle described above can clearly transform what would be considered a perpendicular temperature increase to an increase in average outward ion speed. It is worth mentioning here that the term "temperature" is often used rather loosely in that the velocity distributions are often not the maxwellians assumed in the original gas-dynamic definition. However, if one generalizes the meaning to describe a moment of the distribution function involving the kinetic energy in the integral (see Chapter 3 for a discussion of moments), the basic idea of a measure of particle velocity spread with respect to the average is preserved. Given the many processes that exist to exchange energy between the waves and particles, it is not surprising that different ion species in the solar wind have slightly different average speeds. Where these exchanges occur, however, may be in the corona rather than in the solar wind. In fact, the proton distribution functions also show suprathermal, field-aligned features that may be a portion of the original coronal source flux. While this ion subpopulation is not of primary interest, we shall see that a similar feature in the solar-wind electron distribution function receives considerable attention.

The solar wind's electron distributions tell a different story than those of the ions. The low-mass electrons act as a charge-neutralizing fluid and do not noticeably affect the bulk plasma parameters. They move through the ion fluid with much higher thermal speeds, changing their densities to match those of the ions that contain essentially all the solar-wind mass and thus momentum. The electrons exert their control by transporting heat and by creating charge-separation electric fields that help the ions resist the Sun's gravitational pull. One interesting fact to think about is that, although the solar-wind fluid picture (Figure 5.2(a)) approximately describes the actual microscopic ion motions, it is not a very good description of the individual electron motions. Electrons are more tightly tied to the magnetic field by their small gyro-radii, making them excellent tracers of prevailing field topology for the time at which they are measured and of coronal conditions at their field-line footpoint. While their average or bulk speed is nearly the same as that of the ions, they can reach 1 AU via their thermal speeds in hours instead of the several days typical for the solar-wind ions. The examples of solar-wind electron velocity distributions in Figure 5.14, where the plasma bulk speed is indicated, show this important difference from the proton distribution in Figure 5.13(a). Another major difference is the prominent field-aligned feature above ~50 eV. This field-aligned electron beam within the distribution, known as the **strahl**, is particularly clear and narrowly focused in the high-speed streams from the centers of large coronal holes. It may be a direct observation of electrons heated near the Sun that stream outward relatively unaltered by Coulomb collisions along the way. They are considered carriers of the coronal heat flux, whose beam-like appearance at 1 AU is enhanced by mirror force focusing in the diverging coronal and heliospheric magnetic fields. There is also a more isotropic suprathermal "halo" population of the electron spectrum that likely results from the scattering of some strahl electrons out of their beam en route to the observer. The strahl or heat flux electrons have turned out to be an important vehicle for studying local magnetic field connections to the corona. For example, because the electron heat flux nominally points away from its solar source, heat flux reversals with respect to the magnetic field vector can be used as a means of distinguishing interplanetary field current sheet crossings from bends and folds in the field. As described in Section 5.3, they are also especially useful for studying the topologies of coronal transients in the solar wind.

Several types of waves are important in the physics of the solar wind and also diagnostics of its kinetic properties. In addition to MHD modes associated with the bulk behavior of the plasma in its fluid-like limit, there are numerous microscopic modes associated with details of the plasma particles' distribution functions (see Chapter 13). The corona is where the solar-wind plasma distribution functions are initially set, after which they are modified en route to other heliospheric locations. All outgoing ions and electrons are subject to the focusing by the mirror force in the diverging coronal and interplanetary fields, but there are other effects en route associated with current sheets – both heliospheric and localized, from field compressions producing local mirroring or cross-field drifts and/or from shocks producing local heating. There are also particle interactions with waves that may be locally self-generated by unstable distribution functions, or generated remotely or locally by minor particle populations or plasma processes. Waves are observed *in situ* in the solar wind in parameters whose measurements have sufficient time resolution. Among these are plasma waves seen by electric field antennas at frequencies near the local plasma frequency and gyro-frequencies (see Chapter 3) and

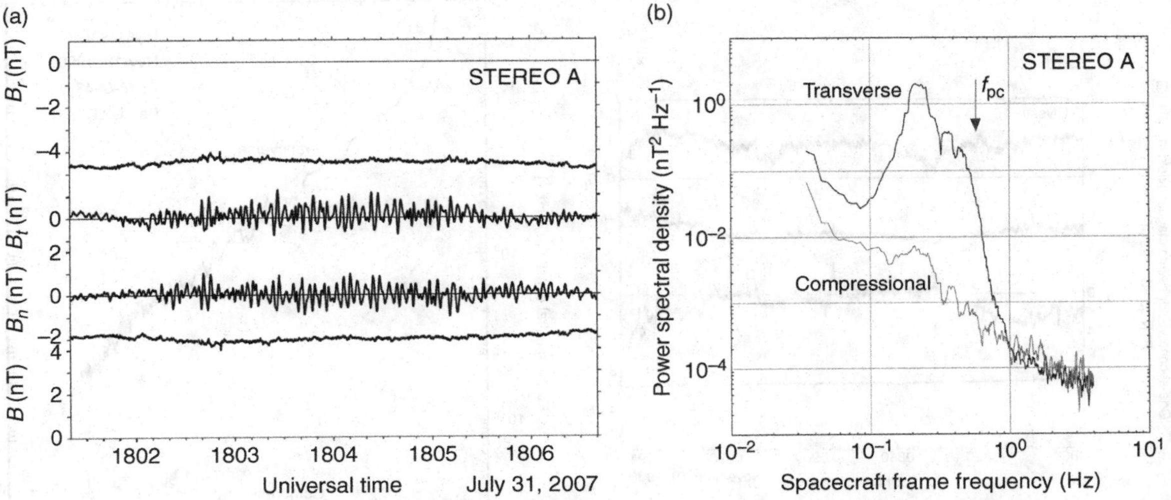

FIGURE 5.16. Five minutes of magnetic field data obtained at 1 AU (a) and its power spectrum (b), showing that the nearly transverse fluctuations have most of their power around the ion cyclotron frequency (f_{pc}).

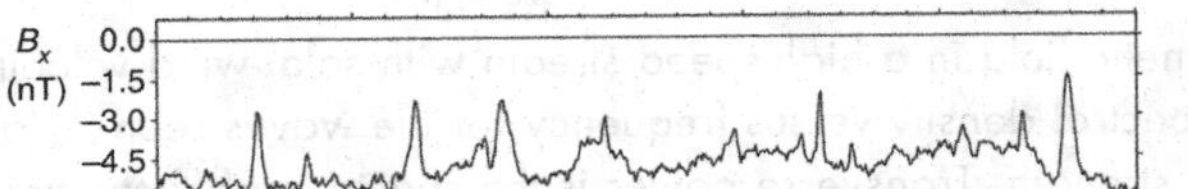

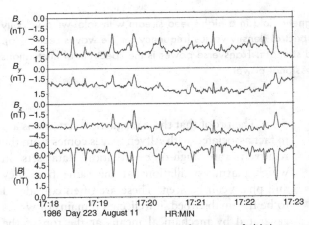

FIGURE 5.17. Sample time series of magnetic field data obtained at 0.7 AU showing fluctuations characteristic of mirror-mode waves. These can appear either as sharp depeletions in the field or as sharp peaks rather than classical waves, depending on the parameter regime and setting.

oscillations of the magnetic field at the ion cyclotron frequencies. Research on these types of heliospheric waves ranges from the diagnostics of coronal and extended solar-wind heating to the interpretation of interplanetary shock acceleration processes. Possibly key wave modes for the ions that have been detected in high cadence *in situ* magnetic field measurements are the ion cyclotron mode and the **mirror mode**, examples of which are shown in Figures 5.16 and 5.17, respectively.

A technique for identifying the nature of the waves is the analysis of the measurements in the frequency and/or wavelength domains for comparison with expected dispersion relations or instability criteria (see Chapter 13). The presence of mirror-mode waves is a signature of perpendicular ion distribution function anisotropy in the generation region. While it is easy to understand how the mirror force in diverging coronal and heliospheric magnetic fields continues to maintain a field-aligned anisotropy in the solar-wind distribution functions, the fact that perpendicular ion anisotropies are sometimes present raises the question of what produces these far from the region of coronal heating. Analyses of the anisotropies in the proton distribution functions within $T_\perp$ and $T_\parallel$ space (obtained from bimaxwellian fits) show that their ranges are bounded by the instability thresholds for the mirror, firehose, and ion cyclotron instabilities. For still unknown reasons, both the ion cyclotron and mirror-mode waves seem to occur in "storms," with the ion cyclotron episodes associated with particularly radial interplanetary fields. Interpreting what produces these different wave signatures can reveal insights into why the

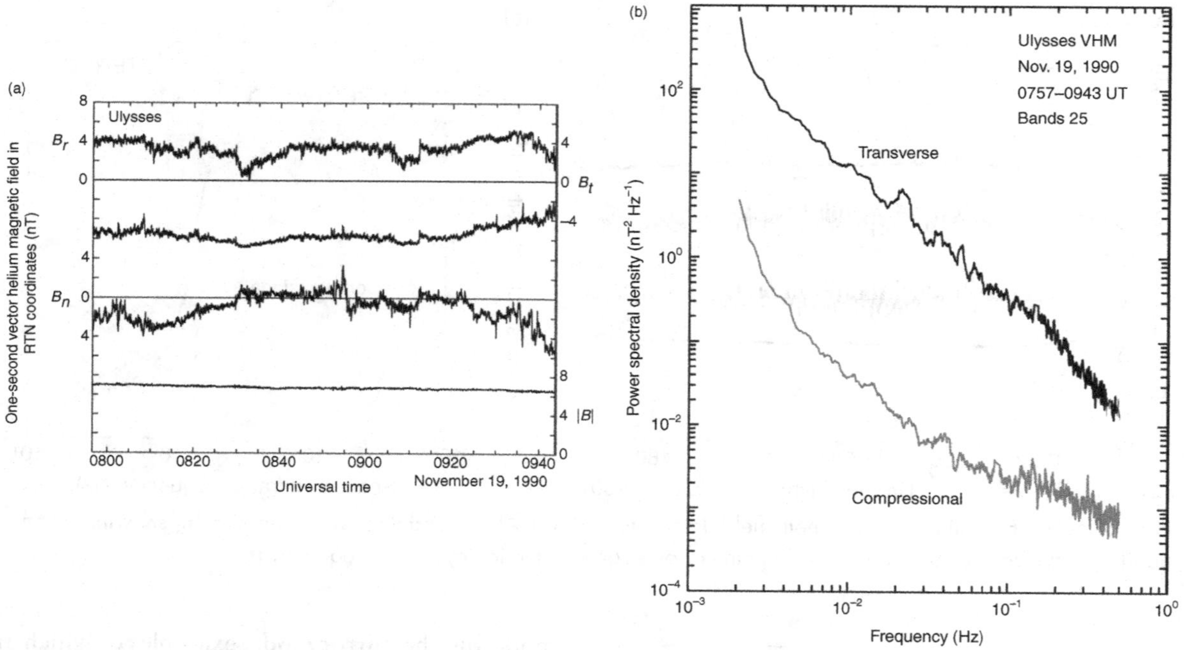

FIGURE 5.18. Measurements of the interplanetary magnetic field in a high-speed stream with solar-wind velocity v_{sw} = 725 km s^{-1}. (a) ~2 hour time series. (b) Power spectral density versus frequency for the waves seen in (a). Compressional power is the spectrum of magnetic field strength. Transverse power is the spectrum of fluctuations summed over three orthogonal sensors less the compressional power.

solar-wind ion distributions remain more nearly thermal than expected for a radially expanding collisionless plasma. In a similar vein, the plasma waves detected at higher frequencies with plasma wave/electric field antennas are attributed to electron distribution function irregularities and related plasma instabilities. The suprathermal strahl "beams," through their available free energy, must maintain a certain background of plasma waves in the solar wind to limit their streaming anisotropy. Studies of waves associated with electron behavior give insights into the physics and remote sensing of interplanetary shocks, which provides a good segue to waves associated with macroscopic attributes of the solar-wind plasma.

Of the fluid or bulk plasma modes (see Chapter 3), the local Alfvén waves are, in principle, readily identifiable by their magnetic oscillations with a corresponding velocity component. A sample time series and spectrum of ~5 min period Alfvén waves seen in the high-speed wind magnetic field is shown in Figure 5.18. The proof that these magnetic fluctuations are in fact associated with Alfvén waves comes from the relatively rare high-time-resolution data sets in which plasma oscillations at the same frequency and phase can be seen. These are often considered to be the unabsorbed, unreflected remains of waves generated by mechanical means at the top of the solar convection zone. Part of the reason for this interpretation is that they are observed most clearly in well-defined high-speed streams from large coronal holes, far from the coronal streamer and coronal hole boundary complications described earlier. Although MHD waves in the corona may undergo mode conversions (Chapter 13) after they are generated, these may be the strongest evidence we have of a role for Alfvén-wave-related coronal heating and post-critical point momentum deposition by these waves. The other MHD mode that is ubiquitous in the solar wind is the magnetosonic mode that includes density and pressure signatures. Compressions like those in stream interaction regions are typical settings for studying magnetoacoustic waves. The process of interplanetary shock formation by magnetoacoustic wave steepening (see Chapter 6) is central to the transfer of energy between different types of solar wind and the formation of suprathermal tails on the solar-wind electron and ion distribution functions. But the energy transfer in the solar wind must also involve other processes, such as fluid instabilities that generate macroscopic bulk plasma variations and contribute to macroscopic solar-wind variations. The Kelvin–Helmholtz instability, for example, transforms velocity shear between streams to vortices and breaking waves, and can introduce so-called turbulent boundary layers.

In solar-wind research, the study of turbulence covers variations on a wide range of spatial scales observed in the heliospheric magnetic field and plasma. Dictionary definitions of the word turbulence include "disorder," "agitation," and "eddying motion." The formal study of turbulence was originally developed in fluid mechanics, where there is much literature on the mathematical treatment of topics like wave steepening and power spectral descriptions of scales of turbulent structures. The role of turbulence in boundary fluid instabilities, and in the dissipation or transfer of flow energy, also applies in the solar-wind plasma. As mentioned above, shear flows at stream boundaries (for example) are found in abundance. In addition, the solar wind contains processes such as reconnection related to magnetic shears and MHD-unstable velocity shears. Thus it is often difficult to determine what fluctuations are waves, dynamic current sheets, rotational discontinuities, or solar-wind turbulence. The magnetic fluctuation spectrum that is observed on a spacecraft includes structures on scales from microphysical (e.g., plasma waves) to fractions of an astronomical unit (e.g., stream interaction regions). Some are simply convected by the solar wind, although they evolve with radial distance, while others are in the process of being actively generated or dissipated, or have additional propagation velocities of their own in the plasma frame. Complete sets of plasma measurements do not usually exist across the whole spectrum, which allows the sorting out of the different contributions: waves, convected structures formed closer to the Sun, or structures generated en route. Yet sometimes a classical "Kolmogorov" turbulence spectrum with a power law slope in wave number or energy of $-5/3$ is found in analyzing time series of solar-wind parameters. Still, turbulence terms and theory must be applied only to specific periods where conditions are well understood. In particular, studies of turbulence in the slow solar wind require an appreciation of its sometimes unsteady source(s), a topic that provides a good segue to the following section.

5.3 INTERPLANETARY TRANSIENTS

Movies of coronal images reveal that the solar-wind plasma typically does not come out in a steady stream. While the solar wind is continuously present in interplanetary space, with the steady-state picture described in the earlier parts of this chapter providing a good working model, it is constantly evolving in response to the evolving coronal magnetic field. Even under quiet coronal conditions, its field can be time varying, and its evolution can sometimes be explosive, with many consequences of interest for planet interactions and astrophysics. These different transient behaviors that, in their extreme form, include the results of coronal mass ejections, are the subject of this section.

5.3.1 The transient component of the solar wind

The expectation of a regular transient component of the solar wind has been discussed for decades, in part because observations of the evolving coronal structure require the existence of such a component. As the solar surface field changes in response to the dynamo-dependent active region flux emergence and ongoing flux transport processes described in the preceding chapter, the coronal field is constantly adjusting its open and closed field geometry to adapt to the new boundary conditions and minimize magnetic stresses. If the required changes are slow enough, the corona can adjust in a quasi-steady manner. Under such circumstances it can be approximated by a succession of potential field source surface models, even though coronal field reconnections and disconnections are occurring to make the necessary modest

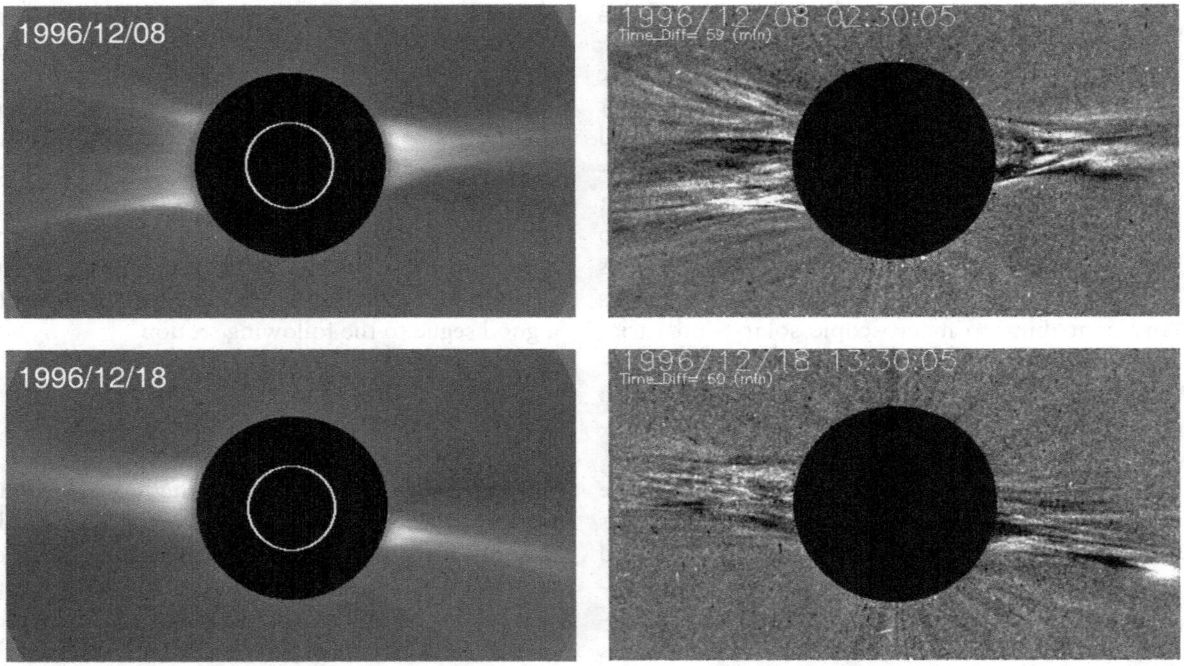

FIGURE 5.19. Examples of SOHO LASCO coronagraph images showing apparent "blobs" of denser coronal plasma departing from the ends and edges of coronal streamers in the difference images on the right. (ESA/NASA images.)

topological changes. These in turn must result in some solar-wind signature, especially at the boundaries of the different streams that connect to the boundaries of the various coronal hole sources.

While the difficulty of both theoretically describing and numerically modeling the evolving solar wind has delayed its detailed study, observational evidence for its contribution has grown with the launch of the twin STEREO spacecraft. In particular, both coronagraph and the Heliospheric Imager images mentioned earlier show the apparent release of "blobs" of coronal plasma from the vicinity of coronal rays (see Figures 5.19 and 5.12B). These are observed to accelerate from low speeds up to the slow solar-wind speed (~300–400 km s^{-1}) within the coronagraph field of view. Given that IPS observations depend on convected small-scale structures for the technique to work as a solar-wind tracer, these observations are not surprising, but their overall impact on the character of the solar wind has not been clear. Investigations now concern the details of the generation and propagation of this blob wind, also called "streamer wind" because of its collocation with the coronal streamers – or, more precisely, their boundaries. Some of these implicitly or explicitly explore the question of whether this transient component is the main source for the slow solar wind, or whether there is also an equally important steady slow solar-wind component associated with the highly diverging open coronal fields near the edges of coronal holes as some solar-wind source mapping investigations and models have suggested. Toward this goal, studies connecting imaged blobs leaving the Sun to *in situ* measurements of features in the solar wind have been carried out with some success, while other studies of inferred *in situ* samplings of coronal hole boundaries show no evidence of transient structures. Not all boundary locations may adjust at once. It is also important to consider, when viewing the coronagraph and heliospheric images of these blobs, that the viewer perspective may give different impressions of a transient's origin, size, shape, speed, and direction. In addition, solar-wind stream interactions and evolution that occur in transit to 1 AU complicate the identification of specific coronal features. Relating what is imaged to *in situ* information is especially challenging in this area of investigation.

Important independent clues to the existence and origins of a transient solar-wind component are the ion composition and charge state measurements mentioned earlier. The notion of hot, magnetically closed coronal loops producing higher charge states (e.g., higher O^{+7}/O^{+6} ratios) and more abundant heavy ion content (e.g., higher Fe/O) in slow solar wind, for example (recall Table 5.1), could be used as part of the overall argument that recently closed/newly opening coronal fields supply much of the slow wind. Another feature sometimes mentioned in related work is the so-called heliospheric plasma sheet, a region of higher than average solar-wind density sometimes found around the heliospheric current sheet but not associated with interstream compressions. Finally, there is evidence in some *in situ* magnetic field observations of small flux-rope-like features that suggest plasmoids (see Chapter 9) are shed from the coronal streamer belt rays. The relationships between all of these different observations of "slow wind" and coronal features are still under investigation. The generation of the transient component of the solar wind is probably best investigated through time-dependent MHD modeling, where evolving surface fields, including flux emergence and evolution, effectively drive related time-dependent behavior in the modeled corona and solar wind. MHD models of coronal streamers have been developed to study streamer properties, and some have even exhibited instabilities, suggesting streamer shedding of material at their boundaries and cusps. But the interested reader who investigates these will find they do not (yet) explore the effects of the lower boundary evolution at the center of our discussion here.

5.3.2 Interplanetary coronal mass ejections

Coronagraph images show a broad spectrum of transients, with the largest and brightest being the CMEs, the **coronal mass ejections** described in Chapter 4. In contrast to the small blobs just discussed, CMEs stand out because they often appear as well-defined expanding loops or jets. Figure 5.20 shows an example of a CME as it appears first in the coronagraph image and later in the Heliospheric Imager picture. While most move outward at speeds that reach values no more than typical solar-wind speeds, others explosively blast through the ambient corona, rapidly accelerating and reaching up to several thousand kilometers per second within a few tens of solar radii. Some of the slower CMEs must merge into the general coronal outflows, as they produce no obvious *in situ* signatures at 1 AU. The faster ones produce interplanetary counterparts known as ICMEs, where the 'I' distinguishes the *in situ* or interplanetary entity from the phenomenon seen in the corona. The classic ICME plasma and field signature, an example of which is shown in Figure 5.21, takes about 1–2 days to pass a stationary observer. It sometimes includes a moderate to strong ($M_{ms} > 3$) leading shock, followed by an interval of shock-heated, compressed solar-wind plasma. This interval is the magnetosheath or sheath of the ICME. It is produced by the compression and deflection of solar wind around the coronal ejecta plowing into the ambient flow ahead. The coronal material is sometimes referred to as the "driver gas" because it acts like a piston. It typically has normal solar-wind density but enhanced and unusually fluctuation-free magnetic fields. The ion temperature is often low relative to the solar wind, and the magnetic field strength is enhanced, making the plasma beta noticeably lower than in typical solar wind. In the preceding chapter it was noted that occasionally a filament is seen erupting together with a CME, giving the appearance of a bright core in the coronagraph image that is otherwise dominated by the higher densities at the leading edge of the departing loop. The filament material, which is chromospheric in origin, is sometimes detected in the ICME as a density increase within or trailing the ejecta. Its special nature is often identifiable by an unusual He^{+} content (most solar-wind helium is fully ionized He^{++}). ICMEs may also show other atypical ion composition signatures, such as enhanced He and Fe abundances, and higher charge states of heavy ions, perhaps related to the occurrence of an associated flare in the low corona.

Roughly one-third of the time, the magnetic fields in the ejecta exhibit slow, smooth rotations

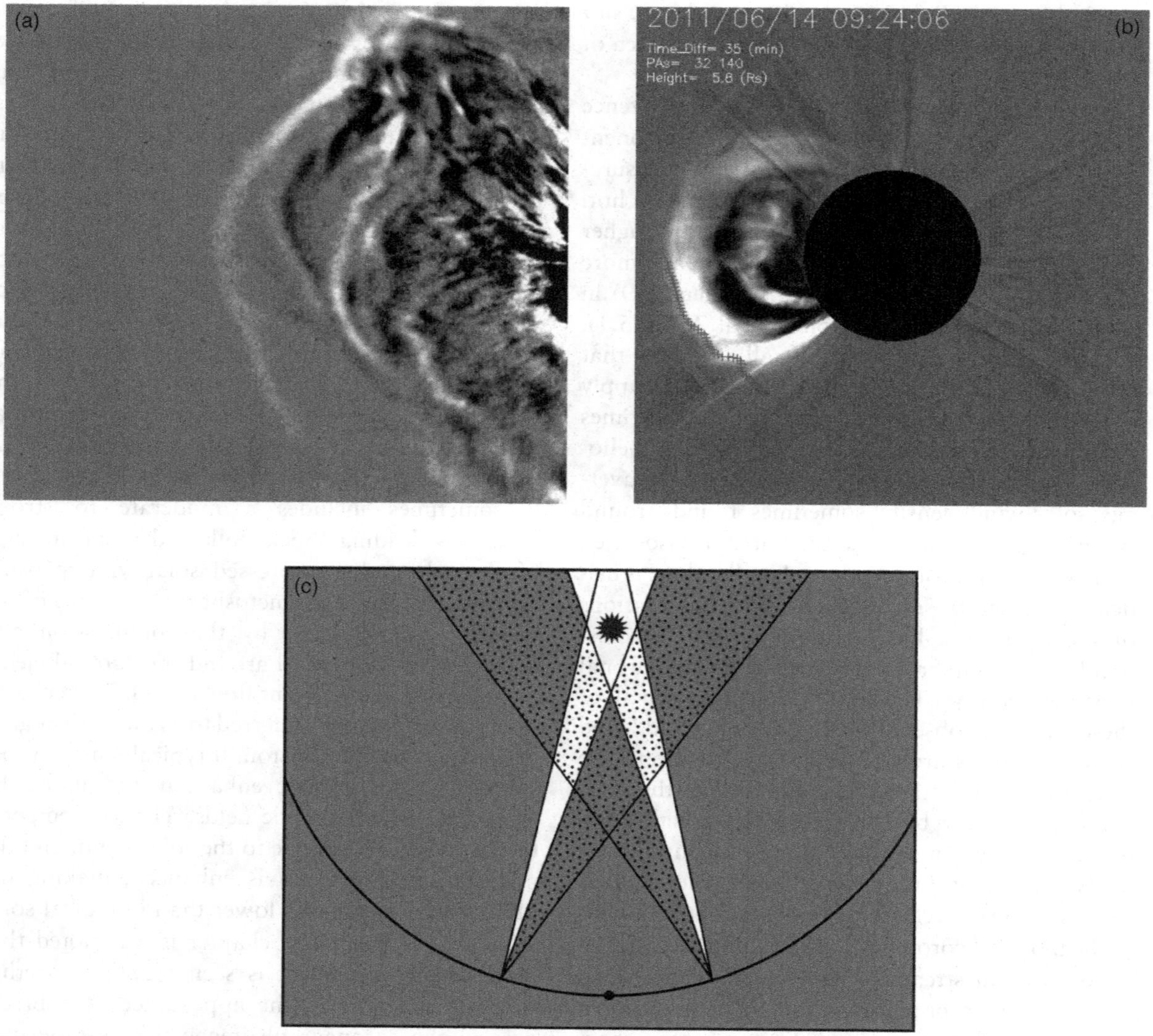

FIGURE 5.20. (a) STEREO-SECCHI heliospheric difference image of the coronal mass ejection seen earlier close to the Sun with its coronagraph (b). (c) Geometries of the two STEREO Heliospheric Imager fields of view (dark stippled) relative to their coronographic counterparts (white fields of view). (NASA.)

with a significant out-of-ecliptic (B_z) component as the structure passes. The observed field rotation may appear to include the entire ejecta or only part of it, or a succession of field rotations may be present. On occasion, especially in high-heliolatitude observations like those from Ulysses, there is also a reverse shock following the ejecta, indicating significant expansion as well as bulk motion of the erupted mass. In many discussions of this phenomenon, the enhanced, smoothly rotating magnetic field is referred to as a **magnetic cloud** (MC), and is interpreted as a large flux rope originating with the coronal CME. This is the origin of the standard illustration of an ICME in Figure 5.22, which suggests the expulsion of a coronal flux rope into the heliosphere that remains magnetically connected to the Sun at both ends. Sometimes intervals of particularly distinct counterstreaming suprathermal electrons are apparent inside the ejecta, consistent with the strahl coming from the

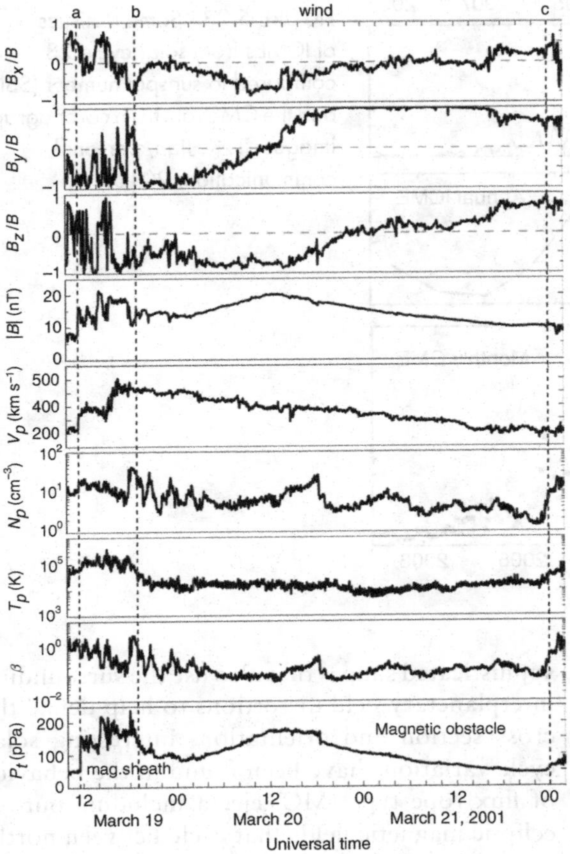

FIGURE 5.21. Examples of a ~2.5 day time series of solar-wind observations at 1 AU obtained during the passage of an ICME. The primary regions of the classic ICME signature – the leading shock, the sheath, and the ejecta or magnetic obstacle – are indicated by vertical lines a, b, c, respectively. (WIND spacecraft data (NASA).)

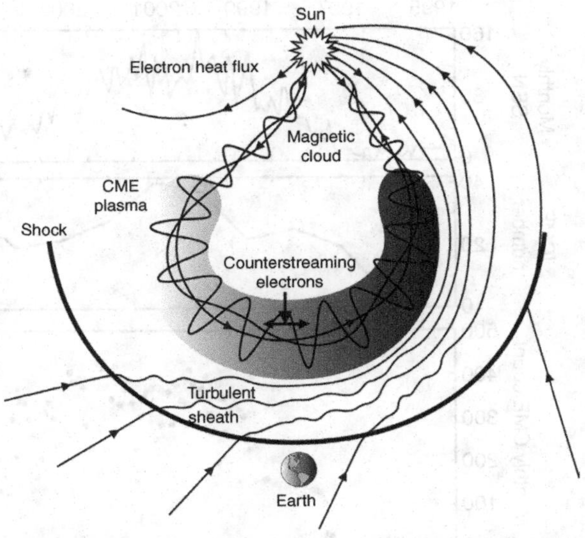

FIGURE 5.22. Illustration often used to explain the ICME signature. The ejecta is shown as a magnetic flux rope called a "cloud" in many descriptions. This illustration is probably over-simplified for most real cases; however, it appears to provide a good first-order description of simple events that are not heavily distorted or eroded in the course of their propagation. Some ICMEs that appear to originate from streamer belt CMEs are observed within the interplanetary field sector boundary.

corona along both "legs". However, one must keep in mind that Figure 5.22 is only a rough idealization of the basic ICME concept, and that counterstreaming electrons are not always present.

In general, not all observed ICMEs fit the above picture well, and there is still discussion of whether all ejecta are flux ropes. The total pressure time profiles for ICMEs in the ecliptic have a distinctively different character than those for SIRs (Figure 5.11(a)). Instead of a symmetric signature, they exhibit different time profiles depending on whether or not the coronal ejecta are encountered. If only the leading shock and its magnetosheath are encountered, and not the driving magnetic cloud, the total pressure signature jumps up at the shock arrival and then gradually diminishes. If the magnetic cloud is encountered, the total pressure often grows to a maximum, following the initial increase at shock arrival, before decreasing. There are, in practice, a number of other ICME properties used in observational analyses, some of which are described below. But it is important to be aware that a particular interplanetary structure rarely has all of them. It will also become apparent how limiting *in situ* spacecraft sampling is for diagnosing the details of these transients.

Figure 5.23 shows the results of a survey of ICME occurrence during the previous solar cycle, based on the *in situ* data. The total pressure signature was used to identify the events in this case. ICMEs clearly follow the sunspot number trend, as expected from CME occurrence rates derived from coronal images. Even though there can be

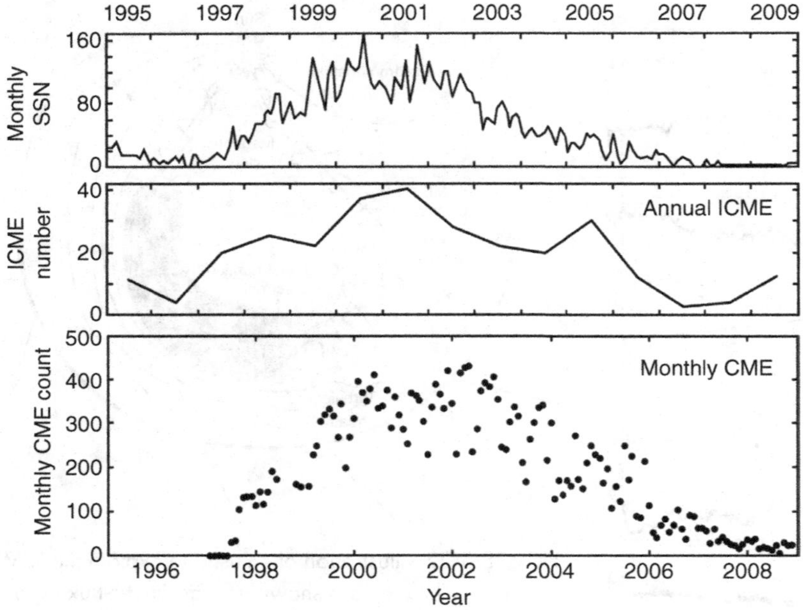

FIGURE 5.23. Some statistics of ICMEs from solar cycle 23 compared to sunspot number (SSN) and the CME rate from coronagraph images. (L. K. Jian, personal communication, 2010.)

several CMEs a day observed by coronagraphs at active times, their finite spatial extent reduces the number of related *in situ* (ICME) signatures at a single location in the ecliptic at 1 AU to only a few per month. In comparison, there are routinely two to four stream interaction regions observed near Earth during each ~27 day solar rotation period throughout the cycle. The influence of the effects on average interplanetary conditions of the few ICMEs contributing per solar rotation, with few exceptions, only modestly influences solar cycle trends in solar rotation-averaged velocity and density (see Figure 5.8). On the other hand, the ICMEs produce the most extreme plasma and field parameters at active times, and their out-of-ecliptic magnetic field contributions are key to their sometimes major geomagnetic effects (see Chapter 9).

Much attention has been focused on the fitting of ICME ejecta *in situ* magnetic fields (e.g., Figure 5.21) with force-free flux ropes (introduced in Chapter 3), in part to gain an understanding of the implications of the spacecraft sampling and also in attempts to relate the ejecta to their coronal source regions (see Chapter 4). These fits are usually applied using axisymmetric circular or elliptical cross section flux ropes with axis orientation as a parameter, or more sophisticated shapes that also use the surrounding interplanetary field distortions to help define the cross section and orientation. Interesting solar cycle variations have been found in the behavior of flux rope type (MC) ejecta including out-of-ecliptic magnetic fields that cycle between northward to southward rotations and southward to northward rotations on alternate cycles (Figure 5.24). Some studies suggest this behavior can be understood if streamer belt or polar crown filament channel CME source locations are assumed (see Figure 4.28). However, while the long-term trends in the ejecta field polarities stand up to scrutiny, the details are not understood to the point where a particular ICME field can (yet) be confidently predicted from its observed coronal source location.

The fact that only about one-third of ICME ejecta fields are fit well by simple flux ropes may in part be due to the limited *in situ* sampling of the passing structure, but also because not all CME ejecta have idealized forms. Some CMEs appear to develop from their onset in complicated ways, with images suggesting rotation or twisting of even initially simple structures before they leave the coronagraph field of view (Figure 5.25). The CMEs that involve a filament in the initiation region may be

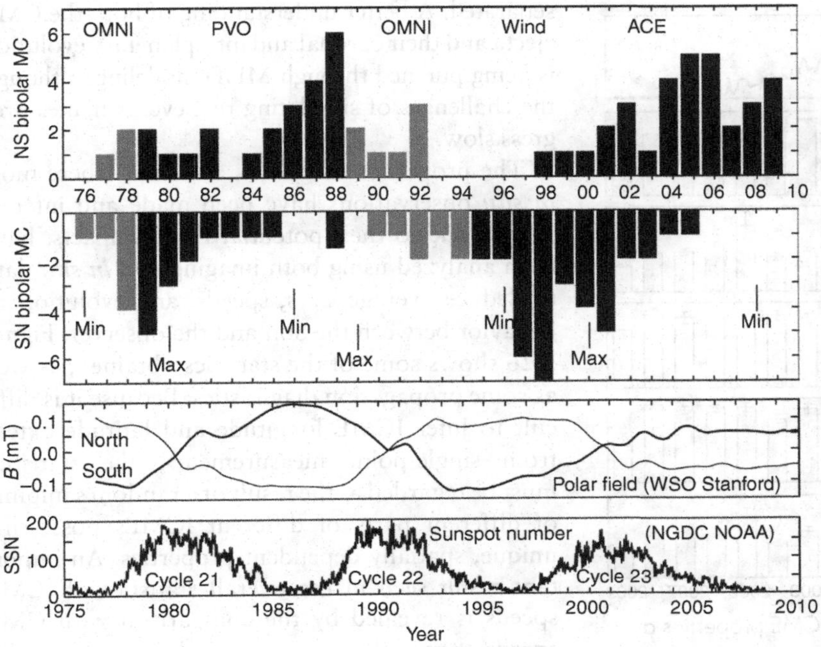

FIGURE 5.24. Solar cycle statistics of ICME magnetic field "polarity" or north–south magnetic field component cloud ejecta time sequences (NS or SN). The solar polar field cycle* and the sunspot cycle are also shown for comparison. Recall that the solar polar field cycle has a different phase than the active region cycle (see Chapter 4). The relationship between all these fields is still under investigation.
* See Wilcox Solar Observatory (wso.stanford.edu). (Adapted from Li *et al.*, 2011.)

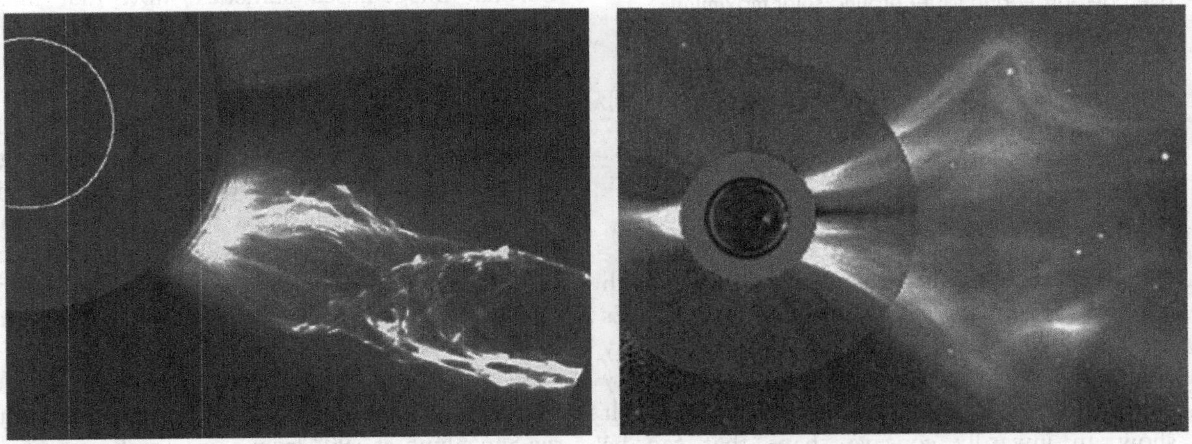

FIGURE 5.25. Illustrations of the distortion of CMEs that is sometimes observed in coronagraph and heliospheric images as they erupt. These changes in the appearance and orientation of the initial erupted coronal structure may be due in part to its interaction with the surrounding coronal structures. They can continue to accumulate as the ejecta move out into the solar wind, where the stream structure presents yet another source of modifications, making it difficult to relate what is observed in the low corona to what is observed driving the ICME at 1 AU. (SOHO LASCO images (ESA/NASA).)

affected if it makes a sufficiently massive addition to the erupting structure. And considering the previously described complications of the surrounding coronal and solar-wind structure, even simple ejections may be significantly altered by their environs as they move outward from their source regions. Especially around solar maximum, when CMEs are frequent, the *in situ* appearance of ICMEs is further complicated by the interaction of eruptions – either in the corona or in transit. Also, recall from

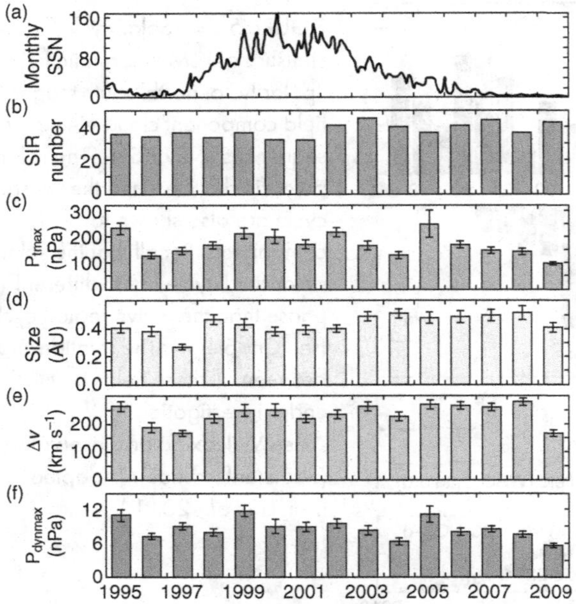

FIGURE 5.26. Annual statistics of ICME properties at 1 AU obtained over the solar cycle. ICMEs tend to be, on average, larger and faster around solar maximum. (a) Average monthly sunspot number; (b) number of stream interaction regions per year; (c) average static pressure at maximum of each SIR (magnetic plus thermal); (d) average radial dimension of ICMEs; (e) average change in velocity across ICMEs. (L. K. Jian, personal communication, 2015.)

Chapter 4 that CMEs are often associated with coronal streamers, which means that many CMEs are launched into and interact with the already complex slow solar wind. Images of particularly wide loop-like CMEs centered on the streamer belt show an inwardly concave shape that can be understood kinematically as structures whose outer parts move at higher speed with the faster coronal hole wind at their edges, while their centers run into the slow wind associated with the streamer head-on. The coronagraph images in Figure 5.25 illustrate the appearance of CME shape changes from interaction with the ambient and coronal solar-wind conditions. Multiple-spacecraft measurements of the same events are used to investigate ICME structures at different locations with mixed results. They often appear quite different at each site, even when the spacecraft are not widely separated. A better understanding of both the CME ejecta and their coronal and interplanetary evolution is being pursued through MHD modeling, although the challenges of simulating real events makes progress slow.

The properties of ICMEs at 1 AU, where most *in situ* observations have been made and interest is high due to their potential Earth impacts, have been analyzed using both imaging and *in situ* data to deduce average sizes, speeds, and evolutionary behavior between the Sun and the observer. Figure 5.26 shows some of the statistics obtained, as well as some propagation diagnostics. Because it is difficult to infer ICME longitude and latitude extent from single-point measurements, the statistics must be regarded as the result of a random sampling of different parts of different ICMEs possessing unique, spatially dependent properties. An important result seen in the overall statistics of ICME speeds is revealed by the comparison with CME speeds over the same period (Figure 5.27). Coronal image-based statistics show that most CMEs are slow when they leave the coronagraph field of view (at up to a few tens of solar radii), with many starting at only a few tens of kilometers per second to solar-wind speeds. At 1 AU, the ICME speed statistics show a maximum near the average solar-wind speeds of ~300–400 km s^{-1}. Both distributions show a high-speed tail, although some CMEs exhibit higher speeds. This difference can be better understood using so-called quadrature observations of CMEs and ICMEs, a technique illustrated in Figure 5.28(a).

In quadrature observations, an observer with a coronagraph observes CMEs departing from near the Sun's limb at ~90° from a spacecraft measuring the related *in situ* ICMEs. This allows a more accurate determination of speed and speed changes in transit. The results (see Figure 5.28(b)) show that CMEs moving slower than the solar wind at the Sun accelerate up to the solar-wind speed by the time they reach 1 AU, while CMEs moving faster than the average solar-wind speeds slow down until they converge on the ambient solar-wind value. In both cases the interaction with the ambient solar wind must influence the CME speed in transit. Identifying the physics that produces these effects, which may include various friction-like forces from fluid plasma

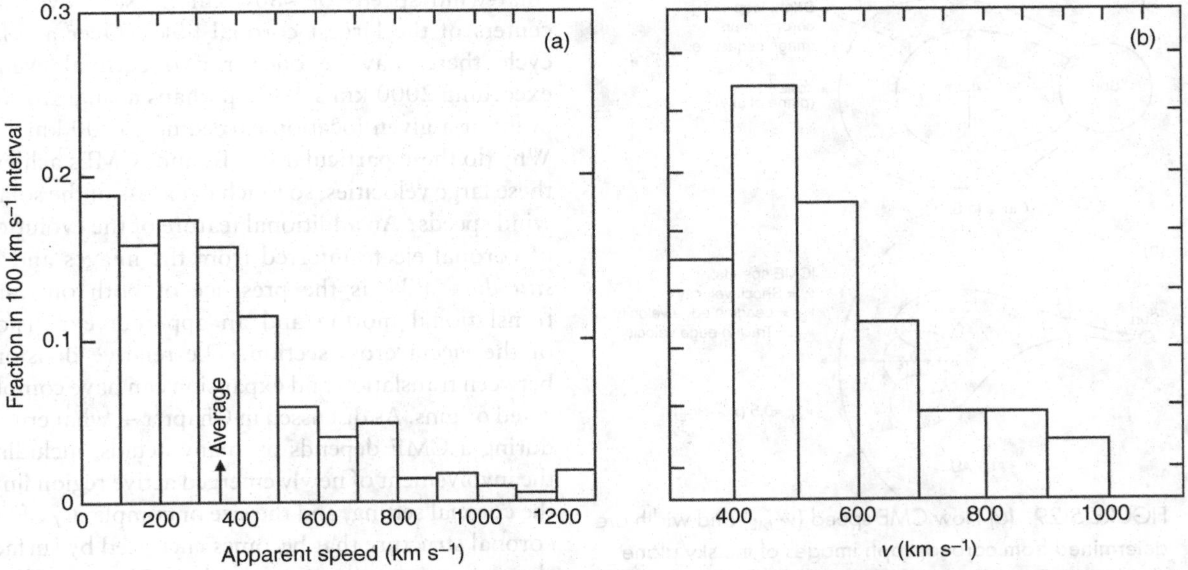

FIGURE 5.27. Comparison of the statistics of CME speeds obtained from coronagraph images (a) and ICME speeds obtained at 1 AU (b), suggesting that the slowest CMEs at the Sun must either vanish from detectability in the local solar wind or be accelerated to average solar-wind speeds as they move outward. ((a) After Hundhausen, Burkepile, and St. Cyr, 1994; (b) after Bothmer and Schwenn, 1998.)

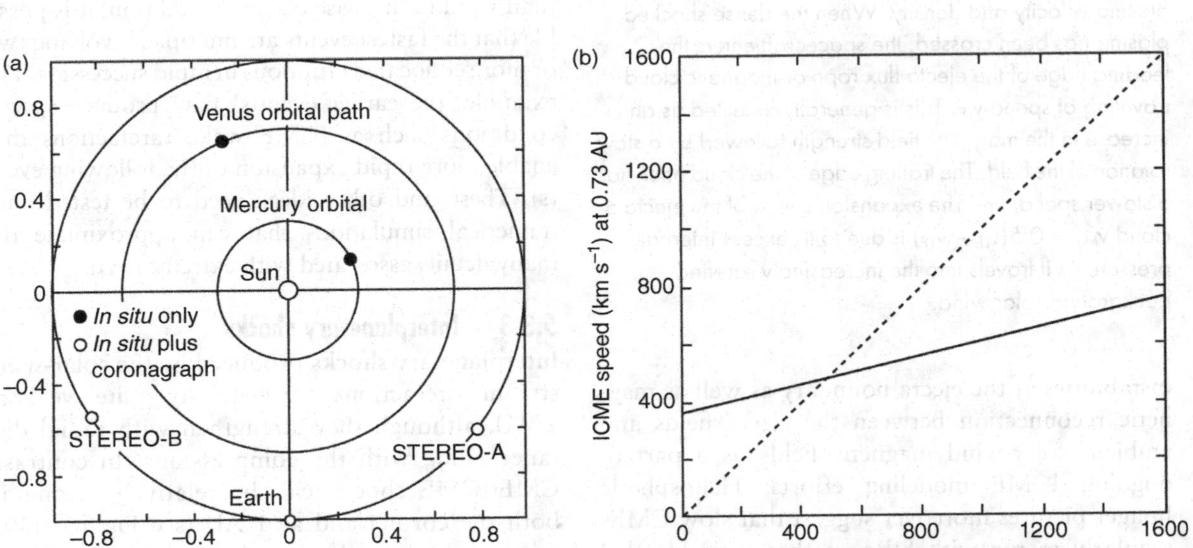

FIGURE 5.28. (a) Illustrations of possible "quadrature" observations to compare CME speeds measured at the solar limb (from images) and *in situ* related ICME speeds, roughly 90° away. (b) The trends in speeds determined from such measurements are indicated by the solid line, while the dashed line represents no difference. This technique suggests that slow CMEs accelerate before they are seen as ICMEs, while fast CMEs decelerate. ((b) Based on results of Lindsay *et al.*, 1999.)

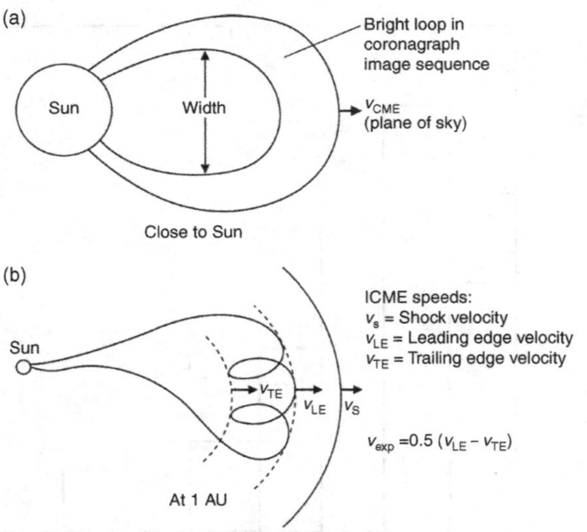

FIGURE 5.29. (a) How CME speed (v_{CME}) and width are determined from coronagraph images of the sky plane close to the Sun. These speeds are typically based on the motion of the front of a bright loop in a sequence of images. (b) Various speeds involved in ICME diagnostics, usually with *in situ* measurements. The speed of any leading shock is labeled v_S. It is detected as a jump in the plasma velocity and density. When the dense shocked plasma has been crossed, the spacecraft enters the leading edge of the ejecta flux rope or magnetic cloud traveling at speed v_{LE}. This is generally detected as an increase in the magnetic field strength followed by a slow rotation of the field. The trailing edge of the cloud often has a slower speed, v_{TE}. The expansion speed of the ejecta or cloud $v_{exp} = 0.5(v_{LE} - v_{TE})$ is due to its excess internal pressure as it travels into the increasingly rarefied background solar wind.

instabilities at the ejecta boundary as well as magnetic reconnection between the ejecta fields and ambient solar-wind magnetic fields, is a part of ongoing ICME modeling efforts. Heliospheric Imager pictures moreover suggest that slow CMEs regularly merge with SIRs so that their blended structures become practically indistinguishable by 1 AU. Recall that the coronal streamer transients also meet this fate, suggesting that a complicated mix of evolving structures of several scales is always present in the solar wind at some level.

The statistics of CME and ICME speeds (Figure 5.27) show how few reach speeds above the fastest solar-wind speeds of ~800 km s^{-1} seen over the centers of the largest coronal holes. Over a solar cycle there may be one or two coronal events exceeding 2000 km s^{-1} and perhaps a single *in situ* event at a given location exceeding ~1200 km s^{-1}. Why do these particular CMEs and ICMEs achieve these large velocities, so much in excess of the solar-wind speeds? An additional feature of the evolution of coronal ejecta inferred from the images and *in situ* data alike is the presence of both outward translational motion and an apparent expansion of the ejecta cross section. The relative divisions between translation and expansion can have complicated origins. As discussed in Chapter 4, what erupts during a CME depends on many details, including the involvement of newly emerged active region flux, the coronal setting, and the size or complexity of the coronal structure that becomes energized by surface shear flows or flux cancelation. However it is decided, the internal pressure of the ejecta relative to its surroundings must determine how much expansion contributes to the creation of a larger and faster ICME with a stronger leading shock than would otherwise occur. In addition, it is possible that the fastest events are multiple, involving two or more collocated eruptions in rapid succession. For example, the earlier event(s) may produce special conditions such as ICME wake rarefactions that enable more rapid expansion of the following event (s). These and other ideas need to be tested with numerical simulations that can approximate the many details associated with extreme events.

5.3.3 Interplanetary shocks

Interplanetary shocks produced by the solar-wind stream interactions are generally quite weak at 1 AU, although they strengthen with radial distance along with the compressions. In contrast, CME/ICME shocks can be relatively strong in both the corona and at 1 AU (see Figure 5.30). Still, Earth's bow shock and planetary bow shocks generally remain the strongest heliospheric shocks because of the speed of the external plasma flow relative to the obstacle. The fact that the CME ejecta are moving in the same direction as the ambient solar-wind plasma impacts their strength, so that only those few with speeds >800 km s^{-1} (Figure 5.27) can make comparable shocks.

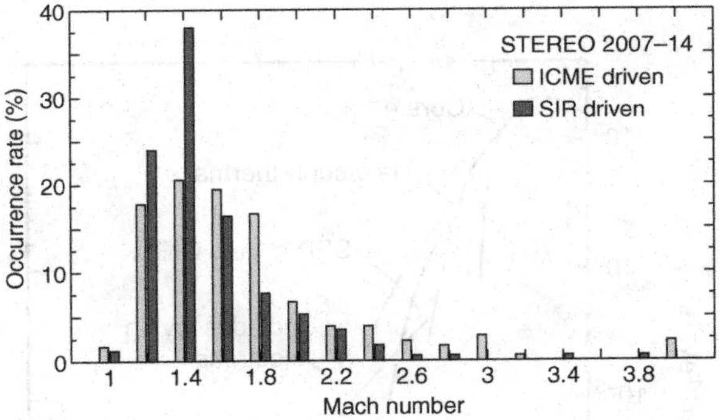

FIGURE 5.30. Comparison of statistics of Mach numbers of observed ICME and SIR/CIR shocks at 1 AU. (L. K. Jian, personal communication, 2015.) The ICMEs generally produce stronger shocks. The shock normals are also different given the different shapes of the shock drivers. (See Figures 5.11(b) and 5.22.)

Nevertheless, by the same physics (see Chapter 6), magnetosonic waves steepen ahead of the CME ejecta to form a leading structure that is much like a planetary bow shock except for its relatively large scale. Whether the ICME shock can be regarded as a continuation of a coronal shock or as a separate shock was discussed earlier, but the answer of course depends on the speed of the particular ejection in question and its path through the coronal domain inside ~$20R_S$ and the heliosphere beyond. Recall that the ejecta often expand as well as move bodily through the solar wind, and so these combined actions determine the interplanetary shock extent and shape, which is modified by the non-uniform surrounding solar wind. One might expect the larger, faster, and higher internal pressure events to preserve more of their coronal attributes when they reach 1 AU than do the more typical modest events. Indeed, analyses of strong ICMEs suggest they expand more nearly self-similarly, with their shock front normals oriented in an approximately radial direction as if the distortion effects described above are minimal.

ICME shocks might be expected to have the equivalent of Earth's foreshock, where waves are generated by a reflected component of incident solar-wind ions – producing anisotropic ion distributions and associated instabilities. However, attempts to organize wave observations around interplanetary shocks are still at a relatively early stage and show less of an obvious pattern. The reason is in part that the large scale makes it difficult to infer the global context of a particular shock crossing, and in part because high enough cadence plasma and field measurements have not been as routine on heliospheric spacecraft as on Earth orbiters. In moving ahead to the next subject, solar energetic particles, it is worth pointing out that the MHD models of CMEs mentioned in Chapter 4 involving "cone-shaped" injections of plasma into realistic solar-wind models produce global pictures of the evolution of an ICME-like shock. These models give a rough idea of how the shock shape and parameters change between the corona and a heliospheric observer, with the caveat that the absence of ejecta/driver descriptions limits their applications.

5.3.4 Solar energetic particles

A discussion of heliospheric shocks naturally segues into the topic of **solar energetic particles** (SEPs). The discussion of coronal particle acceleration and coronal shocks in Chapter 4 already touched on this subject; here we confront the larger picture of what is detected in the heliosphere. In the literature one finds several related acronyms: SEPs, ESPs, and SPEs, which can be confusing. These respectively refer to solar energetic particles (SEPs), energetic storm particles (ESPs), and solar particle events (SPEs). The more widely used terms SEP and ESP events respectively refer to solar energetic particles in general (SEPs) and enhancements in their fluxes that may occur around the time the ICME shock passes the observer (ESP events). Some discussions of SEP events include ESP events as a part of a broader definition. Here we adopt this broader definition and regard ESP events as a feature of some but not all SEP events. SEP events are further broken down into categories based on their duration and time profile appearances as impulsive and gradual events, and also by their inferred origins as flare or ICME (interplanetary) shock-source events. The observations and physics underlying this nomenclature will become clearer in the following paragraphs.

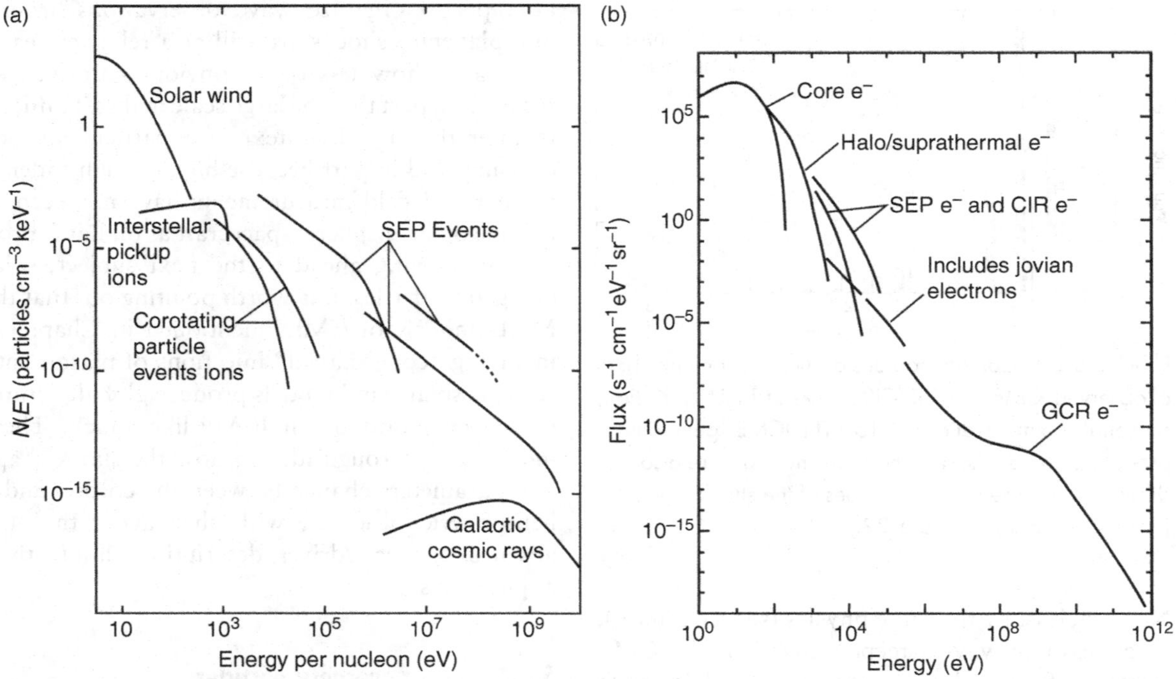

FIGURE 5.31. Heliospheric particle spectra. (a) Protons (at 1 AU); (b) electrons, showing the broad perspective on fluxes and energies of the different populations, including solar energetic particles.

Solar energetic particle (SEP) events are characterized by temporary intensifications of ions or electrons with suprathermal energies exceeding some loosely defined threshold, usually on the order of hundreds of kiloelectronvolts for ions and tens of kiloelectronvolts for electrons. As is the case for all heliospheric ion populations, SEP ions are dominated by protons but have heavy ion contributions roughly in accord with photospheric values. The place SEPs occupy in the overall heliospheric particle energy and flux spectrum at 1 AU is shown in Figure 5.31. Unlike galactic cosmic rays that are the dominant contributor above ~100 MeV for ions and a few megaelectronvolts for electrons and are relatively steady, SEP fluxes increase up to several orders of magnitude in events lasting from less than an hour for the weakest events to several days for strong events. On some occasions, unusually long durations suggest the merging of several events or interplanetary conditions conducive to SEP trapping in the background magnetic fields. In Chapter 4 a few mechanisms for particle acceleration in the corona, including in flare sites or at flare- or CME-driven shocks, were briefly described. SEPs may originate in these settings and propagate outward along open coronal field lines, or they may instead originate at ICME shocks in the solar wind, at heliocentric distances up to and beyond 1 AU. Together, the time profiles, energy spectra, pitch angle distributions, and composition of SEP events provide evidence of their origins. Here we emphasize the dominant proton component of SEPs, with some briefer mention of the accompanying heavier ions and electron characteristics.

The simplest SEP events to interpret are in a group of relatively weak events labeled "impulsive." These often accompany flares and their associated x-ray and radio bursts (see the descriptions of the various solar radio bursts in Chapter 4). In addition to their modest fluxes, these events are characterized by their rapid onset and decay in time profiles lasting hours at most, their steep ("soft") energy spectra, and their relatively high electron to proton ratio relative to the much longer duration "gradual" events discussed below. Figure 5.32(a) shows examples of typical impulsive events. Impulsive SEPs are sometimes

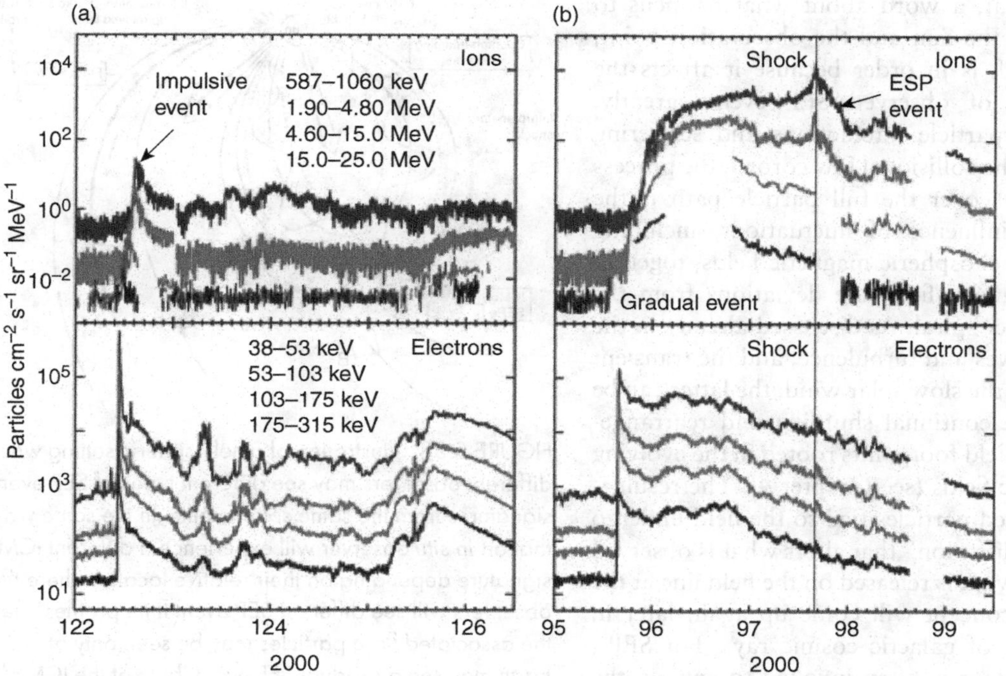

FIGURE 5.32. Illustrations of the main types of observed solar energetic particle (SEP) events as time series or profiles of ion and electron measurements at several energies. SEP events are generally categorized as impulsive, with rapid onsets and decays over several hour time scales, and "gradual," which may have either a sudden or slow onset and can last for several days. Flare events generally fall under the impulsive category, while ICME-related events define the gradual category. The increase in ion flux at the time of the local shock passage in (b) is called an ESP event (see text). (After D. Lario, 2005.)

also referred to as flare SEPs because they are considered to come directly from flare sites on the Sun, some of which may be too small or faint to detect. The onsets of the impulsive events in Figure 5.32 (a) nearly coincide with the x-ray event onset, allowing for the difference in light travel (~8 min) and particle travel (~ tens of minutes to hours) times. As discussed in Chapter 4, acceleration mechanisms at flare sites can include reconnection-related (mainly electron) acceleration, shock-related acceleration, and wave–particle interactions. The dominant mechanisms may be quite different for electrons and ions. The electrons, which also cause the radio bursts, may be most easily energized in reconnection sites, whereas the ions are more often associated with shock acceleration and, especially in the case of specific heavy ion species such as helium-3, may be singled out by resonant wave–particle interaction processes (see the discussion of wave acceleration by gyro-resonance in Chapter 4). Impulsive events also show enhanced abundances of heavy ions such as Fe relative to the averages in their gradual counterparts, as well as higher ionization states of these ions than in the solar wind, further suggesting they originate in an exceptionally heated coronal site. It is generally thought that the observation of an impulsive event depends on the observer having a good magnetic connection to the source at the Sun. Note the focusing action of the mirror force of the diverging open coronal fields turns what may be initially isotropic pitch angle distributions into nearly field-aligned beams. Both ions and electrons in these events are inferred by flare timing and their flux profiles to have had little interaction with irregularities in the solar wind en route, instead traveling with minimal cross-field diffusion from their site of origin at the Sun to the observer.

At this point, a word about what happens to SEPs between the Sun and the observation point, often at 1 AU, is in order because it affects the interpretation of observed SEP events greatly. Whereas interparticle interactions and scattering may occur in the collisional low corona, the process that dominates over the full particle path is the randomizing influence of fluctuations, including waves in the heliospheric magnetic fields, together with any magnetic field line deviations from the expected Parker spiral. As discussed already in the context of waves and turbulence, and the transient component of the slow solar wind, the latter can be caused by the continual shuffling and rearrangement of open field footpoints rooted in the evolving solar magnetic fields (see Chapter 4). The result is that the charged particles tied to the field undergo an effective "diffusion" that alters what is observed compared to what is released on the field line at the source. This concept will come up again later in the discussion of galactic cosmic rays. For SEPs, diffusive behavior is often invoked to explain the impulsive event decay phases in the flux time profiles (e.g., as in Figure 5.32(a)). Similarly, enhanced particle scattering in the corona, especially of ions, is sometimes used to explain extended delays of particle release relative to an x-ray event or delayed onsets from the expected arrival time for the ~1.2 AU Parker spiral path from the Sun to the Earth. But the biggest alteration of a SEP event profile is from the presence of an additional traveling shock source if a CME has also been produced.

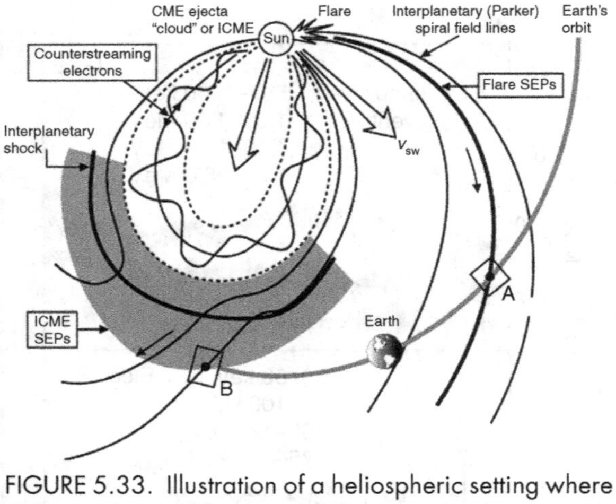

FIGURE 5.33. Illustration of a heliospheric setting where different observers may see different types of SEP event signatures from the same solar events. In the same way that an *in situ* observer will experience a different ICME signature depending on their relative location, here the observers will see different SEP event time profiles. Here the associated flare particles may be seen only at A. Earth may see a gradual SEP event, but not the ICME itself, while B may experience both. (After Luhmann *et al.*, 2008.)

Before proceeding to gradual SEP events, we take a brief diversion back to the ambient solar-wind SIR and CIR compressions. As discussed earlier in this chapter, SIRs are sometimes bracketed by spiral-shaped forward and reverse shocks. Although the compressions become stronger and thus most shocks occur beyond Earth's orbit, they still routinely produce energized particles at 1 AU. The reason is that the accelerated ions and electrons travel back toward the Sun along the heliospheric field lines as well as outward. Their typical contributions to the overall heliospheric energetic particle spectra are also indicated in Figure 5.31. SIR-related events are easy to identify as increases in suprathermal ion and electron fluxes occurring in the few days around an SIR or CIR passage seen in plasma and magnetic field data (see Figure 5.11). Although they are weaker than the SEP events of most interest, SIR/CIR events give insights on particle acceleration in the solar wind, and, because they produce a widespread suprathermal heliospheric population, their contributions may provide a boost to the generation of gradual SEP events.

A useful way to remember the different classes of SEP events is to think of impulsive SEP events as flare source related and the so-called gradual SEP events as primarily interplanetary shock-source events. As sketched in Figure 5.33, the flare source is fixed at the Sun while the shock source is traveling outward ahead of its CME ejecta driver. The observer's magnetic field connection to the traveling shock source, like the connection to the flare site for impulsive events, is of utmost importance to the gradual event appearance at a particular location. However, in this case, there is the added

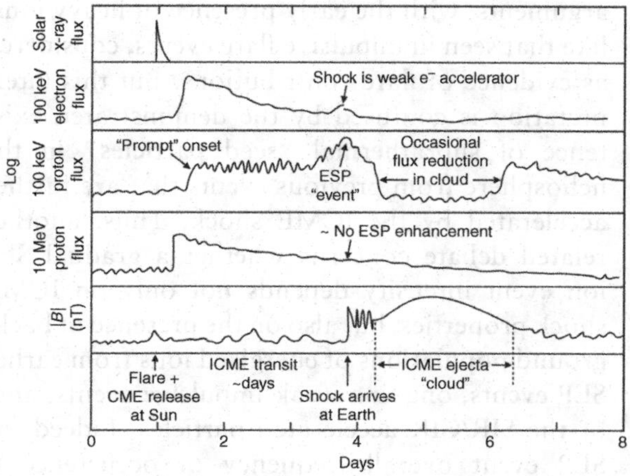

FIGURE 5.34. Sequence of events that may be observed in the course of a single large outburst of solar activity. The nature and timing of these is important to those who seek to forecast space weather. Note that after the flare x-rays and radio bursts, the SEPs are the first energetic output to arrive. However, as mentioned in connection with Figure 5.33, they may not be the harbinger of a geomagnetic storm, which is caused by the ICME.

complication that the source not only moves, but also its details, and the observer's field line connection, are continually changing as the shock moves out. The energetic particle time profile at different heliospheric locations thus depends on the particular shock history sampled. Shock compression ratio (density jump), velocity jump, and shock normal angle (see Chapter 6) all play a role. Since the shock moving ahead of a CME ejecta driver typically takes a few days to reach the observer, the initial part of the gradual event time profile, including its rise time and peak flux, depends on where and when a magnetic field connection to the shock is established.

If the initial observer connection occurs in the low corona at a strong shock, the gradual event onset can be rapid. This can happen when there is a fast CME ejected at low solar latitudes near the west limb of the Sun. Under these circumstances, the observer magnetic connection can be lost as the shock moves out at right angles to the Earth's location. As a result, the event appears impulsive in onset but its decline is related to the way in which the traveling shock leaves the observer's field line. Note this is quite different from the flare impulsive events described earlier where the decay phase is attributed to only the source behavior and interplanetary diffusion. A cautionary note for the reader is the use in the literature of the term "impulsive" to describe both flare and rapid onset, short duration, CME shock-source events. It is worth noting here that this geometrical situation is common for a special set of SEP events called GLEs (ground-level events); GLEs are SEP events that have a harder proton spectrum than most, with significant contributions up to a few gigaelectronvolts. The typical western disk location of the related solar events suggests that the observer connection along the spiral heliospheric magnetic field is probably to the shock of an exceptionally energetic eruption, while it is still low in the corona (~2–5R_S).

In contrast Figure 5.34 illustrates the timing that is typical of an event that occurs near the central disk as seen by an observer (e.g., for a halo CME-related SEP event). In this case, the full sequence of gradual SEP event phenomena and also the ICME ejecta impact are experienced. Gradual events (e.g., Figure 5.32(b)) tend to be ion dominated compared to the impulsive flare-related events, and also have average ion compositions more similar to the solar wind than the flare site (although there is sometimes a mixture of the two, as mentioned in the following). They are sometimes associated with a Type II radio burst, which indicates the presence of a CME shock. Moreover, at the time of the shock passage, an ESP enhancement in flux may occur. ESP "events" are often interpreted as additional flux from submersion in the diffusive shock acceleration region itself, where particles are still trapped by field fluctuations and still actively undergoing acceleration. Indeed, the ESP intervals tend to have softer proton energy spectra than the rest of the gradual event. Fast halo CMEs are particularly likely to produce ESP events at L1 because the "nose" of their ICME shocks squarely impact the Earth. After the ESP enhancement subsides, the energetic particle directions may change from what was an initially antisunward flux, when the shock was inside 1 AU, to sunward. In addition, a SEP flux

Table 5.3. Solar energetic particle (SEP) common classifications/characteristics

Classification	"Impulsive" event	"Gradual" event
Association	flares	CME, ICME shocks
Duration	hours	days
Composition identifier	higher $^3He/^4He$, Fe, ionization states of heavy ions, and electron contributions	solar-wind ion composition
Radio burst type	III	II

decrease may occur coincident with the ICME ejecta passage, suggesting the ejecta constitute a separate magnetic flux system in the post-shock region to which the SEPs do not have easy access. Alternatively, what appear to be flare SEPs are sometimes observed at the onset of a gradual event and/or inside ICME fields that may remain connected to the site of coronal activity. The existence of such combined events is not surprising considering the association of major CMEs and flares (see Chapter 4). Finally, when a CME occurs on the Sun's eastern disk or limb relative to an observer, magnetic connection to the strongest shock source is less likely due to the interplanetary field nominal spiral geometry. As a result, eastern disk events tend to be weak and have slow onsets relative to their western counterparts.

Table 5.3 summarizes the properties associated with the two types of SEP events. It makes this categorization clearer to think of impulsive events as strictly flare related, distinct from gradual events for which the observer shock connection along field lines is key. Gradual SEP events include the spatially diverse flux histories described above that have been sorted out, at least to first order, using decades of SEP observations and their solar event associations. However, even now debate continues about the early gradual event particles that arrive from close to the Sun, with prompt onsets, and whether they are accelerated by (related or background) flare processes as the CME is launched, or from the CME shock when it is just starting out. Investigations of this question often make use of ion composition arguments, with the early presence of heavy ions, like that seen in impulsive flare events, considered as evidence of flare contributions. But this interpretation is confused by the demonstrated existence of suprathermal "seed particles" in the heliosphere from previous events that are further accelerated by the ICME shock. Thus, another related debate concerns whether a gradual SEP ion event intensity depends not only on ICME shock properties, but also on the presence of background populations of energized ions from earlier SEP events, ongoing weak impulsive events, and/or the SIR/CIR accelerated particles. Indeed the SEP event overall frequency of occurrence is highly dominated by weak, impulsive events, which are nearly always present. On the other hand, like fast CMEs, the gradual events indicated by the statistics of energetic particle fluxes in Figure 5.35 have a strong solar cycle dependence. As mentioned above, there are also times when SEP fluxes remain high for times longer than expected for a single interplanetary shock passage. It is speculated that at these times the presence of multiple ICMEs or combinations of ICMEs and stream interaction regions produce GMIRs (global merged interaction regions), which act as both sources and traps of SEPs, thus delaying their departure from the heliosphere. The potential for larger SEP fluxes and longer duration events associated with CMEs are therefore of most interest for observers, theorists, modelers, and solar event forecasters.

5.3.5 Space weather

At this point it is worth stepping back to take a broad view of what we have just covered and its consequences for the solar system, and in particular for the Earth. The term "space weather" is sometimes used to describe practical consequences of the solar activity such as communications disturbances, spacecraft operational anomalies, or astronaut radiation hazards, some of which are indirectly connected to the actual conditions on the Sun. But space weather can also be defined more broadly, like meteorology, as the state of the space environment that has a host of particular causes and impacts. In practice, research relating to space weather explores the connections

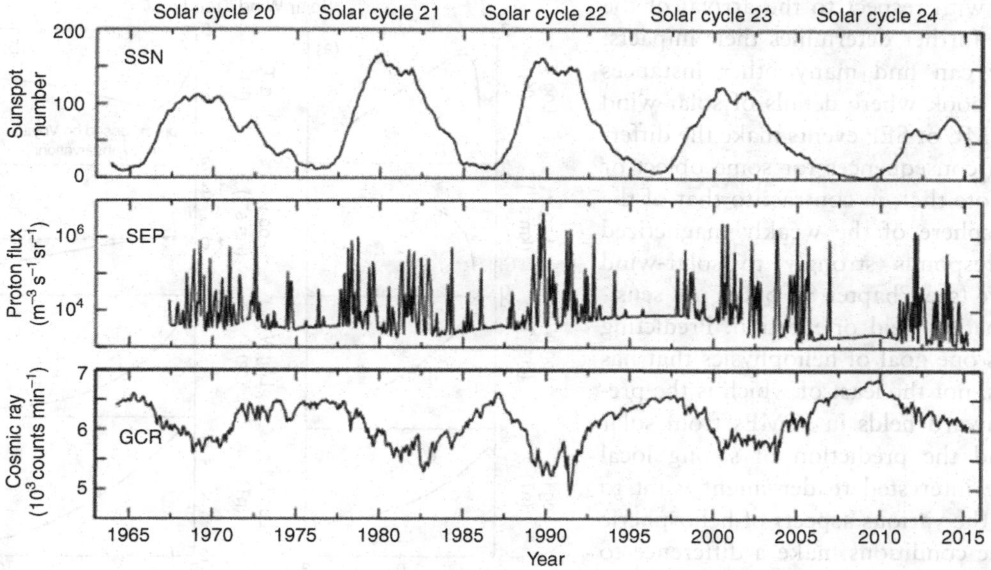

FIGURE 5.35. Recall that solar cycle dependence of the occurrence of ICMEs follows that of CMEs. (See Figure 5.23.) SEP events (identified here in the proton flux at energies > 10 MeV) follow a similar trend, expected given that most large events are associated with ICME-generated shocks. Galactic cosmic rays, also shown here, are also affected by solar cycle changes, but exhibit opposite trends. (L. K. Jian, personal communication, 2015.)

between what is going on at the Sun, in interplanetary space, and in the Earth's magnetosphere and upper atmosphere – a highly coupled system. For example, in Chapter 9 we shall learn about geomagnetic storms, but the severity of these storms and the details of their development depends intimately on the solar-wind "drivers." The injection of the ring current into the magnetosphere occurs primarily during intervals in the passing ICME, where the interplanetary magnetic field has a large southward component. This southward field is sometimes concentrated in the sheath preceding the ejecta and sometimes in the ejecta, but occasionally both are southward. Moreover, interaction of the ICME with the solar wind can compress and thus strengthen the southward fields in the ejecta in some circumstances. The stronger and longer the southward field is, the stronger the ring current injection and its consequences, such as induced currents in conductors on the Earth's surface. The coupling of the solar-wind electric field $\mathbf{E} = -\mathbf{v} \times \mathbf{B}$ to the Earth's polar regions is also much stronger because the magnetosphere's open field geometry during the interval(s) of southward fields greatly enhances that coupling.

Earlier, the "polarity" of the ejecta fields for some simple magnetic cloud type ICMEs was mentioned, along with the observation that the north–south field component often changes in a particular order, from north to south or south to north. A south-first cloud (such as that in Figure 5.21) that follows a sheath, which also has large southward fields, can make a particularly intense geomagnetic storm, whereas a north-first cloud that is followed by a high-speed stream that compresses the trailing southward field portion can cause an unusually strong late storm ring current injection. Thus, understanding the details and patterns of ICME plasma and field behavior both provides insight on the Sun–Earth connections and enables efforts to predict the magnitude, timing, and duration of geomagnetic storm effects. In addition, the entry of SEPs into the magnetosphere, where they can affect mesospheric ozone and high-latitude radiation hazards, depends on the more open polar cap fields that result from the southward interplanetary field orientation. The timing of the southward field with respect to the arrival of the peak SEP fluxes further determines their impacts. And the reader can find many other instances throughout this book where details of solar-wind conditions or ICME or SEP events make the difference in terms of consequences for some object or phenomenon. Note that, in contrast to that of the Earth, the ionosphere of the weakly magnetized planet Venus responds strongly to solar-wind dynamic pressure (see Chapter 8), but is not sensitive to interplanetary field orientation. Predicting space weather is one goal of heliophysics that has many challenges, not the least of which is the prediction of southward fields in ICMEs from solar observations and the prediction of strong local SEP effects. The interested reader might want to investigate how the various aspects of heliospheric field and particle conditions make a difference to Earth as a planet and to the humans who occupy it, as well as broader solar system responses and consequences past and present.

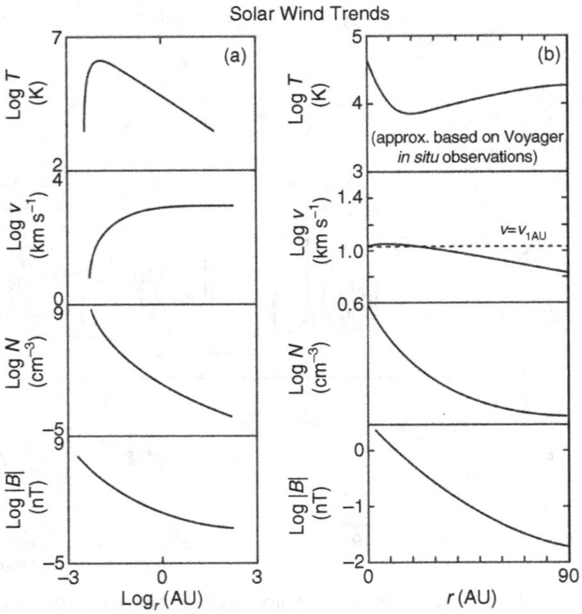

FIGURE 5.36. Variations of basic solar-wind parameters with radial distance from the Sun, as inferred from collected measurements. (After Wang and Richardson, 2001.)

5.4 THE LARGER HELIOSPHERE AND ITS CONTENTS: PICKUP IONS, DUST, ENAS, ACRS, AND GCRS

In the introduction to this chapter, a concise definition of the heliosphere was given as the "volume the solar wind carves out in interstellar space." Solar-wind properties have been measured with *in situ* particles and fields instrumentation from the heliocentric distance of Mercury at ~0.3 AU out to its boundary at ~100 AU. Collected results from several missions have provided an idea of how basic solar-wind parameters vary over heliocentric distance (Figure 5.36). The measurements closest to the Sun are from the Helios twin spacecraft, flown in the mid 1970s through the 1980s. The most distant solar-wind measurements are from the Pioneer 10–11 and two Voyager spacecraft, originally designed for their 1970s reconnaissance of the gas and ice giants. Most of these measurements were in the vicinity of the ecliptic, although the Pioneer and Voyager spacecraft sampled increasingly higher latitudes as they traveled beyond the orbit of Saturn at 10 AU. Near-polar heliospheric measurements were made by Ulysses in its high inclination orbit between 1 AU and 5 AU in the 1990s, complementing the already obtained observational knowledge from the lower-heliolatitude measurements with its unique perspective.

The Helios mission orbited the Sun with a perihelion at Mercury's orbit at ~0.3 AU and aphelion at ~1 AU. Its *in situ* measurements showed the expected denser solar wind and stronger and more nearly radial interplanetary fields in basic agreement with Parker's radially expanding solar wind (r^{-2} dependence) and spiral interplanetary field descriptions. Helios also provided plasma ion and electron distribution functions in different types of solar wind, plus observations of ICMEs at locations where they are fresher, and of SIRs and CIRs still in the process of formation. The interested reader will find details of Helios mission results in the literature, including comparisons with other heliospheric mission data obtained at the same time, as well as models aimed at understanding the radial evolution of different features from distribution functions to ICME shocks. Solar Probe is a new mission in the development phase

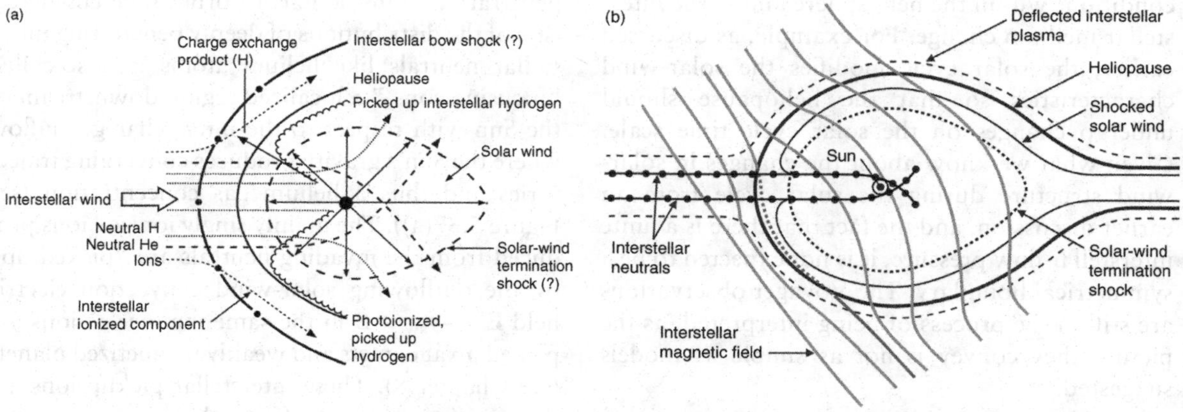

FIGURE 5.37. (a) Illustrated interaction of the flowing interstellar medium with the heliosphere, and (b) some related features, such as the external interstellar field draping in the heliosheath and the solar gravitational focusing of entering interstellar neutral gas atoms. ((b) Adapted from McComas *et al.*, 2009.)

destined finally to make measurements inside the Helios perihelion. It will approach the Sun to within ~10R_S (~0.034 AU), for the first time obtaining *in situ* measurements in a region where the solar wind preserves much stronger evidence of its origins in the corona.

Beyond Earth's orbit, the focus of investigations changes to the increasing influence of the interstellar medium on the outer heliosphere. Speculations on the heliosphere boundary radius from estimates of local interstellar pressure suggested for some time that the primarily dynamic pressure of the solar wind would come into balance with external plasma and field pressure at ~100 AU. The >1 AU observations indicated that, while the solar wind generally behaves as expected, with its density falling off nearly as r^{-2}, its Parker spiral magnetic field evolution, and its initially distinct streams merging as their momentum is exchanged, other effects became significant in its character by the orbit of Saturn (~10 AU). These effects are related to processes not so important for the solar wind in the inner solar system, although, as noted below, they are present in observations there as well.

The Sun and the solar wind do not exist in a vacuum but rather in a spiral arm of the Milky Way Galaxy containing weakly ionized (~10–20%) interstellar gas, with a density of ~0.3 cm^{-3} flowing at ~25 km s^{-1} relative to the Sun, and interstellar magnetic fields of microgauss strength. These properties are inferred from various techniques of remote sensing, such as the polarization of starlight and the Lyman-α sky background. In the absence of *in situ* observations, these inferences from remote sensing were used to estimate the heliocentric distance where the solar wind is expected to come into pressure/force balance with the interstellar plasma and field. This is similar to the way the magnetopause distance is calculated (see Chapter 9), except in this case the internal pressure is not mainly magnetic, but is instead dominated by the solar-wind dynamic pressure. This source-in-a-flow picture of the heliosphere–interstellar medium interaction is illustrated in Figure 5.37. Estimates of the total local interstellar plasma and field pressure indicate that, in another contrast to the solar wind–magnetosphere interaction case, the external flow pressure does not dominate the thermal and magnetic pressures. Thus, a more nearly spherical pressure-balance boundary at ~80–100 AU was expected. Prior to encountering this region on the Voyager spacecraft, it was anticipated that a "termination shock" should occur at which the flow energy of the solar wind was converted to thermal energy prior to its interface with the interstellar plasma and field at ~130–150 AU. The actual solar-wind boundary at this interface is called the heliopause. Note that these boundaries are expected to move in and out as prevailing conditions within the heliosphere and in the interstellar medium change. For example, as discussed earlier, the solar cycle modifies the solar-wind characteristics so that the heliopause should undergo changes on the solar cycle time scale. Given what we know about the changes in solar-wind structure during the solar cycle from an earlier discussion, and the fact that there is a finite interstellar flow pressure, it is not expected to be a symmetrical boundary. The Voyager observations are still in the process of being interpreted, as the picture they convey is not as simple as models suggested.

5.4.1 Heliospheric pickup ions

An important consideration in the heliospheric boundary that differs from other plasma boundaries encountered in this book is that the interstellar medium has a large neutral component. This state of weak ionization is maintained because at >100 AU it is far from the ionizing photon flux horizon of a star like the Sun. Both the ions and neutrals in this mainly hydrogen and helium medium flow at the ~25 km s^{-1} velocity with respect to the Sun mentioned above. But, while the plasmas and magnetic fields from the Sun exclude the ionized component of the interstellar medium at the heliopause, the neutrals can flow into the heliosphere relatively freely. It is estimated that at the orbit of Jupiter at ~5 AU there are about as many interstellar neutrals present as solar-wind ions. These interstellar neutrals in the heliosphere experience their environment in distinctly different ways from the plasma component, which is forced to flow around the heliopause. As they approach the Sun, they are affected by solar gravity and some end their lives as neutrals when they are eventually ionized by photoionization, solar-wind electron impact ionization, or charge exchange with solar-wind ions. The radius at which ionization occurs depends on the gas species, which sets both the energy required for ionization and the process. Thus the different interstellar gases that exist inside the heliosphere each have their own distinctive spatial distributions, with holes of different sizes and shapes carved out around the Sun. In particular, although little interstellar hydrogen makes it in as far as 1 AU without being ionized, helium can penetrate well inside Earth's orbit. One characteristic of the distributions of deeply penetrating interstellar neutrals like helium atoms is a so-called "focusing cone," a localized region downstream of the Sun with respect to the interstellar gas inflow where the Sun's gravity produces converging trajectories and thus a helium gas concentration (see Figure 5.37(a)). The mainly singly ionized ions produced from the invading neutrals are "picked up" by the outflowing solar-wind convection electric field $\mathbf{E} = -\mathbf{v}_{sw} \times \mathbf{B}$ in the same manner as ions are picked up at comets and weakly magnetized planets (see Chapter 8). These interstellar pickup ions are then carried outward, effectively becoming part of the solar wind.

Because ion pickup is an important process in space physics in general, it is worth considering it in more detail here. The concept is most appropriate in contexts where a minor species is involved, such that the ion can be regarded as a test particle in the plasma, moving in the magnetic and electric fields in the background medium as if undergoing simple single-particle motion in the combined **B** and $\mathbf{E} = (-\mathbf{v}_{flow} \times \mathbf{B})$ fields. An easy way to think about the physical consequences of the process for perpendicular flow and magnetic field situations like that in the outer heliosphere is that the ion is accelerated from rest to the flow speed but is forced to gyrate around the frozen-in field. Thus its guiding center (see Chapter 3) moves with the field while it gyrates around it at the flow velocity v_{flow}, with a gyro-radius $M_{ion}v_{flow}/qB$. This combination means that its trajectory in the inertial frame of the Sun is a cycloid (see Figure 8.3(a)), where the ion cycles between zero velocity and twice the flow speed, v_{sw}, with corresponding energies. In the solar-wind frame the pitch angle distribution is a "ring beam" perpendicular to the magnetic field. Thus, pickup ions have very characteristic distribution functions and energy spectra. Of course in reality pickup ions do not necessarily start with zero velocity and they also can be created in locations where the flow and field in the background plasma are not perpendicular or uniform. For example, if $\mathbf{v}_{flow}$ and **B** are parallel the ion is not picked up because there is no electric field. The reader can consider what happens to the ion for other $\mathbf{v}_{flow}$–**B** angles between these two extremes.

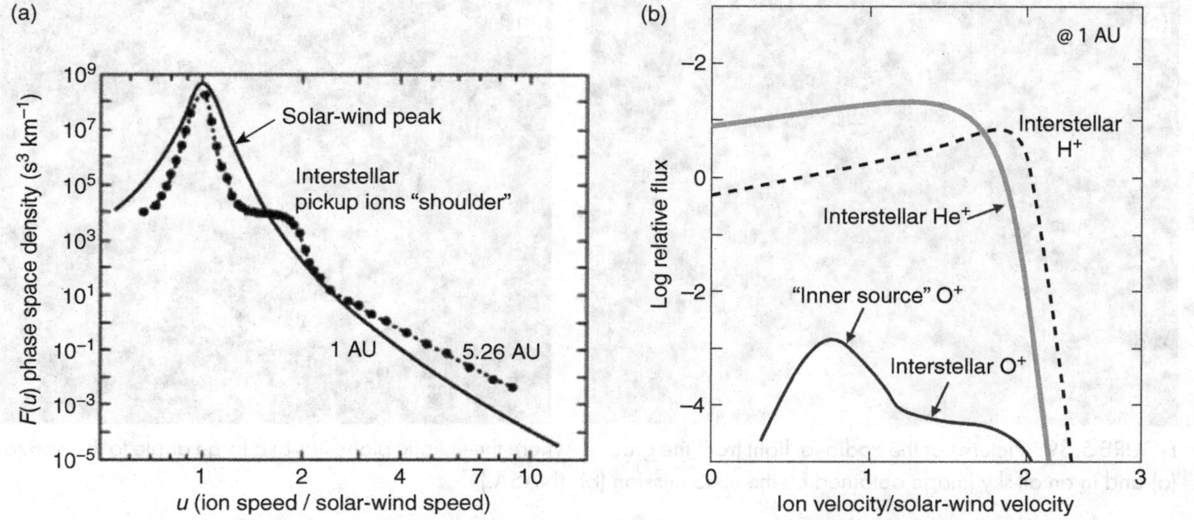

FIGURE 5.38. (a) Illustration of the heliospheric proton spectrum including interstellar pickup H+ and solar-wind protons. Note the relatively sharp cutoff in energy in the pickup ion spectrum at twice the solar-wind speed. (b) Heavier interstellar pickup ion spectra, showing the different shapes that result from the different spatial distributions of their neutral source populations. (Based on figures in Gloeckler and Fisk, 2006.)

The presence of pickup ion production effectively adds mass to the solar wind, with the consequence of a reduction in its flow speed. Observations of the distant solar wind confirm that it undergoes deceleration and modification with increasing heliocentric distance (see Figure 5.36). Moreover, interstellar pickup ions with the composition reflecting the makeup of the interstellar medium have been detected over a wide region, with He^+ an especially detectable contributor at 1 AU. Because most interstellar pickup ions are only singly ionized, in contrast to solar-wind ions, it is relatively easy to distinguish those with mass greater than hydrogen from the background solar-wind ions. In addition, because interstellar pickup ions have the characteristic spectral feature of an energy limit equivalent to about twice the background solar-wind speed, they are clearly seen in the 1 AU spectra of both heliospheric protons and heavier ions (Figure 5.38).

There is also clear compositional and spectral evidence that not all pickup ions detected in the heliosphere come from outside sources. In addition to the known sources at comets and planets (see Chapter 8), there is also a proposed widespread contribution from the solar-wind plasma interaction with dust and other small ice- and rock-derived particles in the heliosphere. In Section 5.4.2 we summarize our current knowledge of the dust sources and distributions, but for the purposes of the present discussion it is sufficient to know that most of it is concentrated around the ecliptic, where the solar wind interacts with it and, by some mechanism, produces another source of pickup ions. This source of pickup ions has several characteristics that distinguish it from the interstellar source. First, its composition favors heavy ions more characteristic of solar system solids, including oxygen, nitrogen, and silicon, for example. Second, the spatial distribution of the source appears to be more uniformly distributed in heliolongitude than its interstellar counterpart. Its radial distribution inferred from the pickup ion energy spectrum locates its peak production around $10–30R_S$. This information comes from the understanding that pickup ions produced where the solar-wind plasma flow and magnetic field are more nearly parallel will have their spectral peak at energies significantly below that equivalent to twice the solar-wind speed (see Figure 5.38(b)). (The reader is encouraged to explore the outcome of $\mathbf{E} = -\mathbf{v}_{sw} \times \mathbf{B}$ acceleration of ions at different heliocentric radii in a

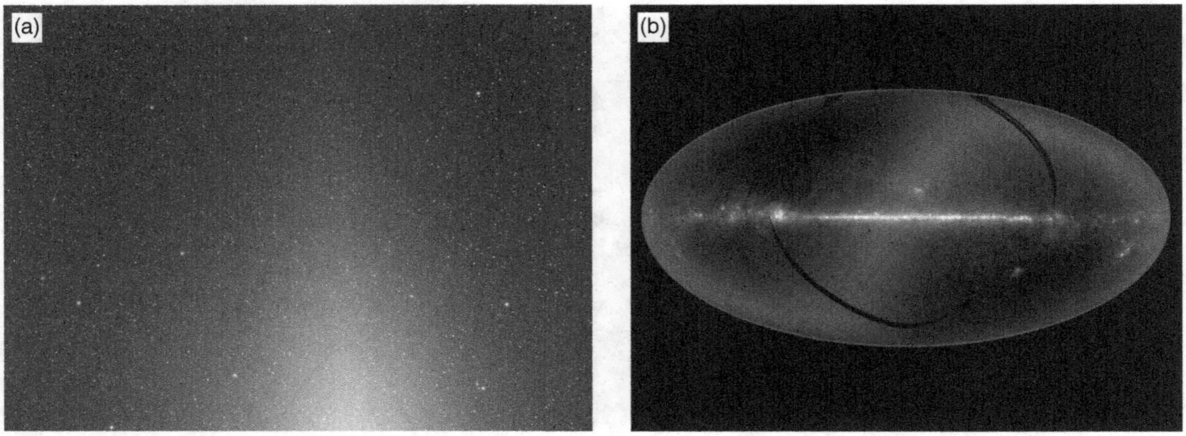

FIGURE 5.39. Pictures of the zodiacal light from the ground, where the ecliptic plane lies at a large angle to the horizon (a) and in an all-sky image obtained by the IRAS mission (b). (NASA.)

Parker spiral field.) These so-called "inner source" pickup ions add to the overall heliospheric populations of oxygen and other ions that also have significant contributions from the interstellar medium source, and can be dominant for selected elements such as carbon.

5.4.2 Dust

The presence of dust in the solar system has been known for a long time from observations of zodiacal light, a background glow in the night sky that is concentrated in the general vicinity of the ecliptic (see Figure 5.39). The detection by coronagraphs of dust near the Sun was also mentioned in Chapter 4. This zodiacal light is from the reflection and scattering of sunlight from submicron to millimeter size particulates concentrated in a disk extending from near the Sun out to the ~5 AU orbit of Jupiter. Scattered light also suggests the presence of another more spherically distributed "halo" population of lower density. In addition, orbit-like dust trails have been inferred from infrared remote sensing and from the occurrences of direct impacts on spacecraft surfaces (creating backgrounds in images) and responses by plasma-wave antennas. Interplanetary dust particles can be detected and collected in the Earth's upper atmosphere, where they are a regular source of incoming material from space. Each of these observational signatures is sensitive to a somewhat different dust size, shape, and composition range. The main sources of the dust in the inner heliosphere (<5 AU) are the debris from collisions in the asteroid belt, located between the orbits of Mars and Jupiter, and comet evaporation. At larger distances, the giant planets Jupiter and Saturn have been found to be dust sources as a result of collisions in their ring systems and outgassing or volcanic activity on their satellites (e.g., as in the cases of Enceladus and Io, respectively – see Chapter 12). However, the dominant source beyond 5 AU is the interstellar medium, whose ~1% dust component comes into the heliosphere together with the interstellar gas. Dust detection and characterization in space is a relatively young area of experimentation, in that techniques that cover the large size range and can identify composition as well are still in development.

The physics of dust in space plasmas is a challenging area of investigation because it involves plasma–solid material interactions at high relative speeds that are not well understood and are difficult to study in the laboratory. In addition, a dust particle immersed in the solar wind will assume an electrical charge that depends on the properties of the dust material, including its composition, size, and shape, as well as the plasma conditions. Moreover, the charge that it assumes can be positive or negative, depending on the circumstances including exposure (or not) to sunlight, and time dependent because the particle is exposed to a constantly changing environment. A number of calculations of dust charging and charged dust dynamics in the heliosphere have been carried out with the aim of understanding the behavior and distribution of the charged dust from a test-particle standpoint, some treating the dust as ultra-heavy pickup ions also subject to solar gravitational forces and photon radiation pressure. Dust masses and charges range from 10^{-20} to 10^{-5} kg (compare the proton mass of ~10^{-27} kg) and up to millions of electron charges, and so their effective gyro-radii in the heliospheric magnetic fields, determined by their ratio of mass to charge (see Chapter 3), have a large range, up to ~1 AU. Note that the entering charged interstellar dust is likely deviated from its interstellar medium flow path when it enters the heliosphere where the fields are stronger.

The dust distribution in the heliosphere is continually evolving as new collisions occur in the asteroid belt, the planets continue to give off particulates, and comets pass through. There are few "sinks" for this dust other than being swept up by larger solar system bodies including planets, evaporation on close approach to the Sun, and expulsion from the heliosphere by radiation pressure and solar-wind interaction processes. In general, its fate depends on the dust size and composition. It is estimated that most of the dust that moves inward is vaporized by solar heating by ~$10R_S$, giving the dust disk an inner boundary there. At larger distances sputtering by energetic ions can erode the dust over time. The smallest grains are transported outward like the pickup ions described above. Effects of the dust on the solar wind, the reverse of what we have discussed here, are still under investigation. But, as previously noted, the dust provides a source of heliospheric pickup ions, which with planets and comets constitutes an "inner source" with particular composition. Thus some broader consequences of the dust's presence are established even though the physical details of the plasma–dust interaction are still not well understood. Ongoing laboratory work may provide an alternative avenue for progress in this area. Meanwhile, the subject of dust in the heliosphere continues to gain attention in part because it provides unique insights for studies of planetary system and planetary ring formation and evolution. We note that when bodies larger than about a meter in diameter are destroyed in a collision, they may provide a sufficiently dense ionized dust cloud or plasma that is acted upon by the collective force of the magnetic field, and the cloud of dust may be accelerated to the solar-wind speed. We discuss the physics of this process in the context of the solar-wind interaction with the larger unmagnetized bodies in the solar system in Chapter 8.

5.4.3 Heliospheric energetic neutral atoms

Note that, as part of the heliospheric pickup ion production and transport scenario in Figure 5.37 (a), the charge exchange process, which is particularly effective for the solar wind interacting with the neutral interstellar hydrogen that has entered, produces a population of "fast" neutralized solar-wind hydrogen. These hydrogen atoms, and a smaller population of hydrogen and helium atoms from H^+ and He^+ pickup ion charge exchange, continue to move on the outward trajectories of the original ions. As a population, they constitute an ENA, or energetic (~1 keV, the original solar-wind proton energy) neutral atom, component of the solar wind. This neutral solar wind starts to accumulate at ~3 AU, where the interstellar neutral hydrogen penetrates, and grows to an estimated ~20% of the heliospheric outflow by the distance of the termination shock at ~100 AU. Efforts to detect ENAs created from the solar wind have encountered numerous technical challenges, but within the last few years the IBEX mission obtained a whole-sky map of ENAs seen from Earth orbit. As the first-generation solar-wind ENAs are expected to pass through the heliopause and into the interstellar medium, the origin of this flux returning to the inner heliosphere remains under investigation. The key observational evidence that any interpretation must explain is the clear pattern in the ENA maps that looks like an elongated "ribbon" or belt, consistent with the location where the inferred local interstellar field is most tangent to the heliopause boundary (see Figure 5.37(b)). One interpretation is that heliospheric pickup ions that reach the outer heliosphere charge exchange with the interstellar neutrals there, producing a second generation of ENAs that have inward velocity components. The interested reader is encouraged to consider the geometry of this process, based on Figure 5.37(b), and to refer to other proposals as to why this location might produce an enhanced return flux of heliospheric ENAs.

5.4.4 Galactic and anomalous cosmic rays

The solar system currently resides deep within the heliosphere due to the combined present states of the solar wind and conditions in the local interstellar medium. The interstellar plasma and field are deflected around the heliospheric plasma and field at ~100 AU even though both the neutral interstellar gas and dust approach the Sun relatively unimpeded (Figure 5.37). However, some charged-particle populations from outside are able to enter the heliosphere due to their high (tens of MeV to GeV) energies. These are the **galactic cosmic rays**, or GCRs, thought to be accelerated to their exceptionally high energies in astrophysical shocks such as those associated with supernova explosions. The observed GCRs are about 90% ions and 10% electrons (for unknown reasons, likely relating to differences in their acceleration and transport processes). About 80% of the ions are protons and ~15% are helium with the rest being heavier species whose relative abundances are used to speculate about broader astrophysical implications. The ability of these ions and electrons to pass through the heliospheric boundary is primarily due to their very large gyro-radii (at least many astronomical units) in the weak (subnanotesla) fields in this region, though this transmission is not complete, as will be further discussed below.

Prior to their measurement in space, GCRs were detected for many decades through their interaction with the Earth's atmosphere. Incoming energetic ions in particular produce "air showers," cascades of secondary muons, pions, electrons, neutrons, and charged fragments of molecules from the primary particles' interactions with the gases of the upper atmosphere. Most of what reaches the ground is this secondary component. However, some primaries can be detected at stratospheric balloon altitudes and even on the ground on high mountaintops. Devices called neutron monitors, which are distributed throughout the world at more modest altitudes, record the secondary components of the most energetic GCRs. These ions can enter into Earth's atmosphere because their gyro-radii, even in the Earth's relatively strong magnetic field, are sufficiently large. They at least partially avoid magnetic deflection back into space with a latitude-dependent energy (or "rigidity," where rigidity = momentum/charge) "cutoff" determined by the Earth's dipole field strength. (The interested reader can look up the term "Störmer orbits.") The access of the GCRs to the magnetosphere is best in the polar regions where the field is most nearly radial in the atmosphere and is stretched out or open to interplanetary space. The GCR electrons, because of their much smaller masses and thus gyro-radii, are especially sensitive to these geomagnetic cutoffs.

The fluxes and spectra of GCR ions and electrons are included in the heliospheric particle spectra in Figure 5.31. The reader will notice that these spectra also have some notable features in the form of diminished fluxes relative to a power law at their lowest energies, below ~100 MeV for ions and ~a few megaelectronvolts for electrons. These are the effects of GCR "modulation" by the solar wind and interplanetary field as seen at the typical measurement radius of 1 AU. Most measurements of GCRs are made at Earth's orbital distance, where the full galactic spectrum cannot be observed because the structure in the heliospheric magnetic fields, including fluctuations and irregularities on all scales from ion gyro-radii to stream interaction regions and ICMEs, essentially sweeps out the particles by pitch angle scattering – "modulating" the particle spectrum at low energies. The solar cycle dependence of the interplanetary fields makes the GCR modulation similarly cycle dependent, with the maximum modulation during the most active phases. On shorter time scales of days, the GCR flux responds to large solar events or event episodes, making so-called "Forbush decreases" of up to ~30% in the normal neutron monitor records that typically have their onset with the arrival of the ICME(s).

GCR modulation is well described to first order by a simple radial diffusion-convection equation for an initially isotropic cosmic ray distribution function $f(r, \mu = \cos(w))$, where w is the particle pitch angle with respect to the heliospheric magnetic field and the incoming distribution is specified at some outer boundary. This equation can be derived from the Vlasov equation (Chapter 3) with many simplifying assumptions along the way. In this approximation, the diffusion coefficient D is assumed to depend on an interplanetary magnetic fluctuation spectrum that is convected outward at the solar-wind speed V, and the diffusion is due to pitch angle scattering only with no significant energy changes. The third term on the left represents the effect of the diverging background heliospheric magnetic field on the distribution function:

$$\frac{\partial f}{\partial t} + \mu V \frac{\partial f}{\partial r} + \frac{V}{r}(1-\mu^2)\frac{\partial f}{\partial \mu} = \frac{\partial}{\partial \mu}\left(D\frac{\partial f}{\partial \mu}\right). \quad (5.14)$$

Figure 5.40 shows examples of numerical solutions to this type of treatment for the GCR proton

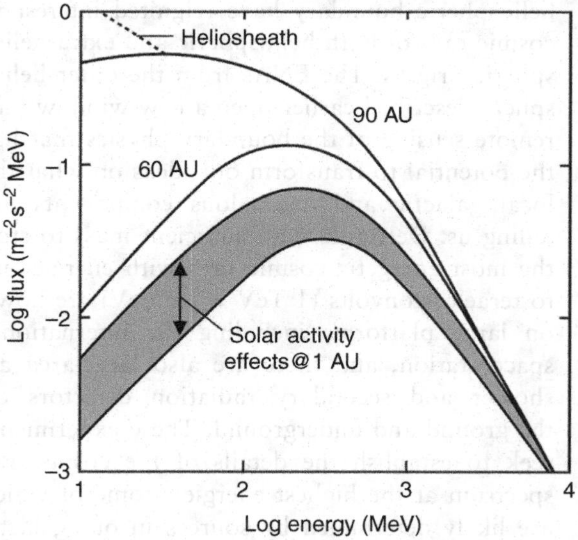

FIGURE 5.40. Sample solutions of the cosmic ray modulation equation, showing the effect of heliocentric distance on the appearance of the energy spectrum of energetic protons entering from outside the heliosphere. The diminishment of flux depends on what is assumed for the diffusion coefficient, under the assumption that perturbations in the interplanetary field moving outward with the solar wind scatter the incoming trajectories of these particles with energy-dependent efficiencies. (After Fisk, 1971.)

flux. Investigations of the physics of cosmic ray transport in the heliosphere today involve much more complicated treatments, starting with use of more general distribution functions and including details of the diffusion processes that occur in both energy and pitch angle. However, the basic understanding of the behavior of GCRs entering the heliosphere remains the same.

One notable insight that ties back in to the larger picture of the global solar wind and heliosphere concerns an additional solar cycle effect seen in the GCR modulation. An alternating shape pattern in the cyclical time series of GCR fluxes inferred from neutron monitors over several consecutive cycles (distinguishable in the middle of the bottom panel of Figure 5.35) suggests that modulation depends on the alternating polarity of the large-scale solar dipolar field component. Entering GCRs undergo different large-scale gradient and curvature drifts depending on whether the north pole of the Sun has a positive or negative polarity. The heliospheric current sheet plays a role in this picture because it acts as a channel enabling GCR entry for only one solar dipole polarity, while for the other polarity entry via the polar regions is favored. The interested student is encouraged to think about how different orientations of the interplanetary field affect the paths of entering protons in a Parker spiral heliospheric magnetic field.

A major goal of GCR research has been to establish the radial gradient of the fluxes in the heliosphere so that the interstellar spectrum can be deduced. However, such efforts have been slowed by the rarity of measurements at large heliocentric distances. Most spacecraft do not carry detectors designed to measure GCRs, in part because this requires instrumentation that is often quite massive. Nevertheless, the Voyager spacecraft has been able to make some key measurements that may at last accomplish this goal. Work on the recent observations obtained at the outer boundaries of the heliosphere should finally establish the true shape of the GCR spectrum unaffected by heliospheric influences.

An additional component of non-solar energetic particles that was a relative latecomer to the big picture of cosmic rays in the heliosphere is the

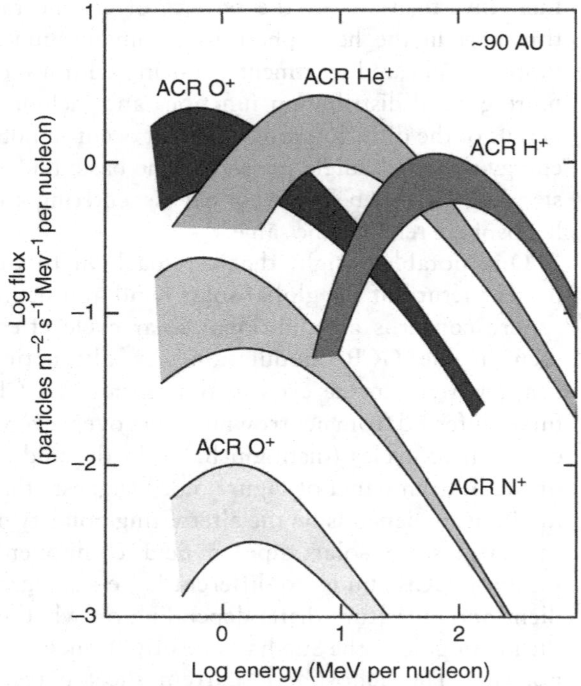

FIGURE 5.41. Illustration of how the energy spectra and composition of anomalous cosmic ray ions might be modified by solar modulation.

anomalous cosmic ray (ACR) component. ACRs contribute to the middle range of cosmic ray energies (see Figure 5.41) and have an unusual ion composition and charge state. Their unusually high abundances of ions heavier than protons, including the same species as the heliospheric pickup ions (He, O, N, C), initially made them stand out. Verification of the single-charge states of these ions cemented the case for interpreting them as heliospheric pickup ions that are accelerated to high energies in the outer heliosphere and redirected inward. Current theories have this acceleration process occurring somewhere in the vicinity of the termination shock. The Voyager mission recently confirmed that the solar wind undergoes a shock-like deceleration at around 100 AU, after which the ACR flux diminishes. However, the question of whether or not the ACR source region was crossed there remains unresolved. Clearly the heliosphere is a complicated place, including at the high-energy end of the particle spectrum. To make matters even more interesting, the energetic electrons have a distinctive added contribution of their own from Jupiter.

When the details of the GCR electron spectrum were being fleshed out, it was found that there was a low-energy component below ~2 MeV that has the appearance of a separate power law superposed on a modulated galactic spectrum (see Figure 5.31(b)). Moreover, it was found that this additional spectral component was not modulated by solar activity in the same way as the rest of the spectrum. Rather, its presence depends on the direct interplanetary field connection between Jupiter and the observer along the Parker spiral field. This observation provided evidence that Jupiter's magnetosphere leaks energetic electrons from its radiation belts that travel toward the Sun along the interplanetary field, at least to the orbit of Earth. Thus Jupiter's influence in the solar system extends even to the cosmic rays.

In the present era, the new results from the heliospheric boundary have reignited interest in cosmic rays of both heliospheric and extra-heliospheric origins. The ENAs from the outer heliosphere described earlier open a new window into remote sensing of the boundary physics that has the potential to transform our ideas on what the local galactic and anomalous components are telling us. Detectors with sufficient mass to stop the most energetic cosmic rays with energies up to teraelectronvolts (1 TeV = 10^{12} eV) are flown on large platforms including the international space station, and there are also large-area air shower and secondary radiation detectors on the ground and underground. These experiments seek to establish the details of the cosmic ray spectrum at the highest energies, some of which are likely determined by sources in our galactic center and outside our galaxy. These highest energy particles of course suffer no significant modulation in the heliosphere, and little if any geomagnetic exclusion, but their fluxes are extremely low and so their detection and characterization are challenging. Thus, while the early work on cosmic rays had a relatively narrow perspective, the field of cosmic ray astrophysics has greatly expanded the possibilities of what can be learned about the universe-at-large from information carried by energetic particles in Earth's neighborhood.

Solar cycle modulation of the effects of GCRs on Earth's atmosphere and surface have been used to infer the history of solar cycles up to at least ~60 000 years ago. The interaction of the GCRs with the atmosphere produces two particular unstable isotopes, carbon-14 and beryllium-10, which decay with half-lives of ~6000 yr and ~1.4×10^6 yr, respectively. The carbon-14 is taken up into atmospheric CO_2 and then absorbed by plants, which preserve a record, through their carbon isotope ratios of ^{12} C/^{14} C, of the time from the creation of the carbon-14. By this means, dating the wood in the annual rings of tree trunks provides an effective means of deducing the history of the annual fluxes of GCRs. Similarly, beryllium-10 produced in the atmosphere precipitates out with rain and snow, the latter of which can accumulate in high-latitude regions over many centuries. While the carbon-14 record is fairly straightforward in that most years a tree ring is produced, the beryllium-10 ice core record is compromised by the mixture of climate-related modifications over time. Annual snowfall depends on atmospheric circulation and the evaporation and precipitation cycle, which are both temporally and regionally variable, and the ice record can be confused by periods of warmth, erosion, or ice migration such as occurs in the formation and evolution of glaciers. Effects of wildfires and volcanoes are particularly problematic. Thus, while the existence of major changes in the solar modulation of GCRs can be inferred for periods such as the Maunder minimum in the 1700s, when the flux of GCRs reaching Earth was exceptionally large and the record is still relatively fresh, the interpretation of ice core records remains challenging. These records also may record large solar energetic particle (SEP) events that have similar effects on the atmosphere, and moreover tend to occur in the periods of minimum GCR flux, thus potentially affecting the inferred minimum in GCRs. All of these issues compound the already difficult task of deriving the history of solar and heliospheric activity before the era of reliable record keeping.

A few words are also in order here regarding the connection of these longer-term effects referred to as "space climate" and the space weather effects described earlier in this chapter. The literature is full of reported correlations between solar cycle proxies such as sunspot number and terrestrial climate variables such as temperature and circulation patterns. These have in turn inspired decades of investigations of physical mechanisms by which other than radiative (solar constant) effects of solar activity can influence Earth's lower atmosphere. One proposal that has received significant attention in the past few decades concerns possible effects of GCR flux on cloud formation. This mechanism involves the related atmospheric levels of radiation that may affect the efficiency of nucleation of cloud droplets. If present, this could alter both albedo and precipitation patterns, thereby influencing key atmospheric circulation drivers and their related thermal balances in the lower atmosphere. Although research is ongoing to determine whether cloud cover responds to GCR flux, the observational evidence at this time of writing does not appear to support the hypothesis. Nevertheless, observations are often incomplete or difficult to interpret, and it is known from climate modeling that regional responses may occur that influence the complex global atmosphere system. Thus, the debate over whether any effects of solar activity on the Earth's space environment alter the path of terrestrial climate (and weather), and if so, how, remains open.

5.5 SUMMARY

In this chapter we focused on the outermost solar atmosphere, the solar wind, and its effects on circumsolar space. We considered the widely used MHD fluid description of the basic coronal outflow that creates what we measure in its larger sphere of influence, the heliosphere. This concept, initially developed by Parker in the early 1960s, provides a relatively straightforward framework for understanding the nature of features such as the interplanetary magnetic field and solar-wind stream structure. Although there are outstanding questions concerning the details of the solar-wind plasma's observed properties, including its velocities, temperatures, composition, and variability, our understanding is good enough to enable the construction of models that can produce realistic facsimilies of the data under quiet Sun circumstances. In particular, coronal models such as those described in Chapter 4 can be used to describe the detailed source(s) of the outflows, including the existence of fast polar coronal hole solar winds, and slower solar winds from low- and mid-latitude coronal holes and coronal hole boundaries. Extended into the heliosphere beyond the corona, these also show the development of the source geometry-dependent solar-wind stream structure, including stream interaction regions that form compressions and sometimes shocks at the interfaces of streams of differing speeds and may corotate for several ~27 day solar rotations near solar minimum (the CIRs).

The existence of solar-wind plasma and field observations from almost four solar cycles allows validation of these models but also brings out their weaknesses, such as the neglect of unsteady or transient components that become especially dominant at active solar times. We learned about the small transients from coronal streamers and coronal hole boundaries, imaged by coronagraph and heliospheric imagers, which provide at least some of the "slow" component of the solar wind observed in the ecliptic, even at times of no solar activity. These can be regarded as a result of the continually evolving photospheric field boundary conditions that determine the coronal source geometry described here and in Chapter 4.

We also learned that the larger extreme of the spectrum of transients, the interplanetary coronal mass ejections, or ICMEs, are the consequences of the CMEs in the corona that evolve as they propagate outward through the ambient solar wind. The stream structure introduces complexities into the evolution of both the ejected CME material and the solar-wind disturbance surrounding it that includes a compressed leading "sheath" region sometimes preceded by a shock. We learned that these ICME shocks, as well as the CIR shocks, accelerate some of the heliospheric particles to higher energies. The ICME shocks in particular can produce major solar energetic particle or SEP events which can last for several days and, together with the ICMEs themselves, produce "space weather" consequences of interest.

We learned about some additional non-solar particle populations in the heliosphere: ions produced from the entering interstellar neutral gases and solar system sources that are accelerated ("picked up") by the solar-wind convection electric field; charged dust particles that similarly act like pickup ions with large masses and variable charges; ENAs (energetic neutral atoms), which are a product of charge exchange processes in the heliosphere and provide a relatively new remote-sensing diagnostic of heliospheric processes; and cosmic rays, both galactic (GCRs) and anomalous (ACRs), which comprise an exceptionally energetic particle population from beyond and from the outer regions of the heliosphere, respectively.

In the near term, heliospheric research is enabled by a Great Heliospheric Observatory of spacecraft that are sampling various regions of its vast ~100 AU space. This observatory is in part a legacy of planetary missions and in part targeted to obtain a broad perspective on the local space environment system and its solar connections. New plans for missions include solar imaging from, and *in situ* measurements of, the high-latitude heliosphere, to fill in information about the important solar polar regions, and near-Sun probing, to determine how the solar source(s) evolve into the solar wind we observe. The interested reader will find by exploring the internet and the literature that it is an exceptional time to be working on the heliosphere.

Additional reading

Discussion of solar-wind missions and the instruments that provide the types of measurements presented in this chapter can be found in the following references.

RUSSELL, C. T., R. A. MEWALDT and T. T. VON ROSENVINGE (1998). The Advanced Composition Explorer, *Space Sci. Rev.*, **86**, 1–4. The ACE mission has been obtaining solar wind and energetic particle data upstream of the Earth since 1997. It provides the primary source for Earth–solar wind interaction studies.

RUSSELL, C. T. (2008). The STEREO mission. *Space Sci. Rev.*, **136**, 1–4. Launched in late 2006, this twin spacecraft mission has provided 1 AU solar-wind and energetic particle data at increasing separations from Earth, together with solar imaging. This has expanded our view of solar and heliospheric phenomena to a more global, solar system wide perspective.

A more physically based discussion of solar-wind phenomena can be found in:

BALOGH, A., L. LANZEROTTI and S. SEUSS (2008). *The Heliosphere through the Solar Activity Cycle*. Dordrecht: Springer. A collection of review papers incorporating the results from the Ulysses mission, which transited the solar poles under different solar activity conditions.

KALLENRODE, M. B. (2010). *Space Physics – An Introduction to Plasmas and Particles in the Heliosphere and Magnetosphere*. Berlin: Springer. An annotated textbook providing an alternative introduction to space physics, with particular attention to the subject of heliospheric energetic particle populations.

SCHRIJVER, C. and G. SISCOE (2010). *Heliophysics, Vol. 2*. Cambridge: Cambridge University Press. A compendium of lectures based on a summer school held in Boulder, Colorado, over a number of years, with contributions from experts in the field. Useful perspectives on the state of knowledge in subareas.

Problems

5.1 Using Eq. (5.9) for the critical radius, find the coronal temperature for which a supersonic solution would be impossible.

5.2 Let us examine the properties of stellar winds on other stars that could be subsonic.

(a) Two Sun–like stars have the mass and size of our Sun, and hydrogen-dominated isothermal coronas with temperatures of 3×10^6 K and 0.5×10^6 K. What are the radial variations of the sound speeds for each of these stars, and at what radius does the sonic critical point lie in stellar (solar) radii? Compare with the Sun, whose coronal temperature is 1.0×10^6 K.

(b) A star has a dipole field magnitude of $B_0 = 10$ gauss (10^{-3} tesla) on the surface at its equator. The coronal hydrogen ion density is n_0 at this point, and has a hydrostatic fall-off with radius $n = n_0\exp(-h/H)$, where $h = r - R_S$, $H = GM_S m_p/(2kT)$, and m_p is the proton mass. Calculate the sonic, Alfvén, and magnetosonic speeds as a function of radius above the surface for both the equatorial plane and the poles, using the radial range $1\text{–}100R_S$. Plot the results separately for the equator and for the pole. What does the relative behavior of these plots imply for the formation of shocks traveling through the corona?

(c) A belt of closed dipolar magnetic fields exists around the equator of the star. Assuming this closed field belt extends approximately to a radial distance where the coronal thermal pressure is equal to the magnetic pressure $P_b = B^2/2\mu_0$, what is that distance r_{bal} for the three coronal temperatures? [Hint: Assume the ideal gas law and plot the thermal pressure $P_{th} = nkT$ and equatorial magnetic pressure against radius to see where they cross.] The "last closed field line" in the corona is often used to define the solar-wind "source surface," but the PFSS model source surface is often assumed to be located at $2.5R_S$. Considering that coronal temperatures can vary, how can this (qualitatively) affect the results for coronal holes determined with this model (e.g., if the more realistic source surface radius is smaller or larger than that assumed)?

5.3 Waves are frequently detected in the solar wind as fluctuating magnetic field and solar-wind velocity. Assume that the solar-wind magnetic field is oriented along the spiral angle.

Linear wave perturbations with magnetic polarization perpendicular to the ecliptic plane are observed. What MHD wave mode is relevant to such perturbations?

5.4 The mean distances from the Sun to Mercury, Venus, Earth, Mars, and Jupiter are 0.39, 0.72, 1.0, 1.5, and 5.2 AU, respectively. The average properties of the solar wind at 1 AU are as follows: density = 7 protons cm^{-3}; velocity = 440 km s^{-1}; proton temperature = 0.9×10^5 K; electron temperature = 1.3×10^5 K; magnetic field strength = 7 nT, lying along a 45° archimedean spiral angle to the flow. Calculate the following quantities at each of the planets: magnetic field strength and spiral angle, proton-number density and temperature, electron temperature, and plasma beta, or ratio of thermal pressure to magnetic pressure. Assume that the proton temperature varies with heliocentric radius as r^{-1}, and electron temperature as $r^{-1/2}$.

5.5 Using the information calculated in Problem 5.3, calculate the Debye length and number of particles in a Debye cube at Mercury and Jupiter. Calculate the electron plasma frequency and proton gyro-frequency at Venus and Mars. Calculate the gyro-radius of a 1 keV proton moving perpendicular to the magnetic field at Earth, Mars, Saturn, and Pluto. If a neutral oxygen atom were ionized in a 440 km s^{-1} solar wind at these four distances, what would its gyro-radius be?

5.6 At its typical velocity of 440 km s^{-1}, how long in days does it take the solar wind to arrive at Mercury (0.39 AU), the Earth (1 AU), Jupiter (5.2 AU), and Pluto (39 AU)? If an ICME leaves the Sun at 1500 km s^{-1} and slows at a rate of 50 km s^{-1} per AU, how long would it take to reach these four bodies? When would the ICME reach the solar-wind speed?

5.7 The location of the heliospheric boundary, where the pressures of the solar wind and the interstellar plasma are in balance, depends on both the solar-wind properties and the external (local interstellar medium, LISM) properties.

(a) Assume a setting where the LISM (mainly hydrogen) gas density is $n_{0\mathrm{gas}} = 1$ cm^{-3} and that it is 1% ionized, has a plasma temperature of $T_0 = 100$ K and a magnetic field strength of $B_0 = 0.01$ nT. Calculate the total pressure (thermal pressure $P_{\mathrm{th}0}$ plus magnetic pressure $P_{\mathrm{b}0}$). Is this LISM a high or low beta medium? Now assume the interstellar medium flows at $v_0 = 10$ km s^{-1} parallel to the magnetic field. What is its dynamic pressure? Which of the three pressures is dominant?

(b) Assume that a star, with a cold, weakly magnetized hydrogen plasma wind and a proton density of $n_i = 10$ cm^{-3} at 1 AU flowing radially outward at $v_i = 200$ km s^{-1}, is within the LISM. Where (in AU) is the *astropause* (the location at which the internal, dynamic pressure-dominated, and external pressures balance as a function of the angle away from the interstellar flow direction)? Describe the general shape of the resulting *astrosphere* (the approximate heliocentric radius, R, in the flow, flank, and wake directions) based on these ideas of internal/external pressure balance. How is this changed if the stellar-wind speed is increased to 600 km s^{-1} and its density at 1 AU is reduced to 0.01 cm^{-3}?

5.8 Flux ropes.

(a) The magnetic field within a coronal mass ejection (CME) often has the appearance of a flux rope. Assuming a force-free configuration, derive the equations determining the magnetic field variation as a function radial distance from the center of the rope for a cylindrically symmetric flux rope. Assume the z-axis is the axis of the flux rope. What is the restriction on the current in order that the rope be force free? [Note: For cylindrical coordinates, $(\nabla \times \mathbf{A})_r = \frac{1}{r}\frac{\partial A_z}{\partial \phi} - \frac{\partial A_\phi}{\partial z}$, $(\nabla \times \mathbf{A})_\phi = \frac{\partial A_r}{\partial z} - \frac{\partial A_z}{\partial r}$, $(\nabla \times \mathbf{A})_z = \frac{1}{r}\frac{\partial}{\partial r}(rA_\phi) - \frac{1}{r}\frac{\partial A_r}{\partial \phi}$.] What assumption can you make for the variation of the current density within the rope that produces a simple familiar structure?

(b) Compare your results with the Bessel function equation:

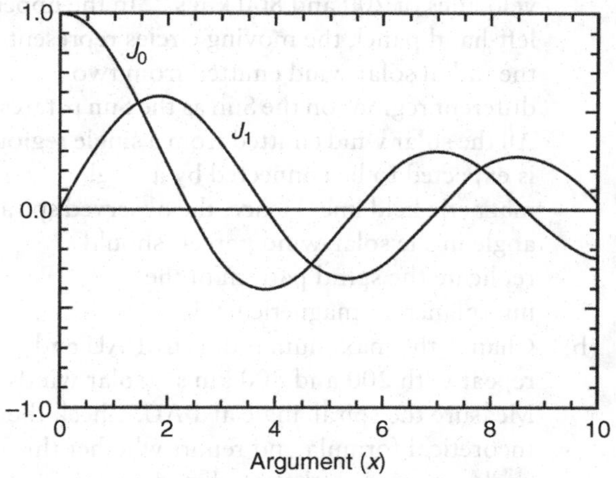

FIGURE 5.42. Bessel functions of order 0 and 1.

$$\frac{\partial^2 \psi}{\partial x^2} + \frac{1}{x}\frac{\partial \psi}{\partial x} + \left(1 - \frac{n^2}{x^2}\right)\psi = 0,$$

with solution $\psi = J_n(x)$ – the Bessel function of the first kind of order n, with argument x. See Figure 5.42.

(c) Given the equations you have derived for the magnetic field variation, determine which component of the field varies as J_0 and which as J_1.

(d) Flux ropes are thought to be generated by twisting the magnetic field at the foot of the flux tube. If that is the case, based on the magnetic field variation with radial distance, does the force-free model become unphysical at larger distances away from the axis? If so, why?

5.9 What is the magnetic field $B_\varphi(r)$ produced by a cylinder of radius a and uniform current density j along the z-direction? Find the resulting pressure $P(r)$ if the plasma is in equilibrium, and sketch both B and P.

5.10 Prove Parker's theorem that the mean square axial field B_z of a flux tube that is confined by a constant uniform pressure P_e and is in equilibrium is unaffected by twisting.

5.11 The field $B_z(r, t)$ of a cylindrically symmetric flux tube is of the form $B_0\exp(-r^2/4\eta t_0)$ at time t_0. Subsequently it diffuses according to the equation

$$\frac{\partial B_z}{\partial t} = \frac{\eta}{r}\frac{\partial}{\partial r}\left(r\frac{\partial B_z}{\partial r}\right).$$

By seeking a solution of the form $f(t)\exp(-r^2/4\eta t_0)$, find $B_z(r, t)$ and sketch it as a function of r for several times. Find the magnetic flux, and show that it is conserved. Show that the rate of change of magnetic energy is negative and comment on the result.

5.12 For a cylindrically symmetric magnetic flux tube of length L in equilibrium under a balance between magnetic and pressure force, find the pressure and azimuthal field as functions of r if $B_z = B_0(1 + r^2/L^2)$ and the twist Φ(through which a given field line is twisted in going from one end of the tube to the other) is uniform.

5.13 Use the Potential Field option of the Space Physics Exercises* to explore the effect of the source surface on the magnetic field of the corona.

(a) Select the source surface (WSO) option; set the latitude steps to 20 keeping the longitude step at 1. Set the source surface radius to one solar radius and all coefficients to zero except g_{10}. Calculate the field line pattern and print the screen. Count the number of "open" field lines that go to the edge of the box. Repeat for 1.5, 2.0, 2.5, 3.0, 3.5, 4.0, 4.5. Print screen for last value. Plot the number of open field lines as a function of source surface radius.

(b) Repeat step (a) but set all coefficients except g_{20} to zero. Print the first and last screens and plot the number of open field lines versus source surface radius.

(c) Repeat step (a) but set all coefficients except g_{30} to zero. Print the first and last screens and plot the number of open field lines versus source surface radius.

* Space Physics Exercises are at http://spacephysics.ucla.edu.

(d) Select the source surface (WSO) option. This option draws the closed magnetic field lines that just graze the source surface radius. Where they reach maximum altitude (the neutral line) is marked with a dot. Leave the source surface at $2.5R_S$. Pick "Sol Min Dipole" and calculate its field pattern. Print the screen. Choose the simplest set of coefficients that you believe will replicate "Minimum Dipole." Be sure to set coefficients you do not want to use to zero. Print the screen. What do you conclude about the magnetic field during the solar minimum chosen in this example?

(e) Repeat part (d) using "Intermediate Quadrupole Pole." Again choose the simplest set of coefficients that replicate the observed field. What do you conclude about the magnetic field during this period of the solar cycle? Rotate your magnetic configuration if needed to demonstrate your goodness of fit.

(f) Repeat parts (d) and (e) but now using the magnetic map (WSO) option. Print both analytic (i.e., your simple approximation) and observed maps for "Minimum Dipole" and "Intermediate Quadrupole." Are these results consistent with those obtained in (d) and (e)?

5.14 Use the Solar Wind/Parker Spiral option of the Space Physics Exercises.* Note the two secondary options.

(a) Change the maximum radius to 6 AU. Run the program (click calculate) for solar-wind velocities of 200 and 800 km s^{-1}. In the upper left-hand panel, the moving circles represent the radial solar wind emitted from two different regions on the Sun as the Sun rotates. All the solar wind emitted from a single region is expected to be connected by a single magnetic field line. Hence, the observed spiral angle in the solar-wind parcels should replicate the spiral pattern of the interplanetary magnetic field.

(b) Change the maximum radius to 1 AU and repeat with 200 and 800 km s^{-1} solar winds. Measure the spiral angle at 1 AU. Check the theoretical formula and report whether this angle agrees with that predicted.

(c) Choose the Solar wind/Parker spiral model/x–y plane option. Change the velocity to 800 km s^{-1} and run the program. Print the screen. Change the latitude to 45° and 80°. Describe and explain the effect of changing the solar-wind velocity and the latitude.

(d) Choose the Solar wind/Neutral sheet model/3D Structure option. Change the magnetic axis tilt angle to 25°. Redraw for solar-wind velocities of 200 km s^{-1} and 800 km s^{-1}. Describe and explain the effect of changing the tilt angle and velocity.

(e) Choose the Solar wind/Neutral sheet model/Stream interaction option. Run it for a magnetic axis tilt angle of 30°. Change the radial distance to 0.1 AU and repeat. Compare the quantities in the right-hand panels for the two runs. How do they differ and why?

* Space Physics Exercises are at http://spacephysics.ucla.edu.

6 Collisionless shocks

6.1 INTRODUCTION

Shocks occur in a gas or liquid when the speed of their interaction with an obstacle exceeds the velocity of the compressional (sound) wave that would ease the deflection of the flow around that obstacle. This process is both non-linear and dissipative. Thus, the phrase "collisionless shock" is at first sight an oxymoron, given that collisions appear to be a necessary part of the shock process. In an ordinary gas, collisions are an effective way to dissipate energy and bend the flow around the obstacle. In a collisionless plasma, where particles do not collide in the usual sense, kinetic processes and collective forces take the place of the collisions that occur in more familiar shocks. Understanding how this occurs under the wide range of parameters found in space and astrophysical plasmas is an ongoing study, but one whose bases we understand through observation, theory, and numerical models.

Collisionless shocks are important in astrophysical, heliospheric, and planetary settings. They are found throughout the heliosphere, wherever the solar wind flows, from the Sun's surface to the heliopause. As we discussed in Chapter 5, collisionless shocks arise when fast streams in the solar wind overtake slow streams, and also when fast ejections of magnetized plasma called coronal mass ejections move swiftly through the ambient solar wind. They are found as bow shocks, standing in the flow in front of all the planets visited to date. They deflect flows around obstacles, heat the plasma, and alter the properties of the flowing plasma and how it interacts with the obstacles. The physical processes that take place at these shocks depend on the strength of the shock, as measured by its Mach number, the speed of the shock relative to the speed of the equivalent linear wave and local plasma parameters. Consequently, the downstream properties of the shocked plasma also depend on its Mach number.

6.2 SHOCK BASICS

6.2.1 Waves

As discussed in Chapter 3, there are three propagating waves in the plasma at low frequencies, well below the proton gyro-frequency. The **fast magnetosonic wave** compresses and deflects the plasma around an obstacle. The slow magnetosonic wave strengthens the magnetic field while the mass density drops. This is akin to the stretching of the field line, since when a straight field line gets longer, keeping the same mass content, the field strength will remain the same (in a cold plasma) while the density must drop. In a warm plasma, the thermal energy density of the plasma is positively correlated with the number density, and, when the number density drops, the energy density must too. Hence, in a **slow magnetosonic wave**, the magnetic field and number density are anticorrelated. Of the three propagating waves, slow-mode and fast-mode waves, as their names imply, travel most slowly and most rapidly in a plasma, respectively, at least in the magnetohydrodynamic (MHD) approximation that is generally applicable. Figure 6.1 shows a polar plot of the phase velocity

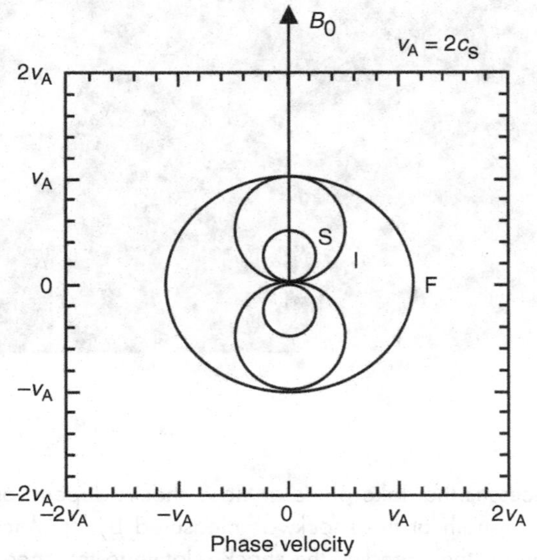

FIGURE 6.1. Polar plot of phase speed of MHD waves in a magnetized plasma in which the Alfvén speed is twice the sound speed. The magnetic field direction is vertical. The fast, intermediate, and slow magnetosonic waves are marked F, I, S, respectively.

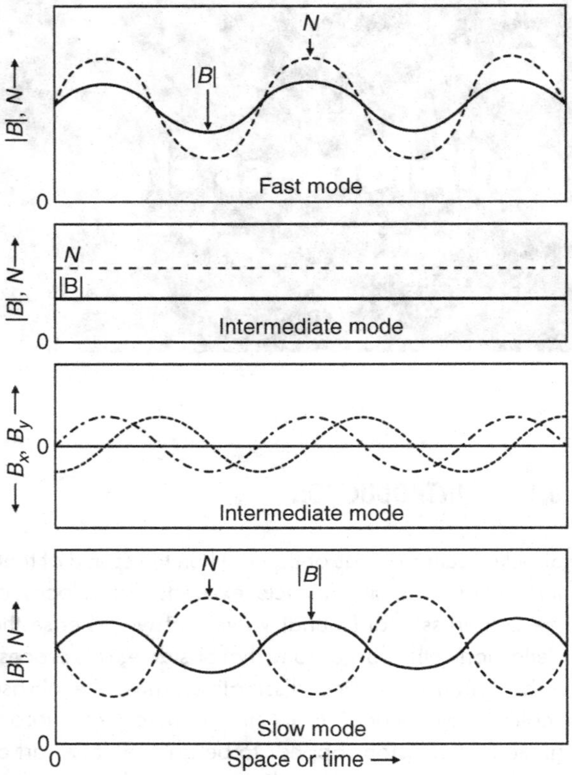

FIGURE 6.2. Waveforms of the fast, intermediate, and slow magnetosonic waves. The signature in the field strength and density are shown for all three waves. Two magnetic components are shown for the intermediate mode because the magnetic field strength and density are constant in this wave.

of these three modes for the case when the Alfvén speed is twice the sound speed.

The third wave mode in the plasma is the **shear Alfvén** or intermediate mode, which is a pure bending wave in which the direction of the magnetic field and the plasma velocity change, but the density and magnetic field strength do not. Each of these three waves can be associated with a collisionless shock, but only in a very limited range of circumstances for the intermediate mode. Figure 6.2 shows the variation of density and magnetic field strength versus space or time for the three modes – fast, intermediate, and slow. For the intermediate mode, two orthogonal components of the magnetic field are shown.

6.2.2 Dependence on upstream conditions

Two other factors affect the processes occurring at the shock: the beta of the plasma, the ratio of the magnetic to the thermal plasma energy density; and the direction of the magnetic field in the unshocked flow relative to the normal to the shock front. This shock normal angle is usually abbreviated as θ_{B_n}. Because of the latter dependence, the geometry of the interaction is very important, and, since most obstacles to the flowing solar-wind plasma are quasi-spherical objects of different sizes, the radius of curvature of the shock becomes important. Thus, while one-dimensional treatments (as we pursue in this chapter) are instructive, we often have to treat a global model of the interaction in order to understand the observed behavior. We defer discussion of such global models until Chapter 7.

Some of the interest in the study of collisionless shocks stems from their associated acceleration of charged particles to large energies. Systems of collisional particles generally evolve with time to maxwellian distributions with very few particles at high energies. However, in collisionless plasmas, a small minority of the particles can gain a

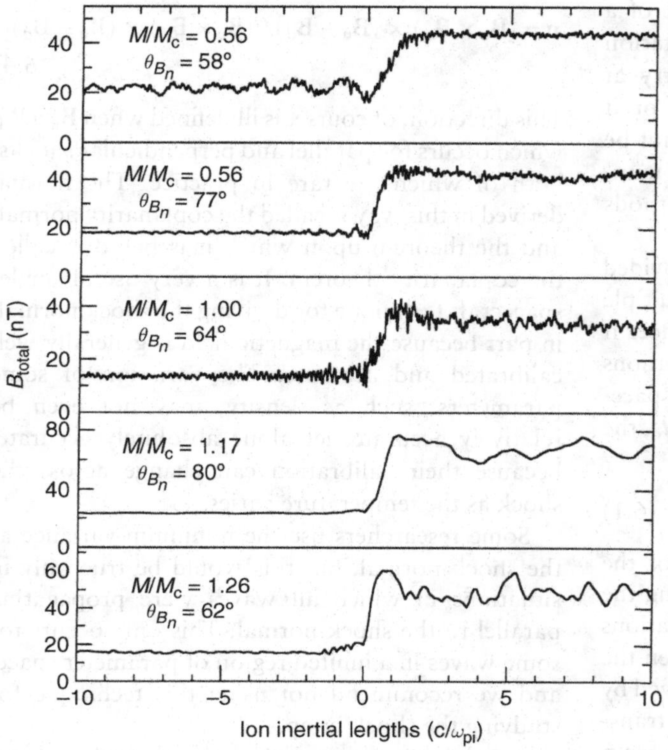

FIGURE 6.3. Observations of magnetic field strength for five shocks of low Mach number, ranging from subcritical to slightly supercritical. All shocks are quasi-perpendicular shocks. The measured temporal profile has been transformed to a distance scale using simultaneous measurements from a companion spacecraft that observed these shocks moments later. The low-frequency upstream waves in the two top panels are standing precursors that move with the shock. The higher-frequency waves are traveling waves in the shock frame. The appearance of these waves is sensitive to the value of θ_{Bn}, the shock normal angle. M_c is the critical Mach number.

disproportionate share of the energy, and their distribution of function becomes non-maxwellian. Frequently, the cause of this favoritism of a minority of the particles is acceleration by a collisionless shock. Such acceleration can occur at bow shocks at planetary obstacles, at interplanetary shocks, and at astrophysical shocks in which particles reach ultra-relativistic energies. When these particles enter our solar system, they are called cosmic rays.

Figure 6.3 shows the magnetic field strength during five crossings of the bow shock by ISEE 1 and 2 under a variety of solar-wind conditions, albeit all at low beta and low Mach numbers when the magnetic field was at a large angle to the shock normal. As can be seen, the shock is thin, comparable to, or less than an ion inertial length, the speed of light divided by the ion plasma frequency. When the Mach number is high, or when the magnetic field upstream is more aligned with the shock normal, the fluctuation amplitudes are much greater. We return to the observed structure of the bow shock after we examine the Rankine–Hugoniot equations that allow us to calculate the change in the average parameters across the shock front. The derivation of these equations from conservation laws and Maxwell's equations is straightforward but lengthy. Since it is important enough and its results used frequently, we include the derivation below. If the reader is only interested in the results, please just skip forward to Eqs. (6.25) and (6.26) near the end of Section 6.3. However, to work with shocks and to use these equations, we must first know how to find the direction of the normal to the shock. This is given in Section 6.2.

6.2.3 The shock normal

As discussed in Section 6.2.2, the angle between the upstream magnetic field and the normal to the shock is one of the most important parameters in determining the nature of the physical processes occurring at the shock. Whereas determination of the orientation of the upstream magnetic field is usually straightforward, determining the orientation of the shock surface may not be so simple. If the shock is standing in front of a relatively hard obstacle, the shock orientation may be obtained from the known geometry at the interaction. If the shock is oscillating, or if it is a propagating shock, this surface will not be geometrically constrained in any obvious manner. In such a case, we must use other methods to determine the shock surface.

An example of such a constraint is provided by the relative timing and separation of multiple observations of the same planar shock surface. If $\delta\mathbf{x}_i$ (i = 1, 2, 3, ...) are the spatial separations between observations of a shock relative to a spacecraft at $\mathbf{x}_o$, and if δt_i are the time delays from t_o, the time of the observation at $\mathbf{x}_o$, then

$$\delta\mathbf{x}_i \cdot \hat{\mathbf{n}} = v\delta t_i, \tag{6.1}$$

where $\hat{\mathbf{n}}$ is a unit vector in the direction of the normal to the surface and v is its speed along the normal. If there are more than four observations (i.e., there are more than three baselines), then the equations are over-determined and can be solved by multiplying each side of the equations by the transpose of the $\delta\mathbf{x}$ matrix and inverting to solve for $\hat{\mathbf{n}}$ and v. However, we often count ourselves lucky to obtain measurements of a shock with one spacecraft, let alone four.

The rate of transport of magnetic flux toward the shock is the tangential component of the electric field. Since magnetic flux does not pile up in the shock layer, our intuition, consistent with Maxwell's equations, tells us that the tangential component of the electric field is conserved across the shock. This, in turn, tells us that the upstream and downstream magnetic field and the shock normal must be coplanar. Thus, the cross product of the upstream and downstream magnetic fields should be perpendicular to the shock normal:

$$(\mathbf{B}_u \times \mathbf{B}_d) \cdot \hat{\mathbf{n}} = 0. \tag{6.2}$$

Because the magnetic field is divergenceless,

$$(\mathbf{B}_u - \mathbf{B}_d) \cdot \hat{\mathbf{n}} = 0. \tag{6.3}$$

With two different vectors perpendicular to the shock normal, we can calculate a vector along the normal and normalize by the magnitude of the cross product:

$$\hat{\mathbf{n}} = (\mathbf{B}_u \times \mathbf{B}_d) \times (\mathbf{B}_u - \mathbf{B}_d)/|(\mathbf{B}_u \times \mathbf{B}_d) \times (\mathbf{B}_u - \mathbf{B}_d)|. \tag{6.4}$$

This direction, of course, is ill defined when $\mathbf{B}_u \parallel \mathbf{B}_d$, which occurs for parallel and perpendicular shocks, both of which are rare in practice. The normal derived in this way is called the coplanarity normal, and the theorem upon which it is based is called the coplanarity theorem. It is a very useful single-spacecraft technique for deriving the shock normal, in part because the magnetic field is generally well calibrated and accurate. The variation of some parameters, such as density, may not even be relatively accurate, let alone absolutely accurate, because their calibration can change across the shock as the temperature varies.

Some researchers use the minimum variance as the shock normal, but this would be true only in situations in which all waves were propagating parallel to the shock normal. This only occurs for some waves in a limited region of parameter space, and we recommend not using this technique for studying the shock jump.

Another constraint that can be used when three-dimensional velocity measurements are available is that

$$(\mathbf{B}_u \times \Delta\mathbf{v}) \cdot \hat{\mathbf{n}} = 0 \tag{6.5}$$

and

$$(\mathbf{B}_d \times \Delta\mathbf{v}) \cdot \hat{\mathbf{n}} = 0, \tag{6.6}$$

where $\Delta\mathbf{v}$ is the change in the flow velocity from upstream to downstream. When a velocity constraint such as this is combined with a magnetic constraint such as $\Delta\mathbf{B} \cdot \hat{\mathbf{n}} = 0$, the resulting normal is called a mixed-mode shock normal. In fact, all these constraints can be combined into one over-determined solution and solved for a best-fit shock normal (Russell *et al.*, 1983).

Finally, we note that, with knowledge of the shock normal $\hat{\mathbf{n}}$ and use of the continuity equation, we can determine the velocity of the shock relative to the measurement frame. It is possible to show that the shock velocity v_{sh} is

$$v_{sh} = (\rho_d - \rho_u)^{-1}(\rho_d u_d - \rho_u u_u) \cdot \hat{\mathbf{n}}, \tag{6.7}$$

where u_d and u_u are the downstream and upstream solar-wind velocities, respectively.

Once the shock velocity is known, it is possible to convert the time series of a shock observation into a spatial profile by converting time into an appropriate distance unit such as ion inertial length (Figure 6.3) or ion gyro-radii.

Now that we know how to find the shock normal, we can proceed to derive the Rankine–Hugoniot equations that are derived in the normal incidence frame, where the flow is along the shock normal.

6.3 RANKINE–HUGONIOT EQUATIONS

With the establishment of a shock normal coordinate system, we can now express the Rankine–Hugoniot conservation relations in terms of the measured values of the plasma. The Rankine–Hugoniot equations allow us to calculate the downstream state from the upstream state as a function of the strength of the shock, as given by its Mach number. For a fast magnetosonic shock, this is the speed of the shock front, relative to the plasma measured along the shock normal, divided by the speed of the fast magnetosonic wave in the direction of the shock normal in the upstream region. This treatment does not tell us how heating and dissipation occur, only that so much dissipation must occur. One of the assumptions of this treatment is that the spatial changes of the plasma properties are exclusively along the shock normal. The equations are expressed in the normal incidence frame, where the upstream velocity is along the shock normal. Therefore, we assume the gradient is defined as

$$\nabla \equiv \frac{\partial}{\partial n}. \tag{6.8}$$

The Rankine–Hugoniot equations use the conservation equations and Maxwell's equations to derive a set of equations that depend on the upstream state of the plasma (magnetic field strength and orientation, plasma density temperature and velocity) to determine the same parameters downstream as a function of shock strength. We start with an expression for the mass continuity equation. Using our expression for the gradient $\nabla \cdot = (\rho u) = 0$, the continuity equation becomes $\partial(\rho u)/\partial n = 0$. If we integrate over n, we can express the above relation as

$$[\rho u_n] = 0, \tag{6.9}$$

where again the bracketed notation indicates the change across the shock boundary. This indicates that the change in mass flux across the shock remains constant.

For the momentum equation

$$\rho u \cdot \nabla u = \nabla P + \mathbf{j} \times \mathbf{B},$$

we get two conservation equations, one for the normal momentum and another for the tangential momentum. For the normal momentum equation, we take only those components which lie along the **n** direction. Similarly, the tangential momentum equation uses only those components which lie along the **l** direction, where **l** is in the plane of the shock along the projection of the upstream magnetic field. Therefore, we obtain for the two equations:

$$\left[\rho u_n + P + \frac{B_l^2}{2\mu_0}\right] = 0, \tag{6.10}$$

$$\left[\rho u_n u_l - \frac{B_n B_l}{\mu_0}\right] = 0. \tag{6.11}$$

The energy equation

$$\nabla \cdot \left[u\left(\frac{\gamma}{\gamma-1}P + 1/2\rho u^2\right)\right] = \mathbf{j} \times \mathbf{E}$$

can be reduced to

$$\left[u_n\left(\frac{\gamma}{\gamma-1}P + 1/2\rho u^2 + \frac{B_l{}^2}{\mu_0}\right) - \frac{u_l B_n B_l}{\mu_0}\right] = 0 \tag{6.12}$$

by using the fact that

$$\mathbf{j} \cdot \mathbf{E} = \mu^{-1} \cdot (\nabla \times \mathbf{B}) \cdot (\mathbf{u} \times \mathbf{B})$$

and providing the proper velocity and magnetic field components. The equation $\nabla \cdot \mathbf{E} = 0$ becomes a tautology (0 = 0) when the gradient is taken. Therefore, this equation does not play an important role in obtaining a solution. The equations $\nabla \cdot \mathbf{B} = 0$ and $\nabla \times \mathbf{E} = 0$ become

$$[B_n] = 0, \tag{6.13}$$

$$[u_n B_l - u_l B_n] = 0. \qquad (6.14)$$

Therefore, we now have a closed set of equations which can be combined and solved to obtain a solution that we can use to predict the downstream state. To achieve our goal of expressing the downstream state, indicated by subscript 2, in terms of the upstream state, indicated by subscript 1, we rewrite the above equations as follows:

$$\rho_1 u_{1n} = \rho_2 u_{2n}, \qquad (6.9a)$$

$$\rho_1 u_{1n}^2 + \rho_1 KT_1/m + B_{1l}^2/2\mu_0$$
$$= \rho_2 u_{2n}^2 + \rho_2 KT_2/m + B_{2l}^2/2\mu_0, \qquad (6.10a)$$

$$B_{1n} B_{1l}/\mu_0 = B_{2n} B_{2l}/\mu_0 - \rho_2 u_{2n} u_{2l}, \qquad (6.11a)$$

$$u_{1n}\left(\gamma(\gamma-1)^{-1}\rho_1 KT_1/m + 0.5\rho_1 {u_{1n}}^2 + {B_{1l}}^2/\mu_0\right)$$
$$= u_{2n}\left(\gamma(\gamma-1)^{-1}\rho_2 KT_2/m + 0.5\rho_2 {u_{2n}}^2 + 0.5\rho_2 {u_{2l}}^2 + B_{2l}^2/\mu_0\right) - u_{2l} B_{2n} B_{2l}/\mu_0, \qquad (6.12a)$$

$$B_{1n} = B_{2n}, \qquad (6.13a)$$

$$u_{1n} B_{1l} = u_{2n} B_{2l} - u_{2l} B_{2n}, \qquad (6.14a)$$

and define dimensionless parameters

$$x = \rho_2/\rho_1, \quad y = u_{2n}/u_{1n}, \quad z = u_{2l}/u_{1n},$$
$$w_1 = (kT_1/mu_{1n}^2)^{1/2}, \quad w_2 = (kT_2/mu_{1n}^2)^{1/2},$$
$$b_n = B_n(\mu_0\rho_1 u_{1n}^2)^{-1/2}, \quad b_{1l} = B_{1l}(\mu_0\rho_1 u_{1n}^2)^{-1/2},$$
$$b_{2l} = B_{2l}(\mu_0\rho_1 u_{1n}^2)^{-1/2}$$

Using these definitions, our Rankine–Hugoniot equations become

$$1 = xy, \qquad (6.9b)$$

$$2(w_1^2 + 1) + b_{1l}^2 = 2x(w_2^2 + y^2) + b_{2l}^2, \qquad (6.10b)$$

$$b_n b_{1l} = b_n b_{2l} - xyz, \qquad (6.11b)$$

$$2\left(b_{1l}^2 + \gamma(\gamma-1)^{-1} w_1^2\right) + 1$$
$$= y\left(y^2 + z^2 + 2b_{2l}^2 - 2b_n b_{2l} + 2\gamma(\gamma-1)^{-1}\, x w_2^2\right), \qquad (6.12b)$$

and

$$b_{1l} = y b_{2l} - z b_n. \qquad (6.14b)$$

The relationship in the first equation can be used to remove x from the following three, and, using the third and fifth equations, we can solve for b_{2l} and z, obtaining

$$b_{2l} = b_{1l}(b_n^2 - 1)(b_n^2 - y)^{-1},$$
$$z = b_{1l} b_n (y-1)(b_n^2 - y)^{-1}.$$

Continuing this procedure, we obtain a fourth-order equation in y. To keep better track of the coefficients for the quartic equation, we introduce four variables, which can be expressed in terms of the original measured parameters, as well as those variables defined above. The four defined variables are

$$a \equiv w_1^2 \equiv KT_1/(mu_{1n}^2), \qquad b \equiv b_n^2 \equiv B_n^2/(\mu_0\rho_1 u_{1n}^2),$$
$$c \equiv b_{1l}^2 = B_{1l}^2/(\mu_0\rho_1 u_{1n}^2), \qquad d \equiv \gamma(\gamma-1)^{-1}.$$

With these defined variables, the quartic equation in y is given by

$$y^4(1-2d) + y^3(2d + 2ad + cd + 4bd - 2b)$$
$$+ y^2(b^2 + bc - 4bd - 4abd - 2bcd - 2b^2 d - 2c - 2ad - 1) + y(2b^2 d + 2ab^2 d + 2bcd + 4abd + 2b - cd) + (2c - bc - b^2 - 2ab^2 d) = 0. \qquad (6.15)$$

We can reduce this equation to a cubic equation in y by noting that this equation has one trivial root, $y = 1$, corresponding to no shock at all. The cubic equation becomes

$$y^3 + py^2 + qy + r = 0,$$

where

$$p = [d(2a + c + 4b) - 2b + 1]/(1-2d),$$
$$q = [-d(4ab + 2bc + 2b^2 - c) + bc - 2c + b^2 - 2b]/(1-2d),$$

and

$$r = b(2abd + b + c)/(1-2d).$$

Depending on the coefficients p, q, and r in the cubic equation, there will be at least one and as many as three real roots. If we define

$$Q \equiv (p^2 - 3q)/9, \quad R \equiv (2p^3 - 9pq + 27r)/54,$$

and the expression $Q^3 - R^2 \geq 0$, there are three real roots, which are

$$y = -2Q^{1/2}\cos((\varphi + n\pi)/3) - p/3, \text{for } n = 0, 2, 4, \tag{6.16}$$

where

$$\varphi = \cos^{-1}\left(R/Q^{3/2}\right).$$

If the expression $Q^3 - R^2 < 0$, there is one real root, which is

$$y = R/|R|\left[(R^2 - Q^3 + |R|)^{1/3} + Q\{(R^2 - Q^3)^{1/2} + |R|\}^{-1/3}\right] - (p/3). \tag{6.16a}$$

It is important to note that, although there may be three real roots, the root for which $n = 2$ is the only physically interesting root for our discussion of planetary bow shocks. Therefore, depending on the sign of the expression $Q^3 - R^2$, we use either Eq. (6.16a) or Eq. (6.16) with $n = 2$ to determine the parameters for the subsonic downstream state.

We can now calculate the downstream state of the plasma from the upstream parameters. Since parameter y is defined to be the ratio between the downstream and upstream velocities along the shock normal,

$$u_{2n} = yu_{1n}. \tag{6.17}$$

The ratio between the downstream and upstream densities is the reciprocal of the above relation, thus

$$\rho_2 = \rho_1/y. \tag{6.18}$$

Using the definitions for x, y, z, etc. and Eq. (6.16), we can obtain additional downstream values of the plasma:

$$u_{2l} = u_{1n}(y - 1)(bc)^{1/2}(b - y)^{-1} \tag{6.19}$$

and

$$B_{2l} = (\mu_0\rho_1 u_{1n}^2 c\delta)^{1/2}, \tag{6.20}$$

where

$$KT_2 = mu_{1n}^2 y\left(a + (1 - \delta)c/2 - y + 1\right), \tag{6.21}$$

$$\delta = (b - 1)^2(b - y)^{-2}.$$

The parameter δ is the square of the tangential component of the magnetic field. From these expressions in Eqs. (6.19)–(6.21), we can calculate the magnitude of the velocity and magnetic field downstream by

$$|u_2| = (u_{2n}^2 + u_{2l}^2)^{1/2}$$
$$= u_{1n}(y^2 + bc(y - 1)^2(b - y)^{-2})^{1/2} \tag{6.22}$$

and

$$|B_2| = (B_{2n}^2 + B_{2l}^2)^{1/2} = (\mu_0\rho_1 u_{1n}^2(b + c\delta))^{1/2}. \tag{6.23}$$

A note should be made about comparing predicted values of the downstream plasma using the above relations with measured values from spacecraft. It is very easy to compare directly the densities, temperatures, and magnetic field strengths measured by spacecraft with the Rankine–Hugoniot predicted values because the absolute values of these parameters are not frame dependent. However, we must remember that certain assumptions were made previously about the direction of the flow across the shock. It is important to remember that a frame transformation was executed so that the direction of the upstream plasma flow is aligned along the shock normal. This fact means that there may be a residual velocity component in the **m** direction along the shock front for the measured velocity not taken into account for the predicted velocity. Therefore, care must be taken when comparing measured and predicted velocities due to the use of a special coordinate system.

We note that this frame is different from the de Hoffman–Teller frame, in which the frame travels along the shock at a speed sufficient to align the flow along the magnetic field (rather than the normal, as shown here). Hence, there is no motional electric field in the de Hoffman–Teller frame. The velocity of the frame is $v_{\mathrm{HT}} = \hat{\mathbf{n}} \times (\mathbf{u}_\mathrm{u} \times \mathbf{B}_\mathrm{u})/(\hat{\mathbf{n}} \cdot \mathbf{B}_\mathrm{u})$. Since the normal component of **B** and the transverse component of **u** × **B** are conserved across the shock, the de Hoffman–Teller velocity is the same on both sides of the shock. In this frame, particles simply spiral around the magnetic field line and move along it. Furthermore, since E is zero, the energies of particles are constant in this frame.

It is also possible to describe the plasma with four dimensionless upstream parameters: plasma beta, β, the fast magnetosonic Mach number, M_{ms1}, the angle of the upstream field to the shock normal, $\theta_{B_{1n}}$, and the ratio of specific heats, γ.

Doing this, we obtain for the upstream version of the four parameters

$$\beta_1 = 2a/(b+c); \theta_{B_{1n}} = \cos^{-1}(b/b+c)^{1/2};$$
$$\gamma = d/(d-1);$$
$$M_{ms1} = (0.5(b+c+ad/(d-1)$$
$$+((b+c+ad/(d-1))^2 - 4abd/(d-1)^{1/2})^{-1/2}, \quad (6.24)$$

where

$$a = \beta_1/(M_{ms1}^2 K_0);\ b = 2\cos^2\theta_{B_n}/(M_{ms1}^2 K_0);$$
$$c = 2\sin^2\theta_{B_n}/(M_{ms1}^2 K_0);\ d = \gamma(\gamma-1);$$

and

$$K_0 = 1 + \gamma\beta_1/2 + (1 + \gamma^2\beta_1^2/4 + \gamma\beta_1$$
$$\times(1 - 2\cos^2\theta_{B_n}))^{1/2}.$$

With these expressions for a, b, c, and d, we can obtain y and the parameters for the downstream state.

Since upstream plasma can be fully described by four dimensionless parameters, the same is true for the downstream plasma. Using our definitions of these dimensionless parameters and the expressions for the downstream values of the plasma given, expressions for the dimensionless parameters downstream can be derived. They become

$$\beta_2 = 2(a + 0.5c(1-\delta) + 1 - y)/(b + c\delta);$$
$$\theta_{B_{n2}} = \cos^{-1}(b/(b+c\delta))^{1/2};$$
$$M_{ms2} = 2(y^2 + bc(y-1)^2/(b-y)^2)/[y(b+c\delta)$$
$$\times (1 + 0.5\gamma\beta_2 + (1 + \gamma^2\beta_2^2/4 + \gamma\beta_2$$
$$\times (1 - 2\cos^2\theta_{B_{n2}}))^{1/2})], \quad (6.25)$$

where the ratio of specific heats γ is assumed to remain constant through the shock.

As the shock is crossed, the beta (in most cases) will go up, the angle θ_{B_n} will increase toward 90°, and the Mach number will be below unity. The jumps in the magnetic field strength, density, temperature, and the plasma beta across the shock can be determined using what we have derived in the preceding paragraphs,

$$\rho_2/\rho_1 = y^{-1};$$
$$|B_2|/|B_1| = (\cos^2\theta_{B_n} + \delta\sin^2\theta_{B_n})^{1/2}$$
$$= [(b+c\delta)/(b+c)]^{1/2};$$
$$\beta_2/\beta_1 = a^{-1}(b+c)(a + 0.5c(1-\delta) + 1 - y)$$
$$\times(b+c\delta)^{-1};$$
$$T_2/T_1 = a^{-1}y(a + 0.5c(1-\delta) + 1 - y). \quad (6.26)$$

With the equations in Eq. (6.24), we are now equipped to determine how parameters change across the shock as a function of Mach number, beta, and θ_{B_n}. Figure 6.4 compares contour plots of the jump in density and the jump in magnetic field strength as a function of upstream θ_{B_n} and Mach number for three different betas. At a θ_{B_n} of 90°, the jump in density and magnetic field are the same, asymptoting to 4 (for a γ of 5/3) at high Mach number. For parallel shocks with θ_{B_n} of 0°, the jump is very different for the density and magnetic field. Generally, the density is less sensitive to changing θ_{B_n} than the magnetic field strength. At a beta of zero, the low Mach number behavior is complicated by the phenomenon known as a switch-on shock, which we discuss in Section 6.4.

Figure 6.5 compares contour plots of the jump in the tangential component of B and in the temperature across the shock as a function of θ_{B_n} and Mach number. Examining first the plots for high beta, we see that, independent of θ_{B_n}, these jumps increase monotonically as the Mach number increases.

For lower beta, the behavior changes markedly for the tangential component, and slightly for temperature. We see a slight change of the jump in temperature for changing θ_{B_n} for both $\beta = 0$ and $\beta = 1$, albeit in opposite senses. For the tangential component of B, there is a very marked change in the θ_{B_n} dependence. At $\theta_{B_n} = 90°$, there is a monotonic increase at $\beta = 0$ and $\beta = 1$, very similar to at $\beta \rightarrow \infty$. However, near θ_{B_n} = zero, the values increase

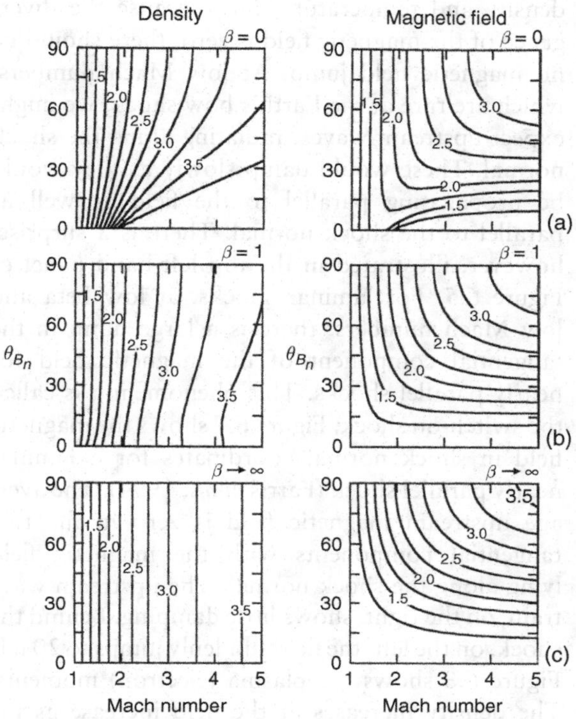

FIGURE 6.4. Contour plots of the jump in density and magnetic field strength as a function of the angle of the upstream magnetic field direction to the shock normal, θ_{B_n}, and the magnetosonic Mach number derived from the Rankine–Hugoniot equations. Contours for (a) $\beta = 0$; (b) $\beta = 1$; (c) $\beta \to \infty$.

and then decrease with increasing Mach number. This is related to the switch-on shock. This behavior is almost counter-intuitive.

Figure 6.6 compares contour plots of the jump in downstream beta and downstream Mach number as a function of θ_{B_n} and upstream magnetosonic Mach number. The downstream Mach number has the simplest behavior. It monotonically decreases with increasing upstream Mach number with almost no θ_{B_n} dependence at high beta, weak dependence for $\beta = 1$, and stronger dependence at $\beta = 0$. The behavior of downstream beta is quite interesting. At $\beta \to 0$, the downstream beta at low Mach number drops across the shock. Eventually it increases. At $\beta = 1$, low Mach number, nearly perpendicular shocks have a drop in beta but then an increase at higher Mach number. At high beta,

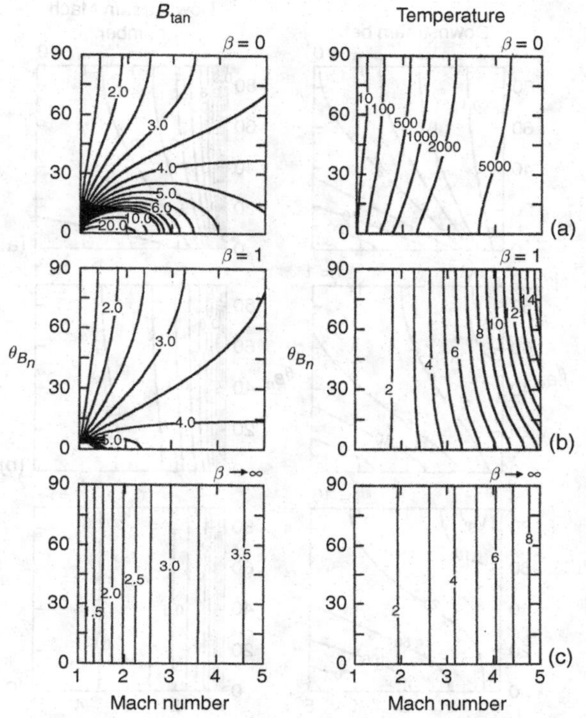

FIGURE 6.5. Contour plots of the jump in B_{tan} and the temperature. Comments of Figure 6.4 apply. Contours for (a) $\beta = 0$; (b) $\beta = 1$; (c) $\beta \to \infty$.

there is an increase in beta across the shock at all angles and Mach numbers.

6.4 OBSERVATIONS OF PARALLEL AND PERPENDICULAR SHOCKS

The generalized MHD Rankine–Hugoniot equations are very powerful as they allow us to predict the downstream state from the upstream state for any conceivable solar-wind condition as long as the MHD approximations hold. However, they tell us nothing about how the dissipation needed by the shock occurs. To understand this dissipation, we must explore the kinetic processes at the shock. For this purpose, high-resolution magnetic field measurements prove to be instructive. Before we explore that territory more generally, it is helpful to examine the behavior of the shock in two end-member states. We look at the nearly parallel

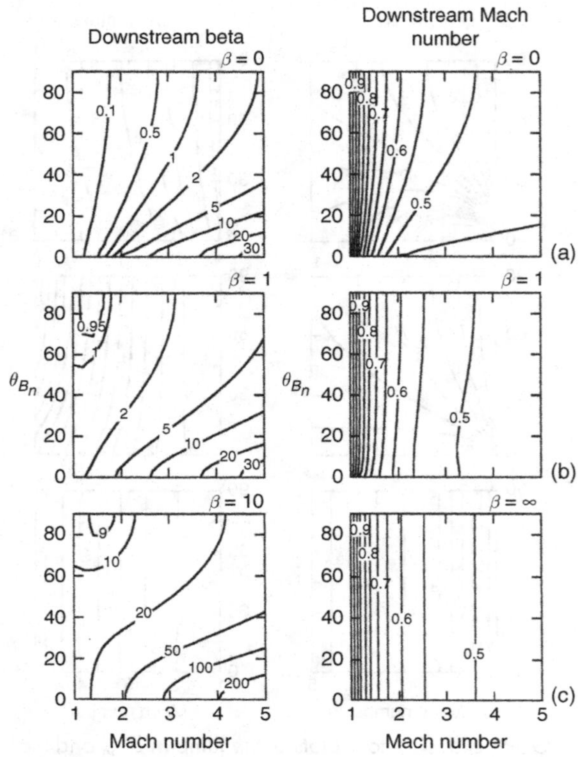

FIGURE 6.6. Contour plots of downstream beta and downstream Mach number. Contours for (a) $\beta = 0$; (b) $\beta = 1$; (c) $\beta \rightarrow \infty$.

shock and its manifestation in the laminar switch-on shock, as well as at the nearly perpendicular shock. Laminar shocks occur when both the Mach number and beta are low, e.g., less than 2 and 0.5, respectively. In the solar wind, shocks are often weak, but the beta is seldom small. The most promising place to find laminar shocks is at the inner planets, where there is always a shock and occasionally the IMF is very large, causing both the Mach number and the beta to drop into the laminar range together.

6.4.1 Parallel laminar shocks

Parallel shocks are very rare because the probability that two uncorrelated vectors will be aligned is very low, much lower than them being orthogonal, for which there are an infinite number of directions. The Rankine–Hugoniot relations tell us that, in general, a parallel shock should produce a jump in density and temperature, but, because the divergence of the magnetic field is zero, there should be no magnetic field jump. At low Mach numbers, which are rare at the Earth's bow shock, we might expect upstream waves, radiating along the shock normal. These would damp slowly as they would be propagating parallel to the field as well as parallel to the shock normal. There is a surprise, however, illustrated in the top left-hand panel of Figure 6.5. For laminar shocks, at low beta and low Mach numbers, there is a large jump in the tangential component of the magnetic field for nearly parallel shocks. This phenomenon is called the switch-on shock. Figure 6.7 shows the magnetic field in shock normal coordinates for a laminar nearly parallel shock (Farris *et al.*, 1994). The average upstream magnetic field is zero in the two tangential components with the magnetic field lying along the shock normal. The upstream wave train, on the right, shows little damping. Behind the shock, on the left, the field suddenly jumps to 20 nT. Figure 6.8 shows the plasma (electron) moments. The density increases at the field increase as the plasma flows through the shock, the velocity drops, and the temperature rises. We note that, even under laminar shock conditions, the quasi-parallel and parallel shocks are quite unsteady.

6.4.2 Perpendicular shocks

The magnetic profiles of strong quasi-perpendicular shocks consist of a turbulent foot, a jump in field across a ramp region followed by an overshoot, and an undershoot in the magnetic field strength. These features are determined by ion dynamics. The foot length is produced by the turn around of specularly reflected ions at the shock and can be used to determine how fast the shock is moving when there is only a single-spacecraft observation. So, too, the thickness of the ramp and overshoot are associated with ion dynamics, with the ramp thickness being about 0.4 ion inertial lengths. Furthermore, simulations of quasi-perpendicular shocks reproduce the shock structure and the ion dynamics quite well.

For perpendicular shocks, finite ion inertia is not expected to play a role in the dispersive structure of the shock. Rather, the shock thickness is expected to be of the order of the electron inertial length,

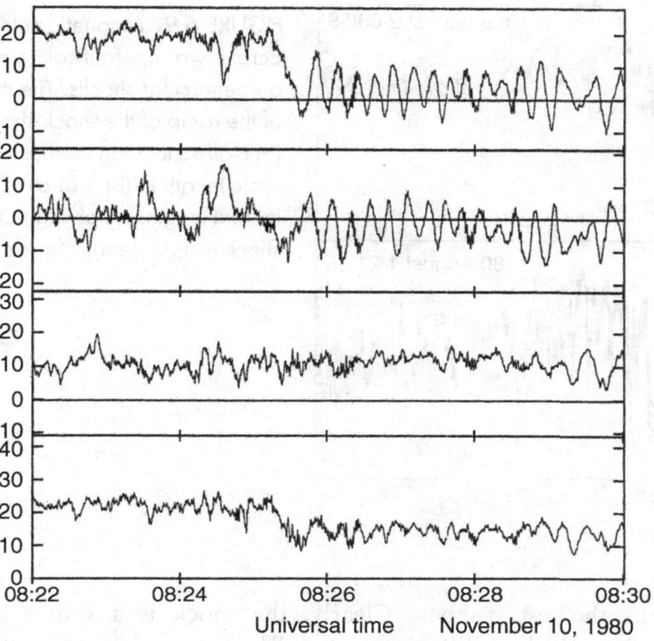

FIGURE 6.7. Magnetic field measurements across a switch-on shock in shock normal coordinates. The n-direction is along the shock normal, the l-direction is parallel to the downstream component of the magnetic field in the shock plane. Mach number = 1.14 ± 0.7; β = 0.2, and θ_{B_n} = 17°.

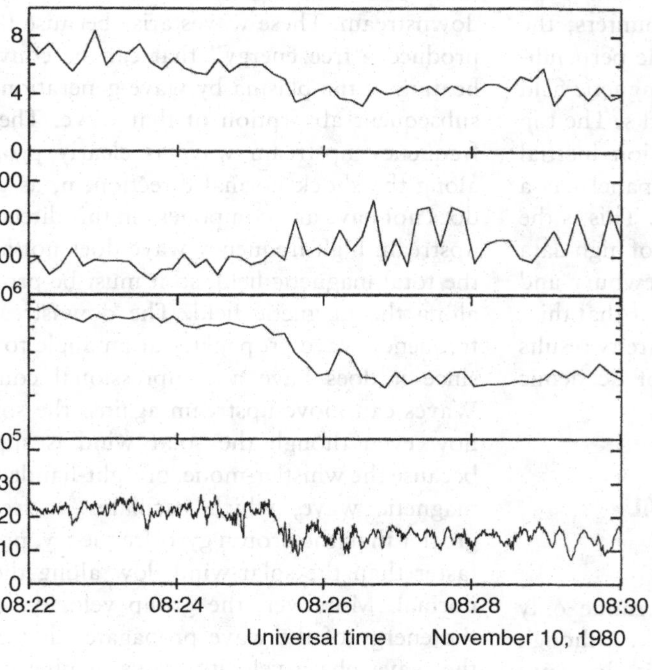

FIGURE 6.8. Electron density, velocity, temperature, and magnetic field strength during the switch-on shock shown in Figure 6.7.

43 times smaller than the ion inertial length. We can only determine the orientation of the normal to the shock to an accuracy of several degrees. Even when we have an accurate average orientation, it will vary with time as the spacecraft encounters it. Since we expect the "perpendicular" shock to exist only in a range of about 1° about the exactly perpendicular orientation, it will be almost as rare as the nearly

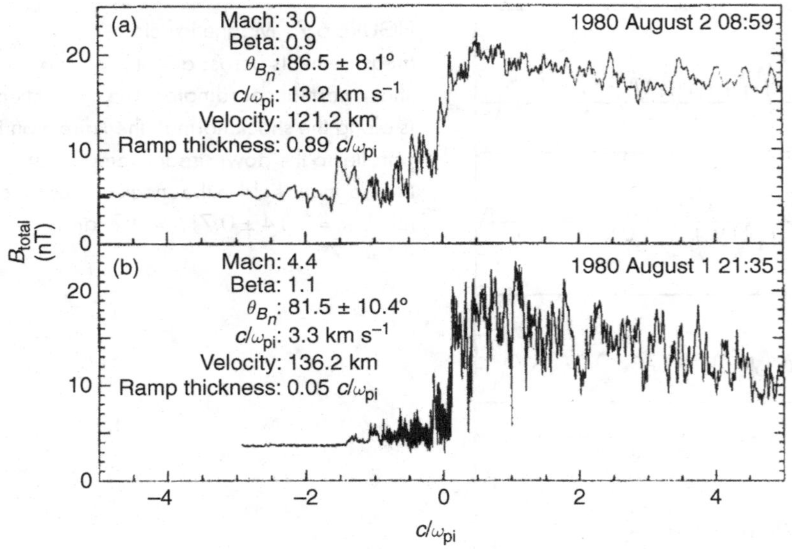

FIGURE 6.9. Magnetic field profile across two supercritical nearly perpendicular shocks. The thickness of the ramp of the shock shown in (a) is close to 1 ion inertial length. The scale length in (b) is about 2 electron inertial lengths, implying that this shock is a perpendicular shock.

parallel shocks we discussed above. Thus, the best we can do observationally is to determine which shocks are very nearly perpendicular and hope that, during one of these shock encounters, the shock will be sufficiently oriented in the perpendicular direction. Figure 6.9 shows the magnetic field through two nearly perpendicular shocks. The top profile has a ramp thickness of 0.9 ion inertial lengths, while the ramp in the lower panel has a thickness of 2 electron inertial lengths. This is the only example found in an examination of high data rate ISEE 1 and 2 shock crossings (Newbury and Russell, 1996). Thus, while it seems clear that thin, electron-scale shocks can exist, their scarcity results in their existence being principally of academic interest in solar system plasmas.

6.5 OBSERVATIONS UNDER TYPICAL CONDITIONS

The Rankine–Hugoniot relations determine only the jump in parameters across the shock. Although this is very important information, it does not inform us how the dissipation, heating, and thermalization occur. These latter processes make the study of shocks extremely interesting. Figure 6.10 shows three components of the magnetic field across a low Mach number quasi-perpendicular shock. Clearly, the shock is a source of many different waves. Waves are growing in the shock ramp and damping as they move upstream and downstream. These waves arise because the shock produces "free energy" that can be converted to heating of the plasma by wave generation and the subsequent absorption of that wave. The lowest-frequency upstream wave is clearly propagating along the shock normal direction, **n**, as the wave does not have any component in this direction. The upstream high-frequency wave does not appear in the total magnetic field, so it must be propagating along the magnetic field. The downstream high-frequency wave propagates at an angle to the field since it does have a compressional component. Waves can move upstream against the solar-wind flow even though the solar wind is supersonic, because the whistler-mode, or right-handed electromagnetic wave, which oscillates at frequencies greater than the proton gyro-frequency, propagates faster than the solar-wind flow along their wave normal. Moreover, the group velocity, at which the energy of the wave propagates, is faster than the wave phase velocity. Thus, a phase standing structure can form as the low frequency wave in Figure 6.10 and be continually provided with energy from the shock. Thus, the wave maintains a temporally steady profile even while damping as the incoming solar wind flows through it.

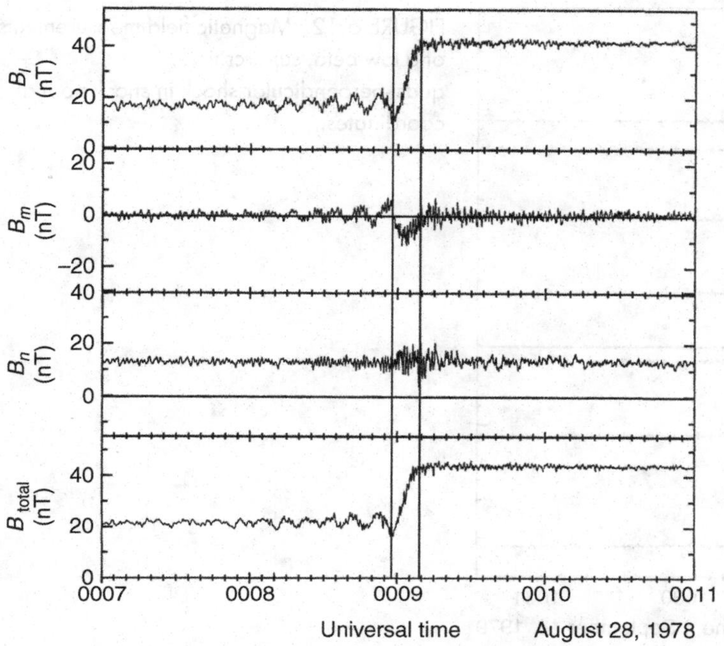

FIGURE 6.10. Magnetic field measurements of a low-beta, subcritical, quasi-perpendicular shock in shock normal coordinates. Except within the shock ramp, the magnetic field is contained within the l–n plane and the shock normal points upstream along the positive n-direction. The two solid lines define the thickness of the shock ramp.

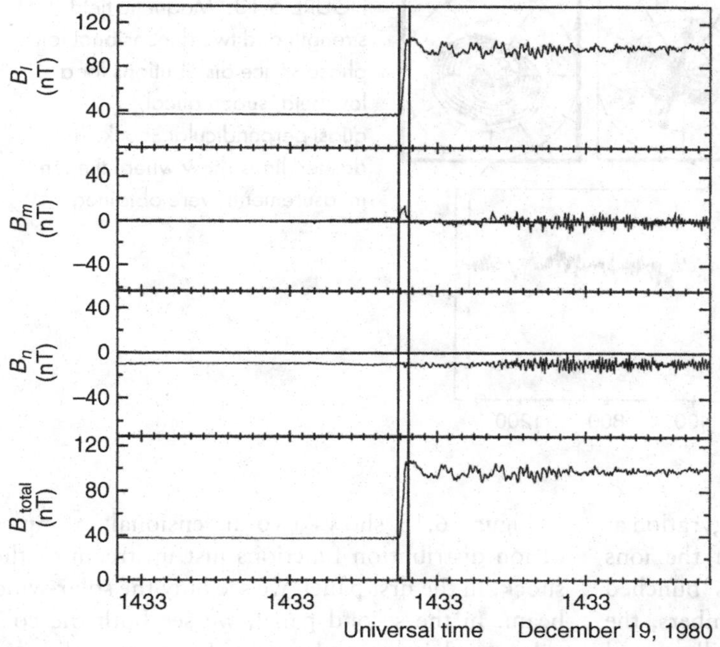

FIGURE 6.11. Magnetic field measurements of a low-beta, marginally critical, quasi-perpendicular shock in shock normal coordinates.

At higher Mach numbers, the waves cannot move upstream and the waves are all carried downstream, as shown in Figure 6.11. Note here the small B_m-component of the field as the shock is crossed and the slow growth of the high-frequency transverse waves downstream. Here the low-frequency downstream "wave" seen in the total field may not be due to a plasma instability but

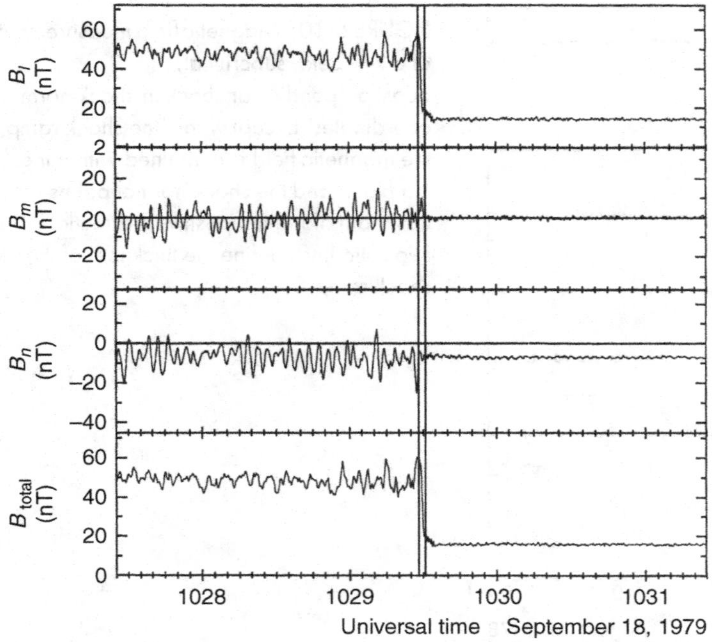

FIGURE 6.12. Magnetic field measurements of a low-beta, supercritical, quasi-perpendicular shock in shock normal coordinates.

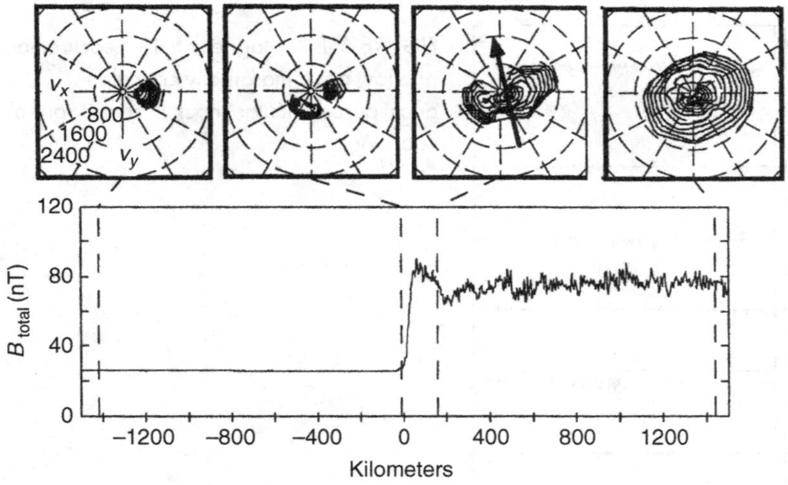

FIGURE 6.13. Magnetic field strength and two-dimensional ion phase space distributions for a low-beta, supercritical, quasi-perpendicular shock. The dashed lines show where the ion measurements were obtained.

rather just due to the bunching of the ion gyration at the thin shock. The thermal spread of the ions allows the ion motion to become less bunched downstream. At even higher Mach numbers, the downstream region is very turbulent, as illustrated in Figure 6.12. The downstream waves serve to thermalize the plasma by growing in a velocity space anisotropy produced as the ions cross the shock. The waves reduce the anisotropy and then in turn damp and heat the plasma.

Figure 6.13 shows two-dimensional examples of ion distribution functions just upstream of the shock. In the first panel, we see only the solar-wind beam. In the second panel, we see both the cold solar-wind beam and reflected ions that did not manage to get through the electric potential and the magnetic ramp of the shock. Panel three shows the downstream ion distribution. The reflected ions have now been swept through the shock and into the magnetosheath. They have been scattered by the

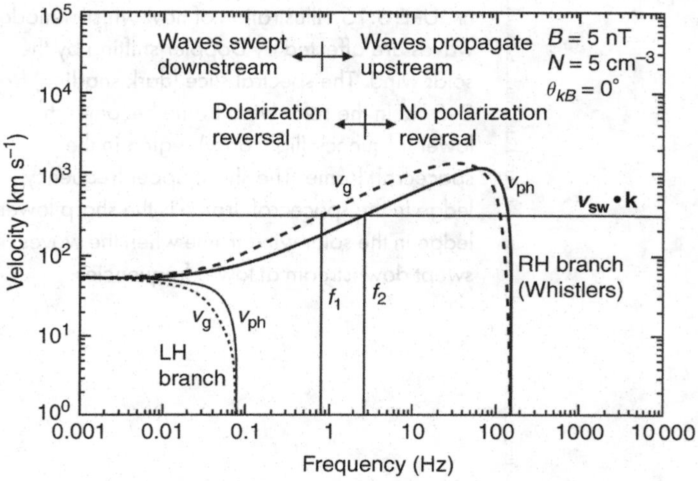

FIGURE 6.14. Phase and group velocities for typical conditions in solar wind near 1 AU, illustrating how whistler-mode waves can group stand (at f_1) or phase stand (at f_2). The velocity of the solar wind along the wave normal, $\mathbf{v}_{sw} \cdot \mathbf{k}$, will sweep wave energy downstream at frequencies below f_1 and reverse their polarization below f_2.

waves but still possess a lot of free energy for the production of waves. This free energy continues to produce waves until it is depleted. The waves damp in the plasma and result in the thermalization of the plasma. The fourth panel shows the particle distribution when it has lost most of its velocity space anisotropy and is essentially completely thermalized. The scale downstream over which these waves exist may be very long.

The upstream whistler-mode waves that are propagating at a small angle to the magnetic field are no less interesting than the low-frequency turbulence. Whistler-mode waves are generally associated with electrons which have little mass and therefore momentum, but move quite rapidly for the same energy or temperature. As Figure 6.14 illustrates, the phase and group velocities of whistler-mode waves can exceed the speed of the solar wind along the wave normal for ordinary solar-wind conditions. There is an interesting frequency range for these waves where the group velocity that is faster than the phase velocity can move upstream while the phase velocity is carried downstream. Thus, the wave appears to be left handed when it is actually right handed in the plasma. Moreover, it produces a very interesting power spectrum, as shown in Figure 6.15. If the wave spectrum in the plasma frame were the dark shaded spectrum from 2–5 Hz, and these were the frequencies corresponding to group velocity standing and phase velocity standing waves in the solar-wind flow, then the waves at 5 Hz would be Doppler shifted to 0 Hz. Lower frequencies can be Doppler shifted to higher frequencies than zero but are polarization reversed. Above 1 Hz in this example, the waves no longer have a group velocity that can transport wave energy upstream, and a sudden drop off of wave energy is seen. The phenomenon reminds us how much Doppler shifting of waves in the solar wind affects their appearance in our sensors. We must always convert the measurements to the plasma frame to understand how these waves are interacting with the plasma constituents.

6.6 THE CRITICAL MACH NUMBER

The concept of a **critical Mach number** arises from the evolutionary model of shock formation in which the shock is formed from a series of wave packets that overtake one another to form a single steepened wave packet. At low Mach numbers, resistive processes can supply the dissipation required by the Rankine–Hugoniot equations. These resistive processes gradually decouple the magnetic and fluid oscillations, and the phase speed of the steepened wave approaches the sound speed. Thus, if another magnetosonic pulse, which steepens in the same evolutionary manner, is launched and the flow speed is greater than the sound speed, the second pulse will not reach the first pulse. For a proper shock transition,

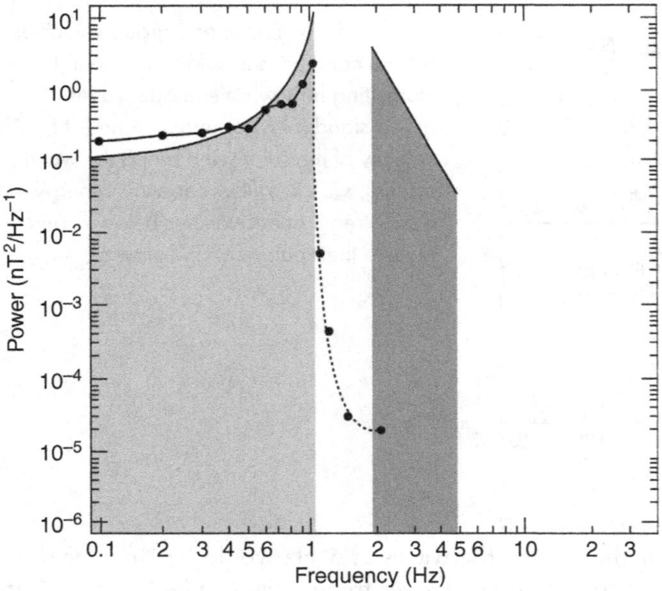

FIGURE 6.15. Illustration of how whistler-mode waves are affected by Doppler shifting by the solar wind. The spectral slice (dark shading) from 2–5 Hz in the solar-wind frame becomes the lower-frequency (light gray) region in the spacecraft frame. The sharp upper-frequency ledge in the spacecraft frame is the sharp lower ledge in the solar-wind frame when the waves are swept downstream at lower frequencies.

dissipation must cause all downstream perturbations to die out away from the shock layer (e.g., Kennel, Edmiston, and Hada, 1985). Thus, the point downstream where the flow speed is equal to the sound speed is the point where resistivity alone cannot provide the required shock dissipation. This is called the critical Mach number. New dissipation mechanisms are required in addition to resistive heating to satisfy properly the predictions made by the Rankine–Hugoniot equations when the upstream Mach number is above this critical Mach number (Kantrowitz and Petschek, 1966; Woods, 1969; Coroniti, 1970). The ratio between the upstream magnetosonic Mach number and the critical Mach number is called the ratio of criticality (M/M_c). When the Mach number is below the critical Mach number, the shock is called "subcritical," and when it is above, it is called "supercritical." One of the phenomena that changes with the ratio of criticality is the size of the overshoot, a rise in the magnetic field strength with a thickness of about an ion gyro-radius immediately downstream of the shock ramp. This behavior is illustrated in Figure 6.16.

As stated above, the critical Mach number is the upstream Mach number for which the downstream flow velocity along the shock normal is equal to the downstream gas-dynamic sound speed, for a particular plasma beta, magnetic field orientation, and ratio of specific heats. Since we have solutions for these two characteristic velocities (see Eqs. (6.17) and (6.21)), we can set them equal to one another and solve for the criterion defining the critical Mach number, recalling that the downstream sound velocity is $(\gamma KT^2/m)^{1/2}$

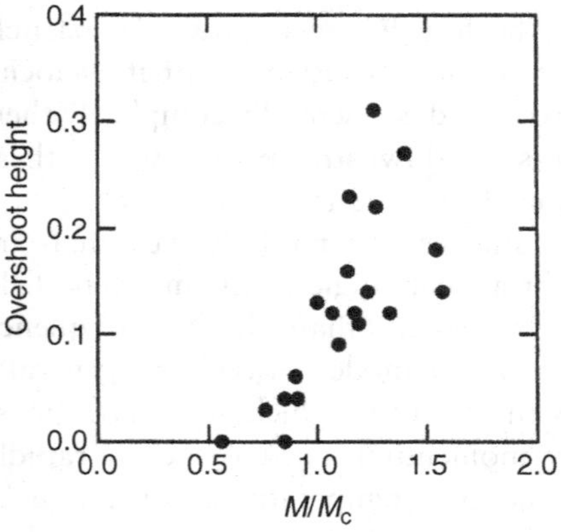

FIGURE 6.16. Overshoot height versus the ratio of the Mach number to the critical Mach number.

$$\gamma(a + 0.5c(1-\delta) + 1 - y) - y^2 = 0. \qquad (6.27)$$

The upstream magnetosonic Mach number which satisfies this relation is the critical Mach

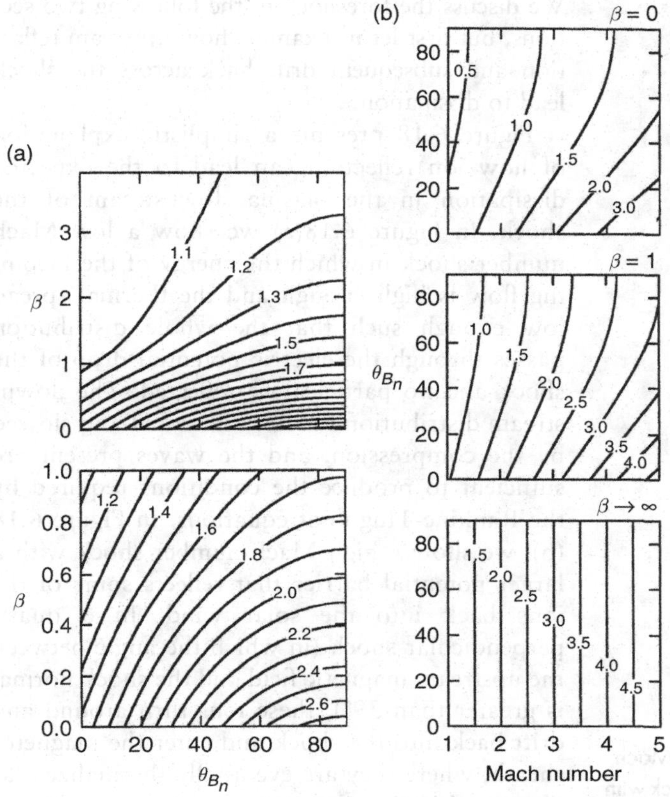

FIGURE 6.17. Contours of (a) the critical Mach number versus beta and θ_{B_n} and (b) the ratio of criticality versus θ_{B_n} and Mach number for $\beta = 0$, $\beta = 1$, and $\beta \rightarrow \infty$. The lower left panel in (a) is an expansion of the low-beta portion of the upper left panel in (a) to display more clearly the θ_{B_n} dependence in this region.

number. It is interesting to see how the critical Mach number varies with plasma beta and θ_{B_n} for a ratio of specific heats of 5/3. Figure 6.17 shows contours of the critical Mach number for varying betas and magnetic field orientations as well as the ratio of criticality versus θ_{B_n} and Mach number for various plasma betas. The critical Mach number is lowest for parallel shocks and increases monotonically as the shock becomes more perpendicular. It drops with increasing beta. The ratio of criticality, in turn, increases monotonically with upstream Mach number and drops somewhat with increasing θ_{B_n}. At high beta, there is very little θ_{B_n} dependence.

6.7 SHOCK DISSIPATION

Ideally, the collisionless shock develops a thin layer in which there is an electric potential drop with an electric field pointing in the direction of the incoming flow so that positive ions are decelerated. In addition, the magnetic field rise results in the deflection of some particles with large pitch angles, even without the deceleration by the electric field. What happens to the reflected particles depends on the direction of the magnetic field relative to the shock normal. In the frame of reference of the solar wind, the reflected ion has a velocity of twice the solar-wind velocity because the particle is now moving backward at its previous forward velocity. However, the upstream magnetic field is acting on this backward-moving particle and turns its motion around. The ion can only move a finite distance upstream until it is turned around and moves downstream through the shocks, unless the upstream field is at an angle of less than 39° to the shock normal. Then, the upstream reflection velocity is too high to be turned around by the gyro-motion to

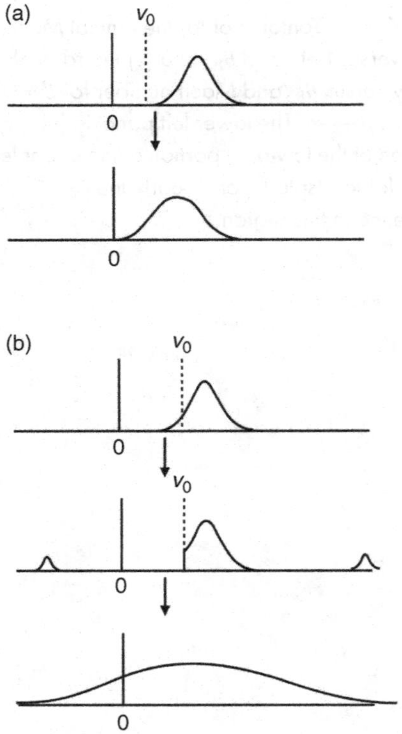

FIGURE 6.18. Variation of shock dissipation with Mach number. (a) The solar wind approaches a weak shock with a low electric potential drop so that the entire ion distribution can pass through without reflection.
(b) A strong shock has a potential drop that turns around part of the distribution so that it returns to the solar wind only to gyrate around the magnetic field and cross the shock at higher speed. The downstream ions are heated in crossing the shock and thermalized with increasing distance downstream.

cross the shock. At angles less than 39°, ions move upstream upon reflection. In practice, this process is very non-time-stationary. Moreover, hot ions created downstream can also move upstream, especially for supercritical shocks. In either case, the result is that ion beams can move upstream. We can simulate this process with hybrid codes, following only ion motion, or with fully kinetic codes. For ions at shocks with radii of curvature of the terrestrial planets, the radius of curvature is an important factor in determining the properties of the foreshock. These simulations will be discussed in Chapter 7. We discuss the foreshock in the following two sections, but first let us examine how upstream reflection and subsequent drift back across the shock lead to dissipation.

Figure 6.18 presents a simplistic explanation of how ion reflection can lead to the expected dissipation in the plasma downstream of the shock. In Figure 6.18(a) we show a low Mach number shock in which the energy of the incoming flow is high enough and the thermal spread low enough such that the whole distribution passes through the electric potential drop of the shock and no particles are reflected. The downstream distribution function is heated and slowed by the compression, and the waves present are sufficient to produce the conditions required by the Rankine–Hugoniot equations. In Figure 6.18 (b), we show a high Mach number shock with a larger potential barrier that reflects some of the ions back into the solar wind. In a quasi-perpendicular shock (in which the angle between the upstream magnetic field and the shock normal is greater than 39°), these ions turn around and drift back into the shock and enter the magnetosheath, where they are eventually thermalized, as illustrated in Figure 6.18(b).

One might expect that the electric potential that slows the ions would accelerate the electrons an equivalent amount, but electrons receive little heating at the shock. The reason that electrons can avoid being accelerated is tied to the magnetic structure of the shock, at least at low Mach numbers where the heating paradox occurs. Due to their small gyro-radii, electrons follow the magnetic field as shown in Figure 6.19, wheras ions move straight across the shock parallel to the normal. The magnetic field rotates as it crosses the shock in just the direction to move the electrons across solar-wind equipotentials to compensate for much of the electric potential drop across the shock normal (Goodrich and Scudder, 1984). This phenomenon is caused by the difference in the masses of electrons and ions. As a result, the electrons, which are generally warmer than ions in the solar wind, are the cooler species by about a factor of seven in the magnetosheath and the plasma sheet after they enter the Earth's tail.

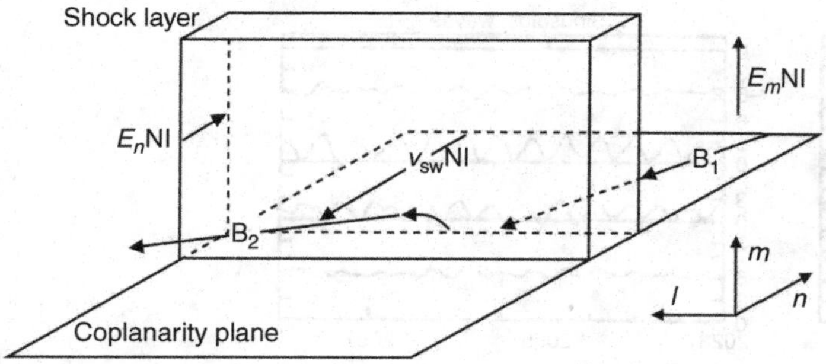

FIGURE 6.19. Schematic of how the electrons move along the magnetic field and are able to avoid the full effect of the potential drop across the shock by moving across interplanetary electric potentials. Quantities labeled "NI" are in the normal incidence frame where the solar-wind velocity is along the shock normal.

6.8 QUASI-PARALLEL SHOCK AND ION FORESHOCK

As mentioned above, the reflected ions can move upstream when the magnetic field makes an angle of less than 39° to the shock normal. Moreover, downstream particles can also cross the shock and move upstream. Ions moving upstream have to pass through the incoming solar wind and thus produce various ion instabilities in which waves grow from the free energy in the beam. These waves can be right handed when the unstable wave is moving slower than the particle and left handed when the particle and wave interact head on. As a result, the region upstream of the quasi-parallel shock, known as the foreshock, is a regular showcase of wave phenomena. Figure 6.20 shows examples of waves seen in this region.

In general, the waves are being carried backward by the solar wind into the shock and are adding to the dissipation. Moreover, they propagate and steepen while they simultaneously alter the particle distributions that produce them. This steepening is illustrated in Figure 6.21. This close coupling and mutual interaction make it difficult to sort out cause and effect, and this underlines the need to use both observation and numerical simulation to determine the physics of these shocks.

A very interesting phenomenon occurs when these waves reach the location of the quasi-parallel shock. The waves suddenly grow and the shock reforms. This process is illustrated in Figure 6.22.

6.9 THE ELECTRON FORESHOCK

Electrons are accelerated back into the solar wind at a much higher velocity and therefore have a foreshock boundary much closer to the tangent field line. They can be accelerated right up to the tangent point, albeit with infinitesimal flux (see Figure 6.23). This beam produces Langmuir oscillations at the plasma frequency of the solar wind.

It is instructive to review briefly the reflection of the electrons, often called fast-Fermi acceleration, to gain insight into the nature of this electron beam. Figure 6.24(a) shows the electron angular distribution in the **de Hoffman–Teller frame,** in which the solar-wind flow is parallel to the magnetic field. The closer the magnetic field is to being tangent to the shock normal, the higher the transformation velocity is, so that the center of the electron distribution moves upward in this diagram. The curved dashed lines indicate that part of the distribution that cannot cross the shock because of the combined action of the magnetic and electric fields. As the de Hoffman–Teller velocity becomes faster and faster, the reflected portion of the distribution moves to higher and higher energy, and the flux reflected becomes less and less. This is made more quantitative in Figure 6.24(b). The result is a Langmuir wave generation region whose extent is controlled by the curvature of the shock. This has relevance to both planetary and interplanetary shocks.

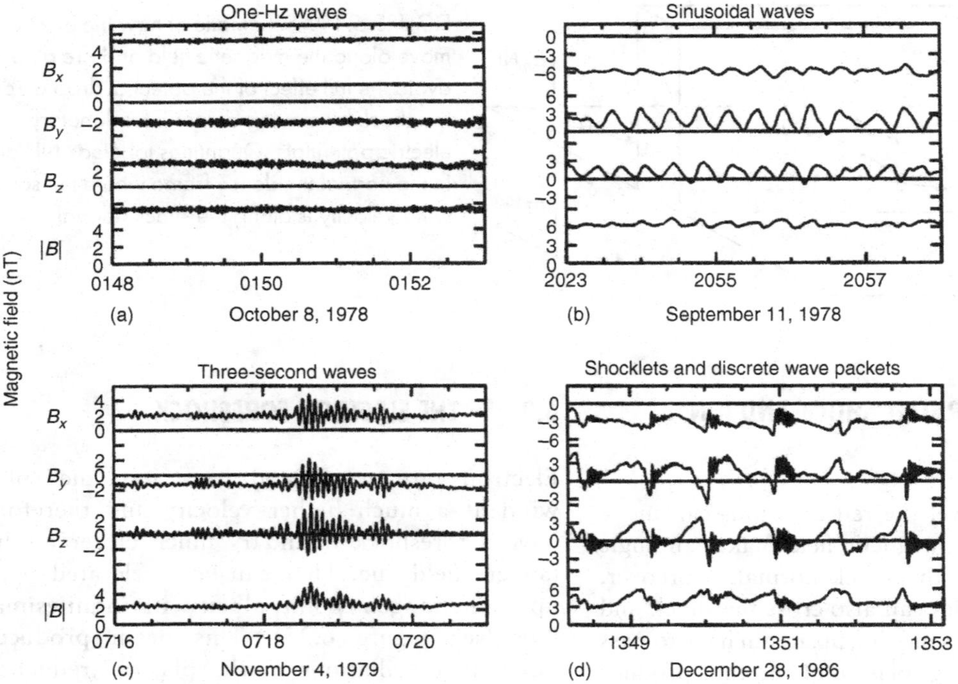

FIGURE 6.20. Examples of upstream waves observed by the ISEE 1 and 2 spacecraft. Magnetic field data are in the geocentric solar ecliptic coordinate system with x to the Sun and z along ecliptic north. (a) Waves commonly called "one-Hz" waves. (c) Three-second waves, whose strength is so large they change the magnitude of the field. Sinusoidal waves (b) are thought to steepen (d), producing shocklets and whistler-mode wave packets.

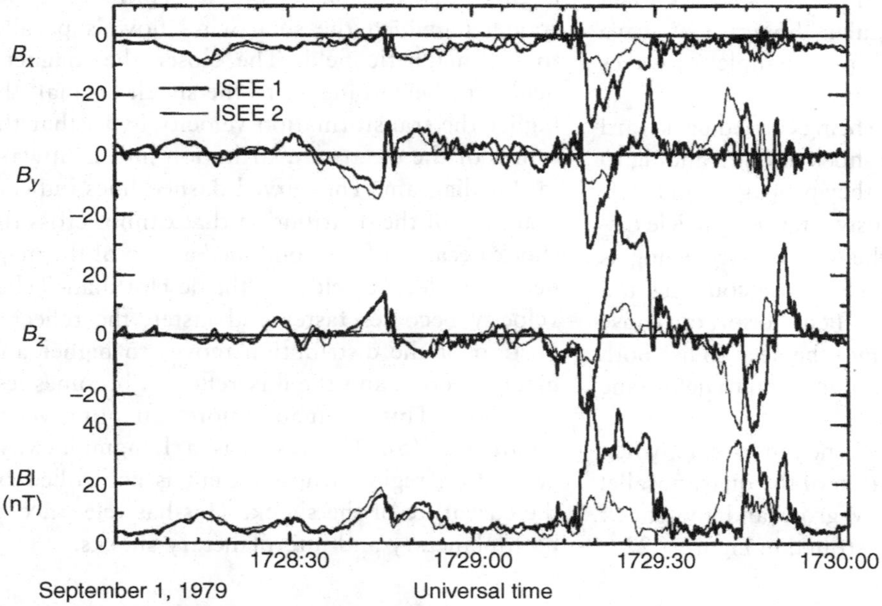

FIGURE 6.21. Shock reformation observed in measurements of the magnetic field by the ISEE 1 and 2 spacecraft as they near the bow shock from the upstream side, with ISEE 2 further upstream than ISEE 1 (heavy line). As the ISEE 1 approaches the shock, the upstream waves both see an increase in strength to shock-like amplitudes.

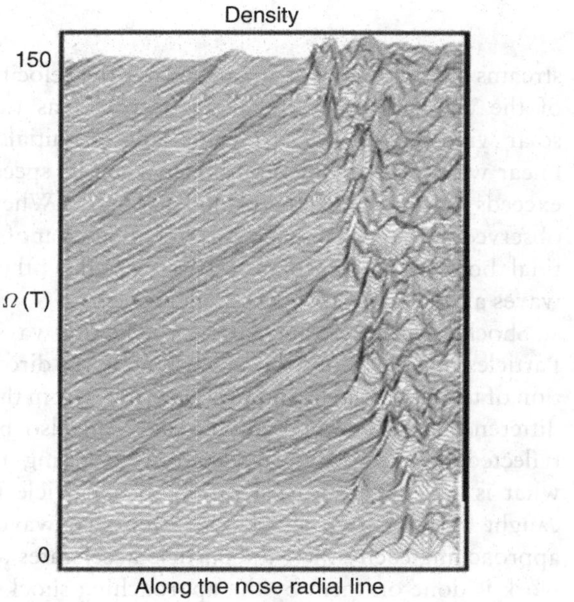

FIGURE 6.22. Two-dimensional hybrid simulation showing upstream waves in the solar-wind density being convected into the shock on the right. As the waves approach the average location of the shock, they strengthen and become part of the shock. (From N. Omidi, with permission; private communication, 2015.)

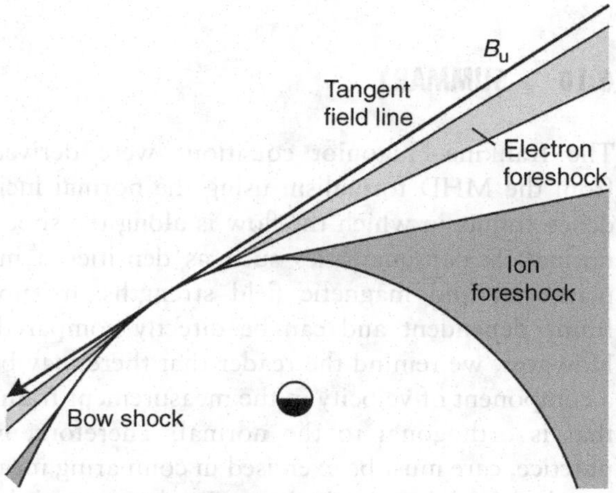

FIGURE 6.23. Schematic of the location of the electron and ion foreshocks in relationship to the magnetic field line that is tangent to the bow shock.

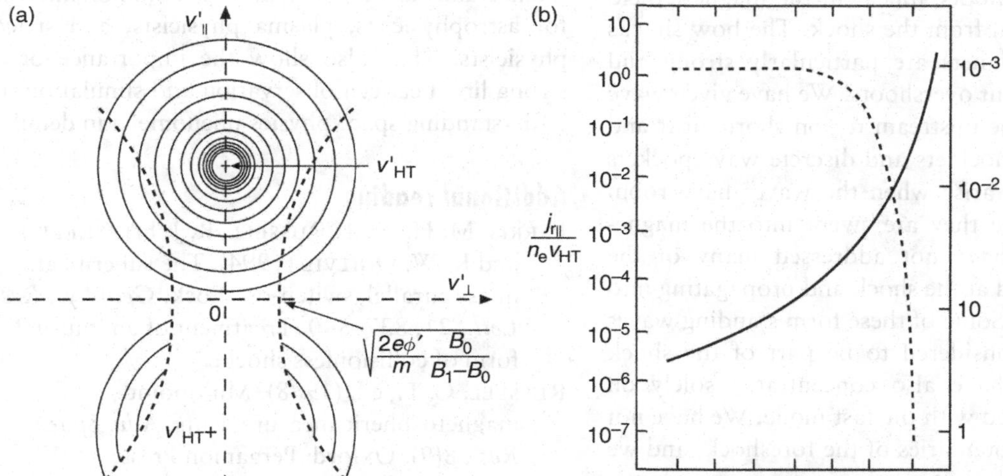

FIGURE 6.24. Schematic of the production of an electron beam at the bow shock by the fast-Fermi acceleration mechanism, shown in the de Hoffman–Teller reference frame. The curved dashed lines in (a) indicate that portion of the electron distribution that cannot pass through the ramp because of the magnetic mirror effect countering the potential drop. The closer the magnetic field line is to tangency, the further is the distribution above the axis, and the smaller is the portion of the distribution reflected, the higher the energy of the reflected ions. B_0 and B_1 are upstream and downstream magnetic fields; v'_{HT} is the de Hoffmann–Teller speed; ϕ' is the shock potential; n_e and T_e are the upstream electron density and temperature; $J_{r\parallel}$ and $E_{r\parallel}$, are the flux and energy per charge of the reflected beam along the magnetic field; and θ_{B_n} is the shock normal angle. (After Leroy and Mangeney, 1984.)

6.10 SUMMARY

The Rankine–Hugoniot equations were derived from the MHD formalism using the normal incidence frame, in which the flow is along the shock normal. Scalar quantities such as densities, temperatures, and magnetic field strengths are not frame dependent and can be directly compared. However, we remind the reader that there may be a component of velocity in the measurement frame that is orthogonal to the normal. Therefore, in practice, care must be exercised in comparing measured and predicted velocities. We also remind the reader that the thickness of the bow shock is of the order of the ion inertial length, and at times of the order of the electron inertia length. Thus the dissipative processes in the shock must be treated at the kinetic scale. MHD at best only describes how much dissipation is required at the shock, not how it occurs.

This chapter provides only a cursory review of the physics of collisionless shocks. Much has been omitted, especially in the area of high Mach number shocks where the magnetic field becomes turbulent and large overshoots appear in the magnetic field just downstream from the shock. The bow shocks of the outer planets are particularly strong and exhibit significant overshoots. We have given wave steepening in the upstream region short shrift and not discussed shocklets and discrete wave pockets (whistlers) that arise when the waves have room to evolve before they are swept into the magnetosheath. We have not addressed many of the waves generated at the shock and propagating into the solar wind. Some of these form standing waves that may be considered to be part of the shock structure. We have also concentrated solely on shocks associated with the fast mode. We have not discussed the boundaries of the foreshock, and we have not discussed a very interesting wave that is generated at the bow shock and convected downstream, the mirror-mode wave that appears to damp only over very long time scales.

Much of our understanding of the collisionless shock arises from the observations, modeling, and theory of bow shocks, but shocks also occur in the solar wind in front of ICMEs and where fast streams overtake shower streams. Since the velocity of the magnetosonic fast mode decreases as the solar wind moves away from the Sun, an initially linear wave can become non-linear when its speed exceeds that of the magnetosonic fast mode. When observed, the steepening process resembles that of a tidal bore in which small waves overtake other waves and make a stronger, single wave.

Shocks can accelerate particles in many ways. Particles can drift along the shock front in the direction of the electric field and pick up energy from the difference in electric potential. They can also be reflected by a moving shock front, resulting in what is called Fermi acceleration. If a particle is caught between two shocks, or even two waves approaching each other, the particle accelerates as work is done on them by the approaching shocks, like a table tennis ball between a descending paddle and the table. This is called second-order Fermi acceleration.

Collisionless shocks are important for many reasons, not only in their role in the energization of charged particles, but also because of their effect on the bulk properties of the flow. They are important intellectually as they are a point of common interest for astrophysicists, plasma physicists, and space physicists. They also show the importance of a strong link between observation and simulation in understanding space physics phenomena in detail.

Additional reading

Farris, M. H., C. T. Russell, R. J. Fitzenreiter, and K. W. Ogilvie (1994). The subcritical, quasi-parallel, switch-on shock. *Geophys. Res. Lett.*, **21**, 837–840. Treatment of an unusual form of collisionless shock.

Russell, C. T., ed. (1988). Multipoint magnetospheric measurements. *Adv. Space Res.*, 8(9). Oxford: Pergamon Press, 464 pp. Collection of research papers on multipoint measurements, many of them on the bow shock and the magnetosheath.

Russell, C. T., ed. (1994). The magnetosheath. *Adv. Space Res.*, **14**. Oxford: Pergamon Press, 135 pp. Collection of research papers on the magnetosheath during the height of collisionless shock research.

RUSSELL, C. T., ed. (1995). Physics of collisionless shocks. *Adv. Space Res.*, 15(8/9). Oxford: Pergamon Press, 544 pp. Collection of research papers during the height of collisionless shock research.

STONE, R. G. and B. T. TSURUTANI, eds. (1985). *Collisionless Shocks in the Heliosphere: A Tutorial Review*. Geophysical Monograph Series, vol. 34. Washington, D.C.: American Geophysical Union, 114 pp. Collection of review papers by senior scientists on the collisionless shock.

TSURUTANI, B. T. and R. G. STONE, eds. (1985). *Collisionless Shocks in the Heliosphere: Review of Current Research*. Geophysical Monograph Series, vol. 35. Washington, D.C.: American Geophysical Union, 301 pp. Research papers accompanying the preceding volume.

Problems

6.1 Sudden changes are detected in the solar wind and interplanetary magnetic field. The radial velocity of the solar wind remains constant, but the density jumps from 5 to 10 cm^{-3}. The proton temperature jumps from 5 eV before the discontinuity to 13.8 eV afterward, but the electron temperature remains constant at 15 eV. The magnetic field of (0, –8, 6) nT before the discontinuity rotates and drops in strength to (0, 3, 4) nT. What type of discontinuity might this be, and why?

6.2 An interplanetary shock crosses the spacecraft Space Physics Explorer, and its magnetometer detects an upstream magnetic field of (6.36, –4.72, 0.83) nT and a downstream magnetic field of (10.25, –9.38, 1.74) nT. Using the magnetic coplanarity assumption, determine the orientation of the normal. The plasma analyzer detects an upstream velocity of (–378, 33.1, 19.9) km s^{-1} and a downstream velocity of (–416.8, 7.3, 51.2) km s^{-1}. Calculate the mixed-mode normal. If the upstream density is 7.5 cm^{-3} and the downstream density is 11 cm^{-3}, calculate the shock velocity.

6.3 A cold solar-wind proton encounters a strong collisionless shock and is reflected back into the solar-wind flow. If the magnetic field is perpendicular to the flow, and the flow and shock normal are aligned, how far backward from the shock does the proton move before its motion is reversed by the solar-wind electric field? This is the extent of the shock foot. Express this distance analytically in terms of the gyro-radius of a proton moving at the solar-wind velocity.

6.4 The dispersion relation for fast magnetosonic waves is $v_{\rm ms}^4 - v_{\rm ms}^2\left(v_{\rm A}^2 + c_{\rm s}^2\right) + v_{\rm A}^2 c_{\rm s}^2 \cos{}^2\theta = 0$. Show that, when the IMF is aligned with the solar-wind flow, the asymptotic Mach cone angle equals $\sin^{-1}(1/M_{\rm c})$, where $M_{\rm c} = M_{\rm A}M_{\rm s}/\left(M_{\rm A}^2 + M_{\rm s}^2 - 1\right)^{1/2}$. Include a diagram showing the asymptotic shock normal angle and its relationship to the flow direction and the wave propagation angle θ.

6.5 Use the MHD/Shocks module of the Space Physics Exercises* and choose the MHD Wave Graphs option. Vary the magnetic field strength while keeping the other parameters at their default values. Capture the screen for magnetic fields of 3.5, 4.5, 5.5, 6.5, 15 nT. Choose the MHD Wave Case Studies option and calculate the Alfvén speed and sound speed for each of your cases. Describe how the phase velocity surfaces and the group velocity surfaces evolve as the ratio $c_{\rm s}^2/c_{\rm A}^2$ changes. For what mode(s) is the phase velocity zero for perpendicular propagation? What mode(s) have a guided group velocity? For which mode is the group velocity most tightly guided by the magnetic field?

* Space Physics Exercises are at http://spacephysics.ucla.edu.

7 Solar-wind interaction with magnetized obstacles

7.1 INTRODUCTION

As correctly foreseen by Chapman and Ferraro (1930) (see Chapter 1), a planetary magnetic field provides an effective obstacle to the solar-wind plasma. The solar-wind dynamic pressure, or momentum flux, compresses the outer reaches of the magnetic field, confining it to a cavity. This magnetic cavity has a long magnetotail consisting of two antiparallel bundles of magnetic flux that originate in the polar regions and stretch in the antisolar direction, as sketched for the Earth in Figures 1.16 and 1.17 in Chapter 1. The pressure of the internal magnetic field and the plasma it contains establishes an equilibrium with the solar-wind dynamic pressure. When the solar wind "blows harder," the magnetosphere shrinks. When the solar-wind pressure abates, the magnetosphere expands, akin to a balloon as it rises in the upper atmosphere.

In the first part of this chapter we describe the mathematical formalism to describe this magnetic obstacle to the solar wind. We begin with a dipole magnetic field, describing how we express the generalized magnetic field arising from currents below (and above) the surface of the Earth and how the solar wind distorts this magnetic field. As discussed in the preceding chapters, the solar wind is highly supersonic when it reaches the planets. For supersonic flow, the wind velocity exceeds the velocity of any pressure wave that could act to divert the flow around the magnetosphere. In the case of a magnetized plasma, this wave is principally the fast magnetosonic wave introduced in Chapter 3. Because the velocity of this wave is too slow to move upstream to divert the flow, a non-linear fast magnetosonic shock forms, standing in the flow, relatively close to the magnetopause. The physics of collisionless shocks was discussed in Chapter 6 and will not concern us here. Rather, here we examine why the shock forms where it does and what the physical processes are that determine where the shock stands in front of the magnetosphere. In a neutral, collisional gas this can be done with the sound wave (albeit non-linear) introduced in Chapter 2. In a magnetized plasma there are three wave modes, not just the single compressional mode of gas dynamics. These three waves are needed both to divert the plasma and to distort the magnetic field as it is carried with the plasma around the obstacle. To describe the deflection process we use the MHD formalism introduced in Chapter 3.

The MHD equations have served us well in the solar wind where the scale size vastly exceeds that of the gyro-motion of the constitutive particles of the plasma. But when we treated the shock in Chapter 6, the kinetic motions of the electrons and ions were found to be important in determining the heating and dissipation in and near the shock. Here too we need to concern ourselves with the kinetic scale. After examining what we learn from the MHD equations alone, we move to an examination of what more we learn when we include ion motion. In particular we find that obstacles with different magnetic moments will create different interactions in the same flowing magnetized plasma. The scale size of the interaction relative to the gyro-scale of the particles is important in determining the physical processes that occur in the interaction.

To treat the entire solar-wind interaction with a magnetic dipole is intractable if we follow the motion of all the electrons. In situations where we have to follow electron motion, we can first treat the large-scale problem with an approximate solution such as MHD and then focus our attention kinetically on a small key region such as the shock transition or the "x-point" in reconnection. We briefly mentioned the former problem in the preceding chapter. We defer the reconnection problem to Chapter 9.

7.2 PLANETARY MAGNETIC FIELDS

Almost two centuries ago, Gauss showed that the magnetic field of the Earth could be described as the gradient of a scalar potential:

$$B = -\nabla\Phi = -\nabla(\Phi^{i} + \Phi^{e}), \tag{7.1}$$

where Φ^{i} is the magnetic scalar potential due to sources inside the Earth and Φ^{e} is the scalar potential due to external sources. Gauss and his colleague Weber founded a chain of magnetic observatories around the world. With the data they collected they demonstrated that the field at the surface of the Earth was almost entirely internal and principally dipolar. This latter fact is very useful in treating the terrestrial magnetosphere, because it allows for the simplification of otherwise complex problems.

The **dipole magnetic moment** of the Earth is tilted about 10.2° to the rotation axis, with a present-day moment of about 7.8 × 10^{15} T m^3 or 30.2 μT·R_E^3. The dipole moment and its direction are not constant but vary greatly, with the sense of the dipole moment reversing on occasion (on time scales of hundreds of thousands of years or more). The tilts of the magnetic dipoles of the other planets are also varied, ranging from less than 1° to over 50°, and their moments span the wide range discussed in Chapter 12. A tilted dipole moment rotates as the planet rotates, and in most cases will vary in orientation with respect to the solar-wind flow in the course of a day. Where the solar wind and magnetic field are in pressure balance, the magnetic field is far from being dipole like. However, near the planet, if the magnetic moment is strong, charged-particle motion will be dominated by the internal planetary field, not by the solar-wind interaction. Thus, the dipole approximation of the magnetic field is often useful.

In spherical coordinates,

$$B_r = 2Mr^{-3}\cos\theta, \tag{7.2a}$$

$$B_\theta = Mr^{-3}\sin\theta, \tag{7.2b}$$

$$|B| = Mr^{-3}(1 + 3\cos^2\theta)^{1/2}, \tag{7.2c}$$

where θ is the magnetic colatitude, as defined in Figure 7.1, and M is the dipole magnetic moment. In this coordinate system aligned with the dipole, there is no B_φ-component.

The magnetic field of a dipole can also be expressed in cartesian coordinates. If we again define a coordinate system with the z-axis along the magnetic dipole axis, then:

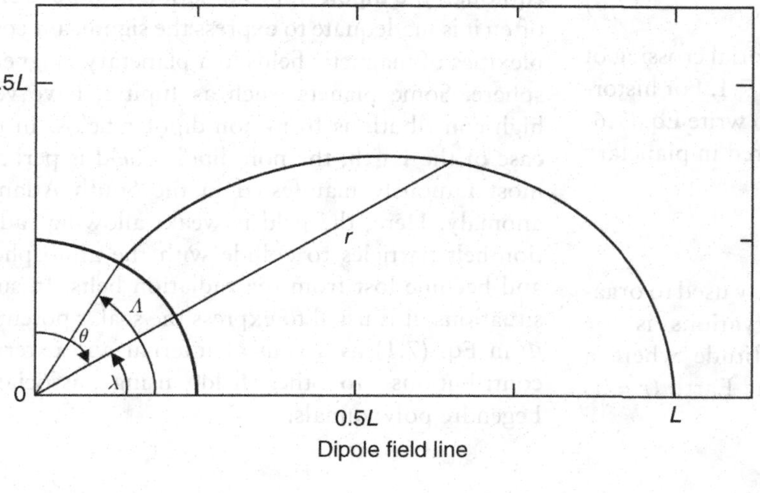

FIGURE 7.1. Dipole magnetic field line. L designates the magnetic drift shell and is equal to the distance in planetary radii from the center of the planet to the point where the field line crosses the equator. The angle λ is the magnetic latitude of a point on the field line; the colatitude is θ and R is the radial distance to that point. The angle Λ is the latitude where the field line intersects the surface of the planet.

$$B_x = 3xzM_z r^{-5}, \tag{7.3a}$$

$$B_y = 3yzM_z r^{-5}, \tag{7.3b}$$

$$B_z = (3z^2 - r^2)M_z r^{-5}, \tag{7.3c}$$

where M_z is the magnetic moment along the z-axis. This representation can be easily generalized to the case for a dipole moment with arbitrary orientation:

$$B = r^{-5}\begin{pmatrix} (3x^2 - r^2) & 3xy & 3xz \\ 3xy & (3y^2 - r^2) & 3yz \\ 3xz & 3yz & (3z^2 - r^2) \end{pmatrix}\begin{bmatrix} M_x \\ M_y \\ M_z \end{bmatrix}. \tag{7.4}$$

Given a series of observations $\mathbf{B}_i$ at locations $\mathbf{r}_i$ (x_i, y_i, z_i), this equation can be solved by standard matrix-inversion techniques.

7.2.1 Magnetic field lines and the L parameter

The spherical coordinate representation of a dipole magnetic field allows us to calculate readily the equation of a magnetic field line. A field line is everywhere, tangent to the magnetic field direction. Thus,

$$rd_\theta / B_\theta = dr/B_r \tag{7.5a}$$

and

$$d\varphi = 0. \tag{7.5b}$$

Integrating Eq. (7.5a), we obtain for the equation of a field line:

$$r = r_0 \sin^2\theta, \tag{7.6}$$

where r_0 is the distance to the equatorial crossing of the field line, as illustrated in Figure 7.1. For historical reasons, it became customary to write Eq. (7.6) in terms of L (with distance measured in planetary radii) and the magnetic latitude λ:

$$r = L\cos^2(\lambda).$$

A related parameter that is frequently used to organize inner magnetospheric observations is the **invariant latitude.** This is the latitude where a field line reaches the surface of the Earth (r = 1) and is given by

$$\Lambda = \cos^{-1}(1/L)^{1/2}.$$

Thus, a dipole field line that extends to $4R_E$ in the equatorial plane of the magnetosphere maps to an invariant latitude of 60° at the Earth's surface. A dipole field line that extends to $10R_E$ in the equatorial plane maps to a latitude of 71.6°.

The most intense portions of the Earth's radiation belts are found near the Earth where the magnetic field is primarily dipolar. In 1961, C. McIlwain realized that particle observations could be organized by using the properties of particle motion in a dipolar field. As we saw in Chapter 3, in addition to their gyro-motion about a magnetic field line, particles bounce between the northern and southern hemispheres. This motion reflects the action of the magnetic mirror force associated with conservation of the first adiabatic invariant (see Chapter 10). If the dipolar magnetic field is sufficiently slowly varying, this bounce motion is maintained until the particle is scattered by some interaction onto a trajectory that takes it into the atmosphere or outside the magnetopause. The gyrocenter of a particle trapped on magnetic field lines will be confined to a surface specified by the distance L to the equatorial crossing of the field line. Moreover, it will drift around the Earth under the gradient and curvature forces (Chapter 10) in a charge-dependent direction. These basic forces allow magnetospheres to possess trapped radiation belts.

7.2.2 Generalized planetary magnetic fields

Although the dipole approximation is quite useful, often it is inadequate to express the significant complexities of magnetic fields in a planetary magnetosphere. Some planets, such as Jupiter, have very high contributions from non-dipolar fields. In the case of the Earth, the non-dipolar field is perhaps most famously manifested in the South Atlantic anomaly. Here, the field is weak, allowing radiation-belt particles to collide with the atmosphere and become lost from the radiation belts. In such situations, it is usual to express the scalar potential Φ in Eq. (7.1) as a sum of internal and external contributions to the field using associated Legendre polynomials:

$$\Phi^{i}(r,\theta,\varphi)=a\sum_{n=1}^{\infty}\sum_{m=0}^{n}[r/a]^{-n-1}P_n^m(\cos\theta)(g_n^m\cos(m\varphi)+h_n^m\sin(m\varphi)) \quad (7.7)$$

and

$$\Phi^{e}(r,\theta,\varphi)=a\sum_{n=1}^{\infty}\sum_{m=0}^{n}[r/a]^{n}P_n^m(\cos\theta)(G_n^m\cos(m\varphi)+H_n^m\sin(m\varphi)), \quad (7.8)$$

where a is the planet's radius, and θ and φ are the colatitude and east longitude, respectively, in planetographic coordinates. $P_n^m(\cos\theta)$ denotes associated Legendre functions with Schmidt normalization:

$$P_n^m(\cos\theta) = N_{nm}(1-\cos^2\theta)^{m/2}d^mP_n(\cos\theta)/d(\cos\theta)^m,$$

where $P_n(\cos\theta)$ is the Legendre function, and $N_{nm} = 1$ when $m = 0$, and $[2(n - m)!/(n + m)!]^{1/2}$ otherwise. The coefficients g_n^m, h_n^m, G_n^m, and H_n^m are chosen to minimize the difference between the model field and observations. The coefficients of the internal magnetic field are monitored very carefully to gain a better understanding of the internal structure of the Earth and the geomagnetic dynamo. A consensus list of the g_n^m and h_n^m coefficients and their temporal (secular) variation is published regularly under the name International Geomagnetic Reference Field (IGRF).

We can see the relation of this to our dipole approximation by examining the $n = 1$, $m = 0, 1$ terms in the series. The dipole moment becomes

$$M = a^3[(g_1^0)^2 + (g_1^1)^2 + (h_1^1)^2]^{1/2}, \quad (7.9)$$

and the **dipole tilt** to the rotation axis is

$$\alpha = \cos^{-1}(g_1^0/M). \quad (7.10)$$

These coefficients are functions of time and have sizable secular or temporal variations. Three variations are worthy of note. The dipole moment was about 9.54×10^{15} T m^3 in 1550, but had decreased to only 7.84×10^{15} T m^3 in 1990. Recently, the pace of this decrease has accelerated somewhat, and it is now about 0.1% per year. The geographic colatitude (tilt) of the dipole axis was close to 3° in 1550, and rose to 11.5°, where it remained from 1850 to 1960. Since then, it has declined, reaching 10.8° in 1990 and 10.2° in 2007. The third principal secular variation of the dipole field is a westward drift. The dipole axis lay at 334° east longitude in 1550, but in 1990 had drifted to 289° east longitude. This averages to a drift of 0.1° per year, but, as with the other properties of the dipole field, this rate varies. This drift explains why there are many reports of auroral phenomena from northern Europe during the middle ages, such as the woodcut shown in Figure 1.1. The current drift rate would put the auroral zone at lowest latitudes over China at the start of the common epoch (CE). We cannot yet successfully predict the temporal variations of the internal magnetic field; it has to be measured.

Although it is common to draw the Earth's magnetosphere with its dipole axis perpendicular to the solar-wind flow, it is seldom in this configuration. In addition to the 10.2° dipole tilt, the Earth's rotation axis is inclined 23.5° to the ecliptic pole. Thus, in the course of its daily rotation and its annual journey around the Sun, the angle between the direction of the dipole and the direction of the solar-wind flow varies between 90° and 56°. Because, in part, the interplanetary magnetic field is ordered in the ecliptic plane (or, more properly, the Sun's equatorial plane), and because, as we discuss in Chapter 9, interplanetary magnetic fields that are opposite in direction to the Earth's field more strongly interact with it, there are annual and semiannual variations in geomagnetic activity.

7.3 THE SIMPLEST MAGNETOSPHERES

To the extent that the magnetized plasmas of the solar wind and magnetosphere are collisionless and dissipationless, they should form an impenetrable boundary between them when they intersect. The boundary between the two field domains is called the **magnetopause.** If the boundary were planar, the resulting magnetic field in the magnetosphere would be that of the Earth's dipole plus induced currents in the planar interface. Mathematically, this is equivalent to the field due to a magnetic dipole and its mirror image, equal in magnitude, of the same orientation, and an equal distance on

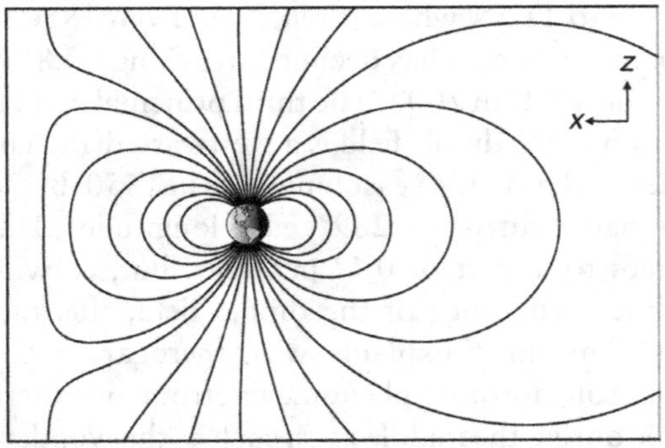

FIGURE 7.2. Magnetic field lines in the noon–midnight meridian of the image dipole magnetosphere. This magnetosphere results from an infinite plane of superconducting material placed upstream of a magnetic dipole to represent the solar-wind plasma. The magnetic field of this magnetosphere can be calculated by placing a second identical dipole an equal distance upstream of the magnetopause.

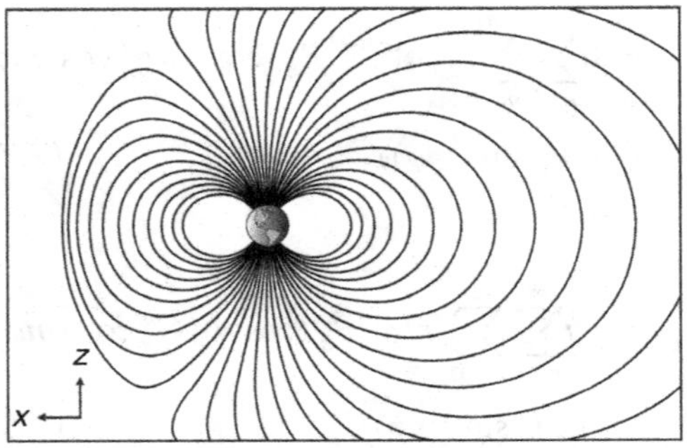

FIGURE 7.3. Magnetic field lines in the noon–midnight meridian of a magnetosphere formed by placing a dipole magnetic field in a vacuum inside a curved superconducting boundary with the observed shape of the boundary. (This model described by Tsyganenko, 1989a.)

the other side of the boundary. This simplest of all magnetospheres in a "flowing" plasma is shown in Figure 7.2. It has several properties reminiscent of the Earth's magnetosphere. At the subsolar point, the magnetic field is double that of the dipole in a vacuum. To the north and south lie neutral points, or **cusps**, where the magnetic field reverses direction as its strength goes through zero. At this point, there is nothing to stop the plasma from entering the magnetosphere, with some flowing all the way down to the surface of the Earth, or more precisely to collide with the upper atmosphere. Above the northern "cusp" and below the southern "cusp" the magnetic field turns in the direction opposite to its low-latitude orientation. Field lines diverge from the southern neutral point and converge at the northern neutral point. On the antisunward side of this magnetosphere the magnetic field is also enhanced, but less than on the dayside, so the resulting configuration is asymmetrically compressed, as is the real magnetosphere.

The real interface between the solar wind and the Earth's magnetosphere is curved. Such a curved boundary can be created by increasing the strength of the mirror field, but a better approach is to make the mirror assume the measured shape of the magnetopause and solve for the field inside the "vacuum" region under the constraint that no magnetic field lines cross the boundary (Tsyganenko, 1989a). As illustrated in Figure 7.3, it provides a good first approximation to a three-dimensional magnetospheric field shaped by the solar-wind interaction. This model can be generalized to include finite tilt (S. M. Petrinec, personal communication, 1992).

The magnetic field at the subsolar magnetopause is now 2.4 times the purely dipolar field at that position. Comparing this value with the doubling at the planar magnetopause illustrates that the field enhancement depends on the radius of curvature of the magnetic field in the neighborhood of the subsolar region. The location of this balance point between the pressure applied by the solar wind and pressure applied to the solar wind by the Earth's magnetic field depends on the strength of the solar-wind pressure and the magnitude of the dipole magnetic moment. We discuss this in the following section.

The real magnetosphere contains warm plasma. The thermal pressure of this plasma adds to the magnetic pressure and alters the magnetic configuration of the magnetosphere, as shown in Figure 7.4. The changes are most noticeable in the magnetic field in the tail region known as the plasma sheet. This empirical model was one of earliest produced by N. Tsyganenko (1989b), based on observational data. The plasma remains implicit in this empirical model of the magnetic field only.

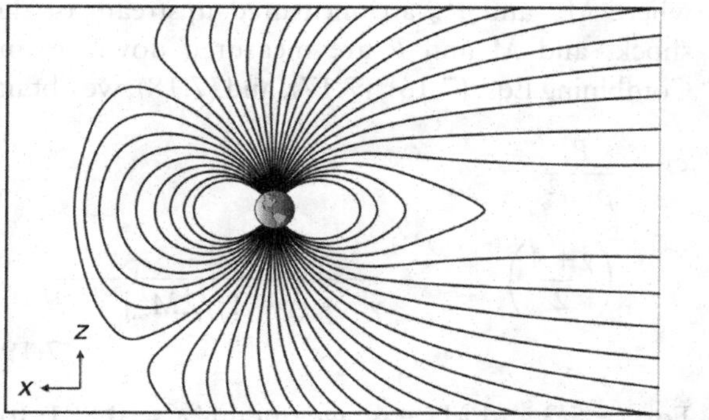

FIGURE 7.4. Magnetic field lines in the noon–midnight meridian of an empirical magnetic model fit to spacecraft observations. (This model described by Tsyganenko, 1989b.)

The most recent versions of the Tsyganenko model also include interplanetary magnetic field effects on the field geometry. Reconnection with the interplanetary field and its effects will be discussed in detail in Chapter 9. However, it is worth mentioning here that these effects produce field lines that penetrate the magnetopause boundary for some field orientations. Dungey (1961) was the first to appreciate that the dipole field of Earth would interconnect with southward interplanetary magnetic fields which oppose the dipole field at the subsolar magnetopause.

7.4 SIZE OF THE MAGNETOSPHERIC CAVITY

In the Chapman and Ferraro model of the solar-wind interaction with the Earth's magnetic field, the boundary of the magnetic cavity was located halfway between the Earth's dipole and its mirror image. The forces that determine where the boundary was to be drawn were not specified. The solar-wind plasma was treated solely as a superconductor. The real solar wind has mass and momentum and a magnetic field. It exerts a force outward from the Sun on every obstacle in its path. The Earth's magnetic field is such an obstacle. Because the magnetic fields of the Earth and of the solar wind are "frozen in" to their respective plasmas by the high electrical conductivity of the plasmas, they do not interpenetrate on short time scales. To first order, the major effect of the magnetized solar wind is to exert pressure, or more precisely normal stresses, on the magnetosphere.

In a steady-state situation, the force of the solar wind against the magnetosphere and the force of the magnetosphere against the solar wind are in balance, and an equilibrium is struck. The forces are exerted by pressure gradients. At the magnetopause, there is a pressure gradient in the magnetospheric magnetic field and plasma exerting an outward force, and a pressure gradient in the magnetosheath plasma and magnetic field exerting an inward force. The location of this equilibrium point is pressure sensitive. If the magnetosheath plasma pushes harder, the magnetopause moves inward to where the magnetic field is stronger and the magnetosphere can exert sufficient outward force to balance the new level of magnetosheath pressure. In order to determine the location of the magnetopause, we must determine how much force the solar wind exerts on the magnetosphere and how the counterbalancing magnetospheric force on the solar wind varies with the size of the magnetosphere over the surface of the magnetosphere.

7.4.1 The pressure exerted by the solar wind on the magnetosphere

The solar-wind pressure consists principally of dynamic pressure or momentum flux, ρu^2, where ρ is the mass density, which includes, on average, a 20% (by mass) contribution from doubly ionized helium, and u is the solar-wind velocity. The magnetic field pressure and thermal plasma pressure typically add about 1% to the total. The balance among these different forms of pressure is altered in passing through the shock and the magnetosheath. At the magnetopause, the flow is tangential to the surface, so that the contribution of the dynamic pressure to the pressure balance across the boundary surface is zero. The pressure here must be totally due to magnetic and thermal contributions. It is proportional to the incident dynamic pressure, but even at the nose of the magnetosphere it is smaller than the incident pressure because of the divergence of the flow around the obstacle. To see this, we consider the momentum flux through unit area in the direction **n**:

$$\rho u(u \cdot \mathbf{n}) + P\mathbf{n}.$$

Integrating over the surface of a stream tube, we obtain the momentum-conservation equation:

$$(\rho u^2 + P)S = \text{constant.} \qquad (7.11)$$

Upstream (at infinity in the solar wind), P_∞ is small, and, at the magnetopause, ρu^2 can be neglected. Thus,

$$\mathcal{K} = \frac{P_s}{\rho_\infty u_\infty^2} = \frac{S_\infty}{S_s}. \qquad (7.12)$$

Here, the subscript s refers to measurements at the magnetopause, and ∞ refers to those in the solar wind. The parameter $\mathcal{K}$ indicates how much the pressure has been diminished by the divergence of the flow. We can calculate $\mathcal{K}$ from **Euler's equation** for an ideal fluid with no viscosity or thermal conduction:

$$\frac{\partial u}{\partial t} + (u \cdot \nabla)u = -\frac{1}{\rho}\nabla P. \qquad (7.13)$$

For an adiabatic fluid,

$$\rho P^{-\gamma} = \text{constant}, \qquad (7.14)$$

where γ is the ratio of specific heat or the polytropic index. Using the identity

$$u \cdot \nabla u = \frac{1}{2}\nabla u^2 - u \times (\nabla \times u)$$

we obtain, in the steady state,

$$\frac{1}{2}u^2 + \gamma(\gamma - 1)^{-1}P/\rho = \text{constant}, \qquad (7.15)$$

which is **Bernoulli's equation** for adiabatic flow. Substituting Eq. (7.14) into Eq. (7.15) and recalling that the sonic Mach number M_s is $u(\rho/\gamma P)^{1/2}$, we can relate the stagnation pressure to the pressure at any point upstream along the same streamline:

$$P_s/P = [1 + (\gamma - 1)M^2/2]^{\gamma(\gamma-1)}. \qquad (7.16)$$

From the Rankine–Hugoniot equations discussed in Chapter 6,

$$P/P_\infty = 1 + 2\gamma(\gamma + 1)^{-1}(M_\infty^2 - 1) \qquad (7.17)$$

and

$$M^2 = [1 + (\gamma - 1)M_\infty^2]/[2\gamma M_\infty^2 - \gamma - 1], \qquad (7.18)$$

where M_∞ and P_∞ are measured upstream of the shock, and M and P are measured downstream. Combining Eqs. (7.16), (7.17), and (7.18), we obtain

$$\mathcal{K} = \frac{P_s}{\rho_\infty u_\infty^2} = \left(\frac{\gamma+1}{2}\right)^{(\gamma+1)(\gamma-1)} \frac{1}{\gamma[\gamma - (\gamma - 1)/2M_\infty^2]^{1/(\gamma-1)}}. \qquad (7.19)$$

For $\gamma = 5/3$ and $M_\infty = \infty$, we obtain $\mathcal{K} = 0.881$; for $M_\infty = 4.5$, $\mathcal{K} = 0.897$. For $\gamma = 2$, which corresponds to a gas with two degrees of freedom, and $M_\infty = \infty$, $\mathcal{K} = 0.844$. Because the effective polytropic index of the magnetosphere is found empirically to be about 5/3, and the typical solar-wind Mach number is around 6 at 1 AU, the pressure exerted by the solar wind on the nose of the magnetopause is about 11% less than that of the momentum flux or dynamic pressure of the solar wind. We note here a difference in the definition of dynamic pressure in space plasma physics and aerodynamics. Aerodynamics defines dynamic pressure to be $\frac{1}{2}\rho u^2$, whereas space physics defines it to be equal to the momentum flux, ρu^2.

7.4.2 Pressure exerted by the magnetosphere on the magnetosheath plasma

If the magnetosphere were a vacuum, then the magnetic field just inside the magnetopause would provide the total pressure to stand off the magnetosheath plasma. In practice, there is a variable contribution from the magnetospheric plasma. At the Earth, and all the other "intrinsic" magnetospheres, the plasma pressure is thought to be less than that of the magnetic field near the nose of the magnetosphere. Thus, a solution that ignores the plasma contribution to the pressure can be instructive. This assumption may not be appropriate at Jupiter and Saturn, whose magnetospheres contain copious amounts of cold corotating plasma that adds to the outward pressure of the magnetic field gradient.

To calculate the pressure at the boundary, we must determine how much the magnetic field at the nose of the magnetopause is compressed in the interaction. An image dipole model would cause a doubling of the field, but, as noted in Section 7.1, a more realistically shaped magnetosphere would produce a larger effect. Moreover, other current systems, such as the ring current, the tail current, and the **field-aligned current** or **Birkeland current**, will also contribute some positively, some negatively, to the magnetic field at the magnetopause. Thus, the magnetopause position depends on the "state of the magnetosphere" as well as the dynamic pressure of the solar wind. Despite this dependence on the state of the magnetosphere, it is instructive to proceed, leaving the compression factor as a free parameter, a, to be determined empirically. Balancing the solar-wind pressure against our assumed magnetospheric pressure, we obtain

$$\mathcal{K}\rho_\infty u_\infty^2 = (aB_0)^2(2\mu_0 L_{\mathrm{mp}}^6)^{-1}, \qquad (7.20)$$

where B_0 is the equatorial surface field of the planet and L_{mp} is the distance to the subsolar magnetopause in planetary radii. We can use this size-dependent pressure-balance equation to solve for the size of the magnetosphere. For the Earth, we obtain

$$L_{\mathrm{mp}}(R_{\mathrm{E}}) = 8.53a^{0.33}(\mathcal{K}\rho_\infty u_\infty^2)^{-0.167}, \qquad (7.21)$$

where the term in parentheses on the right is expressed in nanopascal. For a typical solar-wind momentum flux of 2.6 nPa (see Table 5.1), the magnetopause is observed to lie at about $10R_{\mathrm{E}}$. Solving Eq. (7.21) for a, we obtain 2.44. Thus, the subsolar standoff pressure of the magnetosphere is approximately 2.44 times that expected for the vacuum dipole field at that distance. This value falls, as expected, between the value for the infinite planar magnetopause and that for the spherical magnetopause, and is in accord with the compression factor developed for a vacuum magnetosphere with the observed boundary shape. In more practical units, we can rewrite Eq. (7.21) in terms of the measured solar-wind number density and velocity as

$$L_{\mathrm{mp}}(R_{\mathrm{E}}) = 107.4(n_{\mathrm{sw}}u_{\mathrm{sw}}^2)^{-0.167}, \qquad (7.22)$$

where n_{sw} is the proton-number density in proton masses per cubic centimeter (adjusted for helium content, noting the mass factor of four), and u_{sw} is the solar-wind bulk velocity in kilometers per second. We defer discussing the shape of the magnetopause for planets other than the Earth until we discuss their internal pressure sources, which can be quite different from those of the Earth.

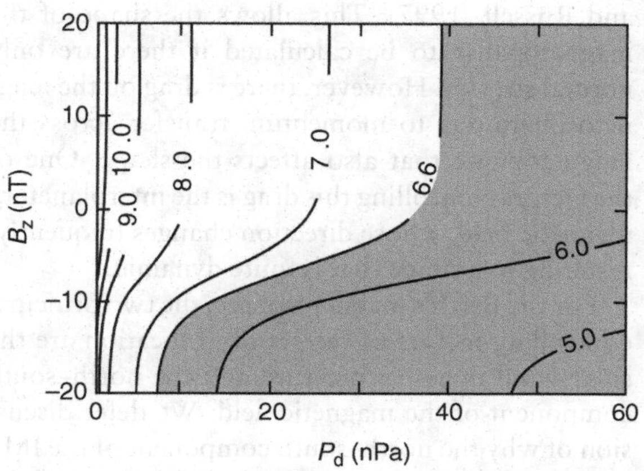

FIGURE 7.5. Location of the magnetopause nose or standoff distance as a function of the north–south solar magnetosphere component of the interplanetary magnetic field. (After Shue *et al.*, 1998.)

7.4.3 Shape of the magnetospheric cavity

The terrestrial magnetopause may be defined as the boundary at which the magnetosheath, as determined ultimately by the solar wind, is balanced by the pressure of the Earth's magnetic field and of its contained plasma. Thus far, we have concerned ourselves with only the location of the nose of the magnetopause. The magnetospheric cavity is a three-dimensional object with a specific shape. It has a rather blunt nose region and an extended tail. The problem in calculating the shape reduces to determining the pressure distribution in the magnetosheath. The newtonian approximation states that the pressure derived above for the subsolar point is modified away from the subsolar point by $\cos^2\Psi$, where Ψ is the angle between the magnetopause normal and the solar-wind flow. This relationship breaks down as Ψ approaches 90° because the exterior pressure as determined by this relation approaches zero. If this were true, then the magnetopause would never reach an asymptotic radius downstream or the total pressure in the tail would decrease to zero, neither of which is observed. To rectify this and to maintain reasonable boundary conditions at the subsolar point, we adjust the pressure $\mathcal{K}\rho_\infty u_\infty^2$ with the additional term, $P_\infty \sin^2\Psi$, where P_∞ is the thermal pressure of the solar wind (Petrinec and Russell, 1997). This allows the shape of the magnetopause to be calculated if there are only normal stresses. However, there is drag on the magnetosphere due to momentum transfer across the magnetopause that also affects the shape. One of the factors controlling this drag is the interplanetary magnetic field, whose direction changes frequently, resulting in a shape that is quite dynamic.

For the Earth's magnetosphere, the two principal controlling factors of the standoff location are the solar-wind dynamic pressure and the north–south component of the magnetic field. We defer discussion of why the north–south component of the IMF has such an effect until Chapter 9 because of the even greater importance of this component of the field on energy and momentum transfer. This is illustrated in Figure 7.5 for extreme solar-wind conditions. The shape of the magnetopause was fit in this study using the functional form

$$r = 2^{\alpha} r_0 (1 + \cos\theta)^{-\alpha}, \tag{7.23a}$$

$$r_0 = \{10.22 + 1.29 \tanh[0.184\,(B_z + 8.14)]\}(P_d)^{-1/6}, \tag{7.23b}$$

$$\alpha = (0.58 - 0.007 B_z)[1 + 0.024 \ln(P_d)], \tag{7.23c}$$

where B_z is the north–south component of the interplanetary magnetic field in GSM coordinates (see Appendix A.3) in nanotesla and P_d is the dynamic pressure in nanopascal. The hyperbolic tangent dependence on the magnetic field results in a weak dependence on B_z for positive B_z values and very negative values. Note that this model does not take into account any change in shape due to the erosion process.

7.4.4 Magnetic flux transport to the tail

Tangential stress, or drag, transfers momentum to the magnetospheric plasma and causes it to flow tailward. This stress can be transferred by diffusion of particles from the magnetosheath into the magnetosphere, by boundary-wave processes that cause motion in the magnetosphere, by the finite gyro-radius of magnetosheath particles, and by the reconnection process, as discussed in detail in Chapter 9. The greatest tangential stress on the magnetopause occurs when reconnection links interplanetary magnetic field lines with planetary ones when the interplanetary and planetary fields lie in opposite directions.

These processes act to transfer magnetic flux and plasma from the dayside to the nightside magnetosphere, and hence have the potential to alter the shape of the magnetosphere. To illustrate the effect of such transfer on the configuration of the magnetosphere, we employ the two-dimensional treatment by Unti and Atkinson (1968) that used the amount of magnetic flux in the tail as a parameter in the model. Figure 7.6 shows how the shape of the

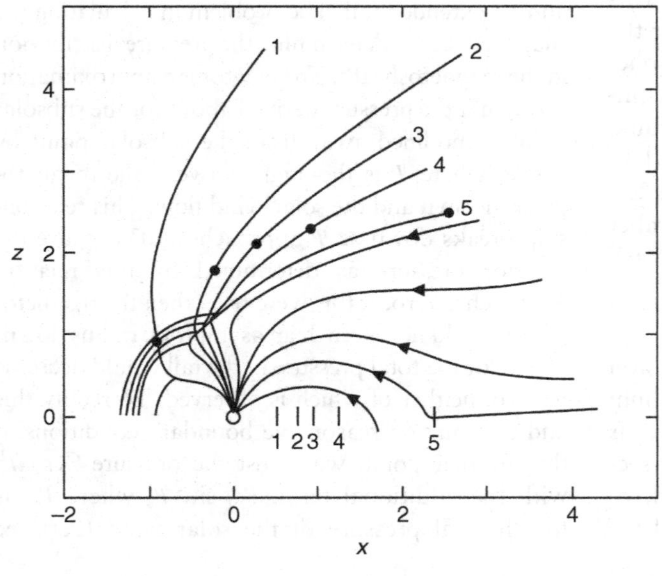

FIGURE 7.6. Two-dimensional boundary shapes for a model magnetosphere in which the tail was represented by an equatorial current sheet, with return boundary currents. The five shapes correspond to different amounts of magnetic flux in the tail (1 contains the most flux and 5 the least) for constant solar-wind conditions.

magnetopause and the position of the inner edge of the tail neutral sheet (labeled 1–5 in the night magnetosphere) varied as the tail flux was altered while the solar wind was held constant. This picture is qualitatively consistent with the idea that the magnetopause shape ultimately depends on how its flux is distributed, which in turn depends on its interplanetary magnetic field connection.

7.4.5 Limiting width of the tail

The tail expands only to an asymptotic diameter, where there are no normal stresses associated with the solar-wind dynamic pressure, only the normal stresses of the solar-wind plasma and magnetic field as expressed in the magnetosheath flow. If we assume that one tail lobe is a semicircle, then the magnetic flux in that tail lobe is given by

$$F_{\mathrm{T}} = \frac{\pi R_{\mathrm{T}}^2}{2} B_{\mathrm{T}}, \tag{7.24}$$

where R_{T} is the lobe radius and B_{T} is the field strength. Because the tail magnetic pressure is in balance with the solar-wind thermal and magnetic pressure, the asymptotic radius of the tail is given by

$$R_{\mathrm{T}}^2 = 2^{1/2} F_{\mathrm{T}} / (\pi^2 \mu_0 P_{\mathrm{sw}})^{1/2}, \tag{7.25}$$

where P_{sw} includes both the thermal and magnetic pressures of the solar wind.

7.5 EMPIRICAL MODELS OF THE MAGNETOSPHERIC MAGNETIC FIELD

In many situations, we need a reference model of the magnetospheric magnetic field that can be calculated quickly with analytic expressions. We mentioned one such model in Section 7.3 by Tsyganenko (1989a) that was based on pure theory with a single assumption about the location and shape of the magnetopause. This has been a very useful and instructive model, but it is for vacuum conditions. The hot plasma in the magnetosphere has significant pressure over much of the magnetosphere, and this influences the magnetic profiles. As previously mentioned, Tsyganenko has extended his analytic models with a series of empirical models, culminating in the model known as Tsyganenko-2004 (Tsyganenko and Sitov, 2007). This model is based on physics-based approximations of magnetospheric current systems and their dependences on solar-wind conditions, including the interplanetary field. It is quite possible that a better approximation to the magnetic field in the magnetosphere can be obtained with these empirical models than with the numerical computations described in Sections 7.6 and 7.7 because of the better spatial resolution of narrow physical features and fewer explicit assumptions in the empirical models. On the other hand, these models may not be self-consistent, and they provide a description of the magnetic field only.

7.6 FLUID SIMULATIONS OF THE SOLAR-WIND INTERACTION

The supersonic nature of the solar wind ensures that the region in which the flow is diverted around the magnetosphere is highly non-linear. In particular, it is bounded on its upwind side by a fast-mode shock. Alternative approaches to modeling the solar-wind interaction have used both the approximations of gas dynamics or magnetohydrodynamics (MHD). Hybrid models (fluid electrons and kinetic (particle) ions) have also been attempted. Each step in this hierarchy comes at some cost but provides new information. These computer models are continually being refined, developed, and compared with data to determine their accuracy. Computers have several very important advantages that are non-trivial. Computer power is increasing at a rapid rate, so that computer models can become more and more precise or sophisticated with time, with little increase in cost. They can also assume global, three-dimensional descriptions that can be steady state or time dependent, and can accommodate details of the physics and couplings at boundaries that are impossible to include with analytical approaches. The results from numerical models also have to be regarded with some caution because they are subject to artifacts from numerical diffusion or dissipation, inadequate spatial and/or temporal resolutions, and the need for parameterizations of sometimes important physical elements

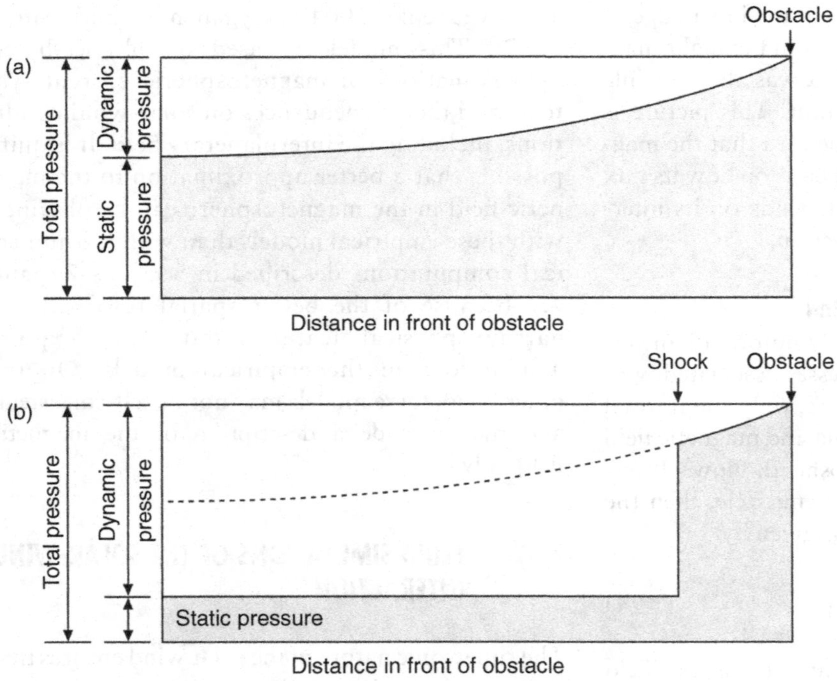

FIGURE 7.7. Schematic variation of kinetic and dynamic pressure as an obstacle to a gas dynamic flow is approached in a subsonic (a) and supersonic (b) interaction. In the subsonic case, the kinetic pressure is sufficient that a pressure gradient can be established that can stop the flow. In the supersonic case, the formation of a shock is needed to create the hot gas from the original cold flow that can stop the flow.

and processes. Nonetheless, it is fair to say they have revolutionized the way we think and work in space physics research.

The numerical simulations we discuss in this section all treat the interaction of supersonic or supermagnetosonic flows with an obstacle to that flow. In these simulations, a standing shock forms ahead of the obstacle. Figure 7.7 illustrates two situations: an approaching subsonic fluid or gas flow, to which the dynamic pressure is able to be fully transformed to the thermal pressure as the obstacle is approached, and a supersonic flow, where the incoming dynamic pressure cannot be transformed without shock dissipation processes. These pictures apply to the stagnation streamline. In three dimensions, away from the stagnation streamline of a blunt obstacle like the magnetosphere, only part of the dynamic pressure needs to be converted to thermal pressure. Behind the shock, in the magnetosheath, the pressure gradient slows and deflects the remaining flow and its associated pressure. The shock is located such that the compressed plasma has sufficient room to flow between the shock and the obstacle.

7.6.1 Gas-dynamic simulations

Gas-dynamic simulations are used for situations in which the magnetic forces can be neglected, so that increased spatial resolution and faster computational speed can be obtained. Gas-dynamic simulations have been employed since the mid 1960s in studying the solar-wind interaction with the magnetosphere and have been very influential in guiding our understanding of this problem.

The most productive application of the gas-dynamic model of the solar-wind interaction has been its use in investigating the properties of the magnetosheath (Spreiter, Summers, and Alksne, 1966). While the model ignores magnetic forces, it calculates magnetic field lines by convecting frozen-in field lines along with the fluid. Thus, this model is often called the convected-field gas-dynamic model. The results of the simulations depend on the specified shape of the obstacle, the Mach number of the flow, and the polytropic index γ. For considerations of computational speed, it is usual to solve for the flow parameters with a shape that is cylindrically symmetric about the Sun–planet line and then convect the three-dimensional magnetic field through the flow field by using these flow parameters. The Mach number of the solar-wind flow that is appropriate for use in this gas-dynamic analog is that of the fast magnetosonic wave, because the bow shock is a fast magnetosonic shock. The fact that the velocity of the magnetosonic wave is anisotropic is a complication, but one that is no more serious than the choice of an axisymmetric obstacle to the flow. For most situations found at 1 AU, the variation of the magnetosonic velocity around the shock front is a few percent. The polytropic index is usually chosen to be 5/3, which is appropriate for an ideal gas with three degrees of freedom.

Figure 7.8 shows the streamlines for a Mach 8 flow with $\gamma = 5/3$ past a magnetosphere. The streamlines show the direction of the flow. Figures 7.9, 7.10, and 7.11 show lines of constant density, velocity, temperature, and mass flux normalized by the upstream solar-wind value. The density ratio immediately behind the shock is close to the maximum permitted by the Rankine–Hugoniot relations discussed in Chapter 6. Here, this limiting value $(\gamma + 1)/(\gamma - 1) = 4$. At the subsolar magnetopause, this ratio has increased to 4.23. In the actual magnetosheath, magnetic effects act to limit this density increase. As the gas expands around the obstacle, it decreases below the upstream value near the magnetosphere, but it is always compressed in the region just behind the shock. The temperature contours are the same as the velocity contours in Figure 6.8, because the temperature ratio is related to the velocity ratio by the expression

$$T/T_\infty = 1 + 0.5(\gamma - 1)M_\infty^2(1 - u^2/u_\infty^2), \quad (7.26)$$

which is obtained by integrating the energy equation (Spreiter *et al.*, 1966). We note that the temperature rise in the magnetosheath is substantial. If the solar-wind temperature were 50 000 K in the solar wind, it would be over 1 000 000 K throughout much of the dayside magnetosheath. We note that because the single gas-dynamic temperature is a proxy for the sum of the electron and ion temperatures, and because the electron temperature in the solar wind is often more than twice that of the ions, but changes only slightly across the bow shock, the actual change in ion temperature in the magnetosheath can be many times that shown in Figure 7.10. In contrast to the case for the solar wind, the magnetosheath ion to electron temperature ratio is rather constant, averaging about 6 everywhere.

The mass flux contours shown in Figure 7.11 were obtained by multiplying the velocity along the streamline by the mass density shown in Figures 7.9 and 7.10. The mass flux is important because it determines the location of the bow shock. All mass

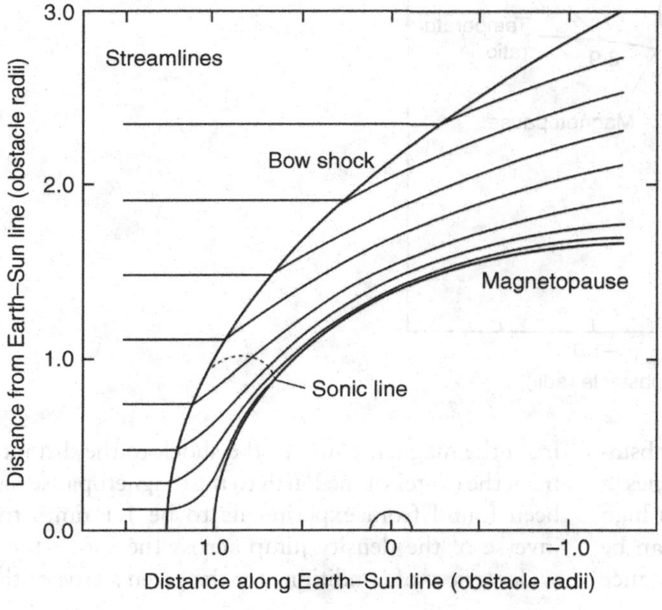

FIGURE 7.8. Streamlines for supersonic flow past the magnetosphere for a Mach number of 8 and a polytropic index of 5/3. The flow-line spacing has been chosen for convenience in the illustration of the magnetosheath flow and is not an indication of mass flux. (After Spreiter *et al.*, 1966.)

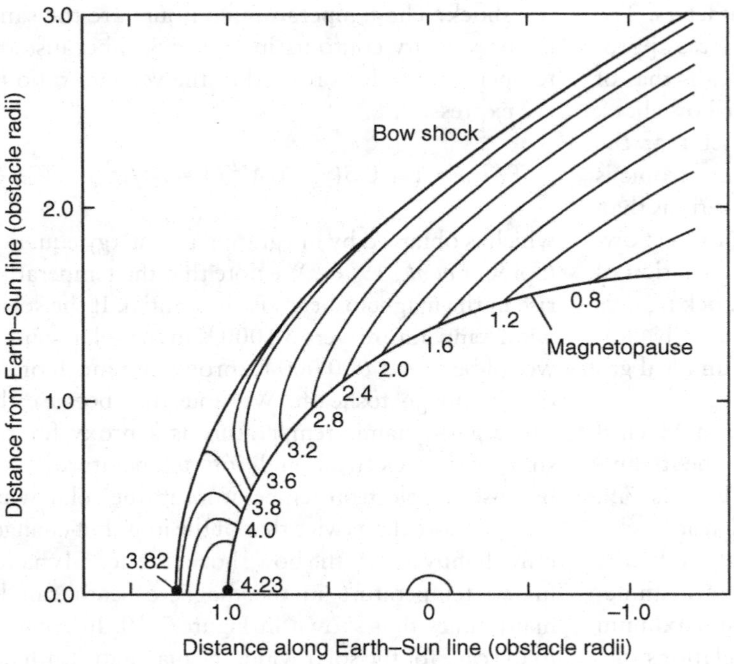

FIGURE 7.9. Density contours for supersonic flow past the magnetosphere for a Mach number of 8 and a polytropic index of 5/3. (After Spreiter *et al.*, 1966.)

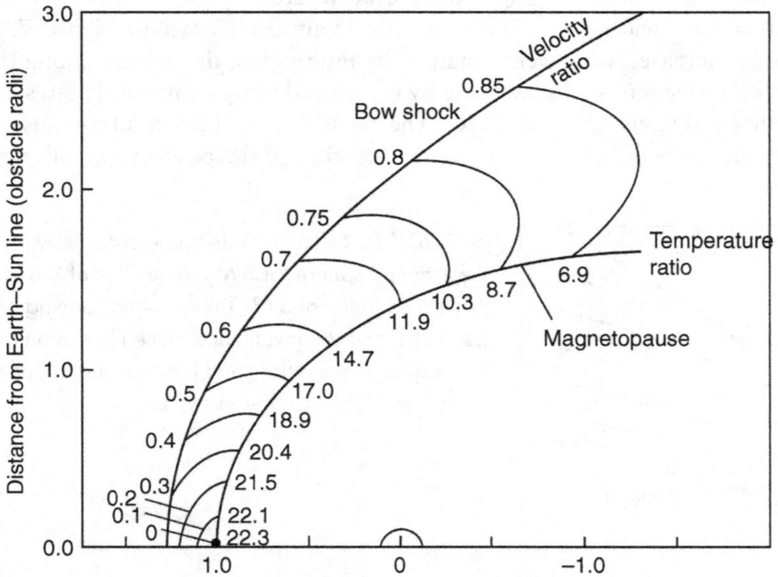

FIGURE 7.10. Velocity and temperature contours for supersonic flow past the magnetosphere for a Mach number of 8 and a polytropic index of 5/3. (After Spreiter *et al.*, 1966.)

flux through the bow shock flows around the obstacle in this model, and the bow shock assumes a position that allows this flow. In particular, at high Mach numbers the subsolar shock location can be determined empirically. The ratio of the distance from the magnetopause to the shock to the distance from the center of the Earth to the magnetopause has been found from experiments to be 1.1 times the inverse of the density jump across the shock for a wide variety of conditions for the given shape of the

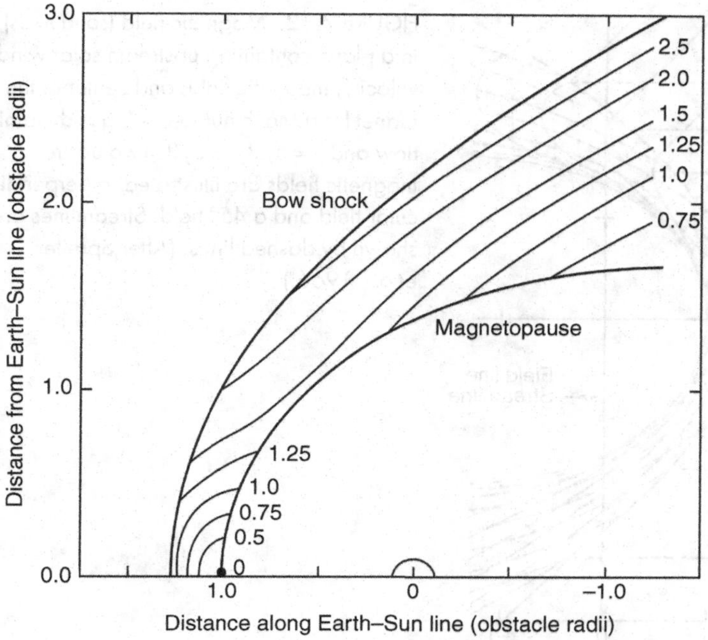

FIGURE 7.11. Mass flux contours for supersonic flow past the magnetosphere for a Mach number of 8 and a polytropic index of 5/3. The mass flux is measured along the streamlines and is numerically equal to the product of the density ratio and the velocity ratios shown in Figures 7.9 and 7.10. (After Spreiter *et al.*, 1966.)

magnetosphere. In this gas-dynamic solution, the density jump is solely a function of Mach number and γ and is equal to $\{(\gamma - 1)M^2 + 2\}/(\gamma + 1)M^2$ (Spreiter *et al.*, 1966). For a Mach number of 8 and γ of 5/3, the bow-shock nose should be 29% farther out than the magnetopause distance. Obviously, at low Mach numbers, this formula does not work, because as the Mach number approaches unity the compression of the plasma approaches zero. The shock should move to infinity, whereas our formula predicts a shock that is at a finite distance from the magnetopause. Furthermore, the standoff distance should depend on the radius of curvature of the subsolar magnetopause. Spreiter *et al.*'s (1966) model applies only to the shape of the obstacle used in that simulation. An improved formula that is a good approximation at both low and high Mach numbers and has an obstacle shape dependent on the radius of curvature is

$$D_{BS} = R_C[D_{OB}/R_C + \{0.8(\gamma - 1)M_1^2 + 2\}/\{(\gamma + 1)(M_1^2 - 1)\}, \quad (7.27)$$

where D_{BS} is the distance from the region to the subsolar bow shock, D_{OB} is the distance from the origin to the subsolar obstacle, and R_C is the radius of curvature (Farris and Russell, 1994). The radius of curvature of the "nose" of a conic section of revolution is equal to the semilatus rectum, $\mathcal{K}$.

Figure 7.12 shows the magnetic field configuration obtained by convecting field lines with the gas-dynamic flow field, with no magnetic-pressure effects considered. This might be expected to be the case over most of the simulation volume for high-Mach-number conditions. Two situations are shown: one with the magnetic field perpendicular to the flow, and one with it at 45° to the flow. In both situations, the magnetic field is seen to pile up in the subsolar region. We do not expect this to happen to the same extent in the actual magnetosheath, because the associated magnetic-pressure gradients should alter the flow pattern there. At the stagnation point in the flow, the sums of the magnetic and thermal pressures of the magnetosheath and magnetosphere should be in balance.

While the gas-dynamic solution is very instructive about the pressure balance between the flow and the obstacle, it does not reproduce some of the important properties of the flow, notably the plasma depletion layer in the subsolar region. Here, when reconnection is not occurring, the plasma density drops, and the magnetosheath magnetic field provides the pressure

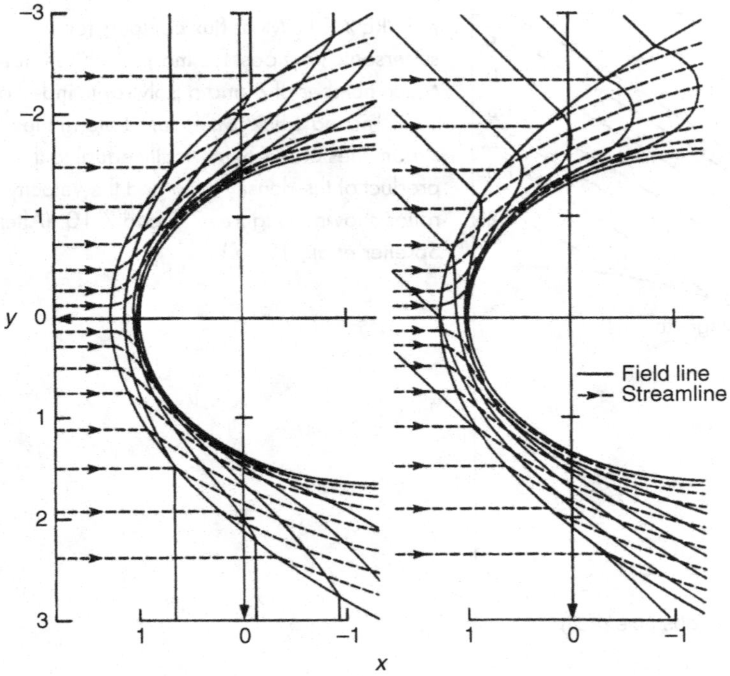

FIGURE 7.12. Magnetic field (solid lines) in a plane containing upstream solar-wind velocity, magnetic field, and center of the planet for a Mach number = 8 gas-dynamic flow and $\gamma = 5/3\gamma = 5/3$. Two upstream magnetic fields are illustrated, a perpendicular field and a 45° field. Streamlines are shown by dashed lines. (After Spreiter *et al.*, 1966.)

balance with the magnetospheric magnetic field. This effect can only be provided by a consideration of the magnetic forces in the equations, which is not part of the gas-dynamic approximation. For this, and for the effects of reconnection at the magnetopause, we must use MHD approaches.

7.6.2 Magnetohydrodynamic simulations

In the MHD approach to modeling the magnetosphere, we solve the equations of motion and Maxwell's equations for a magnetized plasma. The MHD equations were introduced in Chapter 3. The forms of the equations frequently solved in the magnetospheric models are as follows:

$$\text{continuity equation } \partial\rho/\partial t = -\nabla \cdot (\mathbf{u}\rho), \quad (7.28)$$

$$\text{momentum equation } \partial U/\partial t = -(\mathbf{u} \cdot \nabla)\mathbf{u} - (\nabla P)/\rho + (\mathbf{J} \times \mathbf{B})/\rho, \quad (7.29)$$

$$\text{pressure equation } \partial P/\partial t = -(\mathbf{u} \cdot \nabla)P - \gamma P \nabla \cdot \mathbf{u}, \quad (7.30)$$

$$\text{Faraday's law } \partial B/\partial t = \nabla \times (\mathbf{u} \times \mathbf{B}) + \eta \nabla^2 B, \quad (7.31)$$

$$\text{Ampére's law } J = \nabla \times (\mathbf{B} - \mathbf{B}_d), \quad (7.32)$$

where ρ is the plasma density, $\mathbf{u}$ is the flow velocity, P is the plasma pressure, $\mathbf{B}$ is the magnetic field, and $\mathbf{B}_d$ is the Earth's internal field. The polytropic index γ is taken to be 5/3, and η is a magnetic diffusivity set to zero in the ideal MHD equations, although there is always finite numerical resistivity. The boundary conditions on the left-hand side are set to the constant incoming solar-wind flow conditions. A major advantage of this approach, when the Earth's dipole is used as the obstacle, is that the physical boundaries of the problem (i.e., the magnetopause and the bow shock) are calculated naturally without limiting assumptions. As noted above, the limitation of the calculations is that they include numerical dissipation, which is not always of the correct magnitude or location to mimic the physics of the solar wind–magnetosphere interaction.

The gas-dynamic solutions discussed above provide good spatial resolution for the magnetosheath, but have to slow and deflect the flow with solely a compressional wave and without the help of the magnetic field. The MHD solutions add the Alfvén mode, which can bend the magnetic field and the flow, and the slow-mode wave, which can decrease density (and increase field strength) as occurs when field lines stretch. Such field line stretching occurs, for example, on magnetic flux tubes convected to and draped across the subsolar magnetopause. The resulting strong-field, low-density boundary layer is called the plasma depletion layer. We can see the differences between the physics of the gas-dynamic interaction of the solar wind with a magnetopause-shaped obstacle shown earlier and the MHD interaction with the help of Figure 7.13. The pressure force, $-\nabla P$, is shown in Figure 7.13(a); the magnetic force, $\mathbf{j} \times \mathbf{B}$, is shown in Figure 7.13(b), and their sum is shown in Figure 7.13(c). The arrows denoting the pressure gradient force point strongly outward along the shock to slow down the flow in Figure 7.13(a), but point inward downstream from the shock where the thermal pressure drops as the boundary is approached. In the gas-dynamic solution, the pressure gradient force is outward everywhere. In Figure 7.13(b), we see the magnetic force. At the shock, it too acts to slow the flow, but downstream from the flow it continues to push against the incoming flow, unlike the forces in Figure 7.13(a). Thus, the magnetic field is very important in controlling the deflection of the flow near the magnetopause. When the forces are summed in Figure 7.13(c), we see the MHD forces are such as to slow and divert the flow, contributing to the low-density, strong-field region at the subsolar magnetopause, called the depletion layer.

MHD models such as that shown here are very useful for describing the magnetic configuration and the flow around it. However, they have limitations. Because the plasma is assumed to be a magnetized fluid, the models do not simulate kinetic effects and the plasma instabilities that could arise. Often, parameters must be chosen for numerical stability rather than on the basis of physical constraints. Hence, other techniques have been developed to address these problems. The hybrid simulation technique treats ions as particles and electrons as a massless fluid in order to include some of the kinetic effects in a plasma. We should note here that MHD models also describe the region interior to the magnetopause, and that the details of the magnetopause boundary and its interplanetary field connections are very important for the magnetosphere's response to the solar-wind interaction.

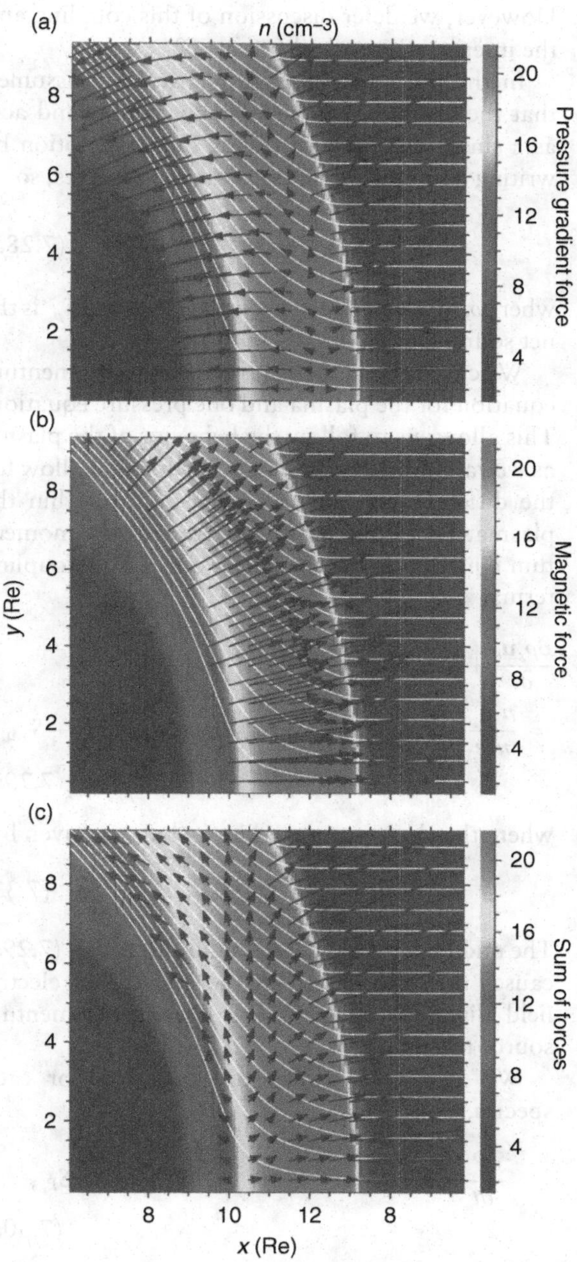

FIGURE 7.13. Forces in the equatorial plane of an MHD simulation of the bow shock and magnetosheath. Arrows show the pressure gradient force (a), the magnetic force (b), and the net force (c). The background contours show the number density. The interplanetary and magnetosheath magnetic fields are northward. Upstream density is 5 cm^{-3}. Streamlines are shown in white. (Simulation described by Wang, Raeder, and Russell, 2004.)

However, we defer discussion of this coupling and the interior region until Chapter 9.

In the preceding discussion, we have assumed that the plasma contains a single species and acts as a single fluid. We can relax this assumption by writing continuity equations for each species, so

$$\frac{\partial \rho_s}{\partial t} + \nabla(\rho_s \mathbf{u}_s) = S_{\rho_s}, \tag{7.28a}$$

where ρ_s is the mass density of species s and S_{ρ_s} is the net source rate of ρ_s.

We can still assume that there is one momentum equation for the plasma and one pressure equation. This allows us to follow the behavior of the plasma as we vary different species, but it does not allow for the different behavior of each species within the plasma. To do this, we need to rewrite the momentum equation for each species and add a coupling term:

$$\frac{\partial \rho_s \mathbf{u}_s}{\partial t} + \nabla \cdot (\rho_s \mathbf{u}_s \mathbf{u}_s + P_s \mathbf{I}) = \frac{n_s q_s}{n_e e}(\mathbf{j} \times \mathbf{B} - \nabla P_e) + n_s q_s (\mathbf{u}_s - \mathbf{u}_+) \times \mathbf{B} + S_{\rho_s \mathbf{u}_s}, \tag{7.29a}$$

where the charge-averaged ion velocity is given by

$$\mathbf{u}_+ = \sum\nolimits_s n_s q_s \mathbf{u}_s / e n_e. \tag{7.33}$$

The middle term on the left-hand side of Eq. (7.29a) causes flow separation in the convection electric field direction. Here $S_{\rho_s \mathbf{u}_s}$ is the net momentum source term.

We also need a pressure equation for each species,

$$\frac{\partial P_s}{\partial t} + \nabla \cdot (P_s \mathbf{u}_s) = -(\gamma - 1) P_s \nabla \cdot \mathbf{u}_s + S_{P_s}, \tag{7.30a}$$

where S_{P_s} is the net rate of change of pressure.

The magnetic induction equation is

$$\frac{\partial \mathbf{B}}{\partial t} - \nabla \times (\mathbf{u}_e \times \mathbf{B}) = \nabla \times \left(\frac{1}{\sigma_0 \mu_0} \nabla \times \mathbf{B} \right), \tag{7.34}$$

where the electron velocity is

$$\mathbf{u}_e = \mathbf{u}_+ - \mathbf{j}/n_e, \tag{7.35}$$

and the pressure is

$$P_e = \sum\nolimits_s P_s. \tag{7.36}$$

This approach allows us to treat plasma with two or more ion species with quite different masses, and even small charged dust particles. The multi-fluid model is needed when there are two or more fluids moving with different velocities (which is always true in a solar wind–planetary interaction) and when the magnetic field is relatively weak (such as for Mars and Titan). In collisionless plasmas, different fluids interact mainly through the electromagnetic force. The effect of the force would be to accelerate the flow that is moving slower than the electrons and to decelerate the flow that is moving faster than the electrons. In other words, electromagnetic forces create flow separation along the electric field direction, in order to keep all the flow moving with the same average bulk velocity as the electrons. When the magnetic field is strong, and the ion gyro-radius is small compared with typical scale size, then the single-fluid assumption is sufficient. Otherwise, the multi-fluid model is a better choice. The multi-fluid model is a useful tool in explaining some of the observed plasma phenomena such as different magnetic stretching on the plus and minus electric field sides of a magnetotail (Zhang *et al.*, 2010), which cannot be resolved by a single-fluid model. The cost of using a multi-fluid model, however, is much longer computation times.

7.7 HYBRID SIMULATIONS: EXPLORING MULTISCALE BEHAVIOR

In the earlier sections of this chapter, we treated as irrelevant the fact that underlying the macroscopic properties of the magnetized plasma were microscopic motions of ions and electrons. In the gas-dynamic approximation, we consider only the conservation of mass, momentum, and energy. The magnetic field does not matter, and the size of the obstacle to the flow does not matter. The solutions are self-similar. If we add magnetic forces and treat an ionized medium under the MHD approximation, we add a rotational Alfvén wave and a slow compressional wave to the fast compressional wave of gas dynamics, but still the solutions remain self-similar. In this approximation, it does not matter how large or how small an object is; to zeroth order, it will still cause the same relative deflection of the solar wind. A shock will always form if the incoming flow is supermagnetosonic, and its location will be determined by the upstream flow parameters and obstacle shape. However, the gyro-radii of the particles compared to the scale sizes of the macroscopic interaction do matter in space plasmas, and are especially important at boundaries between plasma and field domains.

The fluid approaches, moreover, do little to inform us how the needed dissipation at the shock takes place. What happens to the plasma when it is shocked, accelerated by an electric field, or reconnected? To answer these questions, we need to move beyond MHD solutions and follow the individual motions of particles. To follow both electrons and ions is possible in restrictive geometries, but fully kinetic, three-dimensional solutions are not yet available to solve realistic global interaction problems. A compromise approach is called the **hybrid simulation**, in which ion kinetic motion is followed, but electrons are treated as a massless fluid. Hybrid simulations have proven to be useful for many localized problems, especially for wave–particle interactions, but recently computers have become powerful enough to use hybrid simulations to treat some aspects of the global interactions.

A particularly insightful way to utilize these simulations is to examine the interaction of the solar wind with magnetic dipoles of increasing strength, from moments so small that the distance from the center of the dipole magnetic field to the solar-wind pressure balance point is much less than an ion inertial length, to moments whose magnetic field pressure matches that of the solar wind at a distance of a hundred or more ion inertial lengths, about the size of Mercury's magnetosphere, for example. The simulations we use to illustrate this section have two spatial dimensions and three velocity dimensions. This is sometimes called a 2.5D approach. In Figure 7.14, the simulation box and the IMF are in the x–y plane with x along the solar-wind velocity (V_{sw}). The magnetic field is perpendicular to V_{sw}. The magnetized body is represented by a line dipole with its axis along y. The planetary size is one (absorbing) cell at the origin, and the solar wind is continuously injected from the left with an Alfvén Mach number M_A of ~5–8. These simulations are intended to represent the upstream conditions in the solar wind at 1 AU. The boundaries are open so that the plasma leaves the simulation box at the downstream boundaries. The simulation is run until a steady state is reached. Details of these simulations can be found in Omidi *et al.* (2002) and Blanco-Cano, Omidi, and Russell (2004).

7.7.1 A hierarchy of magnetospheres

We can compare the different members of our hierarchy of interaction regions using the parameter, D_p, the distance (in ion inertial lengths, λ_i) from the dipole at which solar-wind dynamic and dipole magnetic pressures are equal; D_p can be interpreted as the effective size of the obstacle. As D_p increases, we find that there are four different types of solar-wind interaction with a magnetized body. When D_p is smaller than the ion scales ($D_p \ll \lambda_i$), the plasma is not modified, but a whistler wake is formed. As discussed in Chapter 13, the whistler wave is an electromagnetic wave in which the perturbed electric and magnetic fields rotate about the background magnetic field in the direction of motion of the gyrating electrons. Larger values of $D_p < \lambda_i$ lead to some plasma modification near the obstacle, and to the generation of three separate wakes corresponding to whistler, as well as fast and slow magnetosonic modes. (A magnetosonic wave has a longer wavelength than a whistler-mode wave, is compressional, and affects ions and electrons similarly.) When $D_p \approx \lambda_i$ the interaction region changes dramatically, the plasma is perturbed, a shock-like structure develops ahead of the dipole, and a hot plasma sheet appears in the tail. A magnetosphere similar to Earth's is formed when $D_p > 20\lambda_i$. In short, the characteristics of the interaction depend on the size of D_p relative to ion scales. The size of the interaction region and complexity increase with D_p. This dependence on the relative size of the obstacle to the flow to the scale size of ion motion is

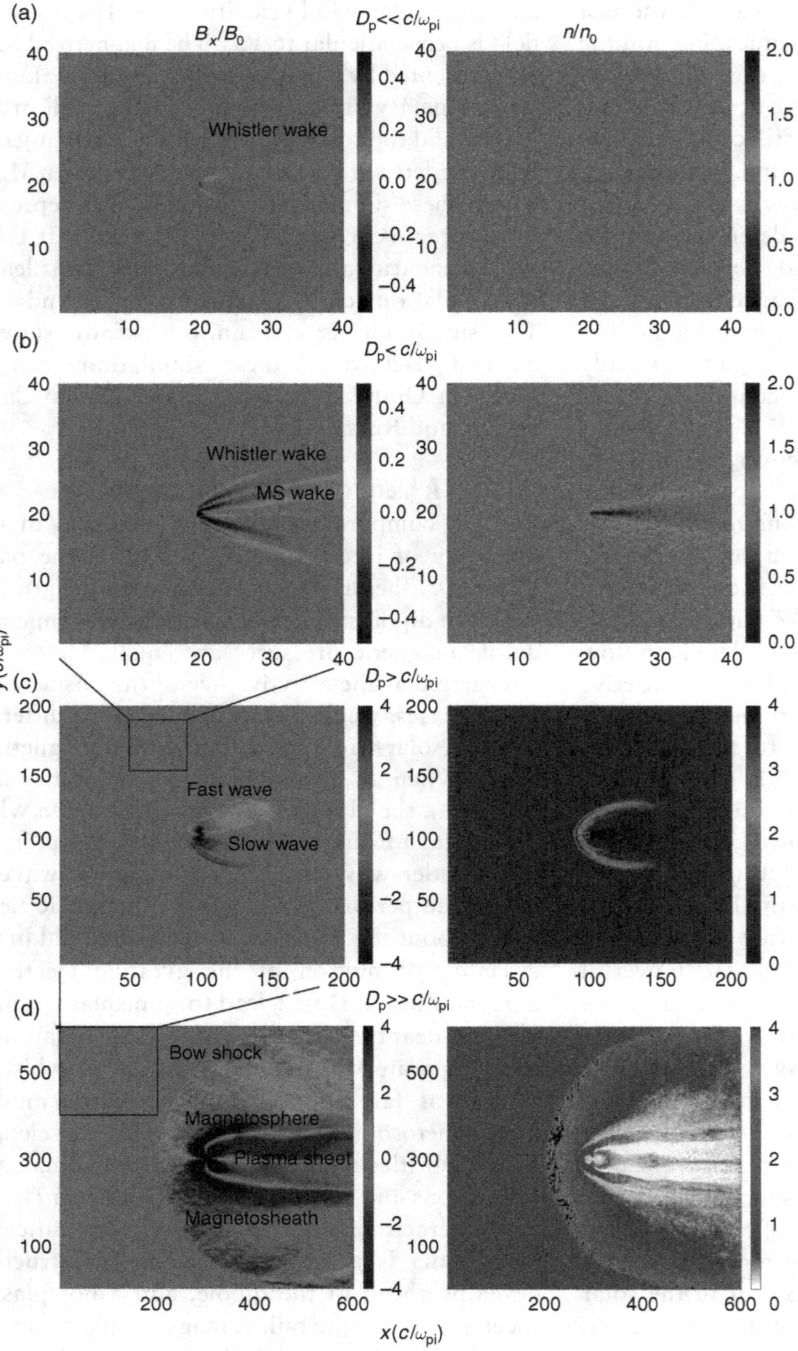

FIGURE 7.14. Magnetic field component along the flow and the plasma density for four different hybrid runs of increasing obstacle size, illustrating the cross-scale coupling of kinetic and global scales. The simulations use a two-dimensional dipole field equivalent to that produced by two current-carrying wires into the page. The box size is longer in parts (c) and (d), as indicated. The interplanetary magnetic field is northward. (After Blanco-Cano *et al.*, 2004.)

demonstrated in Figure 7.14 with a sequence of contour plots using the same flowing plasma conditions at varying magnetic obstacle strengths.

The solar-wind interaction with a weak dipole is like that of interaction with a weakly magnetized non-outgassing body whose effective size is much smaller than the ion scales ($D_p = 0.05\lambda_i$). Only a magnetic signature is generated (as shown in Figure 7.14(a)). In this case, the magnetic dipole is so weak that it does not constitute an impenetrable obstacle to the solar-wind flow, which reaches the cell containing the dipole, and is absorbed. There is no deflection of solar-wind ions, and no density or temperature changes occur. As a result, the perturbation consists only of a non-compressive whistler wake downstream of the body with small B_x and B_z fluctuations, and no changes in field magnitude. A situation in nature where such an interaction occurs is in the solar-wind flow past a small magnetized asteroid (Blanco-Cano, Omidi, and Russell, 2003). Weakly outgassing comets may have a similar signature.

For stronger magnetization and $D_p = 0.2\lambda_i$, the wake still forms downstream of the obstacle (as shown in Figure 7.14(b)). In contrast to the previous case, the plasma is modified close to the nose of the obstacle. The density increases ahead of the dipole, and decreases in the tail behind. There is no pile-up region in front of the dipole because the flow velocity never goes to zero, but some flow deflection occurs. The density and velocity change over a region upstream slightly larger than D_p. In the tail, the temperature and velocity are perturbed. The wake is formed by whistler and magnetosonic fast and slow waves. Non-compressional whistler waves arise in front of the region where the density increases. Closer to the dipole, where the density is enhanced, magnetosonic waves able to compress the flow form. A situation in nature where such an interaction appears to occur is at Saturn's moon, Iapetus, when it is in the solar wind. Also, small outgassing comets will have similar signatures.

When the obstacle size is comparable to ion scales, the dramatic interaction changes in Figure 7.14(c) appear. The density increases, where the flow stagnates ($V_x = 0$), and pile-up occurs at distances $r \approx D_p$ ahead of the dipole. A magnetosonic wave forms upstream, compressing, slowing, heating, and diverting the flow around the obstacle. The magnetosonic wave resembles a fast shock, but spatial scales associated with the wave are comparable to ion gyro-radii and therefore different dissipation processes operate. Ion reflection at the bow wave and their subsequent acceleration leads to asymmetries in the wave structure. The magnetosonic wave, with both plasma and field compression, results from density pile-up. The plasma is heated by particle acceleration at the dipole. Downstream of the dipole, a region of lower magnetic field identified as a slow mode appears. This separates a slower, cooler plasma from a faster and hotter one in the central wake region. Test-particle runs show that this fast and hot plasma results from acceleration in the dipole region. That density pile-up starts to become important when $D_p \approx \lambda_i$ clearly illustrates that, at this scale size for the obstacle, ion gyration has a deep influence on the structure of the interaction region and on the way that the plasma is modified. In contrast to a planetary magnetopause, pile-up occurs inside dipolar field lines. This, in addition to the fact that the region is small compared to ion scales, indicates that the region of the interaction does not have the properties of a planetary magnetosphere or magnetic cavity. A moderately strongly outgassing comet might have this structure. The only interaction reported thus far in nature that resembles this simulation is the lunar limb compressions that form above magnetized regions of the lunar surface which are too small to form shocks (Russell and Lichtenstein, 1975).

The interaction generates an Earth-like magnetosphere when $D_p > 20\lambda_i$ (i.e., the magnetized obstacle is much larger than the ion scales). As illustrated in Figure 7.14(d), in this case the planetary field is strong enough that a distinct magnetosphere, magnetosheath, and bow shock are produced. Plasma parameters show that there is not a pile-up region within a distance $r \approx D_p$ from the planet as before. The interaction is mediated by a shock wave at a large distance upstream ($\approx 100\lambda_i$), where the solar wind is compressed, heated, and completely diverted. This results in a magnetosheath where the density, temperature, and field are enhanced, and the flow is decelerated. The shock leads to ion reflection, which in turn can modify the oncoming plasma. The magnetosphere extends ~$30\lambda_i$ upstream, and a much larger distance downstream. Solar-wind density drops at the distance to which closed dipolar field lines extend, so that the dipole field is an impenetrable obstacle to the flow, and a magnetopause current layer is formed. A thin magnetopause current layer separating solar wind from magnetospheric plasma is possible when the ion gyro-radius becomes much smaller than the magnetopause thickness. The simulated magnetosphere shows the basic features observed at Earth: a cusp, a tail with a plasma sheet bounded by boundary layers, and the existence of energetic ions at the dipolar, magnetopause, and central tail regions. Evidence of magnetic reconnection, in which IMF lines link up with magnetospheric lines (Dungey, 1961), is provided by the magnetic islands, or plasmoids, at the high-latitude magnetopause and in the central tail region. Although one must bear in mind that these features are all produced by a two-dimensional dipole interaction, the results suggest many of the most important processes are at work in this simulation, including the formation of plasmoids.

In summary, the numerical models help us appreciate that the solar-wind interaction with magnetized bodies can occur in many different ways. It is clear from the hybrid models, in particular, that the obstacle size, in relation to plasma ion scales, strongly affects the nature of the interaction region and the characteristics of the waves involved. The formation of a magnetosphere requires that the dipolar field is strong enough so that the dipole field obstacle's effective size is larger ($D_p > 20\lambda_i$) than ion scales. In this case, a bow shock, magnetosheath, and magnetospheric system (with a bow shock and magnetosheath) is driven, and the flow is modified far upstream. Since the magnetospheres of all planets in the solar system have $D_p > 20\lambda_i$ (e.g., 85 for Mercury, 640 for Earth, and 5800 at Jupiter), they have the same basic features. Hybrid simulation results show that even ion kinetic processes, which act on the microscale, are very important to the global structure of the solar-wind interaction region with a magnetized body. One very important additional effect that arises in the hybrid simulation that is absent from gas-dynamic and magnetohydrodynamic simulations is the formation of a particle and wave foreshock region, where ions stream back into the solar wind, generating waves of various sorts, including cavitons and heated current sheets called hot flow anomalies. This region preprocesses the solar wind before it reaches the shock and can play a major role in governing the behavior of the shock.

7.7.2 Upstream waves and particles

The global hybrid simulation gives a particularly clear picture of the origin of upstream particles and waves, even for relatively small magnetospheres. Figure 7.15 shows the results of simulations for $D_p = 64$, which corresponds to a Mercury-size magnetosphere scale. Figure 7.15(a) shows a shaded contour of the density and the panels in Figure 7.15(b) show cuts through the simulation. The IMF is at 45° to the flow, and the upper (lower) portion of the bow shock corresponds to quasi-perpendicular (parallel) geometries. Figure 7.15(b) shows two crossings of the quasi-parallel shock with one (top) closer to the nose and the other (bottom) further in the flank. As can be seen in both cases, the transition from solar wind to the magnetosheath takes place in a number of steps, which is the result of local heating and deceleration of the solar wind by the upstream generated waves. The shock profile is highly turbulent and time varying due to convection of the upstream-generated ultralow-frequency, ULF, waves (see also Chapter 10) into the shock and its cyclic reformation. Examination of the ULF waves generated in this run and the ion distribution functions in the foreshock reveals the presence of both parallel and obliquely propagating waves with the former being generated by field-aligned ion beams and the latter by gyrating ions closer to the quasi-

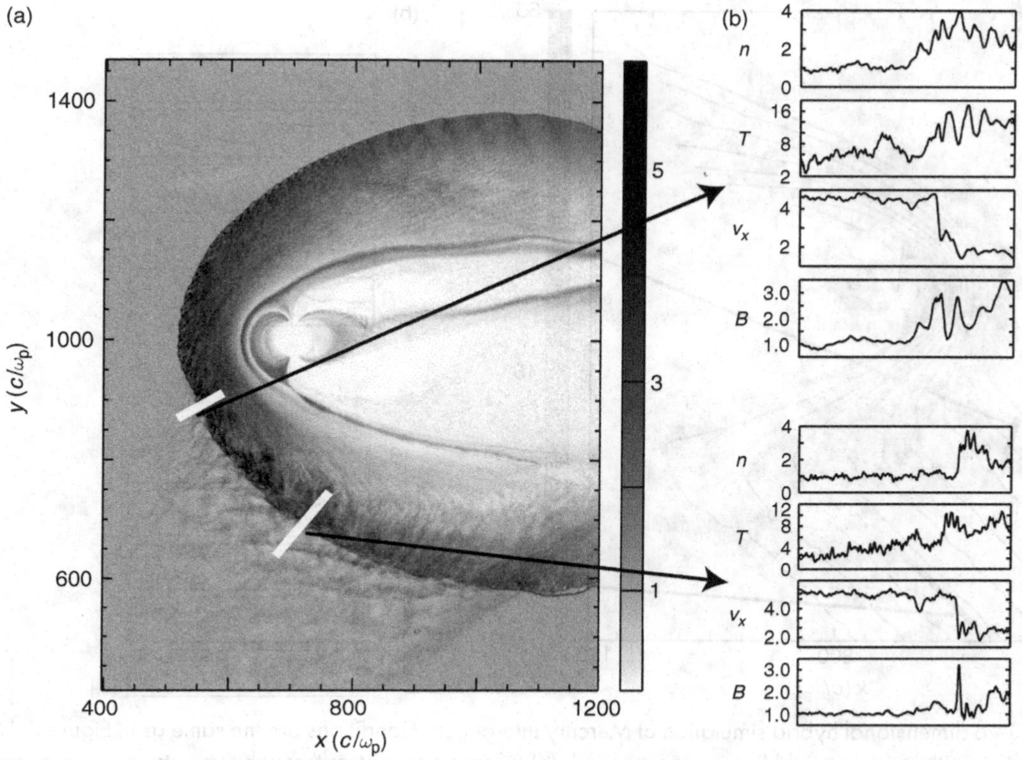

FIGURE 7.15. Two-dimensional hybrid simulation of the solar-wind interaction with the Mercury magnetosphere for which the ratio of the size of the obstacle to the ion inertial length is 64. (a) Contour map showing density. (b) Plasma parameters for two slices, near the nose and further away. The interplanetary magnetic field is at 45° to the flow from lower left to upper right, so that the shock is quasi-perpendicular in the top portion of the simulation. (After Omidi, Blanco-Cano, and Russell, 2005.)

parallel shock. Figure 7.16 shows the ion temperature for this run with the magnetic field lines superimposed. The panels in Figure 7.16(b) show examples of both waves, with the oblique ones (top) having steepened to form shocklets and the parallel ones (bottom) having a sinusoidal wave form. The results demonstrate that generation of shocklets is essential for the formation of the quasi-parallel shocks, as discussed at the end of Chapter 6. One can also see the convection of some of these waves into the downstream magnetosheath, a result that can have an impact on the obstacle boundary and the obstacle itself, as we shall see later in this book. Finally, we note that the curvature of the bow shock and the varying angle of the upstream magnetic field significantly affect the structure of the foreshock. Thus, interplanetary shocks which are more nearly planar can have simpler temporal behavior than bow shocks. However, the spatial variability of the IMF causes much temporal variation as interplanetary shocks pass through this varying structure.

The upstream foreshock region and the magnetosheath downstream from the foreshock can be quite turbulent and dynamic regions, in fact, as well as in simulations. A plethora of different phenomena have been reported such as cavitons, turbulent flux enhancements, filamentary structures, high-speed jets, hot flow anomalies, and more. At the present time, there is no agreed upon nomenclature or clear definition for these features.

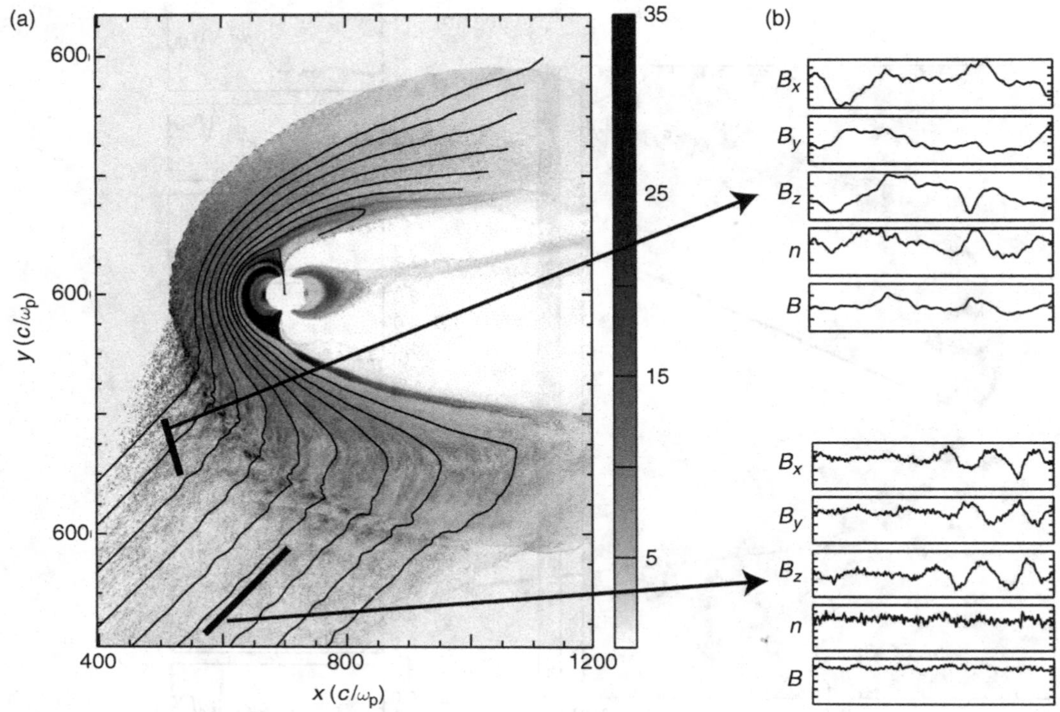

FIGURE 7.16. Two-dimensional hybrid simulation of Mercury interaction. Conditions are the same as in Figure 7.15. (a) Ion temperature with magnetic field lines superimposed. (b) Plasma parameters through two cuts in the upstream waves. The upper panel illustrates compressional waves and the lower panel illustrates transverse waves. (After Omidi *et al.*, 2005.)

7.8 SUMMARY

As discussed in this chapter, the solar-wind interaction with a magnetosphere is a complex process. The magnetosphere acts as a nearly impenetrable obstacle about which the solar wind must flow. The solar wind is supersonic, and so a shock must form and stand in front of the obstacle. We can examine this problem with varying degrees of approximation. The simplest treatment is that the flow is a gas-dynamic one, in which magnetic forces are ignored. This provides a simple and useful description of the magnetosheath that serves many purposes. However, it has many limitations. In particular, it predicts a magnetosheath density buildup and a magnetic field buildup at the magnetopause that are not observed. We also expect the real magnetosheath to include at least weak Alfvén and slow-mode waves in addition to the fast shock, because such waves are required, in general, to generate an arbitrary perturbation in the plasma. It is unlikely that the perturbation in the flow required to divert the solar wind around the magnetosphere could be accomplished with solely a compressional wave. Another limitation of these techniques is their assumption of isotropic pressure. We expect anisotropies in the magnetosheath pressure if the pressure of the magnetic field is a significant fraction of the thermal pressure. These anisotropies will lead to interesting plasma instabilities, which in turn will attempt to restore the isotropy of the plasma. Finally, the MHD approach and, of course, the gas-dynamic approach do not simulate small-scale features of the ion gyro-radius or smaller size, nor do they reproduce the foreshock. At the bow shock, these have been found to be of critical importance in providing the dissipation required by the Rankine–Hugoniot equations. These processes may have equal importance in the magnetosheath and at the magnetopause. By following the kinetic motions of the ions in hybrid models, we can simulate some of the microscale processes plasma as we simulate global interaction, but we are far from fully kinetic global interaction models. At present, when we need to consider the kinetics of the electrons as well as the ions we are generally restricted to the local problems like the shock layer physics or the neutral point physics in the reconnection problem.

Even though MHD and hybrid simulations do couple the flow to the magnetic obstacle, we have minimized any discussion of these coupling processes in this chapter. Rather, because they are so important to the Earth's magnetosphere, we defer the discussion of the coupling across the magnetopause and the resulting dynamics of the magnetosphere and upper atmosphere to the following chapters.

Additional reading

PETRINEC, S. M. and C. T. RUSSELL (1997). Hydrodynamic and MHD equations across the bow shock and along the surfaces of planetary obstacles, *Space Sci. Rev.*, **79**, 757–791. Review of how the solar wind confines the magnetosphere.

SPREITER, J. R., A. L. SUMMERS, and A. Y. ALKSNE (1966). Hydromagnetic flow around the magnetosphere. *Planet. Space Sci.*, **14**, 223–253. Classic gas-dynamic treatment of the flow of the supersonic solar-wind flow with an incompressible magnetospheric obstacle.

Problems

7.1 The magnetic field due to a dipole moment M along the z-axis and centered at (0, 0, 0) is

$$B_x = 3xzM/r^5,$$
$$B_y = 3yzM/r^5,$$
$$B_z = (3z^2 - r^2)M/r^5.$$

If the magnetic moment of the Earth is 31 000nT · R_E^3 and is aligned along the z-axis, calculate the magnetic field at the following points if an infinite flat plane of solar-wind plasma forms a magnetopause at $x = 10R_E$, as postulated by Chapman and Ferraro:

(a) at radial distances of 2, 4, 6, 8, and $10R_E$ along the Earth–Sun line;

(b) at just inside the magnetopause at $(10, 0, 2i)$ R_E $(i = 1, 4)$ and at $(10, 2j, 0)R_E$ $(j = 1, 3)$.

(c) If the neutral point is defined as the place on the magnetopause where the magnetic field is normal to the magnetopause, locate the neutral point.

(d) Graph these results in a manner that suitably illustrates the variation observed.

7.2 Use the image dipole model and a magnetopause distance of $10R_E$ to calculate the increase in surface magnetic field strength on the equator as a function of local time.

7.3 A magnetic field line crosses the magnetic equator of the Earth at $4R_E$. Assuming that the Earth's field is dipolar, where does this magnetic field line intersect the surface of the Earth?

7.4 The spacecraft Space Physics Explorer returns to its South Pole base from its repair mission at synchronous orbit and along the way makes the measurements found in Table 7.1.

What is the magnetic moment of the Earth (in $\mathrm{nT} \cdot R_E^3$), according to these measurements? Do you consider them to be consistent? How could you obtain a best-fit solution that is based on all three observations instead of averaging the solutions for the individual observations?

7.5 If the magnetic moment of Mercury is 3×10^{12} T m^3, what is the subsolar distance to its magnetopause in planetary radii for a solar-wind velocity of 500 km s^{-1} and a density of 20 cm^{-3}? The radius of Mercury is 2440 km. What is the magnitude of the field at the subsolar point? Assume that the shape of the Mercury magnetopause is the same as that empirically found for the Earth.

7.6 Waves generated at the magnetopause can be observed behind the Earth's bow shock. At least one of the MHD wave modes can always travel from the magnetopause to the nose of the bow shock. Which wave mode is it, and how do you reach this conclusion?

Table 7.1.

Position (geomagnetic)	Field (geomagnetic)
$(6.6, 0, 0)R_E$	(0, 0, 111) nT
$(0, 2, 0)R_E$	(0, 0, 4005) nT
$(0, 0, -1)R_E$	(0, 0, −64233) nT

7.7 Determine the equation of a magnetic field line and sketch the magnetic field lines corresponding to $B_x = y$, $B_y = x$, making sure that their relative spacing indicates the field strength. Calculate the magnetic curvature and pressure forces for this field. Include on the diagram several isocontours of magnetic field strength and several vectors indicating the direction and strength of the curvature and pressure forces. [Hint: Magnetic-pressure force is $-\nabla\,(B^2/2\mu_0)$ and curvature force is $\mathbf{B} \cdot \nabla\,(B/\mu_0)$, where $\mathbf{B} = y\hat{\mathbf{i}} + x\hat{\mathbf{j}}$]

7.8 Find the equation of a field line and sketch the field lines for the field $B_x = B_0 \cos kxe^{-kz}$, $B_z = -B_0 \sin kxe^{-kz}$, for $|x| \langle \pi/(2k)$, $z \rangle 0$. Verify that it has zero current. This is a useful model for a coronal arcade.

7.9 Find a solution for a force-free arcade by seeking solutions to $(\nabla^2 + \alpha^2)\mathbf{B} = 0$ and $\nabla \times \mathbf{B} = \alpha\mathbf{B}$ that are separable in x and z.

7.10 Find the equation of a field line and sketch the field lines for $B_x = B_0$, $B_y = 2B_0 x$. What is the magnetic force at a point (1, 0)? Does pressure or curvature force dominate?

7.11 Use the Magnetosphere option of the Space Physics Exercises* to study dipole magnetic fields.

(a) Measure the Earth's magnetic field at three different equatorial locations and calculate the dipole magnetic moment of the Earth. List the radial distance, latitude, and field components at each point. Calculate the mean, median, and standard deviation of the calculated moments. Why might your values of the moment differ? What would you do to improve the measurements? Use the number of significant digits appropriate to the

* Space Physics Exercises are at http://spacephysics.ucla.edu.

accuracy of your measurement in presenting your results.

(b) Measure the magnetic moments of each of the planets using several measurements of each. Units should be T m^3 for dipole moment. See Note below. Normalize the moments by the terrestrial moment.

(c) Plot the magnetic field in the equatorial plane of Jupiter versus radial distance on a log-log plot. Compare with dependence of an inverse cube law. Repeat for a radial path directly above the north pole. Plot the results and compare with the equatorial data.

[Note: The dipole magnetic moment is defined to be the product of the equatorial field strength times the cube of the radial distance from the center of the planet. The radii of the planets are: Mercury, 2440 km; Earth, 6371 km; Jupiter, 71 400 km; Saturn, 60 300 km; Uranus, 25 900 km; and Neptune 25 000 km.]

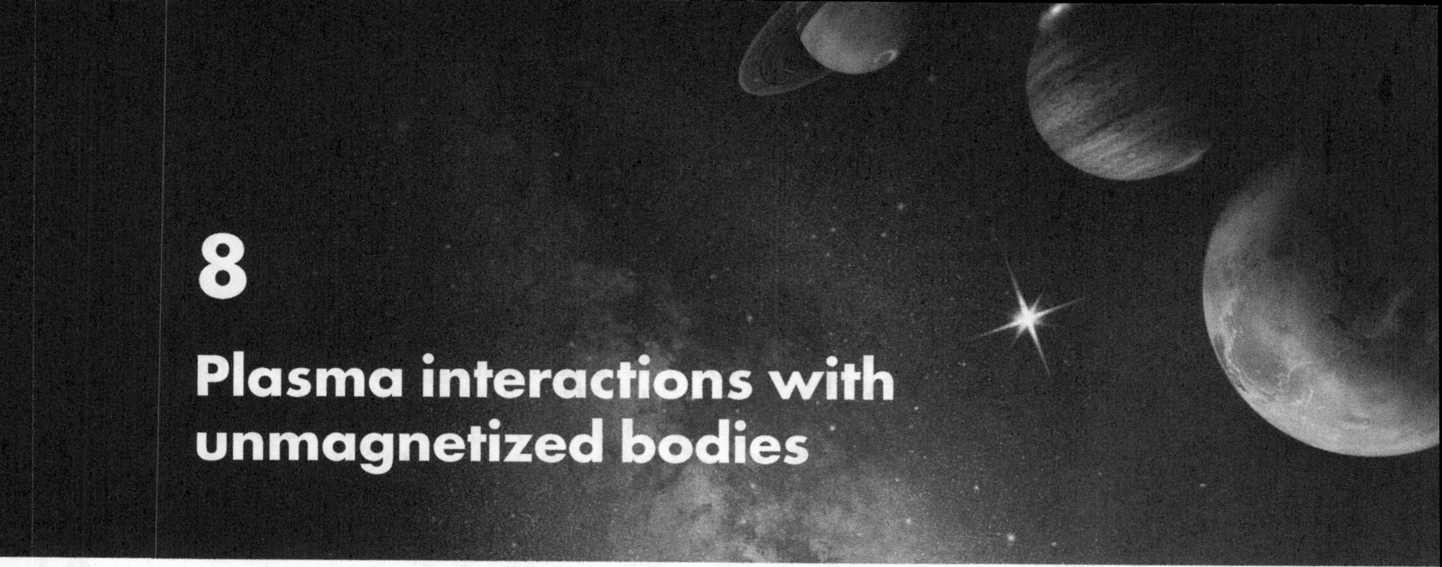

8 Plasma interactions with unmagnetized bodies

8.1 INTRODUCTION

When a planet or satellite has a weak internal magnetic field, or none at all, its interaction with a flowing magnetized plasma can be drastically different from the interaction with an Earth-like magnetosphere (see Chapter 7). In this chapter we shall see that such interactions can vary widely, depending on both the size of the body and whether it has a significant atmosphere. In addition, the body may be rocky or icy, and have layers of different composition and hence conductivity. If an atmosphere is present, the obstacle's characteristics depend on the atmospheric properties. In fact, there are many examples of such bodies in the solar system. These include Venus and Mars, comets and asteroids, Pluto, and many planetary moons. Some of these moons interact with the plasma and field of their host planet's magnetosphere instead of, or sometimes in addition to, interacting with the solar wind. Mars has significant crustal remanent magnetic fields that complicate the basic plasma interaction as we shall learn below. There are also complications introduced by an obstacle's size relative to gyro-radius scales in the external plasma flow and in the local atmospheric pickup ion population (see Chapter 5 to find pickup ions defined in the heliospheric context). In particular, when the ion gyro-radius either in the external medium or of the planetary ions is not small compared to the scale of the obstacle, extra physical processes and features arise. We begin this chapter with the simplest generic case of a spherical body like the Earth's moon, made of generally rocky, insulating material and having a negligibly thin atmosphere, located in the solar wind. We proceed from there to add other features found at the different weakly magnetized bodies to see the various changes that occur, as well as their consequences.

8.2 PLASMA INTERACTIONS WITH MOON-LIKE BODIES

Our moon, in orbit at ~60 Earth radii (R_E), represents the prototypical unmagnetized "planet." Its orbit at ~$60R_E$ means it is often located in the flowing, superfast solar-wind plasma. (The subsolar-terrestrial magnetopause is at ~$10R_E$.) The basic description of the solar-wind interaction with the Moon was envisioned early in the study of space plasmas. Without any substantial magnetospheric or atmospheric shield of its own, its rocky crust absorbs the plasma particles that are incident on the 1738 km radius body. As a result, lunar soil and rocks contain a historical record of the composition and energy flux of the solar wind. A bow shock does not form upstream because the plasma has no way of detecting the lunar obstacle in advance, and thus nothing requiring its deflection – at least to first order. While the incident particles are absorbed, the magnetic field frozen in the solar wind continues on, diffusing into and through the weakly conducting

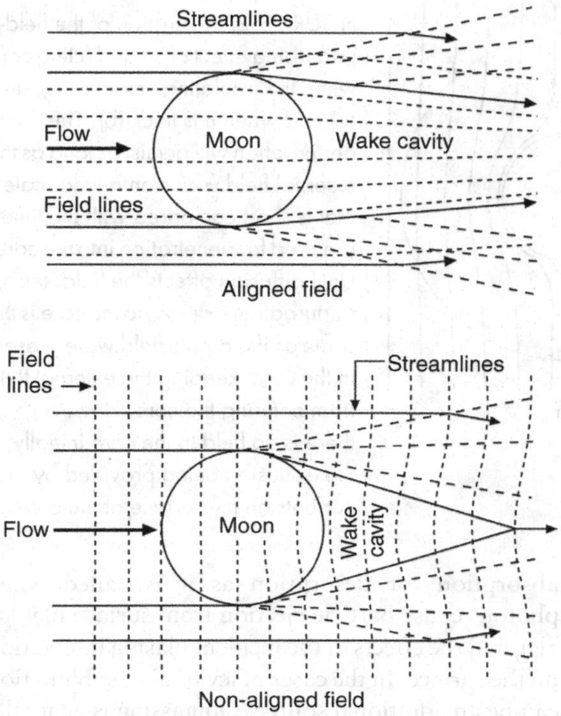

FIGURE 8.1. Illustration of the interplanetary plasma flow and magnetic field perturbation by the Moon, assumed here to be unmagnetized and non-conducting. (After Spreiter *et al.*, 1970.) The wake created by solar-wind absorption closes more quickly when the magnetic field is not aligned with (parallel to) the undisturbed flow.

body at a rapid rate so that it is barely perturbed from its upstream orientation. Figure 8.1 illustrates this basic atmosphereless, insulating body–plasma interaction.

The plasma interactions for magnetic fields parallel and perpendicular to the flow are shown in Figure 8.1. While the upstream is nearly featureless, a significant wake structure is present due to the absorption of the incident plasma. The details of this wake depend on external conditions. If the flow speed is high compared with the plasma thermal velocity, the wake will persist to large distances; but if the flow is slow relative to the thermal speed, thermal motions perpendicular to the flow direction can refill the empty space within a short distance downstream of the body. However, because the plasma particles travel more easily along magnetic field lines than across them, the direction of the magnetic field can either inhibit this refilling of the wake, if the field is nearly parallel to the upstream flow, or have minimal effect, if it is perpendicular. The sketches in Figure 8.1 show two hypothetical configurations for the lunar wake in the solar wind for these two magnetic field orientations. Because the magnetic field in the solar wind is frozen in, the field is slightly perturbed as the flow closes behind the Moon to fill the plasma void created by the absorbing obstacle. Magnetohydrodynamic (MHD) simulations and spacecraft data show that the wake is an MHD **rarefaction** wave standing in the flow that deflects the plasma in the region behind the Moon. Large asteroids, such as Vesta and Ceres, and atmosphereless planetary satellites in the solar wind or in rapidly rotating magnetospheric plasma should have similar plasma interactions.

But various complications apply in the lunar–solar-wind interaction, in spite of its first-order simplicity relative to other planetary bodies. If a planetary body has a conducting core, and the external field varies with time as it does at the orbit of the Moon, currents are generated or induced in that core. These produce a magnetic field that cancels the field internal to the conductor and enhances it outside the core. Eventually, the external field, if steady, will diffuse into the core.

For a spherical core, this field external to the core is the same as if an opposing dipole were present in the core. As illustrated by Figure 8.2, if the core is small, as it is at the Moon (the radius of the core is less than one-third of the radius of the Moon, R_M) these effects are small above the lunar surface. As the core becomes a larger fraction of the body, the perturbations are larger above the surface. While small, the perturbations observed during the Apollo missions were large enough to measure the size of the lunar core accurately. An important fact to remember is that these induced currents decay if the external field is steady and the core is not infinitely conducting. Induced

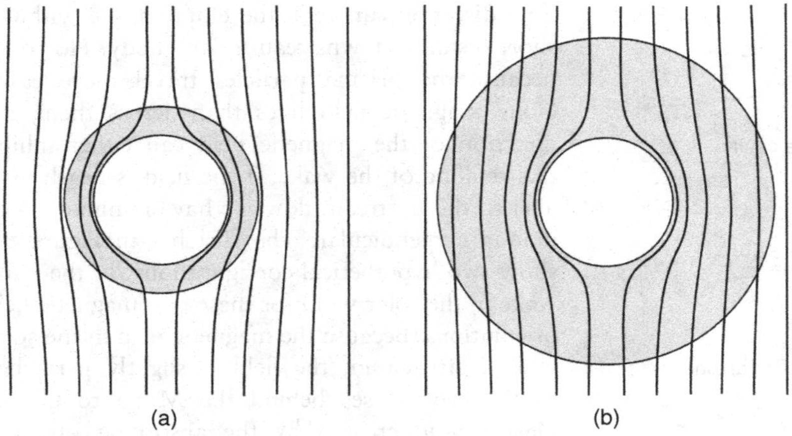

FIGURE 8.2. Illustration of the field-perturbing effect of a conducting core when the insulating "mantle" is thin (a) and when it is thick (b). This perturbation will persist as long as the external field varies on a time scale that is short compared with the time required for penetration into the core. Until diffusion affects the field, the perturbation external to the core is the same as if a dipole field were present in the core, keeping the external field from entering the core. However, there is no field in the core initially, the exclusion being provided by currents on the surface of the core.

fields can be produced by external field variations, such as those the solar wind provides in the ecliptic (see Chapter 5), and by the motion of the Moon into and out of the magnetosphere each month.

How long does it take the external field to migrate into or out of a spherical conductor like the lunar core? Cowling (1957) gives the diffusion time constant as

$$\tau = R^2 \mu_0 \sigma, \tag{8.1}$$

where R is the core radius, σ is its conductivity, and μ_0 is the magnetic permeability. The time constant for the lunar core is expected to be of the order of 1000 years. In the solar wind, the magnetic field changes more rapidly than this time constant, and so the picture in Figure 8.2 generally holds. The situation when the Moon is in the portion of its orbit inside Earth's magnetosphere is somewhat similar to that when it is outside, except that this picture is modified by the different external flow and field conditions there. However, this scenario does not pertain to planetary satellites with cores that have been immersed in a steady magnetospheric field for millennia. (More about such cases later.)

Another complication in the lunar interaction is that the Moon has a very thin atmosphere which can be quite variable. This atmosphere comes from natural outgassing by the lunar crust, solar photon absorption (or desorption as it is called when photons cause particle ejection from surface materials), and the effects of the incident plasma interaction on the surface. In the cases of icy bodies, sublimation can be an additional source. Outgassing is generally controlled by the interior and surface processes, and sometimes in response to collisional perturbations, such as impacting meteors, from outside. The process called photon desorption releases material at the surface when the incident photon flux – often at UV and EUV wavelengths – is sufficient to deposit energy in a way that heats the surface and thereby dislodges an atom or molecule. Incident plasma particles can transfer sufficient energy to the surface layer to "sputter" away material in a similar way, or alternatively produce a chemical reaction on the surface that results in a released particle.

The thin atmosphere produced by these processes at the Moon have little effect on the basic plasma interaction as long as it remains rarefied, but it nevertheless has observable signatures in the form of scattered or absorbed sunlight seen by telescopes with spectrographs, and heavy ions detected *in situ* in the vicinity of the obstacle. Gases such as sodium and potassium are particularly observable in imaging spectrographs due to their resonant scattering of sunlight, although they make up only a small fraction of the lunar atmosphere. In the process analogous to that described in Chapter 5 for heliospheric pickup ions, ions can be produced by

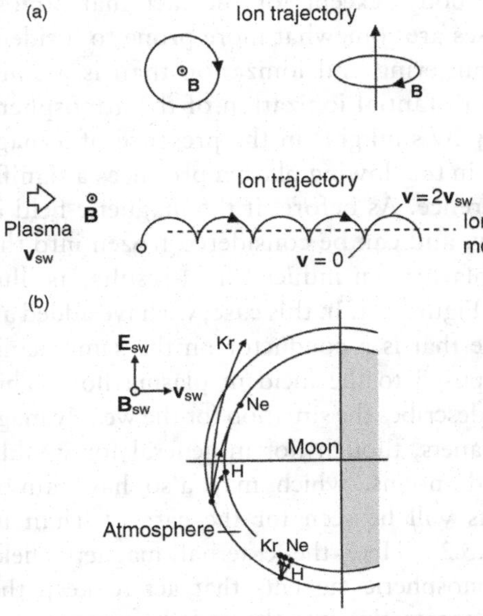

FIGURE 8.3. (a) Illustration of the pickup process and (b) the related fate of ions formed in the lunar atmosphere. (Adapted from Manka and Michel, 1970.)

photoionization, impact ionization (usually by solar-wind electrons), or charge exchange (wherein an ion in the incident plasma passes its charge to an atmospheric atom). As soon as they acquire their usually single positive charge, the atmospheric ions feel the convection electric field **E** = −**v** × **B** in their surroundings (where **v** is the ambient plasma velocity and **B** is the magnetic field) and start to move in the manner illustrated in Figure 8.3. An atmospheric ion in a rarefied atmosphere, in the case where the solar-wind magnetic field is nearly perpendicular to its flow, is thus accelerated as shown in Figure 8.3(a). In the presence of an obstacle the size of the Moon, solar-wind behavior may be fluid like, but the ion gyro-radius, $mv_{\perp}/eB$, of many atmospheric species cannot be neglected, as illustrated in Figure 8.3(b).

What happens to a pickup ion in Figure 8.3 depends on its mass, gyration speed $v_{\perp}$, and the magnetic field strength – as well as where it is produced in the external flow. As suggested here, at the Moon both light (H) and heavy (Ne, Kr, Na, K, etc.) ions are produced, whose composition is determined by the surface materials and the ease of their ejection. If the gyro-radius of the *pickup ion* is small compared to the size of the body, and is produced upstream close to the body, it is carried back into the body with the plasma flow. If the gyro-radius is large compared to the body, the ions can gyrate either over and away from the body or back into it, depending on where they are created. Although these few ions may not be important to the plasma interaction at the Moon, their detection provides a means of studying the weak lunar atmosphere and surface composition from an orbiting spacecraft.

This example also provides our first insight regarding the possible evolutionary effects of a flowing plasma incident on such bodies and their atmospheres. When the atmospheric atoms and molecules are heated by sunlight or non-thermal processes, they may or may not have sufficient energies or the right trajectories to escape the body's gravity near the surface. However, if they are ionized in the external magnetized plasma flow, they can easily acquire sufficient energy from the convection electric field present there. Pickup ions can in some cases represent a significant loss of constituents from the body or its atmosphere. They also introduce the presence of "finite gyroradius" effects in these plasma interactions, where the usual fluid treatments of the plasma interaction fail to capture all the important physics. Finally, we note that in other cases, with denser atmospheres, the ion pickup process can produce significant feedback on the fields and flows in the plasma interaction through momentum exchange. We return to this idea a little later in this chapter.

An additional departure of the Moon from an electrically passive obstacle occurs because remanent magnetism is present. This field, presumably a relic from some past lunar dynamo or other magnetizing events, was detected on the early low-altitude spacecraft as they orbited, and has since been mapped in great detail. Over much of the Moon, this magnetic field is too weak to have much effect on the solar wind. But when one of the more strongly magnetized regions is near the lunar limb, defined here as the edge of the wake caused by solar-wind absorption, it causes a small deflection of the flow and produces a weak compression of the local magnetic field and plasma density. These are referred to as limb compressions or limb shocks. More recent observations with sophisticated plasma spectrometers show modifications in the near-surface particle distributions consistent with loss cones and magnetic mirroring by the crustal fields. There are also hints of reconnection processes occurring as the external and crustal fields come in contact, and of surface features related to possible crustal field shielding of the local surface from the externally incident particles. As these represent only modest modifications of the overall plasma interaction described earlier, we leave it to the interested reader to seek out further discussions of such lunar interaction features in the literature.

As you can tell from the preceding discussion, quantitative treatments of flowing plasma interactions with even insulating bodies like the Moon are not simple exercises. Another class of challenge arises for bodies such as asteroids that are comparable to or smaller than the Debye length for the flowing plasma (~10 m in the solar wind). Under such conditions, the details of the individual plasma particle motions are also important, as is the electrical potential of the body surface, which can charge up if there is an imbalance of positive and negative charges incident on and emitted from the body. MHD fluid approximations become inadequate, and one is forced to deal with additional matters such as the properties of the surface materials. "Space weathering" (as opposed to "space weather"), where solid surfaces are altered by exposure to space particles and fields, is a subfield in itself. Fully kinetic (particle) treatments of flowing plasmas interacting with solid bodies have, up to now, fallen mainly in the realm of spacecraft-charging studies – where the details of the body shape and surface materials' properties are well known. Computational capabilities are now allowing the latest generations of researchers to confront larger-scale problems with such simulation tools, including approaches that include at least the ion kinetics in "hybrid" numerical models (see Chapter 7) of the lunar plasma interaction.

8.3 PLASMA INTERACTIONS WITH BODIES WITH ATMOSPHERES

8.3.1 Unmagnetized planets

An insulating body with an atmosphere residing in the darkness of space represents no more of an obstacle to the external plasma than does a bare insulating body, except for the fact that atmospheric gases are somewhat more prone to incident plasma sputtering and ionization than is a solid surface. Substantial ionization of the atmosphere (especially by sunlight) in the presence of a magnetic field in the flowing plasma produces a significant difference. As before, if the magnetic field is not steady, and can be considered frozen into the external plasma, an induced field results, as illustrated in Figure 8.4. In this case, we have added an ionosphere that is a conductor on the same hemisphere exposed to the incident plasma flow. This geometry describes the situation for the weakly magnetized planets, though not in general for weakly magnetized moons, which may also have atmospheres (as will be seen for the case of Titan in Section 8.3.2). Here the external magnetic field induces ionospheric currents that act to keep the field from penetrating into the body by generating a canceling field – somewhat analogous to the lunar core situation (see Figure 8.2). And, as for the lunar core, these currents persist as long as the magnetic field keeps changing its orientation (as it does in the solar wind). A steady field would eventually diffuse in on a time scale that depends on the ionospheric conductivity. Another possibility in such situations is that there may be partial field diffusion into the planet if the ionosphere is too weak to carry sufficient shielding current. This basic picture applies most directly to the solar-wind interaction with Venus, although it also has much relevance for Mars.

The basic details of the plasma interaction picture shown in the final panel of Figure 8.4 have been validated by observations at Venus. A bow shock is to be expected, because the solar-wind plasma with its frozen field is flowing at supermagnetosonic velocity toward a conducting (impenetrable) obstacle with a scale much larger than the solar-wind proton gyro-radius. (Recall that, for the lunar case, the conductor is deep inside the body so that the solar wind is not deflected before it is absorbed.) This circumstance allows one to calculate the position of this bow shock using fluid models of flow around a blunt obstacle. The results of one such calculation, from a hydrodynamic or gas-dynamic model of Spreiter and Stahara (1980), are shown in Figure 8.5. In this calculation, the position of the shock is sensitive only to the obstacle shape,

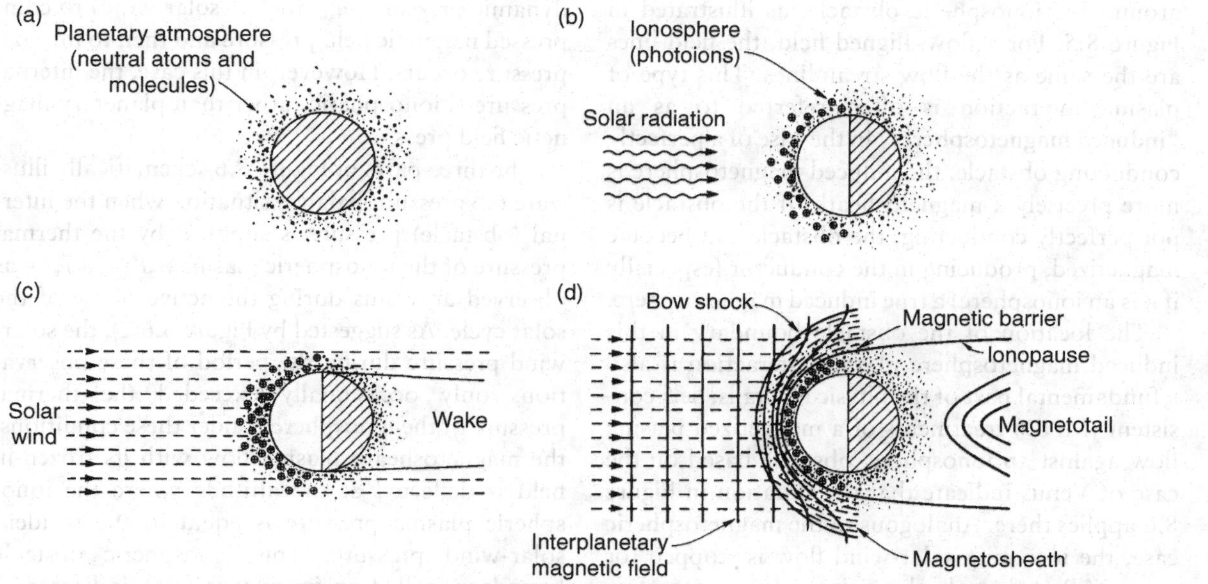

FIGURE 8.4. Illustration of the steps (a) to (d) that lead to the formation of an ionospheric planetary obstacle in a flowing plasma like the solar wind. Ionization by solar radiation, for example, is followed by diversion of the external plasma flow only if that flow is magnetized.

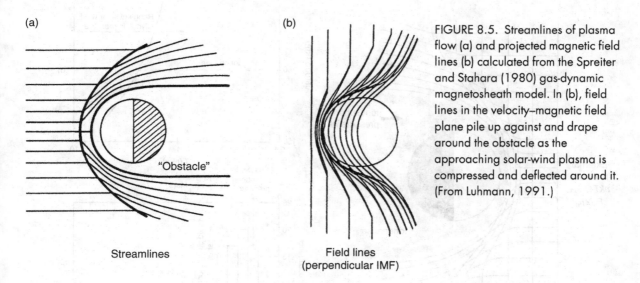

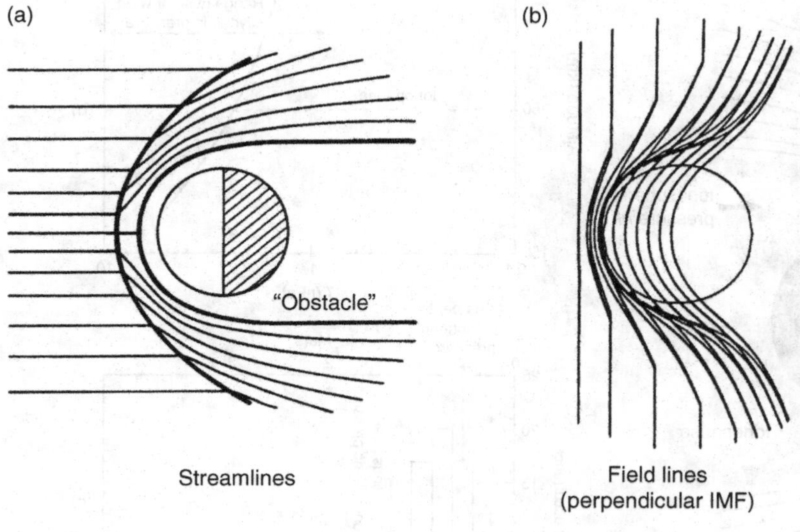

FIGURE 8.5. Streamlines of plasma flow (a) and projected magnetic field lines (b) calculated from the Spreiter and Stahara (1980) gas-dynamic magnetosheath model. In (b), field lines in the velocity–magnetic field plane pile up against and drape around the obstacle as the approaching solar-wind plasma is compressed and deflected around it. (From Luhmann, 1991.)

the upstream Mach number, and an assumed ratio of specific heats for the solar-wind plasma. The fluid equations that are solved are those of continuity, momentum, and energy. The solutions give the plasma density, velocity, and temperature throughout the region shown. The magnetic field in between the bow shock and the obstacle, the same "magnetosheath" as that found in the solar-wind interaction with the Earth's magnetosphere (see Chapter 7), is calculated separately from the velocity in this model, using the assumption that it is frozen in (i.e., that it satisfies $\partial \mathbf{B}/\partial t = -\mathbf{u} \times \mathbf{E} = \nabla \times (\mathbf{u} \times \mathbf{B}) = 0$). The field lines for a perpendicular upstream field drape around the ionospheric obstacle, as illustrated in Figure 8.5. For a flow-aligned field, the field lines are the same as the flow streamlines. This type of plasma interaction is often referred to as an "**induced magnetosphere.**" In the case of a perfectly conducting obstacle, the induced magnetosphere is more precisely a magnetosheath. If the obstacle is not perfectly conducting, the obstacle can become magnetized, producing in the conductor (especially if it is an ionosphere) a true induced magnetosphere.

The location of the obstacle boundary in this induced magnetosphere plasma interaction is also a fundamental part of the physical details. Self-consistent (MHD) treatments of a magnetized plasma flow against an ionospheric obstacle based on the case of Venus indicate that the situation in Figure 8.6 applies there. Analogous to the magnetospheric case, the incoming solar-wind flow is stopped (or stagnated) at the subsolar point, where a transformation of upstream pressure (initially dominated by dynamic pressure ρu_{sw}^2 for the solar wind) to compressed magnetic field pressure and then to internal pressure occurs. However, in this case, the internal pressure is ionospheric rather than planetary magnetic field pressure.

The three parts of Figure 8.6 schematically illustrate the pressure-balance situation when the internal (obstacle) pressure is supplied by the thermal pressure of the ionospheric plasma $n_e K(T_i + T_e)$ – as observed at Venus during the active phase of the solar cycle. As suggested by Figure 8.6(c), the solar-wind pressure during the period of these observations only occasionally exceeded the thermal pressure of the ionosphere. Under these conditions, the magnetosheath plasma flow with its frozen-in field is deflected at the altitude where the ionospheric plasma pressure is equal to the incident solar-wind pressure. This ionospheric obstacle boundary, called an **ionopause** (also indicated in Figure 8.4), has the average solar-maximum

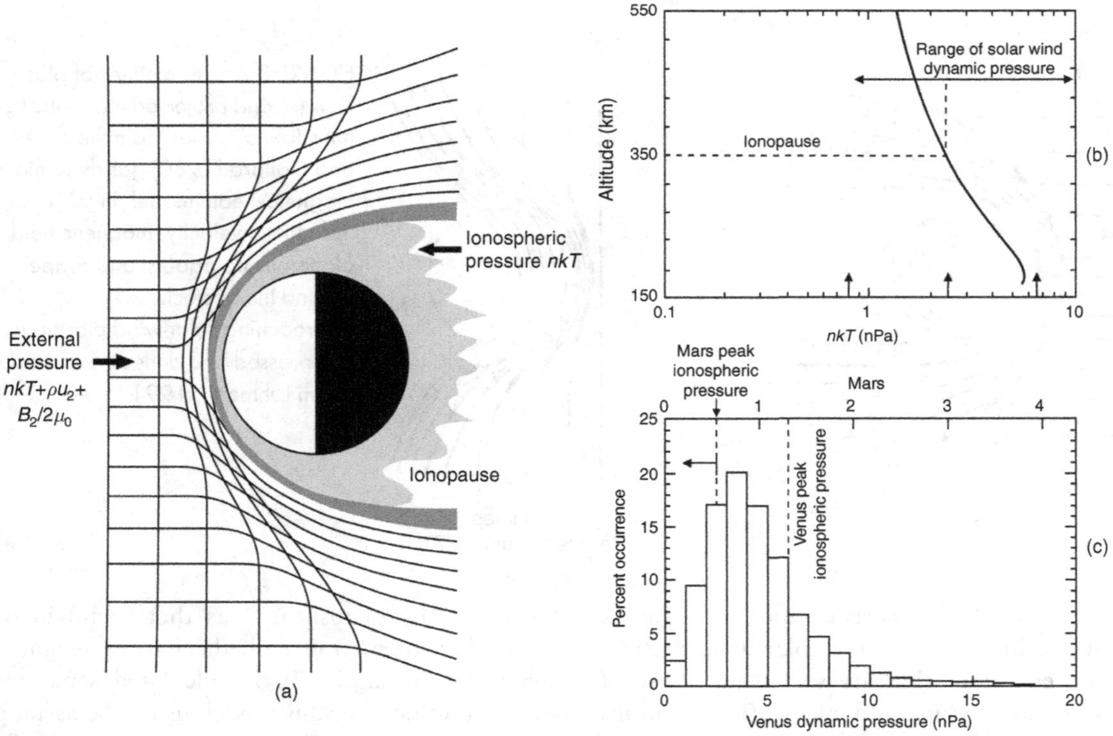

FIGURE 8.6. Illustration of pressure balance between the solar wind and the thermal pressure of the ionosphere which determines the height of the ionopause. The observed variability of the solar-wind pressure is indicated by the histogram in (c). ((a) After Luhmann, 1986; (b) and (c) after Luhmann *et al.*, 1987.)

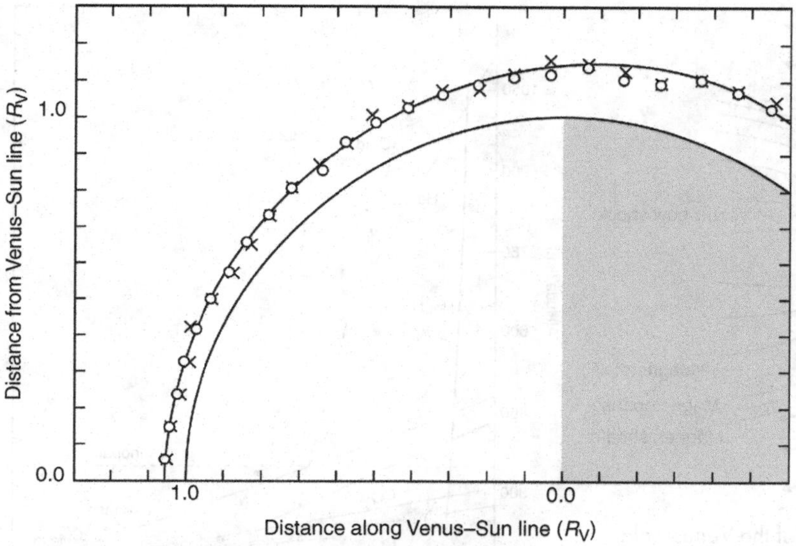

FIGURE 8.7. Venus at solar maximum ionopause location from two different ionospheric particle measurements on the Pioneer Venus orbiter: solid line from Theis, Brace, and Mayr (1980); circles from Knudsen, Miller, and Spenner (1982), compared with pressure balance ionopause (crosses) where plasma and magnetic pressures are equal. (After Phillips, Luhmann, and Russell, 1984.)

location at Venus shown in Figure 8.7. Later, we shall learn that the ionopause term is also used in circumstances where the induced magnetosphere picture is not so classic (Mars), but where induced ionospheric currents still make an important contribution.

Venus is a roughly Earth-sized body with a radius ~6050 km. The Venus ionopause obstacle to the solar wind is nominally at ~300 km altitude at the subsolar point, and flares to ~1000 km near the terminator – beyond which additional physics of the plasma wake must be considered. An important consideration is that this altitude is everywhere above the nominal exobase at Venus at ~200 km. (See Chapter 2 for a description of the exobase.) Thus the pressure balance occurs in a collisionless region of the upper atmosphere of Venus during the periods of moderate to high solar activity when these measurements were obtained. Calculations of the expected ionospheric properties of Venus during solar minimum indicate some changes to this picture that also apply to Mars, which usually has a weaker ionosphere pressure than incident solar-wind pressure. Recall from Chapter 2 that an ionosphere's density (and pressure) depend on the solar EUV flux, which is ~3–4 times higher on average when the Sun is active than when it is quiet. When the ionospheric thermal pressure is too weak or the solar-wind dynamic pressure is too strong, the ideal "collisionless" ionospheric obstacle interaction in Figure 8.6 does not occur. If the projected pressure-balance altitude (see Figure 8.6(b)) lies near or below the exobase, i.e., in the collisional ionosphere, the overlying magnetosheath magnetic field diffuses into the ionosphere and shares in providing the obstacle pressure perceived by the solar wind. In this picture, it is important to remember that the neutral upper atmosphere extends throughout this interaction region and into the solar wind, with additional effects of its own, described in the following.

Figure 8.8 compares the size of the solar-wind interaction region of nearly Earth-size Venus with that of Earth's magnetosphere. A normalization of planet sizes would lead to a similar picture for Mars, which at ~3390 km is about half the size of Venus. The proximity of the effective obstacle boundary at Venus to the planet surface is also apparent in the measured ionopause location (see Figure 8.7). The (moderate to high solar activity) Venus subsolar ionopause altitude is compared to the nominal altitude profile of the dayside Venus neutral ionosphere in Figure 8.9. It is apparent that, whereas the Earth's atmosphere lies protected deep inside its magnetospheric "bubble," the upper atmosphere of Venus (and also Mars) is regularly exposed to the passing solar-wind plasma. As a result, planetary atmosphere ions that are produced

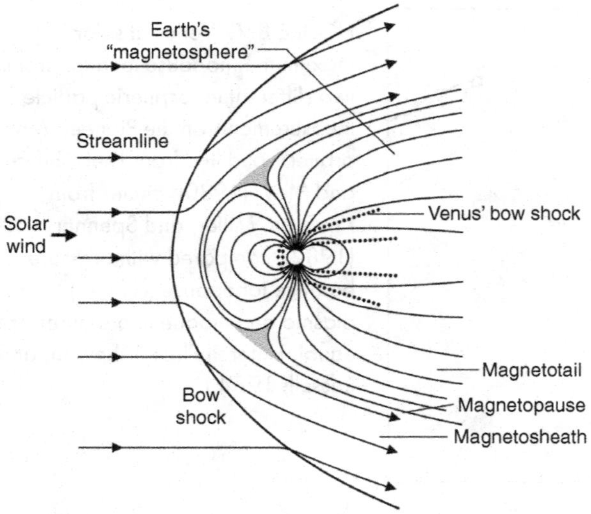

FIGURE 8.8. Comparison of the size of the Venus–solar-wind interaction region with that of the magnetized Earth. (After Luhmann and Brace, 1991.)

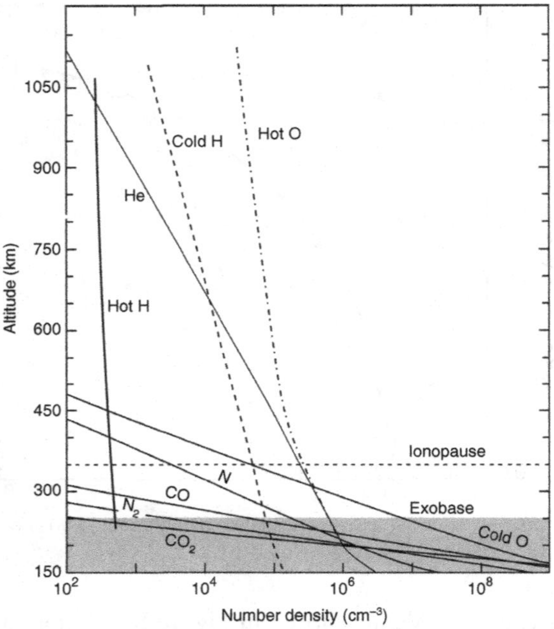

FIGURE 8.9. Altitude profile of the Venus neutral upper atmosphere showing exobase and ionopause relative locations at solar maximum. (Adapted from Nagy *et al.*, 1981.)

by sunlight, solar-wind electron impact ionization, or charge exchange with solar-wind protons in the neutral atmosphere above the ionopause can be picked up in the same manner as described for the lunar atmospheric ions (see Figure 8.3). The difference in this case is that the magnetosheath **u** and **B** must be used to evaluate the accelerating electric field close to the planet. As one can infer from the illustration of the gas-dynamic interaction in Figure 8.6, this modifies the simple picture shown for the lunar pickup ions in Figure 8.3 where, to first order, there is no departure from upstream solar-wind conditions. However, the general nature of the asymmetry of the atmospheric pickup ions is similar.

Oxygen atoms are the main constituent of the Venus upper atmosphere, as shown in Figure 8.9. Both Venus and Mars have mostly carbon dioxide atmospheres that transition to mainly oxygen and hydrogen in the region of their solar-wind interactions. The hydrogen at high altitudes is expected because it is the lightest constituent. In contrast, the existence of atomic oxygen is a special consequence of the photochemistry of carbon dioxide atmospheres that make primarily molecular oxygen ionospheres. The hot O dominating the high-altitude exosphere arises from **dissociative recombination** of electrons and O_2^+ ions. Upon neutralization, the O_2 molecule remains excited (often denoted as O_2^*). This energy is sufficient to split the O_2^* molecule into two suprathermal or "hot" oxygen atoms ($O^* + O^*$), which divide the excess energy between them in the form of motion. Atoms moving upward can extend the oxygen exosphere to thousands of kilometers at Venus. As a result, the pickup ions produced there are mostly O^+, with a small amount of planetary H^+. Figure 8.10 illustrates the typical gyro-radius of the atmospheric oxygen ions produced in the subsolar magnetosheath and upstream relative to the scale of the Venus obstacle. As at the Moon, the escaping O^+ pickup ions are asymmetrically distributed around the obstacle because these heavier ions gyrate into one hemisphere rather than escaping downstream. Associated O^+ **ion cyclotron waves** (see Chapter 12), another signature of pickup ions in addition to their direct detection, have not been observed near Venus by visiting spacecraft. (In contrast, proton pickup produces copious ion cyclotron waves at Mars, whose interaction is discussed later.)

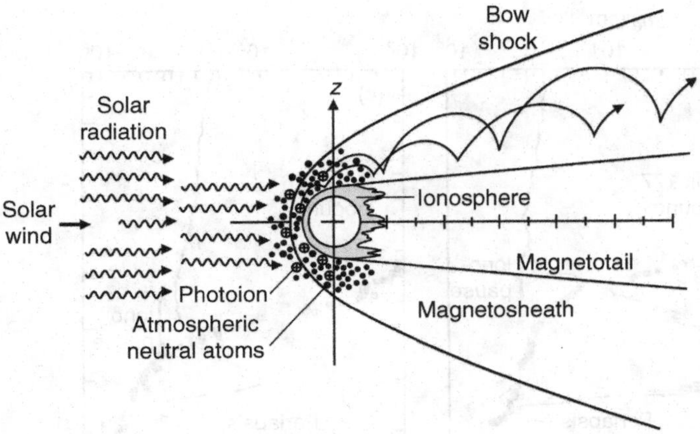

FIGURE 8.10. Illustration of planetary pickup ion trajectories at Venus. The cycloid sizes are approximately scaled for O^+ (oxygen is the main constituent of the Venus upper atmosphere, as shown in Figure 8.9.) (After Luhmann, 1990.)

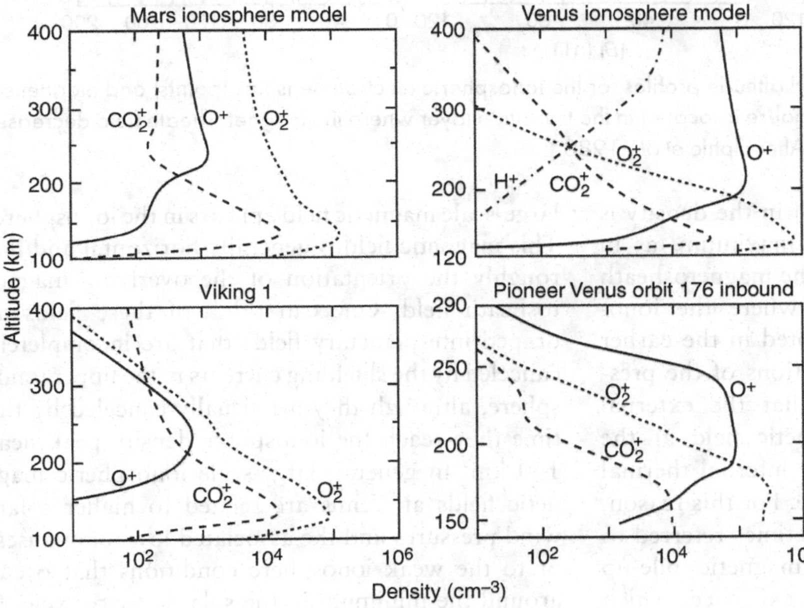

FIGURE 8.11. Comparison of modeled (top) and observed (bottom) ionosphere altitude profiles at Venus and Mars. The difference between observation and theory is due to the loss of the upper ionospheres to the solar wind. (After Shinagawa and Cravens, 1989; Shinagawa, Cravens, and Nagy, 1987.)

The consequence of the high-altitude ion-removal processes can be seen in the measured ionospheric profiles for both Venus and Mars in Figure 8.11, where they are compared with profiles modeled from their neutral atmospheres under the assumption that no ion loss mechanism operates other than the usual ionospheric recombination (see Chapter 2). The observed ionospheric densities are depleted at high altitudes, but they are relatively unaffected below a few hundred kilometers. Integrated over the lifetime of the planet (~4.5 billion years), this scavenging of planetary ions by the solar-wind-related ion pickup process can have an impact on atmosphere evolution at Venus and Mars. Additional loss of atmosphere may also occur by sputtering when precipitating pickup ions collisionally interact with the atmosphere near the exobase, resulting in energy transfer to some neutral atoms that may escape, although the ion itself is retained.

Let us now return to the ionospheric magnetic field, which of all these phenomena probably provides the deepest insight into how induced magnetospheres work. Figure 8.12 shows some altitude profiles of magnetic field and ionospheric electron density obtained at Venus during solar

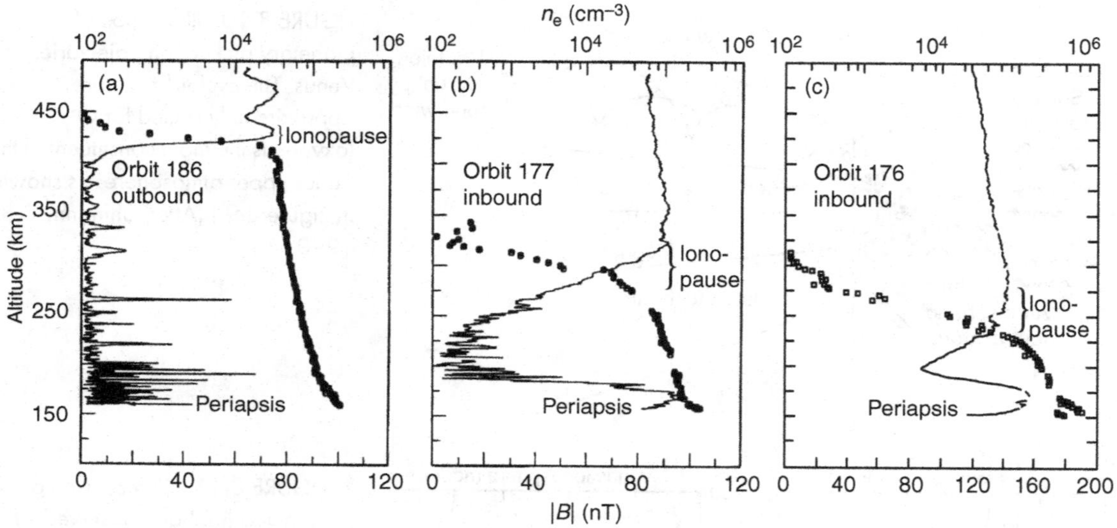

FIGURE 8.12. Examples of observed altitude profiles for the ionospheric electron densities (points) and magnetic fields (solid line) at Venus. The ionopause is located in the transition layer where the magnetosheath field decreases and the plasma density increases. (After Elphic *et al.*, 1980.)

maximum. The ionopause cutoff in the density is often clear and sharp at solar maximum, as in Figure 8.12(a). For such cases, the magnetosheath magnetic field stops abruptly where the ionospheric density rises. As was noted in the earlier discussion of Figure 8.6, evaluations of the pressures on both sides confirm that the external pressure of the piled-up magnetic field at the ionopause is about equal to the internal thermal (ionospheric) pressure just inside. For this reason, the inner magnetosheath is sometimes referred to as a "**magnetic barrier**" or "magnetic pile-up boundary" (see Figure 8.4). Its existence, which is not taken into account in the gas-dynamic magnetosheath model, implies that a slightly larger obstacle exists than the ionopause itself presents. For now, the small-scale structures in the magnetic field inside the ionosphere in Figure 8.12 will be neglected, as they contribute negligibly to its total pressure.

A consequence of the ionopause forming on the dayside, where the thermal pressure is equal to the incident solar-wind dynamic pressure (Figure 8.6), is that it gets lower when the solar-wind pressure gets higher. As mentioned earlier and illustrated in Figure 8.12, when it approaches the exobase at ~200 km altitude the ionopause thickens and a large-scale magnetic field appears in the ionosphere. This magnetic field is generally horizontal and has roughly the orientation of the overlying magnetosheath field. One can think of these fields as draped interplanetary fields that are incompletely canceled by the shielding currents in the upper ionosphere, although they are usually canceled by the time they reach the ionosphere density peak near 140 km. In general, large-scale ionospheric magnetic fields at Venus are related to higher solar-wind pressures and the associated low ionopauses, or to the weak ionosphere conditions that occur around the minimum in the solar activity cycle. It should be noted that the ionopause at Venus is observed to have a minimum altitude of about 225 km, just above the exobase altitude. After this point, any increases in solar-wind pressure at Venus manifest themselves in the compensating increases of the ionospheric magnetic field pressure.

This physical concept has been translated to a straightforward one-dimensional MHD treatment. The production of the ionospheric field at an unmagnetized planet can be thought of most simply as inward diffusion of the overlying magnetosheath magnetic field, but in the case of Venus the downward convection of the dayside ionosphere also contributes. The vertical

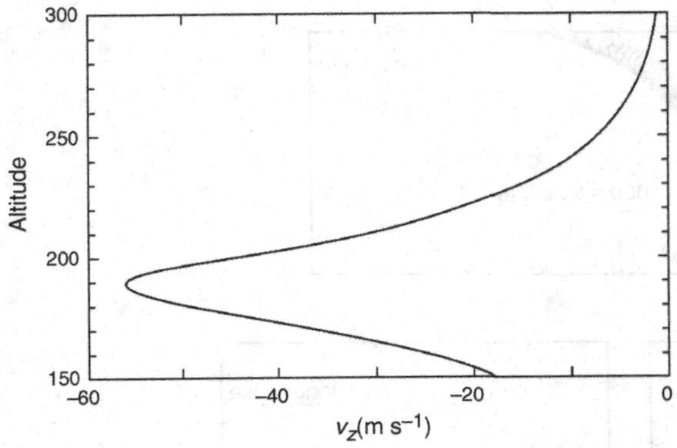

FIGURE 8.13. Ionospheric plasma drift velocity profile at Venus derived from a semiempirical model. (After Cravens *et al.*, 1984.)

ionospheric plasma drift, u_h, can be calculated from observations and the vertical steady-state momentum equation described in Chapter 2:

$$u_{\mathrm{h}} = \frac{1}{n_{\mathrm{e}} m_{\mathrm{i}} \nu_{\mathrm{in}}} \left[\frac{\partial P_{\mathrm{T}}}{\partial h} + n_{\mathrm{e}} m_{\mathrm{i}} g \right]. \tag{8.2}$$

Here the usual notations of Chapter 2 are used: electron density, n_e; ion mass, m_i; ion–neutral collision frequency ν_{in}; total plasma pressure P_T; gravity, g; and h is the altitude.

A result for Venus, from a semiempirical model of the ionosphere (Cravens, Shinagawa, and Nagy, 1984) assuming a negligible magnetic pressure, is shown in Figure 8.13. Although ionosphere is produced at all altitudes, it recombines most readily at lowest altitudes, where the collision frequencies are greatest. Thus the Venus ionosphere in this altitude range drifts downward. The quasi-steady-state ionospheric magnetic field has a boundary condition at the top controlled by the solar-wind magnetic field and dynamic pressure. To a first approximation the (horizontal) field, B, obeys the one-dimensional dynamo, diffusion/convection, or induction equation:

$$\frac{\partial B}{\partial t} = 0 = \frac{\partial}{\partial h} D \frac{\partial B}{\partial h} - (B u_{\mathrm{h}}), \tag{8.3}$$

where the diffusion coefficient D is given by

$$D = \frac{m_{\mathrm{e}}(\nu_{\mathrm{en}} + \nu_{\mathrm{ei}})}{n_{\mathrm{e}} e^2 \mu_0}. \tag{8.4}$$

Equation (8.3) is readily derivable from Maxwell's induction equation, $\partial \mathbf{B}/\partial t = -\nabla \times \mathbf{E}$, the ion- and electron-momentum equations, and Ampère's law (see Chapter 3), where m_e is electron mass and ν_{en}, ν_{ei} are electron–neutral and electron–ion collision frequencies, respectively. The collision frequencies ν_{en} and ν_{ei} for Venus, shown in Figure 2.22, are such that the diffusion coefficient is very small at high altitudes and large at low altitudes. Both convection and diffusion enable the field's downward penetration, and collisional diffusion minimizes its gradients. The collisions also cause significant current (and thus field) dissipation at the lowest altitudes. It is simplest to solve the equation for B numerically, as a time-dependent problem from some assumed initial state (such as zero ionospheric magnetic field with an upper boundary field representing the overlying magnetic barrier), until it converges to a steady solution. Some examples of the evolving solution are illustrated in Figure 8.14 for different upper boundary conditions characteristic of varying altitudes of the dayside ionopause of Venus. The same figure shows some comparable observations. Because the ionopause typically is lowest in the subsolar region, the large-scale ionospheric magnetic field is more frequently present there.

Two caveats relating to this downward diffusive "field conveyor belt" picture must be given here, one of which relates to the neglect of horizontal motions in the ionosphere. The upper ionosphere of Venus around solar maximum is observed to convect antisunward, as illustrated in Figure 8.15, mainly because of the large day to night pressure gradients in the ionospheric plasma. This antisolar flow can also be inferred from the pressure gradients calculated from the global plasma density and temperature measurements. In the one-dimensional model only dissipation removes the field, but with horizontal convection there is another alternative. The full three-dimensional problem of the ionospheric field structure has been solved within the framework of global MHD models of the Venus–solar wind interaction. However, these models have yet to be fully compared against observations, and are also made difficult by the need for high spatial resolution and the dependence of the solutions on the assumed lower boundary conditions on the

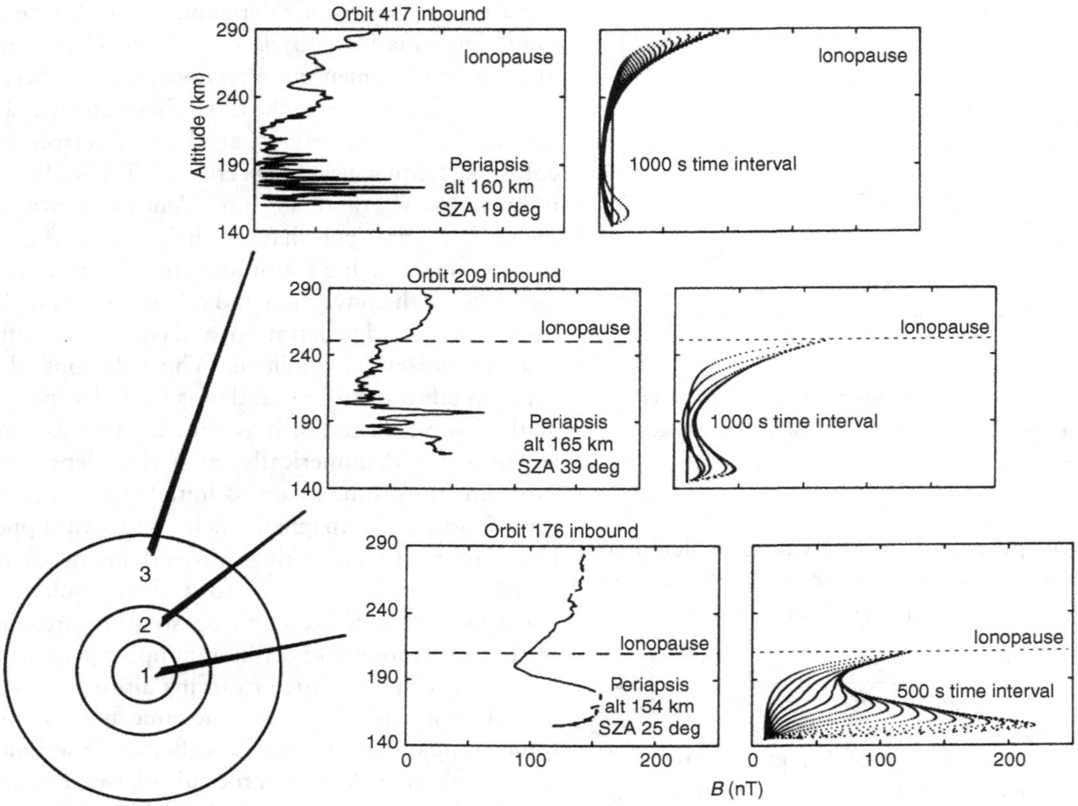

FIGURE 8.14. Examples of solutions of the diffusion/convection equation for the ionospheric magnetic field given in the text, compared to observations. These magnetic profiles are determined by the plasma velocity profile shown in Figure 8.13 and the altitude profile for the collisional diffusion coefficient, together with the upper boundary condition on the field at the ionopause. The "bull's eye" diagram represents the dayside of the planet as seen from the Sun. The ionospheric magnetic field usually is largest in the subsolar region, where the ionopause is lowest and the overlying magnetosheath field is strongest. SZA = solar zenith angle. (After Phillips *et al.*, 1984.)

field. Nevertheless, from the good agreement of even the simple one-dimensional treatment above with observations at Venus, the major physical process of ionospheric magnetization appears to be understood. The same physical process is also relevant at Mars, where the ionosphere is typically in a state of relatively weak pressure, and thus should nearly always be magnetized. However, the effects of Mars' crustal magnetic fields complicate the application of this simple picture to its ionosphere. For Mars, comparisons of models combining the crustal and induced fields to the relevant *in situ* observations are still under way. We briefly return to the Mars case at the end of this section.

A fascinating phenomenon of ionospheric magnetization that is not as well understood is illustrated by the data in the first panel of Figure 8.12(a). The small-scale (less than a few tens of kilometers) magnetic field structures observed inside of the dayside ionosphere of Venus appear when the large-scale field is absent. Their vector properties can be described by field lines that are twisted around an axis like ropes made of smaller strands. One model for the **flux rope** structures, and their hypothetical ionospheric distribution, is illustrated in Figure 8.16. The problem with Figure 8.16 is that the individual rope axes do not appear to be parallel or even horizontal. Especially at low altitudes, the ropes seem to be kinked or twisted. Their detailed characteristics

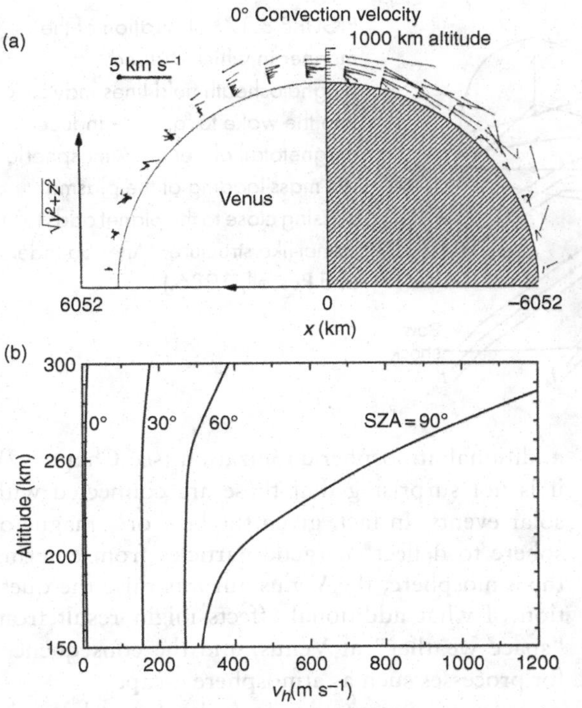

FIGURE 8.15. Observed antisolar flow in the ionosphere of Venus. SZA = solar zenith angle. (After Knudsen *et al.*, 1980.)

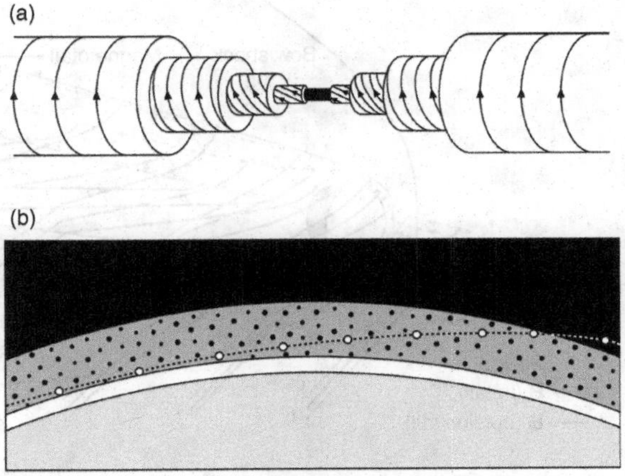

FIGURE 8.16. The inferred configuration of magnetic field lines in a Venus ionospheric flux rope (a) and their distribution in the ionosphere (b). (After Russell and Elphic, 1979.)

also change with solar zenith angle, with the classic structures mainly in the dayside. They may form by an interchange instability near the subsolar ionopause in which a small portion of the magnetic barrier flux is pulled into the ionosphere by the curvature force of the draped field there. Velocity shears due to the horizontal flows in the ionosphere then twist this flux into ropes. While they are clearly affected by ionospheric dynamics, it is not as clear that they affect the dynamics. Since magnetic ropes are a common phenomenon in solar system plasmas, unraveling the physics behind their curious nature is of some interest. We may learn some very basic information about ionospheric plasma interactions and space plasmas in general. Similar features have been observed in the weakly magnetized regions of the martian ionosphere. Some larger twisted field structures are also present in Titan's ionosphere. Crustal magnetic fields may play a role in their formation at Mars, for reasons described below.

When the pickup of ionospheric ions by the solar wind was described earlier, we ignored the reaction of the process on the ambient flowing plasma and magnetic field. Momentum must be conserved so when **mass loading** by these locally produced ions occurs, the plasma slows down and the stressed magnetic field results in a current. In a fully self-consistent treatment of the problem, all of these effects are automatically taken into account. For example, the MHD continuity equation gets a source term, and, if the neutrals are moving before they are ionized, so does the momentum equation.

Observations, and also global numerical simulations with atmospheric ion production, show that the pickup ions contribute to the formation of an **"induced" magnetotail** in the plasma flow wake at Venus, as illustrated in Figures 8.4 and 8.17. A defining characteristic of induced magnetotails is the double-lobed fields (sections with fields pointing toward and away from the Sun, divided by a current or plasma sheet) that are an extension of the draped magnetosheath fields (see Figure 8.17). Furthermore, the geometry of the fields tells us there are $\mathbf{j} \times \mathbf{B}$ forces on the bulk plasma acting in the antisolar direction in the regions of sharp field draping, especially at the "draping poles" at the flow terminator and in the region of the central wake. Note that the draping poles are not necessarily at the geographic poles because the interplanetary field has orientations

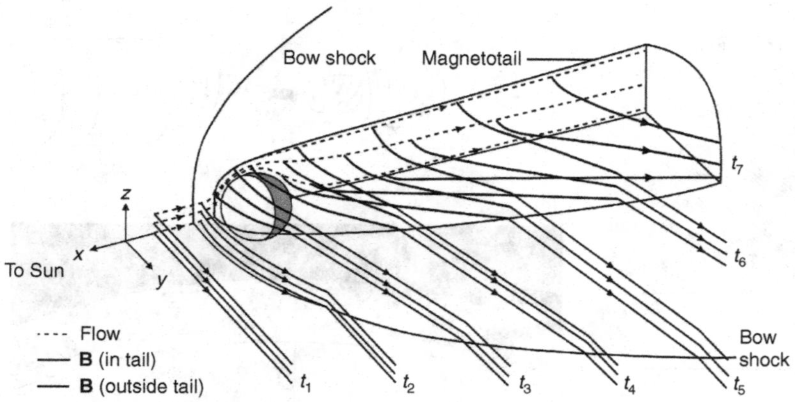

FIGURE 8.17. Illustration of the manner in which draped magnetosheath field lines may sink into the wake to form the induced magnetotail of Venus. Atmospheric ion mass-loading of the plasma passing close to the planet adds to the comet-like structure. (After Saunders and Russell, 1986.)

other than parallel to the equator. Ionospheric plasma is found in the region of the plasma sheet separating the induced magnetotail lobes. This planetary plasma would be expected there if the **j** × **B** forces of the draped polar fields pull ionospheric plasma into the wake. The pickup ions may also contribute to the plasma sheet, but in a more asymmetric way (Figure 8.10 gave an idea of the Venus O^+ gyro-radius scaling for upper atmosphere pickup sites). The tail current sheet in induced magnetotails is also a location where magnetic reconnection or merging can occur. Plasmoid-like structures and x-points (see Chapter 9 for a discussion of these in Earth's magnetotail) have been inferred from the Venus wake observations. Some of the draped ionospheric fields can wrap around the planet like a belt, pinching off the tail portion. It is not at present certain how much additional planetary ion removal occurs in connection with the formation and evolution of these structures. In any case, an analogy can be made between the induced magnetotails of these weakly magnetized planets and cometary tails, to be discussed later in this chapter.

Given the induced nature of the solar-wind–Venus interaction, one would not expect to find auroral activity. However, both patchy nighttime ultraviolet and spatially unresolved auroral green-line emissions have been detected there in association with solar activity – especially the kind that produces solar-wind pressure enhancements and/or energetic particles (see Chapter 5). As auroral emissions are, by definition, excited by suprathermal particle "precipitation" into atmospheres that excites electrons in molecules and produces additional atmospheric ionization (see Chapter 2), it is not surprising that these are connected with solar events. In fact, given the lack of a magnetosphere to deflect energetic particles from entering the atmosphere, the Venus auroras raise the question of what additional effects might result from "space weather" at Venus, and the consequences for processes such as atmosphere escape.

A few comments are in order here about the application of this overall "induced magnetosphere" picture to Mars. At Mars, crustal magnetic fields contribute substantially to the solar-wind pressure balance, especially when the strongest fields – located in the southern hemisphere in a restricted longitude sector – are on the dayside. For this reason, the solar-wind boundary there is sometimes more like a magnetopause. The non-uniform distribution of the crustal fields produces protrusions on the Mars–obstacle boundary that extend out into the magnetosheath above the nominal ionospheric pressure-balance boundary (Figure 8.18). Observations suggest how the Mars plasma interaction changes as the crustal fields rotate with the planet. When on the dayside, the strongest crustal fields seem to protect the ionosphere they cover from solar-wind scavenging processes there. In other locations that are unshielded, penetration of overlying interplanetary fields can be seen, similar to the situation in the ionosphere of Venus. (The larger distance of Mars from the Sun and its thinner atmosphere typically result in an ionosphere thermal pressure that is too weak to hold off solar-wind pressure by itself.) Partial reconnection of the crustal fields with the overlying draped interplanetary fields means that fields connected to

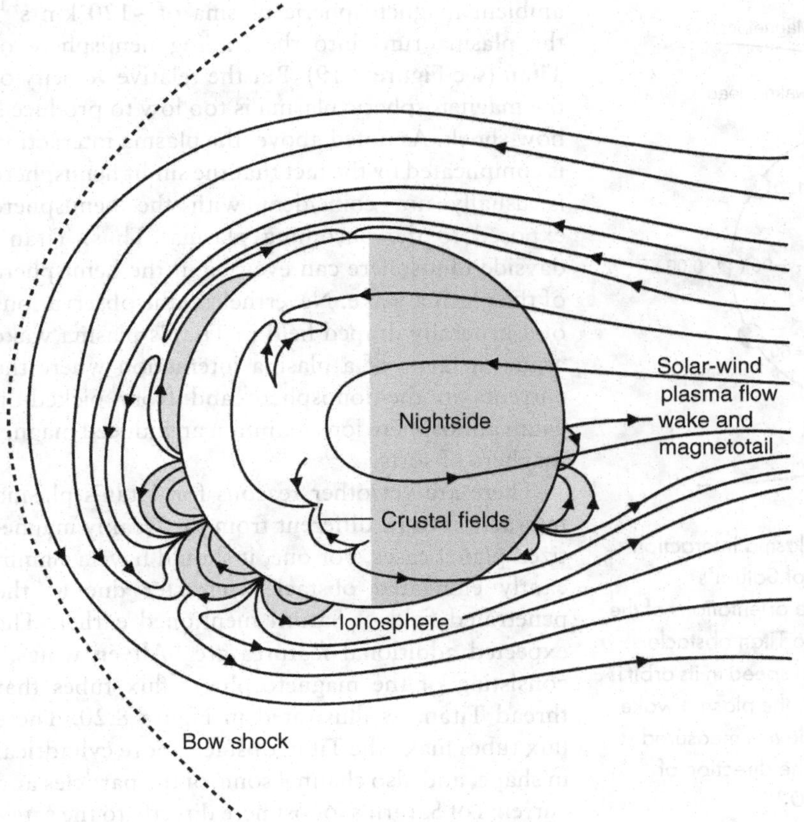

FIGURE 8.18. Illustration of the solar-wind plasma interaction with Mars, where both an ionosphere and remanent crustal magnetic fields define the obstacle. The plasma interaction includes the induced aspects of the Venus interaction, but also has the equivalent of "mini-magnetospheres". The crustal fields can merge or reconnect with the draped interplanetary field where they have antiparallel components – just as for a planetary magnetosphere. This produces regions of magnetically "open" and "closed" fields in the obstacle boundary, and also in the magnetotail in the solar-wind wake. Lunar remanent fields interact with the solar-wind fields in a similar way, but are not additionally affected by induced fields in an ionosphere.

the crustal sources contribute to the field in the inner magnetosheath and martian magnetotail. Just how much of the martian magnetotail is connected to the crustal fields will depend on many factors, including the solar-wind pressure, the ionospheric pressure, the draped external field orientation, and the position of the crustal fields relative to the Sun. Compared to Venus, this complicates the Mars case considerably. For example, like Venus, Mars has auroras that have been seen from orbit at ultraviolet wavelengths. However, at Mars the presence of the strongest crustal fields controls their occurrence, together with season and interplanetary conditions. These weakly magnetized planet auroras are still revealing their causes and consequences. Measurements from Mars Express and now the MAVEN mission are providing major observational resources for this and other Mars plasma interaction topics.

8.3.2 Unmagnetized moons

Saturn's moon, Titan, is another interesting example of the interaction of a magnetized flow with an atmosphere and ionosphere. At ~2575 km radius, in between the Moon and Mars in size, it exhibits some features of the weakly magnetized planet–plasma interactions, but with distinctive elements of its own. Like Venus, Titan has a substantial atmosphere, although it is dominated by nitrogen and methane instead of carbon dioxide. Titan is located at about 20 Saturn radii (R_{Sat}) from its host body, which means it is typically orbiting inside Saturn's magnetosphere. As a result, there is an external field that does not average to zero over time, and the oncoming plasma flow is mainly consistent with nearly corotating magnetospheric plasma. The importance of this combination for the **Titan–plasma interaction** is that the Titan–Sun

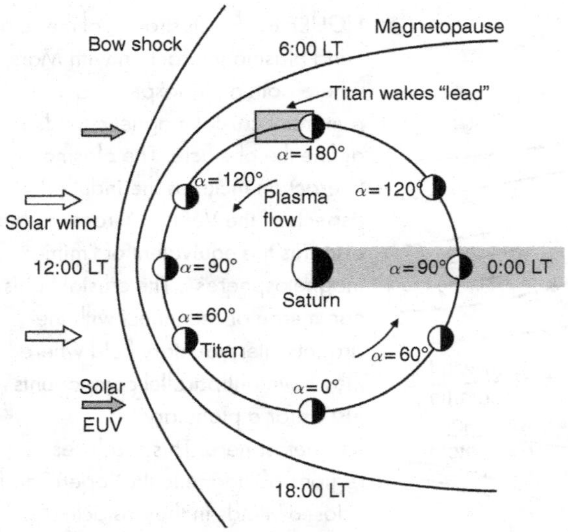

FIGURE 8.19. The setting of Titan's plasma interaction (not to scale) in its $20R_{Sat}$ orbit inside of Saturn's magnetosphere. The changing relative orientations of the sunlit and upstream hemispheres of the Titan obstacle distinguish it. Because Titan's keplerian speed in its orbit is less than the corotating plasma speed, the plasma wake leads Titan in its orbital path. The angle α is measured between the direction to the Sun and the direction of motion of Titan. (After Blanc *et al.*, 2002.)

axis and the plasma flow direction are usually not coincident (see Figure 8.19).

The consequences of Titan's environment are unique compared to the cases discussed here so far. First, there is likely durable penetration of Titan itself by the intrinsic dipole field of Saturn, where it has resided for as long as Saturn's field has been maintained. Even though the magnetospheric field at Titan is distorted into a magnetodisk by Saturn's icy satellite and ring-related water-product plasma sheet (see Chapter 12), and Titan oscillates above and below the related Saturn plasma sheet with season, a Saturn dipole component, locally pointing northward at ~5 nT strength, generally remains. Titan has been found to have an ionosphere of mainly solar photoionization origin, somewhat surprising because of both Saturn's location at ~10 AU from the Sun and the presence of magnetospheric ionizing particles. As its keplerian orbital speed around Saturn is significantly exceeded by the near-corotation speed of the ambient magnetospheric plasma of ~120 km s^{-1}, the plasma runs into the trailing hemisphere of Titan (see Figure 8.19). But the relative velocity of the magnetospheric plasma is too low to produce a bow shock. As noted above, the plasma interaction is complicated by the fact that the sunlit hemisphere is usually not coincident with the hemisphere exposed to the oncoming plasma. Thus, Titan's dayside ionosphere can even be in the hemisphere of the plasma wake. Nevertheless, the observations of a generally draped field in Titan's plasma wake argue in favor of a plasma interaction where the currents in the ionosphere and from picked-up Titan atmosphere ions maintain an induced magnetosphere of sorts.

There are yet other reasons for Titan's plasma interaction to be different from the weakly magnetized planet cases. For one, it should have a significantly elongated obstacle character due to the penetrated field of Saturn mentioned earlier. The expected additional features are "**Alfvén wings,**" consisting of the magnetospheric flux tubes that thread Titan, as illustrated in Figure 8.20. These flux tubes make the Titan obstacle more cylindrical in shape, and also channel some of the particles and currents of Saturn's ionosphere directly to the satellite, and from Titan's ionosphere to Saturn. For another, there are potentially important effects from the often water-group ion-dominated external plasma. The heavy plasma ions in the incident flow have significant energies from their own pickup origins, making their gyro-radii important in the incident flow interaction. Because the magnetic field at the orbit of Titan is only about 5 nT, both the mass 16–17 amu (O^+, OH^+) incident plasma ions and the mass 14–17 amu and 28 amu (N^+, $N2^+$, $CH4^+$) atmospheric pickup ion gyro-radii matter. In addition, Titan is sometimes within the solar-wind plasma in Saturn's magnetosheath or upstream when the solar-wind pressure is high enough to push Saturn's subsolar magnetopause inside Titan's orbit (and Titan happens to be there). Although these are relatively infrequent occurrences, they introduce some interesting time-dependent external conditions. Finally, Titan's nitrogen and methane atmosphere is also particularly extended, with an exobase at around 1200 km altitude, almost half a Titan radius. At the same

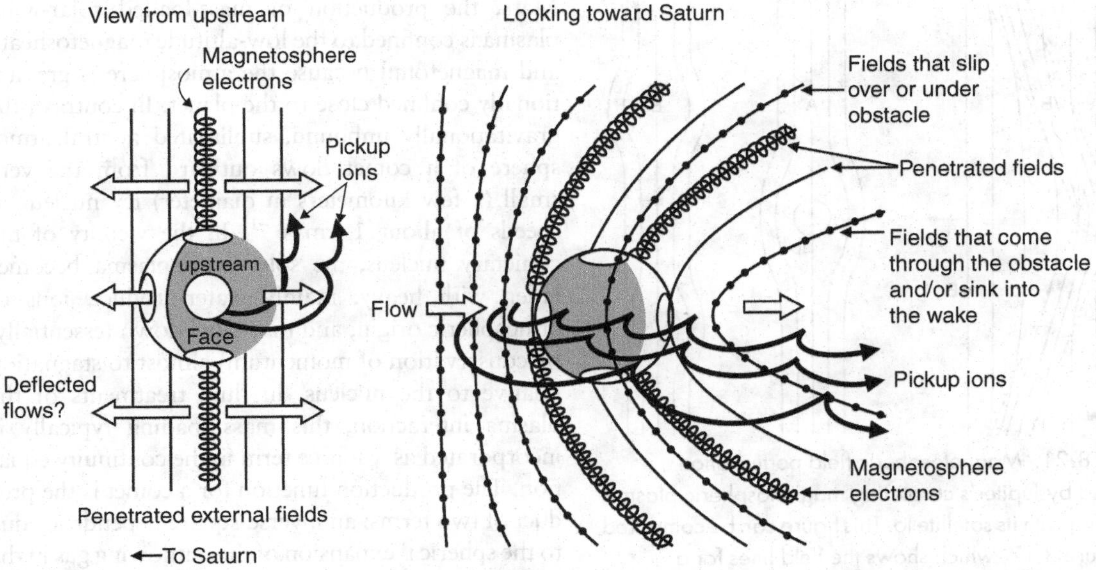

FIGURE 8.20. Saturn magnetospheric plasma interaction with Titan in the case where the external field is that of Saturn's dipole at $20R_{Sat}$ and the sunlit face of the Titan obstacle (including its main ionosphere) is not the incident approximately corotating plasma face – a common occurrence for the Titan case. Titan must also have absorbed Saturn's field over time, giving the obstacle an elongated shape and some effective magnetic field of its own. (After Luhmann *et al.*, 2012.)

time, the Titan ionosphere is also significantly extended, with an ionopause around 1500 km. Thus the obstacle "interior" is somewhat distinctive compared to the cases of Venus and Mars, with the atmosphere/ionosphere occupying a large fraction of its volume. The effects of all these differences are still under investigation with both measurements and models.

Some of the considerations that apply to Titan also apply to Jupiter's moon Io. Io has a much thinner and more variable sulphur dioxide-dominated atmosphere derived from a combination of sublimation of sulfur dioxide frost on its surface, magnetospheric energetic particle sputtering of its surface, and outgassing by its volcanoes. Like Titan's atmosphere, Io's atmosphere is ionized by both sunlight and magnetospheric particle impact. Io also is subjected to a submagnetosonically flowing magnetospheric plasma in its close-in ~6 Jupiter radius orbit within Jupiter's huge magnetosphere (see Chapter 12). In addition, the satellite's outgassing and sputtering rates are sufficient to produce the Io torus (see Chapter 12), a co-orbiting cloud of sulfur, oxygen, and some other gases (e.g., sodium and potassium) that are also ionized, and, like Saturn's water disk, greatly affect the composition of the corotating magnetospheric ions impacting Io. In essence, Io determines its own incident plasma properties.

The Io interaction is somewhat simpler to describe than Titan's because the relatively strong, nearly dipolar field of Jupiter in its inner magnetospheric location makes the ion gyro-radius effects much less important. A simple MHD fluid approximation to the Io interaction treats Io as an absorbing body with a source term to represent the production of heavy ionospheric ions near the body. It is found that the magnetic field can be well reproduced by a model neglecting the weak ionospheric mass source, but assuming a partially conducting body. Figure 8.21 illustrates the solution. Here the Alfvén wings, produced by the external field disturbance caused by Io and traveling along the magnetospheric field lines at the Alfvén

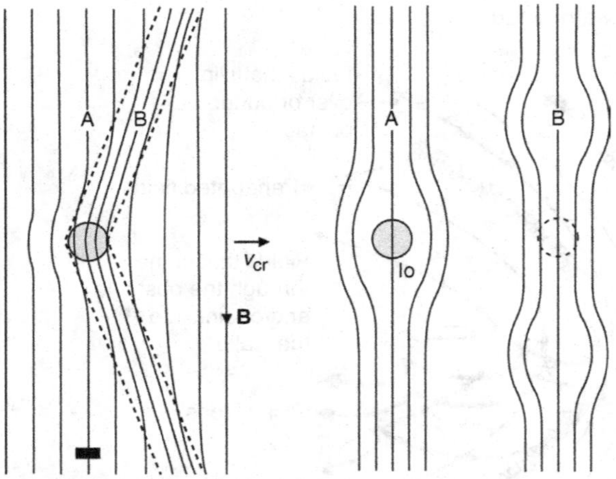

FIGURE 8.21. Magnetospheric field perturbation produced by Jupiter's corotating magnetospheric plasma interacting with its satellite Io. This figure can be compared with Figure 8.17, which shows the field lines for a supersonic interaction with an unmagnetized body with an atmosphere. (After Southwood *et al.*, 1980.)

velocity v_A, provide the main obstacle to the incident plasma flow. The angle that the Alfvén wings make with the ambient flow at the relative speed $(v_{cr} - v_k)$ (where the two speeds are the corotation (v_{cr}) and keplerian (v_k) orbital speeds) is given by $\theta = \arctan(v_A/(v_{cr} - v_k))$. It is expected that any additional field draping from pickup ions at Io will be much less severe than at Titan because of its less extensive atmosphere. It is worth mentioning that obstacles like Venus and Mars (see Figure 8.17) have magnetosheaths and induced tails, not Alfvén wings, because the solar-wind plasma flow speeds greatly exceed the Alfvén and the fast magnetosonic wave speeds in the solar wind.

8.3.3 Comets and other outgassing small bodies

Perhaps the most extreme example of an atmosphere interacting with a flowing plasma occurs in the case of comets. Comets near the Sun can have huge atmospheres compared to their small (~few kilometers scale) solid body, or nucleus. These atmospheres, produced by the increasing sublimation of ices from the cometary surface as it approaches the Sun, undergo much evolution with heliocentric distance. For comets, the term mass loading, introduced earlier, takes on particular importance. At Venus or Mars, the production of mass-loaded solar-wind plasma is confined to the low-altitude magnetosheath and magnetotail because the atmosphere is gravitationally confined close to the planet. In contrast, the gravitationally unbound, sublimated neutral atmosphere of a comet flows outward from the very small (a few kilometers in diameter) icy nucleus at speeds of about 1 km s $^{-1}$. In the vicinity of the cometary nucleus, the solar-wind plasma becomes laden with heavy, mainly water-product, ions of atmospheric origin, and may slow down (essentially, by conservation of momentum) almost to stagnation relative to the nucleus. In fluid treatments of the plasma interaction, this mass loading typically is incorporated as a source term in the continuity equation. The production function for a comet is the product of two terms: an inverse-square dependence due to the spherical expansion of the outflowing gas in the near-vacuum of space, and an exponential decay due to the loss of gas to ionization processes such as photoionization. Thus the source term is given by

$$Q = \frac{Q_0}{r^2} \exp\left(-\frac{r}{u\tau}\right), \tag{8.5}$$

where Q_0 is the gas production rate, u is in this case the neutral gas outflow velocity, r is the distance from the nucleus, and τ is the ionization time.

The huge ionosphere that is produced by the ionization of the cometary gas (one model of which is illustrated, together with a model neutral atmosphere, in Figure 8.22(a)) adopts the outward velocity of the expanding neutrals and may create a planet-sized cavity in the plasma flow around the nucleus (Figure 8.22(b)). The cavity, or obstacle in this case, is created by the dynamic pressure of the outflowing cometary plasma, instead of by the ionospheric thermal pressure. The boundary of this cavity is a tangential discontinuity called the contact surface. However, most of the neutral atmosphere extends outside of this pressure-balance boundary (Figure 8.22(a)). In the planetary ionosphere case, the solar-wind plasma was slowed and deflected close to the obstacle (see Figure 8.7), but in this case the mass loading by the extended atmosphere substantially slows down the solar-wind plasma flow long before the obstacle of the contact surface is encountered. In fact, because of ion pickup and charge exchange along the way, the

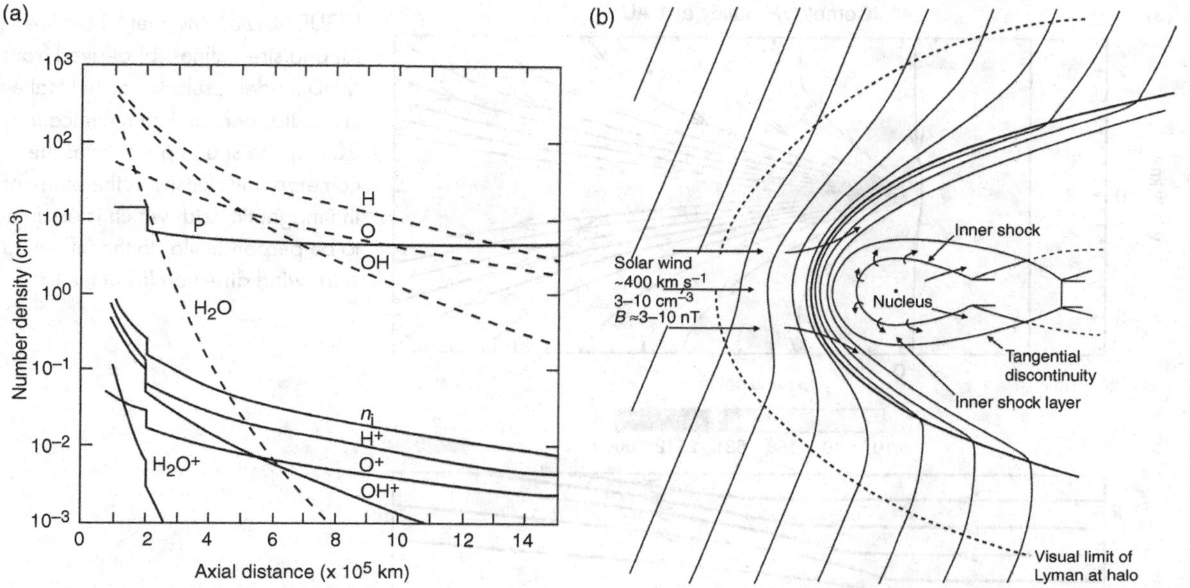

FIGURE 8.22. (a) Altitude profile of a model comet atmosphere and ionosphere. (After Ip and Axford, 1982.) (b) Inner region of the solar wind–comet interaction, illustrating the cometary gas outflow and possible features, such as internal shocks, that may occur depending on the relative speed of the incident flow and the magnetic field strength. In cometary interactions the dynamic pressure of the outflow is an important part of the obstacle (tangential discontinuity or contact surface) that forms when the comet is close to the Sun.

composition of the flowing plasma becomes primarily cometary in origin by the time the incident flow reaches the contact surface. The boundary above the contact surface where this transition takes place has been called the cometopause, following its first observation in a flyby of comet Halley. The extended region where the flow is slowed down by the mass loading is permeated by draped interplanetary magnetic field, which in this case is draped because the overall plasma flow velocity has decreased although there has been little deflection. Figure 8.22(b) illustrates the complement of processes occurring in the solar-wind interaction with a comet. The approximate visual limit of the Lyman-α halo indicates the extent of the hydrogen component of the neutral atmosphere of the comet for comparison. A bow shock is also shown, although it is generally weaker than that found near the planets because it occurs in the mass-loaded plasma where the flow has already slowed down.

Comets have been modeled using self-consistent MHD simulations with a source term in the continuity equation. As at the planetary obstacles, the incident flow is in this case supermagnetosonic, but the solid body of the nucleus itself is considered so small as to be negligible compared to the outflowing cometary atmosphere. Still, the basic equations are the same. The draped magnetic field and nearly undeflected plasma streamlines from a model of the solar-wind interaction with comet Halley are shown in Figure 8.23.

The weak bow shock is practically invisible in these results. Again, note the contrasts with the gas-dynamic planetary magnetosheath model with no mass loading shown in Figure 8.6. In that case, the shock and flow deflection at the obstacle surface are the key features of the interaction and the features that produce the draped magnetic field. As mentioned above, the sources of mass loading by the Mars or Venus atmospheres are largely confined to the inner magnetosheath (and wake, if a more self-consistent global MHD treatment is carried out). For the comet, the extended mass-loading region plays the major role in defining the features of the solar-wind interaction. On the other hand,

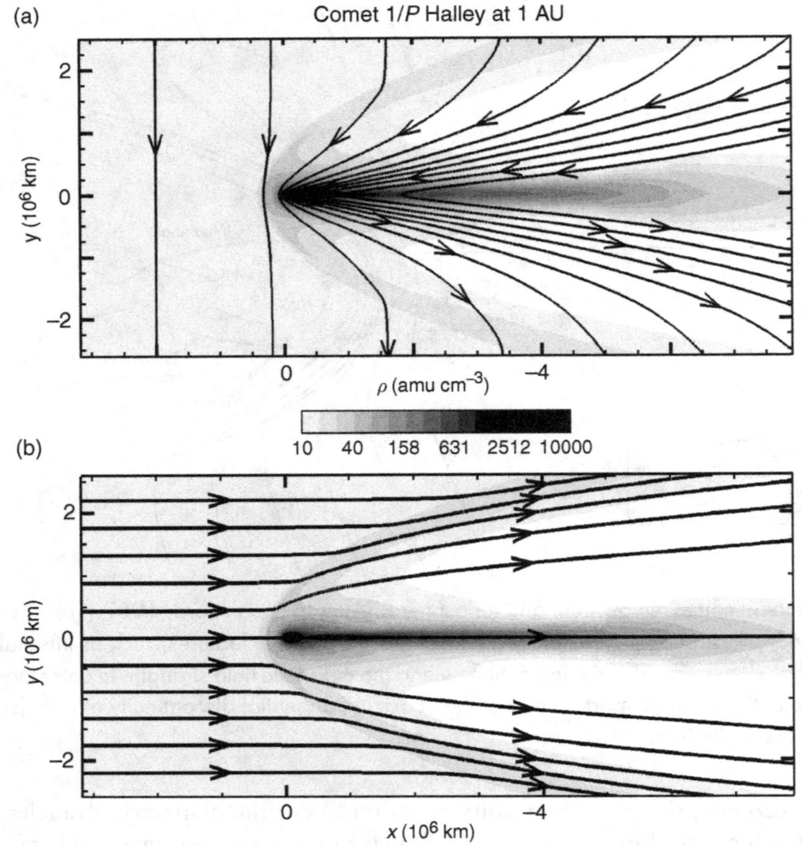

FIGURE 8.23. Magnetic field lines (a) and streamlines (b) derived from MHD model results for comet Halley (Y. D. Jia, personal communication, 2014.) The shading indicates the cometary ion density in the plane of the magnetic field, which is assumed to be perpendicular to the inflowing solar-wind direction (from the left).

the magnetic field of an induced planetary magnetotail inferred from observations (Figure 8.17) closely resembles the draped field of a comet. Future modeling of the planetary cases and comets should allow us to compare the physical processes important for both. One can argue that these fluid models cannot accurately model these systems, because the behavior of individual particles and the details of ion chemistry are neglected. Nevertheless, in the case of both comets and planetary bodies, many aspects of the observed gross properties of the plasma and field behavior are consistent with fluid-model results.

One last cometary topic of relevance here is the well-known duality of observed comet tails. The tail we have been describing so far is the ion tail, whose orientation relative to the Sun is defined by the nearly radially outward solar-wind flow, with some aberration due to the comet's own orbital motion (which may be significant near perihelion). The other comet tail is the gas and dust tail that is subject mainly to solar gravitational and radiation pressure forces, and so more closely follows the comet's orbital path. Both the cometary coma and tail(s) generally brighten as the comet nears the Sun and produces more gas, dust, and ions. An interesting phenomenon called tail disconnection is sometimes observed in the ion tail when a disturbance or current sheet in the solar wind passes through it. Similar transient changes may also occur at Venus and Mars, as a form of space weather response of their plasma interactions.

Several other weakly magnetized bodies with extended exospheres like comets include Enceladus, a small (~250 km radius) icy satellite in the inner magnetosphere of Saturn, and Pluto. Pluto is an icy dwarf planet of the solar system orbiting the Sun at ~33 AU. It has a cold, sublimated nitrogen- and methane-dominated atmosphere that goes through cycles of density due to its highly eccentric orbit. Because of its distance and small size (~1100 km radius, similar to the Moon), many details of its cold atmosphere and its solar-wind interaction are unique. In particular, there are complications associated with changes in the solar wind at this distance from the Sun (see Chapter 5). In addition to the very low densities of the original solar-wind plasma, the locally weak interplanetary field, and the infiltration and ionization of the interstellar gases in its environment make Pluto a case requiring significantly more investigation to describe properly. Recently updated knowledge from the New Horizons mission flyby, coupled with latest modeling capabilities, are available now to make progress.

Enceladus is another small, icy body, probed by *in situ* particles and field instruments during a number of spacecraft close flybys, as well as imaged in detail. It seems that Enceladus is to Saturn as Io is to Jupiter in its influence on the host magnetosphere. Like a comet inside the magnetosphere, Enceladus produces substantial water ice-derived neutral and ion torus populations in its orbit. These become widely dispersed throughout Saturn's magnetosphere – affecting every aspect of its behavior (see Chapter 12). However, what interests us here is Enceladus itself. Like Io, it alters the composition and other properties of its magnetospheric environment. Although the icy surface of Enceladus is sublimated by sunlight and sputtered by the magnetospheric energetic particles in its vicinity, several fissures in the southern polar region of Enceladus are producing dense, geyser-like plumes of water vapor and ice crystals. These fissures are thought to result from the flexing of the body of Enceladus in its ~$4R_{Sat}$ orbit due to the gravitational field of nearby Saturn (a mechanism that may similarly cause Io's volcanism in Jupiter's proximity). The off-center geometry of this stronger "atmospheric" source in the south makes the Enceladus–plasma interaction extremely asymmetric, with the plume providing the main obstacle in the southern part of the interaction, and the icy body and its thinner sputtered/sublimated atmosphere producing the obstacle in the north (see the rough sketch in Figure 8.24). However, the gross features of the plasma interaction are generally much the same as those described in this chapter. Like Titan, the body is threaded by Saturn's dipole field, with some field distortions related to its conductivity, while the plume acts like a cometary source of gas, ions, and dust. In fact, an added consideration here, that pertains to comets as

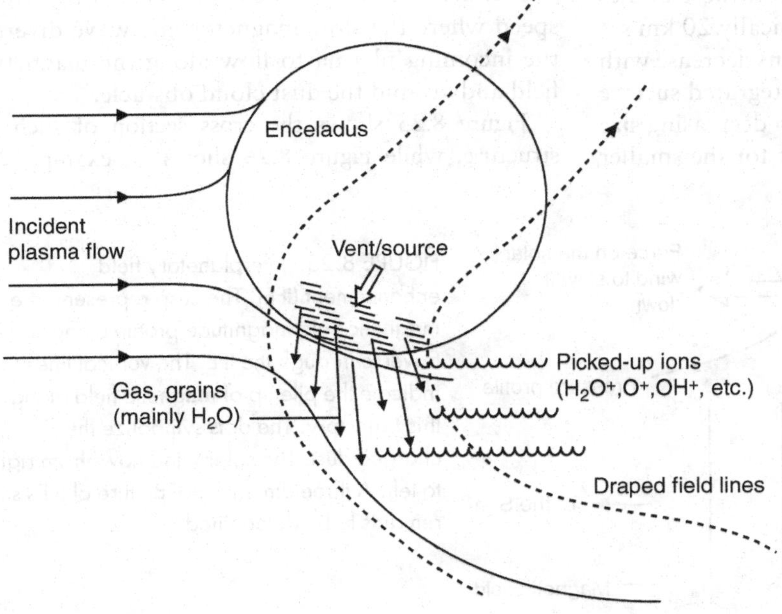

FIGURE 8.24. Saturn magnetospheric plasma and field interaction with Enceladus, inferred from observations. The southern hemisphere plumes of mainly water vapor and dust grains provide a comet-like obstacle "seen" by the plasma, making the interaction highly asymmetric with respect to the body center.

well, is the presence of the large component of dust in the emitted material of the plume. The physics of how this dust interacts with the plasma, including the physics by which it develops a spectrum of electrical charges, is a far-reaching topic (relevant to planetary rings, dust in atmospheres, and protoplanetary disks, for example) for the interested reader to pursue in more depth.

8.4 LOCALIZED DUST OBSTACLES

The subject of solar system dust arose in Chapter 5, where it was discussed from the perspective of a distributed population of small solid particles in the solar wind. However, not all solar system dust is spatially diffuse. For example, infrared images of the sky show many lanes of more concentrated dust associated with cometary and asteroid orbits, as have other *in situ* dust detectors. Thus, when interplanetary magnetic field signatures were found that suggested the presence of an invisible obstacle in the solar wind, dust provided an obvious candidate. However, something more substantial than the broad dust population was indicated.

The space around the Sun is filled with orbiting bodies of varying sizes. Unlike the circular rings of Saturn, these bodies are in orbits that intersect other orbits, and occasionally two of these bodies will collide at very high speed, typically 20 km s^{-1} at the orbit of Earth. While collisions decrease with time, the average body size, the integrated surface area of these bodies, increases with decreasing size so that the collision rate is greater for the smaller bodies. At a relative speed of collision of 20 km s^{-1}, a rock of 1 m diameter can completely disrupt a 100 m diameter rock with a mass of over 10^9 kg. There are a sufficient number of these small rocks or meteoroids flying about the solar system such that collisions with small asteroids of the class of 10–100 m diameter cannot be considered to be rare. Since the particles produced in these collisions are expected to be charged because of solar photons, solar electrons, and the collisions themselves, we must ask what we should see when the solar wind flows through such a cloud. While the center of mass of the cloud will move at the speed determined by the momenta of the colliding bodies, the fastest and smallest particles will be expelled at an energy determined by the specific kinetic energy of the colliders, probably about 10 km s^{-1}, so that in 1000 s (~15 min), they would form a cloud about 10^4 km across. This cloud of charged dust grains would be dense enough to form a dusty plasma, and the solar wind and interplanetary magnetic field would not penetrate it, albeit the cloud would be magnetized as it was formed in the interplanetary field. The resulting magnetic structure should be comet like, anchored in the dust cloud, and, once accelerated to the solar-wind speed, it should be "fronted" by a magnetic barrier similar to that seen at Earth for northward IMF and at Venus for low solar-wind speed where the slow magnetosonic wave diverts the incoming plasma to flow along the magnetic field and around the dust cloud obstacle.

Figure 8.25 shows the cross section of such a structure, while Figure 8.26 shows an example of

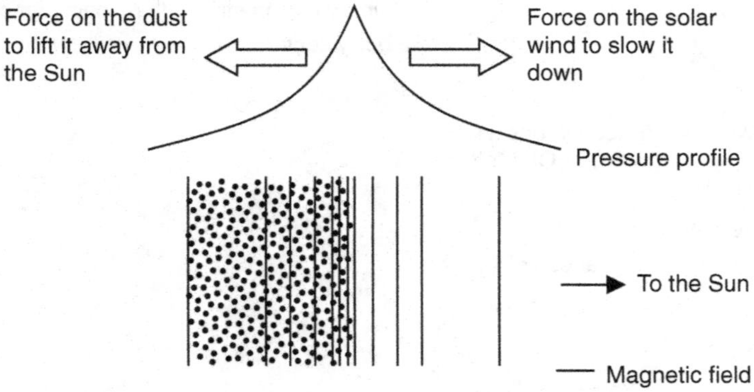

FIGURE 8.25. Interplanetary field enhancement (IFE). The cusp represents the magnetic field magnitude profile along a traverse through the IFE. The vertical lines indicate the pile up of magnetic field along this trajectory. The dots symbolize the charged dust. The solar wind flows from right to left. A three dimensional picture of IFEs still remains to be determined.

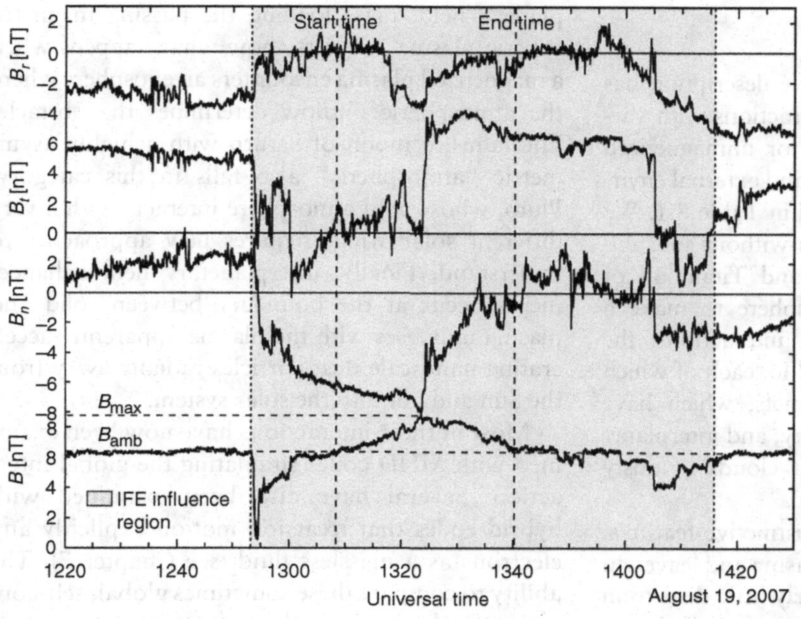

FIGURE 8.26. Magnetic field measurements during the passage of an IFE. The outer vertical dashed lines show the beginning and end of the IFE region of influence. The decreases in field strength just inside these two boundaries are compensated by increases in thermal pressure (not shown). The pressure exerted by the IFE on the dust and on the solar-wind flow from the Sun is purely magnetic. The density and temperature (not shown) do not change. Thin current sheets occur throughout the IFE with usually the largest current near the peak field. The dashed lines marked "start time" and "end time" delimit the region of magnetic-pressure forces.

the magnetic field signature with the boundary of the cloud and the region of magnetic pressure indicated. These magnetic field signatures called **interplanetary field enhancements** have been used to find asteroid debris and follow its evolution with time (over decades). The magnetic dust cloud is on the side away from the Sun and the solar-wind-only side toward the Sun (Figure 8.26). The area perpendicular to the solar-wind flow and strength of the magnetic-pressure-gradient force balanced against solar gravity can be used to weigh the associated dust clouds. The magnetic dust cloud is on the side away from the Sun and the solar-wind-only side toward the Sun. Such structures are seen about seven times a year at 1 AU with strengths great enough to be unambiguously identified. Their durations vary from minutes to many hours, and the masses they contain of fine-scale dust can range from 10^6 to 10^{12} kg. At present, this is the only way to detect these streams.

8.5 SUMMARY

In this chapter, a largely qualitative description has been given of flowing plasma interactions with various types of weakly magnetized or unmagnetized bodies, whose most basic obstacle and external environment attributes are summarized in Table 8.1. We have considered moon-like bodies without substantial atmospheres; Venus, Mars, and Titan, all of which have enough of an atmosphere to make a major difference in their plasma interactions; the small bodies Pluto, Enceladus, and Io, each of which has its own special features; comets, which have atmospheres unrestrained by gravity; and interplanetary field enhancements, apparent clouds of dusty plasma moving with the solar wind.

Each plasma interaction has distinctive features. The Moon absorbs the incident plasma and leaves an empty wake, but produces relatively little distortion in the magnetic field. The weakly magnetized planets with substantial ionospheres deflect the incident plasma, thereby forming a bow shock and magnetosheath, but also produce near-planet mass loading that contributes to the formation of the induced magnetotail in the wake. Mars and the Moon have small-scale crustal magnetic fields sufficiently strong to affect their plasma interactions, for which our basic pictures described in this chapter take on additional complications. The planetary moons with atmospheres, such as Titan and Io, represent a distinct combination of these two types of interactions, in that they are typically embedded in the host planet's field, mass loading the passing magnetospheric plasma. Comets show what happens when a magnetized plasma encounters an exosphere where the atmospheric outflow determines the obstacle. Enceladus, a moon of Saturn with a highly asymmetric "atmosphere," also falls in this category. Pluto, whose cold atmosphere interacts with a very different solar wind, requires new approaches to understand. Finally, interplanetary field enhancements occur at the boundary between solid and plasma universes with the plasma apparently accelerating nanoscale dust particles radially away from the Sun and out into the solar system.

Most of these interactions have now been examined with MHD codes simulating the global interaction. Several have also been examined with hybrid codes that treat ion motion explicitly and electrons as a massless fluid (see Chapter 7). The ability to compare these sometimes global, self-consistent model results with observational data is one key to reaching deeper understanding. As we have learned from past experience, comparisons between physics-based models and observations can best cement our knowledge. Dust in plasmas is a topic where even the most fundamental interaction is still only roughly understood. One feature in common to all of these objects is that they are all sources in a plasma, whose interactions with the plasma lead to material escape from the sources. It is interesting to compare their estimated production rates – thinking of them all as various types of exotic comets perhaps.

Table 8.1. Summary of some key scales and parameters for the obstacles to magnetized plasma flows discussed in this chapter

Body	Typical external B (nT)	Plasma velocity V_{sw} (km s^{-1})	Body radius (km)	Primary pickup ion species	Environment
Moon	7 (sw) 15 (tail)	300–600[a]	1738	He^+, N^+, O^+, Ar^+, K^+, H^+, Na^+	sw, Earth tail
Venus	13	300–600[a]	~6050	O^+, O_2^+, CO^+, CO_2^+	sw
Mars	3	300–600[a]	~3390	O^+, O_2^+, CO^+, CO_2^+	sw
Titan	~5	80–120 200–400 in sheath, sw	~2575	CH_4^+, N_2^+, N^+	magnetosphere sheath, sw
Io	1500	57	1830	SO_2^+, SO^+	magnetosphere
Enceladus	~330	~20	~250	H^2O^+, O^+, OH^+,	magnetosphere

[a] If the solar wind (sw) is disturbed, e.g., by a fast ICME (see Chapter 5), plasma speeds may be up to ~2000 km s^{-1}, and densities and magnetic fields up to about ten times their normal values.

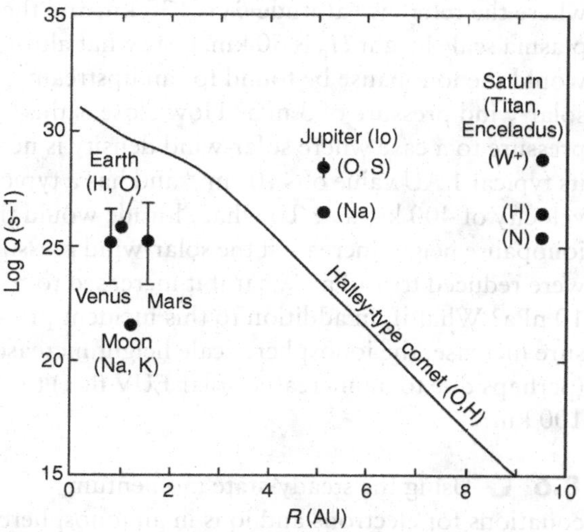

FIGURE 8.27. Comparison of some of the estimated production rates, Q, of the various solar system "sources" discussed in this chapter. The main compositions are also indicated. The Earth is shown for comparison, where the "source" in that case is the polar cap ionospheric outflow (see Chapter 11). The rates for Jupiter and Saturn are essentially those provided by Io and Enceladus, respectively. The solid line indicates the changing rate of production by comet Halley as it approaches the Sun. These rates give an idea of the strengths of planetary sources of ions in the heliosphere.

Figure 8.27 shows such a comparison, based on some numbers in the literature and understood to be gas or ion escape into the external plasma by various combinations of mechanisms, including thermal, photochemical, mechanical (e.g., sputtering), and plasma physical (as in the case of pickup ions, for example). As described above, these objects produce different species depending on their makeup, and they all have variations in production rates that may range over orders of magnitude. When these somewhat average rates are integrated over time, they usually (at least for the planetary bodies) imply only minor consequences for the atmosphere or surface, although a comet can be evaporated at any time depending on its orbit and body cohesiveness.

However, these loss rates may not be the same under all circumstances, and may not have been the same in the early solar system. One goal of planetary research is to understand how these atmospheric escape processes have changed over the 4.5 billion years of solar system history. The planets are constantly losing materials to space, but what has been the time-integrated effect? One might also ask the same question of the Earth and other planets that have magnetospheres (see Chapter 11). Their plasma interaction-related atmospheric loss processes differ, but by how much, and does the absence of a magnetic field make a planetary body more or less exposed to such scavenging? Current rates of ion escape at Earth are similar to those at Venus and Mars (Figure 8.27). Are magnetospheres really "shields" as suggested by Figure 8.8? Readers who find these questions intriguing after working through this chapter may want to follow (or join) the debates on this larger-perspective question.

Additional reading

BOUGHER, S. W., D. M. HUNTEN and R. J. PHILLIPS (Eds.) (1997). *Venus II: Geology, Geophysics, Atmosphere, and Solar Wind Environment.* Tuscon, AZ: University of Arizona Press. Second book summarizing our knowledge of Venus gathered in the space age.

HUNTEN, D. M., L. COLIN, T. M. DONAHUE, and V. I. MOROZ (Eds.) (1983). *Venus*. Tucson: University of Arizona Press. First comprehensive summary of our knowledge of Venus gathered mainly from Pioneer Venus and the Venera missions up to Venera 14.

RUSSELL, C. T. (Ed.) (1991). *Venus Aeronomy.* Norwell, MA: Kluwer Academic Publishers. Summary of knowledge gained up through Pioneer Venus about the upper atmosphere, ionosphere, and solar-wind interaction.

RUSSELL, C. T. (Ed.) (2007). *The Mars Plasma Environment*. Norwell, MA: Springer. Summary of knowledge gained from the Mars Express mission on the upper atmosphere and solar-wind interaction at Mars.

SZEGO, K. (Ed.) (2011). *The Plasma Environment of Venus, Mars and Titan, Space Sci. Rev.*, **162.** New York, NY: Springer. Overview of the plasma interactions at Venus, Mars, and Titan from a modern perspective.

TAYLOR, F. W. (Ed.) (2006). *The Planet Venus and the Venus Express Mission, Planetary Space Sci.*, **54.** Oxford: Elsevier, pp. 1247–1496. Introduction to the goals of the Venus Express mission.

Problems

8.1 Calculate the maximum energy (in keV) of potassium and sodium ions picked up near the Moon if the solar-wind velocity is 400 km s^{-1} (assume the ions miss the Moon). What is the size of their gyro-radius compared with the radius of the Moon if the external magnetic field strength is 5×10^{-5} G?

8.2 There is no bow shock in front of the Moon. Explain why. When the interplanetary magnetic field is aligned with the solar-wind flow, what is the radius of the lunar wake if the upstream beta of the plasma is 3 (R_M = 1738 km)?

8.3 The Moon moves from the magnetosheath into the relatively steady field, near-vacuum conditions of the geomagnetic tail lobe. Assuming that the average magnetosheath field is effectively zero, in the tail-lobe field how large a disturbance would be expected in the magnetic measurements of a satellite in a 1000 km circular orbit if there were an infinitely conducting core with a radius of (i) 200 km, (ii) 400 km, and (iii) 800 km? Assume that the external field lies in the orbit plane of the spacecraft.

8.4 Assume that the conductivity of a spherical body of radius 1000 km is 10^{-3} mho m^{-1} (that of an insulator) and its magnetic permeability is $\mu_0 = 1.26 \times 10^{-6}$ H m^{-1}. How long would it take for an externally imposed magnetic field to diffuse into the body? How long would it take if the conductivity were 10^5 mho m^{-1} (that of a good conductor)?

8.5 If the ionosphere of unmagnetized planet x had a temperature that is constant with height, equal to 10^5 K, and with an exponential subsolar electron density altitude profile given by

$$n_e(h) = 10^5 (\text{cm}^{-3}) \exp\left(\frac{-(h - h_0)}{H_p}\right),$$

where the reference altitude h_0 is 130 km and the plasma scale height H_p is 50 km, near what altitude would the ionopause be found for an upstream solar-wind pressure of 3 nPa? How close is this pressure to a case where solar-wind density is near its typical 1 AU value of ~10 cm^{-3} and has a typical velocity of 400 km s^{-1}? To what altitude would the ionopause height increase if the solar-wind pressure were reduced to 1 nPa? What if it increased to 10 nPa? What if, in addition to this incident pressure increase, the ionosphere scale height increased (perhaps due to an increased solar EUV flux) to 100 km?

8.6 Using the steady-state momentum equations for electrons and ions in an ionosphere (see Chapter 3), Maxwell's equation $\partial \mathbf{B}/\partial t = -\nabla \times \mathbf{E}$, and Ampère's law $\mu_0 \mathbf{j} = \nabla \times \mathbf{B}$, derive the equation describing the evolution of a horizontal ionospheric magnetic field

$$\frac{\partial B}{\partial t} = \frac{\partial}{\partial h} D \frac{\partial B}{\partial h} - \frac{\partial}{\partial h}(B u_{pl}),$$

where u_{pl} , is the vertical velocity and h is the vertical spatial coordinate. Assume that $B = B\hat{\mathbf{x}}$ depends on h only. The equation for the diffusion coefficient D will naturally come out of this derivation. [Hint: Terms multiplied by the electron mass m can be dropped in the latter stage of the derivation.]

8.7 If a water-ion cometary plasma has a density of 10^6 cm^{-3} at a distance of 10 km from the nucleus and expands outward from the nucleus at 1 km s^{-1}, at what distance from the nucleus does the cometary plasma dynamic pressure ρu^2 balance an incident solar-wind pressure of 3 nPa? (In other words, what is the subsolar distance of the contact surface?) If the solar-wind pressure increases to 5 nPa, what is the new distance?

9 Solar wind–magnetosphere coupling

9.1 INTRODUCTION

As our technological society has advanced, we have made greater and greater use of the Earth's magnetosphere to host our increasingly sophisticated instruments on our monitoring, communication, and global-positioning satellites. The magnetosphere is usually a benign host, but it can be energized greatly by the solar wind and become quite inhospitable, affecting not just systems in space, but even those on the surface of the Earth. Fortunately, we have learned much about the magnetosphere and its control by the solar wind since the early days of the space age when James Van Allen and his co-workers discovered that space was "radioactive." This new understanding allows us to predict the response of the magnetosphere to locally measured solar-wind conditions (called nowcasting) and to take protective measures if needed. This knowledge also allows us to design space systems that are less sensitive to the extremes of the space environment. Nevertheless, an April 2010 anomaly on the Galaxy 15 spacecraft that turned it into a rogue communications satellite illustrates well that potentially harmful magnetospheric behavior can occur even during periods of very low solar activity (Connors, Russell, and Angelopoulos, 2011).

In this chapter, we discuss how the solar wind couples to the magnetosphere, the effects of this coupling on the magnetosphere, and how to predict and quantify that coupling. The process most responsible for the coupling, but least understood, is called magnetic reconnection. We discuss how magnetic reconnection leads to two important manifestations of geomagnetic activity: the geomagnetic storm and the geomagnetic substorm. We begin by discussing the boundary conditions applied to the magnetosphere by the solar wind.

9.2 THE OUTER MAGNETOSPHERE

9.2.1 The stress applied by the solar wind

The conditions in the outer magnetosphere are very sensitive to the properties of the solar wind in which it is immersed. As discussed in Chapter 7, the size of the Earth's magnetosphere is determined by the balance between the magnetic pressure of the Earth's magnetosphere and the three components of the solar-wind pressure: the dynamic pressure or momentum flux of the cold stream of ions flowing radially from the Sun; the kinetic or thermal pressure of the plasma that is measured in the frame of the solar-wind flow; and the pressure of the interplanetary magnetic field. These push along the direction perpendicular to the magnetopause surface and are called normal stresses. Chapter 7 also notes that there is drag, also called tangential stress, which changes the shape of the magnetosphere. This stress causes circulation of the magnetospheric plasma and transports magnetic flux from one region to another within the magnetosphere. The magnetosphere can store energy, gradually accumulated over a relatively long period of time, then release it rapidly, reminiscent of the release of energy in a solar flare discussed in Chapter 4. It is often during this rapid release of energy by the Sun that our terrestrial space systems upon which our technological society now relies are endangered, when that energized plasma and the accompanying solar energetic particles reach the Earth.

There are several possible ways to apply tangential stress to the magnetosphere. On the magnetopause, reconnection is a slow and quasi-continuous process, compared to the reconnection we find in solar flares. Even so, it is the most effective means available for coupling with the solar wind. The difference in reconnection rates on the magnetopause and in the solar corona is due to the differing plasma conditions at the point of reconnection. The plasma conditions in the magnetosheath produce a rather low Alfvèn speed, resulting in low speeds of the reconnected plasma. As illustrated in Chapter 1 in Figure 1.20(b), once reconnection occurs on the magnetopause, it links the magnetospheric and solar-wind magnetic fields. This linkage couples the momentum flux from the solar wind to the magnetosphere. The convecting magnetospheric plasma carries magnetic flux and plasma over the polar cap to the magnetotail, a giant energy reservoir that can release its energy rapidly back toward the inner magnetosphere, striking first the night hemisphere.

Despite the efforts of many talented space plasma physicists, the reconnection process is poorly understood. First, it must involve processes acting on characteristic electron scales. In an electron–ion plasma, the electron motion defines the magnetic field, so that, if the electron motion is disrupted – preventing it from being guided by the magnetic field – the identity of a field line will be lost and it can switch partners. While resistive codes can mimic this switching of magnetic partners, and while MHD codes can reproduce the flows seen once reconnection begins, these codes cannot reproduce the physics that takes place at the reconnection point, especially those controlling when reconnection begins. This is very similar to the situation in Chapter 6, where the processes producing dissipation in the collisionless shock ramp are not reproduced by MHD theory, even though MHD can tell us how much dissipation is required at the ramp. The reconnection problem appears to be beyond the reach of our current numerical tools. The reconnection problem on the Earth's magnetopause is not only collisionless and fully kinetic, but also its application requires a global model, not a local solution.

Also shown in Figure 1.20, there is a second region in the magnetotail in which reconnection takes place with a quite different geometry. The different geometry and plasma conditions produce a very different behavior. The magnetotail consists of two large, adjacent magnetic flux tubes with antiparallel fields separated by a magnetized plasma layer doubly anchored to the Earth by its magnetic field. This situation is quite different than that of magnetopause reconnection, where (as we discuss in the following) the rate and location of reconnection are very sensitive to small changes in the direction of the interplanetary field and the plasma conditions. Magnetopause reconnection can take place nearly anywhere, even at multiple locations simultaneously. In contrast, a well-designed multi-spacecraft mission, such as the five-spacecraft THEMIS mission, can easily monitor tail reconnection occurring during substorms. On the magnetopause the equivalent study is much more difficult. To study how solar-wind conditions affect the magnetopause reconnection rate, we need to use a different approach in which we examine the integrated result of the reconnection on the magnetopause. This approach will have to suffice until such a time as we can simulate collisionless reconnection on the electron kinetic scale under space-like conditions for macroscale interactions, or we can afford massive arrays of *in situ* magnetopause probes.

To proceed we have to appreciate first what the "input" to the magnetosphere is. The solar-wind plasma that interacts at the magnetopause is first processed by the shock that converts the bulk motion of the solar wind into thermal energy. Thus the Mach number of the shock is an important parameter in governing the coupling. Furthermore, the plasma evolves as it moves from the shock toward the magnetopause. Boundary layers in the magnetosheath plasma can form just outside the magnetopause, such as in the plasma depletion layer, where the density drops and the magnetic field strength rises. These changes will also vary the rate of reconnection.

Since we cannot monitor plasma flows produced across the entire reconnecting magnetopause, we must devise proxy measures to do so. A proxy that is available is geomagnetic activity in its various guises. Geomagnetic activity is basically the time-varying magnetic field produced by the coupling process. These variations historically have been quantified in numerical geomagnetic indices of various types, some "linear" and some "logarithmic." Some have rapid cadence and others integrate activity over longer periods of time. Geomagnetic activity informs us about the reconnection process in ways impossible from the isolated measurements of our space probes. Thus, after a brief survey of the structure of the magnetosphere, we discuss what we learn from geomagnetic activity about the nature of the magnetospheric coupling with the solar wind. We then examine *in situ* observations of the interface between the magnetosheath and the magnetosphere to determine what constraints *in situ* observations place on the coupling across the boundary and how that coupling leads to energy storage in the magnetosphere. Next we discuss the reconnection process and its consequences for the magnetosphere, one of which is the geomagnetic storm. We close the chapter with a discussion of the magnetotail and magnetospheric substorms.

9.2.2 Structure of the outer magnetosphere

Before we examine how the solar wind couples to the **magnetosphere**, we should say a few words about the magnetosphere itself. Figure 9.1 shows a cut-away sketch of the dayside and near nightside of the magnetosphere. The dayside part of the magnetopause is blunt, and the dayside magnetosphere

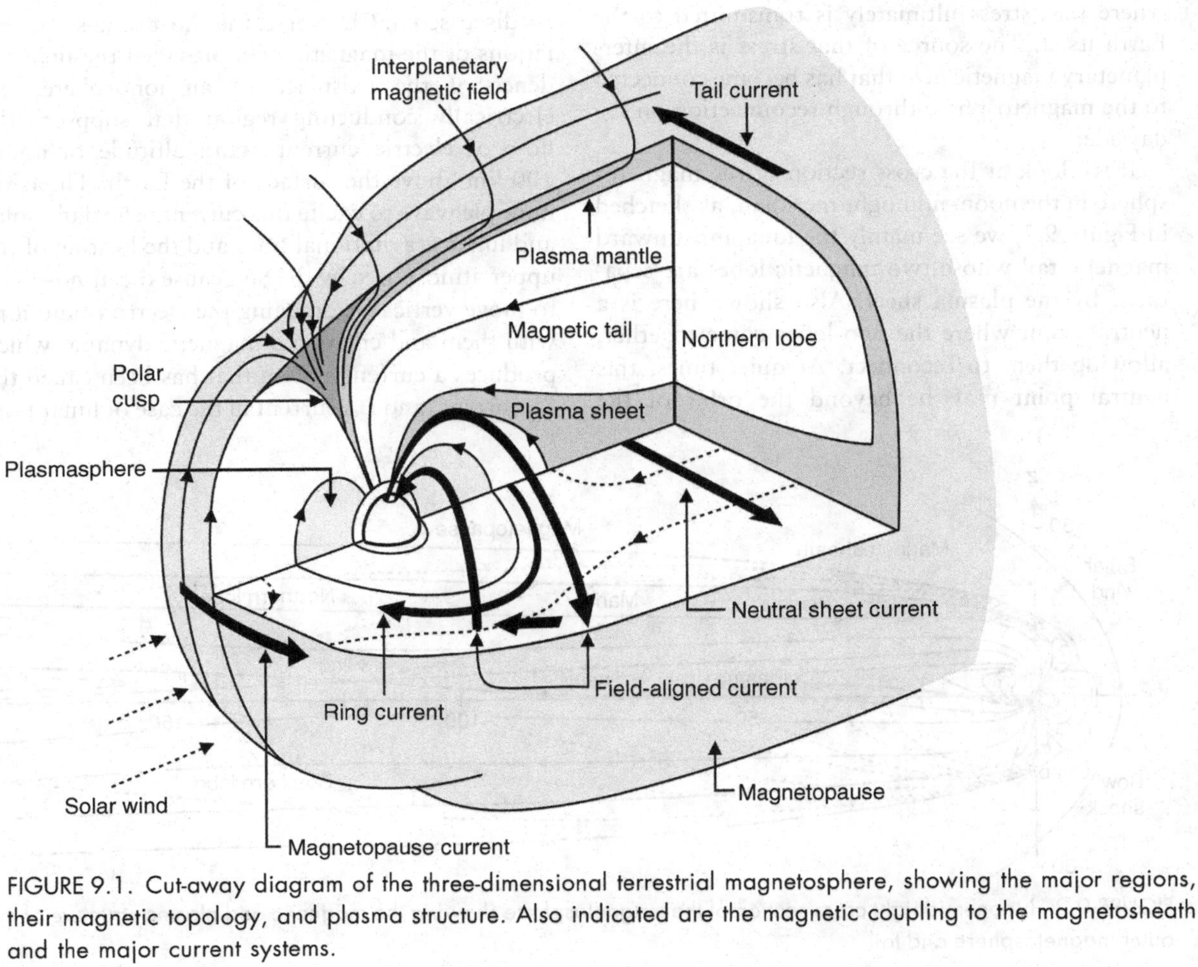

FIGURE 9.1. Cut-away diagram of the three-dimensional terrestrial magnetosphere, showing the major regions, their magnetic topology, and plasma structure. Also indicated are the magnetic coupling to the magnetosheath and the major current systems.

consists of closed magnetic field lines with both feet on Earth. The nightside of the magnetosphere has a swept-back magnetic field forming two lobes of open magnetic field with one foot connected to the Earth. The polar region, where the field changes from bending forward to bending backward, is the polar cusp. Here, magnetosheath plasma extends all the way down to the ionosphere. The two lobes of the tail, the northern lobe and the southern lobe, are separated by a warm dense plasma region called the plasma sheet. This diagram emphasizes the currents: those on the magnetopause that terminate the terrestrial field; those in and on the tail surface that enclose the tail lobes and pass through the plasma sheet between them, labeled here as the neutral sheet current; and the field-aligned currents that transmit the stress of the solar-wind interaction from the outer magnetosphere to the ionosphere, where that stress ultimately is transmitted to the Earth itself. The source of that stress is the interplanetary magnetic field that has become connected to the magnetosphere through reconnection on the dayside.

If we look at the cross section of the magnetosphere in the noon–midnight meridian, as sketched in Figure 9.2, we see mainly the long antisunward magnetic tail whose two **magnetic lobes** are separated by the **plasma sheet.** Also shown here is a neutral point where the two lobes come together, allowing them to reconnect. At quiet times, this **neutral point** may be beyond the orbit of the Moon ($60R_E$) but as close as $10R_E$ at disturbed times. The plasma enters the tail along its boundary. This newly added plasma is called the **mantle.** Other types of boundary layers can form along the magnetopause, the dayside low-latitude boundary layer being one of them. Trapped energetic particles, which form the trapped radiation belts, are found on the more circular closed magnetic field lines. When the energy contained in these bouncing and drifting energetic particles is significant (in comparison with the magnetic energy), we call the resulting flow of electric charge the **ring current.**

9.3 USING GEOMAGNETIC ACTIVITY TO PROBE SOLAR WIND–MAGNETOSPHERE COUPLING

As discussed in Chapter 1, the short time-scale variations of the magnetic field provided the first evidence of the existence of an ionosphere, the electrically conducting region that supports the flow of electric current at an altitude of about 100 km above the surface of the Earth. There are multiple ways to excite this current. The daily solar and lunar gravitational tides and the heating of the upper atmosphere by the Sun cause the atmosphere to move vertically, dragging the electrons and ions with them and creating a magnetic dynamo which produces a current system that has been called the S_Q current (and L_Q current in the case of lunar tidal

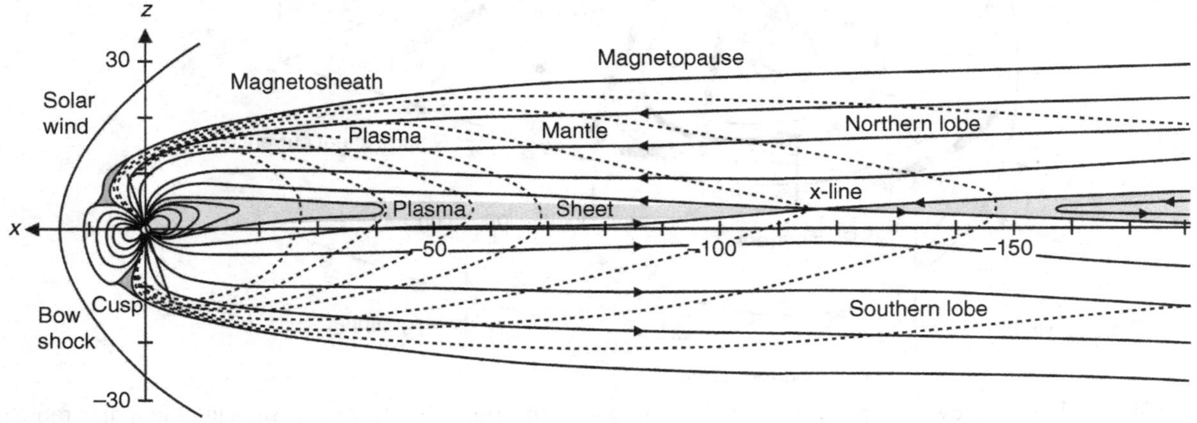

FIGURE 9.2. Noon–midnight cross section of the magnetosphere showing the magnetic and plasma structure of the outer magnetosphere and tail.

action). The solar wind also has a role in driving ionospheric current, and this is called the S_D current. The solar-wind-induced current can have many forms. It can appear as "electrojets" in narrow channels in the auroral zones; it can cross the polar cap; it can appear at mid latitudes. There are many indices that have been created to respond to one or another current system. The AE, AU, and AL indices are used to monitor the intensity of the strongest current in the auroral oval; the polar cap index monitors the current crossing the polar cap; the Dst index monitors the current flowing in the equatorial magnetosphere; and the Ap and Kp indices are designed to monitor the general variability of the magnetic field at mid and low latitudes on a planetary scale. The Kp index is a logarithmic version of the Ap index. There are many versions of the Ap index that attempt to improve one or another aspect of its performance. Here we use the Am index (Mayaud, 1980) because of its careful stewardship by P. N. Mayaud and its long history.

Indices are useful for determining the causes of geomagnetic activity. They reveal solar control through the presence of 22 and 11 year cycles and terrestrial control in the presence of annual, semiannual, and daily cycles. Several different processes have been proposed to produce the observed annual or semiannual cycles, but with some care in interpretation they can help us monitor magnetopause reconnection where we cannot do this with our spacecraft. The point of strongest coupling on the magnetopause moves around and often will be where spacecraft are not. Since geomagnetic indices respond to the integrated global interaction, we can use them to explore the coupling effectiveness of reconnection versus its main competitor.

The main competitor to reconnection for controlling the solar-wind magnetosphere interaction is the process known as the **Kelvin–Helmholtz instability**, which acts similarly to wind over water in producing waves on an interface. Both the Kelvin–Helmholtz instability and reconnection can lead to a semiannual variation because the tilt of the magnetic dipole axis should give the same relative orientation to the solar-wind flow and interplanetary magnetic field in spring and fall. We can simply test which one is responsible for the semiannual variation since the Kelvin–Helmholtz instability does not depend on the sign of the magnetic field and reconnection does. Figure 9.3(a) shows the semiannual variation by subtracting the eight three-hourly values in the winter months from the same values in the summer months. We repeat this in Figure 9.3(b) for various

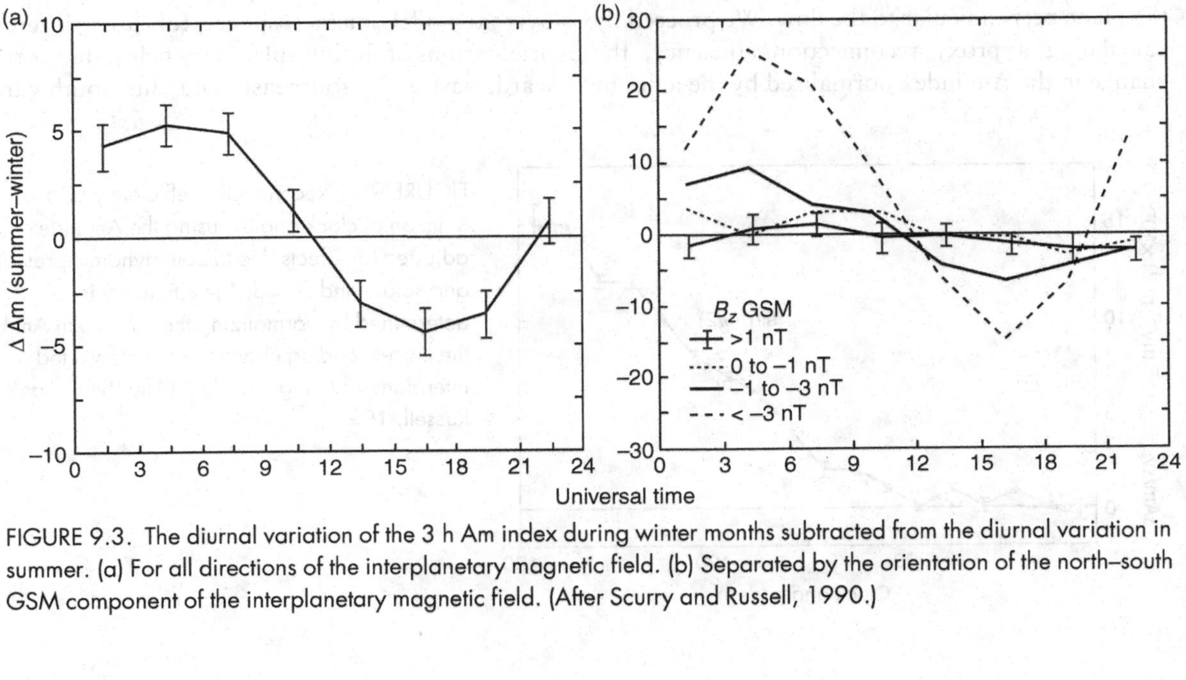

FIGURE 9.3. The diurnal variation of the 3 h Am index during winter months subtracted from the diurnal variation in summer. (a) For all directions of the interplanetary magnetic field. (b) Separated by the orientation of the north–south GSM component of the interplanetary magnetic field. (After Scurry and Russell, 1990.)

southward interplanetary magnetic fields ($B_z < 0$) and all northward fields > 1 nT. We see that the activity is proportional to the southward field component and completely goes away for the northward field. This tells us two very important things about solar wind–magnetosphere coupling. First, the Kelvin–Helmholtz instability plays at most a minor role in solar wind–magnetosphere coupling, and, second, the energy transfer process that the reconnection enables is very finely controlled by the IMF direction. Since many reconnection "theories" and "models" do not have strong directional control of the rate at which plasmas are merging, they cannot easily be used to explain the coupling that we see. In fact, these models should produce continuous, strong reconnection at least somewhere on the magnetopause. This is a strong constraint on any theory of collisionless reconnection.

We can use the same formalism to obtain a functional form for the dependence of reconnection between the solar-wind magnetic field and the closed field lines of the magnetosphere as a function of the direction of the interplanetary magnetic field. Since the bow shock compresses the magnetic field perpendicular to the flow, we examine activity as a function of the angle of the solar-wind magnetic field in the plane perpendicular to the flow direction (clock angle). Here, north is defined to be the direction of the Earth's magnetic dipole axis projected on the plane perpendicular to the flow. We proceed by calculating a proxy reconnection efficiency, the change in the Am index normalized by the amount of magnetic flux convected to the Earth by the solar wind at that time. When the clock angle is near zero (due northward fields), the Am index does not change with the convected magnetic flux to the Earth. As the clock angle turns toward the south, the Am index becomes ever more sensitive to the convected magnetic flux. This dependence increases most rapidly when the interplanetary magnetic field is southward.

This is illustrated in Figure 9.4. To provide an approximate functional form that can be easily remembered, this dependence is compared with $\sin^4(\theta/2)$, where θ is the clock angle: zero for parallel and 180° for antiparallel fields. There is no theoretical basis for any specific functional dependence on clock angle in this problem, nor any particular power law dependence. Geomagnetic activity results from the integrated reconnection over a surface with a complex shape. There is no fixed point on that surface that dominates the reconnection rate, not even the subflow point. The important message provided by Figure 9.4 is that there is very little geomagnetic activity for fields that are even a little bit northward.

This constraint on geomagnetic activity for northward fields places an even stronger constraint on the dependence of collisionless reconnection on the angle between the reconnecting fields. Figure 9.5 shows the region of nearly antiparallel magnetic fields given a draped magnetosheath field over a model magnetosphere for four different orientations of the interplanetary fields: due northward, eastward, southeast, and due southward.

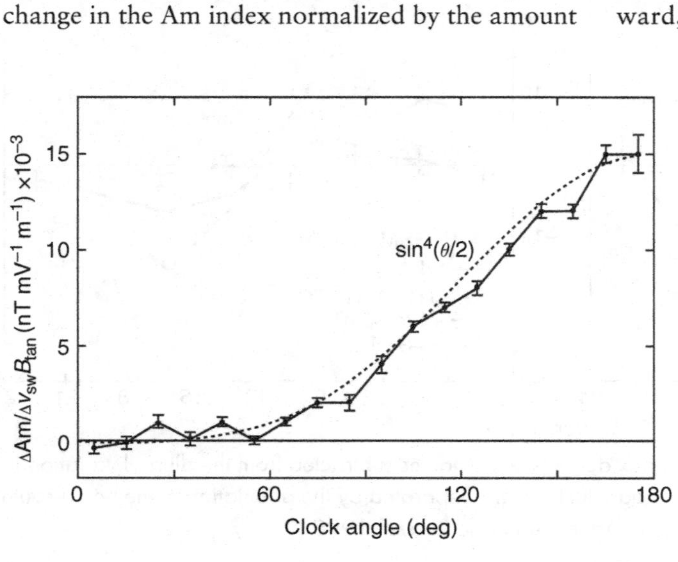

FIGURE 9.4. Reconnection efficiency as a function of clock angle θ using the Am index adjusted for effects due to both dynamic pressure and solar-wind speed. The efficiency is determined by normalizing the change in Am by the corresponding change in the convected interplanetary magnetic flux. (After Scurry and Russell, 1991.)

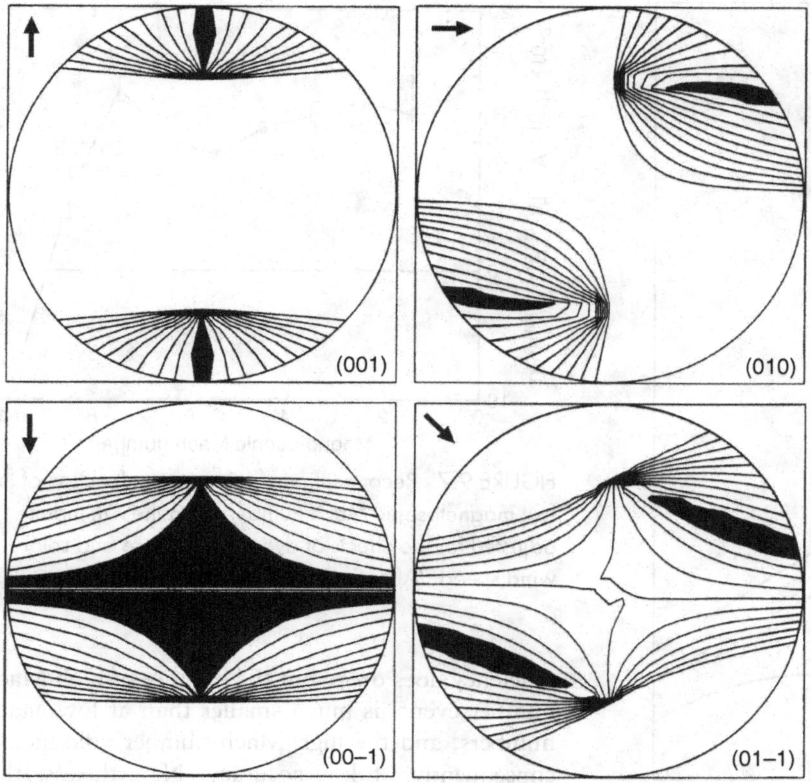

FIGURE 9.5. Blackened regions show where magnetic fields in the magnetosheath and magnetosphere should be nearly antiparallel for the interplanetary magnetic field directions shown in each panel for an idealized magnetospheric field with tilt angle perpendicular to the solar-wind flow. These blackened regions are where reconnection would occur for antiparallel reconnection. Lined areas are where reconnection would occur for guide-field or component reconnection. (Adapted from Luhmann *et al.*, 1984.)

When the field is due northward, the only region of antiparallel magnetic field is in the tail lobe, on open field lines. In this case, there should be no momentum coupling to the closed magnetosphere. In the eastward case, the antiparallel field region moves equatorward close to where the closed field lines could be reconnected, according to an antiparallel reconnection law. For the southeast orientation, the antiparallel reconnecting area is still small, but it occurs in the portion of the magnetopause where we expect closed field lines to lie antiparallel to magnetosheath fields. Thus, reconnection with closed magnetic field lines resulting in flux transfer to the tail can occur here. When the field is due southward, the reconnection region is maximum in size. This exercise illustrates that the half wave antiparallel rectification observed in which northward IMF produces little geomagnetic activity occurs naturally with antiparallel reconnection. Guide-field reconnection is not so successful.

This exercise was calculated for somewhat ideal conditions with the dipole axis perpendicular to the flow. If we allow the dipole to tilt into and away from the solar-wind flow, we find that this tilt also affects the size of the reconnection region. Considering the length of the antiparallel reconnection line on the magnetopause to be the qualitative measure of the expected reconnection rate, we obtain the values given in Figure 9.6. This model is consistent with the observed behavior of the reconnection rate based on geomagnetic activity proxy data, and it helps explain the reason for the significant semiannual variation of geomagnetic activity. Most importantly, it implies that the onset of collisionless reconnection requires antiparallel magnetic fields. This is contrary to what MHD codes with finite numerical resistivity produce.

Finally, we can attempt to extract one more lesson from these proxy data by examining the possible control of the Mach number of the bow shock

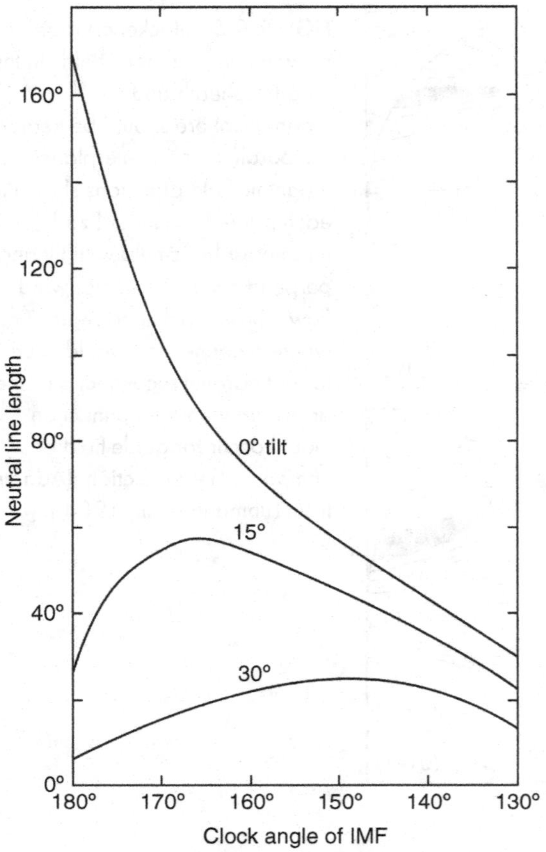

FIGURE 9.6. Length of the neutral line versus the clock angle of the IMF around the flow direction for tilt angles of the terrestrial magnetic dipole axis of 0°, 15°, and 30°. Neutral line is defined to be the line on which the magnetic field is exactly opposite in direction on the two sides of the boundary. (Russell, Wang, and Raeder, 2003.)

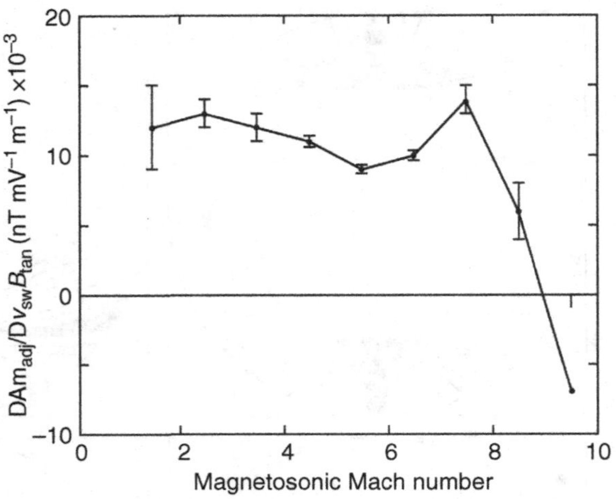

FIGURE 9.7. Reconnection efficiency as a function of the fast magnetosonic Mach number using the Am index adjusted for the effects of dynamic pressure and solar-wind speed. (After Scurry and Russell, 1991.)

on the magnetic reconnection with the magnetosphere. Since magnetic stresses in the reconnected plasma should control the rate at which reconnection can take place, we might expect that conditions downstream of the shock that led to a weak field outside the magnetopause would lead to weaker activity in the magnetosphere. Figure 9.7 shows the reconnection efficiency calculated by the sensitivity of the Am index to transported magnetic flux as a function of the fast magnetosonic Mach number of the solar-wind flow at the subsolar point of the bow shock. When the magnetosonic Mach number approaches 8, the empirical reconnection efficiency does drop, but the number of high Mach number events is much smaller than at low Mach numbers, and the high Mach number efficiency is consequently of less accuracy. Nevertheless, this observation is consistent with the infrequency of reconnection events at the magnetopause of Jupiter and Saturn, whose magnetosheaths are high-beta regions downstream from very strong shocks. We next examine what we have learned from the *in situ* observations of the magnetopause.

9.4 THE OBSERVED MAGNETOPAUSE

In Chapter 2, we discussed the upper atmosphere's various "pauses," the last of which is the magnetopause, the outermost extent of the Earth's magnetic field. This boundary is one between two magnetized plasmas, and, in a collisionless plasma in the absence of any process that links the magnetic fields across the boundary, the magnetized plasmas should not mix. The magnetic fields cannot be exactly parallel for any but a short distance in the geometry of the solar-wind interaction, and, if they were pushed together by an external force, the stresses in the crossed magnetic field would oppose that force and keep the plasmas separate.

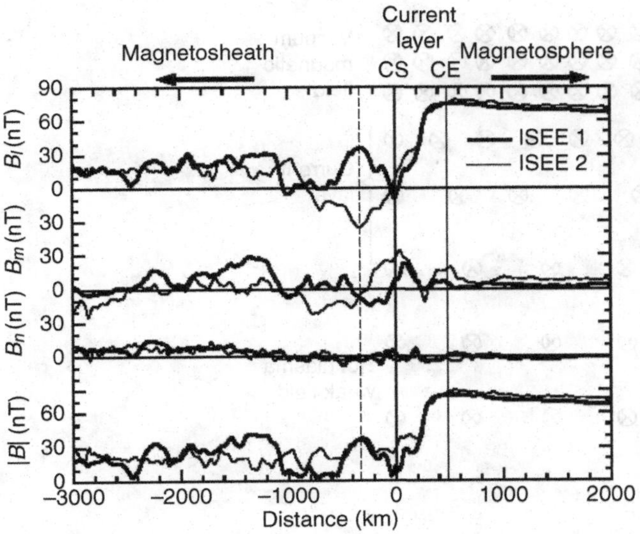

FIGURE 9.8. Magnetopause crossing observed by ISEE 1 and 2 on December 11, 1977, when the magnetosheath magnetic field was unusually weak. Observations are displayed in boundary normal coordinates and are plotted versus distance using the time delay from ISEE 1 to 2 to calculate the speed of the boundary. CS marks the start of the current sheet and CE marks the end. (After Le and Russell, 1994.)

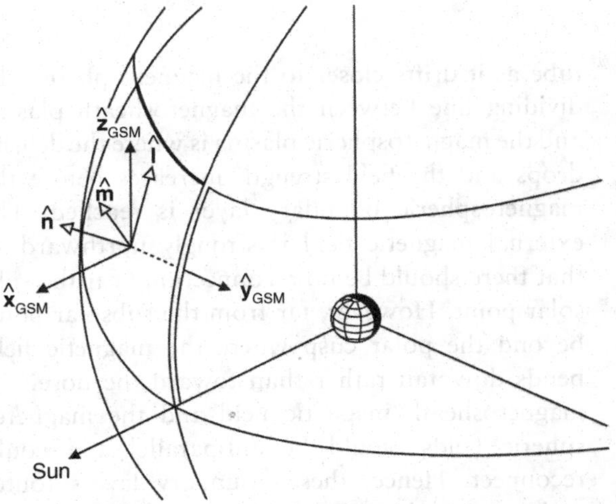

FIGURE 9.9. Definition of boundary normal coordinates. (After Elphic and Russell, 1979.)

The boundary should be at its simplest when there is no magnetic field in the magnetosheath and the force balance is between a plasma pressure gradient force and a magnetic pressure gradient force. As we noted above, this can happen when the Mach number of the bow shock is so high that it produces strong plasma thermal pressure that dominates over the magnetic pressure, i.e., a high-beta plasma. Figure 9.8 gives an example of the magnetic profile across the magnetopause at such a time. In this figure, and in many following it, we use boundary normal coordinates (*l*, *m*, *n*), where *n* is outward along the magnetopause normal and *l* is aligned with the field in the magnetosphere; *m* completes the right-handed set. A defining sketch of this system is shown in Figure 9.9.

The magnetopause carries current. The current arises from two sources. First, the magnetic field direction changes across the magnetopause. Ampère's law requires a current flow in the plane of the discontinuity in the magnetic field in the direction perpendicular to the change in the magnetic field. Second, if there is a change in the plasma pressure across the boundary, then there will be a current flowing in the plane of the boundary, even if the direction of the magnetic field does not change. The source of this current is easy to visualize from the sketch in Figure 9.10. We assume for purposes of illustration that the temperature is constant but the number density is changing to produce the pressure gradient. Close to the inner (magnetospheric) edge of the boundary, there are few ions (or electrons) gyrating in the magnetic field. As one moves away from the magnetosphere, the number of gyrating ions increases, leading to an imbalance in the current where there is a gradient in pressure. When one leaves the region of the number density

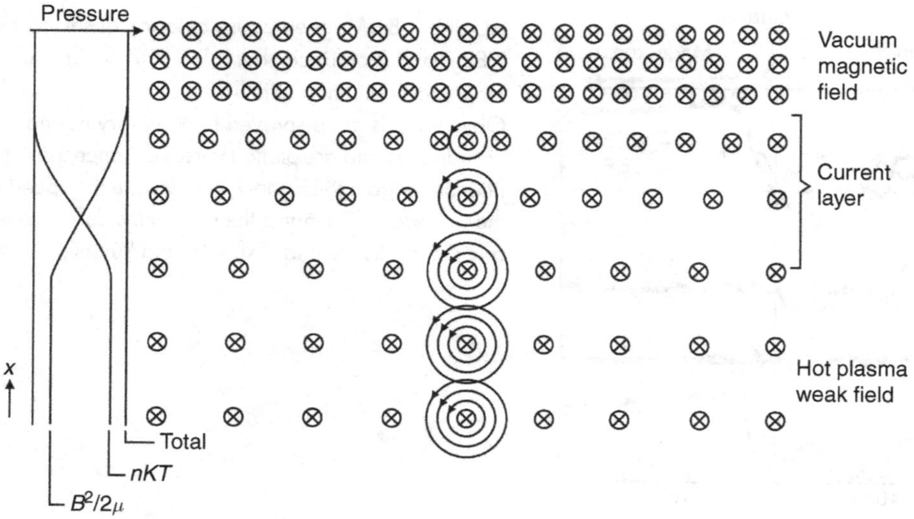

FIGURE 9.10. Pressure balance between a hot weakly magnetized plasma (b) and a vacuum magnetic field (a). The pressure gradient produces a current that acts to reduce the field strength and balances the magnetic pressure with a **j** × **B** Lorentz force.

gradient, there no longer is a current. This current of course is just that needed to reduce the magnetic field to its strength in the magnetosheath.

As noted in Chapter 7, in discussing the MHD simulation of the magnetosheath, when the magnetic pressure in the magnetosheath becomes comparable to the plasma pressure, the magnetic field can help determine the direction of the flow in the magnetosheath. This capability can lead to boundary layers in the magnetosheath such as the magnetic depletion layer, a low-density, high-field-strength layer adjacent to the magnetopause that forms for a northward interplanetary magnetic field. Boundary layers can be also be formed inside the magnetosphere adjacent to the magnetopause. Figure 9.11 shows field and plasma measurements near the subsolar magnetopause from the dual co-orbiting ISEE 1 and 2 spacecraft. The dense cold plasma on the left is the magnetosheath. The oscillating density in antiphase with the magnetic field strength are mirror-mode waves that probably received their start at the bow shock and convected inward to the magnetopause. In the magnetosheath adjacent to the boundary is a transition layer in which the density drops and the magnetic field increases. The plasma presumably is draining from the boundary layer along the magnetic field flux tube as it drifts closer to the magnetosphere. The dividing line between the magnetosheath plasma and the magnetospheric plasma is where the density drops and the field strength increases before the magnetospheric boundary layer is reached. The external magnetic field is strongly northward so that there should be no reconnection near the subsolar point. However, far from the subsolar point beyond the polar cusp where the magnetic field bends downtail rather than toward the nose, the magnetosheath magnetic field and the magnetospheric fields would be antiparallel and could reconnect. Hence, these boundary layers (outer and inner) could be formed by reconnection behind the cusp. Such reconnection does not energize the magnetosphere but rather removes magnetic flux from the geomagnetic tail. This reconnection could provide some momentum transfer to the closed magnetosphere by trapping moving plasma on closed flux tubes.

When the magnetic field is southward, strong flows arise near the subsolar magnetopause, as shown in Figure 9.12. The region around the magnetopause is structured with boundary layers, but these are not stagnant regions like in the northward field case. Here there are dense fast flows that are persistent. The low-latitude "closed" field lines

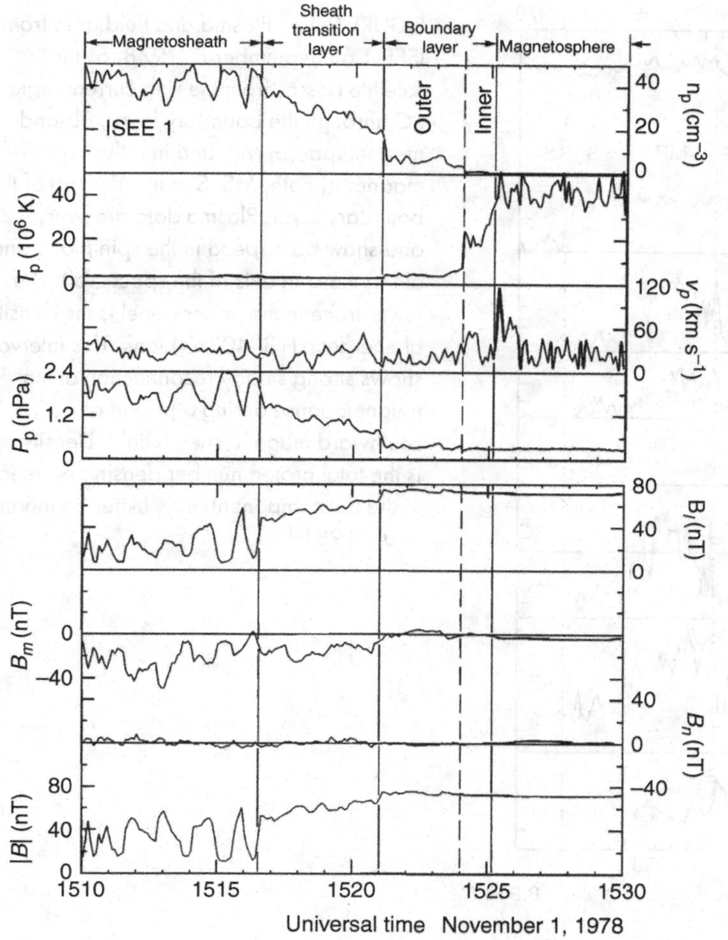

FIGURE 9.11. Magnetopause crossing observed by ISEE 1 on November 1, 1978, near the subsolar point. Plasma data were obtained from the Fast Plasma Experiment at 6 s resolution. Magnetometer data are in boundary normal coordinates. Density n_p is the total proton number density. (After Song *et al.*, 1990.)

have been connected to the flowing magnetosheath. These strong flows, produced by nearly antiparallel magnetic fields, rapidly straighten the field lines and carry them to the polar cusp. Here their curvature changes so that they slow the magnetosheath flow and transfer energy into the magnetotail (a Poynting vector across the magnetopause), storing energy there.

In the two magnetopause examples above, we examined quasi-steady states. In the first, there was no reconnection with the closed magnetospheric field lines, and in the second there was constant strong reconnection with the closed field lines. Intermittent reconnection may also occur as shown in Figure 9.13. Here, short-lived periods of convection are occurring. Both ISEE 1 and 2 see essentially the same twisted magnetic field that is clearly passing over the two spacecraft and being carried away. The appearance of hot electrons characteristic of the magnetosphere shows that these magnetic fields are connected to the "closed" magnetosphere. The magnetosheath has become connected to a patch of the magnetopause. After that patch has been convected past the spacecraft, another patch has become connected. These small patches of connection have been called flux transfer events because they are clearly carrying magnetic flux tubes from the dayside magnetosphere to the tail. They cannot be as important as steady-state reconnection because they carry less magnetic flux. Even though they have been subject to much study, the circumstances involved in their formation are poorly understood. Figure 9.14 shows a schematic of how they might appear if we could image them.

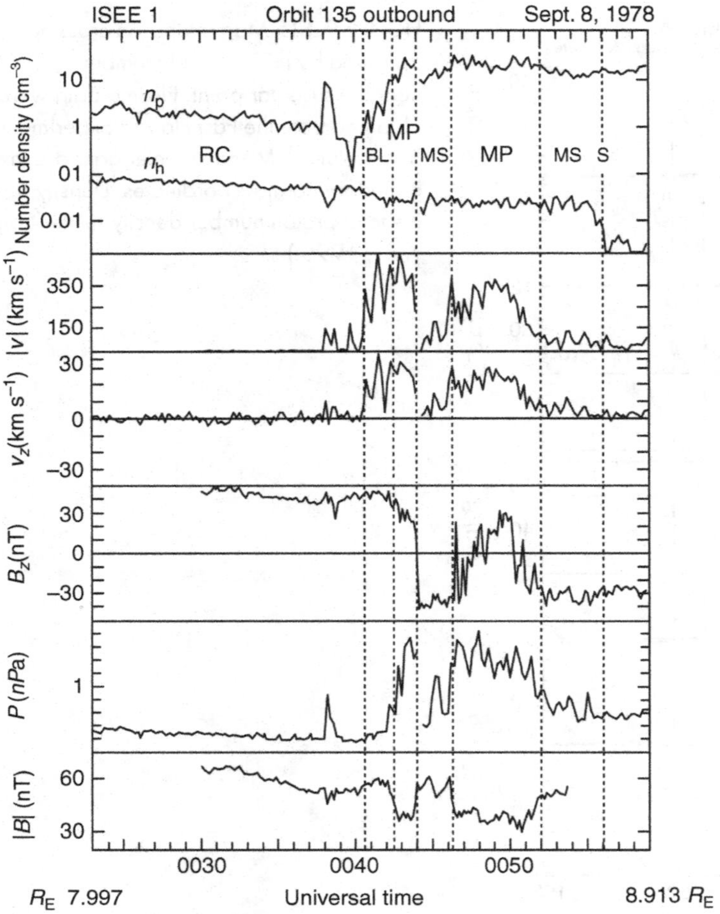

FIGURE 9.12. Plasma and field data from ISEE 1 on September 8, 1978, as the satellite passes from the ring current region, RC, through the boundary layer, BL, and magnetopause, MP, and into the magnetosheath, MS; S marks the end of the boundary layer. Plasma data are every 12 s and show both speed in the spin plane and along the spin axis of the spacecraft. The lower trace in the upper panel is the density of energetic (13–40 keV) ions. This interval shows strong steady reconnection at the magnetopause during a period of southward magnetosheath field. Density n_p is the total proton number density; n_h refers to the hot component only. (After Sonnerup *et al.*, 1981.)

Important clues on the formation of flux transfer events are provided by two planets on either end of the solar system. Mercury, where the solar wind has a small ratio β of plasma to magnetic pressure and a small radius of curvature, produces numerous, small flux transfer events. Saturn, where the solar wind has a high ratio of thermal to magnetic pressure, and dimensions of the system are large, flux transfer events are very rare.

9.5 MAGNETOSPHERIC CONVECTION

That the solar wind drives circulation of the plasma in the magnetosphere has been clear since the earliest days of auroral research. Features in the polar cap tend to flow antisunward and the disturbances in the magnetic fields in the auroral zone are consistent with such a cross-polar cap flow from noon to midnight with a return flow at lower latitudes. Initially it was not clear how the momentum transfer from the solar wind took place. As illustrated in Figure 9.15, several classical mechanisms are possible. For example, particles flowing away from the Sun could diffuse into the magnetosphere across the magnetopause. Waves could be set up on the magnetopause boundary and, if the interior of the magnetopause is dissipative, a flow along the boundary would be established. Observational evidence is that the reconnection process dominates the momentum transfer, as in our above test of the cause of the semiannual variation of geomagnetic activity.

J. W. Dungey was the first to understand how reconnection could lead to this momentum

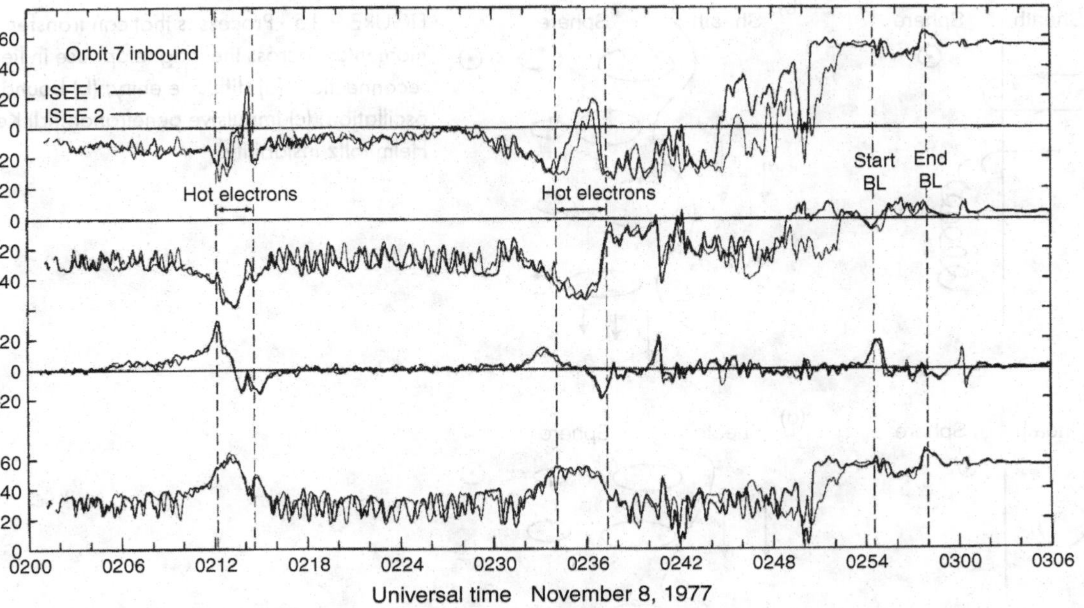

FIGURE 9.13. ISEE 1 and 2 observations of flux transfer events on November 8, 1977, in boundary normal coordinates. The regions between the dashed lines containing hot electrons are the flux transfer events. Boundary layer (BL) plasma is encountered just inside the magnetosphere. (After Elphic and Russell, 1979.)

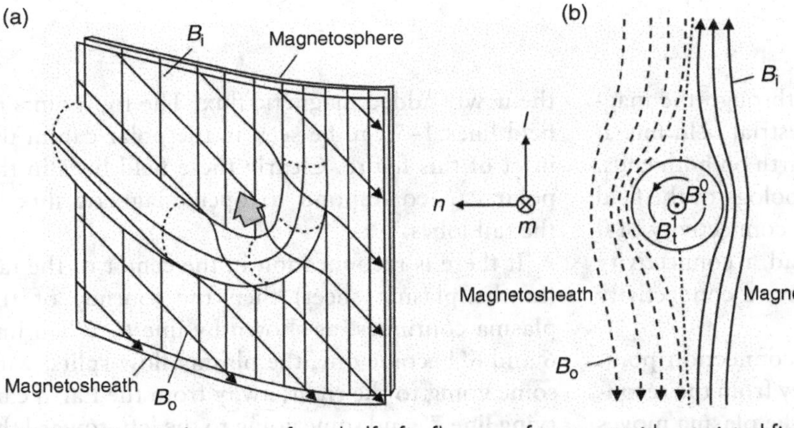

FIGURE 9.14. Schematic of top half of a flux transfer event with a twisted field. (a) Flux rope connected to the magnetosphere through the surface of the magnetopause. (b) Cross section of the magnetopause and the flux rope. (After Elphic and Russell, 1979.)

transfer when he drew the two diagrams shown in Figure 9.16. Figure 9.16(b) is consistent with the flows seen in Figure 9.12 for southward IMF. Figure 9.16(a) is consistent with the boundary layer formation process invoked to explain the observations of the magnetopause for northward IMF in Figure 9.11. Both panels depict a steady state, but in practice, as we discuss later, time variations in reconnection and convection do take place and are very important to magnetospheric processes.

Figure 9.17 is a retelling of the Dungey circulation model including the bow shock and polar cap.

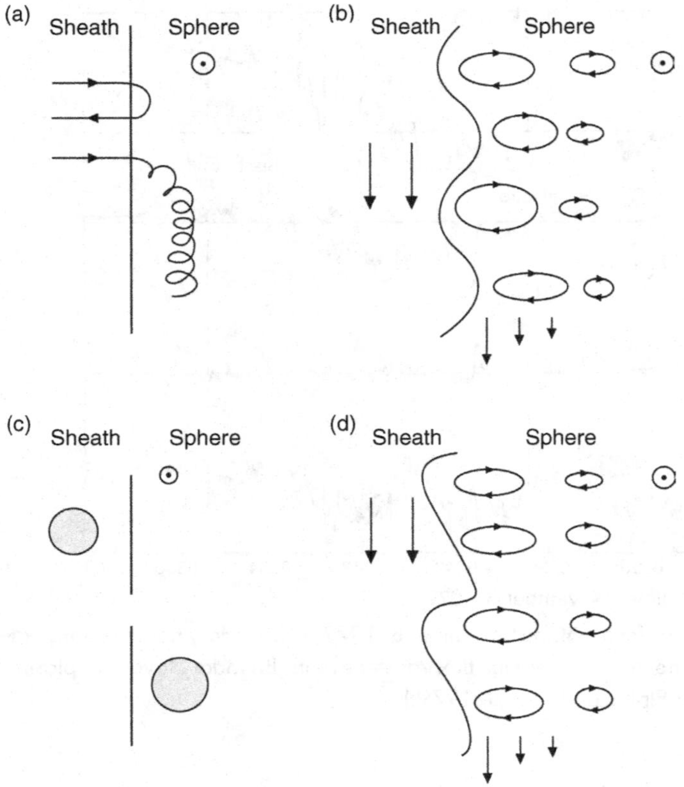

FIGURE 9.15. Processes that can transfer momentum across the magnetopause in lieu of reconnection: (a) diffusive entry; (b) boundary oscillation; (c) impulsive penetration; (d) Kelvin–Helmholtz instability.

A field line labeled 1, convected through the magnetosheath, connects with a terrestrial field line 1′ that had been connected to the Earth on both ends. We refer to this as a change in topology of the field line. Originally field line 1′ had a connectivity of 0 and the terrestrial field line 1 had a connectivity of 2. The connected field lines have a connectivity of 1 with the Earth.

The magnetic stress at the reconnection point for line 1–1′ pulls the plasma away from the reconnection point accelerating it. As the plasma moves poleward and accelerates, the connected magnetospheric magnetic field is relaxing by becoming straighter. Soon after the plasma passes the location of line 3, the bend in the magnetic field is in the opposite direction and the plasma is slowed. Energy is added to the magnetotail as the magnetic flux associated with line 4 is added to the magnetotail. If there is reconnection in the center of the tail, the plasma will sink toward the center of the tail. If not, the size of the tail expands to accommodate the newly added magnetic flux. The footpoints of field lines 1–5 can be seen in the polar cap in the inset of this figure. Clearly these field lines in the polar cap correspond to open magnetic flux in the tail lobes.

If there is reconnection in the center of the tail (in the plasma sheet) then the journey of the plasma continues, as shown by line 6. When line 6 and 6′ become one, the plasma flow splits, with some going to the right, away from the Earth carrying line 7′, and some going to the left, toward the Earth, line 7. In the ionosphere, line 7 continues to connect with the northern (and southern now) auroral zones and the plasma flows toward the dayside. Line 7′ is totally detached from the Earth and has become an "interplanetary" field line, albeit with terrestrial plasma mixed with the solar-wind plasma. Eventually, the plasma and its entrained magnetic flux reaches the dayside, and the circuit repeats if the solar-wind conditions remain the same.

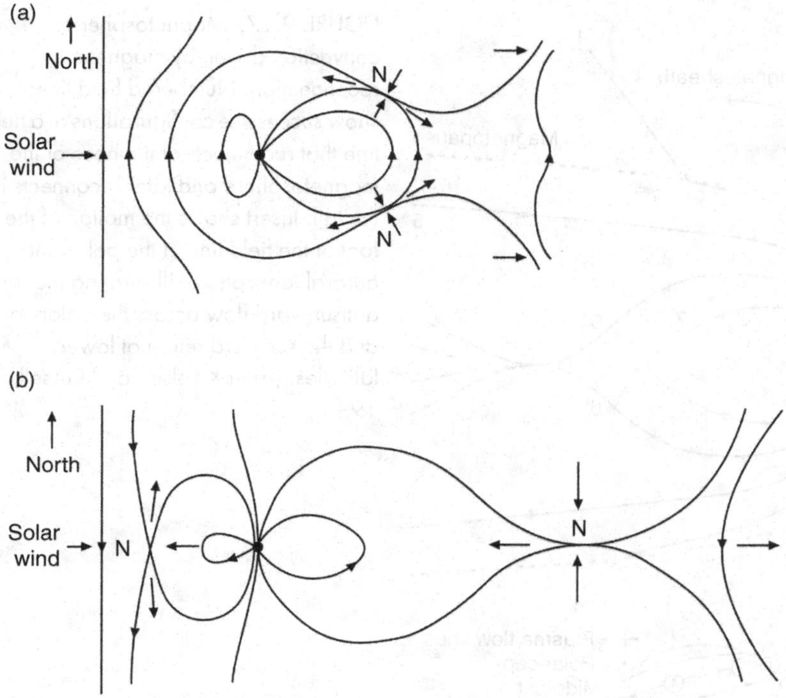

FIGURE 9.16. First models of the reconnecting magnetosphere showing field lines and flows in the noon–midnight meridian.
(a) Northward IMF (after Dungey, 1963). (b) Southward IMF (after Dungey, 1961).

At this point, one might rightly ask why does the footpoint in the ionosphere follow the same circuit in the ionosphere as the plasma in the magnetosphere. While the high-altitude plasma is connected to the solar wind (from 1 to 5 at least), and while the plasma pressure may be pushing the high-altitude flow toward the Sun from 6 to 9, why does the foot have to follow? The footpoint is connected to the Earth's magnetic field through the ionospheric magnetic field and the ionospheric plasma is rotating with the Earth. How does the magnetosphere exert control over the ionospheric plasma? What is the force that overcomes the drag of the ionosphere?

Figure 9.18 shows a bent magnetic flux tube connected at the top to high-altitude plasma and at the bottom to low-altitude plasma. Since the magnetosphere is pulling on the ionosphere, the field line must bend to apply the stress that overcomes the drag of the ionosphere. Shear in the magnetic field as shown in this diagram is equivalent to a field-aligned current. This current closes in the ionosphere and applies a force there. In the magnetosphere it closes along constant pressure surfaces in the region in which the magnetospheric pressure is applied (or solar-wind stress). Most of the interesting physical processes occur in the ionosphere. The Earth's magnetic field lines are frozen into the electrically conducting Earth and the ionospheric field lines are frozen into the magnetosphere. Moving one flux tube, say from the magnetosphere, across others, say from the Earth, in the neutral atmosphere violates no physical laws and applies no force to the Earth. In the electrically conducting ionosphere, this shear in the top and bottom of a flux tube must create parallel electric fields, and these electric fields will accelerate particles along the magnetic field. If these electric fields accelerate electrons toward the ionosphere, electron-excited auroras will result. If electrons go upward, a dark region could result, but these are not as interesting nor as well studied. We defer discussing the physics of auroras until Chapter 11.

If we examine the dawn–dusk plane, and the current systems produced both at dawn and dusk, we see the current pattern shown in Figure 9.19. It flows downward on the morning side from

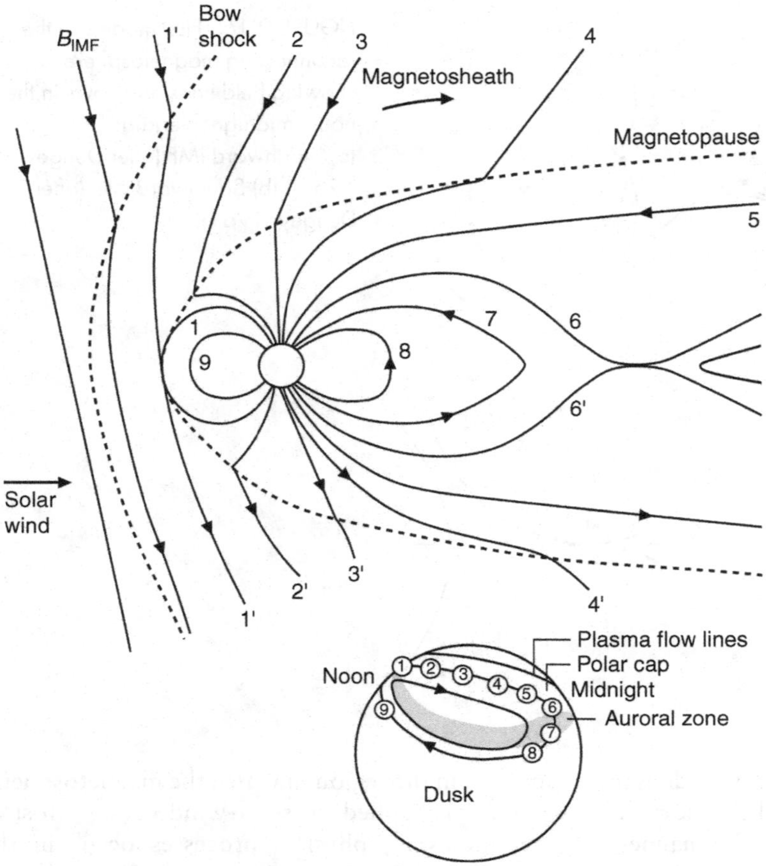

FIGURE 9.17. Magnetospheric convection driven by magnetic reconnection. Numbered field lines show successive configurations of a field line that reconnects at the nose of the magnetosphere and later reconnects in the tail. Insert shows the motion of the foot of the field line in the polar and auroral ionosphere, illustrating the antisunward flow across the polar cap and the sunward return at lower latitudes. (After Kivelson and Russell, 1995.)

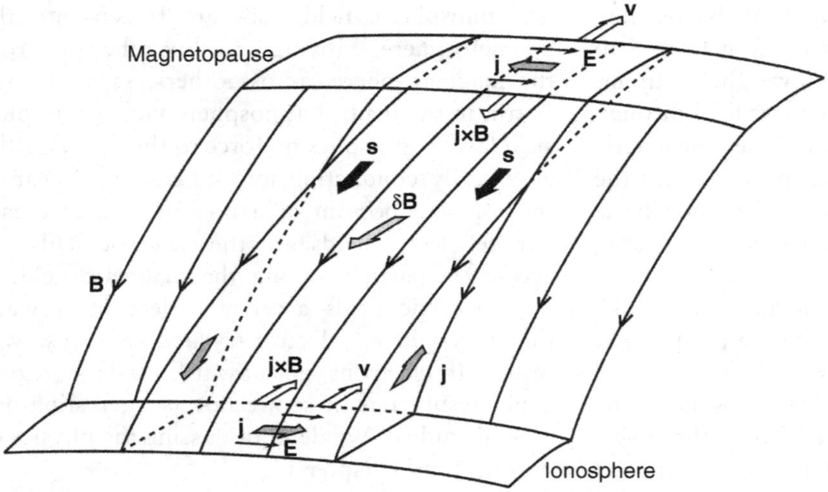

FIGURE 9.18. Magnetic configuration of flux tube being pulled into the page by magnetospheric pressure forces at the top of the flux tube by current **j** across the field. The **j** × **B** forces in the magnetosphere and the ionosphere are shown as well as the magnetic perturbation δ and the pointing vector **S**, the flux tube velocity **v**, and the electric field **E** in the ionosphere. (After Strangeway *et al.*, 2000.)

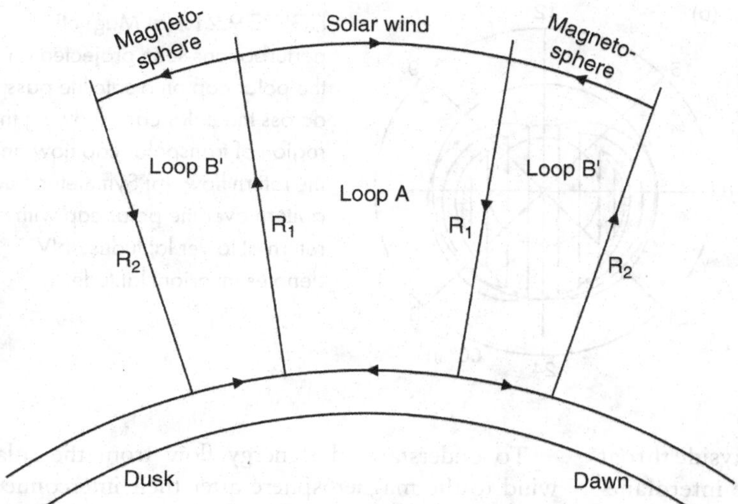

FIGURE 9.19. Region 1 and 2 current systems in the northern hemisphere in the dawn–dusk plane.

the magnetosphere–magnetosheath interface and both across the polar cap and equatorward in the ionosphere. The force in the auroral ionosphere is sunward (eastward), and in the polar ionosphere it is antisunward at dawn. On the dusk side, the current flows upward to the magnetosphere–magnetosheath interface, and exerts a westward (sunward at dusk) force in the auroral ionosphere.

The ionospheric current is carried from dawn to dusk across the polar cap by Pedersen conductivity, which acts in the direction of the component of the electric field that is perpendicular to the magnetic field. There are three loops in each of the northern and southern hemispheres (not shown). Loop A is driven by the solar-wind coupling and loops B and B′ are driven by magnetospheric stresses. Over the long term, we expect loops B and B′ to return all the magnetic flux transported through loop A, but the magnetosphere can store magnetic flux in the magnetotail for later release. During this storage period, the magnetic flux content of the closed dayside magnetosphere decreases and the open flux in the magnetotail increases. When the flux on the dayside decreases, the magnetopause moves inward, even when the dynamic pressure is constant, explaining the behavior seen in Figure 7.5, when the IMF is southward. While we have labeled the loops A and B here, the currents are generally called **region 1** and **region 2**, according to

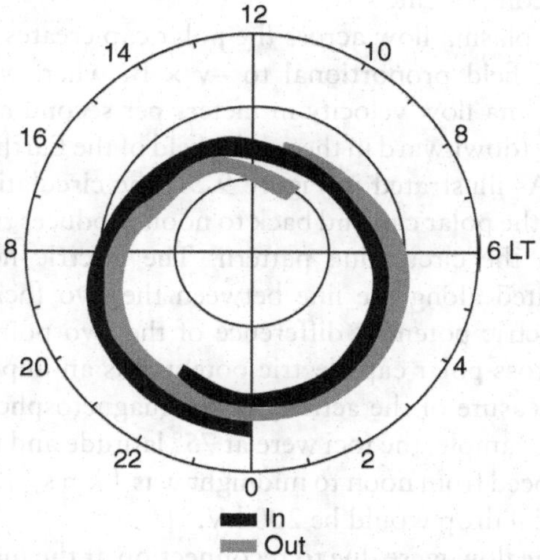

FIGURE 9.20. Current direction into or out of the northern ionosphere for region 1 and 2 currents.

their location relative to the polar cap and not given the physical distinction that is their due. Thus, region 1 (R_1) currents vary with both changes in solar-wind stress and magnetospheric stress, whereas region 2 (R_2) currents respond only to the magnetospheric stress changes.

If we look down on the current systems from above, these field-aligned currents would form a spiral pattern, as shown in Figure 9.20. The region of overlap on the dayside forms a throat for entry of the reconnected flux tubes into the polar cap,

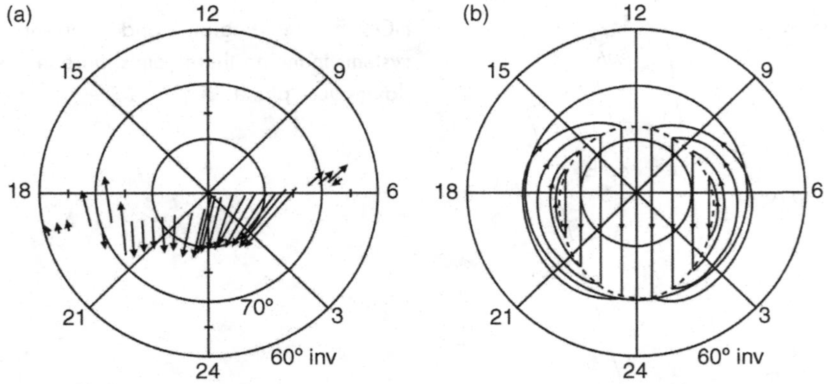

FIGURE 9.21. (a) Magnetic perturbations seen projected on the polar cap on a satellite pass across the polar cap, showing the region of transpolar cap flow and the return flow. (b) Symmetric flow pattern over the polar cap with return at lower latitudes. INV denotes invariant latitude.

and, on the nightside, their exit. The dayside throat is sensitive to the y-component of the interplanetary magnetic field (that opposite the Earth's orbital motion) as well as being affected by the north–south component.

The plasma flow across the polar cap creates an electric field proportional to $-\mathbf{v} \times \mathbf{B}$, where $\mathbf{v}$ is the plasma flow velocity in meters per second and $\mathbf{B}$ is the (downward in the north) field of the Earth in tesla. As illustrated in Figure 9.21, the circulation across the polar cap and back to noon produces two foci in the circulating pattern. The electric field integrated along the line between the two foci is the electric potential difference of the two points. This cross-polar cap electric potential is an important measure of the activity of the magnetosphere. If, for example, the foci were at 75° latitude and the flow speed from noon to midnight was 1 km s^{-1}, the potential drop would be 200 kV.

If the flow were due to reconnection at the magnetopause with a solar-wind flow of 500 km s^{-1} and a southward magnetic field of 10 nT, then a region of solar wind $6.3R_E$ across would have to reconnect with the magnetosphere to power this potential difference. In such a situation, one would say that the solar wind applied 200 kV to the magnetosphere, as sketched in Figure 9.22. However, this description is misleading because there are no capacitor plates here. The electric field results from the flow of the magnetospheric plasma, produced by the drag of the solar wind that has reconnected and by sunward pressure from the tail. The magnetosphere has mass and is controlled by forces that accelerate this mass.

To understand the energy flow from the solar wind to the magnetosphere after their interconnection, let us consider reconnection when the interplanetary magnetic field is 10 nT southward and the solar wind is flowing at 500 km s^{-1}. If $6.3R_E$ is the width of the region in the solar wind that later reconnects with the magnetopause, the potential drop across the polar cap will be 200 kV, a value expected to be associated with significant geomagnetic activity, such as an increase in the strength of the ring current and expansion in the size of the auroral oval. This voltage drop is directly proportional to the reconnection rate, which is 0.2 MWb s^{-1}. If the north lobe of the magnetotail is a semicircle with radius $25R_E$ and a magnetic field strength of 30 nT, it contains 1.2 GWb, so the tail could be replenished by reconnection in 6000 s. The energy stored in $10R_E$ of magnetotail length in one lobe is 900 TJ. The energy stored in the tail is extracted from the kinetic energy of the solar-wind flow. If the solar wind has a proton density of 10 cm^{-3} at this time, the kinetic energy flux is 1 mJ m^{-2} s^{-1}. The energy arrives with the convected magnetic flux, so it too arrives from a region $6.3R_E$ across the solar-wind flow, providing 42 kJ m^{-1} s^{-1}, and, in the 6000 s needed to replenish the tail, the solar wind-provides 250 MJ m^{-1}, where the vertical thickness of the reconnecting layer is a free parameter that was not needed to calculate the flux reconnected but is needed to specify the energy. If we replenish the energy in the tail when we replenish the flux, the needed energy is available from a slab of thickness only $0.6R_E$ thick (or $1.2R_E$ if one includes both lobes). The kinetic energy

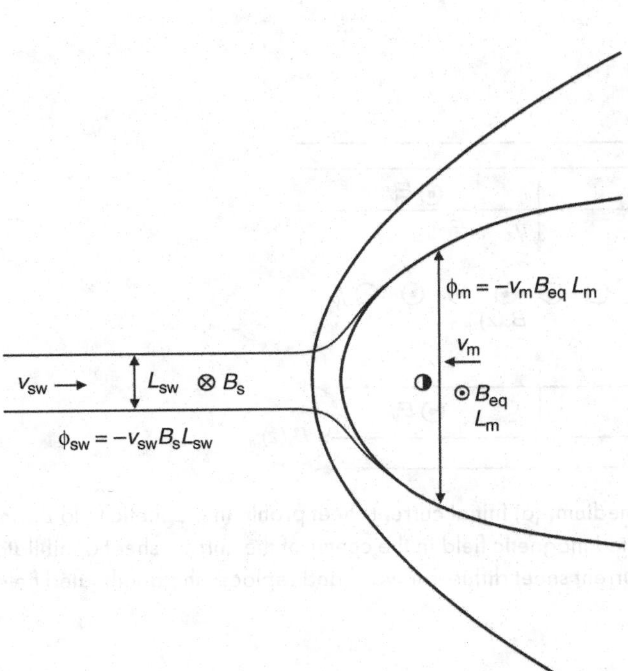

FIGURE 9.22. Mapping a potential drop across a reconnecting slab of solar-wind plasma to the magnetopause and to the day and night equatorial magnetosphere. The plasma flows toward the nose of the magnetopause from both sides, and then flows up and down out of the page and over the polar caps.

from the slab, or more precisely the magnetosheath boundary layer that it forms, flows into the tail via the Poynting vector associated with the small, interconnecting component of the magnetic field.

We can easily calculate how large this normal component needs to be by integrating the Poynting vector over the tail magnetopause, leaving the normal component of the magnetic field across the boundary an unknown, and equating the energy flux to the rate needed to supply the energy in 6000 s. We continue with our example: a $25R_E$ radius tail with a 30 nT field strength, storing its 900 TJ of energy in $10R_E$ length of the tail. The $\mathbf{j} \times \mathbf{B}$ force on the boundary layer plasma slows the flow, and the energy crosses into the tail at a rate equal to the Poynting vector $\mathbf{E} \times \mathbf{B}\mu^{-1}$ integrated over the surface of the tail. We find that our $6.3R_E$ wide by $0.6R_E$ thick solar-wind slab upon reconnection can provide the Poynting vector required with only 1 nT of normal component across the surface of the tail. The amount of magnetic flux that would "leak out" of the lobe as a result of the connectivity would be only 2.5% of its flux content. Again, this is a very reasonable value integrated over the tail magnetopause. The normal component of the magnetic field would be small enough that it would be almost undetectable at a typical magnetopause crossing at disturbed times.

In short, the solar-wind kinetic energy flux can easily supply the energy needs of the magnetosphere at disturbed times through reconnection with the magnetospheric magnetic field. The connectivity extracts momentum from a boundary layer along the tail magnetopause and energy flows from this boundary layer into the tail, providing a store of magnetic energy for possible later release.

9.6 MAGNETIC RECONNECTION

In Section 9.3 we saw how the direction of the interplanetary magnetic field intimately controls processes within the magnetosphere. We determined how the integrated global interaction as monitored by geomagnetic activity varies, as well as how the processes acting locally at the magnetopause vary. We have also examined how the linkage of the solar wind and the magnetosphere through reconnection leads to circulation of the plasma in

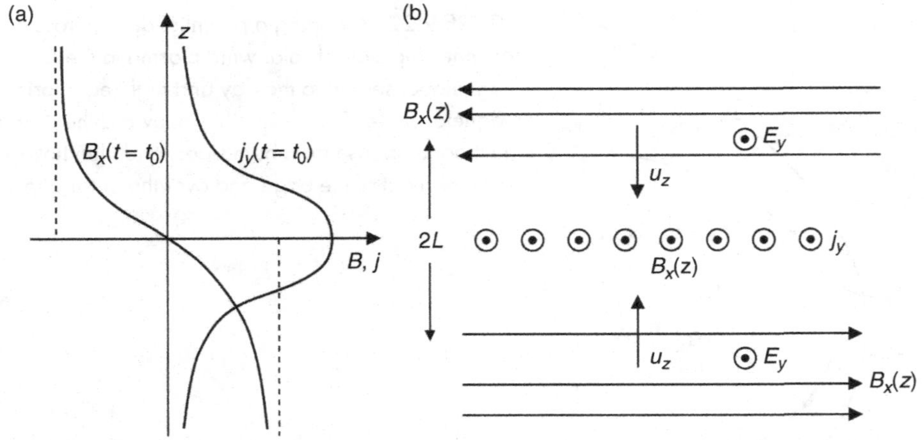

FIGURE 9.23. Current sheet evolution in a resistive medium. (a) Initial current sheet profile in magnetic field B_x and current density j_y. (b) With time, the oppositely directed magnetic field in the center of the current sheet annihilates and heats the plasma. The field on the edge of the current sheet diffuses inward and replaces the annihilated field until all magnetic flux is gone.

the magnetosphere as well as to energy flow into the magnetosphere. However, we have not yet explained how the underlying process, reconnection, works. Since we do not yet understand the physics of the reconnection process as deeply as we should on even the local scale, let alone globally, our discussion will have to be somewhat limited, but we attempt to shed some light on the physics of reconnection that will help the reader assess the various models of reconnection.

9.6.1 The physics of reconnection

If antiparallel magnetic fields lie on either side of a resistive current sheet in the geometry sketched in Figure 9.23, the magnetic fields will diffuse into the sheet, annihilate each other, and convert the magnetic energy into heat, raising the temperature of the plasma. This happens in nature, and it can happen in computer codes, whether the codes are MHD, hybrid, or fully kinetic.

When observations showed that the Sun was magnetic and revealed the Sun's rapid acceleration of energetic particles, scientists of the time turned to collisionless reconnection to explain this acceleration. It was clear that the magnetic geometry mattered, but it was decades later before there was agreement on the geometry. P. A. Sweet made reconnection more rapid by reducing the diffusion to a small region (Sweet, 1958) as sketched in Figure 9.24, but this model is still too slow.

H. E. Petschek introduced the action of MHD waves in the acceleration which increased the rate of reconnection and the energization (Petschek, 1964), but the model still depended on diffusion near the region of reconnection. Others, notably R. G. Giovanelli and his then postdoc J. W. Dungey, emphasized the three-dimensional nature of reconnection, building models of neutral points where the magnetic field crossing the reconnecting current sheet went to zero and not neutral lines, and avoiding planar current layers that slowed reconnection in the geometry of Sweet and Petschek. These latter efforts culminated in the magnetospheric models of Dungey that we discussed earlier.

The geometry of both magnetospheric reconnection regions may be described using the sketch in Figure 9.25. As applied to the terrestrial magnetopause, for which plasma and magnetic field data were presented in Figure 9.12, the Earth and the magnetosphere are on the right and the magnetosheath is on the left. Plasma flows toward the narrow X-shaped region from both sides and is accelerated by magnetic stresses upward and downward in the directions labeled "exhaust." The ($-\mathbf{v} \times \mathbf{B}$) electric field is parallel on either side

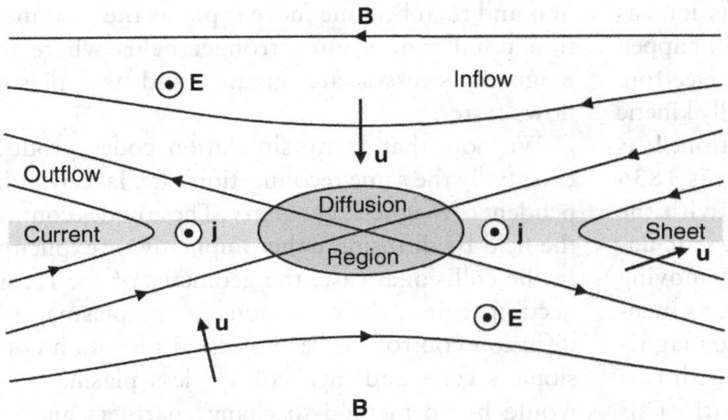

FIGURE 9.24. Current sheet evolution in a resistive medium with outflow channels. If the heated plasma can escape and diffusion can be limited to a small region, the reconnection rate can be increased but the rate is still limited by the rate of diffusion. This process was suggested by Sweet (1958) and improved upon by Petschek (1964).

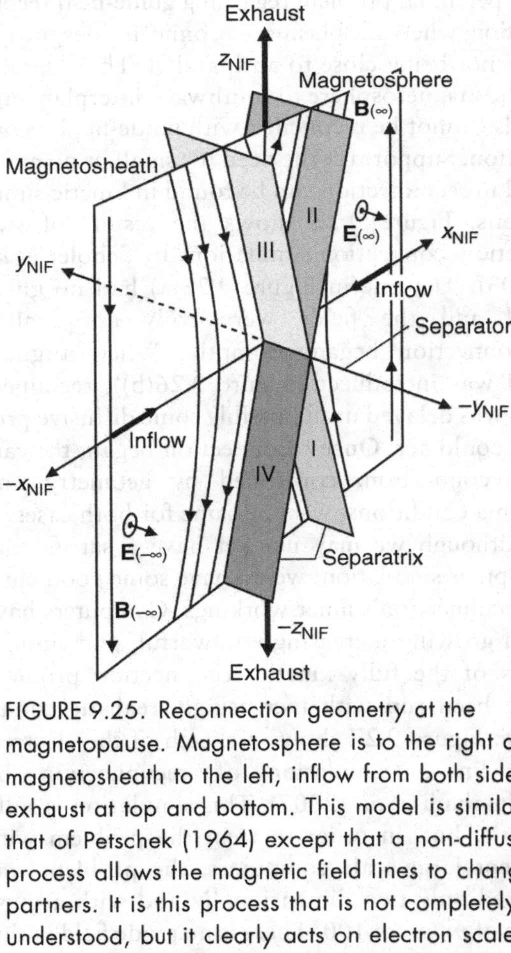

FIGURE 9.25. Reconnection geometry at the magnetopause. Magnetosphere is to the right and magnetosheath to the left, inflow from both sides with exhaust at top and bottom. This model is similar to that of Petschek (1964) except that a non-diffusive process allows the magnetic field lines to change partners. It is this process that is not completely understood, but it clearly acts on electron scales, not MHD or even ion scales. NIF = normal incidence frame.

of the exhaust zone, and, since the magnetic field directions are opposite, this means that the plasma flows inward from both sides. The exhaust regions are critical. If they are both blocked, the reconnection will stop. If one is blocked, the reconnection point must move away from the blockage. The plasma conditions on the two sides of the exhaust region are likely to be different either at the magnetopause or in the tail. They are likely to be moderate- to high-beta plasmas where the plasma stresses are dominated by the plasma. It is only during quiet times that there may be a little hot plasma in the outer magnetosphere, allowing the solar wind to be stood off by predominantly magnetic pressure. Similarly, in the solar-wind/magnetosheath plasma, occasionally there are ICMEs, giant magnetic flux ropes in which the magnetic stresses are dominant. Since magnetic stress depends on the linkage of fields from two sides of the boundary, a weak partner on either side should lower the flow rate out of the reconnection region. This place where the opposing magnetic fields meet is, of course, key to making reconnection work. The magnetic field lines must be readily able to change partners here. The field lines with two feet on the Earth must be able to join with those with no feet on the Earth and each become field lines with one foot on the Earth. If there is no process that can do this, there will be no reconnection. If this exchange process is rapid, reconnection will be rapid as long as the accelerated plasma has an escape path.

In a collisionless plasma, this exchange of partners cannot happen on the MHD scale. Fields are frozen into the flow. In a hybrid code that treats ions as particles and electrons as a fluid, this cannot happen either, because field lines are defined by the electron, not the ion, motion. It might happen in a fully kinetic code that properly treats the electron motion. It is difficult to simulate because the electron is 1836 times less massive than the proton, but with the same quantity of charge, so many more calculations are needed to follow the rapidly moving electrons than the protons. Furthermore, as mentioned earlier, there is a large area on the magnetosphere where closed field lines (those with two feet on the Earth) touch the magnetosheath. This large volume, in which we would need to solve the kinetic reconnection problem and the global interaction problem simultaneously, makes the solution very difficult. The reconnecting regions may also be small and transitory as the magnetosheath magnetic field varies its direction

The situation is simpler in the geomagnetic tail, perhaps even tractable. First, we have symmetry and field lines that should be antiparallel. The approach to an unstable magnetic geometry in the tail is quite the opposite to the approach on the dayside magnetopause. In the tail, the evolution to a reconnection event is generally stretching a connected geometry into a thin sheet where the very weak, connecting field allows reconnection to begin. At the magnetopause, the magnetic fields on either side of the boundary have no connection before the reconnection event begins. The antiparallel geometry occurs when a rotation of the magnetosheath field about a direction normal to the magnetopause aligns it in exactly the opposite direction to the magnetospheric field.

In the tail we turn Figure 9.25 on its side so one side of the figure becomes the north lobe and the other the south lobe. In the region that ultimately becomes unstable to reconnection, the current sheet is stabilized against reconnection by a sufficiently strong normal field across the sheet. This normal component of the field can be diminished by stretching of the tail. In the tail the plasma is not generally being replenished as it evolves to instability as it is at the magnetopause. Thus in the tail the stretching could bring the high-field, low-beta regions closer to each other. Here, if reconnection began in a thin current sheet, it might be slow at first and then become more rapid as the reconnection point moves into stronger fields where the magnetic stresses are greater and the plasma flows faster.

We note that many simulation codes produce essentially the same reconnection rate, largely independent of magnetic geometry. These codes connect the field by diffusion either implicitly or explicitly. In the collisional case, the geometry of the reconnection region, the conditions in the plasma, and diffusion control the reconnection rate. In a collisionless code and in a collisionless plasma there would be no method to change partners and the reconnection would not be expected to happen in the non-antiparallel situation. This problem has led to a perennial problem regarding **guide-field reconnection**, where the plasma "reconnects" despite the field not being close to antiparallel. The response of the magnetosphere to southward interplanetary fields cannot be reconciled with guide-field reconnection. Support for the need for small or no guide field in reconnection can be found in kinetic simulations. Figure 9.26 shows the results of two kinetic reconnection simulations by Scholer *et al.* (2003). The one in Figure 9.26(a) had no guide field and the fields were truly antiparallel. Reconnection began promptly. When a guide field was introduced (Figure 9.26(b)), reconnection was delayed until possibly some diffusive process could act. Once reconnection began, the rate of reconnection, controlled by geometry and plasma conditions, was the same for both cases.

Although we may not yet have a satisfactory computer simulation, we do have some good clues to reconnection's inner workings. Computers have been growing increasingly powerful, and simulations of the fully kinetic reconnection problem have been run with increasingly realistic conditions. Figure 9.27 shows a sketch of the electron flow in a simulation of reconnection by Karimabadi *et al.* (2007). This simulation is fully kinetic, but, in order to treat the problem with current large-scale computers, the problem was treated only two dimensionally and with a mass ratio of m_i/m_e of 100. There is no guide field in this simulation. We see immediately that the electron flow is complex. The electrons flow uniformly into a small region in the center and then neck down

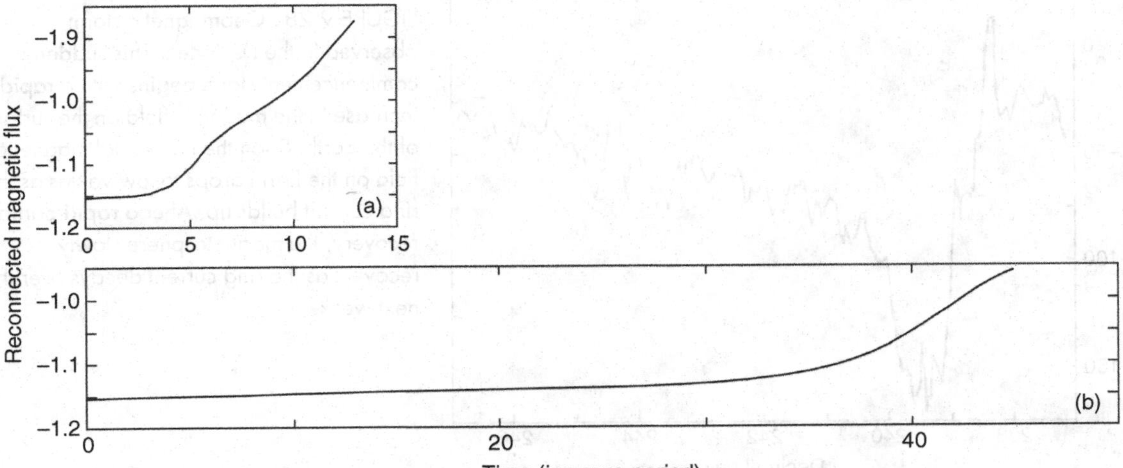

FIGURE 9.26. Reconnection rate versus time in a fully kinetic simulation with no guide field (a) and with a guide field (b). The presence of a guide field significantly delays the onset of reconnection but once started the rates are the same, possibly controlled by geometry and the Alfvén speed. (After Scholer *et al.*, 2003.)

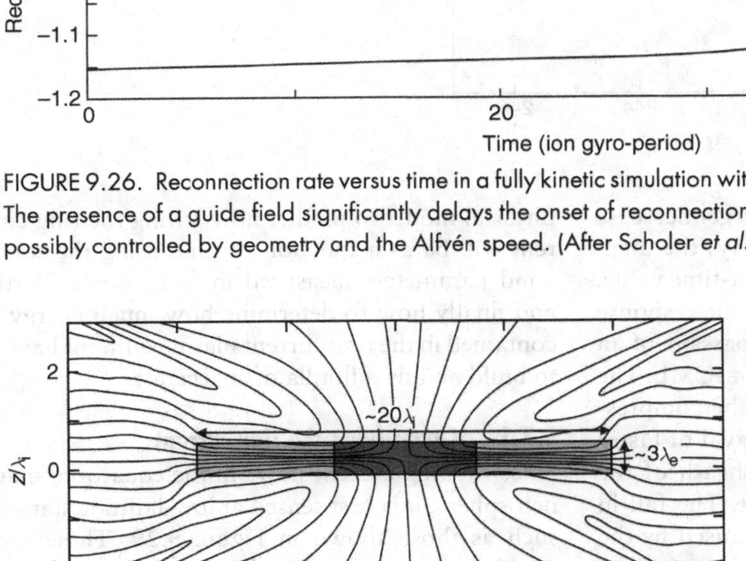

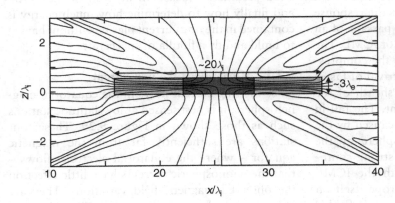

FIGURE 9.27. The multiscale structure of the x-point in a fully kinetic simulation by Karimabadi, Daughton, and Scudder (2007). Electron streamlines are shown. In the center is a smooth uniform inflow heading into strong outflow jets.

into a horizontal electron jet over a larger region. This behavior indicates there are large electric fields here and electron pressure gradients with scale sizes less than an electron gyro-radius or an electron inertial length. These are the features that can demagnetize the electrons and allow field lines to change partners. These simulations are helpful in training our intuition, but it may be some time until such simulations with realistic mass ratios and three-dimensional geometries can be run in the context of a global simulation. Perhaps, too, there will be new observational discoveries in space from the new Magnetospheric Multiscale (MMS) mission that was launched on March 13, 2015.

9.7 GEOMAGNETIC STORMS AND THE RING CURRENT

Prolonged reconnection of the intensity used in our example in Section 9.4 not only stores energy in the tail, but also energizes the plasma on the closed field lines in the outer magnetosphere. This phenomenon is called a **geomagnetic storm** and was recognized early in the study of geomagnetic activity. Figure 9.28 shows the classic pattern exhibited by a geomagnetic storm in low-latitude geomagnetic records. A sudden commencement geomagnetic storm such as this begins with a rapid jump in the horizontal component of the magnetic field

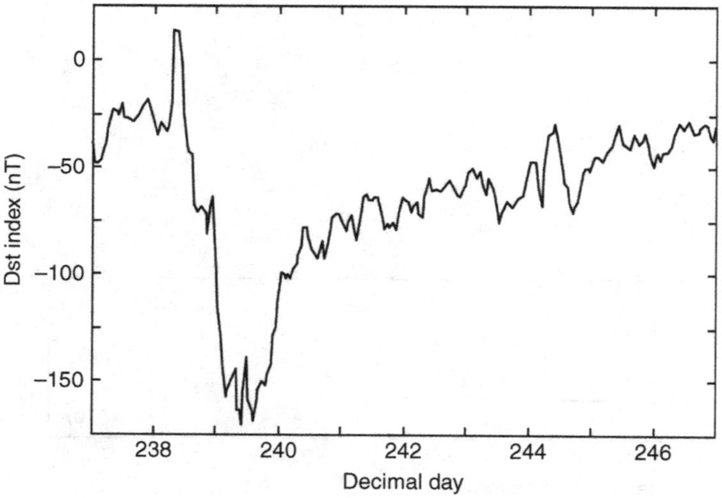

FIGURE 9.28. Geomagnetic storm observed in the Dst index. This sudden commencement storm begins with a rapid increase in the magnetic field on the surface of the Earth. Soon the horizontal magnetic field on the Earth drops to low values as the ring current builds up. After a rapid partial recovery, the magnetosphere slowly recovers as the ring current decays over the next week.

followed some time later by an even larger decrease in the magnetic field. After several days the magnetic field recovers to its previous quiet-time value. We now understand this behavior as the response of the Earth's magnetosphere to the passage of an interplanetary coronal mass ejection or ICME. The jump in the magnetic field is due to the compression of the magnetosphere by the arrival of faster denser plasma, usually that in the sheath of the ICME behind its leading shock front. The fall in the field strength is due to inflation caused by the energy pumped into the magnetosphere by the prolonged reconnection due to the strong southward magnetic field associated with the ICME magnetosheath or the driving flux rope itself on those locations where the ICME magnetic field is southward. This enhanced coupling may be over in about a day and it is possible for energy to pass through the dayside magnetosphere and be lost into the magnetosheath. If it does become trapped, it takes many days for the magnetosphere to lose that energy. One loss process is the generation of ion cyclotron waves that remove energy from the gyro-motion and allow the ions to move along the field line into the atmosphere. Another perhaps more effective way is for the ions to charge exchange with the neutral atmosphere and produce fast neutrals that carry energy out of the system.

In this section we discuss how to create a quantitative index of the ring current, how to use a simple physical model to predict how strong the ring current will be as a function of time using the solar-wind parameters measured in front of the Earth, and finally how to determine how much energy is contained in the ring current plasma without having to build and fly a flotilla of spacecraft.

9.7.1 Quantifying the ring current

Since the ring current flows in the equatorial magnetosphere, it is best sensed at low-latitude stations such as those shown in Figure 9.29. These four stations are sufficiently far from the magnetic equator – where the equatorial electrojet flows – that these ionospheric currents have little effect on the observed magnetic field variations. They are also far from the auroral currents. The quiet days during the year are used to establish the average minimum ring current and the background magnetic field. Then this is subtracted from the records so that only the disturbance from this baseline is recorded. The average worldwide value using the five stations is then taken. This is called the **Dst index**, for disturbance storm time. As we shall see, this simple measurement has a very quantitative interpretation.

9.7.2 Predicting the ring current strength

The passage of an ICME and the subsequent geomagnetic storm is illustrated with interplanetary field and plasma data in Figure 9.30. The dynamic

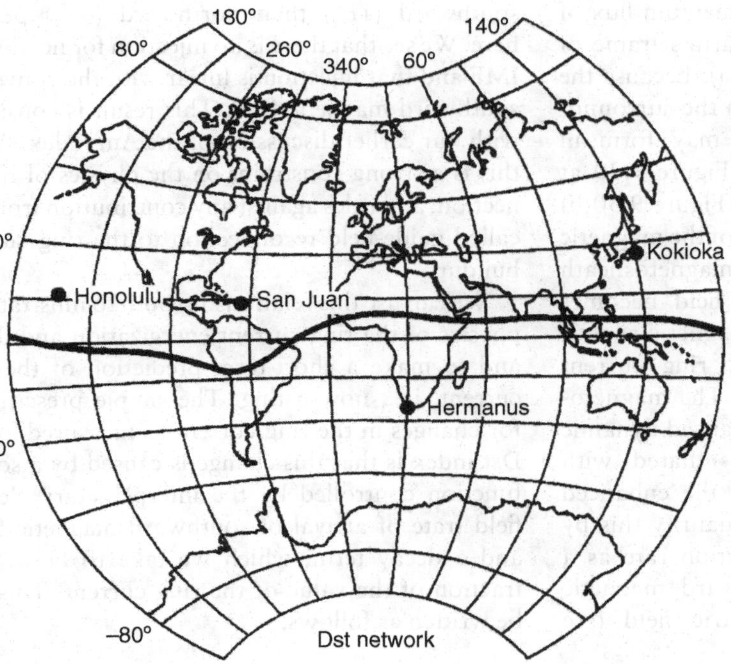

FIGURE 9.29. Location of stations used to calculate the Dst index, which is used to identify and quantify geomagnetic storms. The heavy black line near the rotational equator is the geomagnetic equator, where the equatorial electrojet flows.

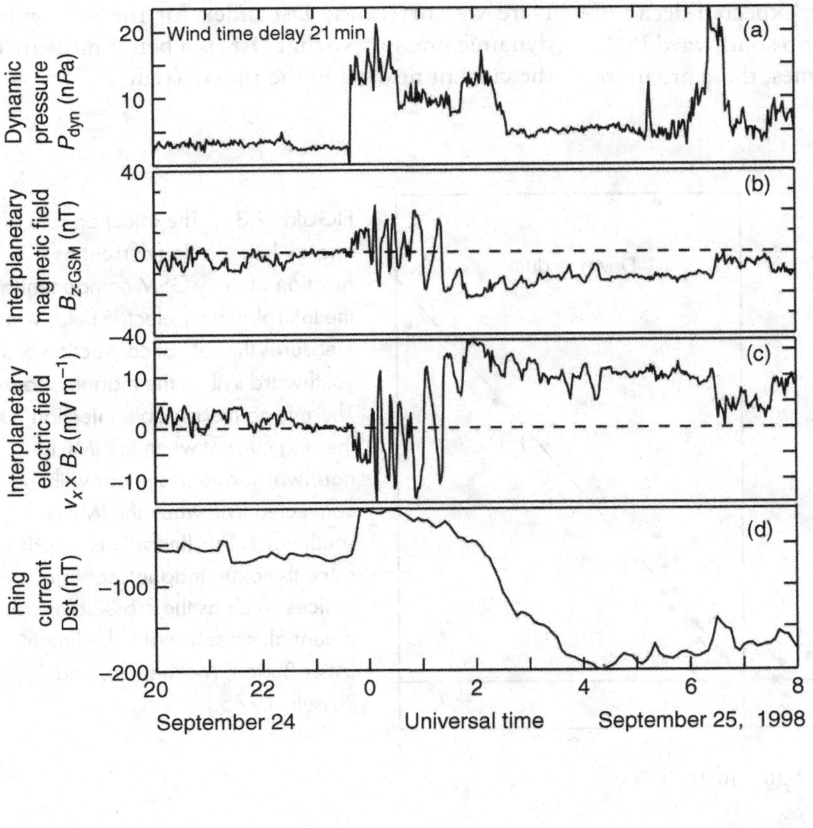

FIGURE 9.30. Solar-wind conditions during a sudden storm commencement. (a) Solar-wind dynamic pressure shifted to the time of arrival of the pressure increase at the magnetosphere. (b) North–south component of the magnetic field in GSM coordinates. (c) Interplanetary electric field, which measures the rate at which southward interplanetary magnetic field is convected to the magnetosphere. (d) Dst index. The sudden commencement is clearly generated by the strong increase in the dynamic pressure, and the main phase begins when sustained strong convection of southward magnetic flux begins.

pressure of the solar wind, the momentum flux of the solar wind measured in the Earth's frame of reference, is shown in Figure 9.30(a). Because the ICME is often traveling faster than the surrounding solar-wind plasma, a shock may form in front of the ICME, as it does in Figure 9.30 at about 23.45 UT on September 24. Figure 9.30(b) shows the north–south component of the magnetic field. It is rather turbulent in the magnetosheath of the ICME, but the magnetic field becomes strong and southward when the magnetic rope is encountered. The magnetospheric ring current reacts as shown in Figure 9.30(d). The magnetosphere is compressed when the enhanced dynamic pressure arrives and it is then inflated with energetic plasma at the arrival of the enhanced flux of southward IMF. We can quantify this by examining the change in the injection rate as a function of the convected southward magnetic field i.e., the interplanetary electric field (see Figure 9.30(c)).

Figure 9.31 shows this calculation. We measure the injection in terms of the change in the ring current per hour, correcting for expected decay of the ring current as if there were no southward IMF. Since we examine only storm times, there are more southward ($+E_y$) than northward ($-E_y$) periods here. We see that there is no injection for northward IMF and that injection is linear with the convected southward magnetic field. This result is consistent with our earlier discussion of the Am index. Again this is a strong constraint on the physics of reconnection, and rules against any contribution from so-called guide-field reconnection to the ring current buildup.

We can use this relationship both to illustrate the physics of the ring current energization and decay and to make a short-term prediction of the ring current, i.e., nowcasting. The simple prescription for changes in the ring current as measured by the Dst index is that this change is caused by a source function controlled by the interplanetary electric field (rate of arrival of southward magnetic field) and a decay term, which we take to be a fixed fraction of the value of the ring current. This can be written as follows:

$$\frac{d}{dt}\mathrm{Dst}_0 = F(E) - a\mathrm{Dst}_0. \qquad (9.1)$$

Here we correct the Dst index for the solar-wind dynamic pressure so that Dst_0 is a better measure of the current flowing in the ring current,

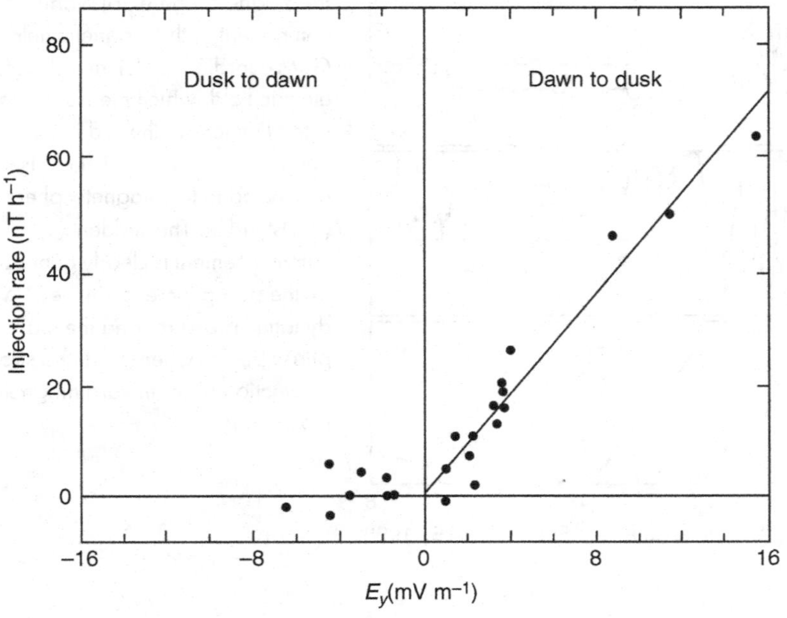

FIGURE 9.31. The injection rate of energy into the ring current as a function of the y-GSM-component of the interplanetary electric field, which measures the rate of convection of the southward IMF to the magnetopause. There is no measurable injection into the ring current when the IMF is northward, and it is linear with the convected IMF when the IMF is southward. This linearity is surprising since the other magnetospheric indices, such as the cross-polar-cap potential, appear not to be linear. (After Burton, McPherron, and Russell, 1975.)

$$\text{Dst}_0 = \text{Dst} - b(P)^{1/2} + C, \tag{9.2}$$

where $b = 5 \times 10^{-4}$ T(N m^{-2})$^{-1/2}$, C is taken to be 20 nT, the value of the field produced by the quiet-time ring current (when the Dst index is defined to be zero), and P is the dynamic pressure in newtons per meter squared.

The function $F(E)$ is obtained from Figure 9.31 and is

$$F(E) = 0, E_y < 0.50 \text{ mV m}^{-1}, \qquad F(E) = d(E_y - 0.5), E_y > 0.50 \text{ mV m}^{-1}, \tag{9.3}$$

where $d = -1.5 \times 10^{-3}$ nT(mV m^{-1})$^{-1}$ s^{-1}.

The parameter a determines how quickly the ring current decays if the energy input drops to zero; $a = 3.6 \times 10^{-5}$ s^{-1} gives an e-folding decay time of 7.7 h.

We calculate the electric field from

$$E = -vB_z \times 10^{-3} \text{mV m}^{-1}$$

and the dynamic pressure from

$$P = 1.67 \times 10^{-15} n_p v^2 \text{ m}^{-2},$$

where n_p is the number of protons per cubic centimeter and v is measured in kilometers per second.

When the energetic particles in the ring current become trapped on the closed field lines they begin to drift around the magnetosphere due to gradients in the magnetic field, and eventually they can drift completely around it. As we discuss in Chapter 10, this gradient drift velocity is given by

$$\mathbf{v}_d = W_\perp \mathbf{B} \times \nabla B / qB^3, \tag{9.4}$$

where ∇B is the gradient of the Earth's field at the orbit of the particle, B is the field strength there, and $W_\perp$ is the perpendicular energy of the particle. If the particle is at a distance LR_E from the center of the Earth, this drift produces a magnetic field at the center of the Earth of

$$\delta B_{\text{dnft}} = \frac{-3\mu_0 W_\perp}{4\pi \, R_E^3 B_0} \hat{\mathbf{z}}. \tag{9.5}$$

The particle also has a gyro-rotational current due to its gyro-motion around the field line. This produces a northward magnetic field at the center of the Earth of

$$\delta B_{\text{gyro}} = \frac{\mu_0}{4\pi} \frac{W_\perp}{R_E^3 B_0} \hat{\mathbf{z}}. \tag{9.6}$$

The total field due to these two sources is

$$\Delta B_{\text{part}} = \frac{-\mu_0}{2\pi} \frac{W_{\text{part}}}{B_0 R_E^3} \hat{\mathbf{z}}. \tag{9.7}$$

The total magnetic energy in the Earth's dipole magnetic field above the surface of the Earth is

$$W_{\text{mag}} = \frac{4\pi}{3\mu_0} B_0^2 R_E^3, \tag{9.8}$$

where W_{part} is the total energy in trapped particles. Thus the ratio of the magnetic field due to the ring current particles to the magnetic field strength at the surface of the Earth is

$$\frac{\Delta B_{\text{part}}}{B_0} = -\frac{2}{3} \frac{W_{\text{part}}}{W_{\text{mag}}} \hat{\mathbf{z}}. \tag{9.9}$$

This equation is called the **Dessler–Parker–Sckopke relationship** after the three scientists who first worked on this problem (Dessler and Parker, 1959; Sckopke, 1966). This equation predicts the field of the ring current at the center of the Earth, but the center of the Earth never sees this transient field. The interior of the Earth is highly electrically conducting and hence this field does not penetrate far below the surface. If we account for this shielding, which enhances the surface effect, we find that

$$\Delta B(\text{nT}) = -W_{\text{ring}}/2.8 \times 10^{13}(\text{J}), \tag{9.10}$$

where W_{ring} is the total energy of the ring current particles in joules. Thus, a ring current of 2.8×10^{15} J would produce a Dst index of −100 nT.

In our discussion of how the solar wind energizes the tail, we used a $10R_E$ long storage region in the tail and were able to store 0.9×10^{15} J in 6000 s using a 10 nT southward IMF with a 500 km s^{-1} solar wind, and a 1 nT normal component of the magnetic field on the tail magnetopause. Returning to the example of a geomagnetic storm onset in Figure 9.30, for which the IMF was about −10 nT and the solar-wind speed 1000 km s^{-1} (double our example), we see that the Dst index changed by about 200 nT in about 9000 s. Using our energy relation to convert the 200 nT to joules, we obtain 5.6×10^{15} J produced at a rate of 6.2×10^{11} J s^{-1}, compared to our earlier example rate of 1.5×10^{11} J s^{-1}. The fact that the dynamic pressure during the September 25, 1998, storm was over three times the value that we used in our example could easily account for the difference.

In summary, the Dst index is more than a qualitative indicator of geomagnetic activity. It is a monitor of the energy in the particles in the magnetosphere. The energy that is stored in the magnetosphere during a storm is very similar to the energy expected to flow into the tail during a period of strong reconnection. In the following section we examine how the tail transports its energy to the magnetosphere.

9.8 THE GEOMAGNETIC TAIL AND SUBSTORMS

In earlier sections of this chapter we have used the tail as a storage place for energy captured by reconnection, assuming that it can store that energy. In contrast, in Section 9.6 it appears that at very active times the energy can pass rapidly through the tail and enter deep into the magnetosphere, where it is stored not in the magnetic field but in the hot plasmas of the magnetosphere. In this section we examine the tail more carefully, studying its quiet-time behavior as well as its dynamics. It is far from being a passive object. In fact it can turn on the magnetosphere in several different meanings of the phrase.

9.8.1 The plasma sheet

As shown in Figure 9.2, the magnetotail consists of two bundles of magnetic flux, one connected to each polar cap. Some of the magnetic flux in the tail is connected to the other lobe of the tail and some of the field is connected to the solar wind. Much of the plasma in the tail derives from the solar wind. This can be deduced from the composition and the electron to proton temperature ratio. At the bow shock, ions are strongly heated as they are slowed by the shock potential drop, receiving thermally much of the energy originally carried in the bulk flow. The electrons can avoid much of this heating because they travel along the magnetic field and not along the shock normal, resulting in a proton to electron temperature ratio of about 8 in the magnetosheath and the plasma sheet compared with a ratio of about 0.9 in the solar wind before encountering the bow shock.

The tail is fairly open to the transport of particles from the solar wind. When the magnetosheath is connected to the solar wind and magnetic flux is transported to the tail, plasma enters with the added magnetic flux. This plasma is called the mantle. Plasma can also drift in from the sides from the sheath and from boundary layers. Thus it is not surprising to find a sheet of plasma separating the two lobes of the tail.

A simple self-consistent analytical model of the plasma sheet is the **Harris current sheet** in which the component in the x-direction (sunward in the anti-sunward solar-wind flow), a function of z (roughly along the rotation axis or the dipole axis), can be written

$$B(z) = B_0 \tanh(z/h)\,\hat{\mathbf{x}}, \tag{9.11}$$

and the plasma pressure is

$$P(z) = P_0 \operatorname{sech}^2(z/h). \tag{9.12}$$

This makes the total pressure in the tail constant:

$$\begin{aligned} B^2(z)/2\mu_0 + P &= P_0 \\ &= B^2/2\mu_0. \end{aligned} \tag{9.13}$$

From Ampère's law we can deduce the current

$$(\nabla \times \mathbf{B})_y = \mu_0 j_y(z) = (B_0/h)\operatorname{sech}^2(z/h). \tag{9.14}$$

The plasma pressure gradient is balanced by the $\mathbf{j} \times \mathbf{B}$ force, where

$$\mathbf{j} \times \mathbf{B} = (B^2/\mu_0 h)\operatorname{sech}^2(z/h)\tanh(z/h)\hat{\mathbf{z}}, \tag{9.15}$$

$$\nabla P = \frac{d}{dz}\left(P_0 \operatorname{sech}^2(z/h)\right)\hat{z} = \mathbf{j} \times \mathbf{B}. \tag{9.16}$$

This current sheet is shown qualitatively in Figure 9.23.

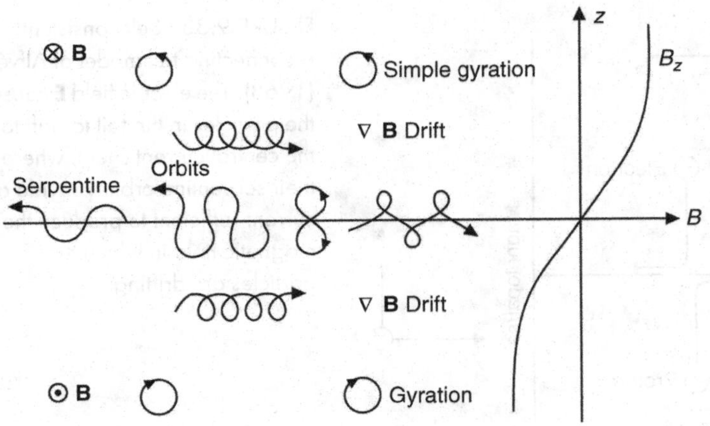

FIGURE 9.32. Proton drift in a simple (e.g., Harris) current sheet. Magnetic field is toward Earth in the upper (northern) lobe. In the absence of field gradients, the motion of the protons is simple gyration. In the region of increasing plasma pressure approaching the current sheet, protons drift to the right (dawn). When they intersect the current sheet, before making a complete circuit they drift to dusk, the direction expected for the current that bounds the tail lobe. The pressure gradient currents also flow duskward, but do not involve particle drift. This is a composite diagram. The left-hand side is the view toward Earth with dawn on the right. The right-hand side of the diagram is a plot of scalar *B* versus *z*, the distance of the observer above or below the $z = 0$ plane.

As noted above, magnetic gradients cause energetic charged particles to drift. As shown in Figure 9.32, particles far from the center of the current sheet simply gyrate and do not drift, but those in the magnetic gradient drift across the tail. The figure here is looking toward the Earth with dawn on the right, dusk on the left, and north up. Ions are left-handed particles and will drift toward dawn to the right, as they have a larger gyro-radius when they are closer to the center of the tail where the field is weakest. Electrons drift in the opposite direction. If we look back at the cut-away magnetosphere in Figure 9.1, we see that the current flows from dawn to dusk as it must to provide the current needed to enclose the northern and southern tail lobes. This is opposite to the gradient drift. The answer to this paradox is hidden in the pressure gradient current that flows from dawn to dusk (see Figure 9.10). This arises because of the increasing density of particles downward so there is a net duskward flux of protons and a net dawnward flux of electrons.

In the center of the tail, where the magnetic field reverses, particles reverse their direction of gyration when they cross the current sheet to the opposite lobe. This produces serpentine orbits in which ions have a net duskward motion and electrons move dawnward. This adds to the "magnetopause" current that crosses the tail and it helps to reverse the field between the two lobes. For gyro-centers farther from the current sheet the serpentine orbits stop and then reverse their drift to that expected for gradient drift motion. We emphasize that the pressure gradient drift current is a real current, which is necessary for force balance, but it does not imply that particle motion is in that direction. The majority of the ions are moving toward dawn in the plasma sheet.

A very complementary model of the tail was published by H. Alfvén in 1968 and is illustrated in Figure 9.33. This model is also self-consistent but with a very different premise. Here an electric field has been applied with capacitor plates on the dawn and dusk flanks of the tail, and newly reconnected magnetic flux is being added at the top (north) and bottom (south) of the tail. These cold electrons and protons are drifting toward the current layer; when they reach it, they execute serpentine orbits to dusk and dawn. The current they produce provides the magnetic field in the lobes.

If the tail boundaries are separated by a distance L then a potential drop Φ requires an electric field $(\phi/L)\hat{y}$. The drifting particles have a density n and a drift speed of E/B, leading to a current of $2enLu$. Ampère's law applied across the current sheet gives a current strength of $2B/\mu_0$. Equating

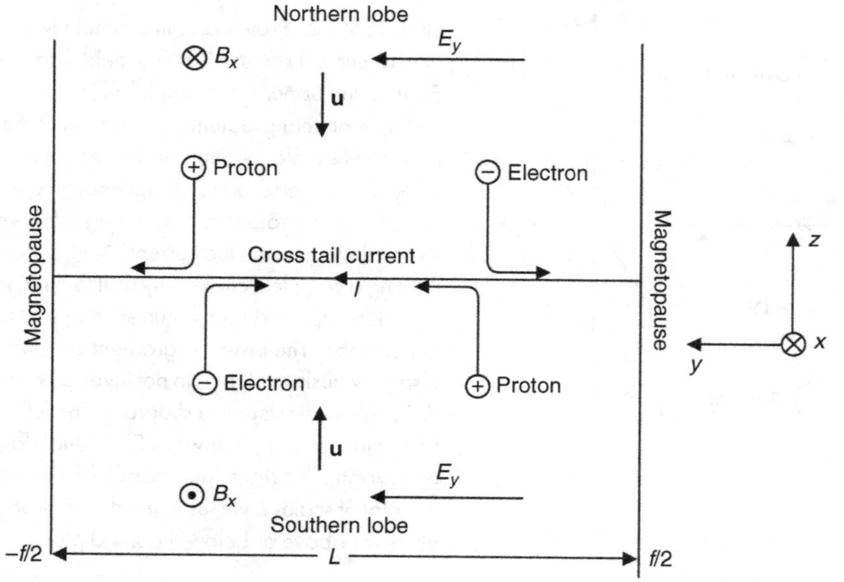

FIGURE 9.33. Self-consistent reconnecting tail model of Alfvén (1968). The electric field **E** causes the particles in the tail to drift to the central current sheet, where their serpentine orbits provide a current sufficient to produce the magnetic field in which the particles are drifting.

these two measures of current, we can solve for the velocity of the plasma $u = B/(\mu_0 neL)$. This drift speed supplies the right number of particles to provide the tail current and to maintain the magnetic field. There is no need for capacitor plates to maintain the electric field; the electric field is provided by the flow between the magnetic flux, added by dayside reconnection on the top (and bottom) of the tail and removed by "reconnection" at the central current sheet.

Of course, both the Harris current sheet and the Alfvén reconnecting tail are idealistic models that are seldom observed in nature. However, they do prepare us for the fact that the tail can exist in quite different states: ones in which there is very little circulation of plasma and ones in which there is much circulation.

Before we leave the plasma sheet we examine one more scenario, intermediate between static and dynamic states discussed above. In Figure 9.34 we show an energetic particle on stretched closed magnetic field lines. The motion in the noon–midnight meridian is shown in Figure 9.34(a). When the particle encounters the current sheet it will drift across the tail, as shown in Figure 9.34(b). At some point the energetic particle will escape the plasma sheet and go north or south as shown. If the tail is convecting inward toward the Earth, there is an electric field across the tail because of the vertical magnetic field. The $-\mathbf{v} \times \mathbf{B}$ electric field is from dawn to dusk, as shown in Figure 9.34(c), and the drifting ions will be energized as they move across the tail. This work was done by T. W. Speiser (1965) under the tutelage of J. W. Dungey and gives good physical insight into how dynamic events in the magnetosphere can energize charged particles.

9.8.2 Quantifying auroral current systems

It is more difficult to devise a simple and adequate measure of the strength of the auroral current systems than for our measure of the ring current because the strong auroral electrojets are at low altitudes (near 100 km) and high latitudes (near 70°). To capture these currents we need a worldwide network at high latitudes over a wide latitude range where people and land area are scarce in both polar regions. Figure 9.35 shows the location of the stations used to calculate the **auroral electrojet index** (AE) and its components AU and AL. We note that the 12 stations, while fairly uniform in longitude, do not provide much latitudinal range, and the degree of activity could move the auroral electrojets poleward or equatorward of the station

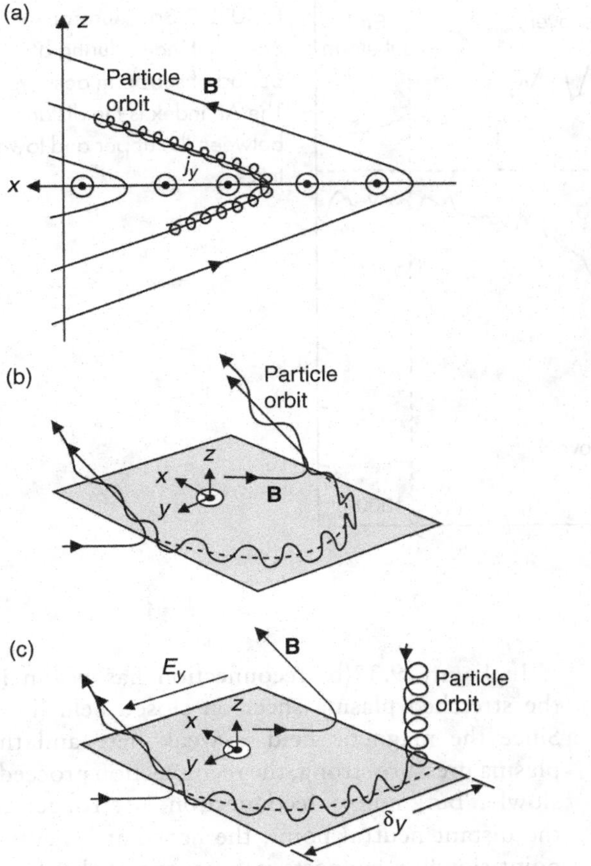

FIGURE 9.34. Acceleration of particles in a moving current sheet. When particles drift across the tail in (b) and the current sheet is stationary, the particles keep their energies. However, if they drift while the current sheet is moving, they cross equipotentials and gain or lose energy. (After Kivelson and Russell, 1995.)

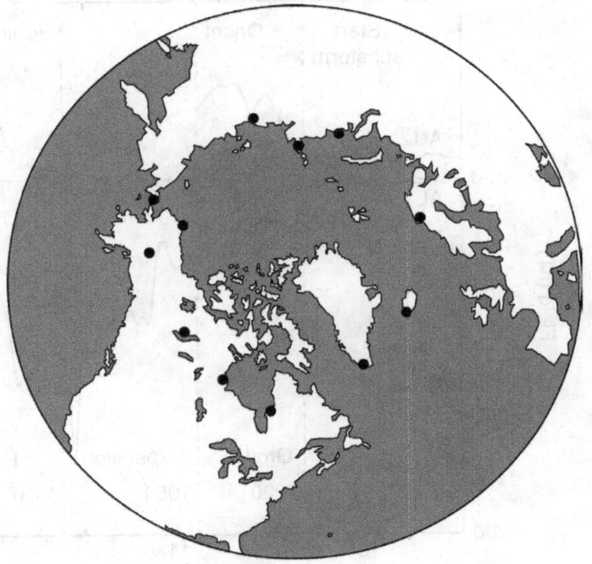

FIGURE 9.35. Location of the stations from which the auroral electrojet index is created.

chain. The objective of the AE index is to determine the greatest current density flowing in the auroral oval at any Universal Time. It does this by measuring the deviation of the magnetogram record from its mean value in the positive (AU) and negative (AL) directions, finding the maximum values in both directions at each time step. The difference between the AU and AL indices is the AE index. The AU and AE indices are usually positive and the AL index is usually negative.

Figure 9.36 shows a typical period of activity including a sequence of substorm growth, expansion, and recovery. The name substorm implies incorrectly that this phenomenon is a small version of the larger geomagnetic storm that we just discussed. In fact, while the substorm can occur during geomagnetic storms, they are a separate phenomenon associated with different conditions in the solar wind, albeit dayside and nightside reconnection are important for both. In this event, during the period in which observations in the solar wind and in the magnetotail indicate that energy should be stored in the tail, the eastward (generally afternoon) and westward (generally morning) electrojets intensify, increasing AU and lowering AL, respectively. At some point the auroras brighten and expand poleward, as do the auroral electrojects. In this substorm, the AU current system did not increase until the recovery phase, when the AL index had substantially weakened.

9.8.3 An empirical model of substorms

The sequence of activity in the auroral oval had been chronicled by auroral all-sky cameras and magnetometer networks well before the more recent period of intensive, *in situ* magnetospheric

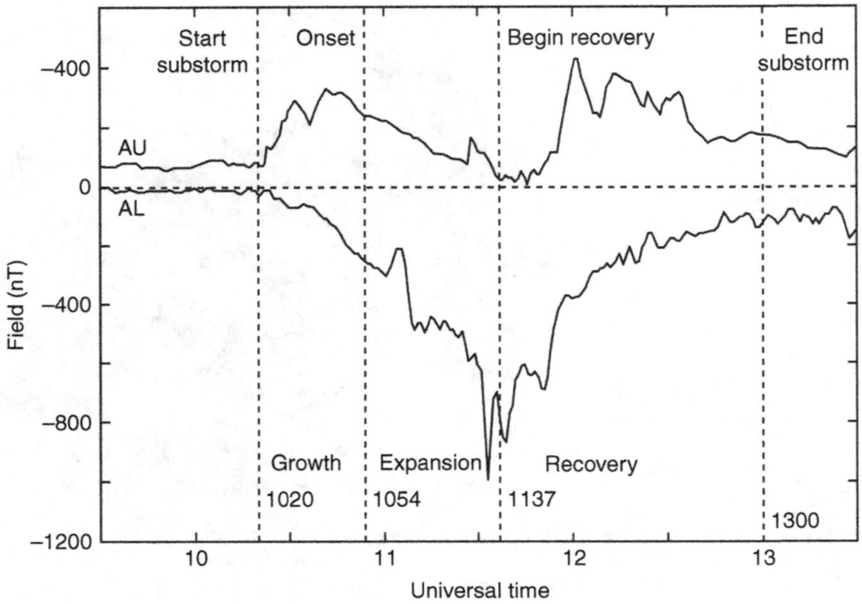

FIGURE 9.36. Auroral electrojet index during a period of substorm activity. The AE index is the distance between the upper and lower traces.

exploration. However, progress in understanding the physics of the **substorm process** awaited the magnetospheric exploration enabled by spacecraft. Most important was the ability to observe simultaneously the solar wind and interplanetary magnetic field in front of the magnetosphere, at the same time as a second spacecraft in the tail lobe, in the plasma sheet, or near the boundary of the magnetosphere was obtaining space plasma observations.

Figure 9.37 shows the phenomenological model that was developed at that time (McPherron, Russell, and Aubry, 1973; Russell and McPherron, 1973) using the OGO 5 spacecraft and other spacecraft in the solar wind and magnetosphere. The dashed line in Figure 9.37(a) shows the initial outlines of the magnetosphere and plasma sheet. When the magnetic field turns southward, the closed magnetic field region on the dayside shrinks and magnetic flux is transported to the polar cap and tail lobe. The magnetopause moves inward at constant solar-wind pressure and the plasma sheet thins. The change in shape of the tail boundary intercepts the solar-wind pressure more, and the strength of the magnetic field in the tail increases.

In Figure 9.37(b) reconnection has begun in the stretched plasma sheet on closed field lines. Since the magnetic field is weak here and the plasma pressure strong, the reconnection proceeds slowly, but, unless reconnection is stronger at the distant neutral point, the near-Earth neutral point slowly reconnects its way out of the dense plasma sheet into the more rarefied plasma sheet boundary layer and then the tail lobe. In the tail lobe the magnetic field is strong, the plasma density low, and the Alfvén speed high. Here, reconnection can proceed rapidly, returning magnetic flux to its closed state much faster than it is being opened on the dayside. The island of closed field lines (called a plasmoid), with no footpoints on the Earth, grows. It resembles in some respects the flux transfer event seen on the magnetopause. When there are no field lines draped over the plasmoid anchoring it to the Earth, it breaks free and moves away from the Earth, as shown in Figure 9.37(c). At this point the magnetosphere returns to its pre-substorm state with the exception of the heating of the auroral and equatorial magnetosphere that has occurred.

It is instructive to look at the three main flux transport rates in the magnetosphere during a

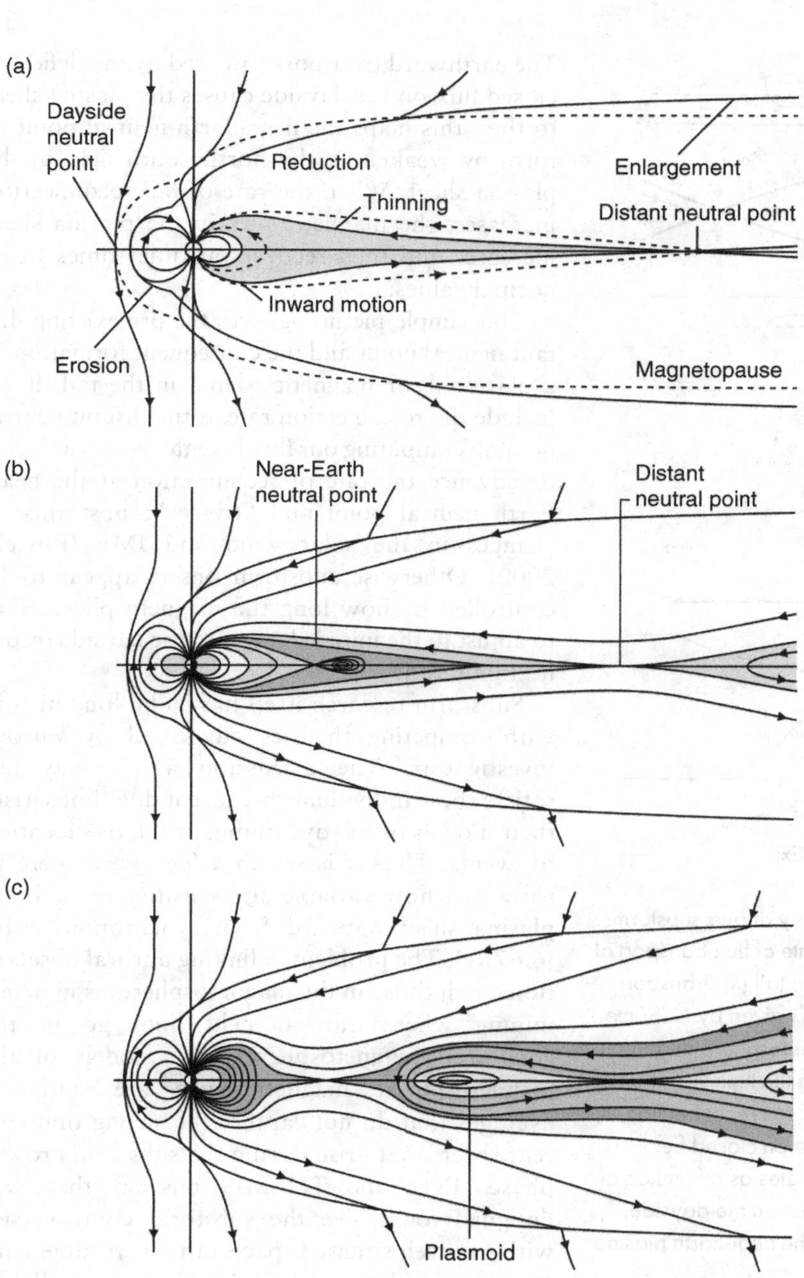

FIGURE 9.37. Phenomological model of a substorm. (After McPherron *et al.*, 1973; Russell and McPherron, 1973.)

substorm and their effects on the inventory of magnetic flux in each of the major regions in the magnetosphere. This is done with the help of Figure 9.38. The rate of merging, M, of the interplanetary magnetic field with the closed magnetic flux on the dayside is controlled by the direction of the interplanetary magnetic field, and turns on and then turns off. The flux is transferred to the tail with the speed of the solar wind. This reduces the closed magnetic flux on the dayside, Φ_{day}, until such a time as convection from the nightside can replenish it. While some convection at rate C can occur as soon as there is a pressure gradient pushing flow toward the dayside, the transport cannot

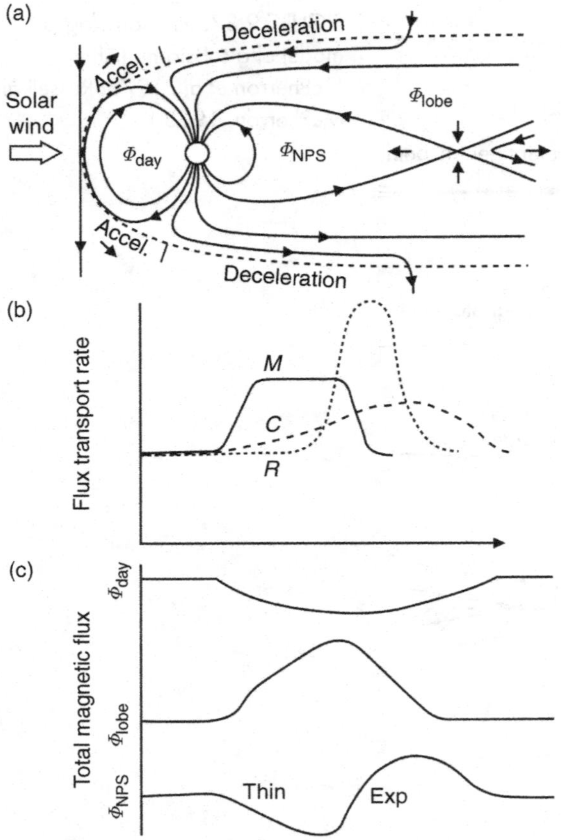

FIGURE 9.38. Magnetic flux inventory during substorms. (a) Magnetospheric geometry. (b) Rate of flux transport at nose (*M*) and at reconnection point in tail (*R*); transport from tail to dayside magnetosphere is given by *C*. Some insight into the behavior of the magnetosphere during a substorm can be gained by following the time variation of the rates of dayside reconnection, *M*; tail reconnection, *R*; and convection from night to dayside on closed field lines, *C*. (c) The differences in these rates as a function of substorm phase alter the magnetic flux on the dayside, Φ_{day}; in the tail lobes, Φ_{lobe}; and in the near-Earth plasma sheet, Φ_{NPS}. (After Russell, 2000.)

reach significant values until the open field lines in the lobe sheet become closed by neutral point reconnection at a rate R. This nightside reconnection depletes the flux reservoir enhanced by the dayside reconnection and Φ_{lobe} decreases back to the quiet-time level and the dayside flux transport can be completed. The flux in the near-Earth plasma sheet Φ_{NPS} is also an important quantity. The earthward transport enforced by the deficit of closed flux on the dayside causes the plasma sheet to thin; this helps the near-Earth neutral point to form by weakening the north–south field in the plasma sheet. When the rate of tail reconnection increases, the magnetic flux in the plasma sheet increases and then recovers at quiet times to its normal values.

This simple picture ignores the pre-existing distant neutral point and the consequent formation of a plasmoid, or magnetic island, in the tail. If we include the reconnection rate at the distant neutral point in computing our flux inventory, we can delay or advance the rate of reconnection at the near-Earth neutral point and "trigger" substorms by changes in the solar wind and IMF (Russell, 2000). Otherwise substorm onsets appear to be controlled by how long the magnetosphere takes to adjust to the initial changes in the dayside reconnection rate.

Substorm research itself has had a long history with competing theories advocated by various investigators. The discussion above may not satisfy these individuals because it does not satisfy their models of relative timing or relative location of events. Thus it is worth a few words here to show just how variable and sensitive is the inner plasma sheet outward from synchronous orbit ($6.67R_E$). The problem in linking auroral observations with those in the magnetospheres is in determining where auroral field lines go in the equatorial magnetosphere. Our models of the magnetospheric magnetic field are statistical averages that do not capture the strong thin current sheets that arise during the substorm growth phase. Even the THEMIS mission that was designed to resolve the substorm controversies with five well situated spacecraft has trouble with this question because its orbit plane is parallel to the current sheet and only crosses it when the current sheet moves. Fortunately we have had a mission that made vertical cuts through the current sheet: the Polar mission, which was sent in a very elliptical orbit high over the north pole to $9R_E$ apogee to study the polar cusp. The oblateness of the rapidly spinning Earth produces a torque on an elliptically orbiting spacecraft such as Polar during every perigee passage, and eventually the line of

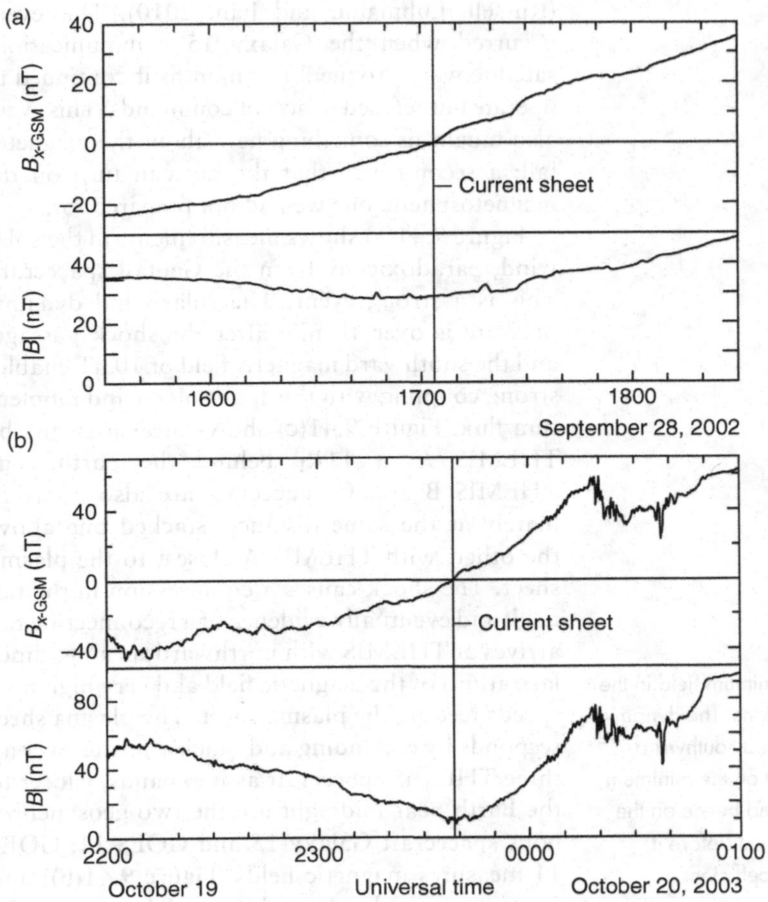

FIGURE 9.39. Two passes of the Polar spacecraft through the equatorial nightside magnetosphere showing the varying magnetic field strength as the spacecraft cuts through the near-Earth current sheet. The spacecraft was (a) at $9.2R_E$ when it encountered the current sheet on September 28, 2002, and (b) at $8.5R_E$ when it encountered the current sheet on October 20, 2003.

apsides moved into the Earth's equator allowing Polar to probe the current sheet in the range $7–9R_E$ in the night hemisphere. This allowed it to determine what controls the magnetic field and current sheet in this region, which some have claimed is the region in which the auroral energization takes place.

Figure 9.39 shows the location of the current sheet by the reversal of the B_x (solar pointing) component of the magnetic field measured by Polar at a quiet period near midnight at $9.2R_E$ and at a disturbed time near $8.5R_E$ just before a substorm onset. The field strengths at the current sheet crossings differ by a factor of three, with the weaker field being the one closer to the Earth.

The current sheet is controlled by both the strength of the solar-wind dynamic pressure and the strength of reconnection before the time of the observation. The control by dynamic pressure occurs because higher dynamic pressure makes the magnetosphere smaller and the root of the tail moves closer to the Earth. The increase in solar-wind connection to the tail stretches the night-time magnetic field, reducing the field strength around a narrow current sheet. As a result of these sensitivities to the solar-wind conditions at the time of a substorm, it is very difficult to infer where the magnetic field through any auroral form goes. A field line that nominally crosses the magnetic equator at $7R_E$ at quiet times can easily cross the current sheet at $20R_E$ if the field strength in the center of the tail drops by a factor of three. Thus, the apparent low latitudes of the auroral arcs involved in brightenings are

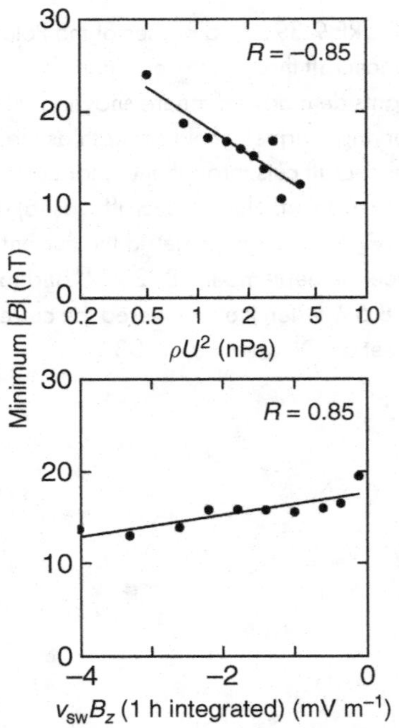

FIGURE 9.40. The dependence of the minimum field in the night magnetosphere on solar-wind conditions. The dynamic pressure of the solar wind and the convected southward magnetic flux both have a strong influence on the minimum field strength near the magnetic equator and hence on the distance to which a field line travels in the tail before it crosses the current sheet. R = correlation coefficient.

not strong evidence against their association with events seen in the tail at distances from 15 to $20R_E$. The night-time magnetic field in the tail current sheet can be very non-dipolar from near synchronous orbit and outward down the tail.

9.8.4 Extreme space weather: overdipolarization

This chapter on the solar wind–magnetosphere coupling has been mainly about energy and momentum exchange and little about the myriad of phenomena that are associated with the energy and momentum exchange we have discussed. We continue to defer much of this discussion to the following chapters and close here with an example of a space weather event, occurring near the solar minimum of 2007–09 that was the weakest since the Dalton minimum (Russell, Luhmann, and Jian, 2010). This event occurred when the Galaxy 15 communications satellite went "rogue." For months it continued to operate but refused to accept commands. This event also taught us something new about the magnetotail: a second way that the tail can turn on the magnetosphere, one we had not prepared for.

Figure 9.41(a) shows measurements in the solar wind, paradoxically from the Geotail spacecraft. This is a strong event. The solar-wind dynamic pressure is over 10 nPa after the shock passage, and the southward magnetic field of 10 nT enables strong coupling with this high solar-wind momentum flux. Figure 9.41(c) shows measurements by THEMIS A at $11R_E$ behind the Earth. The THEMIS B and C spacecraft are also approximately at the same distance, stacked one above the other, with THEMIS A closest to the plasma sheet. The shock causes a compression in the tail field, and eventually evidence of a reconnecting tail arrives at THEMIS with northward turning (dipolarization) of the magnetic field and very high flow speeds feeding the plasma sheet. The plasma sheet responds by expanding and quickly passes over all three THEMIS spacecraft as it expands. Closer to the Earth near midnight are the two geosynchronous spacecraft Galaxy 15 and GOES 11; GOES 11 measures magnetic fields (Figure 9.41(b)) and energetic particles (not shown). We can see the impact of the flows and dipolarization of the magnetic field at THEMIS in the GOES data. The geosynchronous magnetic field becomes dipolar and stronger. One might have expected the dipolarization to have ended when the compressed magnetic field reached 80 nT, the normal quiet-time night-time field at this distance. However, the field continues to climb to well above 140 nT. This is comparable to what we would expect the magnetic field to be on the dayside compressed by a strong solar-wind flow. The tail has effectively turned the solar wind around and directed its momentum flux onto the nightside, compressing the magnetosphere on the back side as it usually is on the front side. A search of geosynchronous data and ground magnetograms shows that these overdipolarizations, although rare, are not unique (Connors *et al.*, 2011).

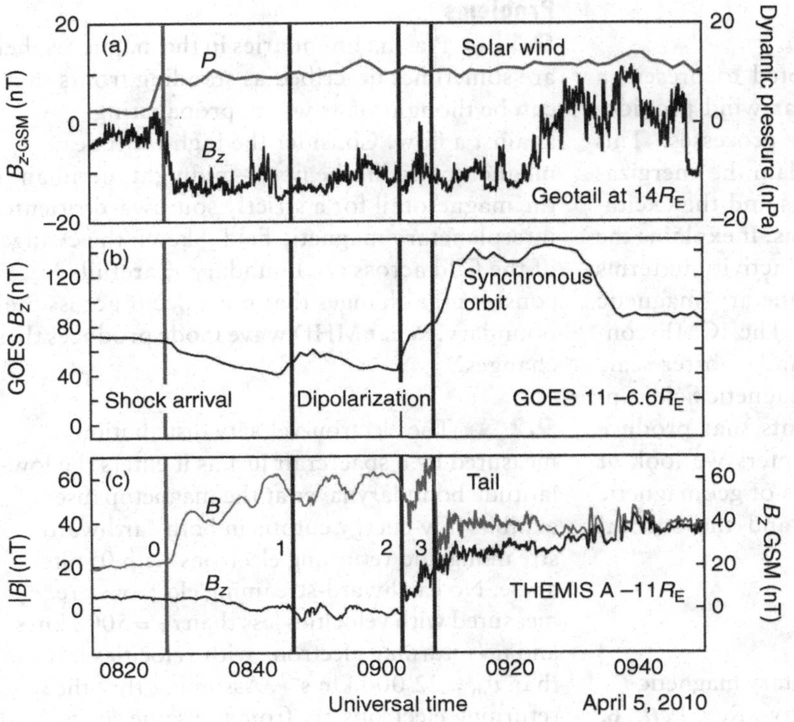

FIGURE 9.41. Measurements in the solar wind, in the tail, and at geosynchronous orbit at midnight during the period surrounding the Galaxy 15 space weather event. (a) Solar-wind dynamic pressure measured by Geotail and the north–south component of the IMF. (b) Magnetic field at geosynchronous orbit near midnight measured on GOES 11. (c) Magnetic measurements at THEMIS A 11 R_E behind the Earth.

While the GOES spacecraft handled this unexpected event well, the Galaxy spacecraft appears not to have. It was in darkness when this event struck, a time when photoelectrons were not available to discharge the spacecraft. Something happened, possibly associated with spacecraft charging if it became immersed in hot electrons. Soon, the Galaxy 15 spacecraft stopped accepting commands. This event reminds us that space weather can affect our technological systems even at solar minimum, and that the tail is not just an energy spigot but also can turn around the dynamic pressure of the solar wind compressing the magnetosphere from the rear.

9.9 SUMMARY

In this chapter we have attempted to present a simple overview of how the solar wind provides the energy for magnetospheric processes. This simple picture enables us to explain the energization of magnetospheric processes and the excitation of the various current systems. It explains the two major types of geomagnetic activity in terms of the behavior of the interplanetary magnetic field that controls the coupling. The ICME controls the geomagnetic storm, and shorter-scale variations in the interplanetary magnetic field supply shorter-period coupling events that produce substorms. In the following chapters we look at the consequences of those drivers of geomagnetic activity on the magnetosphere and the auroral ionosphere.

Additional reading

Dungey, J. W. (1961). Interplanetary magnetic field and the auroral zones, *Phys. Rev. Lett.*, 6, 47–48. This is the paper that revolutionized our understanding of the solar-wind control of geomagnetic activity.

Kivelson, M. G. and C. T. Russell (1995). *Introduction to Space Physics*. Cambridge: Cambridge University Press. The predecessor to this book contains much more material on geomagnetic activity.

McPherron, R. L., C. T. Russell, and M. P. Aubry (1973). Satellite studies of magnetospheric substorms on August 15, 1968. IX. Phenomenological model for substorms, *J. Geophys. Res.*, 78, 3131–3149. This paper used a multi-instrument study of substorms to define the physical processes occurring in a substorm.

Russell, C. T. and R. L. McPherron (1973). The magnetotail and substorms, *Space Sci. Rev.*, **15**, 205–266. This paper put the model derived from OGO 5 data in the context of the existing knowledge of the physics and the structure and dynamics of the magnetosphere from multiple sources.

Problems

9.1 Plasma boundaries in the magnetosphere are sometimes described as standing fronts that can be thought of as waves propagating against a flow. Consider the high-latitude magnetopause in the noon–midnight meridian of the magnetotail for a strictly southward-oriented interplanetary magnetic field. Sketch the change of the field across the boundary. Carefully consider the changes that must occur across the boundary. What MHD wave mode produces these changes?

9.2 The electron-velocity distribution measured by a spacecraft just as it enters the low-latitude boundary layer at the magnetopause exhibits low-energy cutoffs in both earthward-streaming and returning electrons with 0° pitch angle. No earthward-streaming electrons are measured with velocities less than $v_E = 5000$ km s^{-1}, and no returning electrons with velocities less than $v_m = 22\,000$ km s^{-1}. Assuming that the returning electrons are from the same source and that they mirrored close to the Earth, and that the distance from the spacecraft to the mirror point x_m can be obtained from a magnetic field model, derive a formula to estimate the distance from the spacecraft to the acceleration point. If the distance to the mirror point is $12R_E$, use your formula to estimate the distance to the acceleration region, and deduce how long prior to the observation the acceleration began on this field line.

9.3 If the polar cap is defined by the open field lines that enter the tail lobe, and if it can be approximated by a circle of radius 15° centered on the magnetic dipole axis, how much flux is there in a lobe of the geomagnetic tail? If none of this flux crosses the neutral or current sheet in the tail, if no solar-wind plasma enters the tail field lines, and if the strength of the interplanetary magnetic field (IMF) is 6 nT and $\beta = 1$, what is the radius of the distant tail lobe?

9.4 Reconnection can proceed at a fraction of the inflow Alfvén speed, v_A. Calculate the cross-tail potential difference required to convect tail-lobe flux toward the center of the tail at a speed equal to $0.1v_A$ in the lobes. (Use parameter values $n = 8$ cm^{-3}; $B = 15$ nT.) How does this potential drop compare to observed cross-polar-cap potential drops? Discuss any differences.

9.5 During the growth phase of a geomagnetic substorm, the polar cap is observed to expand. If the polar cap were to expand from an initial mean radius of 15° to a mean radius of 20° latitude, how would the geomagnetic tail change? Assume that the tail radius $10R_E$ behind the Earth remains fixed at $R_T = 18R_E$ ($x = 10R_E$), and find how the tail radius, the lobe field strength, and the plasma-sheet pressure change as functions of distance downtail.

9.6 The cross-tail current is carried primarily by plasma-sheet ions, which have a density of $n = 0.3$ cm^{-3}. If the current sheet is $1R_E$ thick, and the lobe magnetic field strength is 20 nT, calculate the average ion velocity required for the ions to carry the current. Compare this velocity to the average ion thermal velocity if the average ion energy is 4 keV. Discuss whether or not a plasma detector on board a spacecraft could measure the current directly.

9.7 Dr. Grant Swinger convinces NASA that he can simulate the collapse of night-time magnetic flux tubes during substorms with the following simple device. A 200 m long solenoid creates a uniform field down its axis. At the ends of the solenoid, extra windings, carrying a very strong current, pinch off the field lines so that ionized particles are trapped in the device. Dr. Swinger injects protons with a parallel velocity of 40 km s^{-1} and a perpendicular velocity of 20 km s^{-1}. After they have executed several bounces, Dr. Swinger moves the pinching coils slowly together to a point only 100 m apart. How fast are his protons moving now? What is their total energy? If they have gained energy, where did it come from? If they have lost energy, where did it go? Dr. Swinger falls asleep at the switch and slowly doubles the background, confining field of the main solenoid. What happens to the protons? Why?

9.8 A unidirectional field $B_y(x)$ vanishes at $x = 0$, and is oppositely directed in the two regions $x < 0$ and $x > 0$. Plasma that is of uniform density and is frozen to the magnetic field moves steadily under the action of an electric field $E_z(x, y)$ with velocity components $u_x = -u_0 x/a$ and $u_y = u_0 y/a$, where u_0 and a are constant. Show that E is uniform, and solve Ohm's law for B_y in the two limits $x \gg l$ and $x \ll l$, where $l^2 = a\eta/u_0$, where η is resistivity. Make a rough sketch of $B_y(x)$, and comment on its behavior. Also, use the steady equation of motion to find the plasma pressure $P(x, y)$. This solution models the annihilation of a magnetic field by a stagnation-point flow.

9.9 Demonstrate the instability of an x-point field $B_x = B_0 y/l$, $B_y = B_0 x/l$ by seeking solutions to the time-dependent equations of motion (with no pressure gradient), induction (with no diffusion), and continuity of the form $B_x = B_0(1 + ae^{\omega t})y/l$, $B_y = B_0(1 + be^{\omega t})x/l$, $u_x = ce^{\omega t}x/l$, $u_y = de^{\omega t}y/l$, $\rho = \rho_0(1 - fe^{\omega t})$. Find the values of the constants a, b, c, d, f, and ω, and suggest what happens ultimately.

9.10 Use the Particle Tracing option of the Space Physics Exercises* to explore particle motion in a Harris current sheet. Use the default field and scale length of 100 nT and 10 km. In a Harris current sheet the magnetic field reverses like a hyperbolic tangent. The weaker field surrounding the current layer is in pressure balance with an enhanced plasma pressure.

(a) Measure the magnetic field strength across the plot by launching particles with a v_x-component only at varying locations in x and measuring the gyro-frequency and its sign. Since the field is entirely along the

* Space Physics Exercises are at http://spacephysics.ucla.edu.

z-direction here, plot the B_z-component versus x (from left to right on the screen). Use small velocities to avoid meandering particles.

(b) With $v_x = 50$ km s^{-1} measure the gradient drift velocity every 5 km from $x = -25$ km to $x = 25$ km using protons. Calculate the expected drift velocity from theory, and plot both observations and theory versus x. Where does gradient drift theory break down? Why does it break down?

(c) Repeat with $v_x = 500$ km s^{-1} electrons from $x = -10$ km to $x = 10$ km. Compare this drift with that of the protons in part (b) on the same graph. Discuss the nature of currents in this current sheet as a function of x if protons and electrons had the same temperature. Is the current (associated with gradient drift) in the direction to weaken and reverse the magnetic field? If not, explain what causes the necessary current.

9.11 Show from Eq. (9.4) that the change in the magnetic field strength at the center of the Earth caused by the gradient drift of a singly positive charged particle q that mirrors in the equatorial plane with energy $W_\perp$ at distance LR_E is given by $-(3/4\pi)(\mu_0 W_\perp)/(R_\mathrm{E}^3 B_0)\hat{\mathbf{z}}$, where R_E is the radius of the Earth and B_0 is the equatorial field strength.

9.12 Show that the total energy in the Earth's dipole field above the Earth' surface is $W_\mathrm{mag} = (4\pi/3\mu_0)B_0^2 R_\mathrm{E}^3$.

10

The terrestrial magnetosphere

10.1 INTRODUCTION

Beginning in Chapter 4, we examined the energy flow starting in the nuclear fires of the Sun that both powers the optical emissions of the photosphere and heats the solar atmosphere producing the solar wind. When it reaches 1 AU, the momentum flux of the solar wind couples to the Earth's magnetosphere. Much of this coupling arises through the process known as magnetic reconnection. Just before the solar wind reaches the Earth's magnetosphere, its properties change dramatically as the solar wind crosses the bow shock that slows, heats, and compresses the solar wind, as discussed in Chapter 6. This alteration, which is controlled by the Mach number of the bow shock, affects its ability to couple the solar-wind momentum to the stationary magnetosphere, as we discussed in Chapter 9. In the present chapter, we discuss the processes that occur within the magnetosphere that result from the boundary conditions that the solar wind and reconnection have imposed upon the magnetopause. These processes stir the magnetosphere, energize the ring current and the radiation belts, and control the loss processes therein. We begin by reviewing the structure of the magnetosphere, including the polar cusp, the plasma mantle, the near-Earth plasma sheet, the ring current, and the plasmasphere. We follow with a discussion of the radiation belts, and close the chapter with a discussion of the waves in the magnetosphere that interact with the radiation belts and the other particle populations in the magnetosphere.

10.2 THE STRUCTURE OF THE MAGNETOSPHERE

The magnetic configuration of the Earth's magnetosphere is captured very simply in Figure 10.1, which shows how the Earth's magnetic field (in the noon–midnight meridian) is confined by the shocked solar-wind flow. After passing through the standing bow shock, the magnetized plasma is deflected about the Earth's magnetosphere, applying a normal stress to the magnetosphere that defines its size and shape. If there is reconnection between the magnetic field in the magnetosheath with that in the magnetosphere, or if there is viscosity (friction) between the flow and the magnetosphere, flow can be induced in the magnetosphere, and the magnetospheric shape can be altered. At high latitudes centered around noon an indentation in the boundary allows the shocked magnetosheath plasma to reach the ionosphere. This indentation is called the polar cusp. On the nightside, an analogous feature is found where the plasma in the plasma sheet can also reach the ionosphere. This plasma sheet plasma can enter from the magnetosheath or be accelerated by reconnection in the central region of the tail. On the outer boundary of the tail is the plasma mantle, consisting of recently reconnected magnetic flux and the plasma it contains. The amount of magnetic flux in the tail is variable depending on the recent history of reconnection on the dayside of the magnetopause and in the tail.

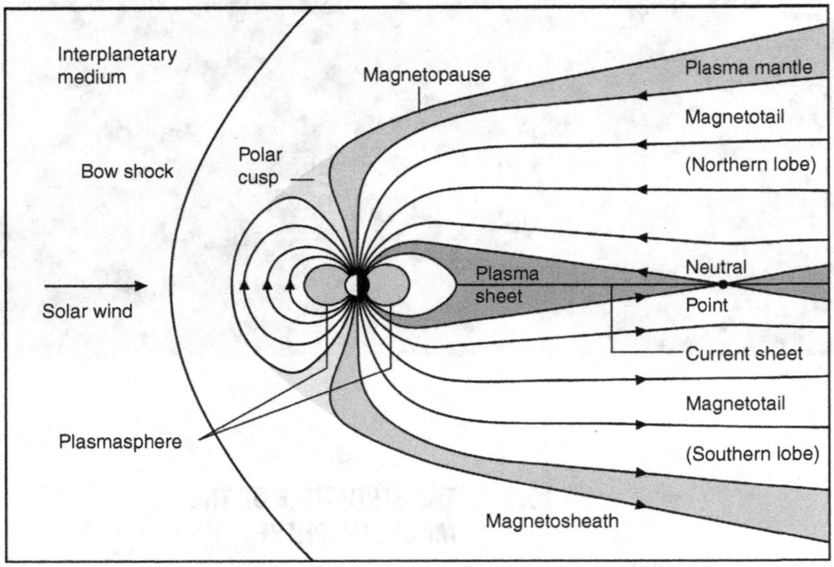

FIGURE 10.1. The terrestrial magnetosphere as understood in the light of early measurements with high-apogee eccentric orbiting spacecraft interpreted in terms of Dungey's (1961) insight. There is a neutral point, not a line, and a current sheet that allows a normal component of the magnetic field to cross it except right at the neutral point. (After Russell, 1972.)

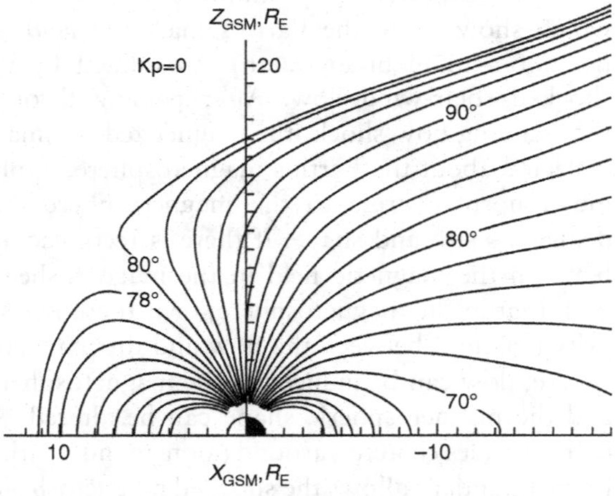

FIGURE 10.2. The magnetic field configuration of the magnetosphere in the noon–midnight meridian. The field lines are drawn from the model of Tsyganenko and Usmanov (1982).

A more realistic picture of the structure of the magnetic field has been produced by Tsyganenko and colleagues in a series of increasingly sophisticated models informed by *in situ* observations. Figure 10.2 shows magnetic field lines drawn in the noon–midnight meridian. The coordinate system here is called geocentric solar magnetospheric (GSM). The x-axis points toward the Sun and the z–x plane contains the magnetic dipole. The model shown is calculated for quiet geomagnetic conditions when the Kp index is zero. The lines are labeled according to their invariant latitudes where they intersect the Earth. Again, the polar cusp is easily identified in this model.

The orthogonal plane contains magnetospheric currents which flow across the field and circle the Earth. Some of the current does not close in the equatorial plane but flows along the magnetic field and closes in the ionosphere, forming a "partial" ring current. This is illustrated in Figure 10.3.

The inner boundary of the magnetosphere is the ionosphere, whose properties vary with altitude and latitude as well as local solar time. The ionosphere (discussed in Chapter 2) is maintained by various sources of ionization including solar EUV and UV radiation, particle precipitation from the radiation belts, and particle precipitation and electron beams in the auroral regions. The resultant ionization is redistributed by transport. This transport makes predictions of ionospheric properties quite complex and difficult at geomagnetically disturbed times.

The ionosphere in turn provides cold plasma to the magnetosphere in the region labeled the plasmasphere in Figure 10.1. Again transport is very important in establishing the plasma content of the plasmasphere and its outer boundary, the plasmapause.

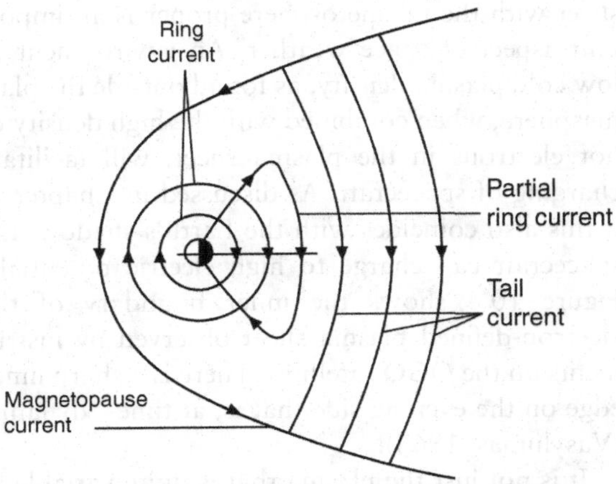

FIGURE 10.3. Magnetospheric currents in the equatorial plane. In the realistic magnetosphere, the currents are not straight across the magnetosphere from dawn to dusk, but adopt some of the curvature of the isocontours of the magnetic field. The innermost currents close in the ionosphere after traveling out of the equatorial plane along the magnetic field, allowing a partial ring current to flow on the night side.

It would be convenient but quite incorrect to assume that the neutral atmosphere only entered this picture as the source of the ionospheric plasma near the surface of the Earth. In fact, the neutral atmosphere has a density greater than that of the magnetospheric plasma over much of the magnetosphere, as noted in Chapter 2. The coupling between the magnetospheric plasma and the atmosphere is weak, but is important in many ways, from the generation of pickup ions and waves, to charge exchange, ring current decay, and the transport of particles across field lines. Thus we should never completely neglect the atmosphere when treating the magnetosphere.

10.2.1 The polar cusp

The magnetopause end of the **polar cusp** is situated where the magnetic field lines switch from closing across the equator to where they join the magnetic flux stretched tailward. This weak point in the magnetic field where the field lines diverge is present in Chapman and Bartels' (1940) image dipole magnetosphere model, discussed in Chapter 1, as well as in Dungey's (1961) reconnection model of the solar-wind interaction with the magnetosphere. Hence, its presence does not depend on the occurrence of reconnection, but its properties certainly do. The polar cusp was initially studied at low altitude by ISIS I and II (Heikkila and Winningham, 1971) and at high altitudes by IMP 5 (Frank, 1971) and by OGO 5 (Russell *et al.*, 1971). Several missions have been devoted to its study, such as ESA's HEOS 2 and Cluster missions and NASA's Polar mission. Of all the regions in the magnetosphere, the polar cusp is in many ways the most sensitive to the solar-wind interaction. When magnetic reconnection is weak, the plasma is quite magnetosheath like, but, when dayside reconnection occurs, the polar cusp moves to lower invariant latitudes, shows evidence of the accelerated plasma, and broadens in local time extent. The polar cusp also reveals the presence of the exosphere through charge exchange between the solar-wind protons and exospheric hydrogen (Le *et al.*, 2001) that leads to the production of proton cyclotron waves. The form of the sensitivity of the polar cusp to the north–south component of the interplanetary field is in accord with all other evidence of the role of reconnection in magnetospheric physics, and forms a consistent picture of magnetic stresses and plasma transport.

10.2.2 The near-Earth plasma sheet

Topologically, as illustrated by Podgorny's (1976) wire model of the solar-wind interaction with the magnetosphere in Chapter 1, the plasma sheet is basically a continuation of the weakened magnetic field around the polar cusp. The magnetosheath plasma that surrounds the magnetotail has direct access to the tail in the weak region of the field between the northern and southern lobes of the magnetotail, except for one problem: the plasma is tied to field lines. Thus the plasma must be able to drift perpendicular to the direction of the magnetic field or there must be reconnection of the magnetic field which could alter the topology of the field lines on which the plasma resides, or annihilate the magnetic field, or a combination of both. The plasma can reach open tail-lobe magnetic field lines by "dayside" reconnection or scattering across the boundary. Dayside reconnection here means reconnection between a solar-wind magnetic flux tube and a closed magnetospheric flux tube with two

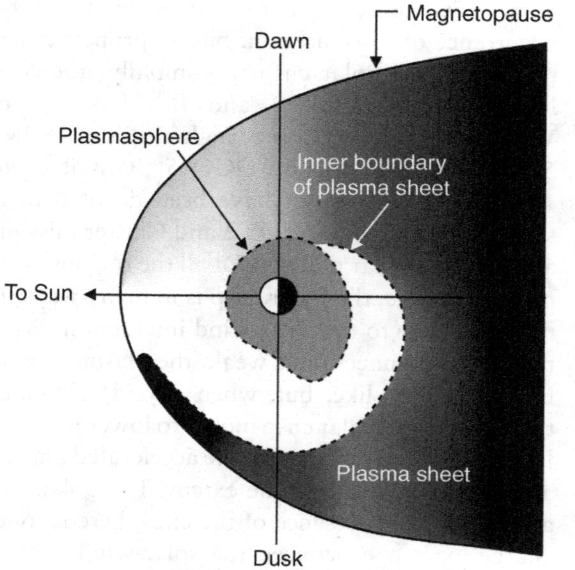

FIGURE 10.4. Schematic of the location of the plasma sheet relative to the plasmasphere as defined by Vasyliunas (1968). The boundary is quite sharp on the evening side and very dynamic near midnight in response to substorm activity, but it is diffuse on the morning side.

feet on the Earth. Once in the tail lobe, convection to the center of the tail is associated with reconnection at the current sheet in the center tail, leading to the topology change and heating that produces dense hot plasma on closed field lines that we call the plasma sheet. We note that reconnection between a solar-wind flux tube and a nightside "open" flux tube can remove flux from the tail and add it to the magnetosphere proper. With the right geometry, it could even reverse the flow over the polar cap.

We have discussed steady-state models of the tail, the Harris current sheet, and the Alfvén steady-state reconnection model in Chapter 9, and we do not repeat these discussions here, except to say that the tail is a vast reservoir of magnetic energy that can be tapped rapidly and deposited in the night atmosphere, the ionosphere, and the radiation belts. This region is the energy storage site that energizes the dynamic auroral phenomena (to be discussed in Chapter 11) as the magnetosphere tries to find a new minimum stress state.

We need to say more about the inner edge of the plasma sheet because the interface of the plasma sheet with the magnetosphere proper is an important aspect of space weather. An environment of low cold plasma density, as found outside the plasmasphere, when combined with the high density of hot electrons in the plasma sheet, will facilitate charging of spacecraft. As discussed in Chapter 9, if this also coincides with the Earth's shadow, the spacecraft can charge to high electric potentials. Figure 10.4 shows the inner boundary of the electron-defined plasma sheet observed by instruments on the OGO satellites. There is a sharp inner edge on the evening side that is, at times, dynamic (Vasyliunas, 1968).

It is not just the plasma that is quite variable in the near-Earth tail; so is the magnetic field. The Polar spacecraft of the Global Geospace mission was launched into an eccentric polar orbit that reached $9R_E$. Initially the orbit was high over the north pole, but eventually it precessed so that it was in the equatorial plane, where it could monitor the dynamic magnetic field of this very important region. Figure 10.5 shows the minimum magnetic field seen near $9R_E$ when Polar was within 2 h of midnight local time. The magnetic field is typically about 17 nT but can be as weak as about 3 nT and, at times, greater than 50 nT. In the period of study, no negative B_z fields during times of low field strength, indicative of reconnection, were found. However, the field strength often became very weak. This suggests that, on occasion, reconnection inside $9R_E$ may occur. Figure 10.5(b) shows a quiet-time pass of magnetic field at midnight measured by Polar. This has a strong north–south (z) component of the field even as the current sheet is crossed (B_x reversal). Figure 10.5(c) shows a pass with a very weak north–south component at the crossing of the current sheet, suggesting the spacecraft was close to the x-point. These weak fields in the near-Earth tail region occur statistically when the solar-wind dynamic pressure is high, causing the magnetosphere to shrink and the plasma sheet to move inward. The reconnection occurs when the IMF is strongly southward (Ge and Russell, 2006).

10.2.3 The ring current

The ring current is also a large reservoir of energy in the form of energized particles trapped in the "dipole" magnetic field. Here energization and

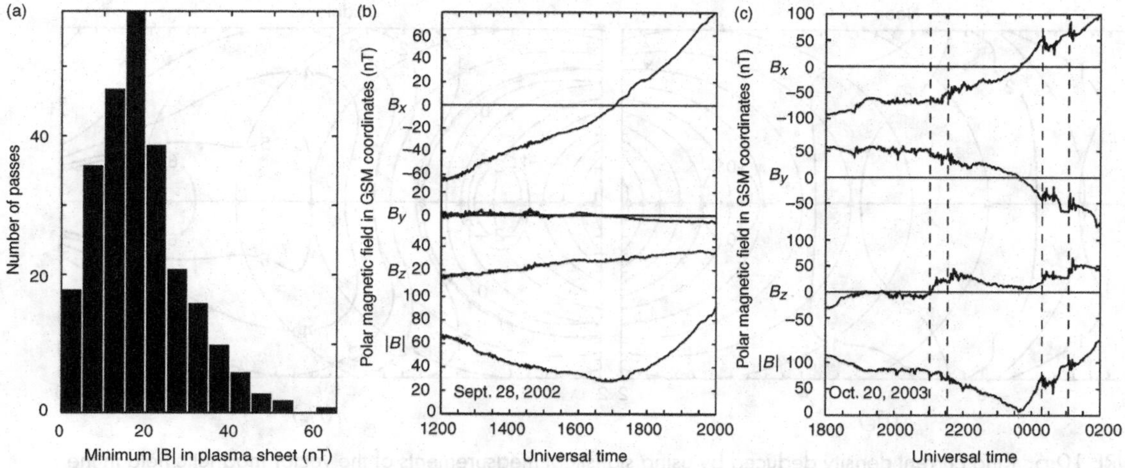

FIGURE 10.5. The magnetic field in the midnight near-Earth plasma sheet. (a) Statistical distribution of the minimum field strength near $9R_E$ from 2200 to 0200 local time. (b) Vector components of the field in GSM coordinates and the total field on a quiet pass. The current sheet is the point where B_x reverses sign. (c) A similar pass when the magnetic field is much weaker. Vertical dashed lines are drawn at plasma sheet expansions. (After Ge and Russell, 2006.)

energy decay are much slower than in the distant regions of the magnetosphere, and the ring current, once developed, can take days to dissipate as the ions charge exchange with the neutral exospheric hydrogen atoms or precipitate into the atmosphere. Since magnetic measurements have now been obtained in all regions of the magnetosphere, and since the electric current can be obtained by calculating the curl of the magnetic field, it is possible to obtain a quantitative empirical model of the ring current by using only spacecraft magnetic measurements. This empirical model of the ring current exists for different phases of a geomagnetic storm, or more precisely for different levels of ring current activity (Le, Russell, and Takahashi, 2004). Figure 10.6 shows contours of the current flowing azimuthally at noon and midnight. The predominant current is in the westward direction, the direction of energetic proton drift in the Earth's magnetic field. This current reduces the strength of the Earth's surface magnetic field at the equator. At low values of latitude (L values), there is a reverse current in the eastward direction. This current occurs where the plasma energy density drops as one moves to lower L values. Only a minor part of the current appears to encircle the Earth completely. We have discussed how we can predict the strength of this ring current using a simple model of solar wind–magnetosphere interaction in Chapter 9. These observations of the magnetic field are consistent with that discussion. Also discussed by Le *et al.* (2004) are the field-aligned currents that carry current to closure paths in the ionosphere and noted in Figure 10.3. These currents enable the nighttime ring currents to be much stronger than those on the dayside, as shown here.

10.2.4 The plasmasphere

Essentially, the **plasmasphere** is just the extension of the ionosphere upwards into the magnetosphere along magnetic field lines. Taking this simplistic picture to its limit, one might expect that the closed magnetosphere would then have cold plasma everywhere and that the ionospheric plasma had sufficient energy to rise along the magnetic field against the force of gravity. Let us hold this thought for a moment, and remove all the cold plasma from the magnetosphere and then let the ionosphere reform itself from photoionization impact ionization and transport. Filling the small volume of magnetic flux tubes at low invariant latitude (L values) would be relatively rapid, but the

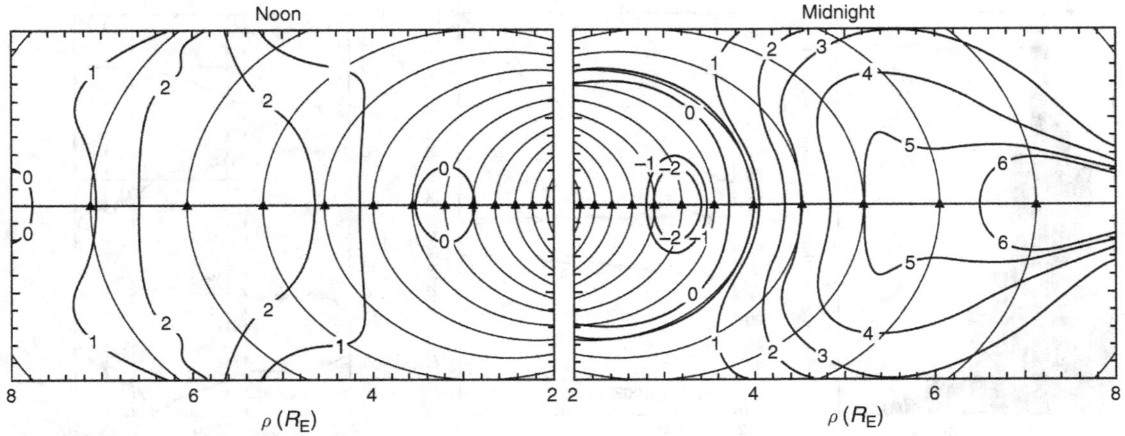

FIGURE 10.6. Ring current density deduced by using statistical measurements of the vector magnetic field in the magnetosphere to calculate the curl of the magnetic field for Dst from −60 to −80 nT adjusted for solar-wind dynamic pressure. Contours of the azimuthal current j_φ are labeled in units of nanoamps per square meter; $\rho(R_E)$ is the distance from the z-axis of solar magnetic coordinates, i.e., from the magnetic dipole axis. (After Le *et al.*, 2004.)

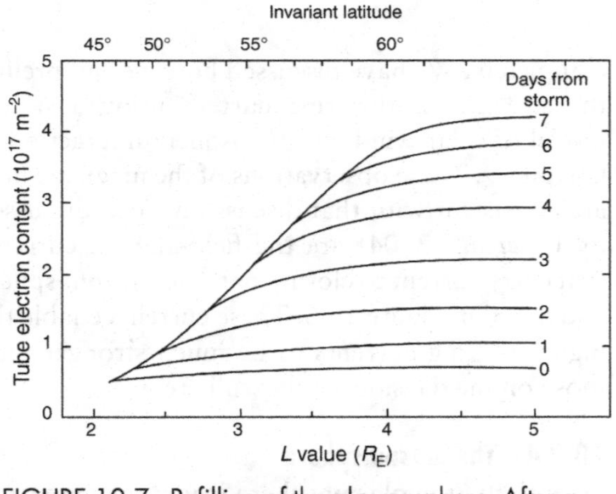

FIGURE 10.7. Refilling of the plasmasphere. After a strong storm, the plasmasphere can be depleted over a wide range of L values. The plasmapheric flux tubes refill at about the same rate until they reach a steady-state value appropriate to their particular L value. This diagram schematically represents refilling after a storm studied by Park (1974).

volume of magnetic flux tubes increases as the fourth power of their equatorial crossing distance since the equatorial cross section of a flux tube is inversely proportional to the magnetic field strength that falls off as L^{-3} and the length of the flux tube is proportional to L. Thus the volume increases by a factor of 10^4 as one moves from near the equator to the cap at an L value of about 10. So, if it took hours to fill up the near-equatorial tube, it would take years to fill the tube at $L = 10$ with the same particle flux. Figure 10.7 shows schematically the filling of the plasmasphere after a strong storm has lowered the density in the tubes to L values as low as $L = 2$. The flux tubes at $L = 4$ are still filling a week later (Park, 1974).

Another complication is the **plasmapause.** Early space measurements by K. I. Gringauz (1969) and ground-based work by D. L. Carpenter (1963), using the lightning-generated whistlers discussed in Chapter 1, found a sharp drop in plasma density in the range $L = 3$–$5R_E$. This sudden drop would not be expected in the simple filling scenario discussed above. The magnetosphere must have a mechanism of dumping plasma at high latitudes. We know no way of quickly pushing the ionospheric plasma back into the ionosphere from whence it came. It can do this slowly at night when the source in the atmosphere is reduced, but never quickly.

The loss of this plasma high in the magnetosphere is caused by the same mechanism that energizes the magnetosphere and creates the ring current. It is the reconnection of the magnetic field of the solar wind with the magnetosphere.

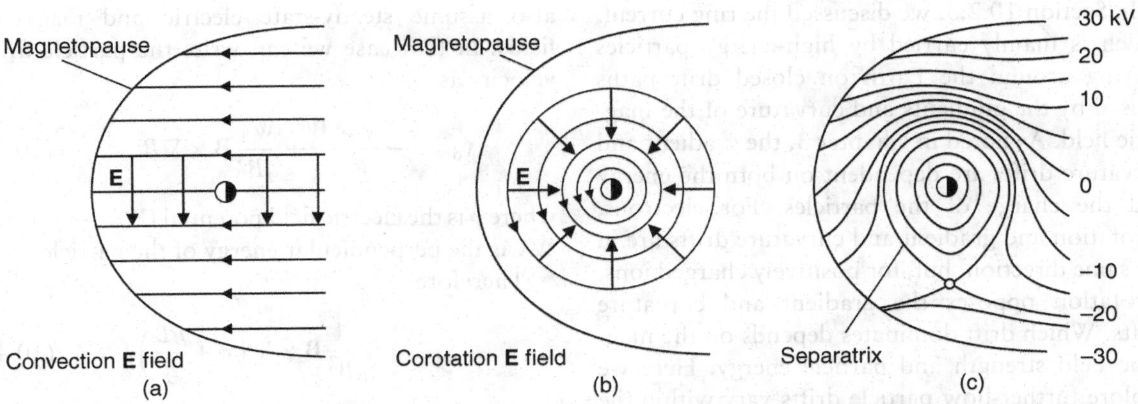

FIGURE 10.8. The formation of the plasmapause. As illustrated in Figure 10.7, the plasmasphere fills from below at a near constant rate, taking many days to fill the flux tube at the higher L values. The depletion mechanism that removes the plasma is convection of plasma to the dayside magnetopause by the electric field of the magnetosphere imposed by dayside and nightside reconnection. In (a), this is assumed to be a uniform electric field. However, the Earth rotates and the ionosphere rotates with the Earth. The corresponding corotating electric field is shown in (b). The motion of the plasma is controlled by the rotation of the Earth and flux transport between the tail and the magnetopause. The electric potential contours corresponding to an ideal steady-state situation are shown in (c). Inside the separatrix, the drift paths are closed and the density can build up to its saturation limit in which the upward and downward fluxes along the flux tube are equal.

The circulation of the plasmaspheric plasma brings it from the nightside to the dayside magnetopause where it is swept antisunward with the solar-wind flow on the now reconnected flux tubes, as illustrated in Figure 10.8. The electric field of the rotating magnetosphere in the equator is $463L$ (m s^{-1}) (where L is the L value in R_E) times the magnetic field strength, B, or $-0.014/L^2$ (V m^{-1}), where the negative sign indicates that the electric field is inward. This is to be compared with a value of 60 kV across the magnetosphere at quiet times for an average electric field of 3.1×10^{-4} V m^{-1}. This would lead to a stagnation point in the flow at $6.7R_E$ in the afternoon sector near geosynchronous orbit, as shown in Figure 10.8. Since B is northward in the Earth's equatorial plane, this electric field is radially inward in the closed low-latitude magnetosphere. When reconnection at the magnetopause drives plasma over the poles, the electric field is from dawn to dusk over the poles.

The cold plasma density in the magnetosphere is thus determined by the combination of photoionization of the atmosphere and upward transport into volumes that vary drastically from equatorial to near-polar footpoints combined with periodic dumping of the high-latitude regions The fact that the geosynchronous orbit where many of our operational spacecraft are stationed is in the region in which this plasma environment is most variable is of concern. The cold dense plasma of the plasmasphere is a benign environment for spacecraft that could be otherwise endangered by spacecraft charging. If a spacecraft is in a low-density region when high fluxes of energetic electrons arise, and especially if the spacecraft is in the Earth's shadow, as happened in the incident described for Galaxy 15 in Chapter 9, it is possible that different parts of a spacecraft charge to different electric potentials and that electric potential discharges could occur from one part of a spacecraft to another.

10.2.5 The dynamics of low-energy plasma

The preceding discussion was concerned with cold plasma that drifts with the electric field, but a fraction of the plasma in the magnetosphere has sufficient energy that its drift motion is affected by the dipolar nature of the magnetic field. In that case, the motions of the plasma are altered from the simple equipotential contours shown above.

In Section 10.2.3, we discussed the ring current, which is mainly carried by high-energy particles moving around the Earth on closed drift paths caused by the gradients and curvature of the magnetic field. As noted in Chapter 3, the gradient and curvature drifts are dependent on both the energy and the charge of the particles. For electrons, corotation and gradient and curvature drifts are in the same direction, but, for positively charged ions, corotation opposes the gradient and curvature drifts. Which drift dominates depends on the magnetic field strength and particle energy. Here we explore further how particle drifts vary within the Earth's magnetosphere.

In Chapter 3 we noted that a particle drift velocity can be generalized to any velocity-independent force **F**:

$$\mathbf{v}_{\rm d} = \frac{\mathbf{F} \times \mathbf{B}}{qB^2}, \tag{10.1}$$

where q is the particle charge.

Furthermore, if we can express the force in terms of a potential, i.e., $\mathbf{F} = -q\nabla\widetilde{\phi}$, similar to the electric potential corresponding to a static electric field, then we can write the drift velocity as

$$\mathbf{v}_{\rm d} = \frac{\mathbf{B} \times \nabla\widetilde{\phi}}{B^2}. \tag{10.2}$$

As a consequence, drift paths follow equipotential contours for this generalized potential.

To explore this further we shall use a simplified model of the Earth's magnetosphere. We shall assume that the Earth's magnetic field can be represented by a dipole, and that the dipole axis is parallel to the Earth's spin axis. We shall further assume that the spin axis is perpendicular to the ecliptic. We define a cartesian coordinate system with the z-axis along the dipole axis and the x-axis pointing toward the Sun. The y-axis therefore points toward dusk, or equivalently opposite to the orbital motion of the Earth around the Sun. We also define a corresponding spherical coordinate system in which the azimuth angle (φ) varies around the z-axis of the cartesian coordinate system, with $\varphi = 0$ pointing to the Sun and $\varphi = \pi/2$ along the y-axis.

We further restrict our discussion to 90° pitch angle particles drifting in the equatorial plane, and also assume steady-state electric and magnetic fields. In that case we can write the particle drift velocity as

$$\mathbf{v}_{\rm d} = \frac{-\nabla\phi \times \mathbf{B}}{B^2} + \frac{W_\perp}{qB^3}\mathbf{B} \times \nabla B, \tag{10.3}$$

where ϕ is the electric field potential ($\mathbf{E} = -\nabla\phi$) and $W_\perp$ is the perpendicular energy of the particle.

Therefore

$$\mathbf{v}_{\rm d} = \frac{1}{B^2}\mathbf{B} \times \nabla\left(\phi + \frac{\mu B}{q}\right), \tag{10.4}$$

where $\mu = W_\perp/B$ is the particle magnetic moment. Consequently we define

$$\widetilde{\phi} = \phi + \frac{\mu B}{q} = \phi + W_\perp, \tag{10.5}$$

and 90° equatorial particle drift paths follow contours of constant $\widetilde{\phi}$. This parameter can be considered to be the total energy of the particles, with the proviso that we have not included the energy associated with the particle drift itself.

In the Earth's magnetosphere there are two primary means by which an electric potential is imposed. The first is corotation. Here the neutral atmosphere forces the ionosphere to corotate through collisions, and this in turn results in a corotation electric field in the magnetosphere. It is frequently stated that corotation is imposed through mapping of the electric field from the ionosphere to the magnetosphere, but this is, strictly speaking, not correct. In the steady state, magnetic field lines are equipotentials; however, changes in convection are transmitted via MHD waves and the associated Maxwell stress ($\mathbf{j} \times \mathbf{B}$). The other source of the electric potential is the convection from the nightside to the dayside of the planet. This convection is driven by reconnection with the IMF, and sunward convection returns magnetic flux that is in turn transported into the lobes via the reconnection process.

For corotation,

$$\mathbf{v}_{\rm cr} = \hat{\boldsymbol{\varphi}} r \frac{2\pi}{\tau_{\rm d}}, \tag{10.6}$$

where $\tau_{\rm d}$ is the number of seconds in a day (86 400). The corresponding electric potential is

$$\phi_{\rm cr} = -\frac{2\pi}{\tau_{\rm d}} B_0 r_0^2 \left(\frac{r_0}{r}\right). \qquad (10.7)$$

In Eq. (10.7) we have assumed a dipole magnetic field of $B = B_0 r_0^3/r^3$. If, as an example, we use the International Geomagnetic Reference Field (IGRF) to specify the dipole magnetic field, then r_0 is the mean Earth radius (6371.2 km).

We can further define a parameter known as the corotation potential:

$$\phi_{\rm c} = \frac{2\pi}{\tau_{\rm d}} B_0 r_0^2. \qquad (10.8)$$

Using the IGRF version 12 magnetic field for 2016, $B_0 = 29816$ nT and $\phi_{\rm c} = 88.01$ kV.

For the reconnection-imposed convection we make the simplifying assumption that the electric field is uniform and points from dawn to dusk (positive y-direction):

$$\mathbf{E}_{\rm conv} = \hat{\mathbf{y}} \frac{\phi_{\rm pc}}{W r_0}, \qquad (10.9)$$

where $\phi_{\rm pc}$ is the cross-polar-cap potential and W is a characteristic width of the magnetosphere, expressed in units of r_0. There are more sophisticated models of the convection electric field, for example the Volland–Stern model, where the convection electric field is shielded from the inner magnetosphere. But for our purposes, the uniform electric field assumption is sufficient.

Again, we can derive a potential corresponding to the electric field in Eq. (10.9). Combining this with the corotation potential, we get the total potential for the sunward convection and corotation:

$$\phi = -\phi_{\rm c}\left(\frac{r_0}{r}\right) - \frac{\phi_{\rm pc}}{W}\left(\frac{r}{r_0}\right)\sin\varphi. \qquad (10.10)$$

At any particular radial distance the potential is maximum at dawn ($\phi = -\pi/2$) and minimum at dusk ($\phi = \pi/2$). The potential lies between these two limits for any other local times.

For equatorial particles, then, from Eq. (10.5),

$$\tilde{\phi} = -\phi_{\rm c}\left(\frac{r_0}{r}\right) - \frac{\phi_{\rm pc}}{W}\left(\frac{r}{r_0}\right)\sin\varphi + \frac{\mu B_0}{q}\left(\frac{r_0}{r}\right)^3 \qquad (10.11)$$

or

$$\frac{\tilde{\phi}}{\phi_{\rm c}} = -\left(\frac{B}{B_0}\right)^{1/3} - \frac{\phi_{\rm pc}}{W\phi_{\rm c}}\left(\frac{B_0}{B}\right)^{1/3}\sin\varphi + \frac{\mu B_0}{q\phi_{\rm c}}\left(\frac{B}{B_0}\right). \qquad (10.12)$$

In Eq. (10.12) the magnetic moment is normalized to $q\phi_{\rm c}/B_0$. This parameter equals 0.2948 MeV G^{-1} (these are clearly non-SI units, but they are used within the radiation belt community since the equatorial magnetic field strength is ~0.3 G at the surface of the Earth). In SI units the parameter equals 2.948 GeV T^{-1}.

From Eqs. (10.2) and (10.11) the equatorial particle drift velocity is given by

$$\mathbf{v}_{\rm d} = \frac{2\pi r_0}{\tau_{\rm d}}\left\{\hat{\boldsymbol{\varphi}}\left[\left(\frac{r}{r_0}\right) - \frac{\phi_{\rm pc}}{W\phi_{\rm c}}\left(\frac{r}{r_0}\right)^3\sin\varphi - \frac{\mu B_0}{q\phi_{\rm c}}\left(\frac{r_0}{r}\right)\right] + \hat{\mathbf{r}}\frac{\phi_{\rm pc}}{W\phi_{\rm c}}\left(\frac{r}{r_0}\right)^3\cos\varphi\right\}. \qquad (10.13)$$

If we consider the azimuthal component of Eq. (10.13), the first term in the square brackets corresponds to corotation and the second corresponds to the φ-component of the sunward convection. The last term in the square brackets is the gradient drift. For electrons, as already noted, this is in the same direction as corotation. For positive ions, on the other hand, the gradient drift opposes corotation. Equation (10.13) shows that the different drifts tend to dominate at different radial distances. At small radial distances, the gradient drift will tend to dominate; at large radial distances, sunward convection dominates; at intermediate distances, corotation is important. But, again, this is complicated by the dependence on the particle magnetic moment.

The radial component of Eq. (10.13) shows that particles move to lower radial distances on the nightside and higher radial distances on the dayside. This is to be expected since it is only the sunward convection that changes the radial distance of the particles. But the other aspect of this is that particles gain energy on the nightside part of their drifts and lose energy on the dayside.

In order to make clear the interplay of the different drifts, we use Eq. (10.5) to determine the types of

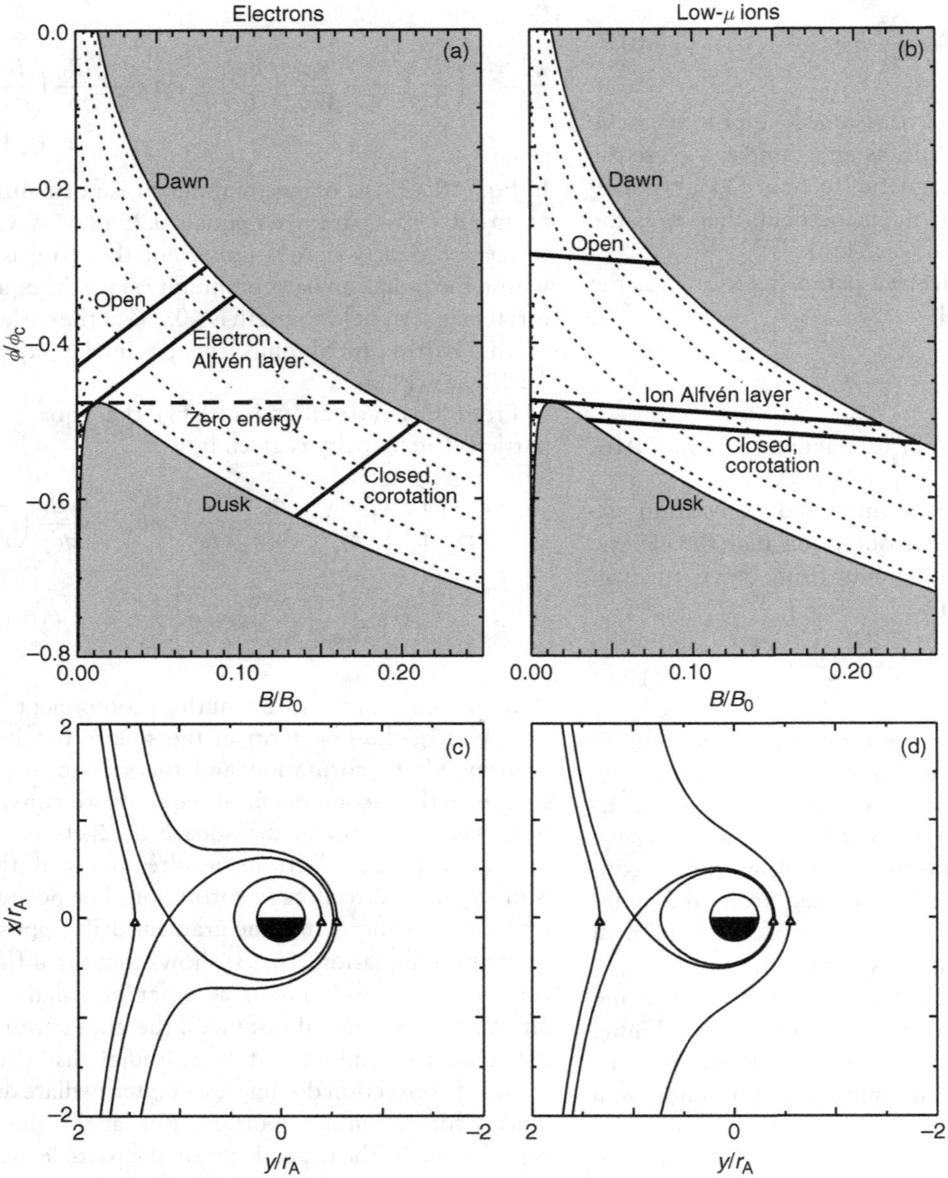

FIGURE 10.9. Potential contours and associated drift paths for electrons and low-μ ions. The use of ϕ and B in the top row to characterize drift paths follows an approach described by Whipple (1978).

drift path we expect for different particles. Figure 10.9(a) shows curves of normalized potential versus normalized magnetic field strength, using Eq. (10.10) to define the potential, again making use of the dipole magnetic field assumption. The trace labeled "dawn" corresponds to the potential at dawn, and the potential is maximum here. The minimum potential is at dusk. The dashed lines show the electric potential for every 2 h of local time (or 30° in azimuth). Because of the symmetry in Eq. (10.10), the potential at noon and midnight is the same, and is shown by the dotted line roughly midway between the dawn and dusk potentials. Similarly, the potential at 4 h local time is the same as at 8 h local time. This holds for all local times when defined relative to dawn and dusk. The difference between the dawn and dusk potentials at $B/B_0 = 0$ is given by $2\phi_{pc}/W\phi$, and we have assumed a relatively strong potential corresponding to 5 kV/R_E (~0.8 mV m^{-1}) so that we can more easily discriminate between the different types of drift path.

The thick solid lines in Figure 10.9(a) correspond to Eq. (10.5) assuming different asymptotic ($B = 0$) potentials for the same magnetic moment. Because we are considering electrons, the slope of these lines is positive. A unit slope corresponds to a magnetic moment $q\phi_c/B_0$. The line labeled "open" corresponds to one type of drift path. This is a drift path that starts for downtail, moves inward to increasing magnetic field strengths, until it passes through dawn, where the potential is maximum. The particle then continues to convect sunward.

The line labeled **"electron Alfvén layer"** is a special case of this type of open trajectory. The line is tangential to the duskside potential contour. This particle starts far downtail until it reaches the dusk meridian; rather than continuing sunward, the particle continues to move inward until it encounters the dawn meridian. The particle continues to drift along a path that, as we shall see, is mirror symmetric with respect to the dawn–dusk meridian. The reason for this reversal in the drift at dusk is because here corotation plus gradient drift exactly balances the sunward convection. The dashed line labeled "zero energy" shows Eq. (10.5) for a zero-energy particle that also reflects at dusk. It is clear that for electrons the radial distance at which this reflection occurs moves to larger radial distances the larger the magnetic moment. The point at dusk where the trajectory changes direction is also referred to as a stagnation point. This is because at this location both the radial and azimuthal gradient in $\widetilde{\phi}$ is zero. From Eq. (10.4), the particle drift velocity is consequently zero.

The "Alfvén layer" appellation is used to mark that drift path that marks the transition from open drifts to closed drift paths that circle around the Earth. One example, corresponding to a closed drift path, is shown by the line labeled "closed, corotation." This is a closed drift path because the particle is constrained to move from dawn to dusk and vice versa, but the particle cannot move to large radial distances, as we are not allowed to extend the line into the gray areas. To understand this better, it may be worth considering the case of a ball rolling along a rippled surface, with the caveat that the ball cannot leave the surface. Thus, the maximum gravitational potential that the ball sees is at the peaks of the rippled surface, and the minimum is at the valleys. The orbit is defined as a "corotation" orbit not because the particle is exactly corotating, but rather that the azimuthal motion is in the same direction as corotation. We know this because the particle is closest to the Earth at dawn (maximum magnetic field), and has therefore gained energy on drifting from dusk to dawn. As noted above, particles gain energy as they drift toward the Earth on the nightside. Thus, the particle drift from dusk to dawn occurs when the particle is on the nightside. This corresponds to an azimuthal drift in the same sense as corotation.

Having used Figure 10.9(a) to determine the different types of drift path for electrons, Figure 10.9(c) shows the drift paths in the equatorial plane. Positions are normalized to the radial distance of the zero-energy stagnation point, and, as already noted, the stagnation point for finite-energy electrons lies outside the zero-energy stagnation point. Each of the drift paths shown in Figure 10.9(c) corresponds to one of the straight lines shown in Figure 10.9(a). The right-most trajectory is open through dawn, and corresponds to the line labeled "open" in Figure 10.9(a). The trajectory immediately to the left is the Alfvén layer trajectory, which first encounters the dusk meridian before continuing to dawn. Inside this teardrop-shaped contour we see a nearly circular drift path. This corresponds to the line labeled "closed, corotation" in Figure 10.9(a). The final trajectory is open through dusk, and corresponds to the short line segment that lies just below the line marking the Alfvén layer in Figure 10.9(a).

The next case we consider is that of low-magnetic-moment (low-μ) ions, as shown in Figures 10.9(b) and (d). Because these ions have low magnetic moment, their drifts are dominated by corotation and sunward convection. As such, we expect the topology of the drift paths to be very similar to that for the electrons.

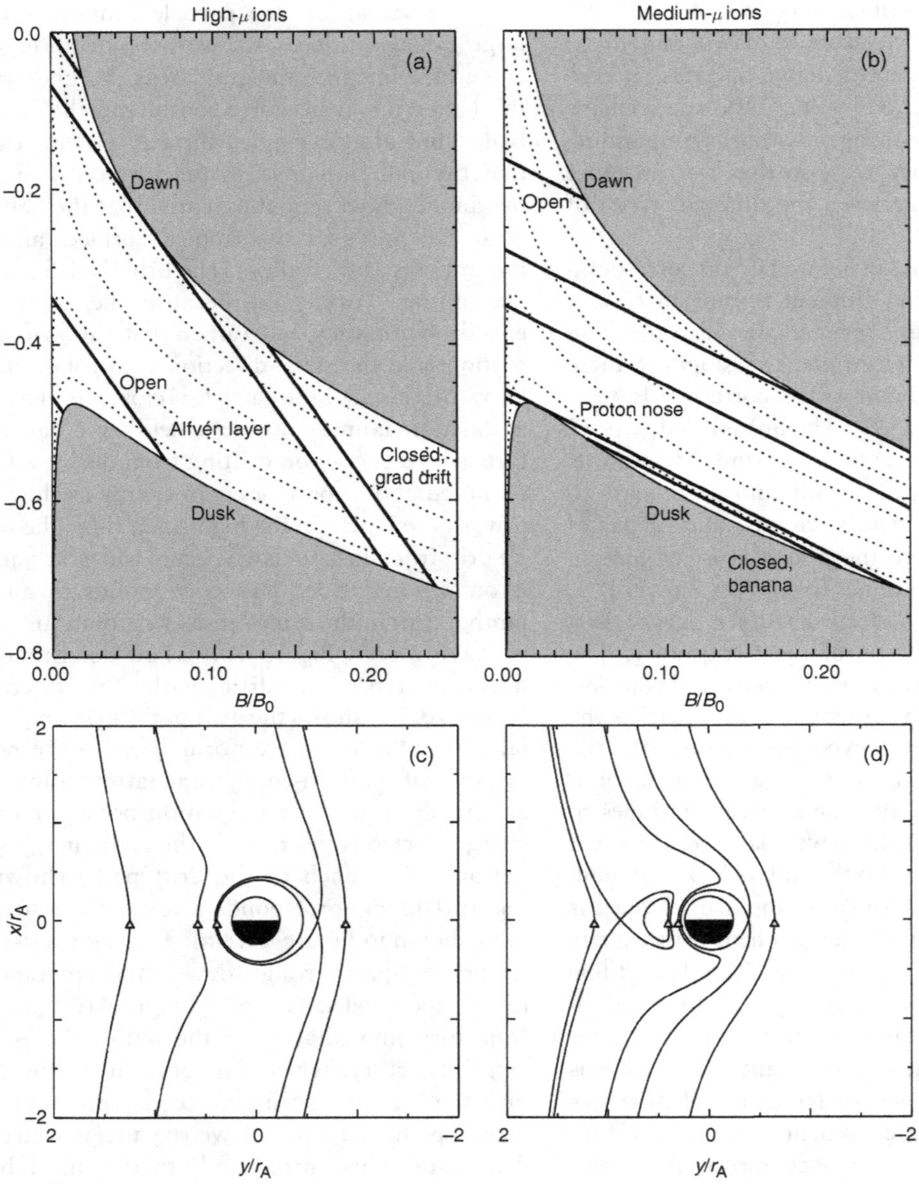

FIGURE 10.10. Potential contours and associated drift paths for high-μ and medium-μ ions.

From Figure 10.9(b) we see similar types of trajectory to those for the electrons as a function of potential and magnetic field strength. For ions, however, the slope of the lines is negative, and the stagnation point for the ion Alfvén layer lies inside the zero-energy stagnation point. Given the similar morphology of Figures 10.9(a) and (b), we expect the low-μ ion drift paths to be similar to those for the electrons, and that is indeed the case on comparing the drift paths in Figures 10.9(c) and (d).

As the ion magnetic moment increases, gradient drift becomes more important. The panels on the left-hand side of Figure 10.10 show the same information as that shown in Figure 10.9, but now for high-μ ions. Figure 10.10(a) shows three open drift paths and the Alfvén layer. In this case, the Alfvén layer stagnation point is at dawn. The trajectory encounters dusk at higher magnetic field strength. For the reasons outlined when we discussed how we determined that the closed electron drift path was in the same direction as corotation, in this case the closed drift path labeled "closed, grad drift" is in the same direction as the gradient drift. The dawnside stagnation point therefore corresponds to where the sunward convection and corotation drifts are balanced by the oppositely directed gradient drift.

The corresponding drift paths are shown in Figure 10.10(c). In many ways, they are the mirror image of the low-μ ion drift paths. Again this is because gradient drift dominates corotation for the high-μ ions.

The final case we consider is that of medium-μ ions. The plot in Figure 10.10(b) now shows the presence of what could be considered to be two Alfvén layers. The trajectory for which the constant-energy line is tangential to the dawn potential is similar to the corresponding case for high-μ ions. We therefore expect the trajectory to show a duskside stagnation point, and, as with the high-μ ions, the stagnation point occurs where gradient drift balances the electric field drifts.

The situation is more complicated for the trajectory that passes through the duskside stagnation point. In this case, the trajectory re-encounters the dusk meridian. As a consequence, this boundary between open and closed drift paths does not encircle the Earth. Furthermore, we see that there is a trajectory marked "closed, banana." This trajectory encounters the dusk meridian twice, and, while the drift path is closed, it does not circle the Earth.

The final trajectory we shall discuss is labeled "proton nose." This type of trajectory is one that is open, but penetrates very close to the Earth. The name derives from particle energy spectra that show an excursion to lower radial distances for particles in the few kiloelectronvolts range. For historical regions this feature has been referred to as the "proton nose," although other species can show a similar feature. The line eventually meets the duskside potential, but at a magnetic field strength larger than the scale used for the plot.

The corresponding trajectories plotted versus position are shown in Figure 10.10(d). The trajectories that pass through the dawn meridian are similar to the high-μ trajectories. But at dusk we see a more complicated pattern. As noted above, there is a region of close drift paths that have a shape reminiscent of a banana (or a kidney bean), hence the name. At the duskside stagnation point the sunward convection and gradient drift are canceled by corotation, but, as the particles drift inward from the stagnation point, away from dusk, gradient drift becomes more important, and eventually the drift paths start drifting back to dusk.

Earthward of these isolated closed drift paths we see that a drift path goes very close to the Earth. This is the proton nose drift path. For this drift path the three different drifts come into play. Initially the particles convect earthward, and, as they come close to the Earth, corotation starts to become more important and the particles start to drift toward dawn. However, as the particles gain energy, gradient drift starts to dominate, and the particles drift toward dusk, eventually passing through the dusk meridian and then drifting toward the dayside magnetopause.

The analysis presented here can be extended to other pitch angles and more complex electric and magnetic field geometries. But, in general, ion drift paths are more complicated than electron drift paths, since gradient and curvature drifts for the ions oppose corotation.

10.3 THE RADIATION BELTS

In Section 10.2 we have given an overview of the structure of the magnetosphere: the polar cusp, the plasma sheet, the ring current, the plasmasphere, and the low-energy plasma. This low-energy plasma is very much affected by the electric fields associated with reconnection in the tail and at the magnetopause. In this section, we move up in energy to examine those energetic charged particles that are less affected by the slowly varying magnetic and electric fields of the magnetosphere because their energies are much greater than the energy they receive or lose when moving through

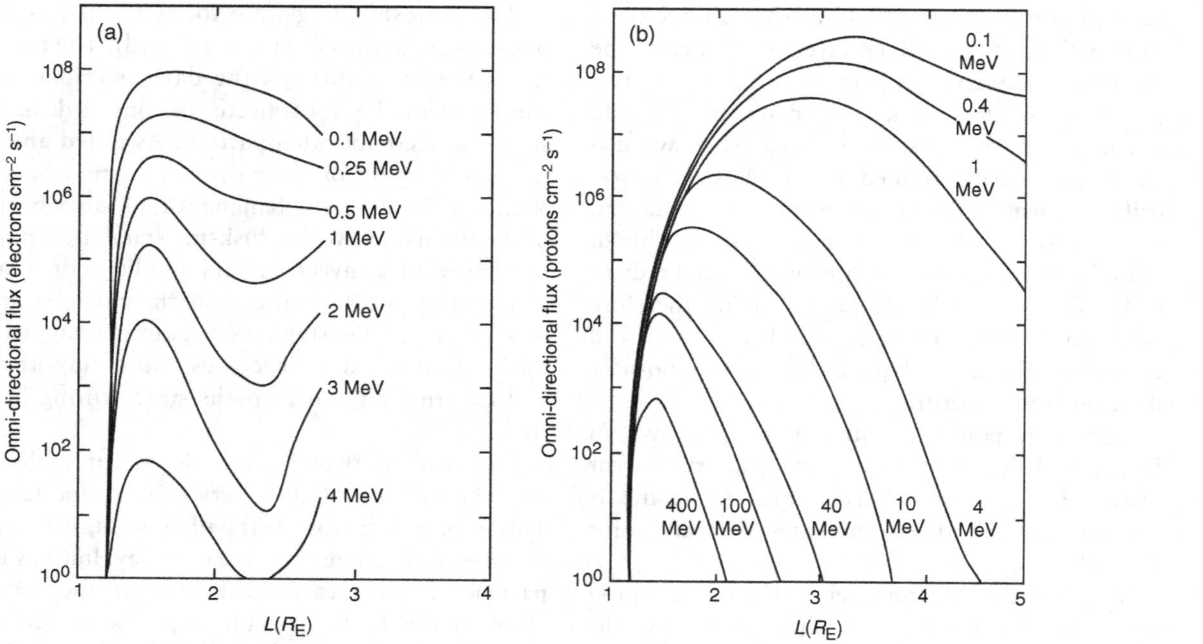

FIGURE 10.11. Omni-directional energetic electron (a) and proton (b) fluxes in the inner magnetosphere at solar minimum. Each curve gives total flux above the threshold value labeled. (After Spjeldvik and Rothwell, 1983.) The dip in the energetic electron fluxes near $L = 2.3R_E$ at energies above 1 MeV is called the "slot."

the magnetosphere. Thus we can draw typical radial profiles of these fluxes of these energetic particles, at least for quiet times, as shown in Figure 10.11. These particles, in contrast, are very much affected by gradient and curvature drifts as they bounce back and forth along magnetic field lines. Understanding these motions is aided by the theory of adiabatic particle drifts.

10.3.1 Time scales

An ion or electron moving through the Earth's magnetosphere executes three motions: (1) gyration or **cyclotron motion** about the magnetic field line, (2) **bounce motion** along the field line, and (3) **drift motion** perpendicular to **B**. For particles below 1 MeV in energy, which are the ones that we shall be considering, these motions take place on three different time scales. The gyro-motion, which occurs at the gyro-frequency Ω (also called the cyclotron frequency), is the fastest. From the definition of Ω given in Chapter 3, we have the following expression for the period of gyro-motion:

$$\tau_g = \frac{2\pi}{|\Omega|} = \frac{2\pi m}{|q|B} \cong (0.66\text{ s})\left(\frac{100\text{ nT}}{B}\right)A, \quad (10.14)$$

where A is the ratio of the particle mass to the mass of a proton ($= 1836^{-1}$ for an electron), and 1 nT = 10^{-5} G (gauss). In writing Eq. (10.14), we assume that the particle is singly charged (i.e., $|q| = e$). Because B averages about 30 000 nT at the Earth's surface at the equator and the field is very roughly dipolar, which implies that the field strength declines like the inverse cube of the geocentric distance r^{-3}, we expect $B \sim 240$ nT at $r = 5R_E$, 30 nT at $r = 10R_E$. For hydrogen ions, τ_c ranges from about 1 ms at low altitudes to about 2 s in the equatorial plane at $r = 10R_E$. The radius of the particle's cyclotron motion (the gyro-radius, see Chapter 3) is given by

$$\rho_g = \frac{mv_\perp}{|q|B} \cong (46\text{ km})A^{1/2}\left(\frac{W_\perp}{1\text{ keV}}\right)^{1/2}\left(\frac{100\text{ nT}}{B}\right), \quad (10.15)$$

Table 10.1. Gyro-radii (km)

	Energy			
L (R_E)	10 eV	1 keV	100 keV	10 MeV
	Electrons			
1.5	0.001	0.012	0.12	3.9
2	0.003	0.028	0.29	9.2
3	0.009	0.095	0.99	31.0
4	0.022	0.220	2.40	74.0
5	0.044	0.440	4.60	140.0
6	0.076	0.760	7.90	250.0
7	0.120	1.200	13.00	400.0
8	0.180	1.800	19.00	590.0
	Protons			
1.5	0.051	0.51	5.1	51
2	0.120	1.20	12.0	120
3	0.400	4.00	40.0	410
4	0.960	9.60	96.0	960
5	1.900	19.00	190.0	1900
6	3.200	32.00	320.0	3200
7	5.100	51.00	410.0	5200
8	7.700	77.00	770.0	7700

where $v_\perp$ is the velocity of the particle's gyro-motion, which is perpendicular to **B**, and $mv_\perp^2/2\,(W_\perp)$ is the energy involved in the cyclotron motion. The gyro-radius is typically several orders of magnitude smaller than magnetospheric dimensions, and it is smaller for electrons than for ions of the same energy.

Table 10.1 lists gyro-radii for electrons and hydrogen ions in the Earth's dipole field for various values of L and $W_\perp$ (the table was computed from the relativistic form of Eq. (10.15) and therefore differs from Eq. (10.15) for very high energies).

The particle's pitch angle is defined to be the angle between **v** and **B**, specifically

$$\alpha = \tan^{-1}(v_\perp/v_\parallel). \tag{10.16}$$

The particle's motion parallel to **B** consists of bouncing between mirror points, as illustrated in Figure 10.12. The period of the bounce motion is given, in order of magnitude, by

$$\tau_b \sim \frac{2l_b}{\langle v_\parallel \rangle} \sim (5\,\text{min}) \left(\frac{l_b}{10R_E}\right) A^{1/2} \left(\frac{1\,\text{keV}}{W_\parallel}\right)^{1/2}, \tag{10.17}$$

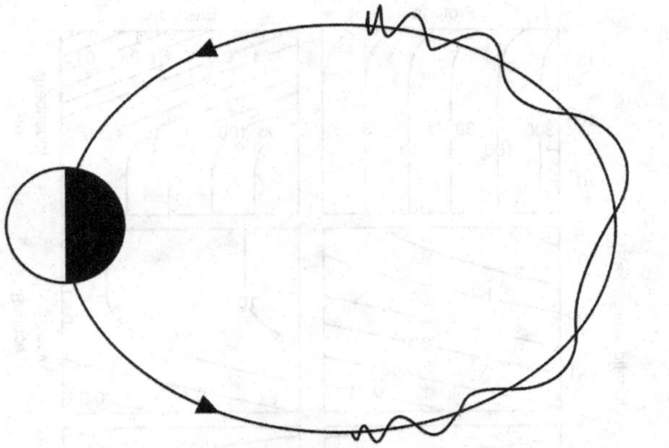

FIGURE 10.12. Sketch of the bounce motion of a charged particle on a dipole field line. Note that the particle spends a large fraction of its time near the mirror points.

where l_b is the distance along the field line between mirror points. Comparison of Eqs. (10.14) and (10.17) indicates that $\tau_b \gg \tau_g$ (i.e., the bounce motion is generally much slower than the cyclotron motion).

The third type of motion that the particles undergo is drift perpendicular to **B**. The drift formula that we shall use is an extension of the drifts as summarized in Section 3.3.7:

$$\mathbf{v}_d = \frac{\mathbf{E}\times\mathbf{B}}{B^2} + \frac{\mathbf{F}_{ext}\times\mathbf{B}}{qB^2} + \frac{W_\perp \mathbf{B}\times\nabla B}{qB^3} + \frac{2W_\parallel \hat{\mathbf{r}}_C \times \mathbf{B}}{qR_C B^2}, \tag{10.18}$$

where $\mathbf{F}_{ext}$ represents an external (non-electromagnetic) force (left unspecified for the moment), R_C is the radius of curvature of the magnetic field line, and $\hat{\mathbf{r}}_C$ is a unit vector outward from the center of curvature. The terms in Eq. (10.18) are called, respectively, **E** × **B** drift, external-force drift, gradient drift, and curvature drift. For a typical particle with pitch angle $\alpha \sim 45°$, the gradient and curvature drifts are of comparable magnitude. Also, in a dipole field, $\hat{\mathbf{r}}_C$ and ∇B are in opposite directions, so that the gradient and curvature drifts are in the same direction. Therefore, we can make an order-of-magnitude estimate of the sum of the gradient- and curvature-drift terms by just estimating one of them. The time scale for gradient/curvature drift for the case of the Earth is therefore given, in order of magnitude, by

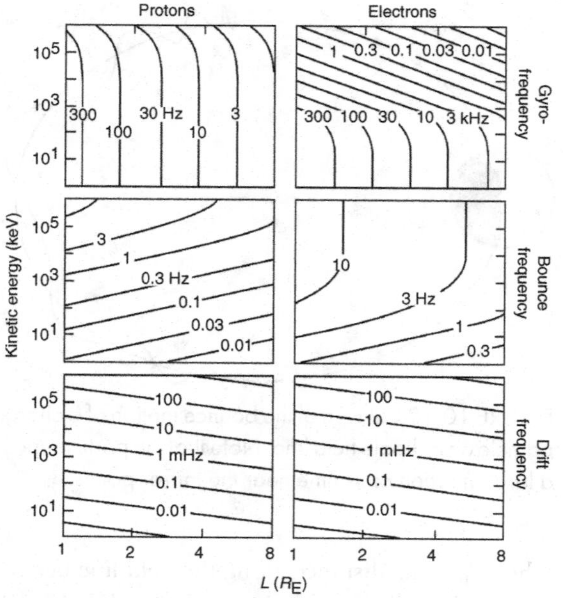

FIGURE 10.13. The gyration, bounce, and drift frequencies for equatorially mirroring particles moving in the Earth's dipole field, as a function of L and particle energy W. (Adapted from Schulz and Lanzerotti, 1974.)

$$\tau_d = \frac{2\pi r}{v_{GC}} \sim \frac{2qBr^2}{W} \sim (56 \text{ h}) \left(\frac{r}{5R_E}\right)^2 \left(\frac{B}{100 \text{ nT}}\right) \left(\frac{1 \text{ keV}}{W}\right). \tag{10.19}$$

Note that for a particle energy of about 1 keV, the three time scales given by Eqs. (10.14), (10.17), and (10.19) are separated by approximately equal factors of about $600A^{-1/2}$, which ranges from more than two orders of magnitude for O^+ to more than four orders of magnitude for electrons. Electrons gyrate and bounce much faster than ions. However, the gradient/curvature drift rate of a particle is independent of mass for a given energy.

Figure 10.13 displays gyro-frequencies (τ_g^{-1}), bounce frequencies (τ_b^{-1}), and drift frequencies (τ_d^{-1}) for equatorially mirroring protons and electrons. (One might think that the bounce period would go to zero as the length of the bounce path $l_b \to 0$, but it does not. The average velocity $\mathbf{v}_\parallel$ goes to zero linearly as $l_b \to 0$ so that, in Eq. (10.17), the bounce period τ_b goes to a finite constant, whose inverse is displayed in Figure 10.13.) The figure shows results for electrons and protons only. To find gyration, bounce, and drift frequencies for O^+, for example, multiply the proton values shown in Figure 10.13 by 1/16, 1/4, and 1, respectively.

10.3.2 Bounce motion and the first and second adiabatic invariants

The wide separation of these time scales allows theoretical separation of the three motions through the use of **adiabatic invariants** (see Section 3.3). The general theory of adiabatic invariants (Landau and Lifshitz, 1960) implies that $\oint p dq$ is conserved under slow changes in a system that exhibits periodic motion in the coordinate q, and p is the corresponding momentum. To apply this to the case of the gyro-motion of a particle in an approximately uniform magnetic field, we let the magnetic field be in the z-direction, $q = x$, and $p = mv_x$. The corresponding adiabatic invariant is then given by

$$\oint p dq = \oint p_x v_x dt = m \langle v_x^2 \rangle \tau_g = \frac{1}{2} m v_\perp^2 \frac{2\pi m}{qB} = \frac{2\pi m \mu}{q}, \tag{10.20}$$

where

$$\mu = \frac{mv_\perp^2}{2B} \tag{10.21}$$

is the magnetic moment of the particle's cyclotron motion. Equation (10.20) specifies the "**first adiabatic invariant**" of the motion of a charged particle in a magnetic field. For the normal situation, where the ratio (m/q) for the particle is constant, Eq. (10.20) implies that the magnetic moment μ is an adiabatic invariant. Specifically, μ is conserved if the magnetic field experienced by the gyrating particle changes slowly (i.e., on a time scale that is long compared with the gyro-period τ_g).

Because the time scale for changes in the magnetic field strength as the particle moves along a field line ($\sim \tau_b$) is in fact very long compared with the cyclotron period, we can use the fact that the magnetic moment is an adiabatic invariant to formulate a simple description of the bounce motion of the particle. We make the further simplifying assumption that there is no electric field parallel to **B**. In that case, because the magnetic field cannot change the kinetic energy of the particle, the kinetic energy W of the particle must remain constant as it moves along a field line in its bounce motion. We can then write

$$W = \frac{1}{2}m(v_\perp^2 + v_\|^2) = \frac{1}{2}mv_\|^2 + \mu B = \text{ constant.} \tag{10.22}$$

Therefore, $\mathbf{v}_\|$ decreases as the particle moves along the field line into regions of stronger field B, and the parallel velocity goes to zero when B reaches a critical value B_m:

$$B_\text{m} = \frac{W}{\mu}. \tag{10.23}$$

In terms of pitch angle α, the kinetic energy associated with the motion parallel to the field line is $W \cos^2\alpha$, and the motion perpendicular to the field line carries energy $W \sin^2\alpha = \mu B$.

Because the bounce motion of the particle on a given flux tube is periodic, we can define a **second adiabatic invariant** $\oint pdq$ associated with that motion:

$$J = \oint p_\| ds = 2\sqrt{2m}\int_{m_1}^{m_2}\sqrt{(W - \mu B(s))}ds, \tag{10.24}$$

where s is distance along the field line, and m_1 and m_2 are the locations of the particle's mirror points. The parameter J should be conserved if the particle experiences magnetic field changes on a time scale that is long compared with the bounce period τ_b. The constant energy W can be pulled out of the integral as follows:

$$J = 2\sqrt{2m\mu}I, \tag{10.25}$$

where

$$I = \int_{m_1}^{m_2}\sqrt{B_m - B(s)}ds. \tag{10.26}$$

Note that I does not depend on the energy of the particle, only on the particle's mirror point and on the field line that is being considered.

10.3.3 Bounce-averaged gradient/curvature drift

Consider a particle that has a non-zero first adiabatic invariant μ, but a zero second adiabatic invariant J. This particle can gradient-drift, but cannot curvature-drift. It is trapped in a magnetic field minimum. Its motion would appear as sketched in Figure 10.14. Assume, for simplicity, that the magnetic field minimum on each field line occurs in an equatorial plane, as would be true, for example, for the configuration shown in Figure 10.2. Substituting Eq. (10.21) into the gradient-drift term of Eq. (10.18) gives

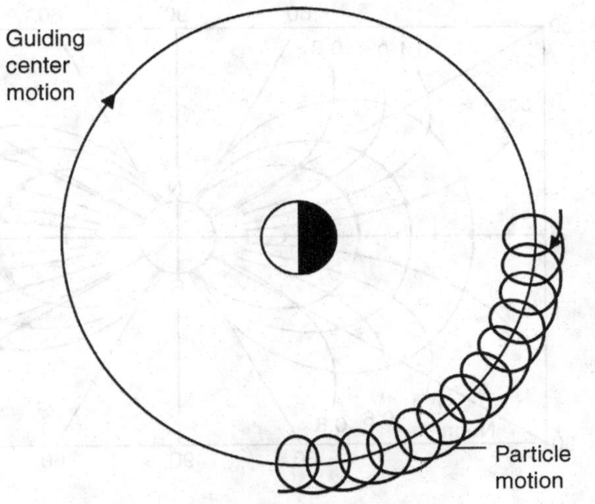

FIGURE 10.14. Gradient drifting charged proton in the Earth's equatorial plane.

$$\mathbf{v}_\text{g} = \mu\frac{\mathbf{B}\times\nabla B}{qB^2} = \frac{\mathbf{B}\times\nabla W}{qB^2}, \tag{10.27}$$

where, in writing the second equality, we use the fact that $W(\mu, \mathbf{x}) = \mu B(\mathbf{x})$ for this equatorially mirroring particle. In taking the gradient of W, we hold μ constant. If the particle is also $\mathbf{E}\times\mathbf{B}$-drifting in a potential electric field $\mathbf{E} = -\nabla\phi$, then we can write the combined $\mathbf{E}\times\mathbf{B}$-drift and gradient-drift velocity as follows:

$$\mathbf{v}_\text{d} = \frac{\mathbf{B}\times\nabla(q\phi + W)}{qB^2}. \tag{10.28}$$

The formula implies that the particle drifts perpendicular to the gradient of its total energy (potential + kinetic), so that Eq. (10.28) is an expression of conservation of energy for the particle. The conservation of energy principle was also invoked in Section 10.2.5; see Eq. (10.4). Because of the requirement of consistency with conservation of energy, we can, in fact, write the general expression

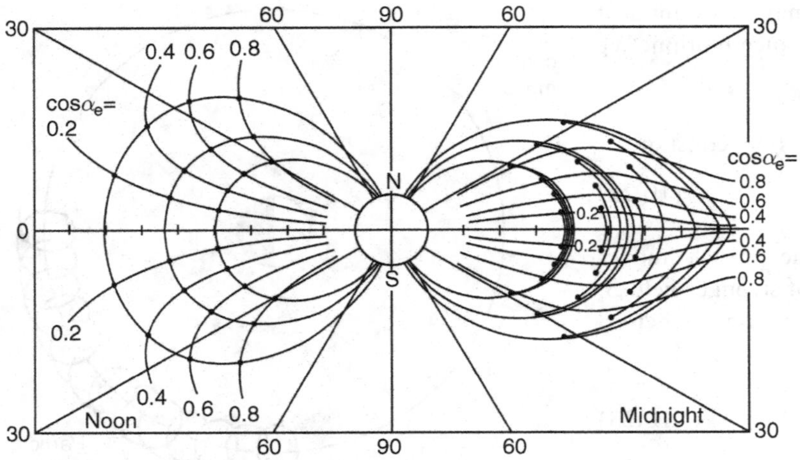

FIGURE 10.15. Computed drift-shell splitting for particles starting on common field lines in the noon meridian. Dots represent particles' mirror points. Curves giving the positions of mirror points for constant equatorial pitch angle α_e are shown. (Adapted from Roederer, 1967.)

for bounce-averaged gradient/curvature drift in the following form:

$$\mathbf{v}_{g+c} = \frac{\mathbf{B} \times \nabla(W(\mu, J, \mathbf{x}))}{qB^2}, \qquad (10.29)$$

where the gradient is now taken at a constant μ and J. (For a more rigorous and general discussion of bounce-averaged gradient/curvature drift, see Roederer (1970) or Wolf (1983).)

10.3.4 South Atlantic anomaly and drift-shell splitting

In a pure dipole field, gradient- and curvature-drift velocities are in the same direction, namely westward for positively charged particles and eastward for negatively charged particles. However, the actual magnetic field in the inner magnetosphere is not exactly dipolar, and the departures from a dipolar configuration have interesting effects on the radiation environment.

For the inner radiation belt, the most important departures from a dipole field result from the higher moments of the Earth's internal field (i.e., the quadrupole, octupole ... moments). The point on the Earth's surface where the field strength is weakest is near the east coast of South America. This feature of the Earth's field is called the "**South Atlantic anomaly.**" Inner radiation-belt particles that mirror near the equator drift on contours of constant magnetic field, according to Eq. (10.27). They have to come closer to the center of the Earth when they pass the weak point in the field that is above the South Atlantic, which means that they move to lower altitudes there and that they have an increased chance of precipitation. A large fraction of the loss from the inner radiation belts is associated with the South Atlantic anomaly.

For outer radiation belts, the most important departures from a dipole magnetic field result from magnetospheric currents. The day–night asymmetry of the magnetic field configuration has particularly interesting effects on the radiation belts. The Chapman–Ferraro current flowing eastward along the magnetopause, which compresses the magnetic field in the inner magnetosphere, has its strongest effect on the dayside. The westward tail current, which tends to expand and weaken the inner magnetospheric field, has its strongest effect on the nightside. The two currents together cause a systematic difference in the shapes of dayside and nightside field lines, as shown, for example, in Figure 10.15. This day–night asymmetry has different effects on particles with different pitch angles. For example, a particle that has $J = 0$, and thus gradient-drifts in the equatorial plane, follows contours of constant equatorial magnetic field strength. Such a particle has to come closer to the Earth on the nightside, where the field tends to be weaker, than on the dayside, where the field is stronger. On the other hand, consider a particle with very small μ and relatively large J. On its bounce path, the field strength is very small compared with the mirror field B_m, except very close to the end of the path, where B is increasing rapidly. Its geometric invariant I is approximately equal to $(B_m)^{1/2}s$, where s is the length of the bounce path. Assuming that electric-field effects are negligible, the particle maintains constant kinetic energy μB_m as it drifts. Because μ also stays constant, it follows that B_m is also a constant along the drift path, and the only way for it to retain constant I is for s to remain constant. In other words, such a particle curvature-drifts on a path of constant field line length s. It is clear from Figure 10.15 that if we compare dayside and nightside field lines that go equally far from the center of the Earth, nightside field lines tend to have shorter overall lengths than dayside field lines. Therefore, particles with near-zero equatorial pitch angles tend to drift farther from the Earth on the nightside than on the dayside, which is opposite from the behavior of particles with 90° equatorial pitch angle. Consequently, the outermost parts of the trapped radiation belts tend to have small equatorial pitch angles on the nightside of the Earth, and to have large equatorial pitch angles on the dayside of the Earth. This phenomenon is called **"drift-shell splitting."**

Figure 10.15 shows the results of a quantitative calculation of the drift-shell splitting effect. As an example of a small-J particle that mirrors near the equatorial plane, consider a particle that passes local noon with cos $\alpha_e = 0.2$, on a field line that crosses the equatorial plane at $8R_E$ geocentric distance. Scaling off the figure, we find that it passes local midnight, bouncing on a field line that extends only to about $7.1R_E$; its bounce path has also been shortened by about a factor of two, a result of the fact that $B(s)$ increases much more rapidly with distance from the equatorial plane on the nightside than on the dayside. As an example of a large-J particle that mirrors near the Earth, consider one that passes local noon with cos $\alpha_e = 0.8$, on the field line that extends to $8R_E$. The figure indicates that the particle passes local midnight on a field line that extends to about $9.6R_E$; its bounce length remains approximately constant, as expected.

10.4 WAVES IN THE MAGNETOSPHERE

In Section 10.3, we learned that energetic particles traveling in the Earth's magnetosphere had adiabatic invariants that controlled their motion. If nothing occurs to disrupt these invariants, the particles remain on their magnetospheric orbits forever. However, something does happen. Seldom is the magnetosphere completely quiet; waves resonate with the motions of the gyrating, bouncing, drifting particles, and they scatter. If they scatter such that their spatial motion leads them into the atmosphere, they will collide with it and be lost. If waves resonate with the drift period of the particle around the Earth, the particles can diffuse inward or outward of the magnetosphere. This could lead to the loss of particles into the atmosphere at low altitudes or into the magnetosheath at high altitudes.

The waves can arise in the atmosphere of the Earth, in the solar wind external to the Earth, or in the magnetosphere itself. The solar-wind interaction is a major source of these waves. The solar wind is unsteady, and that unsteadiness translates to pressure fluctuations on the magnetopause which compresses the magnetosphere. These compressional fluctuations can scatter particles otherwise trapped in the Earth's magnetic field lines.

10.4.1 Classification of magnetospheric waves

We are interested in waves that can interact with the electrons and ions of the magnetosphere. Thus they must have either a fluctuating magnetic or a fluctuating electric field. This field must also have some permanent effect on the particles, so it should not be so slow a variation that the effect is reversible. Waves can be purely compressional, so that the direction of the magnetic field does not change; they can be purely directional changes, in which the strength does not change; and they can be almost anything in between. Sometimes the waves are nearly continuous and narrow banded. Other times, they are short lived or very broadband in their spectrum. In the early days of magnetospheric research, these waves were classified by their discoverers' pet names for them, such as "pearls" for Pc 1 waves, and there was much confusion and difficulty in communication about these phenomena.

Table 10.2. Magnetospheric pulsation classes

Type	Pc 1	Pc 2	Pc 3	Pc 4	Pc 5	Pi 1	Pi 2
Period (s)	0.2–5	5–10	10–45	45–150	150–160	1–40	40–150
Frequency	0.2–5 Hz	0.1–0.2 Hz	22–100 mHz	7–22 mHz	2–7 mHz	0.025–1 Hz	2–25 mHz

In 1963, a meeting of the practitioners in this field was convened in Berkeley, California. They decided to divide the wave types into five classifications, if they were continuous in nature, and two types if they were irregular, as shown in Table 10.2.

These waves are the principal types seen in geomagnetic records, but there are waves also at higher frequencies. These are often recorded with search coil magnetometers and radio antennas. They are usually classified simply by frequency band: 5 Hz–3 kHz as ELF (extremely low frequency) waves; 3–30 kHz as VLF (very low frequency) waves; 30–300 kHz as LF (low frequency) waves; and 300 kHz–3 MHz as MF (medium frequency) waves. These bands are important for electrons in the ionosphere and magnetosphere, whereas the lower-frequency bands are usually more important for ions.

There are many sources of waves that affect the magnetosphere. External sources include terrestrial radio transmitters, solar disturbances, and the solar-wind interaction with the magnetosphere. Internal sources include substorms and the distributions of plasma along and around field lines in the magnetosphere. Often the plasma can move to a lower-energy state by emitting a plasma wave. Generally, this process is self-limiting as the plasma anisotropy is changed to isotropy and the process stops, but occasionally the anisotropy is maintained by some property of the system, for example, loss of the particle by collision with the atmosphere near the "foot" of the field line.

10.4.2 The solar-wind interaction

The solar wind is a conduit for disturbances produced by the Sun and can generate waves as it flows past the magnetosphere. Pressure fluctuations can be carried to the Earth in the changing density of the solar wind, which itself may be balanced in the solar-wind frame in tangential discontinuities that do not propagate but rather apply a varying dynamic pressure to the magnetosphere as they go by. The Sun is a producer of giant outbursts of overpressure in the form of coronal mass ejections, as discussed in Chapter 5. These rapidly compress the magnetosphere, as illustrated in Figure 10.16 (a). The time scale for the passage of a 400 km s^{-1} disturbance to go from the magnetopause to the tail is about 5 min – about the drift time for 30 MeV protons – so we would expect some effects of shock passages on the energetic protons drifting in the magnetosphere. The flow of the solar-wind plasma by the magnetosphere could also generate waves by the Kelvin–Helmholtz instability in the way that waves on the ocean are created by a strong wind. These waves may have a much broader spectrum and longer duration. The velocity shear is smallest at the subsolar point, and grows with angle away from this point. The perturbation in the magnetopause location caused by the instability also grows with distance from the nose of the magnetosphere, as shown in Figure 10.16(b). Waves of opposite polarity are seen on either side of noon.

Another contributor to the spectrum of ultralow-frequency (ULF) waves is the generation of back-streaming particles at the bow shock. These particle streams flow antiparallel to the solar-wind flow, producing a two-stream instability in which the particle energy in the counterstreaming solar-wind and upstream ion beams is isotropized through the production of ULF waves that are then convected against the magnetopause. These compressional waves can cross the magnetopause and enter the magnetosphere, usually in the Pc 3–4 band. These waves can provide a measure of the strength of the interplanetary magnetic field because their period is controlled by the field strength in which the waves are generated (Troitskaya, Plyasova-Bakunina, and Guglielmi, 1971; Russell and Fleming, 1976). These waves, when detected on the surface of the Earth, also provide a diagnostic of the mass density of the plasmasphere, as it is possible to determine what

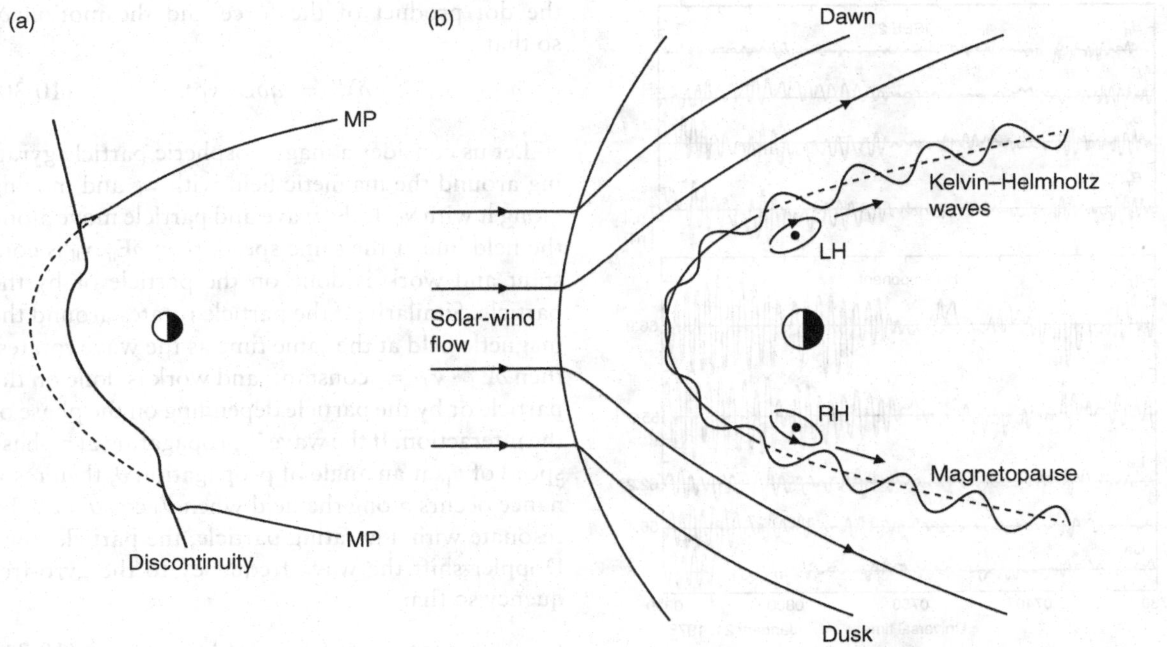

FIGURE 10.16. (a) A discontinuity in density or dynamic pressure encountering the magnetosphere launches a pressure wave into the magnetosphere as the magnetopause is pushed inward. (b) The velocity shear at a viscous boundary between the magnetosheath and the magnetosphere generates a surface wave that grows in amplitude away from the nose, causing left-hand (LH) and right-hand (RH) waves in the morning and afternoon sectors, respectively.

frequencies are standing waves along magnetic field lines (e.g., Russell *et al.*, 1999a). When the length of a field line contains an integral number of wavelengths, so that there is a node in the motion of the field line in the ionosphere at both ends, the field line is in resonance with the wave at this invariant latitude. On either side of this resonance to lower and higher latitudes, the phase of the oscillation shifts so that comparison of the phase of **Pc 3–4 waves** along a north–south chain of stations can identify the resonant points. This can then be inverted to determine the mass density of the plasmasphere at the equator, and then the ion composition of the plasmasphere if the electron density is independently known. A meridian chain enables a daily scan from about dawn to dusk, and a two-dimensional array can take mass density snapshots of much of the dayside plasmasphere. Figure 10.17 shows Pc 3–4 waves seen at a chain of stations near the Greenwich meridian, whilst the ISEE 2 spacecraft detected the same waves in the magnetosphere.

10.4.3 Internal generation

The magnetosphere can generate waves from the energy stored in the fields and particles. The largest reservoir of stored energy is in the geomagnetic tail, and this is released in substorms when the stress state of the tail is suddenly changed so that curved field lines are accelerating plasma to the nightside of the Earth. The sudden rearrangement of stresses causes low-frequency pulses, termed **Pi 2 waves**, to propagate to the surface of the Earth. These waves announce the onset of the substorm. Their arrival time also can be inverted to diagnose the site of reconnection, in a way analogous to the sounding of the plasmasphere by Pc 3–4 waves. Both have been referred to as magnetoseismology (Chi and Russell, 2005).

ULF waves can also be energized by the drift motion of particles in the magnetosphere when the

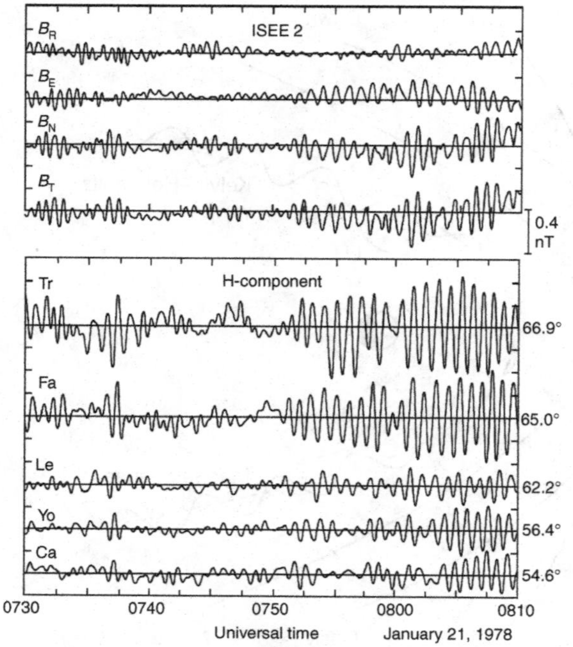

FIGURE 10.17. Pc 3–4 waves seen simultaneously in space and on the ground near the Greenwich meridian. ISEE 2 observations are present in radial, east, north coordinates, while the ground stations from Cambridge (Ca) in the south to Tromso (Tr) in the north are shown in the horizontal (H)-coordinate pointing to the north geomagnetic pole. Other stations are York (Yo), Lerwick (Le), and Faroe Islands (Fa). (After Odera *et al.*, 1991.)

wave resonates with the drifting and bouncing particles. These instabilities may resemble the mirror instability, in which plasma with an excess of pressure perpendicular to the magnetic field spontaneously forms pockets of high-density plasma from an originally, spatially uniform plasma. The mirror-mode instability is found wherever high-beta plasmas are found: in the solar wind, in cometary comas, in the bow shock and magnetosheath, and in the distant night-time magnetosphere. However, since the structure in the mirror mode is close to being in pressure balance, it does not propagate, but it can be convected over large distances.

It is also possible to tap the thermal motions of magnetospheric particles to generate waves. The force on a charged particle due to electric, δE, and magnetic fluctuations, δB, is $q(\delta\mathbf{E} + \mathbf{v} \times \delta\mathbf{B})$. This force does work δW on the particle proportional to the dot product of the force and the motion $\delta \mathbf{s}$ so that

$$\delta W = q\delta\mathbf{E} \cdot \mathbf{v}\delta t. \tag{10.30}$$

Let us consider a magnetospheric particle gyrating around the magnetic field with $\mathbf{v}_\perp$ and moving along it with $\mathbf{v}_\parallel$. If the wave and particle move along the field line at the same speed, then $\delta\mathbf{E}_\parallel \cdot \mathbf{v}_\parallel$ is constant and work is done on the particle or by the particle. Similarly, if the particle rotates around the magnetic field at the same time as the wave rotates, then $\delta\mathbf{E}_\perp \cdot \mathbf{v}_\perp$ = constant, and work is done on the particle or by the particle depending on the phase of the interaction. If the wave is propagating at a phase speed of v_p at an angle of propagation θ, then resonance occurs along the field when $v_\parallel \cos\theta = v_p$. To resonate with a gyrating particle, the particle must Doppler-shift the wave frequency to the gyro-frequency so that

$$v_\parallel \cos\theta = v_p(\omega - \Omega/\Gamma)/\omega, \tag{10.31}$$

where Ω is the signed gyro-frequency of the particle and (see Eq. (10.18)) the relativistic gyro-frequency correction is

$$\Gamma = [1 - (v/c)^2]^{-1/2}, \tag{10.32}$$

which is unity for non-relativistic particles.

Protons are left-handed particles that gyrate around the magnetic field in the direction of one's fingers when the left thumb is along the magnetic field. Electrons are right-handed. These definitions do not depend on the direction of motion of the plasma particle along the field. Similarly, left-handed waves such as ion cyclotron waves do not change polarization when they propagate along or against the direction of the magnetic field. Their properties are determined by their interactions with the ions in the plasma. The same is true for the right-handed waves whose properties are due to their interactions with the plasma electrons.

Energetic particles can interact with waves head-on or by overtaking them. If they overtake them, the particle sees the wave with a reverse polarization. Hence, in an overtaking, resonance can occur between an electron and a left-handed wave and an ion and a right-handed wave. For a head-on resonance, the wave and particle must have the same handedness.

To understand the exchange of energy between a charged particle and a circularly polarized wave, let us examine waves traveling along the magnetic field. If we assume plane monochromatic waves in a uniform medium, we can write

$$\delta\mathbf{E}(x,t) = \delta\mathbf{E}_0\exp(-i\omega t + ikx), \tag{10.33}$$

$$\delta\mathbf{B}(x,t) = \delta\mathbf{B}_0\exp(-i\omega t + ikx). \tag{10.34}$$

From Faraday's law, the ratio of the strength of the electric and magnetic amplitudes is the phase velocity

$$\begin{aligned}\nabla\times\delta\mathbf{E} &= -\partial(\delta\mathbf{B})/\partial t,\\ \mathbf{k}\times\delta\mathbf{E} &= \omega\delta\mathbf{B},\\ |\delta\mathbf{E}|/|\delta\mathbf{B}| &= |\omega|/|k| = |v_{\mathrm{p}}|.\end{aligned} \tag{10.35}$$

The force on a particle moving with the wave phase velocity, $\mathbf{v}_{\mathrm{p}}$, is zero because

$$\begin{aligned}\mathbf{F} &= q\delta\mathbf{E} + \mathbf{v}_{\mathrm{p}}\times\delta\mathbf{B}\\ &= q(\delta\mathbf{E} - \delta\mathbf{E})\\ &= 0.\end{aligned}$$

Hence, the energy of a particle in the frame of reference moving with the wave phase velocity is constant or

$$mv_\perp^2/2 + m(v_\parallel - v_{\mathrm{p}})^2/2 = \text{constant}. \tag{10.36a}$$

For small changes in velocity,

$$mv_\perp\Delta v_\perp + mv_\parallel\Delta v_\parallel - mv_{\mathrm{p}}\Delta v_\parallel = 0. \tag{10.36b}$$

The change in energy in the particle as measured in the rest frame is

$$\begin{aligned}\Delta W &= mv_\perp\Delta v_\perp + mv_\parallel\Delta v_\parallel\\ &= mv_{\mathrm{p}}\Delta v_\parallel.\end{aligned} \tag{10.37}$$

The change of transverse particle energy is

$$\Delta W_\perp = mv_\perp\Delta v_\perp. \tag{10.38}$$

The change in longitudinal energy is

$$\Delta W_\parallel = mv_\parallel\Delta v_\parallel, \tag{10.39}$$

$$\Delta W_\parallel/\Delta W = v_\parallel/v_{\mathrm{p}} \tag{10.40}$$

and

$$\Delta W_\perp/\Delta W = 1 - \Delta W_\parallel/\Delta W = 1 - v_\parallel/v_{\mathrm{p}}. \tag{10.41}$$

When a right-handed particle, most commonly an electron, interacts head-on with a right-handed wave, e.g., a whistler-mode wave, the particle will resonate with the wave when the Doppler shift brings that wave frequency up to the particle's gyro-frequency so it sees the oscillating magnetic and electric field at its gyro-frequency. This resonance can either increase or decrease the perpendicular energy of the particle to the magnetic field depending on the phase of the interaction. Increasing the particle energy decreases the wave energy, and decreasing it increases the wave energy. The pitch angle of the particle's motion becomes more aligned with the magnetic field, and this drives particles into the loss cone. The loss cone is the range of particle pitch angles for which the particle hits the atmosphere and hence cannot return to the equator. Losing the particle maintains a so-called loss cone anisotropy and sustains wave growth in this head-on resonance. Because at any resonant pitch angle, α_0, when the distribution has a loss cone, more particles are just above the resonant point than below the resonant pitch angle so that more particles are scattered toward the loss cone than away. This net scattering to lower pitch angles takes energy from the particle and gives it to the waves. The overtaking resonance is unstable for beam-like anisotropies with more particles along the direction of the magnetic field.

The exchange of energy in cyclotron resonance with right-handed (e.g., whistler mode) waves is summarized in Table 10.3. A similar table can be drawn for the cyclotron interaction with left-handed waves.

If we restrict our discussion to the electron cyclotron head-on resonance with whistler-mode waves, Kennel and Petschek (1966) have shown that the resonant energy, E_{R}, is

$$E_{\mathrm{R}} = B^2/(2\mu_0 n)[(\Omega_{\mathrm{e}}/\omega)(1-\omega/\Omega_{\mathrm{e}})^3]. \tag{10.42}$$

Thus, the resonant energy is approximately proportional to B^3 and inversely proportional to n. Thus, on any field line E_{R} will be least at the equator. Since typically the spectra of energetic electrons are decreasing functions of energy, there will be more resonant particles at the equator and the growth rate will be largest at the equator. If we assume that $\omega \ll \Omega_{\mathrm{e}}$, then the resonant frequency ω_{R} will be

Table 10.3. Energy change in cyclotron resonance with right-handed electromagnetic waves with electrons (head-on) and positive ions (overtaking); the sign of particle energy change depends on the phase of the resonance

Particle type	Resonance type	Particle energy	Wave energy	Parallel energy	Perpendicular energy	Pitch angle
Electron	head-on	↓	↑	↑	↓	↓
Positive ion	overtaking	↓	↑	↓	↑	↑
Electron	head-on	↑	↓	↓	↑	↑
Positive ion	overtaking	↑	↓	↑	↓	↓

$$\omega_R = \Omega_e B^2/(2\mu_0 n E_R). \tag{10.43}$$

Within the plasmasphere, the frequency resonant with a particular electron energy will decrease with increasing L value. However, at the plasmapause, where the plasma density n decreases sharply, the resonant frequency will increase sharply. The parallel resonance energy will be

$$E_R = 2.5(\Omega_e/\omega)(B^2/n), \tag{10.44}$$

where E_R is in electronvolt, n denotes electrons per cubic centimeter, and B is in nanotesla.

The particles in the magnetosphere are often only marginally stable and can be easily pushed to generate waves (such as when an interplanetary shock strikes or a substorm starts). This instability works both for ions (making Pc 1 waves) and electrons (making ELF and VLF waves). ELF waves are usually seen with two quite distinct waveforms, rising tones, called chorus, and broadband hiss. The source of these waves is certainly the energetic electrons trapped on magnetic field lines, but the mechanism by which the coherent rising elements of chorus arise is poorly understood.

Because waves propagate into regions with different plasma properties, they can resonate with a different particle population elsewhere. This complicates our understanding of the particle behavior, but it allows for the removal of particle populations that themselves are not unstable.

As discussed in the first part of this chapter, the varying electric field (or flows) in the magnetosphere due to pressure changes and reconnection causes the plasmas of the magnetosphere to be dynamic. Further, because the energetic particles drift due to their gyro-motions in the magnetic field gradients and their parallel motion along curved field lines, the low- and high-energy plasmas take different paths. Nevertheless, they seem to have some coherent structure despite this disparate motion. Figure 10.18 shows radial profiles of the cold plasma density (plasmasphere), the proton flux from 31–49 keV (ring current), and the $E > 35$ keV electrons (radiation belt) on four cuts through the magnetosphere in 1966 (Russell and Thorne, 1970).

On the first pass (July 3–7, 1966), the ring current is weak and peaks just inside the outer part of the plasmasphere. The energetic particle fluxes are weak. On the second pass, there has been the injection of energetic protons into the ring current. The plasmasphere has been eroded and the ring current and electron radiation belt have been strongly enhanced. Both peak outside the plasmapause. On the following pass, the cold plasma is refilling the magnetosphere and has significant density to $L = 5$. The ring current has disappeared from the inner magnetosphere, quite possibly by the precipitation of the protons by ion cyclotron waves. The fluxes of energetic electrons too have weakened, but not as severely as the protons. On the fourth pass, the plasmasphere continues to refill out to $L = 6$, and the ring current has made a further retreat. The energetic electrons have further decayed.

Figure 10.19 shows another example of motion of a feature of the radiation belts with time, but with a different cause. This is the position of the energetic electron slot at energies near 1 MeV at L values near 3. At quiet times, the slot is near $2.7R_E$ at 3 MeV and $4R_E$ at 70 keV. After a magnetic storm occurred, the slot moved outward at high energies and inward at low energies. In this case, we believe that the change was due to the interplay of enhanced radial diffusion and the loss of electrons due to pitch angle scattering, due to both lightning in the atmosphere and plasma instabilities in

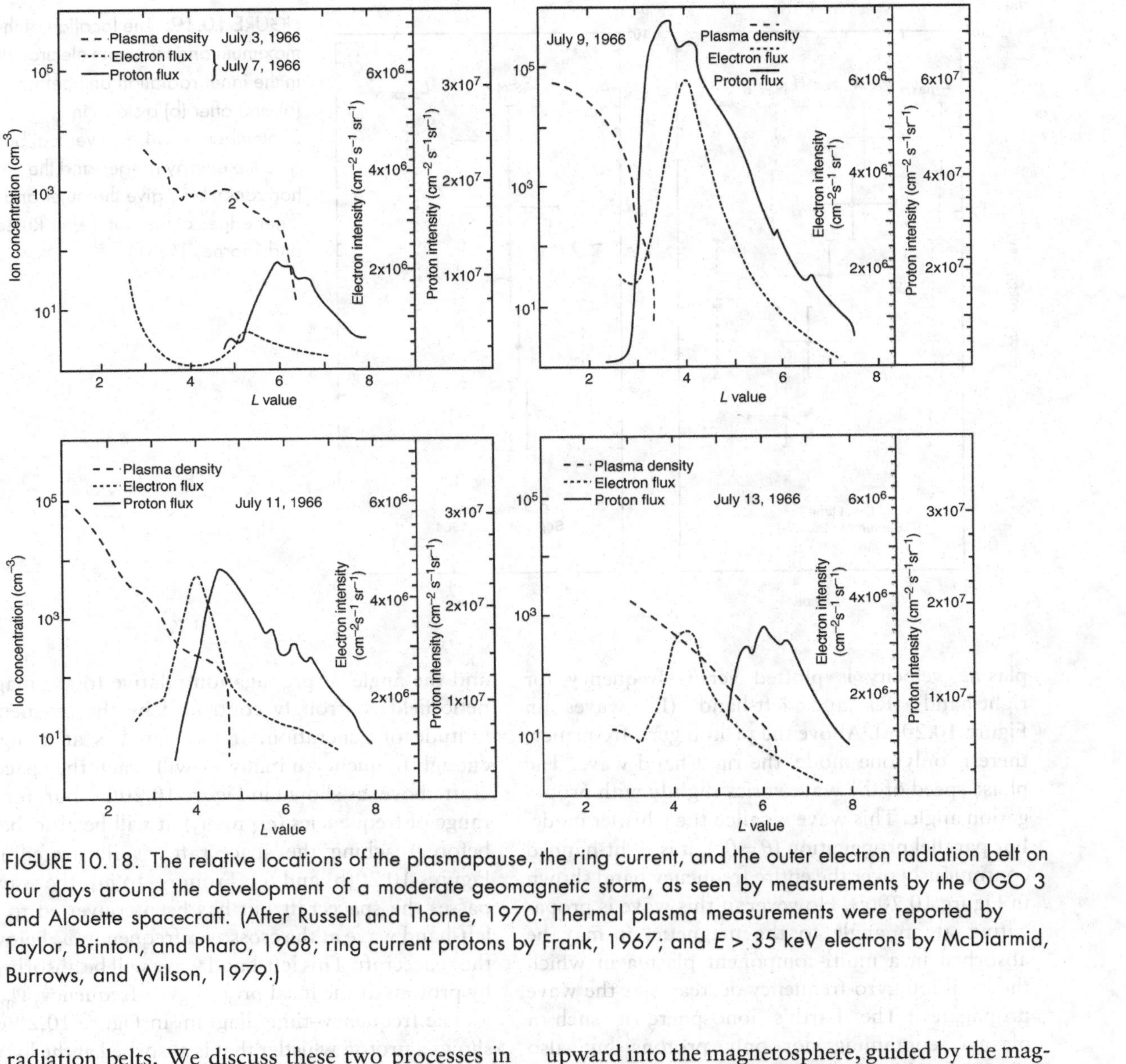

FIGURE 10.18. The relative locations of the plasmapause, the ring current, and the outer electron radiation belt on four days around the development of a moderate geomagnetic storm, as seen by measurements by the OGO 3 and Alouette spacecraft. (After Russell and Thorne, 1970. Thermal plasma measurements were reported by Taylor, Brinton, and Pharo, 1968; ring current protons by Frank, 1967; and $E > 35$ keV electrons by McDiarmid, Burrows, and Wilson, 1979.)

radiation belts. We discuss these two processes in Sections 10.4.4 and 10.5.

10.4.4 Waves from below

In addition to being buffeted by the solar wind and by fluctuations of its own making, the magnetosphere is also disturbed from below. Radio transmitters can cause electrons to precipitate, but a particularly strong radio source is not man-made: it is lightning. Lightning generates a broad band of electromagnetic waves. These waves can propagate upward into the magnetosphere, guided by the magnetic field of the Earth in the ELF and VLF frequency ranges. It is believed that the resonance of the inner zone electrons near $L = 2$ contributes to the loss of electrons from the region, partially explaining the slot seen at energies from 1–4 MeV in the electron radial profiles in Figures 10.11 and 10.19.

Figure 10.20 illustrates how the electromagnetic waves entering the ionosphere above a lightning generation region in the atmosphere are affected by their passage through the ionosphere. The

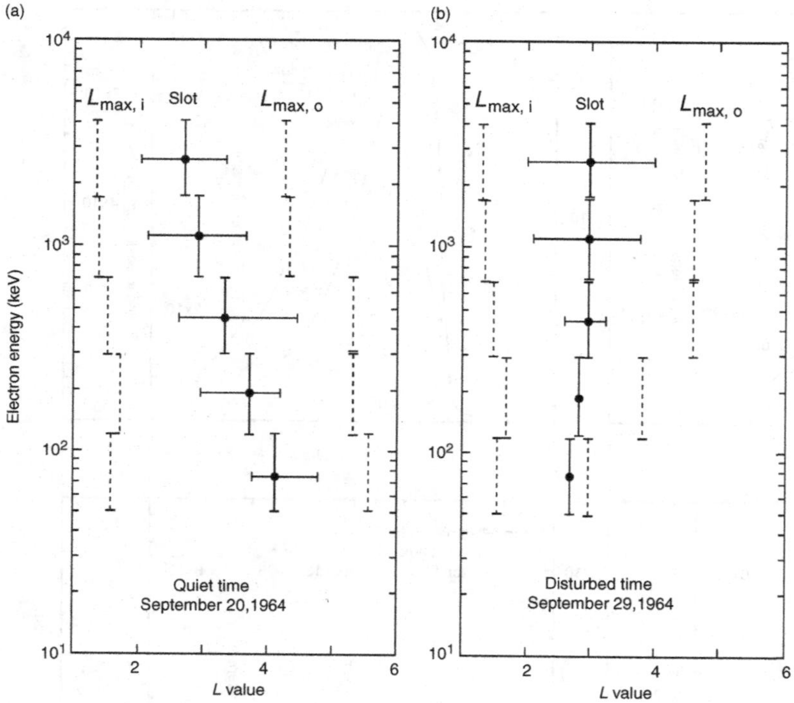

FIGURE 10.19. The location of the maximum and minimum electron flux in the inner radiation belt before (a) and after (b) a storm in September, 1964. The vertical bars give the energy ranges and the horizontal bars give the outer and inner edges of the slot. (After Russell and Thorne, 1970.)

plasma velocity is plotted versus frequency for right-hand (R) and left-hand (L) waves in Figure 10.20(a). Above the proton gyro-frequency, there is only one mode, the right-hand wave. The phase speed of the wave varies slightly with propagation angle. This wave is called the whistler mode. For parallel propagation ($\theta = 0°$), it is right-handed continuously over the entire frequency band shown in Figure 10.20(a). However, if this wave is propagating at an angle to the magnetic, it may be absorbed in a multi-component plasma in which the local ion gyro-frequency decreases as the wave propagates. The Earth's ionosphere is such a plasma, containing not only protons but also significant fractions of helium and heavier ions depending on altitude. Moving from left to right in Figure 10.20(a), as one would if moving upward in the Earth's magnetic field, the wave changes from right-hand to left-hand propagation at the crossover frequency, ω_x, and is then absorbed at the proton gyro-frequency after slowing down considerably. Because of the altitude variation of the index of refraction with altitude, the waves have all refracted vertically upon entry into the ionosphere, and the angle of propagation relative to the magnetic field is strongly controlled by the magnetic latitude of generation. If the signal is at a high enough frequency initially, it will reach the spacecraft above, as shown in Figure 10.20(b), but, for a range of frequencies (ω_4 to ω_5), it will be absorbed before reaching the spacecraft, as illustrated in Figures 10.20(b) and (c). From ω_3 to ω_4, the wave passes the spacecraft but has been converted to a left-hand wave at the crossover frequency ω_x below the spacecraft. This left-hand wave will be absorbed by protons at the local proton gyro-frequency, Ω_p.

The frequency–time diagram in Figure 10.20(c) shows a **proton whistler** that is absorbed at the local proton gyro-frequency. The group velocity of the proton whistler slows drastically as Ω_p is approached. Depending on the height of the detection point above the bottom of the ionosphere, the absorption illustrated from ω_4 to ω_5 will be narrow or broad. In short, the passage of the whistler-mode energy through the ionosphere is difficult at the lowest frequencies. Nevertheless, the waves can get through over a fairly wide band at the higher frequencies, and can be seen on the far side of the

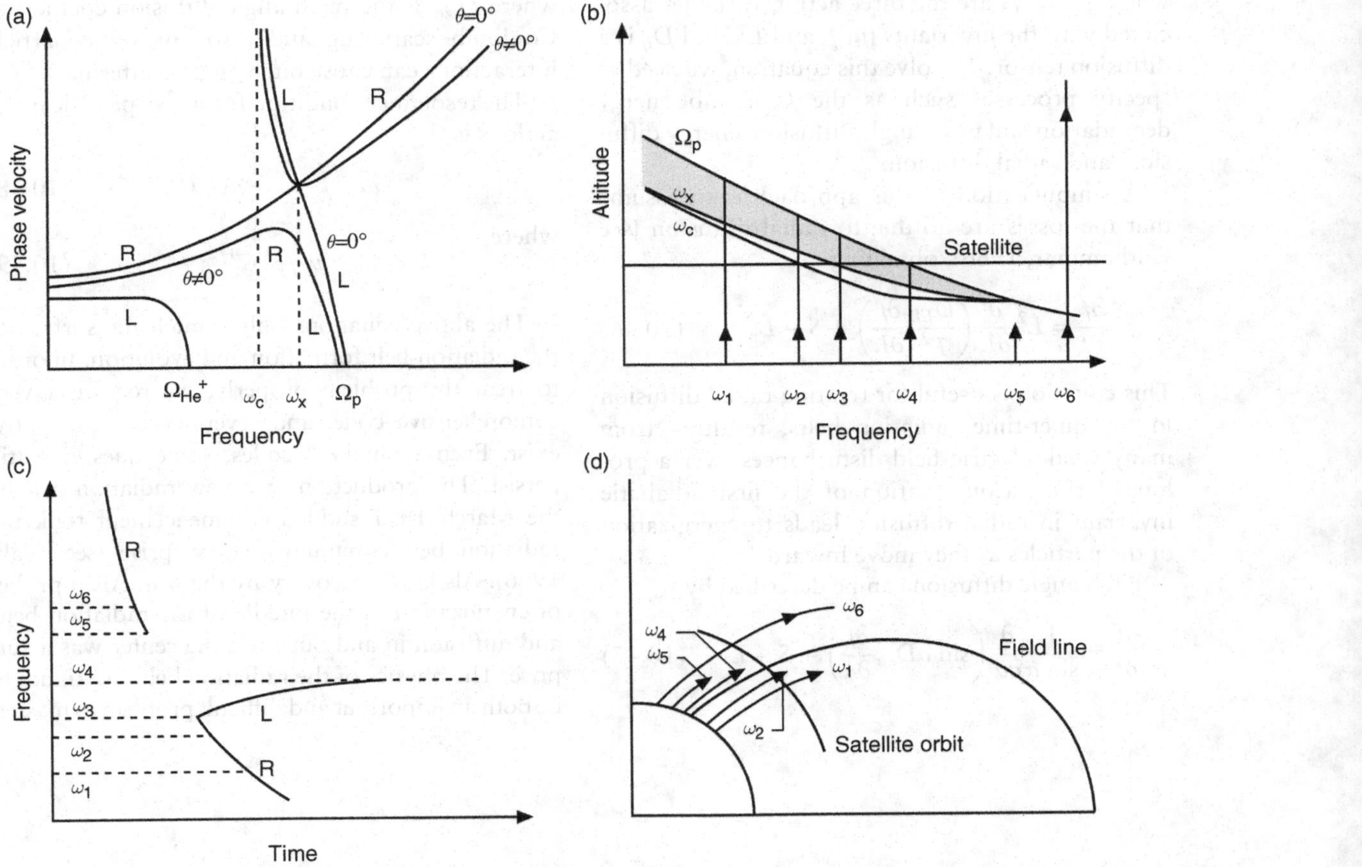

FIGURE 10.20. Formation of a proton whistler. (a) Phase velocity as a function of frequency in a multi-component plasma for parallel and non-parallel propagation. (b) Variation of the cutoff frequency ω_c, the crossover frequency ω_x, and the proton cyclotron frequency with altitude. (c) Frequency–time signature of a proton cyclotron frequency whistler that is converted from right-hand to left-hand polarization and then absorbed. (d) The ray paths of the waves shown propagating upward. (After Russell *et al.*, 1971.)

globe when they return to Earth. Thus they may well be responsible for some of the decay of electrons in the slot region of the radiation belts.

10.5 RADIATION BELT FORMATION AND LOSS

Earlier in this chapter, we discussed the adiabatic constants of the motion of the trapped particles in the radiation belts. These constants of the motion help us not only to understand the long-term constancy of the radiation belts, but also how they form and evolve. To do this, we express the particle phase space density f as a function of μ, J, and L, where μ and J are the first and second adiabatics and L replaces the third. If we describe the evolution of this phase space density by the Fokker–Planck diffusion equation with a source term S and a loss term L, we obtain

$$\frac{\partial f}{\partial t} + \sum_{i=1}^{3} \frac{\partial}{\partial J_i}\left[\left(\frac{\partial J_i}{\partial t}\right)_{\text{fric}} f\right] = \sum_{i=1}^{3}\sum_{j=1}^{3} \frac{\partial}{\partial J_j}\left[D_{ij}\frac{\partial f}{\partial J_j}\right] + S - L, \qquad (10.45)$$

where J_1, J_2, J_3 are the three action variables associated with the invariants (μ, J, and L), and D_{ij} is a diffusion tensor. To solve this equation, we need to specify processes such as the Coulomb energy degradation and pitch angle diffusion, energy diffusion, and radial diffusion.

A simplification to this approach is to assume that the losses are all due to radial diffusion (see Falthammar, 1966), obtaining

$$\frac{\partial f}{\partial t} = L^2 \frac{\partial}{\partial L}\left(\frac{D_{LL}}{L^2}\frac{\partial f}{\partial L}\right) + S - L. \quad (10.46)$$

This equation is useful for treating radial diffusion in the quiet-time radiation belts, resulting from many weak electric field disturbances over a prolonged time. Conservation of the first adiabatic invariant in radial diffusion leads to energization of the particles as they move inward.

Pitch angle diffusion can be described by

$$\frac{\partial f}{\partial t} = \frac{1}{\sin\alpha}\frac{\partial}{\partial \alpha}\left(\sin\alpha D_{\alpha\alpha}\frac{\partial f}{\partial \alpha}\right) + S - L, \quad (10.47)$$

where $D_{\alpha\alpha}$ is the pitch angle diffusion coefficient. Coulomb scattering and resonant wave–particle interactions can cause pitch angle scattering.

The resonance condition for wave–particle interactions is

$$\omega - k_{\|}v_{\|} = n\Omega_e/\Gamma, \quad (10.48)$$

where

$$\Gamma = (1 - v^2/c^2)^{-1/2}. \quad (10.49)$$

The above equations only scratch the surface of the radiation belt formation and evolution. In order to treat the problem properly, we require a very comprehensive code, and several such codes now exist. Even with these codes, some questions still persist. The production of a new radiation belt by the March 1991 sudden commencement took the radiation belt community by surprise (see Walt, 1996). Also, the discovery by the Van Allen probes of energization in the middle of the radiation belts and diffusion in and out from the center was a surprise. The physics of the radiation belts continues to be both an important and difficult problem to master.

10.6 SUMMARY

The terrestrial magnetosphere is of critical importance to the inhabitants of this planet. It is both a shield against the solar wind and galactic energetic particles, and a giant electrical connector to the energy in the solar wind that powers the magnetotail, the ring current and the auroras, and that supplies the radiation belts with very energetic particles. While these particles are "trapped" on the magnetic field lines, they can also be scattered by waves associated with the solar wind, and the particle distributions can be unstable and lead to scattering themselves. Waves can also come from the atmosphere below and scatter the particles in the radiation belts. The various particle populations, including the neutral particles, interact in complex ways. The atmosphere is a source of plasma and not just the solar wind. The net result is a richness of plasma physical processes that are a challenge for even the most adroit plasma physicist.

Additional reading

CAROVILLANO, R. L. and J. M. FORBES (Eds.) (1983). *Solar Terrestrial Physics*. Dordrecht: Reidel. This book contains early ideas that led to many of the current concepts of magnetospheric behavior.

SCHULZ, M. and L. J. LANZEROTTI (1974). *Particle Diffusion in the Radiation Belts*. Berlin: Springer-Verlag. This book includes a good treatment of particle motion and diffusion in the radiation belts.

TAKAHASHI, K., P. J. CHI, R. E. DENTON, and R. L. LYSCK (Eds.) (2006). *Magnetospheric ULF Waves: Synthesis and New Directions*. Geophysical Monograph Series, vol. 169, Washington, D.C.: American Geophysical Union. This book contains a modern compendium of ULF wave processes and generation.

Problems

10.1 In regions of low plasma β in a dipole field, which represents much of the dayside magnetosphere, the cold-plasma approximation is appropriate.

(a) Use your knowledge of the properties of a dipole magnetic field ($B_{eq} \propto L^{-3}$; length of field line proportional to L; volume of flux tube proportional to L^4; equation of field line $r = LR_E\cos^2\lambda$) to explain why the fundamental excitations of field lines at large L occur at lower frequencies than do the fundamental excitations of field lines at small L. Assume that the density is uniform (1 electron cm^{-3}) throughout the magnetosphere and that the fundamental frequency at $6.6R_E$ is 14 mHz. Make a rough plot to show how the fundamental frequency varies with L.

(b) Actually, the magnetospheric plasma density often varies inversely with the flux-tube volume over large parts of the outer magnetosphere. Make a rough plot of the fundamental frequency of field-line excitations normalized to 14 mHz at $6.6R_E$ in a dipole field, assuming this type of variation for the density.

(c) Although the density variation used in part (b) is a good approximation, the magnetospheric density actually drops by a factor of 100 or more across the plasmapause. Allow for a plasmapause at $L = 5$, and assume that the density jumps by 100 inside the plasmapause. Again provide a rough plot of the fundamental frequency versus L.

(d) Where on the surface of the Earth would you expect to find pulsations of 50 mHz for the assumed conditions of part (c)?

10.2 Suppose that a standing Alfvén wave is established on a field line at $L = 5$, where the magnetosphere is approximately cylindrically symmetric. The ambient particle population is taken to include both energetic and cold plasma. Near the equator, the density is ρ in kilograms per cubic meter. Locally, near the equator, the uniform-field approximation is valid; the ambient field is B_0/L^3 and is oriented along the z-direction. The standing wave is a superposition of waves with **k** parallel and antiparallel to $\hat{\mathbf{z}}$.

(a) Assume that the magnetic perturbation **b** is radial. Determine the wave electric field and the fluid velocity perturbations as functions of **b**, ρ, and L. Pay attention to the vector character of the perturbations to determine their directions. Identify the direction of plasma displacement.

(b) The wave oscillations displace the plasma. The rate of displacement is slow enough that the plasma responds adiabatically. Show why this is true, using nominal dipole field values.

(c) Explain why you must consider the variations of particle flux with both L and W (particle energy) if you wish to determine how the particle flux measured at a spacecraft is modulated by a wave.

(d) Assume that only cold electrons and ions are present ($W \sim 0$). Show that the magnitude of the density variations takes the form

$$\delta n = \frac{b(\delta n/\delta L)}{R_E \omega \sqrt{\mu_0 \rho}}.$$

10.3 Consider the bounce motion of a charged particle on a dipole magnetic field line that extends to a maximum distance LR_E from the dipole. Consider specifically a particle that mirrors close to the equatorial plane.

(a) Applying the equation for the strength of the dipole magnetic field, show that, for points near the equatorial plane, the magnetic field magnitude is given by an expression of the form

$$B(L, s) \approx \frac{B_0}{L^3}\left[1 + \left(\frac{\xi}{2}\frac{s}{LR_E}\right)^2\right],$$

where s is the distance from the equatorial plane and ξ is a numerical constant. Find the value of ξ.

(b) By substituting this result in Eq. (10.9), show that, for $|s| \ll LR_E$, the conservation of energy expression for motion along the field line is like that of a harmonic oscillator. Find an expression for the frequency of the oscillation in terms of particle energy W, shell parameter L, and atomic weight A. Compare with Figure 10.13.

10.4 Consider the integral $\int ds/B$, where the integral is taken along a magnetic field line from the southern ionosphere to the northern ionsphere. Because $1/B$ is the area corresponding to one unit of magnetic flux, the integral $\int ds/B$ represents the volume of a magnetospheric magnetic flux tube that contains one unit of magnetic flux.

(a) Show that, for a dipole field and $L \gg 1$,

$$\int \frac{ds}{B} \approx \frac{32 R_E L^4}{35 B_0}.$$

For $L \gg 1$, only a small fraction of the flux-tube volume lies below the ionosphere. Therefore, one can extend the integration all the way to the point dipole, rather than stopping at the ionosphere. [Hint: Remember that $ds^2 = r^2 d\theta^2 + dr^2$.]

(b) Consider an isotropic distribution of particles in the flux tube, each particle having velocity v. The number density of these particles is n. Show that the number of particles hitting a unit area at the end of the tube is $nv/4$. [Hint: Consider a cylindrical tube intersecting a unit circle, such that the angle between the axis of the cylinder and the normal to the unit circle is θ. Over all azimuthal angles, integrate the number of particles hitting the circle as θ goes from 0° to 90°.]

(c) Show that the loss rate $1/\tau$ from a dipole flux tube, specifically the number of particles lost from the flux tube per unit time divided by the number of particles in the tube, is given by

$$\frac{1}{\tau} = \frac{35 v}{128 L^4 R_E}.$$

This is the loss rate for the limit of strong pitch angle scattering (i.e., pitch angle scattering that is so rapid that the loss cone remains full). Plasma-sheet electrons frequently run close to this limit.

(d) Evaluate the loss rate numerically for a 1 keV plasma-sheet electron at $L = 10$.

10.5 Quasi-neutrality is an important property of space plasmas, where the difference in density between the ions and electrons responsible for any space charge is much less than the number density. As an example to show this, consider the electric field associated with a corotating plasma. For simplicity, assume the magnetic field is represented by a dipole and the dipole is oriented along the spin axis.

(a) Sketch the dipole field lines and show that, for the Earth, the corotation electric field requires a negative space charge near the equatorial plane and a positive space charge above the magnetic pole.

(b) Estimate the electric field in the equatorial plane and show that, for the Earth, the electric field is ~$14(R_E/R)^2$ mV m^{-1}, assuming an equatorial magnetic field of 30 000 nT at the surface of the Earth. (Here, R_E is the radius of the Earth, 6371.2 km, and R is the radial distance.)

(c) Assuming that the scale size for variation of the electric field is of order the radial distance, show that the corotation electric field requires a density of ~1×10^{-7} electrons cm^{-3}. Compare this to typical magnetospheric plasma densities.

(d) Determine the electron density exactly for an axisymmetric dipole, and compare this to the estimate given in part (c).

10.6 Use the Magnetic Mirror portion of the Particle Tracing option of the Space Physics Exercises* to study the magnetic mirroring of charged particles. With the particle in the center of the magnetic bottle ($x = y = z = 0$) and $v_x = 0$, trace the motion of protons and alpha particles with equal v_y and v_z velocities ($v_y = v_z = 1, 2, 5$, etc.). Does the mirror distance depend on the absolute value of velocity or on the mass of the particle? Change the ratio of the mirror field from 100 to 50. What happens to the trajectories of the mirroring particles used above? Repeat with different ratios of v_y and v_z, both greater than and less than unity. Explain your results.

10.7 Use the Dipole Magnetic Field portion of the Particle Tracing option of the Space Physics Exercises* to follow particle motion in a realistic model of a planetary magnetosphere. In this model there is mirror trapping along the magnetic field and gradient drift across the magnetic field. Curvature drift arises only when there is a velocity component of the particle parallel to the magnetic field.

(a) Launch protons at right angles to the magnetic field with $v_y = 30$ km s^{-1}, $v_x = v_z = 0$ at three distances $x = 25$, 30, and 35 km, $y = z = 0$. Measure their drift velocity, i.e., the motion of the center of gyration. The magnetic field falls off as the inverse third power of the distance. How does this affect the drift velocity? Use the formula for drift velocity to show what dependence is expected theoretically in a dipole field.

(b) Launch protons at right angles to the magnetic field with $v_y = 30$, 40, and 50 km s^{-1}, and $v_x = v_z = 0$ at $x = 30$ km and $y = z = 0$. Measure their drift velocity. How does the drift velocity depend on the perpendicular energy of the particle, i.e., v_y^2?

(c) Launch protons with $v_y = 30$ km s^{-1}, $v_x = 0$, $v_z = 15$, 30, 45, and 60 km s^{-1}, and fixed starting location. Measure the drift velocity around the dipole and then the mirror latitude. The mirror latitude can be measured at the first reflection point in the x–z plane. How does the drift velocity depend on the total energy, i.e., $v_y^2 + v_z^2$? How does the mirror latitude depend on the ratio of $v_x/\left(v_y^2 + v_z^2\right)^{1/2}$?

* Space Physics Exercises are at http://spacephysics.ucla.edu.

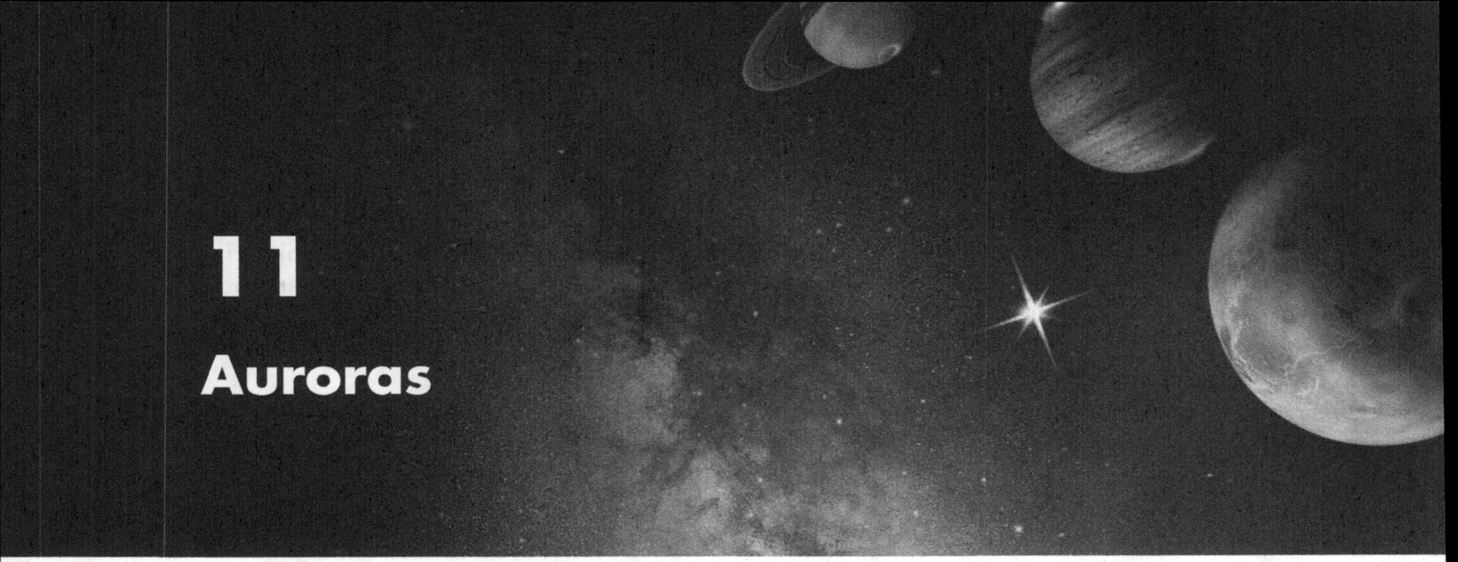

11 Auroras

11.1 INTRODUCTION

As discussed in Chapter 1, auroras may have been mankind's first introduction to solar-terrestrial physics and the plasma universe, but early inhabitants of this planet did not have an understanding of the physical processes that could produce them. Dancing lights in the sky cannot just be ignored. Thus auroras became omens. Religious interpretations and myths had to take the place of physical explanations. This was particularly necessary in the polar regions where this phenomenon was quite common. A good review of these beliefs, especially those in the Nordic countries, can be found in the book by Brekke and Egeland (1983), *The Northern Light: From Mythology to Space Research*.

The founding father of auroral research is widely taken to be Kristian Birkeland, who was the driving force behind the first International Polar Year in 1882–3. Although ships played a major role in exploration in those days, our modern ships are not just balloons and airplanes that can rise above the clouds, but rather artificial satellites that can look down upon the auroras from above and study the electric currents and energetic particle fluxes that create the auroras.

Auroras are still poorly understood despite the work of many auroral physicists armed with sophisticated equipment and the most powerful computers. The auroras are very dynamic, and they occur in a region that is both complex and difficult to monitor as it is above the altitude accessible from balloons and below the altitude commonly accessible with spacecraft. It is complex physically because it is in a partially ionized region of the atmosphere where the plasma is strongly coupled to the neutral atmosphere. It is complex chemically as well, and, because the region has a low collision rate, the excited state of an atom or molecule can last a long time before it loses its energy in a collision with another particle. Further, this region is the site of a tug of war between the atmospheric circulation dominated by the Earth's rotation and weather systems and the plasma flows driven by the solar-wind interaction with the magnetosphere. In this chapter, we first explain the proximate cause and the morphology of the auroral emissions, and then we examine the root cause of the auroras and the coupling to the solar wind through the ionospheric and auroral current systems.

11.2 AURORAL EMISSIONS

The auroras and the auroral ionosphere are the last stop in the energy flow from the solar wind, through the magnetosphere and into the atmosphere. While the visible auroras are the simplest to detect, modern instrumentation can sense the auroras in many ways. Radio signals can reflect from the enhanced ionization associated with visible auroras. They can also be absorbed in the auroral ionization, allowing this absorption to be monitored with simple instrumentation. Although the x-rays created in the auroras do not reach the ground, they can be monitored by balloons that fly in the lower densities at high altitudes. Finally, and most importantly, the auroras are usually accompanied by increases in the intensity of electric currents flowing in the

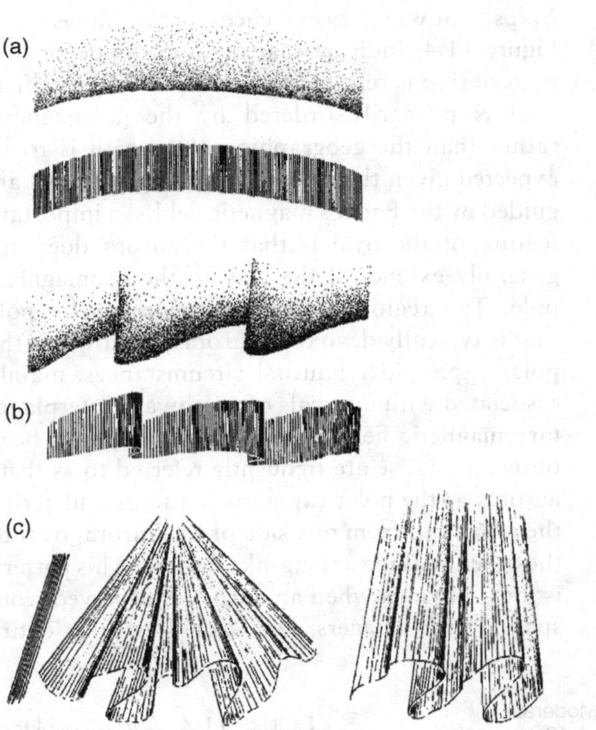

FIGURE 11.1. Examples of discrete auroras. (a) Homogeneous band and rayed band. (b) Folded homogeneous and rayed bands. (c) Complex forms of rayed bands.

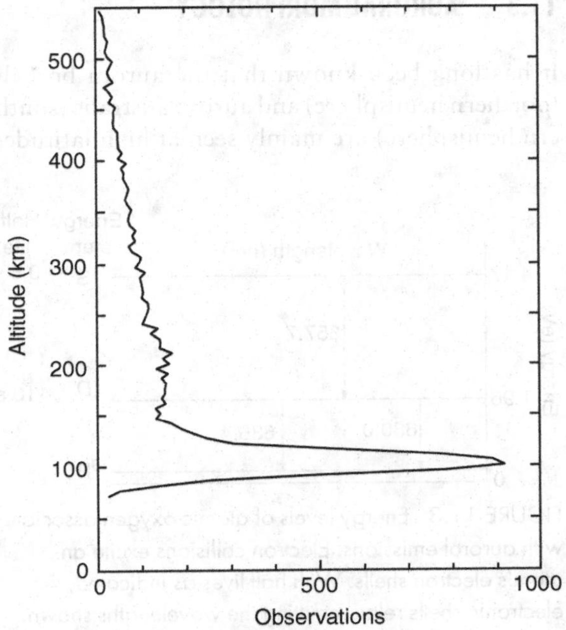

FIGURE 11.2. Height distribution of auroras based on over 12 000 observations by Størmer and colleagues. (After Egeland and Burke, 2013.)

ionosphere. The magnetic fields associated with these currents can be easily monitored with chains of magnetometers. The strength of these electric currents in turn can be used to quantify the strength of the geomagnetic activity during an auroral "storm" or "substorm."

A visible aurora is often quite striking in appearance. It can be dynamic, and it can be quite structured. Its complexity almost defies explanation, even though many explanations have been offered. Diffuse auroras can cover the sky from horizon to horizon, but discrete auroras prove to be the more intriguing form of auroral emissions. Figure 11.1 illustrates two forms of what are called discrete auroras: the homogeneous arc and the rayed band. These emissions arise at altitudes of about 100 km and above, as illustrated in Figure 11.2, and are oriented along magnetic field lines.

In an average sense, auroral arcs are east–west, but they can bend in any direction. Discrete arcs may be very thin, less than 100 m, but at times diffuse auroras can exist over 100 km in the north–south direction. They can extend 1000 km or more in the east–west direction and hundreds of kilometers in height along the magnetic field. They are rich in color. Oxygen atoms can emit in the yellow–green (557.7 nm) and the red (630.0 and 636.4 nm); N_2 molecules can emit in the dark red (650–680 nm); and N_2^+ ions can emit in the blue–violet (391.4 nm).

The physics of the auroral emission is basically quite simple, as illustrated by the energy diagram of the excited state of the oxygen atom in Figure 11.3. If a precipitating electron were to excite the atom to the 4.17 eV state (^1S), it would most probably decay to the 1.96 eV level (^1D) in a time of about 0.8 s, giving off a 557.7 nm photon. The next transition, to the ^3P state, is a "forbidden" transition which requires a long time (typically 110 s). High in the atmosphere, there may be enough time between collisions for this emission to occur, but at low altitudes the atom would be de-excited by a collision before the photon could be emitted.

11.3 AURORAL MORPHOLOGY

It has long been known that the aurora borealis (northern hemisphere) and aurora australis (southern hemisphere) are mainly seen at high latitudes.

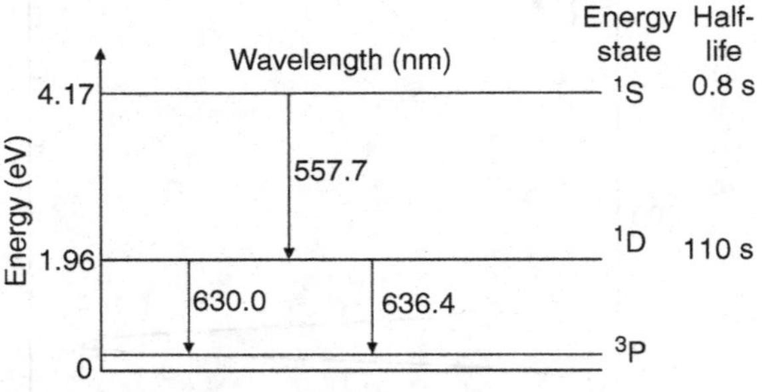

FIGURE 11.3. Energy levels of atomic oxygen associated with auroral emissions. Electron collisions excite an atom's electron shells. With half-lives as indicated, the electronic shells relax, emitting the wavelengths shown.

Maps showing occurrence probabilities (see Figure 11.4) indicate that the auroras occur in a pattern that is referred to as the auroral oval. The oval is primarily ordered by the geomagnetic rather than the geographic pole, which is to be expected given that the precipitating particles are guided by the Earth's magnetic field. An important feature of the oval is that the aurora does not generally extend all the way to the geomagnetic pole. The region around the geomagnetic pole that is typically devoid of auroras is known as the polar cap. Under unusual circumstances, mainly associated with intervals of northward interplanetary magnetic field, polar cap auroras have been observed. These are frequently referred to as theta auroras as the polar cap auroras can extend across the polar cap from one side of the auroral oval to the other, making a theta-like pattern. This pattern is most obvious when an aurora is observed from space, with imagers that can view the entire

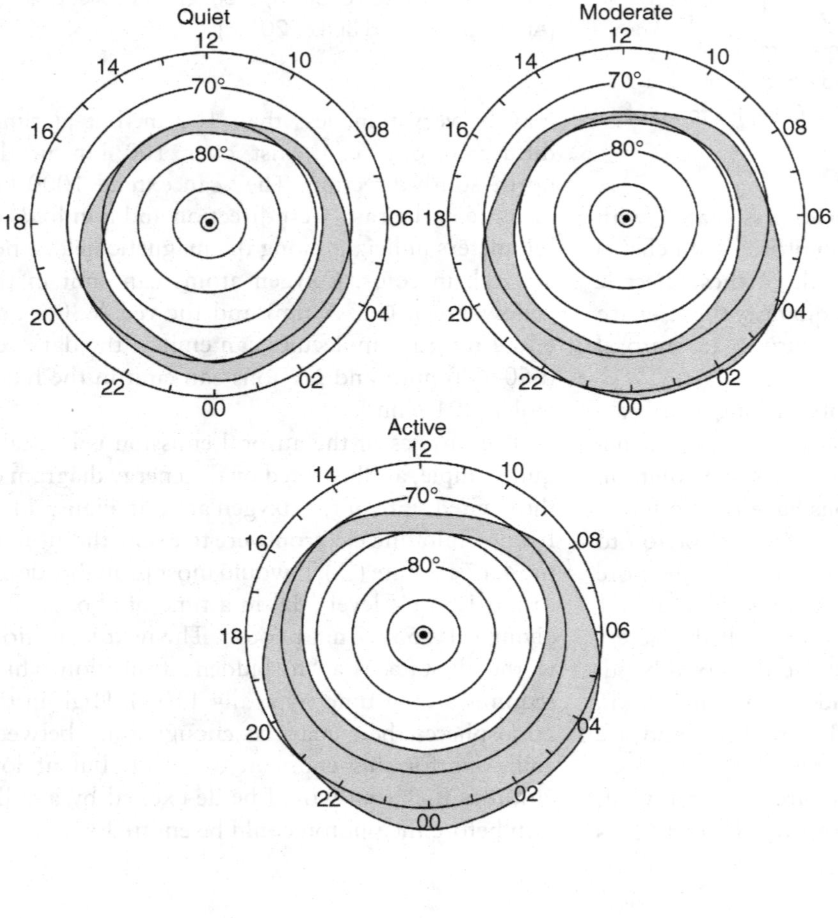

FIGURE 11.4. Auroral oval for different levels of geomagnetic activity. (After Feldstein and Starkov, 1967.)

auroral oval. However, the disadvantage with space-based images is that much of the fine structure of the aurora is lost. Ground-based imagers are usually much better in capturing fine structure, but have the disadvantage of not having continuous coverage, as the aurora can be obscured by clouds or by moonlight.

Nevertheless, the ground-breaking work of S.-I. Akasofu, in the early 1960s, relied entirely on ground-based observations of the auroras. Based on his painstaking observational work, he was able to show how auroras evolve during what became known as the auroral substorm. The results of his work are shown in Figure 11.5.

The **auroral substorm** consists of a relatively quiet interval, at least in terms of the aurora, with a few latitudinally extended arcs. These arcs are not very intense. Suddenly, one of the arcs, usually the most equatorward arc, brightens. This is the substorm onset. At that time the arcs begin to move rapidly toward the pole, and the region of intense arcs also starts to move westward. This is the westward traveling surge. The interval during which the auroral activity increases is known as the expansion phase. After some time, typically a few tens of minutes, the auroral activity begins to decrease, and the substorm enters the recovery phase.

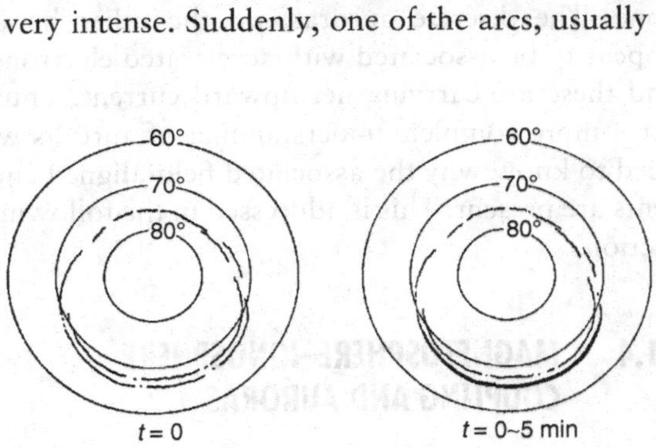

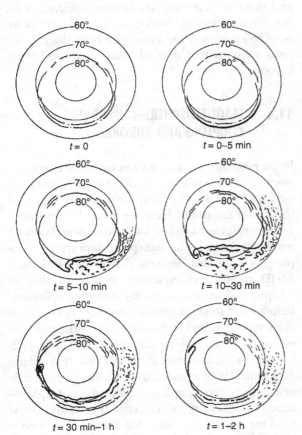

FIGURE 11.5. Auroral substorm temporal evolution. (After Akasofu, 1968.)

There are significant magnetic field perturbations, as measured by ground-based magnetometers. The largest perturbations tend to be in in the north–south direction, with a northward perturbation being associated with an eastward current in the ionosphere, and southward perturbations with a westward current. These currents are referred to as electrojets.

The electrojets flow as **Hall currents** in the ionosphere. Hall currents were introduced in Chapter 2, and we discuss them further in Section 11.4. As noted in Chapter 2, and later in this chapter, Hall currents flow in the $-\mathbf{E} \times \mathbf{B}$ direction, where $\mathbf{E}$ is an electric field present in the ionosphere and $\mathbf{B}$ is the ambient magnetic field. Hall currents are present because the electrons are effectively collisionless in the ionosphere, while the ions are strongly affected by collisions. In addition to the Hall current there is a Pedersen current, where the ions move along the electric field.

We discuss this in more detail later in the chapter, but, when the magnetic field is not time varying, the electric field can be described by equipotentials, and the plasma flow in the ionosphere follows these equipotentials. In the absence of conductivity gradients, the associated Hall currents are divergence free. In particular, without conductivity gradients, Hall currents do not close with field-aligned currents. On the other hand, **Pedersen currents** usually have field-aligned currents associated with them. This point should become clearer in the following section.

One important aspect of the closure of Pedersen currents by field-aligned currents is that the resultant current system tends to be solenoidal in nature. In that case, most of the magnetic field perturbation is interior to the current system, and the exterior perturbation is relatively small. This leads to **Fukushima's theorem** (Fukushima, 1969, 1976), which states that the net perturbation associated

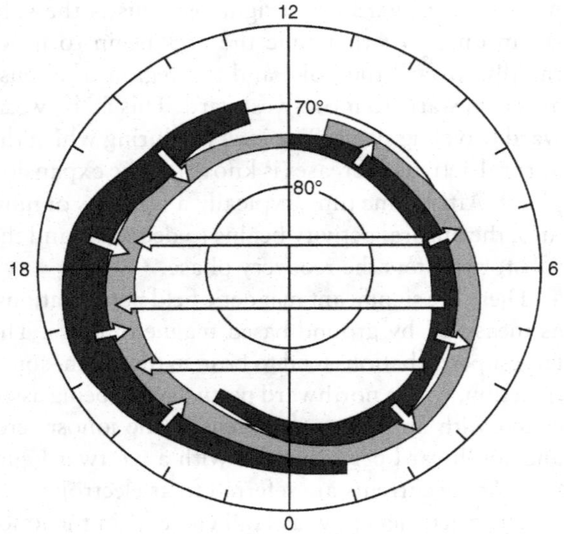

FIGURE 11.6. The large-scale field-aligned current system. The higher-latitude field-aligned current is known as region 1, with the lower-latitude current known as region 2. Near noon and midnight the current system becomes more complicated. (After Iijima and Potemra, 1978.)

with the field-aligned currents flowing into and out of the ionosphere and the Pedersen closure currents cannot be measured on the ground. On the other hand, space-based observations of magnetic field perturbations mainly measure field-aligned currents and the Pedersen currents, rather than the Hall currents.

In the 1970s it was found that there is a large-scale field-aligned current system flowing into the Earth's ionosphere that is almost always present. This is shown in Figure 11.6. This figure is adapted from Iijima and Potemra (1978), where we have added the Pedersen currents that provide closure for the field-aligned currents. Because the Pedersen current is parallel to the perpendicular electric field, the resultant j × B force is in the direction of the plasma flow. Thus the Pedersen current provides the force that overcomes the drag from collisions with the neutral atmosphere, as we shall demonstrate in more detail in the following section. Moreover, the resultant plasma flow pattern corresponds to the two-cell convection pattern driven by reconnection with the IMF. In two-cell convection the plasma convects antisunward over the polar cap. The return flow lies in the region bounded by the two field-aligned currents system.

The final step here is to make the connection between the field-aligned currents and the auroras. A comparison of Figure 11.4 and 11.6 shows that the auroral oval and the field-aligned currents are closely related, at least in terms of latitude. Furthermore, since the auroras are mainly associated with precipitating electrons, we might expect the auroras to be associated with upward current. This is not always the case: diffuse auroras in particular appear to correspond to precipitation associated with scattering of particles into the loss cone. The discrete auroras, on the other hand, appear to be associated with accelerated electrons, and these are carrying net upward current. Thus, for a more complete understanding of auroras we need to know why the associated field-aligned currents are present. This is addressed in the following section.

11.4 MAGNETOSPHERE–IONOSPHERE COUPLING AND AURORAS

In the preceding sections we discussed the characteristics of auroras in terms of the structure, morphology, and collisional excitation processes that lead to auroral emissions. Here we place the auroras in a magnetospheric context and discuss how they are a manifestation of magnetosphere–ionosphere coupling. To do this we first derive the single-fluid MHD momentum equation, similar to that in Chapter 3, but now explicitly including collisions. Before using this equation to show how field-aligned currents are related to flows and pressure gradients within the plasma, we repeat the results of Chapter 2, leading to an anisotropic Ohm's law that relates perpendicular currents to perpendicular electric fields. It is the coupling of the magnetosphere–ionosphere system via currents that leads to the need for auroral processes. But a word of caution is in order. It is tempting to think of the currents as driving the auroras, but the currents themselves are a consequence of dynamical processes within the magnetosphere–ionosphere system.

In order to discuss magnetosphere–ionosphere coupling and the auroras, we start from the **electron- and ion-momentum equations**, including collisions,

$$n_i m_i\left(\frac{\partial}{\partial t}+\mathbf{u}_i\cdot\nabla\right)\mathbf{u}_i = n_i e(\mathbf{E}+\mathbf{u}_i\times\mathbf{B}) - \nabla P_i - n_i m_i \nu_{in}(\mathbf{u}_i-\mathbf{u}_n) - n_i m_i \nu_{ie}(\mathbf{u}_i-\mathbf{u}_e), \quad (11.1a)$$

$$n_e m_e\left(\frac{\partial}{\partial t}+\mathbf{u}_e\cdot\nabla\right)\mathbf{u}_e = -n_e e(\mathbf{E}+\mathbf{u}_e\times\mathbf{B}) - \nabla P_e - n_e m_e \nu_{en}(\mathbf{u}_e-\mathbf{u}_n) - n_e m_e \nu_{ei}(\mathbf{u}_e-\mathbf{u}_i), \quad (11.1b)$$

for ions and electrons, respectively. In these two equations, the subscript "i" indicates ions, "e" indicates electrons, and "n" indicates neutrals. The collision frequencies are given by ν_{in}, etc., with ν_{in}, for example, being the ion–neutral collision frequency. The rest of the terms have their usual meaning, and we have assumed isotropic pressure (P) for simplicity. In addition, we assume only one ion and neutral species. Finally, for the Coulomb collision terms, $n_i m_i \nu_{ie} = n_e m_e \nu_{ei}$, based on conservation of momentum and quasi-neutrality ($n_i \approx n_e$).

We can combine Eqs. (11.1a) and (11.1b) to make a single-fluid momentum equation, as was done in Chapter 3. As before, we shall assume quasi-neutrality, $n_i = n_e = n$, and define the advective derivatives and pressure with respect to the center of mass velocity, $\mathbf{u} = (n_i m_i \mathbf{u}_i + n_e m_e \mathbf{u}_e)/\rho$, where $\rho = n_i m_i + n_e m_e$ is the mass density. Because $m_i \gg m_e$, $\rho \approx n m_i$ and $\mathbf{u}_i \approx \mathbf{u}$. The total plasma momentum equation becomes

$$\rho\frac{D\mathbf{u}}{Dt} = \mathbf{j}\times\mathbf{B} - \nabla P - n(m_i\nu_{in} + m_e\nu_{en})(\mathbf{u}_i-\mathbf{u}_n) + m_e\nu_{en}\frac{\mathbf{j}}{e}, \quad (11.2)$$

where $D/Dt = \partial/\partial t + \mathbf{u}\cdot\nabla$ is the total time derivative using the center of mass velocity.

We shall use Eq. (11.2) when we discuss the role of field-aligned currents in magnetosphere–ionosphere coupling. One important aspect of Eq. (11.2) is that the electric field has vanished, even though it is present in Eqs. (11.1a) and (11.1b). Therefore, it may not be obvious how Eq. (11.2) can be used to discuss ionospheric currents, as these are usually related to the electric field through an Ohm's law (see Chapter 2). For completeness, we show how the Ohm's law is derived from Eqs. (11.1a) and (11.1b).

First, we shall assume that the left-hand side of Eqs. (11.1a) and (11.1b) (i.e., the inertia terms) can be ignored, as well as the pressure terms. Then, after some re-arrangement, we obtain

$$e(\mathbf{E}+\mathbf{u}_i\times\mathbf{B}) = m_i\nu_{in}(\mathbf{u}_i-\mathbf{u}_n) + m_i\nu_{ie}\mathbf{j}/ne, \quad (11.3a)$$

$$e(\mathbf{E}+\mathbf{u}_e\times\mathbf{B}) = -m_e\nu_{en}(\mathbf{u}_e-\mathbf{u}_n) + m_e\nu_{ei}\mathbf{j}/ne. \quad (11.3b)$$

As noted earlier, the last term in Eq. (11.3a) equals the last term in Eq. (11.3b) because of conservation of momentum and quasi-neutrality.

Defining the flow velocities and electric field in the neutral species frame of reference, i.e., $u'_i = \mathbf{u}_i - \mathbf{u}_n$, $\mathbf{u}'_e = \mathbf{u}_e - \mathbf{u}_n$, and $\mathbf{E}' = \mathbf{E} + \mathbf{u}_n\times\mathbf{B}$, we find

$$\mathbf{u}'_i = \frac{e}{m_i\nu_{in}}\left(\mathbf{E}' - \frac{m_e\nu_{ei}}{ne^2}\mathbf{j} + \mathbf{u}'_i\times\mathbf{B}\right), \quad (11.4a)$$

$$\mathbf{u}'_e = -\frac{e}{m_e\nu_{en}}\left(\mathbf{E}' - \frac{m_e\nu_{ei}}{ne^2}\mathbf{j} + \mathbf{u}'_e\times\mathbf{B}\right). \quad (11.4b)$$

These two equations can be solved for the ion and electron velocities, respectively, with the current density obtained from $\mathbf{j} = ne(\mathbf{u}'_i - \mathbf{u}'_e)$. The complication is the Coulomb collision frequency-dependent term in both equations. We shall retain this term when discussing the parallel conductivity, but for simplicity we drop this term when considering perpendicular currents. Formally, we could keep the Coulomb collision term, and show a posteriori that this corresponds to a correction of the order ν_{ei}/Ω_e for the perpendicular current, where $\Omega_e = eB/m_e$ is the electron gyro-frequency. The term ν_{ei}/Ω_e is small in the ionosphere and the magnetosphere.

From the parallel components of Eqs. (11.4a) and (11.4b), given that in general $m_e\nu_{en} \ll m_i\nu_{in}$, we find

$$j_{\parallel} = \frac{ne^2}{m_e(\nu_{en} + \nu_{ei})} E_{\parallel} = \sigma_{\parallel} E_{\parallel}, \quad (11.5)$$

again noting that we have kept the Coloumb collision frequency for this component only.

For the perpendicular current density, on neglecting the Coulomb collision terms, we find

$$\mathbf{j}_{\perp} = \sigma_P \mathbf{E}_{\perp}' - \sigma_H \frac{\mathbf{E}_{\perp}' \times \mathbf{B}}{B}, \quad (11.6)$$

where σ_P and σ_H are the Pedersen and Hall conductivities, respectively, given by

$$\sigma_P = \frac{ne}{B} \left(\frac{\nu_{en}/\Omega_e}{1 + \nu_{en}^2/\Omega_e^2} + \frac{\nu_{in}/\Omega_i}{1 + \nu_{in}^2/\Omega_i^2} \right) \quad (11.7)$$

and

$$\sigma_H = \frac{ne}{B} \left(\frac{1}{1 + \nu_{en}^2/\Omega_e^2} - \frac{1}{1 + \nu_{in}^2/\Omega_i^2} \right), \quad (11.8)$$

where $\Omega_i = eB/m_i$ is the ion gyro-frequency. For completeness we have included the electron terms, but it should be noted that, in the Earth's E- and F-region ionosphere, $\nu_{en}/\Omega_e \ll 1$. On consideration of Eq. (11.3b), this condition also implies that the electron frozen-in condition,

$$\mathbf{E} + \mathbf{u}_e \times \mathbf{B} = \mathbf{E} + \mathbf{u} \times \mathbf{B} - \frac{\mathbf{j} \times \mathbf{B}}{ne} = 0, \quad (11.9)$$

is a good approximation for the ionosphere. We shall make use of this later.

The frozen-in electron approximation also makes it clear that the electrons move with the $\mathbf{E} \times \mathbf{B}$ drift velocity. The Pedersen conductivity is therefore mainly associated with ion motion along the direction of the perpendicular electric field, where the acceleration in the direction of the electric field is balanced by the ion–neutral collisional drag. The Hall current is opposite to the $\mathbf{E} \times \mathbf{B}$ drift because the ions are again slowed down by the ion–neutral collisions, and drift more slowly in the $\mathbf{E} \times \mathbf{B}$ direction. In the Earth's ionosphere, this scenario begins to break down in the D region, as electron–neutral collisions start becoming significant.

11.4.1 Magnetospheric sources of field-aligned currents

For the magnetosphere we can ignore collisions, and Eq. (11.2) reduces to the single-fluid MHD momentum equation as given in Chapter 3. In that chapter we used the MHD momentum equation to derive an expression for the field-aligned current, given as Eq. (3.204) and repeated here:

$$(\mathbf{B} \cdot \nabla) \left(\frac{\mathbf{j} \cdot \mathbf{B}}{B^2} \right) = \frac{\mathbf{B}}{B^2} \cdot \left[2 \left(\nabla P + \rho \frac{D\mathbf{u}}{Dt} \right) \times \frac{\nabla B}{B} + \nabla \times \left(\rho \frac{D\mathbf{u}}{Dt} \right) \right]. \quad (11.10)$$

The term on the left-hand side of this equation is the field-aligned gradient of the field-aligned current density per unit magnetic flux. As such, this takes into account the increase in current density associated with the change in the flux tube area, with $j_{\parallel}/B$ constant if the right-hand side of Eq. (11.10) is zero.

As noted in Chapter 3, the terms on the right-hand side of Eq. (11.10) correspond to thermal pressure gradients, flow braking, and **vorticity**. The vorticity term also corresponds to the Alfvén, or shear, mode. As we shall see, the shear mode is an essential feature of dynamical magnetosphere–ionosphere coupling.

In the inner magnetosphere, the plasma thermal pressure, P, is usually much larger than the dynamic pressure, leading to an approximation to Eq. (11.10) known as the slow-flow approximation. In that case, the thermal pressure gradient term dominates, and the field-aligned currents occur in regions where $\nabla P \times \nabla B \neq 0$. Since the ambient magnetic field is more dipole like in the inner magnetosphere, this usually requires azimuthal pressure gradients. Furthermore, as discussed in Section 11.3, region 2 currents are found at lower latitudes. It is therefore often stated that region 2 currents are driven by pressure gradients. Stating that pressure gradients "drive" field-aligned currents is convenient shorthand, but we have not specifically addressed how these currents are generated, in that we have not identified a dynamo. In particular, from the frozen-in condition and the MHD momentum equation,

$$\mathbf{j}\cdot\mathbf{E} = \mathbf{u}\cdot(\mathbf{j}\times\mathbf{B}) = \mathbf{u}\cdot(\rho D\mathbf{u}/Dt + \nabla P). \quad (11.11)$$

For a dynamo, $\mathbf{j}\cdot\mathbf{E} < 0$. Thus, either the plasma slows down, losing the energy associated with bulk flow, or work is done by the thermal pressure. In passing, it should be noted that Eq. (11.11) can be derived from the electron frozen-in condition (11.9), since $\mathbf{j}\cdot(\mathbf{j}\times\mathbf{B}) = 0$, and is therefore a more general result than would be supposed if Eq. (11.11) were based instead on the ion frozen-in condition.

In comparison to region 2 currents, region 1 currents flow at higher latitudes, often on field lines that connect to the outer magnetosphere, and even the magnetopause. At the magnetopause, in particular, dynamic pressure dominates, and region 1 currents tend to be associated with gradients in the flows. Near midnight, however, this picture is much more complicated, as the geomagnetic field can be severely stretched, which affects the mapping between the ionosphere and the magnetosphere. Furthermore, during substorms, regions of rapid plasma flows are frequently observed, and these can penetrate to quite low latitudes. Understanding the relationship between thermal pressure, dynamic pressure, and vorticity is one aspect of the Time History of Events and Macroscale Interactions during Substorms (THEMIS) mission, and is an active area of research.

11.4.2 Ionospheric currents

In Eq. (11.2) we gave the total plasma momentum equation, including collisions. As written, Eq. (11.2) requires knowledge of the neutral species flow velocity, $\mathbf{u}_n$. In discussing magnetosphere–ionosphere coupling, we often make the simplifying assumption that $\mathbf{u}_n = 0$, implying that Eq. (11.2) is written in a frame moving with the neutral fluid. But $\mathbf{u}_n$ can change, either because of atmospheric forcing from below, or because of the ion–neutral collisions, whereby the neutrals are accelerated. Indeed, a "three-fluid" formalism (Song, Vasyliunas, and Ma, 2005), which includes a simplified neutral momentum equation where the only force acting on the neutrals is the drag force from collisions with the ions, makes it clear that at long time scales the neutral species ultimately provides the inertia in the ionosphere–atmosphere system. At this stage, however, we ignore the effect on the neutrals.

At its simplest, in the ionosphere the dominant terms in Eq. (11.2) are the $\mathbf{j}\times\mathbf{B}$ term and the ion–neutral collision term, in which case

$$\mathbf{j}\times\mathbf{B} = \rho\nu_{in}(\mathbf{u}_i - \mathbf{u}_n). \quad (11.12)$$

We have ignored the change in plasma momentum and the thermal pressure terms, and further assumed the electron–neutral collisions are vanishingly small. These assumptions are similar to the ones we used for the frozen-in electron condition, except that we are assuming that the ion inertia and pressure can be ignored, in addition to the electron inertia and pressure. Because the Earth's ionosphere is a low-beta plasma, neglecting the pressure is reasonable, but neglecting the ion inertia should be treated with caution. For example, the presence of Alfvén waves, with frequencies $\omega \geq \nu_{in}$, would require the inclusion of ion inertia in the momentum equation.

If we take the curl of Eq. (11.12), then, assuming $\nabla\cdot\mathbf{j} = 0$,

$$(\mathbf{B}\cdot\nabla)\mathbf{j} - (\mathbf{j}\cdot\nabla)\mathbf{B} = \rho\nu_{in}(\boldsymbol{\omega}_i - \boldsymbol{\omega}_n) - (\mathbf{u}_i - \mathbf{u}_n)\times\nabla\rho\nu_{in}, \quad (11.13)$$

where $\omega = \nabla\times\mathbf{u}$ is the vorticity.

Taking the dot product of Eq. (11.13) with $\mathbf{B}$,

$$\mathbf{B}\cdot(\mathbf{B}\cdot\nabla)\mathbf{j} - B\mathbf{j}\cdot\nabla B = \rho\nu_{in}\mathbf{B}\cdot(\boldsymbol{\omega}_i - \boldsymbol{\omega}_n) - \mathbf{B}\cdot[(\mathbf{u}_i - \mathbf{u}_n)\times\nabla\rho\nu_{in}]. \quad (11.14)$$

The first term on the left-hand side of Eq. (11.14) gives the change in the field-aligned current, similar to Eq. (11.10) but neglecting the effect of changing flux tube area. Furthermore, the second term on the left-hand side is generally small in the Earth's ionosphere. Equation (11.14) shows that, in the ionosphere, field-aligned currents are associated with vorticity in the plasma flow and gradients in $\rho\nu_{in}$. If, instead, we use the Ohm's law as given in Eq. (11.6), we would state that the divergence of $\mathbf{j}_\perp$ is mainly associated with the divergence of the electric field and gradients in the conductivity. From the frozen-in theorem, divergence in the electric field corresponds to a flow shear, and a flow shear has vorticity. The conductivity depends on the collision frequency, and the last term on the right-hand side of Eq. (11.14) would correspond to a conductivity gradient. But Eq. (11.14) has the advantage of relating the field-aligned currents directly to the plasma flow, similarly to Eq. (11.10) for the magnetosphere.

11.4.3 The coupled system

Equations (11.10) and (11.14), together with the current continuity condition, $\nabla \cdot \mathbf{j} = 0$, allow us to consider the coupled magnetosphere–ionosphere system in terms of forces and flows. This can provide more insight into the coupling that results in auroras than does the more traditional circuit element approach. The latter assumes quasi-static current systems, and does not take into account the coupling between the plasma and the fields or dynamics. But the momentum equations do not form a complete set, in that Eq. (11.10) assumes a prescribed plasma pressure and flow, wheras Eq. (11.14) assumes an ionosphere conductivity profile. Determining the distribution of the plasma pressure and flows within the magnetosphere requires a more complete set of equations. MHD can be useful in that regard, as this automatically includes time-varying fields, and, more specifically, Alfvén waves. Thus, global MHD simulations can be used to indicate where Alfvénic auroras could occur, but the simulations do not include the processes responsible for the auroral particle acceleration. MHD simulations model the solar wind–magnetosphere coupling well, although they tend to underestimate the role of plasma pressure in the inner magnetosphere, resulting in weaker region 2 currents. Simulations that explicitly include particle drifts, known as convection models, yield larger region 2 currents, which shield the sunward convection from lower latitudes. But the particle drift simulations also suffer from the limitation of not fully modeling the region 1 currents, which tend to map to higher latitudes than that included in the simulation domain. Neverthless, much progress is being made in developing models that couple MHD simulations to inner magnetosphere convection models. These models also include modules that mimic the auroral acceleration process, allowing for enhanced ionospheric conductivies because of auroral precipitation.

11.5 AURORAL FIELD-ALIGNED CURRENTS

In the preceding section we discussed why field-aligned currents are expected to flow between the magnetosphere and the ionosphere. In this section we discuss the structure and variability of field-aligned currents, the nature of the current carriers, and the other consequences of field-aligned currents, in addition to current closure.

To set the stage for this discussion, Figure 11.7 presents particle and field data acquired by the Fast Auroral Snapshot (FAST) Small Explorer. This spacecraft was designed to provide high-resolution data acquired within the region where the electrons that create auroras are accelerated. The figure shows a representative pass through the auroral zone, from high latitudes at the left of the figure to low latitudes at the right. The top panel of the figure shows the $\log_{10}$ of the differential energy flux for the electrons as a function of energy and time, with the greyscale at the right showing the range. White corresponds to the most intense differential energy fluxes. Differential energy flux has the units of $\mathrm{eV\,cm^{-2}\,s^{-1}\,sr^{-1}\,eV^{-1}}$, where eV is energy expressed as electronvolts. The energies in the differential energy flux do not cancel, since the energy in the numerator corresponds to the energy of the particles being measured, while the energy in the denominator gives the width in energy over which the measurement is being made. The figure shows regions of very high energy electron fluxes, especially to the right of the figure, labeled "inverted V" electrons, and also intense lower-energy fluxes, labeled "low-energy, bi-directional electrons," mainly to the left of the figure. These correspond to different regions of field-aligned current.

The next panel down shows the differential energy flux as a function of pitch angle (pitch angle is defined in Chapter 3). For the northern hemisphere, where the magnetic field points toward the Earth, 0° pitch angle corresponds to precipitating particles, 180° to upgoing particles, and ±90° to particles that are locally mirroring.

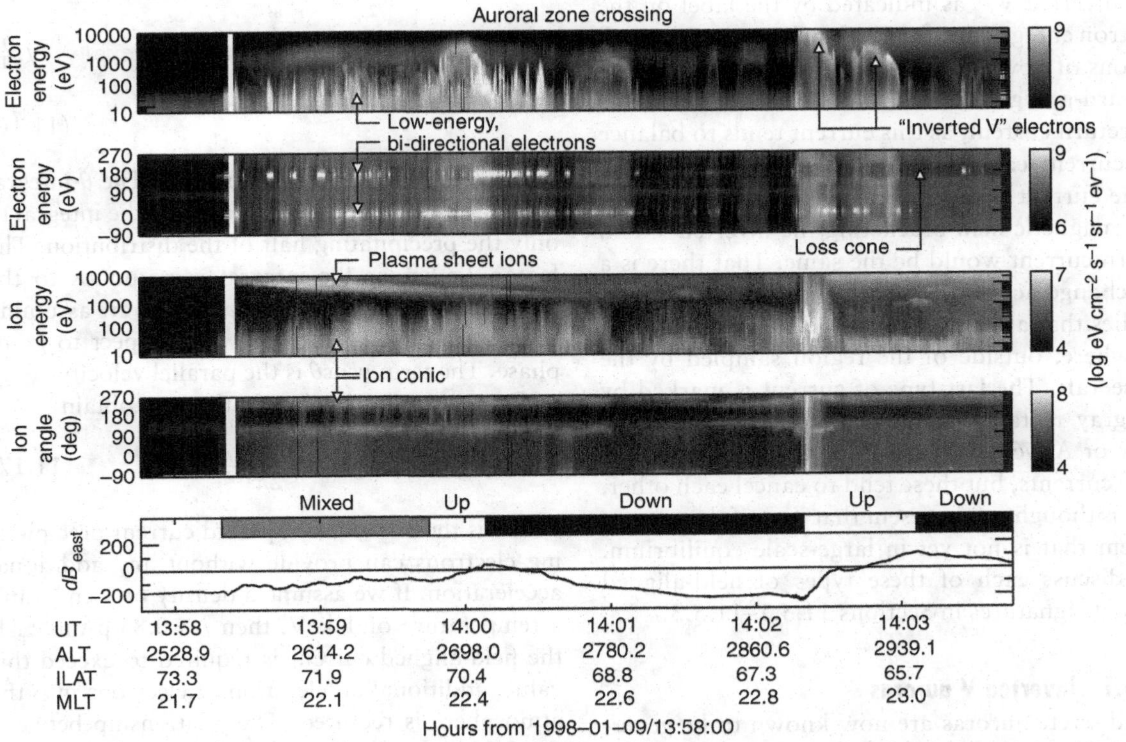

FIGURE 11.7. Observations of particles and fields taken by the Fast Auroral Snapshot (FAST) Small Explorer. The rectangles at the bottom of the figure show the different current regions, with white corresponding to upward current, black to downward current, and gray to a region of mixed upward and downward current.

At the right-hand side of the figure there is a narrow pitch angle from about 160° to 200° that has generally reduced fluxes. This is the loss cone.

The next two panels show the corresponding data for ions. In the energy spectrum there are clearly two populations of ions: one with energies above 1 keV and one with energies around 100 eV. The former consists of trapped and precipitating plasma sheet ions. The low-energy distribution is known as an ion conic. The reason for the name can be seen in the ion pitch angle plot, where there is a differential energy flux maximum just outside the loss cone. These ions therefore form a cone in velocity space with respect to the magnetic field direction. These ions are a source of plasma for the magnetosphere.

The bottom panel shows the eastward component of the magnetic field. This is a perturbation field in that the ambient magnetic field as specified by the International Geomagnetic Reference Field (IGRF) has been subtracted. For simplicity, the figure only includes the eastward component, while the spacecraft is moving from north to south. Thus the gradient along the spacecraft trajectory of this component of the magnetic field, which is known as the along-track gradient, can be used to determine an equivalent current density. In essence it is assumed that the field-aligned currents are carried in current sheets that extend along the magnetic field and perpendicular to the spacecraft trajectory.

The rectangles at the bottom of the figure correspond to the three different types of field-aligned currents typically observed in the Earth's auroral zone. The white rectangles mark regions of upward current carried by energetic precipitating electrons. In early observations of these electrons, mainly on rockets, the characteristic signature in an energy–time spectrogram had the appearance of an "inverted V," as indicated by the label on the electron energy spectra. The black rectangles mark regions of downward current, which is carried by upward-going electrons. It is often referred to as the return current, as this current tends to balance the current carried by the inverted V electrons. If the current balance were perfect, the perturbation magnetic field bracketing the inverted V and return current would be the same. That there is a net change across the white and black rectangles implies that at least some of the current is closing elsewhere, outside of the region sampled by the spacecraft. The last type of current is marked by the gray rectangle. This is known as boundary-layer or Alfvénic auroras. There are many small-scale currents, but these tend to cancel each other. This is thought to represent that part of the current system that is not yet in large-scale equilibrium. We discuss each of these types of field-aligned current signatures in Sections 11.5.1–11.5.3.

11.5.1 Inverted V auroras

The discrete auroras are now known to be generated by precipitating electrons that have been accelerated along the magnetic field by a parallel electric field. In general a magnetized plasma cannot maintain a parallel electric field because the ions and electrons respond in such a way as to short out the electric field. There are exceptions to this statement, however, the auroral acceleration process being one of them. In this case a parallel electric field is set up to enable the precipitating electrons to carry an upward current.

In order to demonstrate this, we consider a maxwellian plasma, with phase space density given by

$$f = \frac{n_0}{\pi^{3/2} v_T^3} \exp(-v^2/v_T^2), \tag{11.15}$$

where n_0 is the electron density, v is velocity, and v_T is the thermal velocity, given by $\frac{1}{2} m_e v_T^2 = KT$, with m_e being the electron mass and T the temperature. For simplicity, we have assumed an isotropic distribution.

Assuming this distribution is present at the foot of a field line, the current density associated with the part of the distribution that is precipitating is given by

$$j_0 = \frac{n_0 e}{\pi^{3/2} v_T^3} 2\pi \int_0^{\pi/2} \sin\theta d\theta \int_0^{\infty} v^2 dv [v\cos\theta \exp(-v^2/v_T^2)], \tag{11.16}$$

where θ is the cone (or pitch) angle. The θ-integral limit is given by $\pi/2$, so as to restrict the integral to only the precipitating half of the distribution. The factor 2π before the integral corresponds to the integration over gyro-phase, φ, but we are assuming the distribution is isotropic with respect to gyro-phase. The term $v\cos\theta$ is the parallel velocity.

On performing the integration, we obtain

$$j_0 = \frac{n_0 e v_T}{2\pi^{1/2}}, \tag{11.17}$$

which is the amount of upward current precipitating electrons can provide without any additional acceleration. If we assume a density of 1 cm^{-3} and a temperature of 1 keV, then $j_0 \approx 0.85\ \mu\text{A m}^{-2}$. If the field-aligned current is required to exceed this value, additional acceleration of electrons into the atmosphere is required. The relationship between the field-aligned current and the net potential of a parallel electric field that accelerates the electrons was developed by Knight (1973), resulting in what has become known as the **"Knight relation."**

In order to derive this relationship, we shall make use of the Liouville theorem, which states that the distribution function is constant along a particle trajectory in velocity and configuration space. Furthermore, as the electrons are accelerated into the atmosphere, they must conserve their total energy and the magnetic moment. Thus, using the subscript "m" (for magnetosphere) to denote the high-altitude end of the acceleration region, where we arbitrarily set the electric potential ϕ to zero, the total energy of the electrons must satisfy

$$v_{\parallel}^2 + v_{\perp}^2 = v_{m\parallel}^2 + v_{m\perp}^2 + 2e\phi/m_e, \tag{11.18}$$

and magnetic moment conservation gives

$$v_{\perp}^2/B = v_{m\perp}^2/B_m, \tag{11.19}$$

where we have separated the velocity into components parallel and perpendicular to the magnetic field.

Thus,

$$v_{||}^2 + v_{\perp}^2(1 - B_m/B) = v_{m||}^2 + 2e\phi/m_e. \quad (11.20)$$

Equation (11.20) describes an ellipse in velocity space, since $B_m<B$, and any accelerated particles must lie in that portion of velocity space outside of the ellipse defined by setting $v_{m||}$ to zero:

$$v_{||}^2 + v_{\perp}^2(1 - B_m/B) = 2e\phi/m_e. \quad (11.21)$$

This is known as the "acceleration ellipse."

Similarly, using the subscript "I" to denote the ionosphere, or bottom of the acceleration region, conservation of energy and magnetic moment gives

$$v_{||}^2 + v_{\perp}^2(1 - B_I/B) = v_{I||}^2 - 2e(\phi_I - \phi)/m_e \quad (11.22)$$

as the relationship between a particular point in $v_\perp$, $v_{||}$ velocity space and the corresponding parallel velocity at the ionosphere, $v_{I||}$. Any electron with $v_{I||}>0$ is lost to the atmosphere. Thus, the other boundary in phase space is given by the **"loss cone hyperbola,"**

$$v_{\perp}^2(B_I/B - 1) - v_{||}^2 = 2e(\Phi_I - \Phi)/m_e. \quad (11.23)$$

The phase space boundaries given by Eqs. (11.21) and (11.23) and an electron distribution measured by the FAST spacecraft in the Earth's auroral zone are shown in Figure 11.8. The combined effect of the downward acceleration by the electric field and the upward acceleration of the magnetic mirror force results in the characteristic **"horseshoe" distribution.** The precipitating particles are at the right of the figure, in the region bounded by the acceleration ellipse and the loss cone hyperbola. The corresponding region to the left of the figure has a much reduced phase space density, filled mainly with backscattered secondary electrons. Outside of the loss cone hyperbola, the phase space density contours are mirror symmetric about $v_{||} = 0$, as is to be expected for particles that are reflected by the magnetic mirror force before entering the atmosphere. Because of the reduced phase space density of upgoing electrons, within

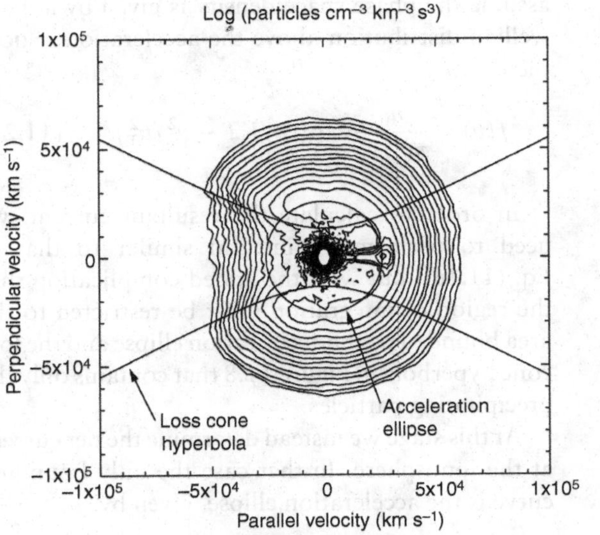

FIGURE 11.8. Phase space density contours for an electron distribution as measured in the auroral acceleration region by the FAST spacecraft. The phase space boundaries defined by Eqs. (11.21) and (11.23) are also shown.

the loss cone there is a net downward flux of electrons, i.e., upward current.

Given data such as those shown in Figures 11.7 and 11.8, we can make direct measurements of at least some of the electron current (a large part of the downward current is carried by electrons with energies below the instrument energy threshold). Although not shown here, there is usually reasonable agreement between the upward current density carried by the electrons and that deduced from the gradient in the magnetic field. Furthermore, the fact that the phase space distribution in Figure 11.8 has features that are consistent with acceleration by a parallel electric field provides strong support for the theoretical formalism of Knight (1973), which we now present.

First, we use the Liouville theorem to specify the phase space density within the acceleration region. If we denote the phase space density at the point of interest as $f(v)$ and the phase density at the top of the acceleration region as $f_m(v_m)$, then the Liouville theorem states that $f(v) = f_m(v_m)$, with v and v_m satisfying Eq. (11.18). In other words, if we again assume the phase space density is given by a maxwellian distribution above the acceleration region, then

$$f(v) = \frac{n_0}{\pi^{3/2} v_T^3} \exp\left(e\phi/KT - v^2/v_T^2\right). \tag{11.24}$$

In order to calculate the resultant current, we need to perform an integral similar to that in Eq. (11.16), but with the added complication that the region of integration must be restricted to the area bounded by the acceleration ellipse and the loss cone hyperbola in Figure 11.8 that contains only the precipitating particles.

At this stage we instead determine the net current at the ionosphere. In that case the only bounding curve is the acceleration ellipse, given by

$$v_{I\|}^2 + v_{I\perp}^2 (1 - B_m/B_I) = 2e\phi_I/m_e \tag{11.25}$$

since all the downgoing particles are assumed to be lost. In Eq. (11.25) B_I is the magnetic field strength at the ionosphere and ϕ_I is the total accelerating potential.

The velocity space integral becomes

$$j = j_0 \left\{ \frac{4}{v_T^4} \left(\int_0^{v_{L\|}} v_\| dv_\| \int_{v_{L\perp}}^{\infty} v_\perp dv_\perp + \int_{v_{L\|}}^{\infty} v_\| dv_\| \int_0^{\infty} v_\perp dv_\perp \right) \right.$$

$$\left. \left[\exp\left(e\phi_I/KT - v^2/v_T^2\right) \right] \right\}, \tag{11.26}$$

where we have split the integral into two parts to emphasize how the bounds to the integral are handled. The first integral covers that region of velocity space where the integral over perpendicular velocity is restricted to values that lie outside of the acceleration ellipse given by Eq. (11.25), i.e., $v_\perp \geq v_{L\perp}$, with

$$v_{L\perp}^2 = \left(2e\phi_I/m_e - v_\|^2\right)/(1 - B_m/B_I), \tag{11.27}$$

where we perform the integral over perpendicular velocity first. The parallel velocity limit ($v_{L\|}$) is given by the requirement that Eq. (11.27) cannot be negative, i.e.,

$$v_{L\|}^2 = 2e\phi_I/m_e. \tag{11.28}$$

In Eq. (11.27), $v_{L\perp} = 0$ when $v_\| = v_{L\|}$.

The second velocity space integral in Eq. (11.26) covers that region of velocity space where $v_\| \geq v_{L\|}$ and the lower limit on $v_\perp$ is 0.

On performing the integral over perpendicular velocity, Eq. (11.26) becomes

$$j = j_0 \left\{ \frac{2}{v_T^2} \left(\int_0^{v_{L\|}} v_\| dv_\| \exp\left[-\frac{\left(v_{L\|}^2 - v_\|^2\right)/v_T^2}{(B_I/B_m - 1)} \right] \right. \right.$$

$$\left. \left. + \int_{v_{L\|}}^{\infty} v_\| dv_\| \exp\left[\left(v_{L\|}^2 - v_\|^2\right)/v_T^2 \right] \right) \right\}, \tag{11.29}$$

where we have used Eq. (11.28) to replace the ϕ_I-dependent term. Clearly, Eq. (11.29) can be simplified through a suitable change in variables, giving

$$j = j_0 \left\{ \frac{2}{v_T^2} \left(\int_0^{v_{L\|}} v_\| dv_\| \exp\left[-\frac{v_\|^2/v_T^2}{(B_I/B_m - 1)} \right] \right. \right.$$

$$\left. \left. + \int_0^{\infty} v_\| dv_\| \exp\left[-v_\|^2/v_T^2 \right] \right) \right\}. \tag{11.30}$$

On performing the parallel velocity integration, we get

$$j = j_0 \left\{ (B_I/B_m - 1) \left(1 - \exp\left[-\frac{e\phi_I/KT}{(B_I/B_m - 1)} \right] \right) + 1 \right\}, \tag{11.31}$$

where we have again made use of Eq. (11.28).

Rearranging terms gives the final form of the Knight relation:

$$j = j_0 B_I/B_m \left\{ 1 - (1 - B_m/B_I) \exp\left[-\frac{e\phi_I/KT}{(B_I/B_m - 1)} \right] \right\}. \tag{11.32}$$

As ϕ_I increases, the current given by Eq. (11.32) reaches an asymptotic value given by $j = j_0 B_I/B_m$. This limit corresponds to accelerating the entire downgoing electron distribution present above the acceleration region into the atmosphere; B_I/B_m is known as the mirror ratio. The inverse of the mirror ratio gives the change in flux tube area, and the asymptotic limit can be understood in terms of flux conservation. Indeed Eq. (11.32) can be used to specify the current density at any height within the acceleration region, simply by using the local mirror ratio as the multiplying factor outside of the curly brackets. Note that the factor inside the curly brackets is unaltered, i.e., this term keeps the ionosphere–magnetosphere magnetic field mirror ratio.

The other limit is given when the term in the exponent is small,

$$j \approx j_0\{1 + e\phi_I/KT\}. \qquad (11.33)$$

This is the limit frequently invoked when deriving a characteristic precipitation energy from the field-aligned current density in global MHD simulations. MHD simulations can give values for the field-aligned current, and the source population temperature, from the plasma pressure, but the simulations cannot self-consistently derive the accelerating potential. Solutions of Eq. (11.32) are shown in Figure 11.9 for different values of the mirror ratio B_I/B_m.

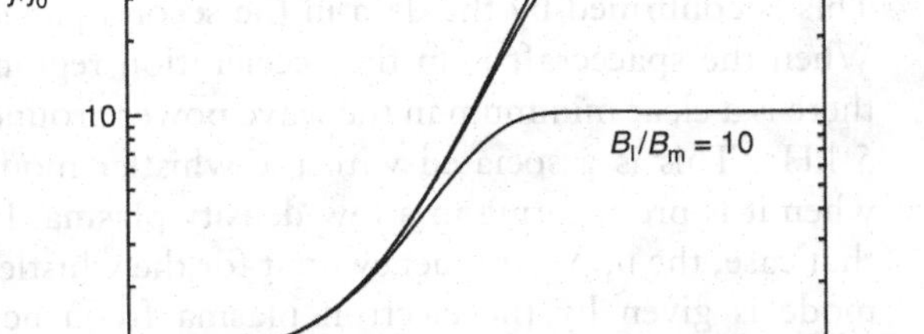

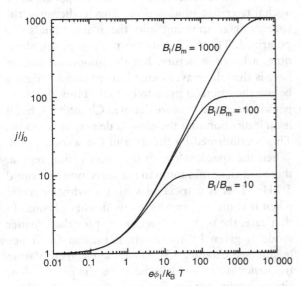

FIGURE 11.9. Solutions of Eq. (11.32) for different mirror ratios.

One aspect of the Knight relation not yet discussed concerns constraints on how the accelerating potential varies with the magnetic field strength. There are two constraints, $d\phi/dB > 0$ and $d^2\phi/dB^2 \le 0$. Two see how these arise, we rewrite Eq. (11.20) using the magnetic moment:

$$\frac{1}{2}m_e v_{\|}^2 = \frac{1}{2}m_e v_{m\|}^2 + \mu(B_m - B) + e\phi. \qquad (11.34)$$

Clearly, the requirement that $d\phi/dB>0$ ensures that the parallel electric field accelerates the electrons into the ionosphere, otherwise the mirror force and the electric field both accelerate the electrons away from the Earth. The second requirement is more subtle. This requires that $d\phi/dB$ decreases for increasing magnetic field strength. This ensures that once Eq. (11.33) becomes zero, corresponding to reflection of the particle, the parallel energy at any larger magnetic field strength (i.e., lower altitude) is also zero or negative. Thus, if an electron has positive downward velocity at the ionosphere, then the electron had positive downward velocity everywhere along its trajectory. In other words, if $d^2\phi/dB^2 \le 0$ is not satisfied, then, even if the parallel energy of the particle is positive above and below the acceleration region, the parallel energy of the particle may have become zero somewhere in the acceleration region, and the particle would have been reflected at that point. This would result in additional empty regions in the phase space distribution besides those given by the loss cone hyperbola and acceleration ellipse shown in Figure 11.8.

There are two consequences of the parallel acceleration, besides the primary one of increasing the field-aligned current density. The first is the enhanced energy flux of the precipitating electrons. The second is the generation of radio waves, known as auroral kilometric radiation (AKR). We shall discuss both further here.

On the dayside of the Earth, the primary source of ionization is solar EUV radiation. But precipitating particles can also increase ionization through collisions. The ionization rates depend on the flux and energy of the precipitating particles (both ions and electrons). There are several models of the resultant conductivities, with varying degrees of sophistication. Here we cite one of the most frequently used relations, that due to Robinson *et al.* (1987). In this paper, the height integrated conductivities from precipitating electrons are given as

$$\Sigma_P = \frac{40\langle W \rangle}{16 + \langle W \rangle^2} Q_0^{0.5} \tag{11.35}$$

and

$$\frac{\Sigma_H}{\Sigma_P} = 0.45 Q_0^{0.85}, \tag{11.36}$$

where Σ_P and Σ_H are the height-integrated Pedersen and Hall conductivities, respectively, expressed in siemens; $\langle W \rangle$ is the average energy of the precipitating electrons, in kiloelectronvolts; and Q_0 is the energy flux of the electrons in milliwatts per square meter. It should be noted that $\langle W \rangle$ is the average energy given by the ratio of the energy flux to the number flux. For a maxwellian, $\langle W \rangle = 10^{-3}(2KT/e)$, where again $\langle W \rangle$ is begin expressed in kiloelectronvolts.

The conductivities in Eqs. (11.35) and (11.36) can be determined directly from measurements of the precipitating electrons, but they can also be used in the context of global numerical simulations. In particular, one of the parameters available in a simulation is the field-aligned current density. The other parameters required are the number density and temperature of the magnetospheric electrons. These can be determined from the mass density and pressure as specified by the simulation, albeit with some assumptions concerning the mass composition and partition of temperature between species with the simulation (single-fluid MHD makes no distinction between ion and electron temperature; see Chapter 3). Equation (11.33) can then be used to determine the potential drop required for the given current density. In that case, if the current density is j, the energy flux $Q_0 = j\phi_I$ and $\langle W \rangle = \phi_I$ when expressed in the appropriate units (if ϕ_I is given in kiloelectronvolts and j in μA m^{-2}, then Q_0 and $\langle W \rangle$ are in the correct units).

One aspect of the changes in conductivity associated with the auroral particle precipitation is that the auroras, which are a consequence of magnetosphere–ionosphere coupling, can in turn modify the coupling.

The other aspect of auroral precipitation that we shall discuss briefly is the generation of AKR. Figure 11.10 shows an example of wave data acquired by the FAST spacecraft in the acceleration region. The figure spans roughly 2.5 min. The top panel shows the AKR spectrum, from 300–500 kHz. The thick white line is the local electron gyro-frequency. The next panel shows VLF wave data, from 20 Hz to 16 kHz. The third panel shows the electron energy spectra, with the fourth panel showing the ions. The eastward component of the magnetic field is shown at the bottom of the figure. The slope of the magnetic field is negative, with the spacecraft moving to higher latitudes. This corresponds to upward current. In the middle of the figure, from ~06:43:45 to ~06:44:45, both the ions and electrons show relatively narrow energy spectra, and we have labeled these as electron and ion beams. This is an indication of electrons being accelerated downward, and ions upward, as is expected when the spacecraft is within the auroral acceleration region. We should point out that, although the electrons are referred to as a beam, this is because they have a narrow energy spectrum; actually, the electrons occupy a large range in pitch angles, cf. Figure 11.7. The ions, on the other hand, are also narrow in pitch angle.

The wave data shown in the top two panels show a characteristic modulation. This is because the spacecraft is spinning and the wave signals are polarized. The AKR data, in the top panel, show quite a lot of structure, but the important feature here is that the waves extend down to and slightly below the electron gyro-frequency. How this happens is discussed in more detail in Chapter 13, but it is an indication that the plasma density is very low. This is confirmed by the data in the second panel. When the spacecraft is in the acceleration region, there is a clear minimum in the wave power around 5 kHz. This is associated with the whistler mode when it is propagating in a low-density plasma. In that case, the upper frequency limit for the whistler mode is given by the electron plasma frequency (again, see Chapter 13). Since the electron plasma frequency is $9n^{1/2}$ kHz, where n is the plasma density per cubic centimeter, the density in the acceleration region is 0.3 cm^{-3}. The AKR propagates in a

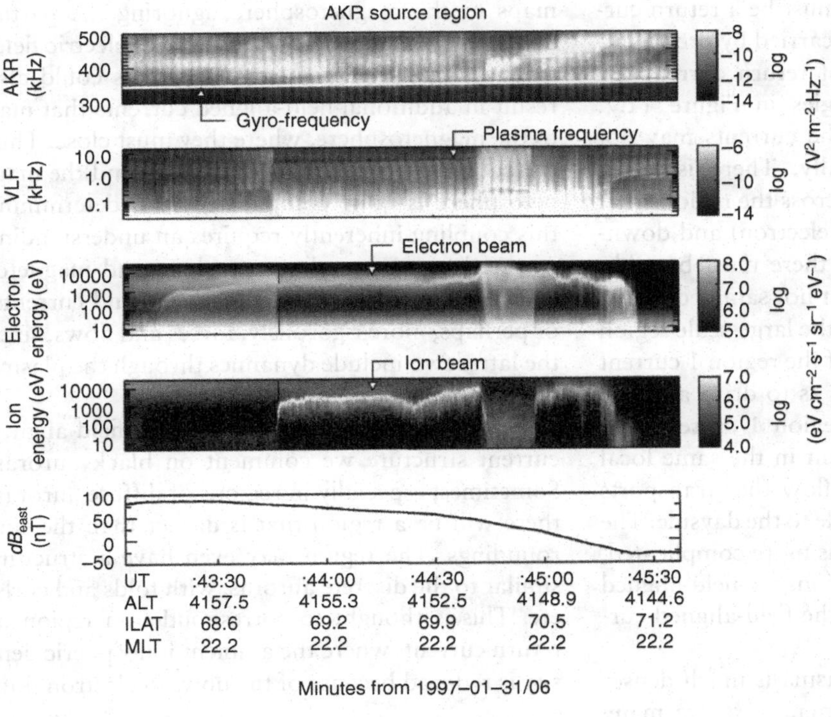

FIGURE 11.10. Particle and field data acquired in the auroral acceleration region.

wave mode known as the R-X mode (see Chapter 13). When the plasma density is this low, the cutoff frequency for the R-X mode can drop below the non-relativistic gyro-frequency, even for ~1 keV electrons, where the Lorentz factor is ~1.002. In a low-density plasma the R-X mode cutoff is given by $\omega_R = \Omega_e\left(1/\Gamma + \omega_{pe}^2/\Omega_e^2\right) \approx \Omega_e\left(1 - v^2/2c^2 + \omega_{pe}^2/\Omega_e^2\right)$, where Ω_e and ω_{pe} are the electron gyro-frequency and electron plasma frequency using the electron rest mass, m_e. As discussed in Chapter 13, when $\omega_R<\Omega_e$, the gyro-resonance instability can couple electrons directly to the R-X mode. Furthermore, relativistic corrections to the resonance condition allow the R-X mode waves to take energy from the perpendicular velocity gradients in phase space distribution functions, such as that shown in Figure 11.8. In passing, we note that the distribution function shown in Figure 11.8 was derived from the electron measurements shown in Figure 11.10, when the spacecraft was in the acceleration region.

As a last comment on AKR, we note that the gas giant planets Jupiter and Saturn emit radiation from high latitudes near the planet. These emissions are known as jovian decametric radiation (DAM) and saturnian kilometric radiation (SKR). It is reasonable to invoke a process similar to that associated with the generation of AKR at Earth, i.e., electron acceleration in a region of field-aligned currents flowing away from the planet. In the case of the gas giants, however, the planetary ionosphere appears to be the source of the currents, as these currents couple to regions of the magnetosphere where the moons are depositing large amounts of material that becomes ionized and must therefore flow at the corotation velocity, rather than the initial orbital velocity of the neutral particles ejected from the moons. This analogy may also extend to other astrophysical objects with intrinsic magnetic fields and where there is differential motion between different regions of the associated magnetosphere.

11.5.2 Return current

Except on short time scales, $\nabla \cdot \mathbf{j} = 0$, and there can be no net current into or out of the ionosphere. Since the currents do not flow out of the bottom side of the ionosphere, there must be a return current that balances the current carried by precipitating electrons. Such regions of return current are marked by the black rectangles in Figure 11.7. This figure also shows that the currents may not be completely balanced locally. There is a net change in the magnetic field across the regions that carry the upward (inverted V electron) and downward (return) currents. Thus there must be additional regions of return current not sampled by the spacecraft. This is the case for the large-scale region 1 and 2 currents, where part of the region 1 current closes across the polar cap so as to drive antisunward convection, while the region 1 current that closes with the region 2 current in the same local time sector drives the return flow that transports magnetic flux from the nightside to the dayside. The current system in Figure 11.7 is more complicated, with more than just two regions of field-aligned current. But again, not all of the field-aligned current closes locally.

Because the ionospheric plasma is much denser than the magnetospheric plasma, there are many more electrons available to carry downward field-aligned current, by accelerating electrons out of the ionosphere. Frequently the acceleration energy is below the low-energy cutoff of the electrostatic analyzer measuring the electrons, but occasionally the acceleration is sufficient for the electrons to be detectable. This is the case around 14:00:30 UT in Figure 11.7, at which time upward flowing electrons are observed. These electrons have a narrow pitch angle, and are in the middle of the loss cone. Some of the electrons appear to be reflected at higher altitudes, present as field-aligned electrons, but the net electron flux is upward. These low-energy electrons can be in Landau resonance (see Chapter 13) with the obliquely propagating whistler-mode waves, observed as VLF saucers in wave spectra.

Another consequence of ionospheric electrons being extracted to generate downward current is the evacuation of the plasma itself. Just as precipitating electrons tend to enhance conductivity, removing electrons tends to reduce the conductivity. The conductivity gradients in turn change the electric field and associated Pedersen and Hall currents. In steady state, the ionospheric electric field maps to the magnetosphere, ignoring the partial decoupling resulting from the parallel electric field. Changes in the perpendicular currents could also result in additional field-aligned currents that map to the magnetosphere, where they must close. Thus the coupling between the ionosphere and the magnetosphere is quite complicated, and determining this coupling inherently requires an understanding of the dynamics as the ionosphere and magnetosphere adjust to match electric fields and currents, or perhaps, more rigorously, forces and flows, since the latter two include dynamics through the plasma momentum equation.

Before addressing the third type of field-aligned current structure we comment on black auroras. Sometimes, especially in regions of diffuse auroras, there will be a region that is darker than the surroundings. The region may even have a structure similar to the discrete auroras, with folds and curls, etc. This is thought to correspond to a region of return current, where the ambient ionospheric density is reduced because of the upward electron flux.

11.5.3 Alfvénic auroras

The grey rectangle in Figure 11.7 marks a region that contains what has come to be known as alfvénic auroras. This is because the field-aligned currents are often very structured, with small-scale upward and downward currents in close proximity, such that the total current across the region is relatively small. The name "**alfvénic auroras**" is meant to indicate two features of this region. First, observationally, the electric and perturbation magnetic fields observed above the ionosphere have the characteristics of Alfvén waves, i.e., the electric to magnetic field ratio is given by the local Alfvén speed, rather than the height-integrated Pedersen conductivity. Second, these auroras are frequently observed at higher latitudes, near the polar cap boundary. This is the case in Figure 11.7. These flux tubes map to the plasma sheet boundary layer (Chapter 9), and alfvénic auroras are likely to be a signature of the magnetosphere and ionosphere coming to equilibrium. It should be noted, however, that alfvénic auroras are not restricted to just the plasma sheet boundary layer.

We now show how the shear mode can carry a parallel electric field. For the wave to carry a parallel electric field it is clear that we must use a different approximation than the ideal MHD approximation $\mathbf{E} + \mathbf{u} \times \mathbf{B} = 0$, as this implies that $E_{\|} = 0$. We shall therefore investigate the generalized Ohm's law (Eq. (3.137)):

$$\mathbf{j} = \sigma\left(\mathbf{E} + \mathbf{u} \times \mathbf{B} - \frac{\mathbf{j} \times \mathbf{B}}{ne} - \frac{\nabla \cdot \mathbf{P}_e}{ne} - \frac{m_e}{ne^2}\frac{\partial \mathbf{j}}{\partial t}\right). \tag{11.37}$$

In Eq. (11.37) we have assumed quasi-neutrality ($n \approx n_e$).

If we further assume that the conductivity is infinitely large, that the electron pressure is isotropic, and only consider the component of Eq. (11.37) parallel to the magnetic field, we get

$$E_{\|} = \frac{\nabla_{\|} P_e}{ne} + \frac{m_e}{ne^2}\frac{\partial j_{\|}}{\partial t}. \tag{11.38}$$

The first term on the right-hand side of Eq. (11.38) leads to what is often referred to as kinetic Alfvén waves, and the second term, which depends on the electron mass, gives the inertial Alfvén wave. It should be noted that the electron pressure term also leads to an ambipolar electric field in the context of the topside ionosphere.

Equation (11.38) shows that a parallel electric field may exist if there is a parallel current. This is the case for the shear mode, whereas the fast mode only carries perpendicular current. Furthermore, in Chapter 3 we saw that for the fast mode the wave vector (**k**), the ambient magnetic field (**B**), and the perturbation magnetic field (**b**) are coplanar. Faraday's law ($\mathbf{k} \times \mathbf{E} = \omega \mathbf{b}$) therefore requires that the wave electric field (**E**) must be perpendicular to the ambient magnetic field. Consequently, the fast mode cannot have an $E_{\|}$-component. For the shear mode, however, **E**, **k**, and **B** are coplanar, and **b** is perpendicular to this plane. The shear mode can have a parallel electric field and still satisfy Faraday's law.

We can now estimate the size of the parallel electric field associated with the electron inertia. For a harmonic perturbation, from Faraday's law and Ampère's law, including displacement current,

$$-i\omega\mu_0 \mathbf{j} = \mathbf{k}(\mathbf{k} \cdot \mathbf{E}) + \left(\frac{\omega^2}{c^2} - k^2\right)\mathbf{E}. \tag{11.39}$$

(The displacement current is included since the plasma density can be very low in the auroral zone and the displacement current ensures that the Alfvén mode does not propagate faster than the speed of light.)

The parallel component of Eq. (11.39) gives

$$-i\omega\mu_0 j_{\|} = k_{\|}(\mathbf{k}_{\perp} \cdot \mathbf{E}_{\perp}) + \left(\frac{\omega^2}{c^2} - k_{\perp}^2\right)E_{\|}. \tag{11.40}$$

From (11.38), neglecting the electron pressure term,

$$-i\omega j_{\|} = \frac{ne^2}{m_e}E_{\|} = \omega_{pe}^2 \varepsilon_0 E_{\|}, \tag{11.41}$$

where ω_{pe} is the electron plasma frequency.

Since **E** and **k** are coplanar, from Eqs. (11.40) and (11.41),

$$|E_{\|}| = \frac{k_{\|}}{k_{\perp}} \frac{\left(c^2 k_{\perp}^2/\omega_{pe}^2\right)}{\left(c^2 k_{\perp}^2/\omega_{pe}^2 + 1 - \omega^2/\omega_{pe}^2\right)} |E_{\perp}|. \tag{11.42}$$

For a plasma density of 1 cm^{-3}, $c/f_{pe} \approx 33$ km, where f_{pe} is the electron plasma frequency in hertz. This is a small transverse wavelength, but it is not unreasonable for the auroral zone. Thus we expect the shear mode to develop a parallel electric field in regions of low plasma density. Because this is a wave field, the field oscillates, and electrons are accelerated both toward and away from the ionosphere. These bi-directional electrons can be seen in Figure 11.7.

The perpendicular electric fields in the Alfvén waves may also result in transverse acceleration of the ions, and a characteristic feature of alfvénic auroras is the presence of ion conics, as can be seen in the fourth panel of Figure 11.7, which shows the ion pitch angle spectra. There is a peak in the differential energy flux just outside the loss cone. This is the ion conic. Ion conics are often observed in the other auroral current regions, indicating transverse heating is also occurring there. This transverse heating causes ions to escape from the ionosphere, providing a source of plasma for the magnetosphere. Unlike the solar wind, this plasma often consists of heavy oxygen ions, and these outflows change the composition of magnetospheric plasma.

11.5.4 Summary of the different types of auroral currents

As shown in Figure 11.7, there are essentially three different types of auroral currents, which we refer to as: inverted V auroras, the return current, and alfvénic auroras. Inverted V auroras correspond to regions of upward field-aligned current carried by electrons that have been accelerated through an electrostatic potential. These electrons typically have energies in the few kiloelectronvolts range, and are responsible for discrete auroras. In addition, these accelerated electrons are unstable to electromagnetic wave generation, and are the source for auroral kilometric radiation (AKR) – the terrestrial counterpart to other planetary radio emissions such as jovian decametric radiation and saturnian kilometric radiation.

As the name implies, the return current region is a region of downward current frequently observed near the inverted V electron current, and often balances most of the upward current. As Figure 11.7 shows, however, the balance between upward and downward current need not be exact. Some of the current may be returned elsewhere. The return current region can manifest itself as "black auroras" because of the associated reduction in plasma density.

The third type of field-aligned current is now known as alfvénic auroras. The field-aligned currents in this type of aurora are often structured on small time scales. The precipitating electrons associated with alfvénic auroras are more typically around 100 eV in energy and are likely to have been accelerated by the parallel electric field that results from electron inertia in the presence of small perpendicular scale shear-mode Alfvén waves. Alfvénic auroras appear to be a signature of the coupled magnetosphere–ionosphere system coming to equilibrium.

11.6 CONSEQUENCES OF AURORA-INDUCED CONDUCTIVITY CHANNELS

In Section 11.5.1 we noted that auroral electron precipitation can introduce horizontal gradients in the conductivity of the ionosphere. This in turn will affect the electric field and currents, and the flows imposed on the ionosphere by the magnetosphere may in turn be modified because of the resultant structure in the ionospheric currents and electric fields. We also note that Eq. (11.14) indicates that gradients in the ionospheric conductivity could require field-aligned currents to provide the current closure. Here we address two aspects of conductivity gradients: flow diversion and Cowling conductivity. The flow diversion is mainly associated with gradients in the Pedersen conductivity, while gradients in the Hall conductivity give rise to the enhanced Cowling conductivity.

11.6.1 Conductivity gradients and flow diversion

Figure 11.11 shows a sketch of a flow channel in the pre-midnight auroral zone. In this figure the x-axis points to the north, the y-axis points to the west, and the flow, **u**, points to the west. The circles indicate upward field-aligned current at the northern side of the channel ($x = x_0$) and downward field-aligned current at the southern side ($x = 0$). In the northern hemisphere the ambient magnetic field ($\mathbf{B}_0$) is down, and the electric field (E) and height-integrated Pedersen current ($\mathbf{J}_P$) are both northward. If we now assume that the height-integrated Pedersen conductivity (Σ_P) varies as a function of x, then horizontal gradients must be introduced in either E, or $\mathbf{J}_P$, or both. What is not clear is which of E or $\mathbf{J}_P$ changes.

One approach is to consider the forces imposed on the ionosphere. Equation (11.2) shows that an increase in the collision frequency or plasma density (both of which would correspond to an increase in Pedersen conductivity) requires an increase in the current density if the flow velocity is held constant (i.e., E is fixed). On the other hand, if $\mathbf{J}_P$ is fixed then the flow velocity must decrease. (Note that Eq. (11.2) is written in terms of current density, **j**, but the same argument applies for the height integrated Pedersen current $\mathbf{J}_P$.) Intuitively, we might expect

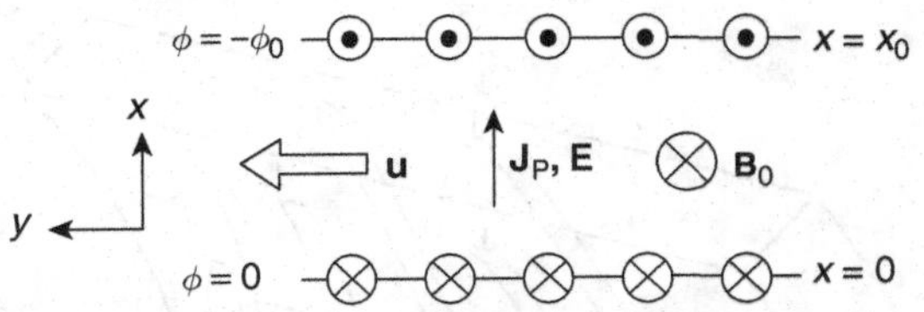

FIGURE 11.11. Sketch of a flow channel in the auroral ionosphere.

that the response of the system is for the flow to be diverted around the region of enhanced conductivity, but we can provide a more rigorous argument, based on minimizing the Joule dissipation in the flow channel.

In steady state the electric field can be expressed in terms of a potential, and

$$\int_0^{x_0} E dx = \phi_0 = \int_0^{x_0} vB_0 dx = \delta\Phi/\delta t. \quad (11.43)$$

Here, Φ is the magnetic flux, and $\delta\Phi/\delta t$ corresponds to the flux transport rate. This rate of flux transport is determined by the convection within the magnetosphere, and we assume that this is fixed. In other words, assume that at $x = 0$ ϕ remains 0 and ϕ_0 is fixed.

We define the energy dissipated per unit length along the flow channel as

$$W_D = \int_0^{x_0} J_P E dx, \quad (11.44)$$

and further assume that

$$J_P = J_0 + \Delta J, \quad (11.45)$$

where J_0 is assumed to be constant. We can specify an electric field corresponding to J_0 as

$$E_0 = J_0/\Sigma_P. \quad (11.46)$$

If we assume that Σ_P varies as a function of x, then one solution is to assume $\Delta J = 0$. In that case the electric field is given by Eq. (11.46). But Eq. (11.43) tells us that

$$\int_0^{x_0} E_0 dx = J_0 \int_0^{x_0} \frac{dx}{\Sigma_P} = \phi_0. \quad (11.47)$$

Again we have assumed that ϕ_0 is constant. From Eq. (11.47) we can determine J_0 given ϕ_0 and Σ_P.

For the more general case, where J_P is given by Eq. (11.45), the corresponding electric field is given by

$$E = E_0 + \Delta E = E_0 + \Delta J/\Sigma_P, \quad (11.48)$$

but, because of Eq. (11.43),

$$\int_0^{x_0} \Delta E dx = 0. \quad (11.49)$$

We can now rewrite Eq. (11.44) as

$$W_D = \int_0^{x_0} (J_0 + \Sigma_P \Delta E)(E_0 + \Delta E) dx$$
$$= \int_0^{x_0} J_0 E_0 dx + 2\int_0^{x_0} J_0 \Delta E dx + \int_0^{x_0} \Sigma_P \Delta E^2 dx. \quad (11.50)$$

The second integral in the bottom row of Eq. (11.50) vanishes because of Eq. (11.49), while the last integral is positive definite because $\Sigma_P > 0$. Thus the minimum dissipation rate is found if $\Delta E = 0$. Hence $\Delta J = 0$, and the lowest dissipation rate is found for $J_P =$ constant.

This implies that the system is likely to respond to the presence of conductivity gradients by adjusting the flow and, hence, the electric field so that the Pedersen current is constant. But clearly the current system sketched in Figure 11.11 is a simplification. In the real magnetosphere the field-aligned currents are not infinitely narrow current sheets but can extend over several degrees in latitude (see Figure 11.7). Furthermore, the magnetosphere may be constrained from relaxing to the minimum dissipation state. For example, the flows may be restricted to higher latitudes because of pressure gradients associated with the ring current. Nevertheless, the arguments presented here give additional justification for assuming that the system adjusts to conductivity gradients by modifying the flows so that the flows tend to occur in regions of lower conductivity.

11.6.2 Cowling conductivity and auroral currents

The concept of Cowling conductivity was introduced in Chapter 2. In the equatorial ionosphere, where the magnetic field is horizontal, if the ionosphere is moving vertically then the Hall currents are also vertical. It is possible to set up a secondary current system with a vertical electric field such that the secondary Pedersen current cancels the primary Hall current. But the secondary Hall current adds to the primary Pedersen current, resulting in a Cowling current given by

$$j_C = \sigma_P\left(1 + \frac{\sigma_H^2}{\sigma_P^2}\right)E_1 = \sigma_C E_1, \qquad (11.51)$$

where E_1 is the horizontal electric field corresponding to the vertical flow. If the Hall conductivity is much larger than the Pedersen conductivity, the Cowling conductivity $\sigma_C \gg \sigma_P$.

We have written Eq. (11.51) in terms of conductivities, but this equation is often written using height-integrated conductivities, i.e., $\Sigma_C = \Sigma_P(1 + \Sigma_H^2/\Sigma_P^2)$. This should be viewed with caution, however. In particular,

$$\int_{h_1}^{h_2} \frac{\sigma_H^2}{\sigma_P} dh \neq \frac{\Sigma_H^2}{\Sigma_P}, \qquad (11.52)$$

where the height integral is from height h_1 to height h_2.

In addition, it is frequently assumed that the height-integrated form of the Cowling conductivity can be applied to the auroral zone. This is also suspect, again because the Pedersen and Hall conductivities very with altitude, with the Pedersen conductivity larger than the Hall conductivity at higher altitudes in the upper E region and F region, while the Hall conductivity is larger in the lower E region.

Nevertheless, since the Cowling conductivity is often invoked when considering auroral current systems, we shall explore the concepts behind the Cowling conductivity as applied to the auroral ionosphere. To do this we use a somewhat simplifying assumption that the auroral ionosphere can be treated as a two-layer ionosphere, with the lower-altitude layer carrying Hall currents and the upper-altitude layer carrying Pedersen currents. This will allow us to explore some of the issues with the Cowling conductivity model, but in the real ionosphere the two layers do overlap. Our approach follows that outlined by Fujii *et al.* (2011).

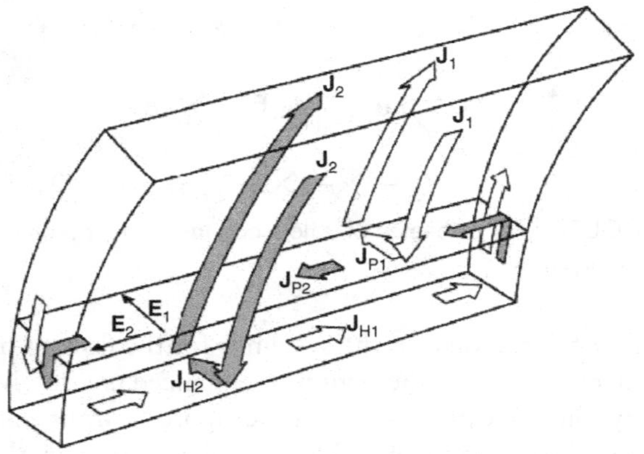

FIGURE 11.12. Sketch of a two-layer auroral Cowling conductivity channel.

Figure 11.12 shows a two-layer Cowling conductivity channel (this is often simply abbreviated to "**Cowling channel**"), corresponding to the auroral current system shown in Figure 11.11. In the figure the channel is assumed narrow in the north–south direction, and is extended in the east–west direction. The open arrows show the primary current system, with the upward current carried by precipitating electrons. The primary field currents are labeled $\mathbf{J}_1$, with the corresponding Pedersen current $\mathbf{J}_{P1}$ (= $\mathbf{J}_1$) and the corresponding electric field given by the thin arrow labeled $\mathbf{E}_1$. We only show a small portion of the primary field-aligned current system, but we assume that these currents and the associated Pedersen current extend uniformly in the east–west direction. It should be noted that, although we only show the electric field in the Pedersen current layer, Faraday's law requires that there are no vertical gradients in the electric field, and $\mathbf{E}_1$ is also present in the Hall current layer, where primary Hall current $\mathbf{J}_{H1}$ flows. At the ends of the region, the Hall current must either flow vertically, as shown, or possibly flow horizontally, depending on the presence or absence of a gradient in the Hall conductivity. If we make the assumption that the Hall conductivity is zero outside of the region shown, then the Hall current must close with vertical currents at the ends of the region.

The gray arrows show the currents for a secondary current system, generated by what is frequently referred to as a polarization electric field $\mathbf{E}_2$. In the Pedersen current layer the corresponding Pedersen current $\mathbf{J}_{P2}$ opposes the primary Hall current $\mathbf{J}_{H1}$. The secondary Pedersen current also closes with the primary Hall current at the ends of the channel. If $\mathbf{J}_{P2}$ exactly balances $\mathbf{J}_{H1}$, then this is referred to as a full Cowling conductivity channel, and the primary Hall current closes within the ionosphere, i.e., the open arrows at the ends of the channel vanish. The opposite case is, of course, when there is no secondary current system and the Hall current closes within the magnetosphere.

One consequence of establishing a secondary current system is that this system also has a secondary Hall current $\mathbf{J}_{H2}$. Again, if the Hall conductivity exterior to the channel is very small, then $\mathbf{J}_{H2}$ must close within the magnetosphere via field-aligned currents $\mathbf{J}_2$. As with the primary field-aligned currents, these currents extend uniformly to the east and west. These field-aligned currents add to the primary field-aligned current.

We can define a parameter

$$\alpha = J_{P2}/J_{H1} \tag{11.53}$$

that specifies the "completeness" of the Cowling conductivity channel, with $\alpha = 0$ corresponding to no secondary current system and $\alpha = 1$ giving complete closure of the primary Hall current through the secondary Pedersen current. We can then define the Cowling current as follows:

$$J_C = \Sigma_P\left(1 + \alpha\frac{\Sigma_H^2}{\Sigma_P^2}\right)E_1. \tag{11.54}$$

It is sometimes argued that this Cowling channel (where $\mathbf{J}_C$ flows) provides a source of Poynting flux that is returned to the magnetosphere. This is based on considering the amount of dissipation associated with the electric fields $\mathbf{E}_1$ and $\mathbf{E}_2$ separately.

Denoting the dissipation per unit area as w_D, then

$$w_{D1} = J_1E_1 = J_CE_1 = \Sigma_P\left(1 + \alpha\frac{\Sigma_H^2}{\Sigma_P^2}\right)E_1^2 \tag{11.55}$$

and

$$w_{D2} = J_2E_2 = (J_{P2} - J_{H1})E_2 = (\alpha - 1)\alpha\frac{\Sigma_H^2}{\Sigma_P}E_1^2, \tag{11.56}$$

where, from Eq. (11.53), $E_2/E_1 = \alpha\Sigma_H/\Sigma_P$.

Unless $\alpha = 0$ or 1, $w_{D2} < 0$, and the secondary electric field appears to be a source of Poynting flux. This is not correct. In terms of dissipation all that matters is the scalar dot product $\mathbf{J} \cdot \mathbf{E} = J_1E_1 + J_2E_2$. It is possible that one of the two constituent terms in the dot product is negative, but the sum will always be positive for the ionosphere, in the absence of a neutral wind dynamo.

Indeed, the net dissipation per unit area is given by the dissipation associated with the Pedersen currents $\mathbf{J}_P \cdot \mathbf{E}$, since the Hall currents, by definition, have $\mathbf{J}_H \cdot \mathbf{E} = 0$. Hence

$$w_D = \Sigma_P(E_1^2 + E_2^2) = \Sigma_P\left(1 + \alpha^2\frac{\Sigma_H^2}{\Sigma_P^2}\right)E_1^2$$
$$= w_{D1} + w_{D2}. \tag{11.57}$$

Although the dissipation is larger if $\alpha \neq 0$, this does not mean that the Cowling channel may not be present in the auroral zone. The problem becomes a question of how complete the channel is. If $\alpha = 0$ then the Hall currents flowing out of the ends of the channel may require additional closure currents within the magnetosphere. If, on the other hand, $\alpha = 1$, then the magnetospheric flows and currents must adjust. This is because in steady state $\nabla \times \mathbf{E} = 0$ and $\nabla \cdot \mathbf{J} = 0$. Thus the flow associated with the electric field $\mathbf{E}_2$ must also be present in the magnetosphere, and Eq. (11.10) shows that any additional sources of field-aligned current require changes in the flow and pressure distribution within the magnetosphere. Thus the problem becomes a question of how the coupled magnetosphere–ionosphere system adjusts so that steady state is achieved. This will determine the parameter α. But the energy principle invoked earlier, where the dissipation is minimized, suggests that α will be as small as possible, as this minimizes Eq. (11.57). In addition, for a Cowling channel in which $E_2/E_1 = \alpha\Sigma_H/\Sigma_P > 1$, the flow associated with the secondary electric field is larger, requiring the plasma flow to be across rather than along the channel. This would constitute a significant change in the flow pattern from the originally assumed flow. This again suggests that α should be small since, in general, $\Sigma_H/\Sigma_P > 1$.

One issue with assuming $\alpha = 0$, however, is that the Hall currents at the end of the channel must be closed through an additional current system. If we assume that the length along the channel in the east–west direction is l and the north–south width across the channel is w, then the total current flowing the additional current system is given by

$$I' = \frac{w}{l}\frac{\Sigma_H}{\Sigma_P}(1-\alpha)I_1, \qquad (11.58)$$

where I_1 is the total primary current, and I' is the total additional current. The additional current is maximum for $\alpha = 0$. But this current should be compared with the secondary field-aligned current that closes the secondary Hall current,

$$I_2 = \alpha\frac{\Sigma_H^2}{\Sigma_P^2}I_1. \qquad (11.59)$$

Thus

$$I' + I_2 = \frac{\Sigma_H}{\Sigma_P}\left[1 + \alpha\left(\frac{\Sigma_H}{\Sigma_P} - \frac{w}{l}\right)\right]I_1. \qquad (11.60)$$

Since, in general, $w/l < 1$ and $\Sigma_H/\Sigma_P > 1$, the minimum total extra current associated with the closure of the Hall currents is minimum for $\alpha = 0$. This again suggests that the Cowling channel may be weak.

11.6.3 Summary of consequences of aurora-induced conductivity channels

The fundamental issue raised in this section is how conductivity gradients affect the coupled magnetosphere–ionosphere system. In the absence of particle precipitation the primary source of the E- and F-region ionosphere is solar EUV. But if we assume that the magnetosphere imposes flows on the ionosphere, which in turn require field-aligned currents, then the conductivity will change because of the enhanced precipitation or reduced density. This will in turn require the currents or electric fields to change so as to ensure current continuity and curl-free electric fields for steady state. The two examples presented here show possible ways in which the system could adjust, but also demonstrate the open-endedness of the coupling. Both examples show that the final state of the coupled system is not well defined, since the magnetosphere is only partially included in describing the coupled system, with the magnetosphere providing a boundary condition but no information of how that boundary condition may evolve as the coupled system evolves.

Nevertheless, some insight can be gained in considering the role of conductivity gradients. First, for the flow channel, we showed that, from an energy principle where the energy dissipation is minimized, the effects of a gradient in the Pedersen conductivity cause flow diversion. That is, the flow imposed by the magnetosphere adjusts so that the flow is largest where the conductivity is lowest, and there is no additional diversion of the Pedersen current to field-aligned current associated with the conductivity gradient.

Second, for the Cowling channel, we explored in detail the role of a secondary (or polarization) electric field in generating a Pedersen current to cancel the Hall current that would otherwise close in the magnetosphere because of Hall conductivity gradients. Again, there is an issue in how the magnetosphere adjusts to accommodate the changes in flow and currents that result in the generation of a secondary electric field. While the answer of how the system evolves ultimately requires coupled models, we can at least suggest that the minimum impact on the system is where the Cowling channel is a minimum and any secondary electric field is small.

11.7 PLANETARY AURORAS

Much of the discussion so far has been related to the auroras at the Earth, where we have assumed that the magnetosphere, whose dynamics is largely governed by the solar wind and the IMF, imposes convection on the ionosphere. This requires currents to flow that provide the $\mathbf{j} \times \mathbf{B}$ force to move the ionospheric plasma through the neutral atmosphere.

Auroras are generated in regions where strong current densities are required and the current carriers (usually electrons) are accelerated into the ionosphere by field-aligned electric fields, as discussed in Section 11.5. It seems reasonable to assume that any magnetized body in which part of the system imposes flows on another part of the system could be expected to have auroras. This is indeed the case.

In particular, within the solar system, the gas giants Jupiter and Saturn have auroras. For these planets, however, much of the auroras are associated with internal processes. Because the solar-wind density and magnetic field strength are so much smaller at 5 AU (~ orbit of Jupiter) and 10 AU (~ orbit of Saturn), reconnection is much less efficient than at the Earth. On the other hand, the spin periods of these planets are less than half that of the Earth (~9.9 h for Jupiter and ~10.7 h for Saturn), and, given the size of their respective magnetospheres, the corotation velocity is much larger. Furthermore, the moons orbiting the gas giants are a source of neutral particles that are subsequently ionized, mainly through charge exchange. This process is referred to as mass loading. Because these newly ionized particles have velocities close to the orbital velocity of the moon from which they come, currents must flow so that the plasma comes up to corotation speed. The planetary atmosphere is the driver of the system, and auroras are observed on flux tubes that connect the planetary surface magnetic field to the orbits of the respective moons.

There are other sources of auroras within the magnetosphere of the gas giants. These seem to be mainly associated with internal processes occurring in the middle to outer magnetosphere. In particular, because of the centrifugal force, mass-loaded flux tubes tend to move outwards, and in doing so lag corotation. Again currents flow so as to provide the $j \times B$ force that at least partially imposes corotation. Substorm-like processes also occur where plasmoids are ejected down the tail. And lastly, auroral forms are also observed at very high latitudes. It is not clear if these auroras are also internally driven, or are associated with the albeit small amount of direct solar-wind driving.

It has also been suggested that auroral processes explain why many astronomical objects emit radio waves. At the Earth, radio waves with wavelengths of kilometer order are generated within the auroral acceleration region. These waves are known as auroral kilometric radiation (AKR); see Section 11.5.1. The gas giant counterparts are jovian decametric radiation (DAM) and saturnian kilometric radiation (SKR). Just as with the Earth, the frequency of the emitted radiation is close to the electron gyro-frequency at the foot of the flux tubes on which the auroras occur. Thus the radio waves emitted by astronomical objects can be used to infer the presence of magnetic fields at these objects.

Chapter 8 discusses auroras as observed at Venus and Mars. Venus is essentially unmagnetized, whereas Mars has crustal magnetic fields. If auroras occur at Venus, they are very weak. At Mars there have been suggestions that auroral acceleration could occur where the solar wind interacts with the crustal magnetic fields. But given the size of the magnetic anomalies, the expected acceleration potentials are small, in comparison to the kiloelectronvolt potentials observed at the Earth (see Section 11.5).

11.8 SUMMARY

In the opening section of this chapter we reviewed the historical context of auroral observations, with emphasis on the pioneering work of Kristian Birkeland. We should note that throughout this chapter we have emphasized the relationship between auroras and field-aligned currents. Because of Birkeland's work, field-aligned currents are also referred to as Birkeland currents.

In Section 11.2 we summarized the primary emission processes. These are usually caused by electrons, most notably the green-line emission. The long-lived red-line emission is also generated by electrons. But these electrons need not be auroral primaries. At lower latitudes, low-intensity red-line emissions are observed. These are thought to be generated by secondary electrons caused by ion precipitation from the plasma sheet. These red-line emissions can therefore be used to place the more intense auroral emissions within a magnetopsheric context.

Following the discussion of the emission process, Section 11.3 described the auroral morphology. We emphasized that auroras tend to occur in a restricted latitude range, known as the auroral oval. This latitude range maps from near the inner edge of the plasma sheet to the polar cap boundary. We also discussed Akasofu's sketch of the auroral substorm, emphasizing the dynamic nature of auroras.

Section 11.4 then placed auroras within the context of magnetosphere–ionosphere coupling. This section emphasized that any coupled system, such as the Earth's magnetosphere–ionosphere system, will have different processes occurring in different regions. For example, the dynamics of the Earth's magnetosphere is mainly controlled by reconnection with the IMF. Internal processes (e.g., the substorm) can also play a role. The ionosphere, on the other hand, is strongly collisional. Thus, there is an interplay between these two parts of the system. If we consider the coupled system in terms of forces and flows, then we can see that a $\mathbf{j} \times \mathbf{B}$ force must be applied to the ionosphere to overcome the collisional drag. These are the two dominant forces in the ionosphere. For the magnetosphere, collisions are largely absent, but other forces are acting, specifically forces related to either thermal or dynamic pressure. The coupling of these two systems leads to the presence of field-aligned currents.

The different types of field-aligned currents were discussed in Section 11.5. These can be split into three types: inverted V or downward accelerated electron currents; the return current; and fluctuating or alfvénic currents. The first two are mainly quasi-steady currents, while the alfvénic currents appear to be a signature of the equilibration of the magnetospheric and ionospheric flows and forces. We also noted that the acceleration process for the inverted V electrons results in free energy for the generation of radio waves, known as AKR, that make the Earth a radio-astronomical object.

Because auroral particle precipitation can modify the underlying conductivity of the ionosphere, we discussed the effects of conductivity gradients in Section 11.6. Based on a minimum dissipation argument, we showed why flows within the ionosphere might tend to be altered by the presence of conductivity gradients in such a way as to reduce the flow in high-conductivity regions, thereby reducing the need for additional field-aligned currents associated with the gradient itself.

Closely related to this is the concept of the Cowling channel, where gradients that confine the Hall current appear to require the establishment of a secondary current system that cancels the primary Hall current. Although the Cowling channel is frequently invoked within the literature, there are reasons for suggesting it may not be important. At a minimum, if a Cowling channel is established, then the flows and currents within the magnetosphere must in turn be modified because of the currents and electric field associated with the secondary current system.

We closed the chapter, in Section 11.7, with a brief discussion of auroras at other magnetized planets.

Additional reading

Banks, P. M. and G. Kockarts (1983). *Aeronomy, Parts A and B*. New York: Academic Press.
These two books provide a comprehensive description of the processes that occur within the ionosphere. Of particular interest to this

chapter are the ionization processes through photoionization and collisions and ionosphere frictional heating.

BURCH, J. L. and V. ANGELOPOULOS (Eds.) (2009). *The THEMIS Mission*. Dordrecht: Springer. This book, which is a reprint of articles published in *Space Science Reviews* (also published by Springer), describes the mission design, science objectives, instrumentation, and preliminary results from the Time History of Events and Macroscale Interactions during Substorms (THEMIS) mission. This mission explores the dynamic coupling of the magnetosphere–ionosphere system, one consequence of which is auroras, as discussed in this chapter.

KEILING, A., E. DONOVAN, F. BAGENAL, and T. KARLSSON (Eds.) (2012). *Auroral Phenomenology and Magnetospheric Processes: Earth and Other Planets*. Geophysical Monograph Series vol. 197. Washington, D. C.: American Geophysical Union. This book extends the discussion in Paschmann, Haaland, and Treumann (2003) (see next reference in this list), including auroral phenomena at other planets.

PASCHMANN, G., S. HAALAND, and R. TREUMANN (Eds.) (2003). *Auroral Plasma Physics*. Dordrecht/Boston/London: Kluwer Academic Publishers. This is a comprehensive book that provides much more detail on auroral processes than can be included in a single chapter.

Problems

11.1 Discuss how the Biot–Savart law can be used to estimate the magnetic disturbance at the ground (in nanotesla) for a line current at 120 km. Assume different values for the conductivity and the electric field.

11.2 Explain how auroral spectrometer measurements could be used to determine the composition and height distribution of the upper atmosphere between 95 and 300 km.

11.3 Choose the Auroral Electrojet portion of the Currents option of the Space Physics Exercises.* This portion has a variable ionospheric current and variable conducting layer under the surface of the Earth to allow you to see how these affect the Earth's magnetic field. Vary the width of the current layer: 0.5°, 10°. How does B_x change? How does B_z change? Turn on the induced currents in the conducting layer. Calculate the magnetic field for depths of the conducting layer of 200, 100, 50, and 0 km for 0° width. How does the conducting layer affect B_x and B_z on the surface of the Earth? Contrast with the case with no conductor.

* Space Physics Exercises are at http://spacephysics.ucla.edu.

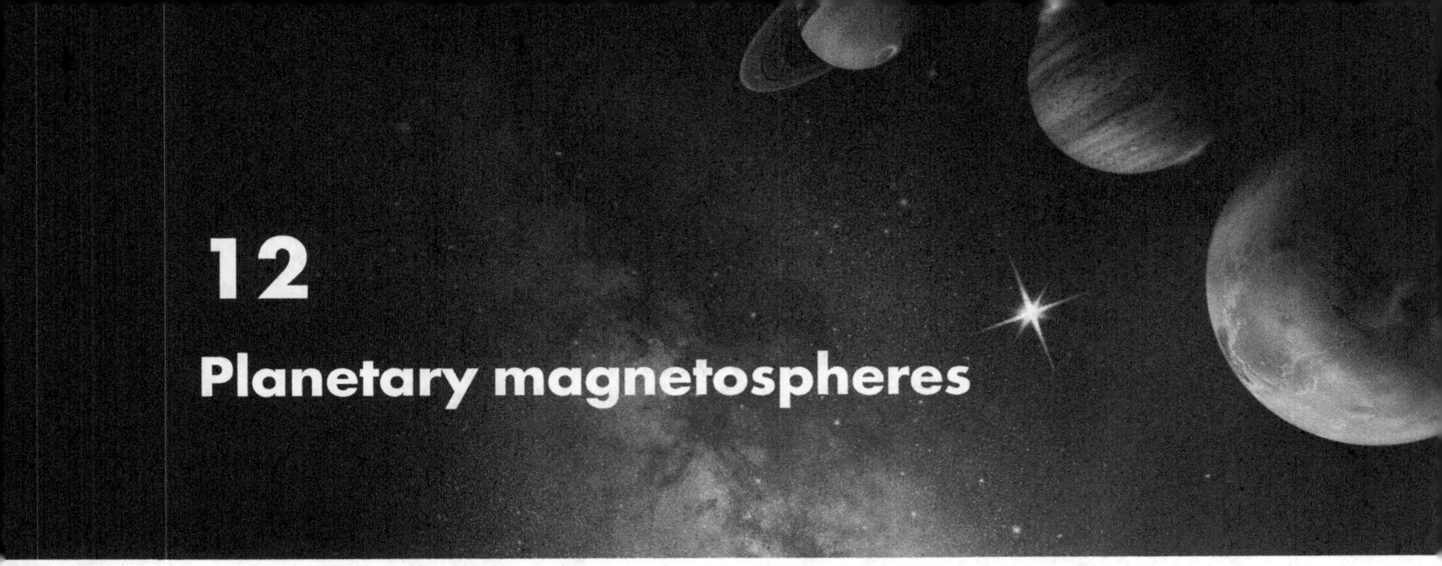

12 Planetary magnetospheres

12.1 INTRODUCTION

In the laboratory, we can modify the conditions in our experimental apparatus and record how the process changes. We would like to be able to make such modifications in studying the solar-terrestrial system, but we cannot. By and large, solar-terrestrial physics is an observational rather than an experimental science. There are few active experiments we can perform to determine how the system works, either in the magnetosphere or in the laboratory. Except by using computers, we cannot create magnetospheres with quantitatively scaled parameters to see how they work. Even in our computer models, we make approximations and do not simulate correctly all the physical processes that occur. Furthermore, our observations are restricted to the magnetospheres that our spacecraft have explored. Fortunately, these magnetospheres differ sufficiently that comparisons among them can give some insight into the governing processes. Mariner 10's three flybys have now been supplemented by over four years of orbital measurements by the MESSENGER mission. Our investigations of the jovian and saturnian magnetospheres are now quite comprehensive: we have measurements of the jovian magnetosphere from the flyby missions Pioneer 10 and 11, Voyagers 1 and 2, and Ulysses, and from the Galileo orbiter. We have surveyed the saturnian magnetosphere with the flybys Pioneer 11, Voyagers 1 and 2, and the Cassini orbiter. In contrast, our knowledge of Uranus and Neptune is based solely on the flybys of Voyager 2.

Mercury provides an important contrast to the terrestrial magnetosphere because it has a small-scale size and has no significant ionosphere or atmosphere. The solar wind exerts strong control throughout the Mercury magnetosphere, and it has direct access to the surface in the polar cusps. It may also contribute to Mercury's weak exosphere by ion sputtering of the surface. The outer planets provide us with other comparisons. First, as we move outward in the solar wind, the properties of the solar wind change. This may affect the coupling of energy flux from the solar wind into the outer planet magnetospheres. The sizes of the magnetospheres of the "gas giants" are much larger than those of Earth and Mercury. The combined rapid rotation and size of the gas giants lead to important centrifugal forces that far exceed those in the Earth's magnetosphere. In the jovian and saturnian magnetospheres, there are also significant sources of plasma and dust from the moons Io and Enceladus. As a result, processes that are minor players at the Earth, such as mass loading, charge exchange, ion cyclotron wave growth, and the interchange instability, become much more important than at Earth.

The available Mercury *in situ* observations have suggested to some researchers that Mercury has substorms analogous to those on Earth, but Mercury's close coupling to the solar wind, its lack of significant ionosphere, and short communication times may limit that analogy. Both Jupiter and Saturn have dynamical magnetotail behavior that resembles terrestrial substorms in many ways, but they put their particular spin on the problem. Since the magnetospheric plasma approaches corotational velocities and these velocities are large, the rapid inward and outward motions of the substorm-accelerated plasma are affected by the need to preserve angular momentum. These substorms also appear to achieve a higher rate of reconnection than terrestrial substorms, causing more dramatic signatures than seen at Earth.

Finally, the outer planets have rings. Some of these are quite tenuous, such as Jupiter's, and have little effect on the magnetosphere or the radiation belts. Thus, the jovian radiation belts reach high intensities. In contrast, the rings of Saturn are thick and extensive, absorbing the trapped radiation efficiently. Mercury has yet a different reason for the absence of a radiation belt. It has only a small and distorted "dipolar" region in its magnetosphere. Thus, it can trap neither the cold particles from its tenuous atmosphere nor the hot particles from interplanetary fluxes that may "leak" in. Hence, it has neither a plasmasphere nor a radiation belt.

12.2 THE RADIAL VARIATION OF THE SOLAR WIND

As discussed in Chapter 5, the solar-wind number density and the radial component of the interplanetary magnetic field (IMF) both vary inversely as the square of distance from the Sun. The tangential component of the magnetic field, however, varies inversely as the first power. This means that the spiral angle of the IMF becomes increasingly tighter and tighter, approaching 90° to the radial direction near the orbit of Saturn. This change in spiral angle is not expected to have a major effect on the interaction with the outer planets, but it will change the foreshock geometry.

The electron and ion densities all decrease with distance from the Sun with an inverse square dependence. This decreases the dynamic pressure and would allow the magnetospheres of the outer planets to expand to enormous scales even if their magnetic dipole moments were similar to those of Earth. Electron and ion temperatures also decrease with distance from the Sun, but, because of heat conduction and dissipation, the radial fall-off is slower than that of an adiabatic process. This has important consequences for two parameters that control aspects of the solar-wind interaction with planetary magnetospheres: the fast magnetosonic Mach number and the beta (β) of the plasma. The fast magnetosonic Mach number is the ratio of the

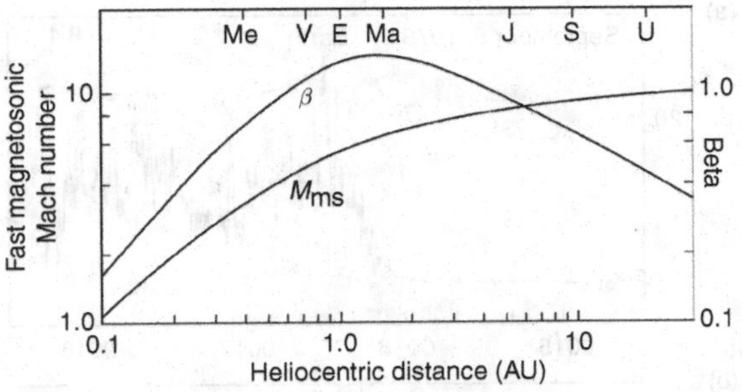

FIGURE 12.1. Expected variations of the fast magnetosonic Mach number and plasma β as functions of heliocentric distance. The locations of the planetary orbits are shown at the top of the figure. (After Russell, Lepping, and Smith, 1990.)

velocity of the solar wind to the speed of compressional waves in the solar wind, as discussed in Chapter 6. It controls the strength of the bow shock, which in turn controls the properties of the shocked plasma in the magnetosheath that bathes the planetary magnetopause. The β of the plasma is the ratio of the thermal pressure in the plasma to the magnetic pressure. The solar-wind β also helps to control the properties of the magnetosheath plasma. As illustrated in Figure 12.1, the expected magnetosonic Mach number increases from about 6 at Earth to about 10 at Saturn, whereas β maximizes at about Mars and then declines slightly in the outer solar system, under our assumptions of radial fall-off. The major implication of this behavior is that, on average, the bow shocks of the outer planets are stronger than those in the inner solar system.

Because of the high β values behind these strong shocks, the magnetic field in the magnetosheath is relatively weak. This will diminish the importance of the magnetic field in the interaction. The flow in the magnetosheath will more resemble that from the gas-dynamic simulation, and the importance of dayside magnetopause reconnection in driving the circulation of plasma in the magnetosphere will be lessened. Figure 12.2 shows a comparison of the magnetic field strengths measured through the bow shocks of Earth, Jupiter, and Uranus. The large overshoot in magnetic field strength just downstream of the shock ramp is a signature of the strengths of these shocks.

Strong shocks, in turn, are accompanied by strong fluxes of particles streaming back toward

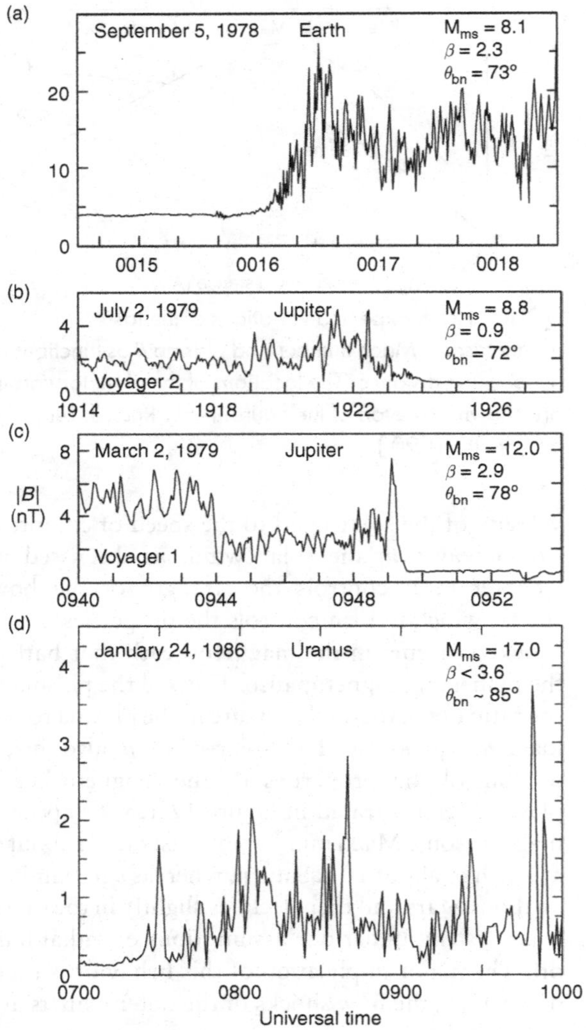

FIGURE 12.2. Magnetic profile of high Mach number shocks at Earth, Jupiter, and Uranus. The quiet fields to the left in (a) and (d) and to the right in (b) and (c) are those of the pre-shock solar wind. (After Russell *et al.*, 1990.)

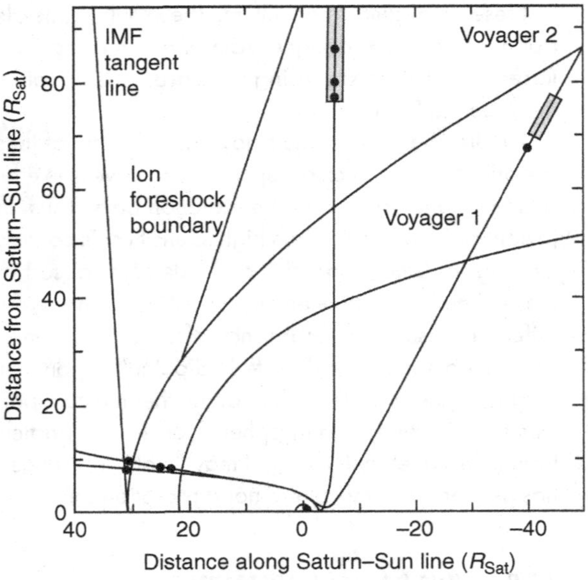

FIGURE 12.3. Geometry of the Saturn foreshock, as deduced from Voyager 1 and 2 measurements. The trajectories of the two spacecraft are shown, as are shock encounters and regions of upstream waves studied by Orlowski *et al.* (1992). Solid circles denote shock crossings, and bars mark the regions of upstream waves. Distances are given in saturnian radii (R_{Sat}).

the Sun. The two mechanisms that can lead to particle streaming along the interplanetary field from planetary bow shocks are leakage of the hot downstream magnetosheath particles and reflection of solar-wind particles. At the outer planets, both mechanisms should be stronger than at 1 AU because of the size of the **overshoot** and the expected temperature of the magnetosheath.

The ever-tightening spiral of the IMF alters the geometry of the foreshocks of the outer planets, as illustrated in Figure 12.3 for Saturn (Orlowski, Russell, and Lepping, 1992). The tangent field line, approximately the boundary of the electron foreshock, is nearly perpendicular to the solar-wind flow, and the ion foreshock is swept back on both sides of the magnetosphere so that the strong ULF waves associated with back-streaming ions are seen only over the terminator regions. This geometry should be compared with the terrestrial foreshock in Figure 1.14.

Another variation that occurs with increasing distance from the Sun is the change in the gyro-radius of the reflected solar-wind ions. However, because the sizes of the magnetospheres of the outer planets are always much larger than the expected size of this gyro-radius, this increasing size is expected not to have any major effects on the interaction of the solar wind with the magnetospheres of the outer planets.

Reconnection plays a significant role in the internal dynamics of Earth's magnetosphere,

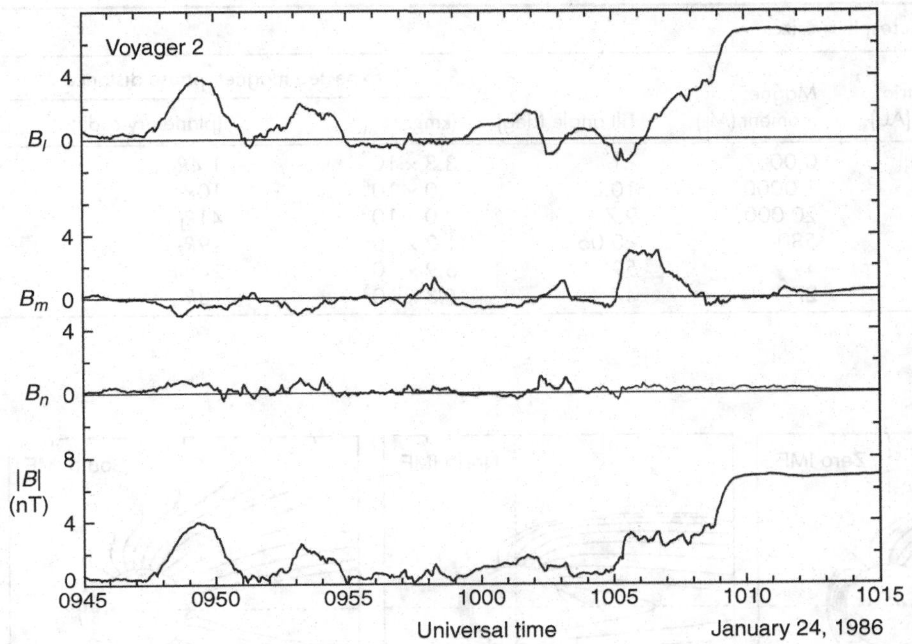

FIGURE 12.4. Magnetic field strength of the time of the Voyager 2 crossing of the uranian magnetopause. The last increase in the field strength into a relatively quiet region is interpreted to be the magnetopause crossing. The preceding activity may be mirror-mode waves excited in the high-β magnetosheath. (After Russell, Song, and Lepping, 1989.)

controlling the occurrence of substorms and storms. This is certainly true at Jupiter and Saturn, as well. However, in the outer solar system, the IMF may not be as important in coupling solar-wind momentum and energy to the magnetosphere. The weakness of the coupling occurs because the magnetic pressure provides a smaller contribution to the total pressure in the magnetosheath at the outer planets than it does at Earth. This is illustrated in Figure 12.4, which shows Voyager 2 magnetic field measurements across the magnetopause at Uranus. The ratio of field strengths across the magnetopause is close to 20. Because the sum of the plasma thermal and magnetic pressures is constant across the magnetopause, and because we expect the major contribution to the pressure in the magnetosphere to be that of the magnetic field, we can conclude that the value of β in the uranian magnetosheath is close to 400 at this time. If reconnection at the magnetopause for Uranus and Neptune is much less efficient than at Earth, we would expect a rather quiescent magnetosphere at each planet. We find that the energetic particle flux and the ULF wave intensity at both planets are reduced compared with the levels found closer to the Sun. However, there is a suggestion of a substorm in the uranian magnetotail data, and there are many plasma waves present in these magnetospheres.

The outer planet magnetospheres again stand in stark contrast to those of Mercury, where the large interplanetary field and solar-wind pressure together with Mercury's modest magnetic moment and weak ionosphere ensure that the solar wind and IMF dominate its behavior. Figure 12.5 illustrates the expected changes in the configuration of Mercury's magnetosphere for different IMF orientations. These models, which appear to match the limited observations from the Mariner 10 flyby, suggest that the entire configuration of the system changes drastically every time the interplanetary conditions vary. With such a highly solar-wind-coupled, externally controlled, dynamic magnetosphere, it is difficult to identify exclusively internal processes. Outside of the magnetosphere

Table 12.1. Relative magnetospheric sizes

Planet	Heliocentric distance (AU)	Magnetic moment (M_E)	Tilt angle (deg)	Expected magnetopause distance (km)	Expected magnetopause distance (planetary radii)
Mercury	0.39	0.0007	<1	3.3×10^3	$1.4R_{Me}$
Earth	1.00	1.0000	10.8	7.0×10^4	$10R_E$
Jupiter	5.20	20 000	9.7	3.0×10^6	$41R_J$
Saturn	9.54	580	<0.06	1.2×10^6	$19R_S$
Uranus	19.2	49	59	6.9×10^5	$24R_U$
Neptune	30.1	27	47	6.3×10^5	$23R_N$

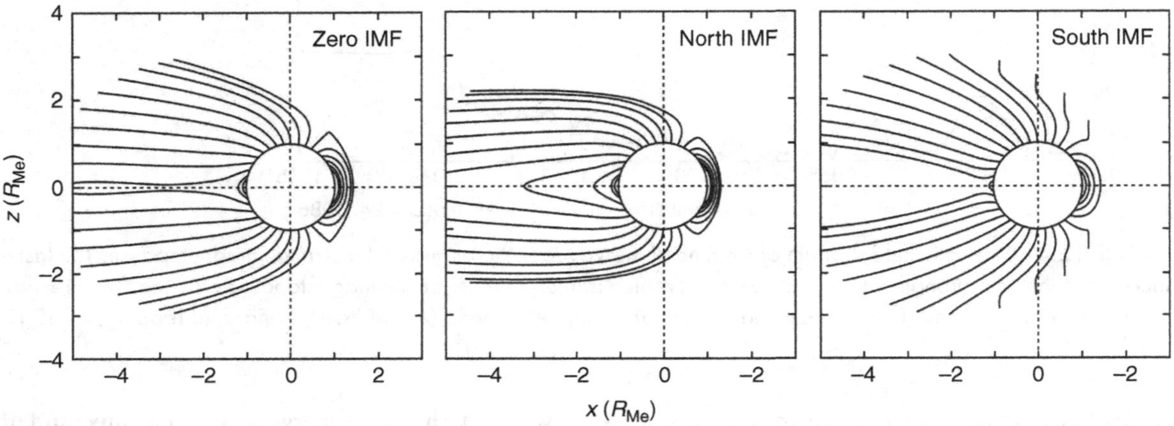

FIGURE 12.5. The expected configuration of Mercury's magnetosphere for varying orientation of the IMF, assuming rapid coupling with the IMF. (After Luhmann, Russell, and Tsyganaenko, 1998.)

proper, Mercury had a well-defined magnetosheath at the time of the Mariner 10 flyby. However, it has been speculated that, for periods of very high solar-wind pressure, the Mercury subsolar magnetopause may be pushed down to the surface.

12.3 MAGNETOSPHERIC SIZE AND COMPRESSIBILITY

As discussed in Chapter 7, the size of the terrestrial magnetic cavity is expected to be proportional to the sixth root of the ratio of the square of the magnetic moment, divided by the dynamic pressure of the solar wind. Because the solar-wind dynamic pressure varies inversely as the square of the heliocentric distance, the magnetospheres of the outer planets should be considerably larger than the terrestrial magnetosphere. The size is further increased because the magnetic moments of the outer planets are significantly larger than the terrestrial magnetic moment. Table 12.1 lists the relevant parameters for these six planets.

In this table, magnetic moments are given in terms of the terrestrial magnetic moment of 8×10^{15} T m^3. The last two columns show the expected distance of the nose, or subsolar point, of the magnetopause, in kilometers and planetary radii, as derived from simple pressure-balance arguments. Figure 12.6 shows the relative sizes of these magnetospheres, including that of Mercury. The expected subsolar distances are all much greater than the terrestrial distances. However, the distances expressed in planetary radii are more similar: 11 for Earth, 45

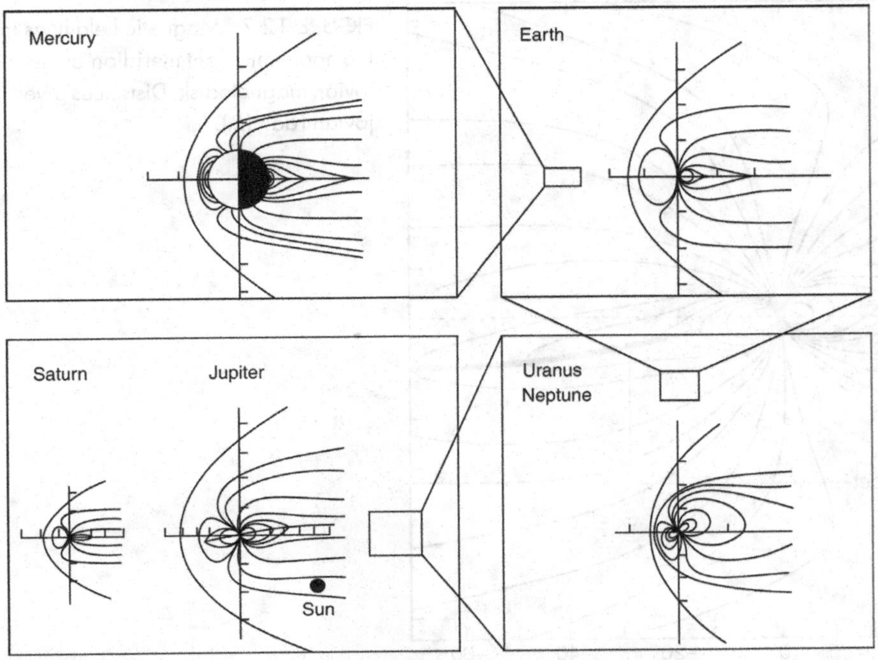

FIGURE 12.6. Comparison of the sizes of planetary magnetospheres.

for Jupiter, and about 22 for the outer three magnetospheres. Mercury is all alone at the other end of the scale, with a small magnetosphere at both absolute and relative scales. Mercury's intrinsic dipole magnetic field is also significantly offset along the rotation axis northward from the planet's center. This is also the case for Saturn's magnetic field, but by a relatively smaller amount.

The tilts of the dipole moments to the rotation axes vary over a large range, from less than 1° for Saturn to close to 60° for Uranus. The effects of these dipole tilts are varied. The axial alignment of the dipole moment with its higher moments results in a rotationally symmetric inner magnetosphere at Saturn. The roughly 10° tilt of the jovian dipole results in a ±10° motion of Io with respect to the magnetic equator, since Io orbits Jupiter in its rotational equatorial plane. Thus, the material that is lost from Io's atmosphere to the magnetosphere, and which is subsequently ionized by the intense radiation in the magnetosphere, becoming the Io torus, is spread over ±10° in magnetic latitude.

For Jupiter and Saturn, our theoretical expectations for the pressure balance with the incident solar wind based on a vacuum magnetosphere are not met. The subsolar magnetopause of Jupiter is found to range from the expected 42 jovian radii (R_J) to over 110 R_J and for Saturn from 15–30R_{Sat}. The reason for these discrepancies, in addition to the variation of the solar-wind dynamic pressure, is that Jupiter's moon Io and Saturn's moon Encaladus provide a significant source of mass to their magnetospheres in the form of neutral particles. Once these neutrals are ionized by solar EUV or by impact with ambient charged particles, the rapidly rotating magnetosphere "picks up" the ions and they become part of the mass of the system, concentrated near the centrifugal equator. The resulting centrifugal force of the plasma pushes out against the solar wind and creates a more distant standoff distance than would be the case if the magnetopause location were determined solely by magnetic pressure. Figure 12.7 shows the magnetic field lines in the noon–midnight meridian in a model of this "**magnetodisk.**" Though quite distorted by the centrifugal force, this magnetodisk still retains the general configuration of the terrestrial magnetosphere.

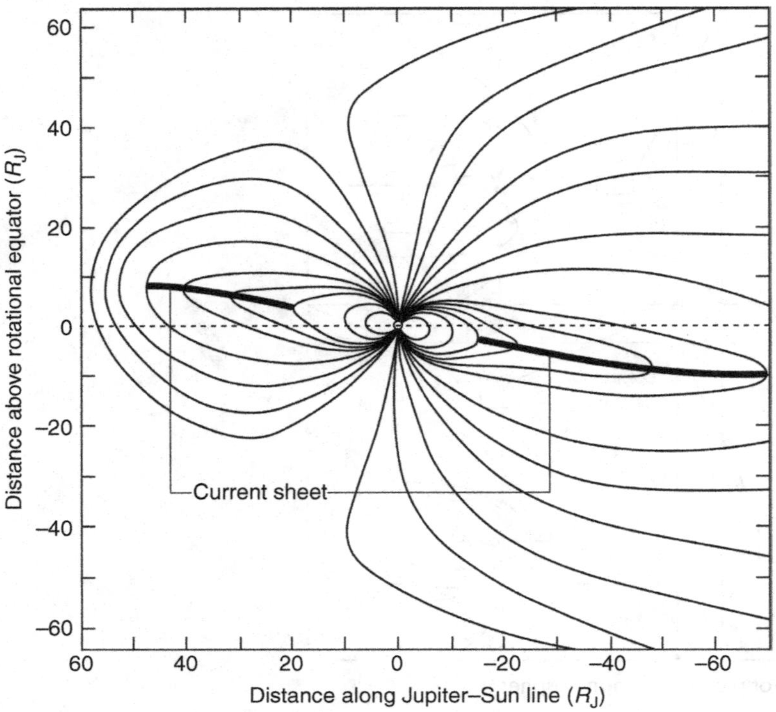

FIGURE 12.7. Magnetic field lines in the noon–midnight meridian of the jovian magnetodisk. Distances given in jovian radii (R_J).

The location of the Earth's magnetopause is affected by magnetic reconnection with the interplanetary magnetic field. This process is sometimes called erosion by the southward interplanetary field. As noted above, the magnetopause of Mercury is quite sensitive to the solar-wind pressure and to the interplanetary magnetic field. In contrast, the magnetospheres of the outer planets appear to be more sensitive to the solar-wind dynamic pressure than to the direction of the solar-wind magnetic field. The magnetic reconnection phenomenon known as a flux transfer event (see Chapter 9) occurs frequently at Earth, but is even more prevalent at Mercury, when it creates numerous small (about 1 s long) events in rapid succession. At Jupiter, in contrast, only a few flux transfer events have been reported, and, at Saturn, none. This may be associated with the higher-beta values found in the magnetosheaths of the outer planets. Since reconnection is a magnetically controlled process, it is less important when the plasma stresses are controlled mainly by the plasma pressure.

The disk-like shape of the jovian magnetosphere has an effect on the location of the bow shock. As discussed in Chapter 7, the bow shock is located at a distance from the magnetopause sufficient to allow the shocked solar wind to flow around the magnetosphere. If the nose of the magnetosphere has a large radius of curvature, this distance will be larger than if the nose has a small radius of curvature. In the analogous situation in aviation, a shock wave will be detached from an aircraft by a distance that is dependent on its shape. A needle-nosed supersonic plane may have the shock attached to the nose of the plane, while a re-entry vehicle slowing down in the upper atmosphere is blunt and has a relative shock standoff distance comparable to an Earth-like magnetosphere. The jovian magnetosphere is sharp enough to cause a measurable reduction in this distance. At Jupiter, the nose of the bow shock is about 20% farther out than the subsolar magnetopause, whereas at Saturn and Earth the nose of the bow shock is about 30% farther out than the magnetopause.

The enormous sizes of the outer planet magnetospheres also have some important consequences for the convection of solar-wind plasma past the planets. At Earth, a solar-wind disturbance can travel the distance from the nose to the terminator in 2–3 min. However, at the outer planets, as shown in Table 12.2, this time is much longer, ranging from about 25 min at Uranus and Neptune, to about 200 min at Jupiter.

The velocity of the plasma just inside the magnetopause is also quite different for the outer planets. If the plasma were to corotate, then it would be moving at close to 1000 km s^{-1} at the jovian magnetopause. In practice, it moves at about half that value, around 500 km s^{-1}. On the dawn flank of the magnetopause, this velocity should help destabilize the Kelvin–Helmholtz instability, whose growth rate is proportional to the velocity shear. On the afternoon flank, this could be a stabilizing effect by reducing the velocity shear. Similar but smaller effects occur at Saturn, Uranus, and Neptune.

Finally, at Jupiter and Saturn, the centrifugal force changes the radial variation of pressure in the magnetosphere so that the sensitivity to changes in the solar-wind dynamic pressure is altered. The two magnetospheres become much more compressible, as shown for Jupiter in Figure 12.8. A very striking feature of the jovian magnetopause is its bimodal distribution of nose or standoff distances. Since the solar wind does not have a bimodal distribution of dynamic pressures, this bimodal distribution, shown in Figure 12.9, must be associated with a bimodal internal structure, such as two most probable mass-loading states. Saturn also has a slightly bimodal magnetopause standoff distance. The solar-wind pressure variations at Jupiter and Saturn are qualitatively similar and not bimodal, so the cause of the bimodality is not external.

We have already noted that the small size of Mercury's magnetic moment and its location relatively close to the Sun result in the potential for its magnetopause occasionally to reach the planetary surface at times of high solar-wind dynamic pressure. The frequency of this occurrence may be estimated from the statistics of the dynamic pressure at Mercury. However, since the magnetosphere is very responsive to the IMF direction, even for modest pressures of the solar wind, much of the surface can be contacted. Thus, a sputtered atmosphere may occasionally appear at Mercury and modify the solar-wind interaction.

Table 12.2. Characteristic flow times and velocities

Planet	Rotation rate (Hz)	Corotation velocity at magnetopause (km s^{-1})	Solar-wind flow time (min)
Mercury	1.96×10^{-7}	~0	0.1
Earth	1.26×10^{-5}	4	2
Jupiter	2.8×10^{-5}	923	200
Saturn	2.6×10^{-5}	196	45
Uranus	1.6×10^{-5}	65	25
Neptune	1.6×10^{-5}	62	23

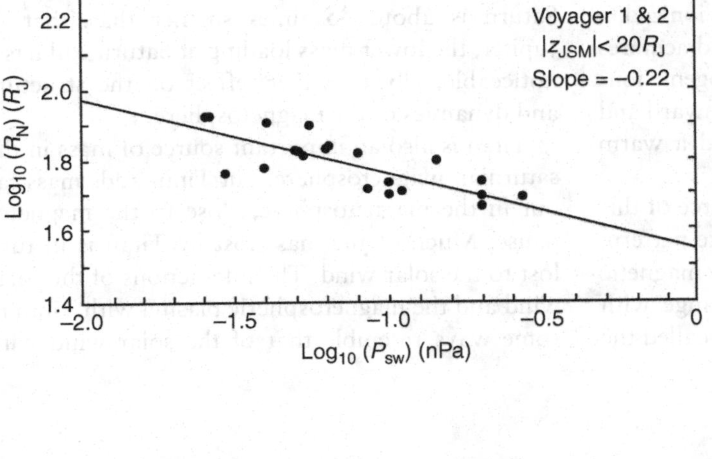

FIGURE 12.8. The location of the nose of the jovian magnetopause as a function of the solar-wind dynamic pressure using Voyager 1 and 2 data. The best-fit line has a slope of −0.22 on this log-log plot compared to the expected −0.16 for a dipole field in a vacuum. Subscript JSM refers to Jupiter solar magnetospheric coordinates; R_N is the distance to the subsolar point (nose) of the magnetosphere. (After Huddleston *et al.*, 1998.)

12.4 MASS LOADING AND CIRCULATION

Of all the cases in which material is added to the interiors of magnetospheres, the source of mass is

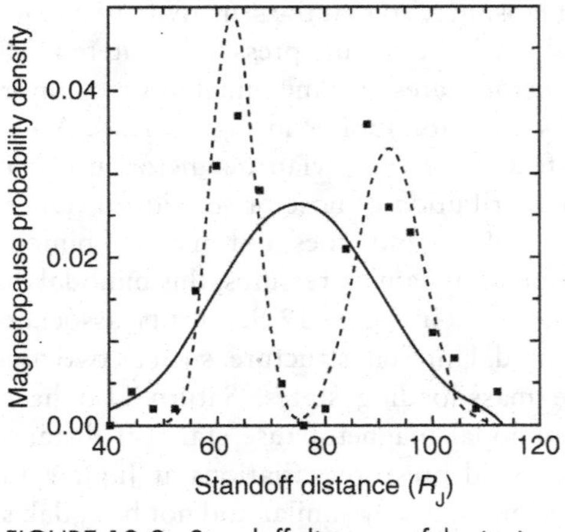

FIGURE 12.9. Standoff distance of the jovian magnetopause derived from Galileo observations (Joy *et al.*, 2002). The bimodal distribution function clearly fits the data better than the single gaussian. Such bimodal behavior is not seen in the solar wind and must be produced by internal magnetospheric processes.

greatest at the jovian moon Io, whose weak atmosphere is maintained by continual volcanic eruptions. The radiation-belt particles collide with atoms and molecules in that atmosphere and with the surface of Io, sputtering surface atoms that can then escape from Io and form an exosphere about the planet. Neutral atoms in the atmosphere become ionized, and the electric field of the corotating magnetosphere accelerates these particles into a torus about the planet. Corotating ions in the Io torus undergo charge exchange with neutral atoms. Any ionized particles that are neutralized by charge exchange will fly out of the torus (along straight lines tangent to their original circular orbits) and spread material throughout the equatorial plane of Jupiter. When this material again becomes ionized, it is accelerated to the local flow velocity and acquires significant thermal energy. In this way, iogenic ions are spread throughout the Io torus both inward and outward, creating a cold inner torus and a warm outer torus.

As described earlier, the centrifugal force of this material stretches the field into a disk-like pattern. The centrifugal force can also cause heavy magnetic flux tubes of the plasma torus to exchange with lighter tubes outside the torus in what is called the interchange instability. The source of greatest mass in the saturnian magnetosphere is the moon Enceladus that adds dust and gas to the E-ring with a plume emanating from its southern polar region. While the number of ions added to the magnetosphere by Enceladus is many fewer than by Io at Jupiter, the weaker saturnian field is significantly stretched in a similar way.

The stress balance in a thin current sheet was first applied to Jupiter by Vasyliunas (1983), and was later used to measure the mass of the jovian current sheet by Russell *et al.* (1999b) and that of the saturnian current sheet by Arridge *et al.* (2007). In a rotating frame, the time-stationary momentum equation can be written as follows:

$$\boldsymbol{\rho}\boldsymbol{\Omega} \times (\boldsymbol{\Omega} - \mathbf{r}) - \mathbf{j} \times \mathbf{B} + \nabla P = 0,$$

where $\boldsymbol{\rho}$ is the mass density, Ω is the angular velocity of the "corotating" plasma, $\mathbf{B}$ is the magnetic field, $\mathbf{j}$ is the current density, and $\mathbf{P}$ is the pressure tensor. Here, we assume that centrifugal and plasma forces far exceed the force of gravity. The stress balance in the radial direction becomes

$$\rho\Omega^2 r = -\frac{B_z}{\mu_0}\frac{\partial B_r^{\mathrm{cs}}}{\partial z} + \frac{\partial P}{\partial r},$$

where the centrifugal force is being balanced by the combination of the vector cross product of the current in the sheet, times the field strength perpendicular to the sheet, plus the radial gradient of plasma pressure in the sheet. This can be solved for the mass density and integrated to obtain the mass of the current sheet. At $20R_J$, Jupiter has about 20 000 tons of plasma per jovian radius in the current sheet, and at Saturn at $20R_{Sat}$ there are about 50 tons. Because the magnetic moment of Saturn is about 35 times smaller than that of Jupiter, the lower mass loading at Saturn still has a noticeable, albeit weaker, effect on the structure and dynamics of the magnetosphere.

Titan is also an important source of mass in the saturnian magnetosphere, but Titan adds mass far out in the magnetosphere, close to the magnetopause. Much of the mass lost by Titan is in turn lost to the solar wind. The interactions of the solar wind and the magnetospheric plasma with Titan in some ways resemble that of the solar wind with

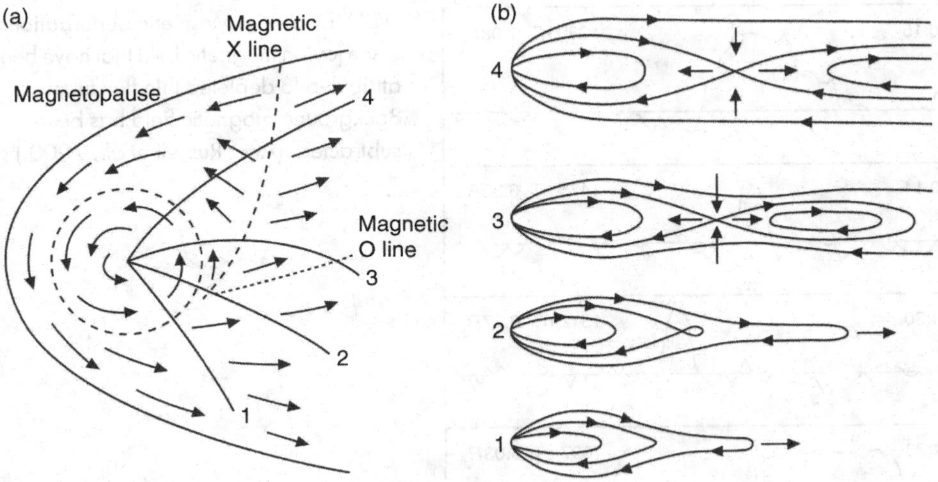

FIGURE 12.10. Schematic of the escape of plasma down the jovian tail via reconnection. The magnetic meridians in (b) show the sequence of neutral-point and plasmoid formation. This process can proceed in a steady state, as envisioned here. The equatorial plane view (a) shows the locations of the four meridian planes and the locations of the X and O neutral lines. (After Vasyliunas, 1983.)

Venus, albeit with slightly different Mach numbers and dynamic pressures. We discuss these moon–magnetosphere interactions in greater detail in Section 12.9.

The moons of the gas giants not only provide mass, but also absorb energetic radiation. Radiation-belt electrons and ions drift relative to the moons and spiral into them to be absorbed. This absorption leaves narrow gaps in the radial distribution of radiation-belt fluxes. The gaps are well defined at Saturn, where the dipole field is centered on the planet and is aligned with the spin axis, but less well defined in other magnetospheres. The speed with which the gaps are filled places constraints on the rate of radial diffusion in the magnetosphere. The rings of Saturn are also efficient absorbers of the radiation-belt particles and limit the buildup of intense fluxes of energetic particles in the inner radiation zone of Saturn.

Finally, we must address the question of what becomes of all that mass that is added to the jovian and saturnian magnetosphere. The mass density cannot build up forever. It must be dumped from the system. In the Earth's magnetosphere, hot plasma is lost through precipitation by the scattering of particles into the loss cone, where the particles can collide or charge exchange with atmospheric particles, thereby depositing energy in the atmosphere. Some may charge exchange with exospheric particles high in the magnetosphere. Cold plasma can sink into the atmosphere and recombine, but much of it in the outer plasmasphere is convected to the reconnecting dayside magnetopause and is lost. At Jupiter and Saturn, we find similar processes but with critical differences. The plasma that is picked up is relatively cold and cannot easily be scattered into the loss cone. It can charge exchange after acceleration when it is in a moon's atmosphere, but generally there is not much neutral density elsewhere. Thus, much of the plasma must be transported by convection. Figure 12.10 shows the convective path followed. Once the plasma reaches the distant tail, reconnection occurs, forming mass-loaded magnetic islands. These islands are lost from the tail once the island is disconnected from the magnetosphere. This leaves us with a dilemma because magnetic flux has been transported to the magnetotail. Now it must make its way back against the outflowing mass-loading plasma. It can do this in three different ways: an inward flow at a different longitude than the outward flow, by interchanging adjacent "full"

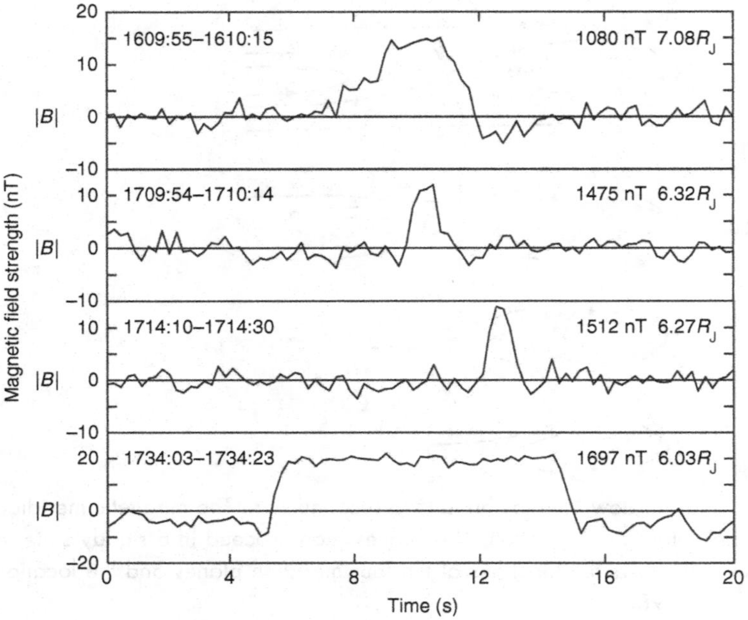

FIGURE 12.11. Magnetic perturbations in the jovian magnetic field that have been attributed to depleted thin flux tubes. Background magnetic field has been subtracted. (After Russell *et al.*, 2000.)

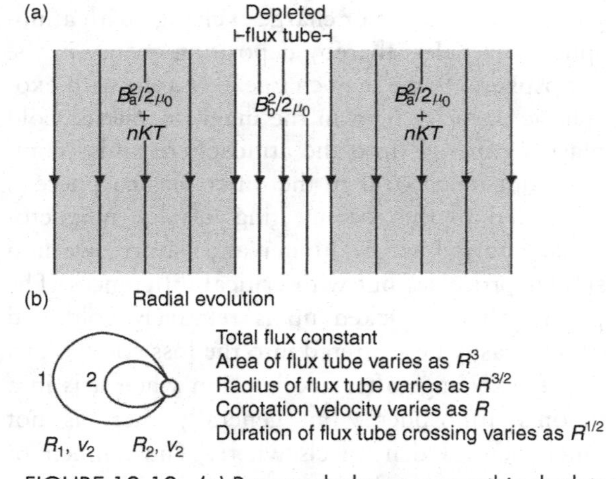

FIGURE 12.12. (a) Pressure balance across thin depleted flux tubes. (b) "Dipolar" magnetic flux tube at times 1 and 2, as it moves buoyantly inward through the Io torus. Subscripts "a" and "b" on B denote regions outside and inside depleted flux tube, respectively.

and "empty" flux tubes, or by the rapid inward transport of thin tubes penetrating the outward flow. An example of these thin tubes in shown in Figure 12.11. Their physical makeup is explained in Figure 12.12.

12.5 SUBSTORMS

As discussed in Chapter 8, a terrestrial substorm is a sequence of events in which energy is stored in the magnetosphere and then released. The energy storage process involves dayside reconnection of the interplanetary magnetic field with the terrestrial magnetic field. In contrast to the Earth, where the IMF orientation is key, at Jupiter and Saturn the major driver of convection in the magnetosphere is the mass-loading process. As we have seen in Section 12.4, internal reconnection appears to play a role in removing ions from the magnetosphere, and releasing them down the tail and returning magnetic flux to the inner magnetosphere for further mass loading. As sketched in Figure 12.10, the picture is steady state, analogous to Dungey's (1961) model of the reconnecting terrestrial magnetosphere. Could the mass-loading process also lead to substorm-like behavior?

Figure 12.13 shows two examples of strong reconnection events in the tail of Jupiter and at Saturn. These events represent extremely strong magnetic reconnection, as the magnetic field more than doubles and becomes vertical and not radial. Figure 12.14 shows our interpretation

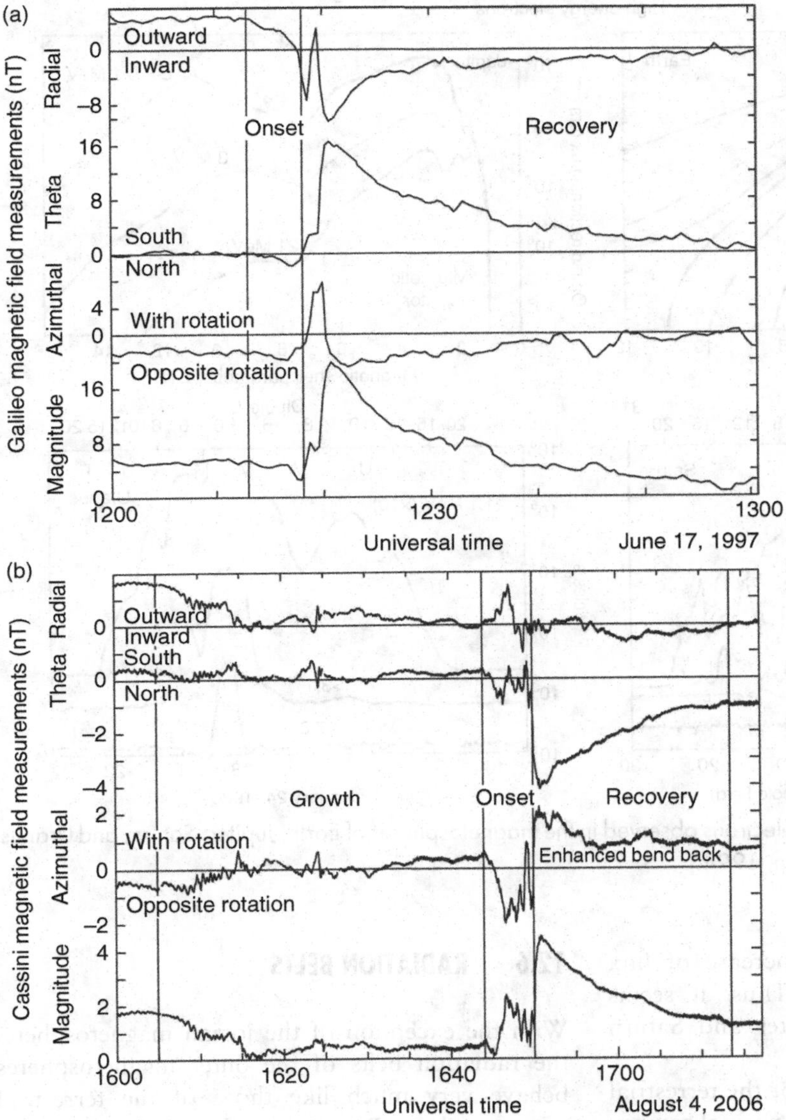

FIGURE 12.13. Example of a strong reconnection event in the tail of Jupiter (a) and Saturn (b) on either side of the neutral point.

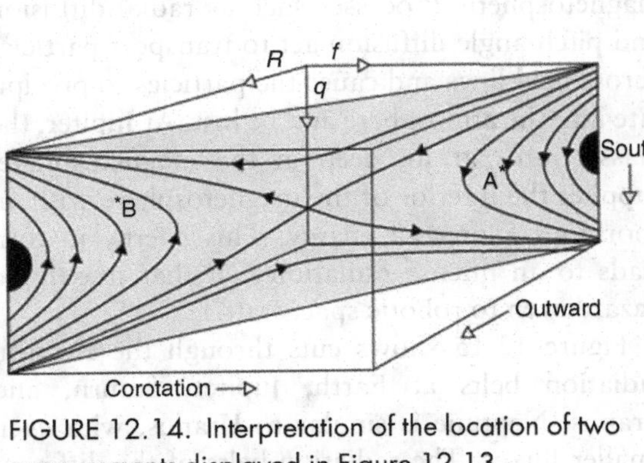

FIGURE 12.14. Interpretation of the location of two measurements displayed in Figure 12.13.

of these two events. The saturnian event occurred outside the neutral point, and the jovian event occurred inside. The magnetic field records in this and other examples indicate that, as the field snaps outward and inward, the conservation of angular momentum bends the field forward and back from the reconnection site. Also, there is evidence that the magnetic field strength in the lobes decreases. This decrease suggests that the magnetic flux content of the tail is lowered by the reconnection activity.

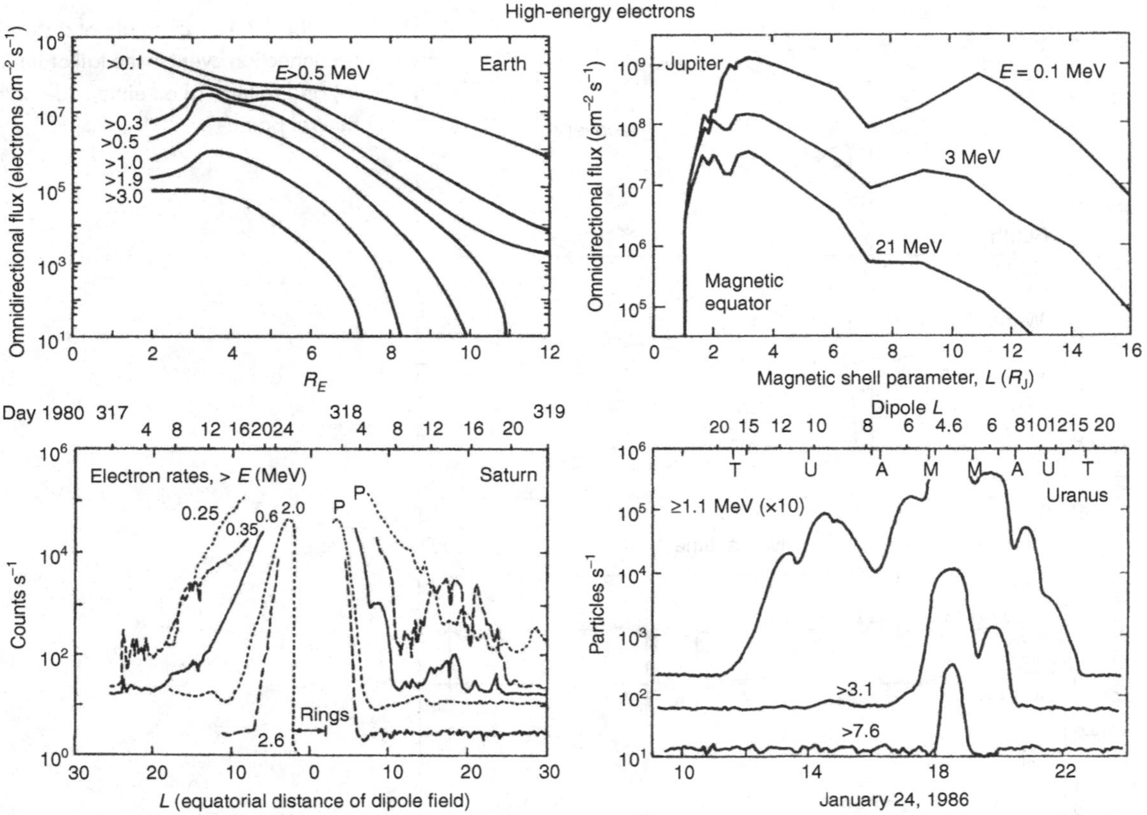

FIGURE 12.15. Fluxes of high-energy electrons observed in the magnetospheres of Earth, Jupiter, Saturn, and Uranus. (D. J. Williams, personal communication, 1994.)

There is also evidence for an increase of flux before the reconnection event. Thus, it seems safe to conclude that both Jupiter and Saturn have substorms.

These events provide a lesson for the terrestrial magnetosphere. The strength of the normal component change indicates that the reconnection rate is controlled by the plasma conditions, the near vacuum condition in the flux lobe. Thus, in the terrestrial substorm when reconnection begins in the center of the plasma sheet, we would not expect reconnection to occur rapidly. However, as the plasma sheet thins due to this reconnection, and therefore the plasma becomes less dense and the magnetic field becomes stronger, the reconnection rate should increase, leading to an explosive onset of the substorm, even though the initiation of reconnection occurred some minutes earlier.

12.6 RADIATION BELTS

With the exception of the jovian magnetosphere, the radiation belts of the outer magnetospheres behave very much like those of the terrestrial magnetosphere. Processes such as radial diffusion and pitch angle diffusion act to transport particles across field lines and cause the particles to precipitate into the atmosphere and be lost. At Jupiter, the mass source at Io, deep in the magnetosphere, supplies the interior of the magnetosphere with an enormous source of energy. This energy in turn leads to an intense radiation belt that presents a hazard even to robotic spacecraft.

Figure 12.15 shows cuts through the electron radiation belts at Earth, Jupiter, Saturn, and Uranus. Neptune is similar to Uranus, with even smaller fluxes. The radiation belts of the different planets are similar. The fluxes are most intense just above the atmosphere (except at Saturn, where the fluxes maximize just outside the rings). At lowest altitudes, the spectrum is harder; that is, the flux decreases less sharply with increasing energy than at high altitudes. However, when one considers the peak fluxes, there is a great disparity. As shown in Table 12.3, the peak electron flux at Jupiter is about 1000 times greater than that at Earth (>3 MeV), and the peak flux at Uranus is an order of magnitude less than that at Earth.

Table 12.3. Peak energetic particle fluxes

	Electrons		Protons	
Planet	Flux (cm^{-2} s^{-1})	Energy (MeV)	Flux (cm^{-2} s^{-1})	Energy (MeV)
Earth	10^5	≥3	10^4	≥105
Jupiter	10^8	≥3	10^7	≥80
Saturn	10^5	≥3	10^4	≥63
Uranus	10^4	≥3	<10	≥63

A similar story holds for the protons, as shown in Figure 12.16. The radiation belts look grossly similar, but the fluxes shown in Table 12.3 reveal an excess of three orders of magnitude at Jupiter and a deficit of three orders of magnitude at Uranus. It is clear that particles in the magnetospheres of Uranus and Neptune do not become highly energized, even when compared with the terrestrial radiation belts, which are formed in a much smaller magnetosphere. Perhaps the reason for this difference is the inefficiency of magnetopause reconnection in the outer heliosphere, about which we speculated earlier.

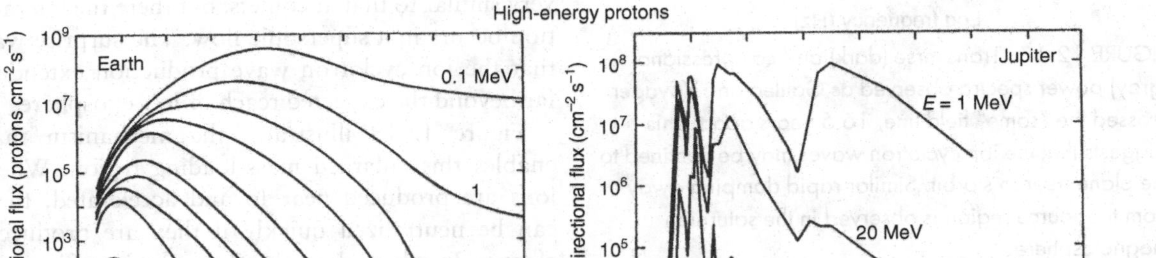

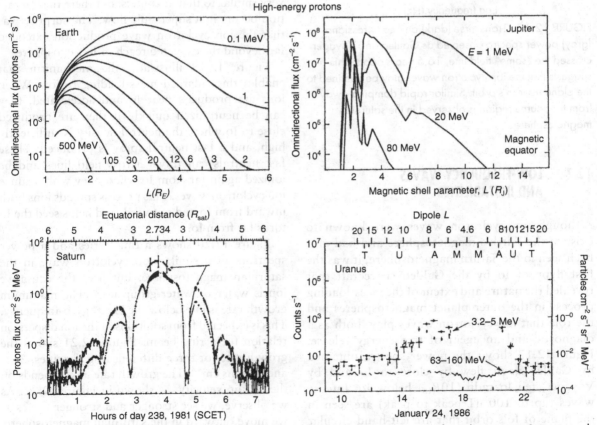

FIGURE 12.16. Fluxes of high-energy protons observed in the magnetospheres of Earth, Jupiter, Saturn, and Uranus. (D. J. Williams, personal communication, 1994.)

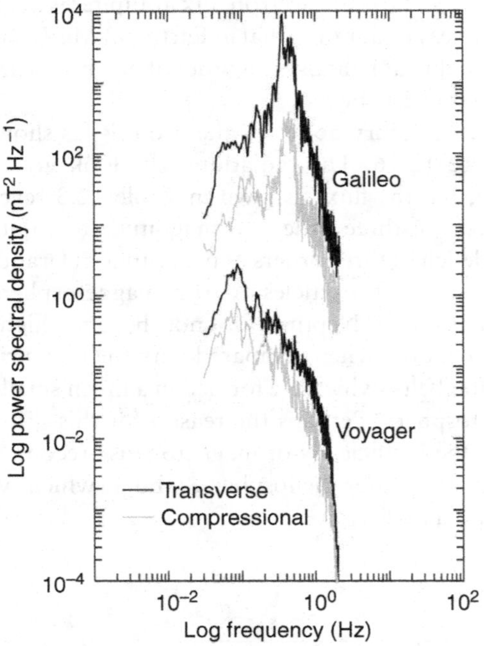

FIGURE 12.17. Transverse (dark) and compressional (gray) power spectra observed as Galileo and Voyager crossed the "same" field line, 16.5 years apart. This suggests that the ion cyclotron waves may be confined to the plane near Io's orbit. Similar rapid damping away from the source region is observed in the saturnian magnetosphere.

12.7 LOW-FREQUENCY WAVES AND INSTABILITIES

Although ion cyclotron waves were known to exist in the Earth's magnetosphere and had also been seen in the Saturn magnetosphere, it was the first flyby of Io by the Galileo spacecraft that revealed the nature and extent of the mass-loading process in the outer planet magnetospheres and the role that ion cyclotron waves play, both as a diagnostic and an agent of free energy release. Figure 12.17 shows the power spectra obtained by Galileo as it flew by close to Io and by Voyager, ten Io radii ($10R_{Io}$) below Io. Intense waves (up to 100 nT peak to peak) are seen in the plane of Io's orbit that are left-hand circular at the gyro-frequency of SO_2^+ ions. At Voyager no such waves are seen. This difference is not just a question of distance as the ion cyclotron waves were observed at Galileo for a radial extent of $30R_{Io}$; rather, the pickup process is nearly two dimensional and restricted to a plane.

Figure 12.18 illustrates how this takes place. The moon Io moves in its orbit around Jupiter at its keplerian velocity of 17 km s^{-1}. The magnetospheric plasma rotates at 74 km s^{-1}, corresponding to the angular velocity of the ionosphere to which it is strongly coupled by the jovian magnetic field. When ions are produced by impact ionization, photoionization, or charge exchange with a neutral in the Io atmosphere, the new ion feels the electric field of the corotating plasma and is accelerated into a ring in velocity space. This particle may have very little velocity along the field. With time, this free energy is liberated in the form of electromagnetic ion cyclotron waves, and the velocity–space distribution becomes more maxwellian. This behavior is very similar to that at comets, but there the interaction occurs in a supersonic flow. The surprise was that the ion cyclotron wave production extended far beyond the expected reach of Io's exosphere.

Figure 12.19 illustrates the mechanism that enables this enlarged mass-loading region. When ions are produced near Io and accelerated, they can be neutralized quickly if they are produced close to Io where the neutral density is sufficiently high, and a fast neutral spray is produced. These fast neutrals can travel across field lines and be ionized again far from Io where they will produce ion cyclotron waves. This process spreads ions both inward from Io and outward, and helps seed the Io torus far from Io.

Figure 12.20 shows a time series and a power spectrum of a similar ion cyclotron wave in the saturnian magnetosphere, but here the causative ion is water or water-group ions. The maximum growth rate is just below the H_2O^+ gyro-frequency. This is expected from solutions of the ion dispersion relation for a ring beam. Figure 12.21 shows the growth rate for three different ring densities: 3, 5, and 10 ions cm^{-3}. The growth rate also depends on the pickup velocity, as shown in Figure 12.22. As we observe, we see stronger and stronger waves as we move outward in the saturnian magnetosphere, past Enceladus, where most of the added ions are produced because the free energy per ion is greater.

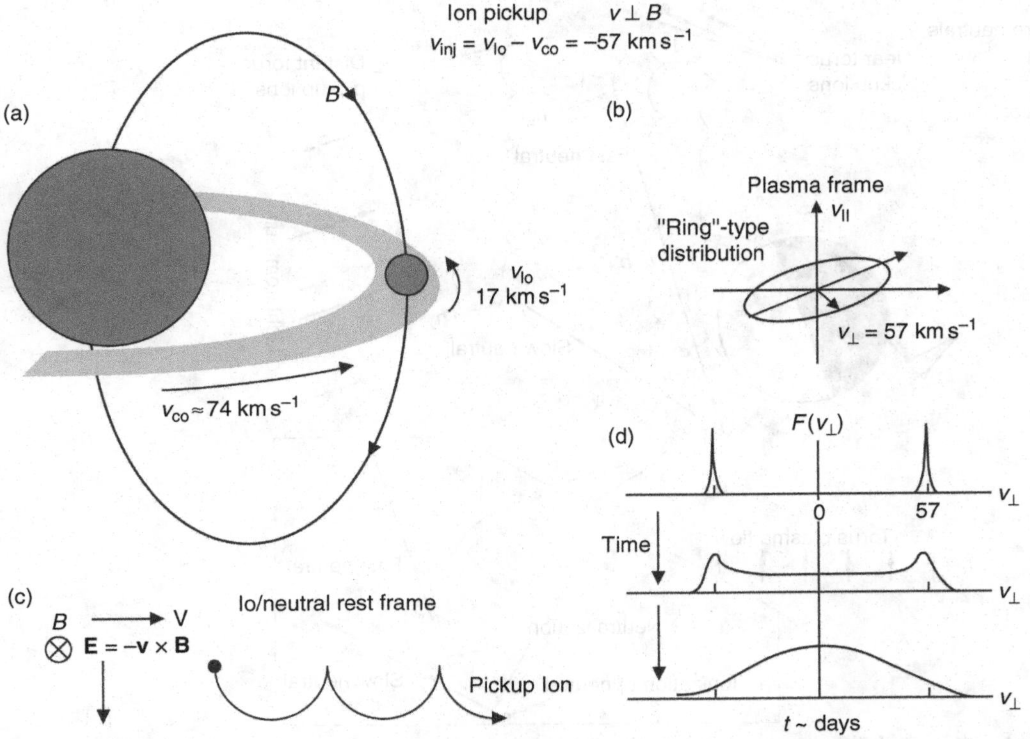

FIGURE 12.18. The ion pickup mechanism at Io. (a) Geometry of the interaction of the corotating plasma with Io. (b) Motion of the particle in velocity space. (c) Motion of the particle in configuration space. (d) Temporal evolution of the distribution function of the particles plotted versus the perpendicular velocity.

The same anisotropy can lead to **mirror-mode waves,** but the growth rate is lower. Figure 12.20 shows a weak mirror-mode wave, labeled MM. Surprisingly, this weak wave growth can lead to a large wave amplitude because of the different directions of propagation of the waves. The ion cyclotron waves move along the magnetic field and eventually damp. The mirror-mode waves are convected outward in the equatorial plane and continue growing. Figure 12.23 shows an example of such waves beyond the region in which ion cyclotron waves are growing.

The radial extent of ion cyclotron wave growth is about 50 000 km at Jupiter and about 120 000 km at Saturn. Since there is a density threshold for wave growth, the actual transport distance of the underlying fast neutrals that make the ring beams may be much greater. In fact, it is quite likely that most of the saturnian neutrals escape from the planet before reionizing, lessening the need for convection to the tail and subsequent dumping of the ions that have been produced in the equatorial plane.

12.8 PLASMA WAVES AND RADIO EMISSIONS

In the generation of plasma waves slightly above the proton gyro-frequency to slightly above the electron plasma frequency, the magnetospheres of the outer planets behave much as the terrestrial magnetosphere does. The distribution functions of the trapped plasma and the various beams and currents flowing in these magnetospheres are not unlike those for the terrestrial magnetosphere. Thus, these magnetospheres produce many, if not all, of the same plasma instabilities as that of Earth. Plasma processes in the jovian magnetosphere seem most intense overall. These waves are responsible for the pitch angle diffusion of energetic

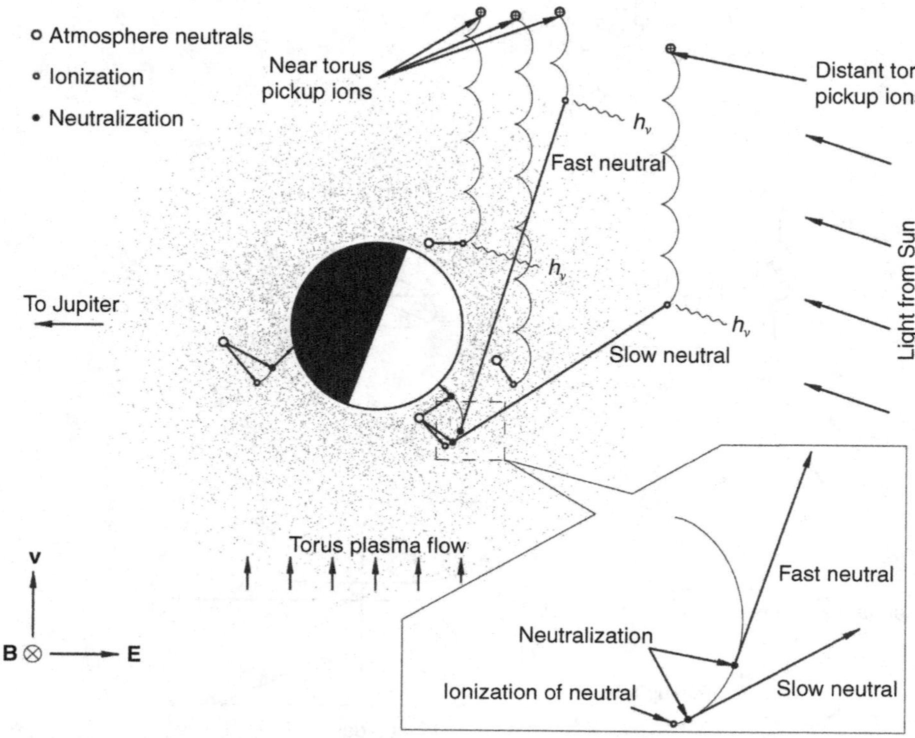

FIGURE 12.19. Mechanism for forming a thin mass-loading disk that enables the production of pickup ions far to the side
of Io.

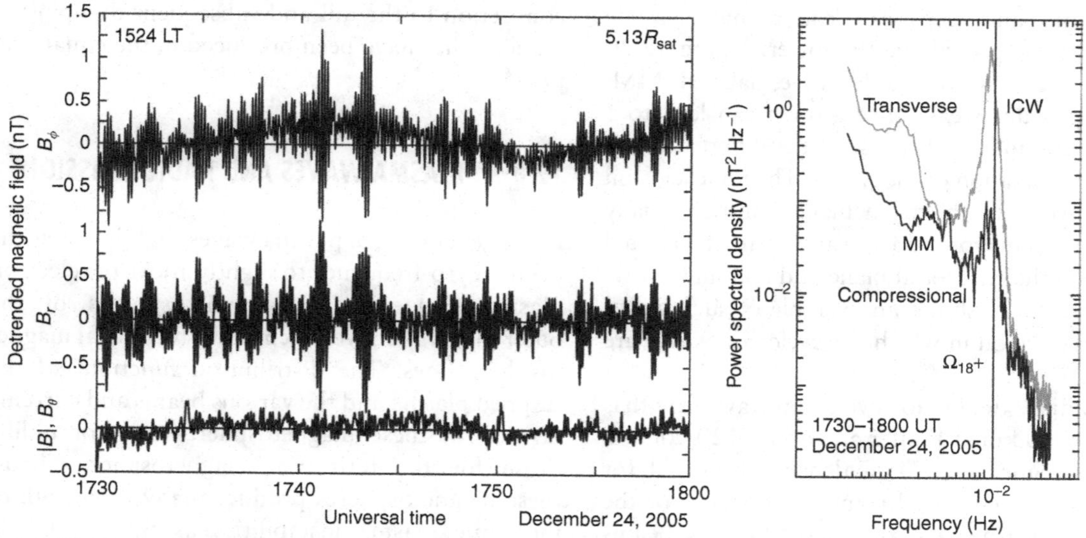

FIGURE 12.20. Detrended 1 s magnetic field measurements in Saturn's E-ring region showing transverse ion cyclotron waves (ICW) due to water-group ion pickup. (a) Time series; (b) power spectrum. Ω_{18^+} denotes the singly ionized water molecule gyro-frequency.

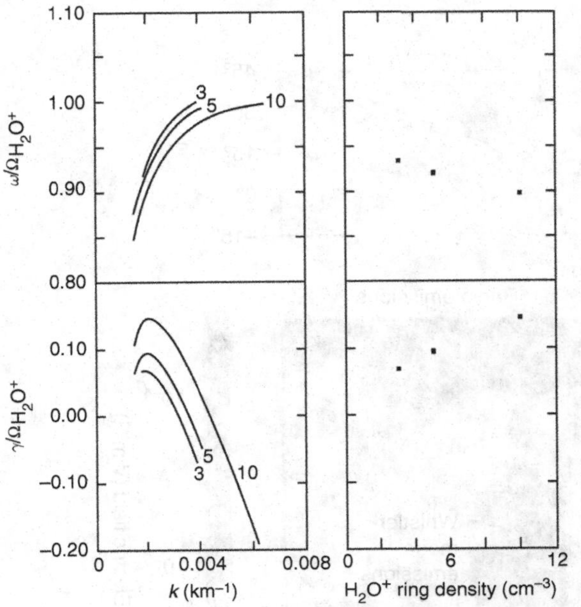

FIGURE 12.21. Growth rate of ring-beam distributions of varying density for conditions appropriate to the E-ring torus. The WHAMP dispersion solver (Ronnmark, 1982) was used for this calculation.

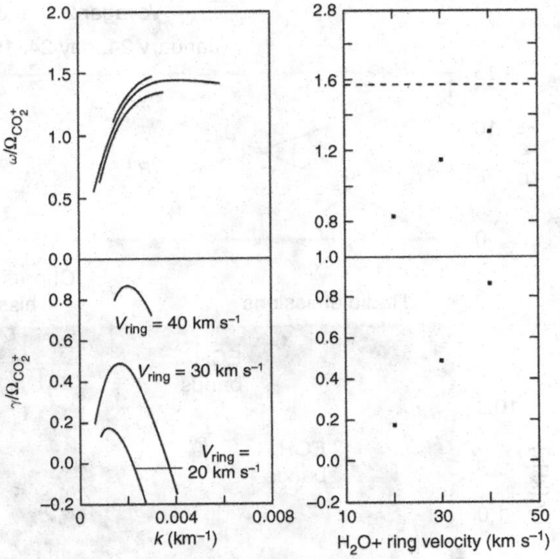

FIGURE 12.22. Growth rate of ring-beam distributions of varying pickup speeds for conditions appropriate to the E-ring torus. The WHAMP dispersion solver (Ronnmark, 1982) was used for this calculation.

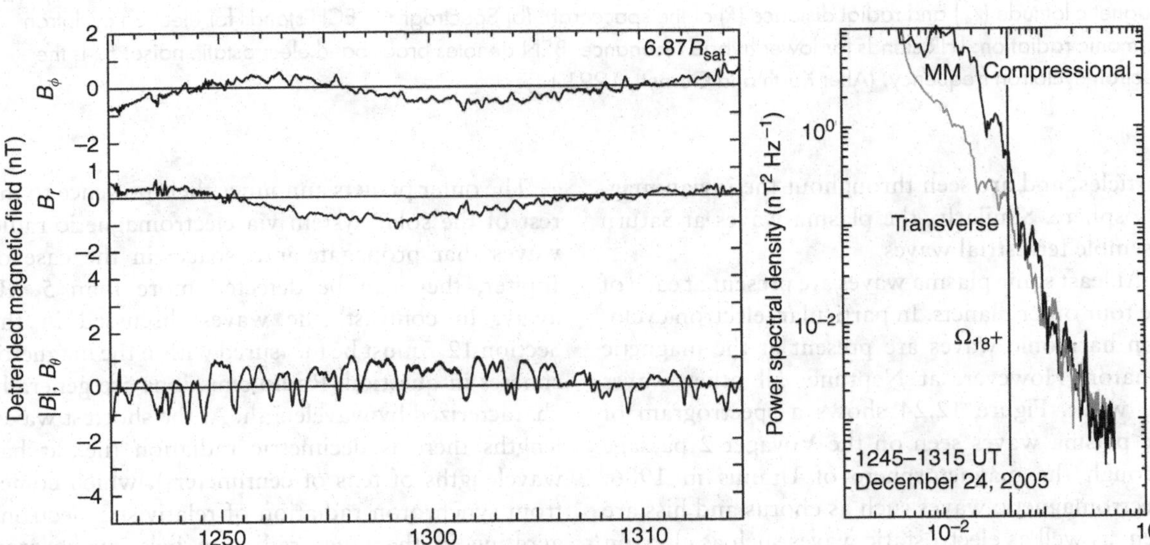

FIGURE 12.23. Detrended 1 s magnetic field measurements in Saturn's E-ring region showing compressional mirror-mode fluctuations. (a) Time series; (b) power spectrum. Ω_{18^+} denotes the singly ionized water molecule gyro-frequency.

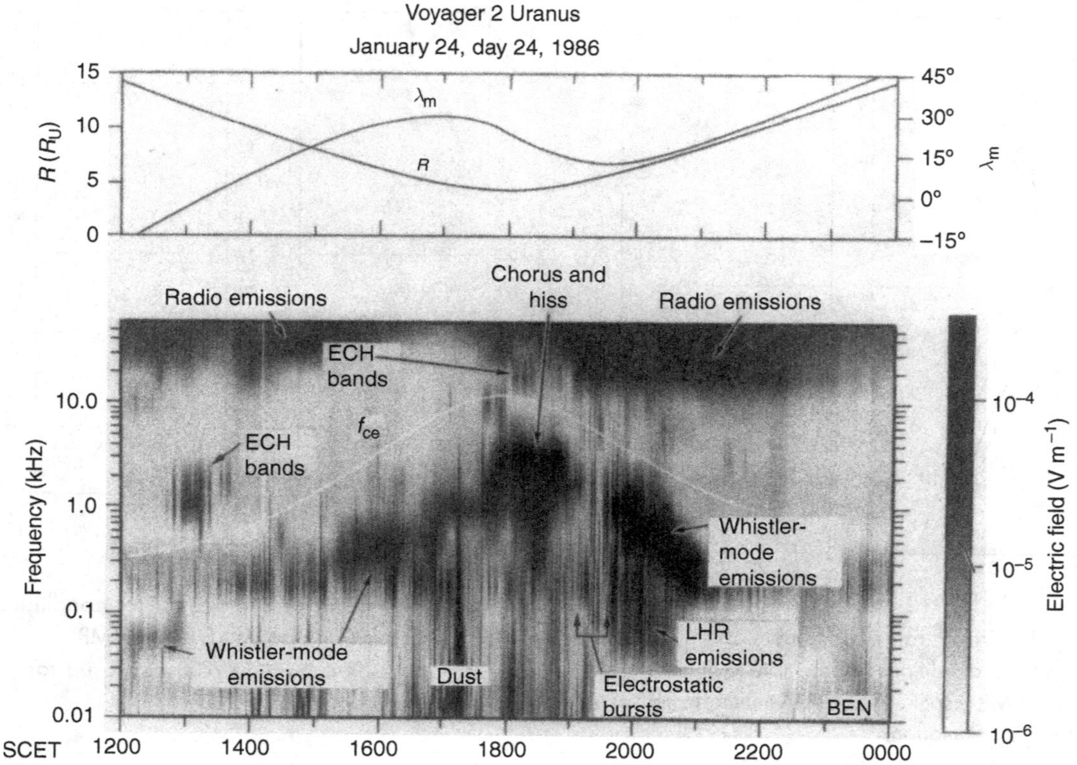

FIGURE 12.24. Overview of the plasma-wave spectrum observed by Voyager 2 in the magnetosphere of Uranus. (a) Magnetic latitude (λ_m) and radial distance (R) of the spacecraft. (b) Spectrogram. ECH stands for electron cyclotron harmonic radiation; LHR stands for lower hybrid resonance. BEN denotes broadband electrostatic noise; f_{ce} is the electron cyclotron frequency. (After Kurth and Gurnett, 1991.)

particles, and are seen throughout the jovian magnetosphere. Similarly, the plasma waves at Saturn resemble terrestrial waves.

At least some plasma waves are present at each of the four outer planets. In particular, electron cyclotron harmonic waves are present at the magnetic equator. However, at Neptune, all other waves are weak. Figure 12.24 shows a spectrogram of the plasma waves seen on the Voyager 2 passage through the magnetosphere of Uranus in 1986. Electromagnetic waves such as chorus and hiss are seen, as well as electrostatic waves such as electron cyclotron harmonic waves and **lower hybrid resonance** emissions. The generation of these waves is thought to be due to processes that are similar to those in Earth's magnetosphere, as discussed in Chapter 13.

The outer planets announce their presence to the rest of the solar system via electromagnetic radio waves that propagate into space; in the case of Jupiter, they can be detected more than 5 AU away. In contrast, the waves discussed in the Section 12.7 must be measured within the magnetospheres in question. Radio emissions are generally characterized by wavelength. At the shortest wavelengths there is decimetric radiation (i.e., it has wavelengths of tens of centimeters), which comes from synchroton radiation of relativistic electrons gyrating in the inner radiation belt. Synchroton radiation led to the discovery of the magnetosphere of Jupiter in the late 1950s, well before spacecraft were sent to the planet.

Decametric and hectometric radiations have wavelengths from 10–100 m. These emissions

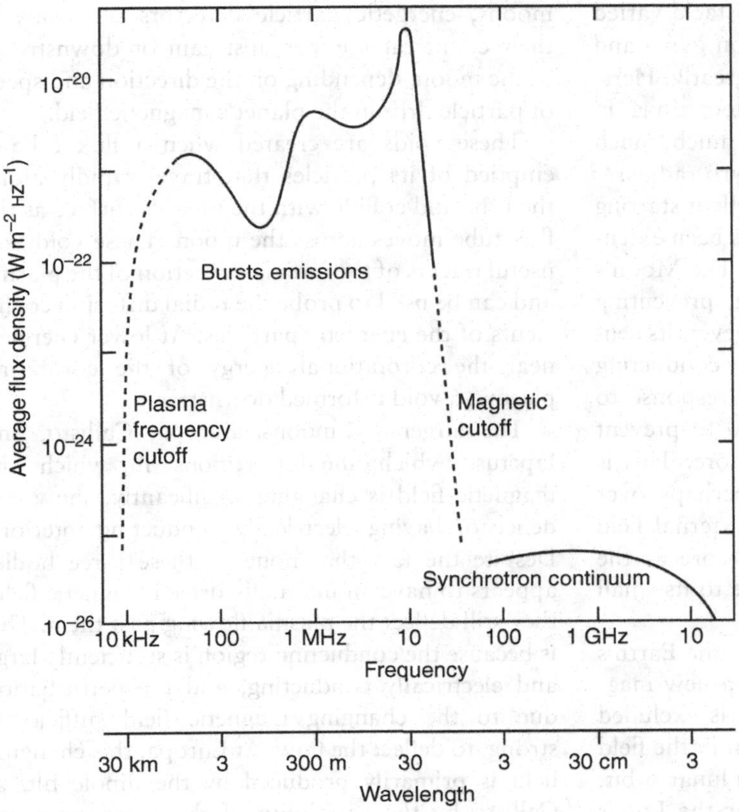

FIGURE 12.25. Average power flux density spectrum of Jupiter's non-thermal magnetospheric radio emissions. The instantaneous spectrum may be quite different. (After Carr, Desch, and Alexander, 1983.)

have much temporal and frequency structure and seem to be associated with instabilities in the plasma. Voyager 1 and 2 showed that this radiation extends to even longer (kilometer) wavelengths that cannot be detected at Earth. Figure 12.25 shows a spectrum of the jovian radio-wave flux versus frequency, illustrating the great intensity of decametric and longer-wavelength emissions.

The decametric emissions from Jupiter have been found to vary in frequency and in occurrence rate, and their variations have been analyzed extensively. Suffice it to say that the radiation received at Earth depends on the longitude of Jupiter of the Jupiter–Earth line and also on the location of Io in its orbit about Jupiter. It is suspected that instabilities in field-aligned currents connecting Io to the jovian ionosphere are responsible for many of these waves.

12.9 MOON–MAGNETOSPHERE INTERACTIONS

In Section 12.4 we discussed the effects that mass loading of the jovian and saturnian magnetospheres by Io and Enceladus have on the structure and dynamics of these magnetospheres. Basically, these moons are the engines that power the circulation and energetic phenomena in these magnetospheres. However, the addition of ions to the magnetospheres is only one aspect of the interaction. There are other aspects that depend on the properties of the moons and the properties of the magnetospheres and which can be used to learn more about the interiors of the moons. This is important because, until we can place landers on the surfaces of these moons with seismometers, we have only gravity and magnetic measurements to probe the interiors. In Chapter 7, we discussed the role of size in determining the nature of the interaction and found that, as the size of the obstacle varied from well below to well above the ion gyro- and inertial scales, the interaction varied greatly. Here, we examine only plasma–moon interactions in which the scale size of the obstacle is much, much larger than the ion inertial length or gyro-radius.

The Earth's moon serves as an excellent starting point for our discussion, because it has been extensively studied and is well understood. The Moon's surface is not electrically conducting, preventing current flow through the surface. However, its central region, or core, is highly electrically conducting and currents can flow in the core in response to changes in the external magnetic field to prevent the changed field from entering the core. This is known as Lenz's law. Eventually, perhaps over thousands of years, a change in the external field could penetrate to the center of the core as the currents on and in the core decay, due to its small finite resistivity.

When the Moon enters the lobes of the Earth's magnetotail, the Moon finds itself in a new magnetic environment. The external field is excluded from the core and a small perturbation in the field can be sensed from a magnetometer in lunar orbit, such as on the Apollo subsatellites or the Lunar Prospector missions, as illustrated in Figure 8.2. These perturbations are consistent with a core of about 400 km, or about 25% of the lunar radius.

Even though today there is no active magnetic dynamo in the lunar core, there may once have been a dynamo, and parts of the crust, when raised above the Curie temperature or blocking temperature, may have become magnetized as the crust cooled. Thus, even a body with quiescent core may have regions of weak magnetic field on the surface. If these fields are stronger than the ambient magnetic field, energetic charged particles moving along the field toward the Moon are reflected before they are absorbed by the surface. By measuring the range of pitch angles over which the reflection occurs, the ratio between the surface field strength and that at the spacecraft can be determined. This has been used to measure the strength of the remanent magnetic field of the surface of the Moon.

Returning to the magnetospheres of Jupiter and Saturn, we can begin now to interpret much of what we see. For example, at the smaller icy moons, energetic particle detectors see voids in their count rates either upstream or downstream of the moon, depending on the direction and speed of particle drift in the planet's magnetic field.

These voids are created when a flux tube is emptied of its particles that travel rapidly along the tube and collide with the moon's surface as the flux tube moves across the moon. These voids are useful tracers of the radial convection of the plasma and can be used to probe the radial diffusion coefficients of the energetic particles. At lower energies, near the corotational energy of the corotating plasma, a void is formed downstream.

The larger icy moons, Europa, Callisto, and Iapetus, which inhabit regions in which the magnetic field is changing significantly, show evidence of having electrically conducting interiors. Despite the fact that none of these three bodies appears to have an internally driven magnetic field, they still deflect the plasma flowing past them. This is because the conducting region is sufficiently large and electrically conducting, and the perturbation due to the changing magnetic field sufficiently strong, to deflect the flow. At Europa, this changing field is primarily produced by the dipole tilt, at Callisto by the variability of the outer magnetosphere, and at Iapetus by the varying interplanetary magnetic field. Figure 12.26 shows the interaction of Europa with the dense plasma at the center of the **jovian plasma torus**, where the dynamic pressure compresses the field against the magnetic field excluded from the conducting one. The Mach number of the flow here is close to unity, and weak shocks are forming. Away from the current sheet, the dynamic pressure is less and the interaction more similar to that discussed previously for the Earth's moon. These signatures have been used to determine the size of the conducting regions within these bodies, which must be global oceans nearly to the surfaces of the bodies.

Before we turn to the moons with atmospheres, we should discuss Ganymede, at present the only moon with its own global intrinsic magnetic field. Figure 12.27 shows a model of this magnetic field. This magnetosphere within a magnetosphere is large enough to have a tail and a radiation belt. Perhaps this is similar to the magnetic state of the Earth's moon when it had self-sustaining dynamo,

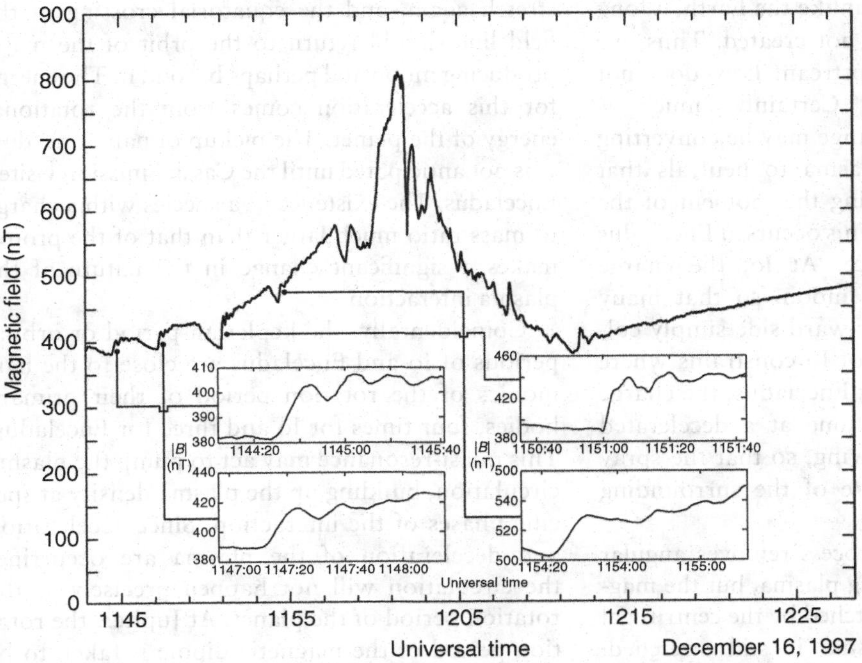

FIGURE 12.26. Interaction of Europa with the dense plasma at the center of the jovian torus/plasma sheet. Insets show four reformations of a weak bow shock.

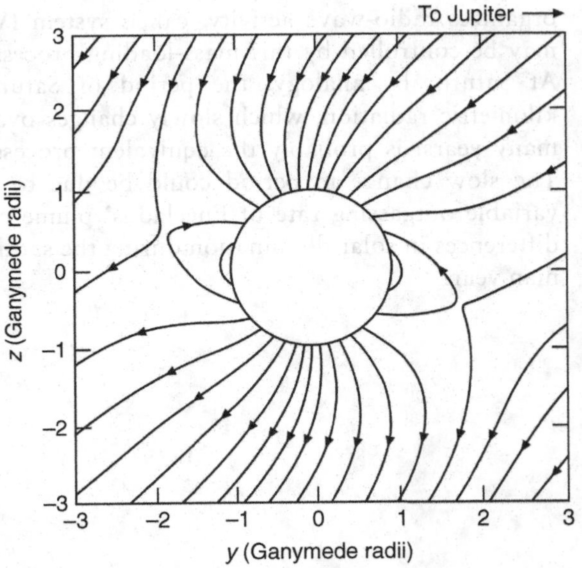

FIGURE 12.27. The interconnection of Ganymede's intrinsic magnetospheric with that of Jupiter.

but possibly not. Jupiter imposes a constant external field on Ganymede, and this field could be amplified by the fluid motions inside Ganymede, even if the dynamo were not self-sustaining.

Io, Enceladus, and Titan have volcanically supplied, plume-driven, and massive primordial atmospheres, respectively, and they exist in quite different plasma and magnetic field environments. As noted above, Io and Enceladus are the engines for their magnetospheres, but we did not describe how their pistons work. Titan is not an engine because of its position near the outer limits of the Saturn magnetosphere; nevertheless, it has interesting effects on the dynamics of the magnetosphere. It appears to help stand off the solar-wind flow when it is near the subsolar region, and it appears to control the timing of substorm onsets in the Saturn magnetotail when Titan is at midnight. To zeroth order, the Titan interaction is like a subsonic version of the Venus and Mars solar-wind interactions, and we do not discuss it further herein.

The Io interaction is subsonic and also very low beta, so that the magnetic field is only slightly distorted by the interaction. Flux tubes are convected across the polar caps of Io, much like the magnetic flux tubes of the Earth are convected across the poles stirred by magnetic reconnection with the solar wind. Like on Earth, the flow over the polar cap is slow, but unlike the Earth, a long tail parallel to the flow is not created. Thus, we know that most of the upstream flow does not cross the polar region. Certainly, much is deflected, but charge exchange may be converting much of the incoming plasma to neutrals that move away from Io carrying the moment of the incoming flow with them. This occurs at Enceladus too, but with a difference. At Io, the charge exchange occurs near the moon, so that many fast neutrals on the Jupiter-ward side simply collide with Io. The shadow of Io constrains where the fast neutrals can go. At Enceladus, the charge exchange occurs in the plume at a decelerated bulk speed with no shadowing, so that the spray of neutrals can reach more of the surrounding magnetosphere.

The change exchange process removes angular momentum from the flowing plasma, but the magnetic field that has been stretched by the centrifugal force exerted by the plasma is still stretched. Further, it may pick up nanoscale dust as well as plasma. The flow may be turned toward the planet and a cold dense plasma region formed interior to each of these moons. At Io, this region is called the ribbon, but at Enceladus it does not have a name. The cold dense plasma is initially traveling only slightly faster than the keplerian orbital velocity, but, since the surrounding plasma and the ionosphere on either end of the field lines are rotating at the planetary rotation period, the flux tube speeds up; as it does, the field line should begin to stretch again, and the equatorial crossing of the field line should return to the orbit of the mass-producing moon and perhaps beyond it. The energy for this acceleration comes from the rotational energy of the planet. The pickup of nanoscale dust was not anticipated until the Cassini mission visited Enceladus. The existence of a species with a charge to mass ratio much larger than that of the proton makes a significant change in the nature of the plasma interaction.

Coincidentally, the keplerian period or orbital periods of Io and Enceladus are close to the harmonics of the rotation period of their primary bodies, four times for Io and three for Enceladus. This quasi-resonance may act to pump the plasma circulation, building up the plasma density at specific phases of the interaction. Since acceleration and deceleration of the plasma are occurring, the circulation will not happen precisely at the rotation period of the planet. At Jupiter, the rotation period of the magnetic dipole is taken to be the rotation period of the interior and is called system III; however, another longer period that organizes radio-wave activity, called system IV, may be controlled by this mass-loading process. At Saturn, by analogy, the period of Saturn kilometric radiation, which slowly changes over many years, is probably the equivalent process. The slow change in period could be due to a variable outgassing rate of Enceladus' plume or differences in solar illumination during the saturnian year.

12.10 SUMMARY

We have learned much about the magnetospheres of Jupiter and Saturn from the Galileo and Cassini missions, and we continue to do so. However, further exploration of Uranus and Neptune seems a long way off. The New Horizons mission has now flown by Pluto, but since it did not carry a magnetometer, mysteries remain about the nature of the solar-wind interaction with this body.

Perhaps such exploration seems moot and of little practical importance. However, each of these magnetospheres is different, and their differences enable us to identify the various possible mechanisms for their various phenomena. We cannot experiment in the usual sense, but by observing different systems we can achieve the same end. Thereby, we learn the general processes by which magnetospheres operate, and in turn learn more about how the terrestrial magnetosphere works.

Additional reading

DESSLER, A. J. (Ed.) (1983). *Physics of the Jovian Magnetosphere*. Cambridge: Cambridge University Press. This book summarizes our understanding of the jovian magnetosphere after Voyager.

DOUGHERTY, M. K., L. W. ESPOSITO, and S. M. KRIMIGIS (2009). *Saturn from Cassini–Huygens*. New York: Springer. This book contains an early summary of the results of the Cassini–Huygens mission to Saturn.

PROBLEM

12.1 If the polar cap of Jupiter can be approximated by a circular cap of extent 10° in colatitude from the dipole axis and is open, calculate the flux content of the jovian tail. If this flux does not close across the central tail current sheet, and if the asymptotic field strength in the tail is 1 nT, calculate the tail cross-sectional area and its radius. Illustrate with a sketch.

13 Plasma waves

13.1 INTRODUCTION

Throughout the book, we have seen how waves play an important role in space plasmas. They are critical in heating the solar wind, in the dissipation occurring at bow shocks, in transferring the energy in gyrating pickup ions into the heat of the thermal ions, and in diffusing the energetic particles in radiation belts so that they have finite lifetimes.

Determining the different waves that occur within in a plasma requires that we derive a wave **dispersion relation**. To do this we use **Maxwell's equations** to relate the wave electromagnetic fields to the current and charge density induced in the plasma by the wave fields. By way of introduction to the basic approach used to derive the dispersion relation, we first consider **electrostatic waves** in an unmagnetized warm plasma. We shall find that this results in a variety of **acoustic modes**.

A large fraction of the chapter will be dedicated to the **electromagnetic wave** modes in a **cold plasma**. First we consider parallel propagation and introduce the **R and L modes**. Next, we derive the dispersion relation for the perpendicularly propagating **ordinary (O) and extraordinary (X) modes**. For oblique propagation we derive the **Appleton–Hartree dispersion relation**. This form of the dispersion relation is particularly useful for radio waves propagating through the ionosphere, and for **whistler**-mode dispersion. Finally, when discussing wave dispersion in a cold plasma, we consider waves in a plasma consisting of multiple ion species. We also show that the low-frequency limit gives the classical MHD results, but in the cold plasma limit. For completeness we also derive the MHD wave dispersion relation for a warm plasma.

Having investigated the dispersive properties of a plasma, we next explore instabilities, whereby waves grow. First we consider streaming instabilities. These can have quite large growth rates, and as a consequence we might expect the source plasmas to be heated by the waves. This then leads us to consider kinetic effects. Up to this point we have used **fluid theory** to determine the wave dispersion characteristics. In fluid theory the details of the particle phase space distribution are ignored, with a particle species being represented by bulk parameters such as density, flow velocity, and pressure. **Kinetic theory**, on the other hand, takes into account the details of the distribution function.

Wave dispersion using kinetic theory can be quite complicated, and we present only an outline of how to approach this topic. We consider three limiting cases: parallel propagating electrostatic waves, parallel propagating electromagnetic waves, and perpendicularly propagating electrostatic waves. The first of these will lead to the concept of **Landau resonance**, the second to **gyro-resonance**, and the third to **Bernstein modes**. The last of these topics will also show, in principle, how to derive a more general dispersion relation using kinetic theory.

The first step in deriving a wave dispersion relation is to establish the equations that govern how a plasma responds to electromagnetic fields, which we show in the following section.

13.2 HARMONIC PERTUBATION – MAXWELL'S EQUATIONS

Here we set out the basic approach to be used in deriving the wave dispersion relation for plasma waves. We start from Maxwell's equations, but we combine Faraday's law (3.1) and the time derivative of Ampère's law (3.2) to give

$$-\nabla \times (\nabla \times \mathrm{E}) = \mu_0 \frac{\partial \mathrm{j}}{\partial t} + \frac{1}{c^2}\frac{\partial^2 \mathrm{E}}{\partial t^2}. \qquad (13.1)$$

Making use of the vector identity for expanding the triple cross product on the left-hand side, we get

$$\nabla^2 \mathrm{E} - \nabla(\nabla \cdot \mathrm{E}) - \frac{1}{c^2}\frac{\partial^2 \mathrm{E}}{\partial t^2} = \mu_0 \frac{\partial \mathrm{j}}{\partial t}. \qquad (13.2)$$

Thus, as a general statement, the task in deriving the wave dispersion relation reduces to determining the current induced in the plasma by the wave electric field. Depending on the nature of the waves, this may be simple or it may be complicated. There are cases where it is useful to recast the governing equations in terms of parameters other than E and j. As an example, in Section 3.7.1, we derived the dispersion relation for MHD waves in a warm plasma using the plasma flow velocity as the fundamental quantity. This was done to elucidate the physics of the different wave modes.

Equation (13.2) relates the current density in the plasma to the wave electric field, but, given that $\nabla \times \mathrm{E} = -\partial \mathrm{B}/\partial t$, the wave also has a magnetic field (unless $\nabla \times \mathrm{E} = 0$), and consequently Eq. (13.2) is the general form used for electromagnetic waves. There is a subset of plasma waves, known as electrostatic waves, that corresponds to setting $\nabla \times \mathrm{E} = 0$. These waves are not completely static (otherwise there would not be any wave to consider), but rather the induction electric field, which is associated with time varying magnetic fields, is small and can be ignored. For electrostatic waves we use Gauss's law:

$$\nabla \cdot \mathrm{E} = \frac{\rho_{\mathrm{q}}}{\varepsilon_0}. \qquad (13.3)$$

This also indicates why we refer to the waves as electrostatic; there is no explicit time dependence in Eq. (13.3). The time variation enters through the charge density ρ_{q}.

The distinction between electromagnetic and electrostatic waves is based on the presence or absence of a wave magnetic field. For the electrostatic approximation, the wave magnetic field becomes vanishingly small. Waves in a plasma are also labeled by whether the wave electric field is longitudinal or transverse. For a **longitudinal wave,** $\nabla \times \mathrm{E} = 0$, which is the same condition for a wave being electrostatic. For a **transverse wave,** $\nabla \cdot \mathrm{E} = 0$. Thus a transverse wave is electromagnetic, but an electromagnetic wave can also have a longitudinal component to the electric field. We should also point out at this stage that the use of longitudinal and transverse is ambiguous when we consider a magnetized plasma. As we shall see later, when considering the direction of propagation of a wave with respect to the ambient magnetic field, **longitudinal propagation** is sometimes used as a synonym for wave propagation parallel to the ambient magnetic field, with **transverse propagation** corresponding to perpendicular propagation. Unless we explicitly state that longitudinal and transverse refer to the direction of propagation, it should be assumed that these refer to the orientation of the wave electric field relative to the wave vector.

There are two additional steps we take in deriving a wave dispersion relation, both of which have been introduced in Chapter 3. The first is **linearization** of the governing equations. At first sight it might not be obvious why linearization is required, since Eqs. (13.2) and (13.3) are linear. But the current density and charge density are derived from a set of equations that are generally not linear, as we shall see below. Linearization is the approximation that assumes the wave fields are first-order perturbations.

The second approximation is to assume that the waves can be taken to be plane waves that vary with space and time as

$$\mathrm{E} = \mathrm{E}_0 \exp[i(\mathbf{k} \cdot \mathbf{r} - \omega t)], \qquad (13.4)$$

where **k** is the wave number, with magnitude given by $2\pi/\lambda$, and ω is the angular wave frequency, $2\pi/f$. The wave amplitude, E_0 in this case, can be a complex quantity. While the formalism developed here will use Eq. (13.4) to describe the waves, it should be remembered that the wave manifests itself as the real part of Eq. (13.4). The term $\mathbf{k}\cdot\mathbf{r}-\omega t$ gives the phase of the wave, and the minus sign in Eq. (13.4) means that the wave phase fronts move in the direction of $\mathbf{k}$. We could have written the phase angle as $\omega t-\mathbf{k}\cdot\mathbf{r}$, and this convention is frequently used. We chose the form in Eq. (13.4) because the frequency can be complex, corresponding to wave growth or damping. If we write $\omega=\omega_r+i\gamma$, then, with our convention, positive γ corresponds to wave growth. Because of the process of linearization, once we specify the electric field as a plane wave, all the associated first-order quantities are also plane waves.

The assumption that the waves are first-order perturbations and vary as Eq. (13.4) is frequently referred to as the **harmonic perturbation** assumption. With this assumption we can replace the differential operators in Eqs. (13.2) and (13.3) with

$$\nabla \equiv i\mathbf{k} \tag{13.5}$$

and

$$\partial/\partial t \equiv -i\omega, \tag{13.6}$$

reducing Eqs. (13.2) and (13.3) to algebraic equations:

$$k^2\mathbf{E}-\mathbf{k}(\mathbf{k}\cdot\mathbf{E})-\frac{\omega^2}{c^2}\mathbf{E}=i\omega\mu_0\mathbf{j} \tag{13.7}$$

and

$$i\mathbf{k}\cdot\mathbf{E}=\rho_q/\varepsilon_0. \tag{13.8}$$

The next step is to specify the current density, $\mathbf{j}$, or charge density, ρ_q, in terms of the wave electric field. In general, the relationship between the current density and the wave electric field will have the form

$$\mathbf{j}=\sigma\cdot\mathbf{E}, \tag{13.9}$$

where σ is the **conductivity tensor**. From charge conservation, the corresponding charge density, which we would use for electrostatic waves, is

$$\rho_q=\mathbf{k}\cdot\sigma\cdot\mathbf{E}. \tag{13.10}$$

Another approach, which is frequently used for high-frequency waves, where the displacement current can be large, is to treat the effect of the plasma as a change in the permittivity of the medium. In that case we rewrite Eq. (13.7) as

$$k^2\mathbf{E}-\mathbf{k}(\mathbf{k}\cdot\mathbf{E})=\frac{\omega^2}{c^2}\mathbf{E}+i\omega\mu_0\mathbf{j}=\frac{\omega^2}{c^2}\boldsymbol{\kappa}\cdot\mathbf{E}, \tag{13.11}$$

with $\boldsymbol{\kappa}$ being the relative permittivity (or dielectric constant) of the medium, which, like the conductivity, is a tensor quantity. The two tensors are related by

$$\boldsymbol{\kappa}=\mathbf{I}+\frac{i}{\omega\varepsilon_0}\sigma, \tag{13.12}$$

where $\mathbf{I}$ is the identity tensor.

The next step is to decide how we determine the conductivity tensor (or the equivalent permittivity tensor). In Chapter 3 we first discussed the kinetic properties of a plasma through the **Vlasov equation**, which describes the behavior of a plasma in terms of the distribution function. We then derived the fluid description, where the plasma is characterized in terms of fluid parameters, such as density, momentum, and pressure. Thus, in order to establish when we could consider a plasma as a fluid, we had to first explore the kinetic properties of a plasma. Having established when a plasma can be considered to be a fluid, we find that it is simpler to determine the response to waves treating the plasma as a fluid and then consider kinetic effects.

13.3 WAVES IN A PLASMA FLUID

In order to determine the wave dispersion, we make use of the fluid continuity equations given in Chapter 3, Section 3.5.2. In doing so we use the assumption of a harmonic perturbation. This means that any terms that are second order or higher with respect to the perturbation quantities can be ignored.

We consider the momentum equation first because this equation relates the plasma perturbations to the wave field perturbations. For a harmonic perturbation, the momentum equation (3.128) becomes

$$-i\omega n_s m_s\mathbf{u}_s+i\mathbf{k}\delta P_s=n_s q_s(\mathbf{E}+\mathbf{u}_s\times\mathbf{B}_0), \tag{13.13}$$

where the subscript denotes species s. The first-order quantities in Eq. (13.13) are the species velocity, $\mathbf{u}_s$, the plasma pressure perturbation, δP_s, and the wave electric field, $\mathbf{E}$. All the other quantities are zero order. We have also assumed that the pressure is a scalar quantity. As noted above, we have ignored the second-order term $(\mathbf{u}_s \cdot \nabla)\mathbf{u}_s$ in the momentum equation.

To first order, the continuity equation (3.119) becomes

$$-i\omega\delta n_s + in_s\mathbf{k}\cdot\mathbf{u}_s = 0, \qquad (13.14)$$

where δn_s is the first-order number density.

To close this set of equations, we need to specify the pressure perturbation, δP_s. This comes from the energy equation (3.127), which appears complicated at first sight. But, again making use of the assumption of a first-order perturbation, this reduces to a familiar form (cf. Eq. (3.167)):

$$\delta P_s = \frac{\gamma_s P_s}{m_s n_s} m_s \delta n_s = \gamma_s K T_s \delta n_s, \qquad (13.15)$$

where $\gamma_s P_s / m_s n_s$ corresponds to a sound speed for species s, γ_s gives the ratio of specific heats for the species, $P_s = KT_s$ is the zero-order pressure, T_s is the corresponding temperature, and K is the Boltzmann constant.

Combining Eqs. (13.13), (13.14), and (13.15), we get

$$\omega^2\mathbf{u}_s - \frac{\gamma_s K T_s}{m_s}\mathbf{k}(\mathbf{k}\cdot\mathbf{u}_s) = i\omega\frac{q_s}{m_s}(\mathbf{E} + \mathbf{u}_s \times \mathbf{B}_0), \qquad (13.16)$$

which provides a general form that relates the perturbation flow velocity for species s to the wave electric field $\mathbf{E}$.

Once we have derived an expression for $\mathbf{u}_s$ in terms of $\mathbf{E}$, we close the set of equations through the current density

$$\mathbf{j} = \sum_s n_s q_s \mathbf{u}_s. \qquad (13.17)$$

Further, if we are only considering electrostatic waves, we can either use Eq. (13.14) to determine δn_s and calculate the charge density explicitly, with

$$\rho_{\mathrm{q}} = \sum_s \delta n_s q_s, \qquad (13.18)$$

or make use of charge conservation:

$$\mathbf{k}\cdot\mathbf{j} = \omega\rho_{\mathrm{q}}. \qquad (13.19)$$

At this stage we can use different approximations to explore further the different wave modes that are present in a plasma. We make three different approximations: electrostatic waves in an unmagnetized plasma; electromagnetic waves in a cold plasma; and low-frequency waves in a warm plasma. The latter are the MHD wave modes discussed in Chapter 3. Our purpose in revisiting these wave modes is to make the connection between the approach used here to derive the plasma-wave dispersion relation and the MHD wave modes.

13.3.1 Electrostatic waves in an unmagnetized plasma

If we assume the plasma is unmagnetized, then the $\mathbf{u}_s \times \mathbf{B}_0$ term in Eq. (13.16) vanishes, and, on taking the dot product of $\mathbf{k}$ with Eq. (13.16) without the magnetic field dependent term, we obtain

$$\left(\omega^2 - k^2\frac{\gamma_s K T_s}{m_s}\right)(\mathbf{k}\cdot\mathbf{u}_s) = i\omega\frac{q_s}{m_s}(\mathbf{k}\cdot\mathbf{E}) \quad (13.20)$$

or

$$q_s\delta n_s = \omega_{\mathrm{ps}}^2 \frac{\varepsilon_0 i\mathbf{k}\cdot\mathbf{E}}{\left(\omega^2 - k^2\frac{\gamma_s K T_s}{m_s}\right)}, \qquad (13.21)$$

where ω_{ps} is the species plasma frequency, $\omega_{\mathrm{ps}}^2 = n_s q_s^2/\varepsilon_0 m_s$. Summing Eq. (13.21) over species yields the perturbation charge density, which in turn is related to $\mathbf{k}\cdot\mathbf{E}$ through Eq. (13.8).

If we now consider a two-species plasma, consisting of electrons and singly charged positive ions, we obtain the dispersion relation for electrostatic waves in an unmagnetized two-species plasma as

$$1 = \frac{\omega_{\mathrm{pe}}^2}{\left(\omega^2 - k^2 c_{\mathrm{se}}^2\right)} + \frac{\omega_{\mathrm{pi}}^2}{\left(\omega^2 - k^2 c_{\mathrm{si}}^2\right)}, \qquad (13.22)$$

where c_{se} and c_{si} are the electron and ion sound speeds, respectively.

If we make the cold plasma approximation, and both sound speeds vanish, we get the classical result that

$$\omega^2 = \omega_{pe}^2 + \omega_{pi}^2 = \omega_p^2, \tag{13.23}$$

i.e., the waves correspond to plasma oscillations, also known as Langmuir waves.

In general, the ion and electron temperatures are not very different, such that $c_{se}^2/c_{si}^2 \approx O(m_i/m_e) \gg 1$. We can use this to explore how the solutions of Eq. (13.22) vary as a function of frequency or wave number, or, perhaps more clearly, as a function of phase velocity. Thus, at highest phase velocity both terms on the right-hand side of Eq. (13.22) are positive, but as the phase velocity decreases the denominator in the first term approaches zero, whereas the denominator in the second term is still such that $\omega^2 \gg k^2 c_{si}^2$. In that case,

$$1 \approx \frac{\omega_{pe}^2}{\left(\omega^2 - k^2 c_{se}^2\right)} + \frac{\omega_{pi}^2}{\omega^2}, \tag{13.24}$$

or, on rearranging terms,

$$\omega^2\left(\omega^2 - k^2 c_{se}^2 - \omega_p^2\right) + k^2 c_{se}^2 \omega_{pi}^2 \approx 0. \tag{13.25}$$

At high frequencies, $\omega^2 \gg \omega_{pi}^2$, the solution of Eq. (13.25) requires the term in parentheses to be small. If we make use of the definition of the species Debye length (Eq. (3.83)), then the high-frequency solution of Eq. (13.25) is

$$\begin{aligned}\omega^2 &= \omega_p^2 + k^2 c_{se}^2 \\ &= \omega_p^2 + \gamma_e k^2 \lambda_{De}^2 \omega_{pe}^2 \approx \omega_p^2\left(1 + \gamma_e k^2 \lambda_{De}^2\right)\end{aligned} \tag{13.26}$$

since $\omega_p \approx \omega_{pe}$.

Equation (13.25) also suggests a low-frequency solution, but some care should be taken, as the assumption that we neglect the ion sound speed in going from Eq. (13.22) to Eq. (13.24) may not be valid.

Before considering additional solutions, we should discuss the meaning of γ_e. In a truly unmagnetized plasma there are three degrees of freedom and $\gamma_e = 5/3$. But the formalism for electrostatic waves propagating along the magnetic field in a magnetized plasma is the same as the unmagnetized case, since the $\mathbf{u}_s \times \mathbf{B}_0$ term does not contribute to the field-aligned component of Eq. (13.26). In the case of parallel propagation in a magnetized plasma there is only one degree of freedom and $\gamma_e = 3$.

Returning to Eq. (13.22), the next solution to consider is one where the phase velocity is low enough that the first term on the right-hand side is negative. In general this will be large as the frequency becomes much smaller than that given by Eq. (13.26). Thus Eq. (13.22) becomes

$$0 \approx -\frac{\omega_{pe}^2}{k^2 c_{se}^2} + \frac{\omega_{pi}^2}{\left(\omega^2 - k^2 c_{si}^2\right)} \tag{13.27}$$

or

$$\omega^2 \approx k^2 \frac{(\gamma_e K T_e + \gamma_i K T_i)}{m_i} = k^2 c_s^2. \tag{13.28}$$

In the present context, the speed c_s is referred to as the ion acoustic speed. It becomes the sound speed of MHD if we assume $\gamma_e = \gamma_i = 5/3$.

The last solution we shall consider is the limit where k becomes sufficiently large that the first term on the right-hand side of Eq. (13.22) can be ignored. In that case

$$\omega^2 \approx \omega_{pi}^2 + k^2 c_{si}^2 \approx \omega_{pi}^2\left(1 + \gamma_i k^2 \lambda_{Di}^2\right). \tag{13.29}$$

The solutions of Eq. (13.22) are shown in Figure 13.1. This figure shows the wave frequency as a function of wave number; it is frequently referred to as an $\omega - k$ diagram or a wave dispersion plot. Figure 13.1(a) shows the high-frequency mode, given approximately by Eq. (13.25). The frequency is normalized to the electron plasma frequency, and the wave number is normalized to the electron Debye length. The gray line shows the asymptotic phase velocity given by c_{se}. In specifying c_{se} we have assumed $\gamma_e = 3$.

Figure 13.1(b) shows the ion mode. For demonstration purposes we have assumed that the electron temperature is ten times larger than the ion temperature and $\gamma_i = 3$. For equal temperature species $\lambda_{Di} = \lambda_{De}$, but in this case $\lambda_{Di} = \lambda_{De}/10^{1/2}$. At low frequencies the wave dispersion is given approximately by Eq. (13.28), which is indicated by the

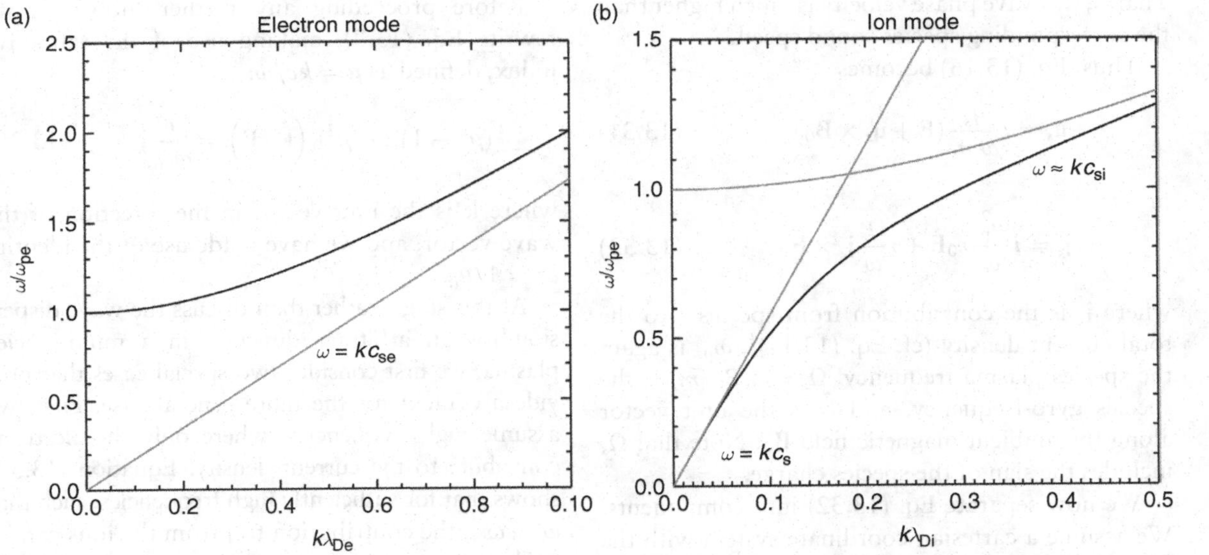

FIGURE 13.1. Wave dispersion plot for the electron (a) and ion (b) modes given by Eq. (13.22). These modes can all be considered to be acoustic modes.

gray line with constant slope. As the frequency increases, the wave approaches the asymptote given by Eq. (13.29), indicated by the other gray line.

Figure 13.1 also allows us to discuss some fundamental properties of waves in a plasma. First, the slope of a line connecting the origin to a point on the $\omega - k$ contour gives the phase velocity. The slope of the dispersion relation, $\partial\omega/\partial k$, gives the group velocity. Thus, for example, for low wave number the phase velocity of the electron mode becomes infinite, while the group velocity is zero. This is an important feature of the dispersion, as wave energy propagates at the group velocity and should not propagate faster than the speed of light. As the wave number increases, both the phase speed and the group speed asymptote to the electron sound speed. For the ion mode, on the other hand, both the phase velocity and group velocity are given by the ion acoustic speed at low frequencies, and asymptote to the ion sound speed at large wave numbers.

The relationship between phase velocity and group velocity becomes more complicated when we consider a magnetized plasma, as the dispersive properties change with the direction of propagation. The phase velocity propagates in the direction given by the wave vector **k**, but the group velocity is given by $\partial\omega/\partial\mathbf{k}$, which need not be parallel to **k**.

Before closing this section, we should point out that the discussion of the electrostatic modes in an unmagnetized plasma is not simply pedagogical: the solar wind can be considered to be weakly magnetized, at least for high-frequency waves, and plasma oscillations and ion acoustic waves are observed in the foreshock region of planetary bow shocks. These waves appear to be associated with particles reflected at the bow shock. We discuss how the reflected particles generate plasma waves in Section 13.5.2.

13.3.2 Electromagnetic waves in a cold magnetized plasma

The next approximation we shall make is the case of electromagnetic waves in a cold magnetized plasma. Taking the dot product of Eq. (13.16) with **k**, we can ignore the effects of plasma pressure (i.e., assume the plasma is cold) provided

$$\frac{\omega^2}{k^2} >> \frac{\gamma_s K T_s}{m_s}. \qquad (13.30)$$

That is, the wave phase velocity is much higher than the corresponding species sound speed.

Thus, Eq. (13.16) becomes

$$\mathbf{u}_s = i\frac{q_s}{\omega m_s}(\mathbf{E} + \mathbf{u}_s \times \mathbf{B}_0), \tag{13.31}$$

or

$$\mathbf{j}_s = i\frac{\omega_{\text{ps}}^2}{\omega}\varepsilon_0\mathbf{E} + i\frac{\Omega_s}{\omega}\mathbf{j}_s \times \hat{\mathbf{b}}, \tag{13.32}$$

where $\mathbf{j}_s$ is the contribution from species s to the total current density (cf. Eq. (13.17)), ω_{ps} is again the species plasma frequency, $\Omega_s = q_s B_0/m_s$ is the species gyro-frequency, and $\hat{\mathbf{b}}$ is the unit vector along the ambient magnetic field $\mathbf{B}_0$. Note that Ω_s includes the sign of the species charge.

We now separate Eq. (13.32) into components. We assume a cartesian coordinate system with the z-axis along the ambient magnetic field $\mathbf{B}_0$. In that case, the components of $\mathbf{j}_s$ are given by

$$j_{sx} = i\frac{\omega_{\text{ps}}^2}{\omega}\varepsilon_0 E_x + i\frac{\Omega_s}{\omega}j_{sy}, \tag{13.33}$$

$$j_{sy} = i\frac{\omega_{\text{ps}}^2}{\omega}\varepsilon_0 E_y - i\frac{\Omega_s}{\omega}j_{sx}, \tag{13.34}$$

$$j_{sz} = i\frac{\omega_{\text{ps}}^2}{\omega}\varepsilon_0 E_z. \tag{13.35}$$

We can combine Eqs. (13.33) and (13.34) to express the components of $\mathbf{j}_s$ in terms of the wave electric fields:

$$j_{sx} = i\frac{\omega_{\text{ps}}^2}{\left(\omega^2 - \Omega_s^2\right)}\varepsilon_0\left(\omega E_x + i\Omega_s E_y\right), \tag{13.36}$$

$$j_{sy} = i\frac{\omega_{\text{ps}}^2}{\left(\omega^2 - \Omega_s^2\right)}\varepsilon_0\left(\omega E_y - i\Omega_s E_x\right). \tag{13.37}$$

Equations (13.36), (13.37), and (13.35) express the current density per species induced in the plasma by a harmonic perturbation solely in terms of the wave electric field. The terms multiplying the different components of the electric field therefore constitute the different elements of the conductivity tensor σ. We can next use Eq. (13.17) to give the total current density in terms of the wave electric fields, and Eq. (13.7) then closes the set of equations.

Before proceeding any further, however, we rewrite Eq. (13.7), making use of the refractive index, defined as $\mu = kc/\omega$:

$$(\mu^2 - 1)\mathbf{E} - \mu^2\hat{\mathbf{k}}\left(\hat{\mathbf{k}} \cdot \mathbf{E}\right) = \frac{i}{\omega\varepsilon_0}\mathbf{j}, \tag{13.38}$$

where $\hat{\mathbf{k}}$ is the unit vector in the direction of the wave vector, and we have made use of the identity $c^2 = 1/\mu_0\varepsilon_0$.

At this stage, rather than discuss the wave dispersion for an arbitrary direction in a multi-species plasma, we first consider two special cases that provide a context for the more general case. First, we assume high frequencies, where only the electrons contribute to the current density. Equation (13.32) shows that for sufficiently high frequencies, such that $\omega \gg \omega_{\text{pi}}$, the contribution to $\mathbf{j}$ from the ions can be neglected, since the electron contribution is $O(m_{\text{i}}/m_{\text{e}})$ larger. Second, we consider waves propagating parallel or perpendicular to the ambient magnetic field.

13.3.3 Cold plasma wave modes – parallel propagation

For parallel propagation, i.e., $\hat{\mathbf{k}}||\mathbf{B}_0$, the components of Eq. (13.38) become

$$(\mu^2 - 1)E_x = \frac{i}{\omega\varepsilon_0}j_x, \tag{13.39}$$

$$(\mu^2 - 1)E_y = \frac{i}{\omega\varepsilon_0}j_y, \tag{13.40}$$

$$E_z = -\frac{i}{\omega\varepsilon_0}j_z. \tag{13.41}$$

Finally, we use Eqs. (13.36), (13.37), and (13.35) evaluated for electrons only to specify the current density in Eqs. (13.39), (13.40), and (13.41):

$$(\mu^2 - 1)E_x = -\frac{\omega_{\text{pe}}^2}{\left(\omega^2 - \Omega_{\text{e}}^2\right)}\left(E_x - i\frac{|\Omega_{\text{e}}|}{\omega}E_y\right), \tag{13.42}$$

$$(\mu^2 - 1)E_y = -\frac{\omega_{\text{pe}}^2}{\left(\omega^2 - \Omega_{\text{e}}^2\right)}\left(E_y + i\frac{|\Omega_{\text{e}}|}{\omega}E_x\right), \tag{13.43}$$

$$E_z = \frac{\omega_{\text{pe}}^2}{\omega^2}E_z, \tag{13.44}$$

where $|\Omega_e| = eB_0/m_e$ gives the magnitude of the electron gyro-frequency.

The set of equations (13.42), (13.43), (13.44) only contains the different components of the wave electric field. We can rewrite these equations using matrix representation as follows:

$$\begin{pmatrix} (\mu^2-1)(\omega^2-\Omega_e^2)+\omega_{pe}^2 & -i\omega_{pe}^2\dfrac{|\Omega_e|}{\omega} & 0 \\ i\omega_{pe}^2\dfrac{|\Omega_e|}{\omega} & (\mu^2-1)(\omega^2-\Omega_e^2)+\omega_{pe}^2 & 0 \\ 0 & 0 & \omega^2-\omega_{pe}^2 \end{pmatrix} \times \begin{pmatrix} E_x \\ E_y \\ E_z \end{pmatrix} = 0. \tag{13.45}$$

The solution of Eq. (13.45) is found by requiring that the determinant of the matrix be zero. However, since there is only one term in the z-column and z-row, we can immediately see that one solution has E_x and E_y equal to zero, and $E_z \neq 0$. For this solution

$$\omega^2 = \omega_{pe}^2. \tag{13.46}$$

This is the same result we found earlier, for electrostatic waves in an unmagnetized cold plasma, Eq. (13.23), ignoring the ion contribution to the plasma frequency. This is not surprising as the ambient magnetic field does not affect parallel particle motion.

We now turn to the other solution of Eq. (13.45). This requires $E_z = 0$ and

$$\begin{vmatrix} (\mu^2-1)(\omega^2-\Omega_e^2)+\omega_{pe}^2 & -i\omega_{pe}^2\dfrac{|\Omega_e|}{\omega} \\ i\omega_{pe}^2\dfrac{|\Omega_e|}{\omega} & (\mu^2-1)(\omega^2-\Omega_e^2)+\omega_{pe}^2 \end{vmatrix} = 0, \tag{13.47}$$

or, since the determinant has the form $a^2 = b^2$, we can take the square root, and

$$(\mu^2-1)(\omega^2-\Omega_e^2)+\omega_{pe}^2 = \pm\frac{\omega_{pe}^2|\Omega_e|}{\omega}. \tag{13.48}$$

On rearranging terms, this becomes

$$\mu^2 = 1 - \frac{\omega_{pe}^2}{\omega(\omega \pm |\Omega_e|)}. \tag{13.49}$$

This mode has a resonance, where the refractive index goes to infinity, when we choose the negative sign in the denominator and $\omega = |\Omega_e|$. Choosing the negative sign in Eq. (13.49) corresponds to choosing the negative sign in Eq. (13.48). From Eq. (13.45) this implies that

$$E_x + iE_y = 0. \tag{13.50}$$

This means that $|E_x| = |E_y|$. Furthermore, if we assume

$$E_x(t) = E_0\exp[-i(\omega t - kz)] \tag{13.51}$$

then

$$\begin{aligned} E_y(t) &= iE_x(t) = iE_0\exp[-i(\omega t - kz)] \\ &= E_0\exp[-i\big(\omega(t-\tau/4) - kz\big)] = E_x(t-\tau/4), \end{aligned} \tag{13.52}$$

where τ is the wave period. Thus, the y-component of the electric field lags the x-component by one quarter wave period. This corresponds to a right-hand circularly (RHC) polarized wave, where the wave electric field rotates around the ambient magnetic field in a right-handed sense. If, instead, we choose the positive sign in Eq. (13.49), then the electric field is left-hand circularly (LHC) polarized. For this reason the dispersion relation given by Eq. (13.49) results in two wave modes – one that is RHC polarized, corresponding to the minus sign in Eq. (13.49), and the other mode that is LHC polarized. These modes are therefore known as the R-mode and the L-mode. The plasma oscillations, given by Eq. (13.46), are also known as the P-mode. The R- and L-modes are electromagnetic modes that are also transverse, with no wave electric field in the direction of the wave vector, while the P-mode is electrostatic and therefore also longitudinal.

The splitting into R- and L-modes also explains why there is a resonance when $\omega = |\Omega_e|$. Electrons gyrate in a right-handed sense about the magnetic field at the gyro-frequency. At the resonance the electrons see a wave field that rotates in the same sense with the same frequency. This is gyro-resonance, and we would expect strong coupling between the electrons and the wave. We shall return to this later, when we discuss wave instabilities.

The other feature of (13.49) is that the refractive index can become zero. This is known as a cutoff.

Setting $\mu^2 = 0$ in Eq. (13.49) and restricting the solutions to positive frequencies, we obtain

$$\omega_R = (\omega_{pe}^2 + \Omega_e^2/4)^{1/2} + |\Omega_e|/2 \qquad (13.53)$$

and

$$\omega_L = (\omega_{pe}^2 + \Omega_e^2/4)^{1/2} - |\Omega_e|/2, \qquad (13.54)$$

where ω_R and ω_L are the cutoff frequencies for the R- and L-modes, respectively. The cutoff frequencies have the following properties: $\omega_R = \omega_L + |\Omega_e|$, $\omega_R > \max(\omega_{pe}, |\Omega_e|)$, and $\omega_L < \omega_{pe}$.

The solutions of Eq. (13.49)] are shown in Figure 13.2. For these solutions we have assumed $\omega_{pe} = 2|\Omega_e|$.

In Figure 13.2 there are two branches of the R-mode: one with $\omega > \omega_R$ and the other with $\omega < |\Omega_e|$. The latter is also known as the whistler mode. Between these there is a stop band where the R-mode cannot propagate. In this band $\mu^2 < 0$, and the wave vector is imaginary, corresponding to spatial decay, or evanescence. The stop band for the L-mode is where $\omega < \omega_L$. At very low frequencies we must include the ions, and the L-mode reappears as a propagating mode, as we shall see later.

The high-frequency branches of the R-mode and the L-mode are both **superluminous** modes that have phase velocities faster than the speed of light, with the phase speed approaching the speed of light for high frequencies. The group velocity, however, given by the slope of the $\omega - k$ curve, is less than the speed of light for both these wave modes. Wave energy propagates with the group velocity, and the waves satisfy the requirements of special relativity. Furthermore, these wave modes have no high-frequency cutoff or resonance. This was discussed in Chapter 11, in the context of the planetary radio emissions: **auroral kilometric radiation** (AKR) at the Earth, jovian decametric radiation (DAM), and saturnian kilometric radiation (SKR). All these waves are generated in the auroral zones of the respective planets, where the radiation primarily couples to the R-mode and so can escape the planetary magnetosphere. The waves are initially generated near the local electron gyro-frequency. As the waves propagate away from the planet, they move to regions where $\omega/|\Omega_e|$ increases. This corresponds to moving vertically in Figure 13.2. Since there is no upper-frequency limit to the R-mode, the waves escape from the planetary magnetosphere.

This is not the case for the whistler mode. This mode only propagates below the electron gyro-frequency. We show later that this mode is guided by

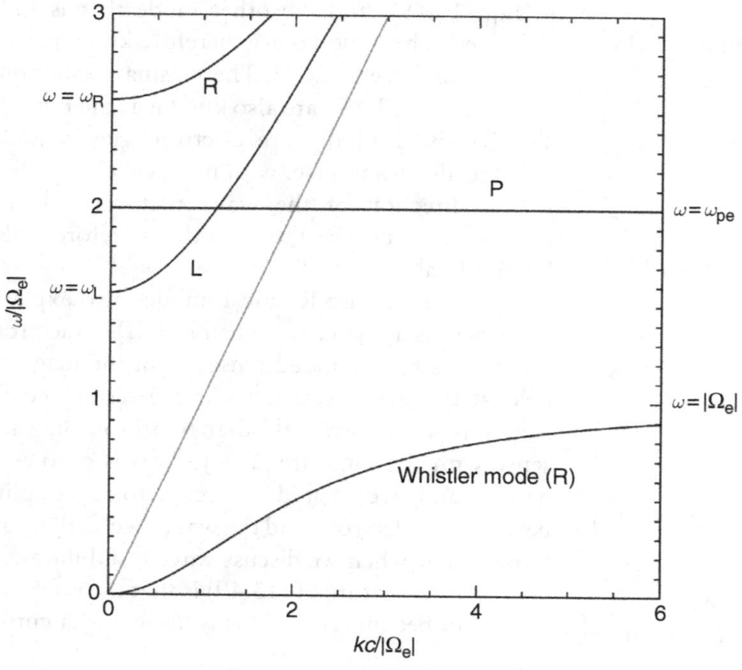

FIGURE 13.2. Wave dispersion plot for high-frequency waves that are propagating parallel to the magnetic field. The gray line of constant gradient shows the speed of light.

the magnetic field. Because the mode is guided by the magnetic field, electromagnetic waves generated by lighting in the ELF range (300 Hz–3 kHz) and VLF range (3–30 kHz) can pass through the ionosphere and propagate along the magnetic field to the conjugate hemisphere. The ELF and VLF bands include frequencies that are audible to the human ear. Figure 13.2 shows that for $\omega < |\Omega_e|/2$ both the phase velocity and group velocity increase for increasing frequency. Thus, the higher frequencies propagate more quickly to the opposite hemisphere, and the signal detected at the conjugate point has a characteristic falling tone. In the nineteenth century, as telegraphy became an important means of communication, these lightning-generated waves would induce signals in the telegraphy wires that telegraphy operators could hear. Because they generally consisted of falling tones, they became known as whistlers. However, it wasn't until the 1950s, when Owen Storey (Storey, 1953) demonstrated that whistlers must be generated by lightning in the conjugate hemisphere, that an explanation was found. Technically, the waves can bounce or "hop" back and forth between the conjugate ionospheres. Odd hops are detected at the conjugate point. The first hop shows the least dispersion, with subsequent hops becoming more and more dispersed, and generally weaker.

13.3.4 Cold plasma wave modes – perpendicular propagation

Having discussed parallel propagation, we now consider the high-frequency wave solutions for perpendicular propagation. We assume, without loss of generality, that the wave vector defines the x-axis of our cartesian coordinate system. Then Eq. (13.38) becomes

$$E_x = -\frac{i}{\omega\varepsilon_0} j_x, \tag{13.55}$$

$$(\mu^2 - 1)E_y = \frac{i}{\omega\varepsilon_0} j_y, \tag{13.56}$$

$$(\mu^2 - 1)E_z = \frac{i}{\omega\varepsilon_0} j_z. \tag{13.57}$$

As we did when considering parallel propagation, we use Eqs. (13.36), (13.37), and (13.35) to specify the current density. Again we assume that only the electrons contribute to the current density, and we write the resultant set of equations in matrix form:

$$\begin{pmatrix} \omega^2 - \omega_{UHR}^2 & i\omega_{pe}^2 \frac{|\Omega_e|}{\omega} & 0 \\ i\omega_{pe}^2 \frac{|\Omega_e|}{\omega} & (\mu^2 - 1)(\omega^2 - \Omega_e^2) + \omega_{pe}^2 & 0 \\ 0 & 0 & (\mu^2 - 1)\omega^2 + \omega_{pe}^2 \end{pmatrix} \times \begin{pmatrix} E_x \\ E_y \\ E_z \end{pmatrix} = 0, \tag{13.58}$$

where we have defined the **upper hybrid resonance** frequency as $\omega_{UHR}^2 = \omega_{pe}^2 + \Omega_e^2$. This is known as a hybrid frequency because, as can be seen from Eqs. (13.33) and (13.34), the particle motion consists of both plasma oscillations from the wave electric field and gyrational motion.

As before, the solution of Eq. (13.58) is found by requiring that the determinant of the matrix be zero. As was the case for parallel propagation, there is only one term in the z-column and z-row, and one solution has E_x and E_y equal to zero and $E_z \neq 0$. For this solution,

$$\mu^2 = 1 - \frac{\omega_{pe}^2}{\omega^2}. \tag{13.59}$$

This is the dispersion relation for the ordinary mode, or O-mode. This mode is called the O-mode because the dispersion relation does not include the electron gyro-frequency, and is the same as would be found for an unmagnetized plasma (although in the latter case there is no restriction on the direction of propagation). The dispersion relation for this mode is shown as an $\omega - k$ curve in Figure 13.3, with the label "O." Similarly to the high-frequency R- and L-modes, this mode is superluminous, but the group velocity is less than the speed of light. The O-mode is both electromagnetic and transverse.

We now turn to the other solution of Eq. (13.58). This requires $E_z = 0$ and

$$\begin{vmatrix} \omega^2 - \omega_{UHR}^2 & i\omega_{pe}^2 \frac{|\Omega_e|}{\omega} \\ i\omega_{pe}^2 \frac{|\Omega_e|}{\omega} & (\mu^2 - 1)(\omega^2 - \Omega_e^2) + \omega_{pe}^2 \end{vmatrix} = 0, \tag{13.60}$$

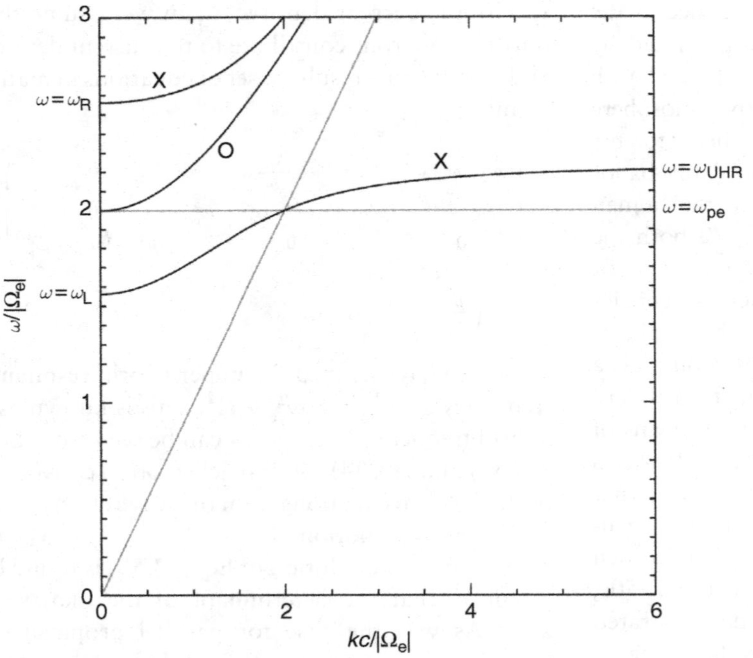

FIGURE 13.3. Wave dispersion plot for high-frequency waves that propagate perpendicularly to the magnetic field. The gray line of constant slope shows the speed of light.

or

$$[(\mu^2-1)(\omega^2-\Omega_e^2)+\omega_{pe}^2](\omega^2-\omega_{UHR}^2)+\frac{\omega_{pe}^4\Omega_e^2}{\omega}=0. \tag{13.61}$$

On rearranging terms this becomes

$$\mu^2 = 1-\frac{\omega_{pe}^2\left(\omega^2-\omega_{pe}^2\right)}{\omega^2\left(\omega^2-\omega_{UHR}^2\right)}. \tag{13.62}$$

This is the dispersion relation for the extraordinary mode, or X-mode. This mode depends on the presence of the magnetic field. The dispersion curve for this mode is also shown in Figure 13.3, labeled "X." There are two X-mode branches.

When discussing the R-mode, we noted the presence of the gyro-resonance of electrons ($\omega = |\Omega_e|$), where $\mu = \infty$. From Eq. (13.62) we see that the X-mode also has a resonance, when $\omega = \omega_{UHR}$. At the upper hybrid resonance, Eq. (13.58) requires that $E_y = 0$. Since we assume the wave vector is in the x-direction, the wave electric field is aligned with the wave vector, and the X-mode becomes electrostatic at $\omega = \omega_{UHR}$.

Similarly to the O-mode, the X-mode has a cutoff when $\mu = 0$. Setting the left-hand side of Eq. (13.62) to zero and using $\omega_{UHR}^2 = \omega_{pe}^2 + \Omega_e^2$, the cutoff occurs when

$$\left(\omega^2-\omega_{pe}^2\right)^2-\omega^2\Omega_e^2=0 \tag{13.63}$$

or

$$\omega^2 \pm \omega|\Omega_e| - \omega_{pe}^2 = 0. \tag{13.64}$$

If we restrict the solutions to $\omega > 0$, then, defining ω_X as the as the X-mode cutoff frequency,

$$\omega_{X\pm} = \left(\omega_{pe}^2+\Omega_e^2/4\right)^{1/2} \pm |\Omega_e|/2, \tag{13.65}$$

which is the same as ω_R (Eq. (13.53)) for the positive sign and ω_L (Eq. (13.54)) for the negative sign. Since the cutoff frequencies are the same, going forward we use ω_R and ω_L to specify the two X-mode cutoff frequencies.

As with the parallel propagating modes, the O- and X-modes have stop bands, where the wave evanesces. For the O-mode the stop band is for $\omega < \omega_{pe}$. For the X-mode there are two stop bands, $\omega_{UHR} < \omega < \omega_R$ and $\omega < \omega_L$. The X-mode above the first stop band is superluminous, similar to the O-mode. The X-mode between the stop bands has a phase velocity less than the speed of light, but becomes superluminous for $\omega < \omega_{pe}$.

Finally, before considering propagation at an arbitrary angle, we have already noted that the wave electric field for the O-mode is parallel to the ambient magnetic field. The X-mode electric fields are perpendicular to the magnetic field and the wave fields are elliptically polarized. From the top row of Eq. (13.58),

$$E_x = -i\frac{\omega_{pe}^2|\Omega_e|}{\omega(\omega^2 - \omega_{UHR}^2)}E_y. \tag{13.66}$$

Since the wave vector defines the x-axis in our chosen coordinate system, X-mode waves include both longitudinal (E_x) and transverse (E_y) electric fields. We have already pointed out that the wave becomes electrostatic at the upper hybrid resonance, where $E_y = 0$. When discussing the R- and L-modes, we noted that $E_x = -iE_y$ (Eq. (13.52)) denoted a wave that was RHC polarized. Because the magnitude of the term multiplying E_y in Eq. (13.66) is generally not unity, $|E_x| \neq |E_y|$, but the wave fields are still in quadrature. The wave electric field therefore traces out an ellipse as it rotates around the ambient magnetic field. For $\omega > \omega_{UHR}$ the sense of rotation is right handed, while for $\omega < \omega_{UHR}$ the sense is left handed. This is to be expected since the higher-frequency branch of the X-mode has the same cutoff frequency as the R-mode, and the lower-frequency branch has the same cutoff frequency as the L-mode.

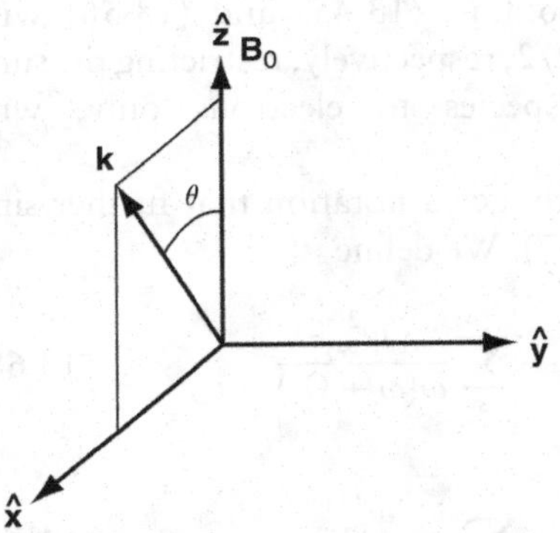

FIGURE 13.4. Coordinate system used for considering waves propagating at an angle to the ambient magnetic field.

13.3.5 Cold plasma wave modes – oblique propagation

Having discussed at some length the methodology used to derive the wave dispersion relation for a magnetized plasma for the two special cases of parallel and perpendicular propagation, we now consider an arbitrary angle of propagation with respect to the magnetic field. Since we also want to consider lower-frequency solutions later in this chapter, we include both ion and electron contributions to the current density.

For perpendicular propagation we assumed that the wave vector defined the x-axis. We now assume that the wave vector lies in the x–z plane, making an angle θ with respect to the magnetic field. The geometry is shown in Figure 13.4. The components of the wave vector are consequently $k(\sin\theta, 0, \cos\theta)$.

Having established the approach used to derive the set of coupled equations, we find from Eqs. (13.35), (13.36), (13.37), and (13.38) that the wave electric fields satisfy

$$\begin{pmatrix} \mu^2\cos^2\theta - 1 + \sum\limits_s \dfrac{\omega_{ps}^2}{(\omega^2-\Omega_s^2)} & i\sum\limits_s \dfrac{\omega_{ps}^2}{(\omega^2-\Omega_s^2)}\dfrac{\Omega_s}{\omega} & -\mu^2\sin\theta\cos\theta \\ -i\sum\limits_s \dfrac{\omega_{ps}^2}{(\omega^2-\Omega_s^2)}\dfrac{\Omega_s}{\omega} & \mu^2 - 1 + \sum\limits_s \dfrac{\omega_{ps}^2}{(\omega^2-\Omega_s^2)} & 0 \\ -\mu^2\sin\theta\cos\theta & 0 & \mu^2\sin^2\theta - 1 + \sum\limits_s \dfrac{\omega_{ps}^2}{\omega^2} \end{pmatrix} \begin{pmatrix} E_x \\ E_y \\ E_z \end{pmatrix} = 0. \tag{13.67}$$

This reduces to Eqs. (13.45) and (13.58) with $\theta = 0$ and $\theta = \pi/2$, respectively, restricting the summation over species to electrons only, with $\Omega_s = -|\Omega_e|$.

We now introduce a notation that further simplifies Eq. (13.67). We define

$$R \equiv 1 - \sum_s \frac{\omega_{ps}^2}{\omega(\omega + \Omega_s)} \tag{13.68}$$

and

$$L \equiv 1 - \sum_s \frac{\omega_{ps}^2}{\omega(\omega - \Omega_s)}. \tag{13.69}$$

From this we define

$$S \equiv \frac{1}{2}(R + L) = 1 - \sum_s \frac{\omega_{ps}^2}{(\omega^2 - \Omega_s^2)}, \tag{13.70}$$

$$D \equiv \frac{1}{2}(R - L) = \sum_s \frac{\omega_{ps}^2}{(\omega^2 - \Omega_s^2)} \frac{\Omega_s}{\omega}, \tag{13.71}$$

and, for the term that does not include the gyrofrequency,

$$P \equiv 1 - \sum_s \frac{\omega_{ps}^2}{\omega^2}. \tag{13.72}$$

It is useful to note that $S + D = R$ and $S - D = L$.

Equation (13.67) now becomes

$$\begin{pmatrix} \mu^2\cos^2\theta - S & iD & -\mu^2 \sin\theta\cos\theta \\ -iD & \mu^2 - S & 0 \\ -\mu^2 \sin\theta\cos\theta & 0 & \mu^2\sin^2\theta - P \end{pmatrix} \begin{pmatrix} E_x \\ E_y \\ E_z \end{pmatrix} = 0. \tag{13.73}$$

For completeness, from Eq. (13.11), the dielectric tensor corresponding to Eq. (13.73) is

$$\mathbf{\kappa} = \begin{pmatrix} S & -iD & 0 \\ iD & S & 0 \\ 0 & 0 & P \end{pmatrix}. \tag{13.74}$$

We can immediately recover our previously derived results from Eq. (13.73). For parallel propagation we find $\mu^2 = R$, $\mu^2 = L$, $P = 0$. These should be compared with Eqs. (13.49) and (13.46), but we have now extended the solutions to include the ions through our definitions of R, L, and P. In particular $\mu^2 = L$ now includes the ion gyro-resonance. For perpendicular propagation we find two solutions, $\mu^2 = RL/S$ (X-mode) and $\mu^2 = P$ (O-mode). If we consider electrons only, then the X-mode solution reduces to Eq. (13.62), albeit after some algebra, while the O-mode solution clearly reduces to Eq. (13.59).

Setting $\mu^2 = 0$ in Eq. (13.73) allows us to determine the various cutoff frequencies. Since R, L, and P do not depend on the direction of propagation, then, for any θ, we find that the cutoffs are given by $R = 0$, $L = 0$, and $P = 0$. Again, at high frequencies, where we neglect ions, these cutoffs are given by ω_R, ω_L, and ω_{pe}, respectively. But the more general result indicates the possibility of additional cutoffs in the wave dispersion at lower frequencies, where ions become important. If we assume only positive ions, we find that the low-frequency cutoffs are given by $L = 0$, since L changes sign through infinity at each ion gyro-resonance, but we also see that the existence of additional cutoffs requires the presence of more than one ion species.

The other limit to consider is where $\mu^2 = \infty$. For parallel propagation this corresponds to the gyro-resonances. For perpendicular propagation this is given by $S = 0$. At first sight, since $\mu^2 = RL/S$ for perpendicular propagation, it might be thought that $R = \infty$, e.g., could also result in a resonance, but, if $R = \infty$, then $S = \infty$ also, and μ^2 remains finite. If we consider a two-species plasma and define $\eta = m_e/m_i$, then $S = 0$ becomes

$$1 - \frac{\omega_{pe}^2}{\omega^2 - \Omega_e^2} - \frac{\eta\omega_{pe}^2}{\omega^2 - \eta^2\Omega_e^2} = 0, \tag{13.75}$$

or

$$\omega^4 - \omega^2[\omega_p^2 + (1 + \eta^2)\Omega_e^2] + \eta\Omega_e^2\left(\omega_p^2 + \eta\Omega_e^2\right) = 0, \tag{13.76}$$

where, again, $\omega_p^2 = \omega_{pe}^2 + \omega_{pi}^2 = (1 + \eta)\omega_{pe}^2$.

In general $\eta \ll 1$, and the last term on the left-hand side of Eq. (13.76) is small. The two solutions of Eq. (13.76) become

$$\omega^2 = \omega_p^2 + (1 + \eta^2)\Omega_e^2 \approx \omega_{pe}^2 + \Omega_e^2 = \omega_{UHR}^2 \tag{13.77}$$

and

$$\omega^2 = \frac{|\Omega_e\Omega_i|\left(\omega_p^2 + |\Omega_e\Omega_i|\right)}{\omega_p^2 + \Omega_e^2} = \omega_{LHR}^2 \tag{13.78}$$

i.e., the upper and lower hybrid resonance frequencies. We have already discussed the upper hybrid frequency in relation to Eq. (13.58). The lower hybrid resonance only appears when we include the ions in the dispersion relation, and it is a hybrid frequency in the sense that it mixes the electron and ion responses to the waves. In many texts the lower hybrid resonance frequency is given by

$$\omega_{\rm LHR}^2 \approx |\Omega_{\rm e}\Omega_{\rm i}|. \qquad (13.79)$$

Inspection of Eq. (13.78) shows that this is the limit found when $\omega_{\rm p}^2 \gg \Omega_{\rm e}^2$. This condition applies in many space plasmas, for example in the Earth's ionosphere or in the solar wind. But, as discussed in Chapter 11, the auroral acceleration region is characterized by a very low plasma density and strong magnetic field. Here $\omega_{\rm p}^2 \ll \Omega_{\rm e}^2$, and for this case we find

$$\omega_{\rm LHR}^2 \approx \omega_{\rm pi}^2 + \Omega_{\rm i}^2. \qquad (13.80)$$

Thus, when $\omega_{\rm p}^2 \ll \Omega_{\rm e}^2$, $\omega_{\rm LHR} \approx \omega_{\rm pi}$ except for an extremely rarefied plasma for which $\omega_{\rm pi}^2 \ll \Omega_{\rm i}^2$. In that case, $\omega_{\rm LHR} \approx \Omega_{\rm i}$.

We now present solutions of Eq. (13.73) for an arbitrary propagation angle. The determinant of the matrix in Eq. (13.73) results in a biquadratic equation

$$A\mu^4 - B\mu^2 + C = 0, \qquad (13.81)$$

with

$$A = P\cos^2\theta + S\sin^2\theta, \qquad (13.82)$$

$$B = PS(1+\cos^2\theta) + RL\sin^2\theta, \qquad (13.83)$$

and

$$C = PRL. \qquad (13.84)$$

Thus

$$\mu^2 = \frac{B \pm (B^2 - 4AC)^{1/2}}{2A}, \qquad (13.85)$$

using the standard result for the roots of a quadratic equation.

Figure 13.5 shows solutions of the wave dispersion relation. This combines the solutions shown in Figures 13.2 and 13.3 for 0° and 90°, respectively, together with a solution for 45°. The shading shows the allowed regions for propagation in the $\omega - k$ space. The dispersion curves are shown for two values of the ratio $\omega_{\rm pe}/\Omega_{\rm e}$, one where $\omega_{\rm pe}/\Omega_{\rm e} = 2$, and the other where $\omega_{\rm pe}/\Omega_{\rm e} = 0.5$. The faster than light modes are labeled as R-X, L-O, and L-X, corresponding to the polarization for parallel propagation, and whether or not the refractive index depends on the magnetic field for

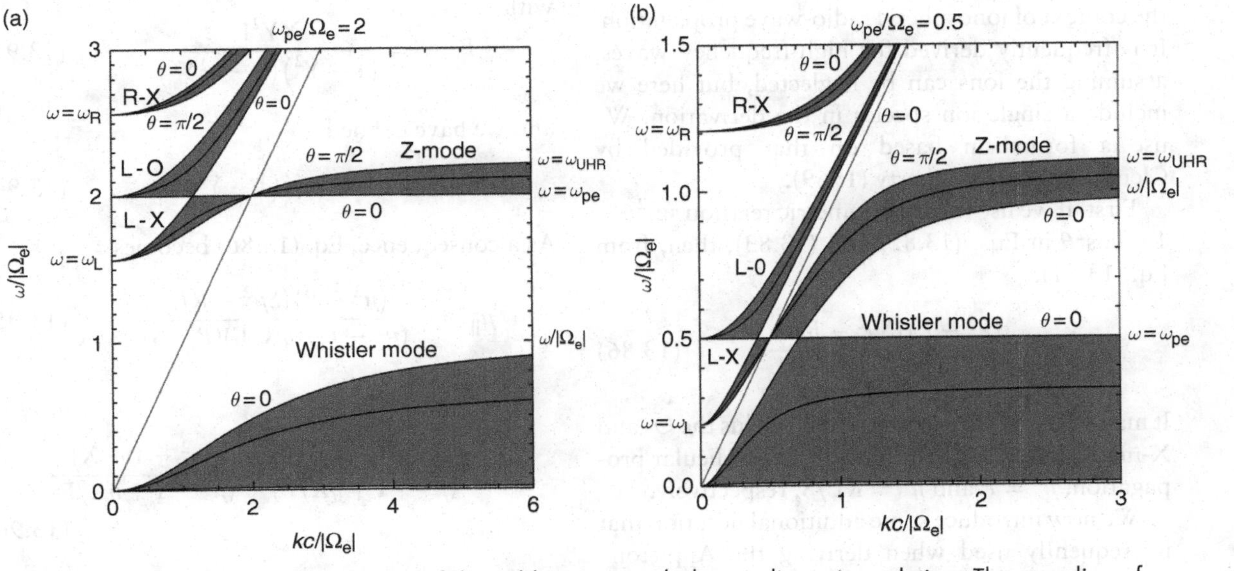

FIGURE 13.5. High-frequency solutions of the cold magnetized plasma dispersion relation. The gray line of constant gradient shows the speed of light.

perpendicular propagation. The two wave modes with phase speeds less than the speed of light are labeled as the Z-mode, with a resonance at the upper hybrid resonance, and the whistler mode at lower frequencies. The main difference between the dispersion relation as shown in Figure 13.5(a) versus that in Figure 13.5(b) is for the wave modes with phase velocities less than the speed of light, given by the gray line. The Z-mode extends from the plasma frequency up to the upper hybrid resonance. However, when $\omega_{pe} < \Omega_e$ the low-frequency limit for this mode is given by the R-mode dispersion relation for $\omega > \omega_{pe}$. Similarly, the role of the plasma frequency changes when $\omega_{pe} < \Omega_e$, becoming the upper limit of the whistler mode instead of the lower limit of the Z-mode. We should also note that we have restricted the appellation "Z-mode" to the waves with phase speed slower than light, whereas the L-X mode corresponds to the faster than light mode. In some texts, and in the context of ionosphere sounding, the signal associated with the L-X mode is also referred to as the Z-trace.

Clearly, we could use Eq. (13.85) to determine the wave dispersion for a cold plasma consisting of several ion species. But before doing so, we shall derive an alternative form of the the dispersion relation known as the Appleton–Hartree relation. This form of the dispersion relation was derived in the context of ionospheric radio-wave propagation. It is frequently derived for high-frequency waves, assuming the ions can be neglected, but here we include a single ion species in the derivation. We use a formalism based on that provided by Clemmow and Dougherty (1969).

First, if we use the trigonometric relation $\sin^2\theta = 1 - \cos^2\theta$ in Eqs. (13.82) and (13.83), then, from Eq. (13.81),

$$\mu_{\|}^2 = -\frac{(\mu^2 - P)(S\mu^2 - RL)}{(P - S)\mu^2 - PS + RL}. \tag{13.86}$$

It may be noted that setting $\mu_{\|}^2 = 0$ yields the O- and X-mode dispersion relations for perpendicular propagation, $\mu^2 = P$ and $\mu^2 = RL/S$, respectively.

We now introduce some additional notation that is frequently used when deriving the Appleton–Hartree relation. We define

$$X = \frac{(1+\eta)\omega_{pe}^2}{\omega^2} \tag{13.87}$$

and

$$Y = \frac{|\Omega_e|}{\omega}, \tag{13.88}$$

where, again, $\eta = m_e/m_i$ is the electron to ion mass ratio. The high-frequency approximation is equivalent to assuming $\eta = 0$.

With this notation, Eq. (13.68) becomes

$$R = 1 - \frac{X}{(1 - Y)(1 + \eta Y)}, \tag{13.89}$$

Eq. (13.69) becomes

$$L = 1 - \frac{X}{(1 + Y)(1 - \eta Y)}, \tag{13.90}$$

with S still being given by Eq. (13.47), and Eq. (13.72) becomes

$$P = 1 - X. \tag{13.91}$$

Further, it can be shown that

$$PS - RL = \left(1 - \frac{\eta X}{\Gamma}\right)(P - S), \tag{13.92}$$

with

$$P - S = \frac{XY^2\Gamma}{(1 - Y^2)(1 - \eta^2 Y^2)}, \tag{13.93}$$

and we have defined

$$\Gamma = 1 - \eta + \eta^2(1 - Y^2). \tag{13.94}$$

As a consequence, Eq. (13.86) becomes

$$\mu_{\|}^2 = -\frac{(\mu^2 - P)(S\mu^2 - RL)}{(\mu^2 - 1 + \eta X/\Gamma)(P - S)}. \tag{13.95}$$

We now define

$$\lambda = \frac{(\mu^2 - P)}{(\mu^2 - 1 + \eta X/\Gamma)} = \frac{(\mu^2 - 1 + X)}{(\mu^2 - 1 + \eta X/\Gamma)}. \tag{13.96}$$

Hence,

$$\mu_{\|}^2 = -\lambda \frac{(S\mu^2 - RL)}{(P - S)}. \qquad (13.97)$$

We can replace μ^2 in Eq. (13.97) by rearranging Eq. (13.96):

$$\mu^2 = 1 - \eta X/\Gamma - \frac{(1 - \eta/\Gamma)X}{1 - \lambda}. \qquad (13.98)$$

After some algebra, Eq. (13.97) becomes

$$\mu_{\|}^2 = \frac{\lambda}{(\lambda - 1)\frac{Y^2}{1-X}\left(\frac{\Gamma}{1-\eta}\right)^2} \left\{ \frac{\Gamma(\Gamma - \eta X)}{(1-\eta)^2} \frac{Y^2}{1-X} - \lambda \right\}. \qquad (13.99)$$

We can make use of Eq. (13.98) again to derive a quadratic equation for λ from Eq. (13.99):

$$\lambda^2 - \lambda \frac{\Gamma(\Gamma - \eta X)}{(1-\eta)^2} \frac{Y^2}{1-X} \sin^2\theta - \left(\frac{\Gamma}{1-\eta}\right)^2 Y^2 \cos^2\theta = 0; \qquad (13.100)$$

substituting the roots of Eq. (13.100) into Eq. (13.98) gives the Appleton–Hartree relation, including ions. The inclusion of the ions makes the relation appear complicated, so, for the purposes of discussion, we now assume that the ions can be neglected. In this case, we set $\eta = 0$ and $\Gamma = 1$. The Appleton–Hartree dispersion relation then becomes

$$\mu^2 = 1 - \frac{X}{1 - \frac{\frac{1}{2}Y^2 \sin^2\theta}{1-X} \pm \left[\left(\frac{\frac{1}{2}Y^2 \sin^2\theta}{1-X}\right)^2 + Y^2 \cos^2\theta\right]^{1/2}}. \qquad (13.101)$$

The Appleton–Hartree dispersion relation is frequently given as Eq. (13.101), i.e., neglecting ions. But we can quickly derive Eq. (13.101) including ions by inspection, starting from Eq. (13.98), i.e., including the $\eta X/\Gamma$ terms and replacing the respective $\sin^2\theta$ and $\cos^2\theta$ terms in Eq. (13.101) with the corresponding terms from Eq. (13.100).

At first sight it is not obvious why Eq. (13.101) has any advantages over Eq. (13.85). However, as already noted, Eq. (13.101) was derived for radio-wave propagation through the ionosphere and it therefore shows how the refractive index departs from unity. Furthermore, there is a useful approximation to Eq. (13.101) that applies when the second term in the square root is much larger than the first. This is known as the **quasi-longitudinal** or **quasi-parallel approximation** (in this case "longitudinal" applies to the direction of propagation with respect to the magnetic field, not the electric field polarization relative to the wave vector). Thus if

$$Y\cos\theta \gg \left|\frac{\frac{1}{2}Y^2\sin^2\theta}{1-X}\right|, \qquad (13.102)$$

$$\mu^2 \approx 1 - \frac{X}{1 - Y\cos\theta} \approx 1 - \frac{\omega_{pe}^2}{\omega(\omega - |\Omega_e|\cos\theta)}. \qquad (13.103)$$

When the inequality is reversed in Eq. (13.102), the approximation is known as the quasi-perpendicular or quasi-transverse approximation. Again, in this context quasi-transverse refers to the direction of propagation relative to the ambient magnetic field.

The dispersion relation (13.103) corresponds to the whistler mode when $\omega_{pe} \gg |\Omega_e|$, which is implied by the inequality in Eq. (13.102). Consequently, for the whistler mode, we can normalize wave numbers to the plasma skin depth, c/ω_{pe}, and normalize wave frequencies to $|\Omega_e|$. Solutions of Eq. (13.103) for two different frequencies are shown in Figure 13.6, plotted as a function of perpendicular and parallel wave number. Figure 13.6 (a) shows the solution for $\omega/|\Omega_e| = 0.75$. The diagonal line gives the asymptotic limit to Eq. (13.103), $\cos\theta = \omega/|\Omega_e|$, which is known as the resonance cone. Figure 13.6(b) shows the solution when $\omega/|\Omega_e| = 0.25$. In the figure we also show the direction of the phase velocity, v_p, and the group velocity, v_g. Because the solution shows a contour of fixed frequency in the wave vector space, the group velocity direction is given by the normal to the constant frequency contour. Near the resonance cone, the phase velocity and group velocity are almost perpendicular. The group velocity is parallel to the magnetic field for parallel propagation. But for $\omega/|\Omega_e| < 0.5$ there is another point where the

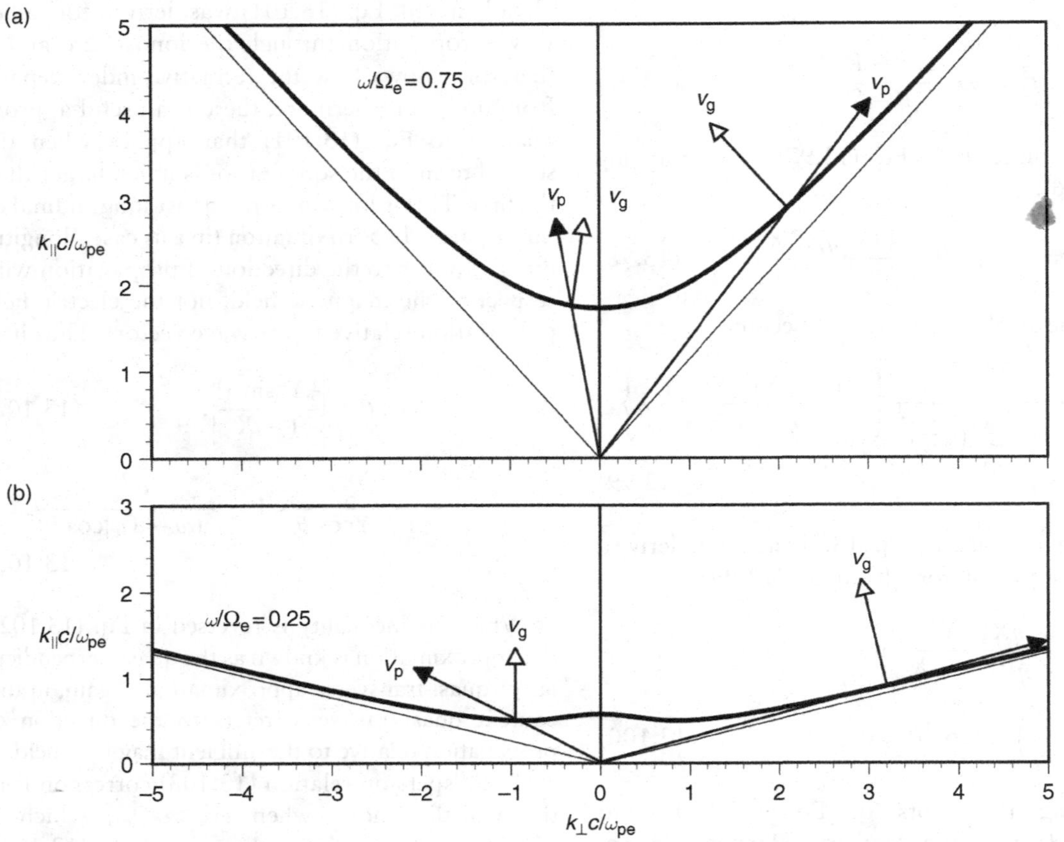

FIGURE 13.6. Whistler-mode dispersion curves assuming the quasi-longitudinal approximation, Eq. (13.103).

group velocity is parallel. This is given by $\cos\theta_G = 2\omega/|\Omega_e|$, and θ_G is known as the Gendrin angle (Gendrin, 1961). As a consequence, whistler-mode waves tend to be guided by the ambient magnetic field, although the presence of field-aligned plasma density structures also contributes to the ducted nature of whistler-mode propagation. The waves tend to refract into regions of higher plasma density for $\omega/|\Omega_e| < 0.5$.

As a final comment on the Appleton–Hartree relation, an important form of this relation includes collisions. This is because the ionosphere is collisional. At low altitudes, collisions with the neutral atmosphere dominate, while at higher altitudes, in and above the F region, electron–ion collisions become more important. Again, for simplicity, we consider electrons only, and include collisions by noting that, from the momentum equation (13.13),

$$-i\omega\mathbf{u}_e = -\frac{e}{m_e}(\mathbf{E} + \mathbf{u}_e \times \mathbf{B}_0) - \nu\mathbf{u}_e, \quad (13.104)$$

where we have assumed that the electrons are cold. We have included collisions in the last term on the right-hand side of Eq. (13.104), where ν is the electron collision frequency. Rearranging terms,

$$-i\omega\mathbf{u}_e = -\frac{e}{m_e(1 + i\nu/\omega)}(\mathbf{E} + \mathbf{u}_e \times \mathbf{B}_0), \quad (13.105)$$

and we can include the effect of collisions in the current density given by Eq. (13.32) by replacing the electron mass with $m_e(1 + iZ)$, where $Z = \nu/\omega$. Consequently, Eq. (13.101) becomes

$$\mu^2 = 1 - \frac{X}{1 + iZ - \frac{\frac{1}{2}Y^2\sin^2\theta}{1-X+iZ} \pm \left[\left(\frac{\frac{1}{2}Y^2\sin^2\theta}{1-X+iZ}\right)^2 + Y^2\cos^2\theta\right]^{1/2}}. \tag{13.106}$$

This is the Appleton–Hartree equation including collisions. Collisions can be important for radio waves propagating in and through the ionosphere. Indeed, waves transmitted in the amplitude-modulated (AM) medium wave band (300 kHz–3 MHz) can propagate large distances at night because the radio waves reflect at higher altitudes, where the collision frequency is lower.

13.3.6 Cold plasma wave modes – multi-ion wave dispersion

In the preceding section we gave two forms of the Appleton–Hartree dispersion relation: one including electrons only in Eq. (13.101), and one with collisions in Eq. (13.106). We also showed how to derive this dispersion relation with the inclusion of one ion species, from Eqs. (13.98) and (13.100), and formally this could be extended to include multiple ion species. Instead, however, we shall return to the dispersion relation given by Eq. (13.52), as this can be more easily expanded to include several ion species. Inclusion of multiple ion species is important for waves in the Earth's magnetosphere, especially the inner magnetosphere, as the plasma consists of protons, singly and doubly charged helium, and oxygen ions. Solutions of the dispersion relation (13.52) are shown in Figure 13.7, using a similar format to that used in Figure 13.5. In this case, however, both the wave frequency and wave number are plotted on a logarithmic scale. Furthermore, the shading indicates the polarization of the waves, given by

$$\frac{iE_x}{E_y} = \frac{\mu^2 - S}{D} \tag{13.107}$$

(from Eq. (13.73)). Light gray corresponds to left-hand elliptical polarization, and dark gray corresponds to right-hand elliptical polarization. In addition to the limiting cases of $\theta = 0°$ and $90°$, we include a dispersion curve for $\theta = 75°$.

The high-frequency portion of the plot, where $\omega > \Omega_p$, corresponds to the whistler mode, as shown in Figure 13.5 (Ω_p is the proton gyro-frequency). For parallel propagation, the dispersion relation is given by $\mu^2 = R$. Below the proton gyro-frequency, the parallel propagating L-mode reappears at high wave number. As the wave number decreases, this mode crosses the R-mode. The frequency at which this occurs is referred to as the **crossover frequency**. For the wave propagating at 75°, the wave polarization changes from left handed to right handed, since $R = L$, and $D = 0$ at the crossover frequency. As the wave number decreases further, we encounter a cutoff where $L = 0$.

For perpendicular propagation, we again see a resonance at $\omega = \omega_{LHR}$. But now there are additional resonances that occur between the different ion gyro-frequencies. These are referred to as **bi-ion hybrid resonances.** The combined effect of the ion gyro-resonances and the bi-ion hybrid resonances is to split the wave dispersion into separate bands. Thus, at large wave number, there is a band between the proton gyro-frequency and the proton–helium bi-ion frequency, and between the helium gyro-frequency and the helium–oxygen bi-ion frequency, with the bands moving to lower frequency for lower wave numbers. Waves observed in these bands within the terrestrial magnetosphere are frequently referred to as electromagnetic ion cyclotron (EMIC) waves. Sometimes the high-frequency portion of the left-hand polarized branch below Ω_O is also included in the definition of EMIC waves. For observations elsewhere in the solar system, these waves are called ion cyclotron waves.

We note that the inclusion of multiple ion species means that only the parallel propagating R-mode dispersion is continuous from $\omega = |\Omega_e|$ down to $\omega = 0$. Whistler-mode waves encounter the $L = 0$ cutoff between the proton and heavier ion (in this case helium) gyro-frequencies for all other angles of propagation. As we shall see, the low-frequency limit for the wave modes corresponds to the fast and shear MHD modes. Because the R-mode becomes the fast mode at low frequency, the low-frequency portion of the whistler mode, i.e., the waves with $\omega < \omega_{LHR}$, are also known as magnetosonic waves. But, again, in a multi-ion species plasma only the R-mode itself is coupled to the low-frequency fast mode. In the case where there is only one ion species in the plasma the whistler

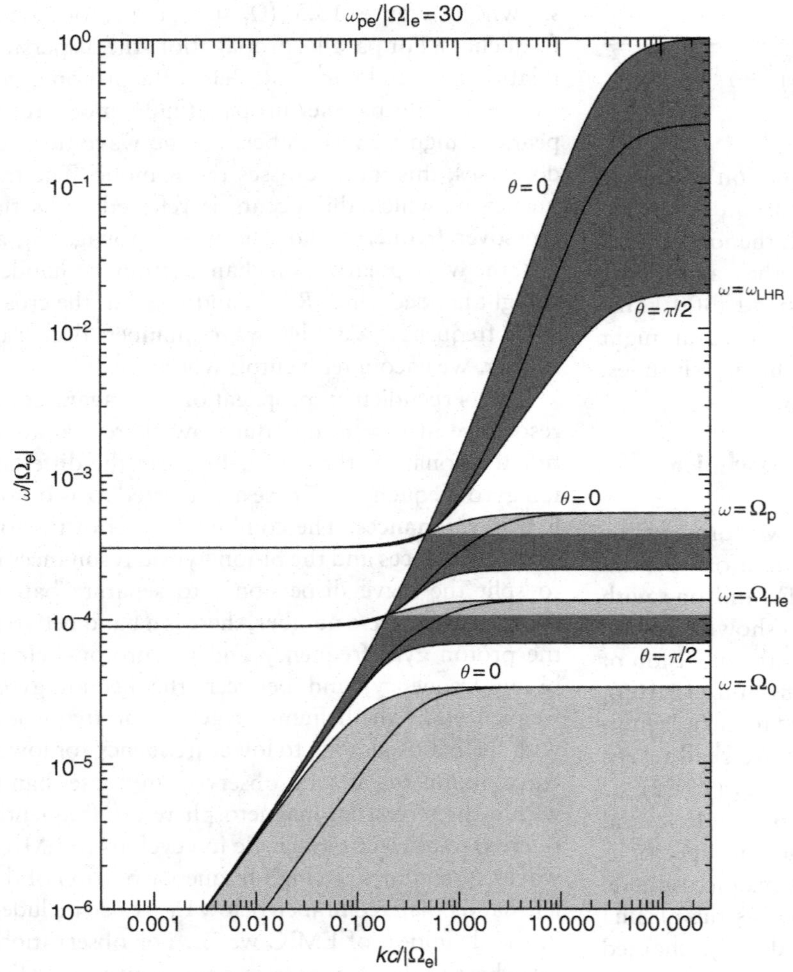

FIGURE 13.7. Multi-species wave dispersion relation. The plasma is assumed to consist of 50% protons, 20% helium, and 30% oxygen ions, where the percentages are given with respect to the number density.

mode continues to low frequencies for all propagation angles, since the $L = 0$ cutoff below the proton gyro-frequency vanishes.

At very low frequencies, where $\omega \to 0$, Eq. (13.72) shows that $P \to \infty$. From the bottom row of Eq. (13.73),

$$\frac{E_z}{E_x} = \frac{\mu^2 \cos\theta \sin\theta}{\mu^2 \sin^2\theta - P} \tag{13.108}$$

and $E_z \to 0$ as $\omega \to 0$. Furthermore, for the low-frequency limit, we find that Eq. (13.68) becomes

$$R \approx 1 - \sum_s \frac{\omega_{ps}^2}{\omega \Omega_s}\left(1 - \frac{\omega}{\Omega_s}\right), \tag{13.109}$$

where we have Taylor expanded the denominator in Eq. (13.68).

From definitions of the plasma frequency and gyro-frequency,

$$\sum_s \frac{\omega_{ps}^2}{\omega \Omega_s} = \sum_s \frac{n_s q_s}{\omega \varepsilon_0 B_0} = \frac{\rho_q}{\omega \varepsilon_0 B_0}, \tag{13.110}$$

where, again, ρ_q is the total charge density. From quasi-neutrality $\rho_q \approx 0$, and we shall ignore this term in Eq. (13.109).

For the last term in Eq. (13.109) we find

$$\sum_s \frac{\omega_{ps}^2}{\Omega_s^2} = \sum_s \frac{n_s m_s}{\varepsilon_0 B_0^2} = \frac{\rho_0}{\varepsilon_0 B_0^2} = \frac{c^2}{v_A^2}, \tag{13.111}$$

where ρ_0 is the total mass density. Thus,

$$R \approx 1 + \frac{c^2}{v_{\rm A}^2}. \qquad (13.112)$$

Similarly,

$$L \approx 1 + \frac{c^2}{v_{\rm A}^2}. \qquad (13.113)$$

Consequently, from Eq. (13.70), $S = R = L$, and, from Eq. (13.71), $D = 0$. From Eq. (13.73) we therefore have

$$(\mu^2 \cos^2\theta - S)E_x = 0 \qquad (13.114)$$

and

$$(\mu^2 - S)E_y = 0. \qquad (13.115)$$

Thus, we have two wave modes, one with $E_x = 0$, and

$$\frac{\omega^2}{k^2} = \frac{v_{\rm A}^2 c^2}{v_{\rm A}^2 + c^2} = \tilde{v}_{\rm A}^2, \qquad (13.116)$$

the other with $E_y = 0$, and

$$\frac{\omega^2}{k^2} = \tilde{v}_{\rm A}^2 \cos^2\theta. \qquad (13.117)$$

These correspond to the fast mode Eq. (13.116) and the Alfvén mode Eq. (13.117), but with a modified Alfvén velocity, $\tilde{v}_{\rm A}$. This is because in deriving the dispersion relation we included the displacement current, and this has the effect of ensuring that the wave phase speed does not exceed the speed of light even when the plasma mass density is very low and $v_{\rm A}^2 \gg c^2$.

Equations (13.114) and (13.115) show that, even for parallel propagation, the wave electric fields are linearly polarized. This appears to contradict the statement that for electromagnetic waves in a cold plasma the wave fields are RHC and LHC polarized for parallel propagation. However, the wave dispersion is degenerate when $\theta = 0$. Thus we can combine the R- and L-mode waves. In particular, if we combine an RHC wave with an LHC wave such that the x-components of the wave electric field are in phase, then the y-components cancel, corresponding to the polarization for the shear mode. Similarly if the x-components are in antiphase, then the y-components are in phase, resulting in a linearly polarized wave corresponding to the fast mode.

In closing this section, we note that we have explored the wave dispersion for electromagnetic waves in a cold magnetized plasma extensively. This is because this allows us to introduce many of the concepts concerning wave propagation in a plasma without other complications, such as the inclusion of thermal pressure or kinetic effects. In the following section we include thermal pressure for low-frequency waves, complementing the discussion in Chapter 3, where we introduced the MHD wave modes.

13.3.7 Low-frequency electromagnetic waves

In Chapter 3 we derived the dispersion relation for low-frequency (MHD) waves. In doing so we used the plasma flow velocity to characterize the wave perturbations. Given the discussion so far in this chapter, we show we can get the same result using **j** and **E**.

Because we are considering low-frequency waves, we use the linearized momentum equation (3.169) from Chapter 3,

$$\omega\rho_0 \left[\mathbf{u} - c_{\rm s}^2 \frac{\mathbf{k}(\mathbf{k}\cdot\mathbf{u})}{\omega^2} \right] = i\mathbf{j} \times \mathbf{B}_0, \qquad (13.118)$$

where $\mathbf{u}$ is the bulk flow velocity induced in the plasma by the wave perturbation and, again, $c_{\rm s} = \sqrt{\gamma P_0/\rho_0}$ is the sound speed. Using the same cartesian coordinate system as shown in Figure 13.4, the different components of Eq. (13.118) become

$$\omega\rho_0 \left[u_x - c_{\rm s}^2 \frac{k_\perp(k_\perp u_x + k_\parallel u_z)}{\omega^2} \right] = ij_y B_0, \quad (13.119)$$

$$\omega\rho_0 u_y = -ij_x B_0, \qquad (13.120)$$

and

$$u_z - c_{\rm s}^2 \frac{k_\parallel(k_\perp u_x + k_\parallel u_z)}{\omega^2} = 0. \qquad (13.121)$$

Using Eq. (13.121) to specify u_z in terms of u_x, we can rewrite Eq. (13.119) as

$$\omega \rho_0 u_x \frac{\left(\omega^2 - k^2 c_s^2\right)}{\left(\omega^2 - k_{||}^2 c_s^2\right)} = i j_y B_0. \tag{13.122}$$

We shall now make use of the frozen-in condition $\mathbf{E}_\perp = -\mathbf{u} \times \mathbf{B}_0$. From Eq. (13.122),

$$j_y = -iE_y \frac{\omega}{\mu_0 v_A^2} \frac{\left(\omega^2 - k^2 c_s^2\right)}{\left(\omega^2 - k_{||}^2 c_s^2\right)}, \tag{13.123}$$

while, from Eq. (13.120),

$$j_x = -iE_x \frac{\omega}{\mu_0 v_A^2}. \tag{13.124}$$

These two equations can now be substituted into Eq. (13.38) to give the dispersion relation for the MHD waves. There is an important caveat, however. The parallel component of the current density is indeterminate, in that j_z appears in Eq. (13.38), but we do not yet have a second equation, similar to Eq. (13.123) or Eq. (13.124), that expresses j_z in terms of the wave electric fields. To overcome this we make use of Eq. (13.35), which on summing over species becomes

$$j_z = i \frac{\omega_p^2}{\omega} \varepsilon_0 E_z. \tag{13.125}$$

Substituting the derived current densities into Eq. (13.38) we find

$$\left[\mu^2 \cos^2\theta - 1 - \frac{c^2}{v_A^2}\right] E_x - \mu^2 \sin\theta \cos\theta E_z = 0, \tag{13.126}$$

$$\left[\mu^2 - 1 - \frac{c^2}{v_A^2} \frac{\left(\omega^2 - k^2 c_s^2\right)}{\left(\omega^2 - k_{||}^2 c_s^2\right)}\right] E_y = 0, \tag{13.127}$$

$$-\mu^2 \sin\theta \cos\theta E_x + \left[\mu^2 \sin^2\theta - 1 + \frac{\omega_p^2}{\omega^2}\right] E_z = 0. \tag{13.128}$$

The set of equations (13.126), (13.127), and (13.128) has some terms that are not usually kept in the MHD approximation. The first of these is the plasma frequency term in Eq. (13.128). We introduced this term in Eq. (13.125) to provide a closed set of equations. In general, however, if $\omega \approx 0$, Eq. (13.125) suggests that, for finite j_z, $E_z \approx 0$. This is indeed the standard approximation used for MHD. However, sometimes, even though the wave frequency may be small, the plasma frequency is also small. This is the case for the Earth's auroral region, and the Alfvén mode can have a finite parallel electric field. These are known as inertial Alfvén waves because the parallel electric field comes from the electron inertia term in the momentum equation, since $\omega_p \approx \omega_{pe}$. These waves are thought to result in the ~100 eV bi-directional electrons observed in what has become known as the alfvénic aurora, as discussed in Chapter 11. We should emphasize that it is the shear mode that can have inertial corrections. The fast and slow modes do not since they do not carry field-aligned current. Going forward, we drop the inertia term and assume $E_z \approx 0$. This removes the E_z term in Eq. (13.126), and we immediately recover the shear-mode dispersion relation (13.117). If we further neglect the displacement current, we get the classical form of the shear-mode dispersion relation:

$$\omega^2 = k_{||}^2 v_A^2. \tag{13.129}$$

This mode requires $E_y = 0$.

The fast- and slow-mode dispersion relation comes from Eq. (13.127). Again neglecting the displacement current, we find

$$\omega^2 = \frac{k^2}{2} \left[\left(v_A^2 + c_s^2\right) \pm \left[\left(v_A^2 + c_s^2\right)^2 - 4 v_A^2 c_s^2 \cos^2\theta\right]^{1/2}\right]. \tag{13.130}$$

These modes require $E_x = 0$ and $E_z = 0$, even if we have included the electron inertia. Thus, as already noted above, only the shear mode can have a parallel electric field, and Eq. (13.123) shows that E_y only results in a j_y current, again demonstrating that the fast and slow modes do not carry field-aligned current.

Solutions of Eqs. (13.129) and (13.130) are shown in Figure 13.8, with $c_s = 0.7 v_A$, and Figure 13.9, with $c_s = 1.2 v_A$. The left-hand plot in

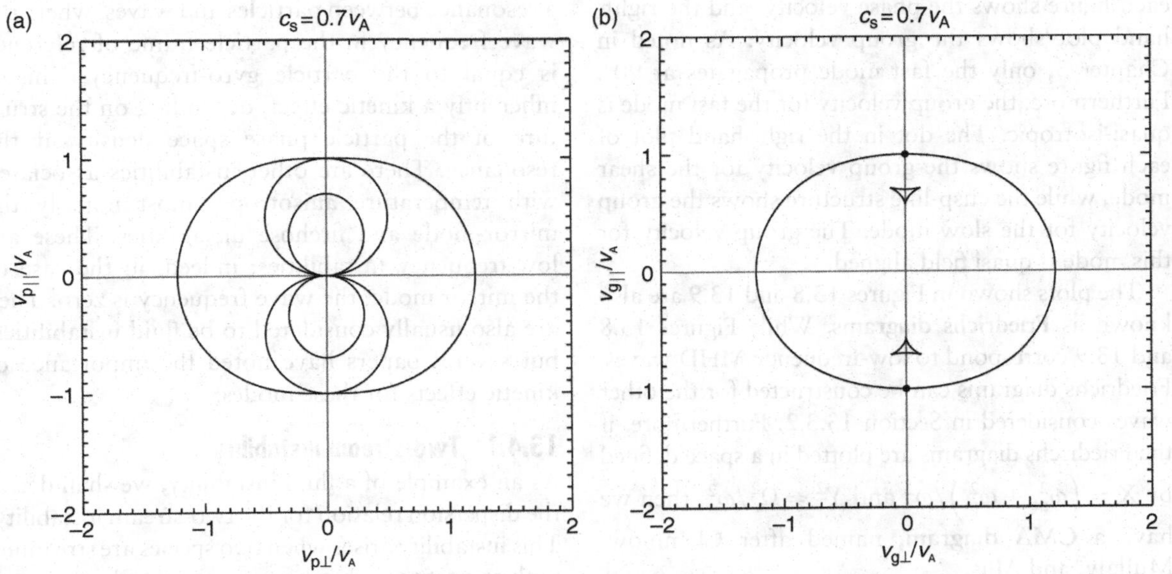

FIGURE 13.8. Phase velocity (a) and group velocity (b) plots for the MHD waves when $c_s < v_A$. The vertical axis corresponds to propagation along the ambient magnetic field. Wave speeds are normalized to the Alfvén speed.

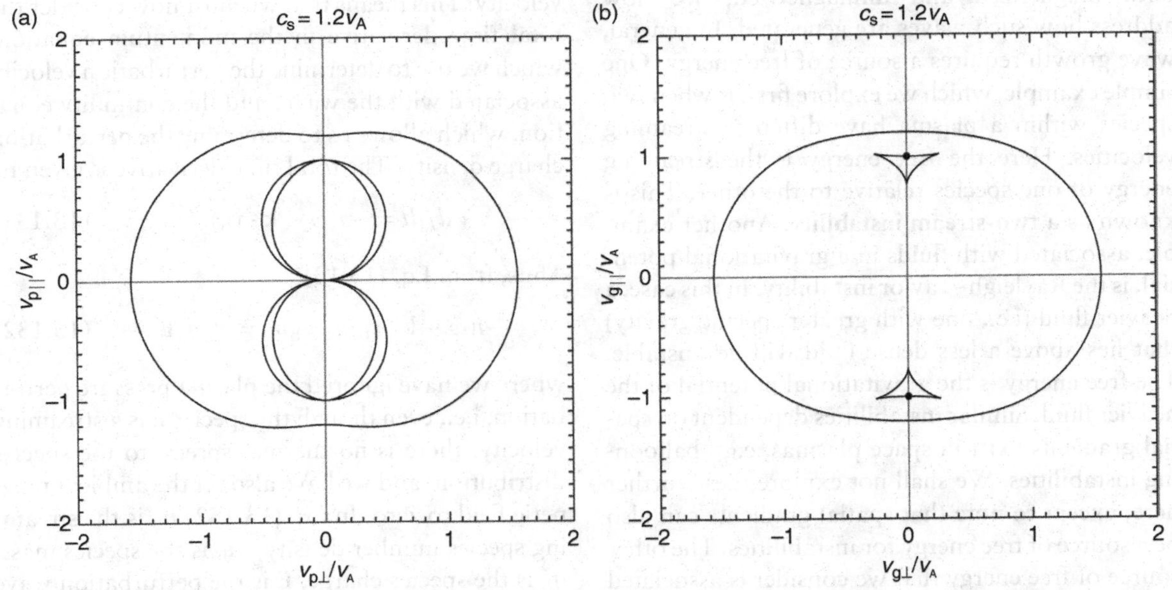

FIGURE 13.9. Phase velocity (a) and group velocity (b) plots for the MHD waves when $c_s > v_A$. The figure has the same format as Figure 13.8.

each figure shows the phase velocity, and the right-hand plot shows the group velocity. As noted in Chapter 3, only the fast mode propagates at 90°. Furthermore, the group velocity for the fast mode is quasi-isotropic. The dot in the right-hand plot of each figure shows the group velocity for the shear mode, while the cusp-like structure shows the group velocity for the slow mode. The group velocity for this mode is quasi field aligned.

The plots shown in Figures 13.8 and 13.9 are also known as **Friedrichs diagrams.** While Figures 13.8 and 13.9 correspond to low-frequency MHD waves, Friedrichs diagrams can be constructed for the other waves considered in Section 13.3.2. Furthermore, if the Friedrichs diagrams are plotted in a space defined by $X = \left(\omega_{pe}^2 + \omega_{pi}^2\right)/\omega^2$ and $Y^2 = \Omega_e^2/\omega^2$, then we have a **CMA diagram**, named after Clemmow, Mullaly, and Allis.

13.4 STREAMING INSTABILITIES

Having explored the wave dispersion for a plasma, both magnetized and unmagnetized, we now address how such waves are generated. In general, wave growth requires a source of free energy. One simple example, which we explore first, is when two species within a plasma have different streaming velocities. Here the free energy is the streaming energy of one species relative to the other. This is known as a **two-stream instability**. Another example, associated with fluids in a gravitational potential, is the Rayleigh–Taylor instability. In this case, a heavier fluid (i.e., one with greater specific gravity) that lies above a less dense fluid will be unstable. The free energy is the gravitational potential of the heavier fluid. Similar instabilities dependent on spatial gradients exist in space plasmas, e.g., ballooning instabilities. We shall not explore these further here, except to note that spatial gradients can also be a source of free energy for instabilities. The other source of free energy that we consider is associated with gradients of the distribution in velocity space. One example is where the distribution has a temperature anisotropy. In Section 13.5.3 we show how a temperature anisotropy results in the growth of whistler-mode waves, as this depends on a resonance between particles and waves, where the wave frequency in the particle frame of reference is equal to the particle gyro-frequency. This is inherently a kinetic effect, depending on the structure of the particle phase space density at the resonance. There are other instabilities associated with temperature anisotropy, most notably the mirror-mode and firehose instabilities. These are low-frequency instabilities; indeed, in the case of the mirror mode, the wave frequency is zero. They are also usually considered to be fluid instabilities, but several papers have noted the importance of kinetic effects for these modes.

13.4.1 Two-stream instability

As an example of a fluid instability, we shall derive the dispersion relation for the two-stream instability. This instability arises when two species are streaming with respect to each other, and the simplest approximation to make is to assume that the waves are electrostatic and the plasma is unmagnetized.

We again assume a harmonic perturbation, but in this case we also allow one of the species to have a streaming velocity, $\mathbf{v}_b$, also referred to as the beam velocity. This means that we must now consider the total time derivative in the momentum equation, which we use to determine the perturbation velocity associated with the wave, and the continuity equation, which allows us to determine the perturbation charge density. The total time derivative is given by

$$d/dt = -i(\omega - \mathbf{k} \cdot \mathbf{v}_b). \tag{13.131}$$

Thus, from Eq. (13.13)

$$-i(\omega - \mathbf{k} \cdot \mathbf{v}_b) n_b m_b \mathbf{u}_b = n_b q_b \mathbf{E}, \tag{13.132}$$

where we have ignored the plasma pressure perturbation, i.e., even though the species has a streaming velocity, there is no thermal spread to the species distribution, and we have also set the ambient magnetic field to zero. In Eq. (13.132) n_b is the streaming species number density, m_b is the species mass, q_b is the species charge, $\mathbf{E}$ is the perturbation wave electric field, and $\mathbf{u}_b$ is the associated perturbation velocity.

From the continuity equation (13.14),

$$(\omega - \mathbf{k} \cdot \mathbf{v}_b)\delta n_b = n_b \mathbf{k} \cdot \mathbf{u}_b. \tag{13.133}$$

Again we have used Eq. (13.131) because the species has a zero-order streaming velocity. Here δn_b is the first-order perturbation to the number density.

To derive the dispersion relation, we combine Eqs. (13.133) and (13.132) to relate the perturbation electric field to the perturbation number density, which, on being multiplied by the species charge, gives the perturbation charge density per species. For simplicity, we assume only two species; on summing over species, from Poisson's equation,

$$s(\omega, \mathbf{k}) = \frac{\omega_{pa}^2}{\omega^2} + \frac{\omega_{pb}^2}{(\omega - \mathbf{k} \cdot \mathbf{v}_b)^2} = 1, \qquad (13.134)$$

where ω_{pa} corresponds to the non-streaming (or ambient) species, and $s(\omega, \mathbf{k})$ is the negative of the electrical susceptibility, i.e., the relative permittivity $\kappa = 1 - s(\omega, \mathbf{k})$.

In general there are four solutions of Eq. (13.134) for the wave frequency ω. Since Eq. (13.134) has two positive infinities at $\omega = 0$ and $\omega = \mathbf{k} \cdot \mathbf{v}_b$, and the asymptotic value of $s(\omega, \mathbf{k})$ for $\omega = \pm\infty$ is 0, there are two real solutions of Eq. (13.134), for $\omega < 0$ and $\omega > \mathbf{k} \cdot \mathbf{v}_b$, where for convenience we have assumed $\mathbf{k} \cdot \mathbf{v}_b > 0$. Between $\omega = 0$ and $\omega = \mathbf{k} \cdot \mathbf{v}_b$, $s(\omega, \mathbf{k})$ has a minimum, and, if the minimum falls below unity, then the remaining two roots are real; otherwise the roots are complex. One of these roots corresponds to a growing wave.

The frequency for which the minimum in $s(\omega, \mathbf{k})$ occurs can be found from $\partial s(\omega, \mathbf{k})/\partial\omega = 0$, giving

$$\omega \left(1 + \left(\frac{\omega_{pb}^2}{\omega_{pa}^2} \right)^{1/3} \right) = \mathbf{k} \cdot \mathbf{v}_b. \qquad (13.135)$$

At that frequency,

$$s_{min} = \frac{1}{(\mathbf{k} \cdot \mathbf{v}_b)^2} \left(\omega_{pa}^{2/3} + \omega_{pb}^{2/3} \right)^3. \qquad (13.136)$$

In order to explore further the streaming instabilities, we first consider $\left(\omega_{pb}^2/\omega_{pa}^2\right)^{1/3} \ll 1$ in Eq. (13.135). This would be the case when a relatively low-density beam of electrons is streaming through a higher density background of cold electrons. We could include the background ions, but in general $m_i \gg m_e$, and we shall assume $\omega_{pa} = \omega_{pe}$, i.e., the plasma frequency for the ambient electrons. Equation (13.135) suggests that solutions of Eq. (13.134) should be found for

$$\omega = (\mathbf{k} \cdot \mathbf{v}_b)(1 + \alpha e^{i\theta}), \qquad (13.137)$$

where $\alpha \ll 1$. The term in the exponent allows the frequency to be complex, and the growth rate for waves is given by

$$\gamma = (\mathbf{k} \cdot \mathbf{v}_b)\alpha \sin\theta. \qquad (13.138)$$

Substituting Eq. (13.137) into Eq. (13.136), and making use of Eq. (13.138), we find for the imaginary part of Eq. (13.134)

$$s_i = -\frac{2\omega_{pe}^2 \gamma}{(\mathbf{k} \cdot \mathbf{v}_b)^3} - \frac{\omega_{pb}^2 \sin^2\theta \sin 2\theta}{\gamma^2} = 0, \qquad (13.139)$$

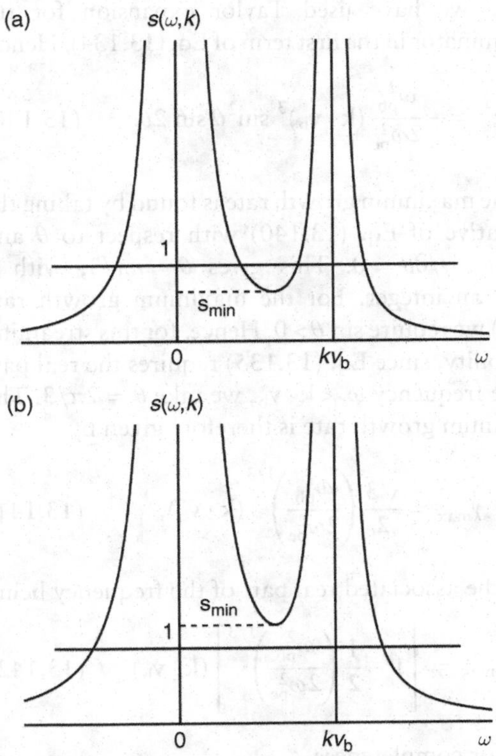

FIGURE 13.10. Plots of Eq. (13.134) for (a) $s_{min} < 1$ and (b) $s_{min} > 1$. The real solutions of the dispersion relation are found where $s(\omega, \mathbf{k}) = 1$. For (a) there are four real solutions, while for (b) two of the solutions are complex.

where we have used Taylor expansion for the denominator in the first term of Eq. (13.134). Hence,

$$\gamma^3 = -\frac{\omega_{\mathrm{pb}}^2}{2\omega_{\mathrm{pe}}^2}(\mathbf{k}\cdot\mathbf{v}_{\mathrm{b}})^3 \sin^2\theta \sin 2\theta. \qquad (13.140)$$

The maximum growth rate is found by taking the derivative of Eq. (13.140) with respect to θ and setting $\partial\gamma/\partial\theta = 0$. This gives $\theta = n\pi/3$, with n being an integer. For the maximum growth rate ($\gamma > 0$) we require $\sin\theta > 0$. Hence, for this streaming instability, since Eq. (13.135) requires the real part of the frequency $\omega_{\mathrm{r}} < \mathbf{k}\cdot\mathbf{v}_{\mathrm{b}}$, we take $\theta = 2\pi/3$. The maximum growth rate is therefore given by

$$\gamma_{\max} = \frac{\sqrt{3}}{2}\left(\frac{\omega_{\mathrm{pb}}^2}{2\omega_{\mathrm{pe}}^2}\right)^{1/3}(\mathbf{k}\cdot\mathbf{v}_{\mathrm{b}}), \qquad (13.141)$$

with the associated real part of the frequency being

$$\omega_{\mathrm{rmax}} = \left[1 - \frac{1}{2}\left(\frac{\omega_{\mathrm{pb}}^2}{2\omega_{\mathrm{pe}}^2}\right)^{1/3}\right](\mathbf{k}\cdot\mathbf{v}_{\mathrm{b}}), \qquad (13.142)$$

and, for completeness,

$$\alpha = \left(\frac{\omega_{\mathrm{pb}}^2}{2\omega_{\mathrm{pe}}^2}\right)^{1/3}. \qquad (13.143)$$

We now substitute Eq. (13.137) into (13.134), with α given by Eq. (13.143), and set $\theta = 2\pi/3$. The imaginary part of Eq. (13.134) is already satisfied, while, from the real part,

$$\mathbf{k}\cdot\mathbf{v}_{\mathrm{b}} = \omega_{\mathrm{pe}}, \qquad (13.144)$$

and we can rewrite the complex frequency at which maximum growth occurs as

$$\omega_{\max} = \omega_{\mathrm{pe}}\left[1 - \frac{1}{2}\left(\frac{\omega_{\mathrm{pb}}^2}{2\omega_{\mathrm{pe}}^2}\right)^{1/3}\left(1 - i\sqrt{3}\right)\right]. \qquad (13.145)$$

The other limit we shall consider is when $\left(\omega_{\mathrm{pb}}^2/\omega_{\mathrm{pa}}^2\right)^{1/3} \gg 1$. This could occur in a two-species plasma where the electrons are streaming and the ions are stationary. Unlike the two-stream instability given by Eq. (13.145), there is no stationary ambient electron distribution. For a singly charged ion species, the limit therefore becomes $(m_{\mathrm{i}}/m_{\mathrm{e}})^{1/3} \gg 1$, which is generally the case for a plasma consisting of ions and electrons. This instability, which is associated with electron beam streaming relative to the ambient ions, is known as the **Buneman instability.**

Following the same approach as used above, we shall now assume

$$\omega = (\mathbf{k}\cdot\mathbf{v}_{\mathrm{b}})\alpha e^{i\theta}, \qquad (13.146)$$

where again $\alpha \ll 1$. Because we are now considering an electron and ion species, we replace ω_{pb} with ω_{pe} and ω_{pa} with ω_{pi}.

Since, from the imaginary part of Eq. (13.146), γ is given by Eq. (13.138), we can use the same analysis to determine when maximum growth occurs. Again $\partial\gamma/\partial\theta = 0$ is given by $\theta = n\pi/3$. In this case, however, we want both $\gamma > 0$ and $\omega_{\mathrm{r}} > 0$, and we therefore choose $\theta = \pi/3$.

We also find that Eq. (13.144) applies for the maximum growth solution, and the complex frequency for the Buneman instability is given by

$$\omega_{\max} = \frac{\omega_{\mathrm{pe}}}{2}\left(\frac{\omega_{\mathrm{pi}}^2}{2\omega_{\mathrm{pe}}^2}\right)^{1/3}\left(1 + i\sqrt{3}\right). \qquad (13.147)$$

It should be noted that Eq. (13.147) indicates that the growth rate for the Buneman instability is actually larger than the real part of the frequency. This implies that non-linear effects become important very quickly. In particular, we expect the wave to scatter the electrons in the beam, thereby increasing the thermal spread of the distribution. We can show that this tends to quench the instability. To do this we include the electron pressure in Eq. (13.132) and make use of Eq. (13.15). Consequently, Eq. (13.134) becomes

$$s(\omega,\mathbf{k}) = \frac{\omega_{\mathrm{pa}}^2}{\omega^2} + \frac{\omega_{\mathrm{pb}}^2}{(\omega - \mathbf{k}\cdot\mathbf{v}_{\mathrm{b}})^2 - k^2 c_{\mathrm{se}}^2} = 1. \qquad (13.148)$$

In Eq. (13.148) c_{se} is the electron sound speed, $c_{\mathrm{se}}^2 = \gamma_{\mathrm{e}} K T_{\mathrm{e}}/m_{\mathrm{e}}$, where γ_{e} is the ratio of specific

Table 13.1. Behavior of $s(\omega, \mathbf{k})$ for four frequency bands; the values for $s(\omega, \mathbf{k})$ correspond to values for ω just above the lower limit of the frequency range to just below the upper limit

Frequency range	$s(\omega, \mathbf{k})$
$-\infty < \omega < k(v_b - c_{se})$	0 to $+\infty$
$k(v_b - c_{se}) < \omega < 0$	$-\infty$ to $+\infty$
$0 < \omega < k(v_b + c_{se})$	$+\infty$ to $-\infty$
$k(v_b + c_{se}) < \omega < \infty$	$+\infty$ to 0

heats for the electron beam and T_e is the beam temperature.

The major difference between Eq. (13.134) and Eq. (13.148) is that the second term goes through infinity for two different frequencies, $\omega = \mathbf{k} \cdot \mathbf{v}_b \pm kc_{se}$. Furthermore, unlike the infinity when $\omega = 0$, where $s(\omega, \mathbf{k}) \rightarrow +\infty$ as $\omega \rightarrow 0$ from above and below, $s(\omega, \mathbf{k})$ changes sign as ω crosses $\mathbf{k} \cdot \mathbf{v}_b \pm kc_{se}$. If $c_{se} > v_b$ we find that $s(\omega, \mathbf{k})$ has the behavior given in Table 13.1.

From Table 13.1 we see that for each of the four frequency bands there must be a point where $s(\omega, \mathbf{k}) = 1$, and consequently where all of the solutions of Eq. (13.148) correspond to a real frequency and there are no unstable solutions. Therefore, once non-linear effects increase the temperature of the streaming distribution such that $c_{se} > v_b$, the Buneman instability is quenched.

It should be noted, however, that the discussion is incomplete. Once the electron distribution has a thermal spread, there are electrons that have the same velocity as the wave phase velocity. As noted earlier, this is a resonance, and we expect a strong interaction between the waves and the resonant particles. But resonant interactions are not included in the fluid formalism presented in Section 13.3. To understand resonant interactions we must resort to kinetic theory.

13.5 KINETIC THEORY AND WAVE INSTABILITIES

As noted in the introduction to Section 13.4, gradients in the particle velocity space distribution can be a source of free energy. The free energy in the distribution is made available for wave growth through wave–particle resonance. Exploring the effect of resonances requires a kinetic approach, and we develop that here.

One type of resonant instability is where some of the particles have the same velocity as the phase velocity of the wave. If the plasma is magnetized, the resonance is when the parallel phase velocity is the same as the particle parallel velocity:

$$\omega = k_{\parallel} v_{\parallel}. \tag{13.149}$$

This is known as Landau resonance. This resonance is equivalent to the statement that the particle sees a wave with zero frequency, i.e., the Doppler shift results in a wave with zero frequency in the particle frame of reference, and so a strong interaction is to be expected.

The other type of resonance is gyro-resonance, where the Doppler-shifted frequency is some multiple of the particle gyro-frequency:

$$\omega - k_{\parallel} v_{\parallel} = n\Omega, \tag{13.150}$$

where n is an integer with $n \neq 0$ and Ω is the particle species gyro-frequency, including the sign of the particle charge, $\Omega = -qB_0/m$. The minus sign is included so that, by convention, Ω>0 corresponds to right-handed gyration about the magnetic field, i.e., electron gyration. In the case of gyro-resonance, the particle senses wave fields that are oscillating at some multiple of the particle gyro-frequency, and, again, a strong interaction is expected.

13.5.1 Linearized Vlasov equation

Plasma kinetic theory was introduced in Chapter 3, and the fundamental equation that forms the basis of kinetic theory is the Vlasov equation:

$$\frac{\partial f}{\partial t} + \mathbf{v} \cdot \nabla f + \mathbf{a} \cdot \nabla_{\mathbf{v}} f = 0, \tag{3.75b}$$

where f is the phase space density or distribution function. The first two terms give the time rate of change and advective derivative following a particle in configuration space (r). The third term, $\mathbf{a} \cdot \nabla_{\mathbf{v}}$, gives the advective derivative following a particle in velocity space (v), since $\mathbf{a} \cdot \nabla_{\mathbf{v}} = (dv/dt) \cdot \partial f/\partial \mathbf{v}$. We also defined the Liouville operator

$$\mathscr{L} \equiv \partial/\partial t + \mathbf{v} \cdot \nabla + \mathbf{a} \cdot \nabla_{\mathbf{v}}. \tag{3.87}$$

One consequence of Eq. (3.75b) is the **Liouville theorem,** which states that f is a constant following a particle trajectory. We shall make use of that theorem here.

Another consequence is **Jeans's theorem**, which states that any phase space distribution that is a function of the constants of the motion is a solution of the Vlasov equation. To see this we note that the constants of the motion are indeed constant following the motion of a particle, i.e., along a particle trajectory. If we denote the constants of the motion as α, then

$$\mathscr{L}f = \frac{\partial f}{\partial \alpha}\mathscr{L}\alpha = 0 \tag{13.151}$$

since $\mathscr{L}\alpha = 0$.

The next step is to expand the Vlasov equation by Taylor expansion, i.e.,

$$\mathscr{L}f = \mathscr{L}_0 f_0 + \mathscr{L}_0 f_1 + \mathscr{L}_1 f_0 = 0, \tag{13.152}$$

where the subscripts 0 and 1 denote zero order and first order, respectively, and $\mathscr{L}_0$ is the zero-order Liouville operator, where the forces in $\mathbf{a}$ correspond to the zero-order forces. Frequently we assume a uniform system with a uniform magnetic field $\mathbf{B}_0$ and

$$\mathbf{a} = \frac{q}{m}\mathbf{v} \times \mathbf{B}_0. \tag{13.153}$$

In this case the constants of the motion (α) are $v_\perp$ and $v_\|$, where perpendicular and parallel are defined with respect to the ambient magnetic field. Since f_0 is assumed to be a function of these constants of the motion, then automatically $\mathscr{L}_0 f_0 = 0$.

We have already noted that $\mathscr{L}$ corresponds to the total time derivative following a particle trajectory, and we can therefore rewrite Eq. (13.150) as

$$f_1(\mathbf{r}, \mathbf{v}, t) = -\int_{-\infty}^{t} \mathscr{L}_1(\mathbf{r}', \mathbf{v}', t') f_0(\mathbf{v}')dt', \tag{13.154}$$

where the integration follows the zero-order particle trajectory that passes through $(\mathbf{r}, \mathbf{v})$ at time t and $(\mathbf{r}', \mathbf{v}')$ at time t'. For a uniform magnetic field, the zero-order trajectory depends on Eq. (13.153), and for that reason $f_0(\mathbf{v}')$ is a function of $\mathbf{v}'$ only.

The next step is to assume that the perturbations are harmonic, i.e., the first-order terms all have form

$$\mathbf{E}(\mathbf{r}, t) = \mathbf{E}e^{-i(\omega t - \mathbf{k}\cdot\mathbf{r})} \tag{13.155}$$

etc. Going forward we use the convention that, for example, $\mathbf{E}(\mathbf{r}, t)$ corresponds to a harmonic perturbation, including the term given by the exponent in Eq. (13.155), while $\mathbf{E}$ is the amplitude of the harmonic perturbation, since the latter does not depend on position or time.

Hence,

$$f_1(\mathbf{v}) = -\frac{q}{m\omega}\int_{-\infty}^{t} e^{-i\omega(t'-t)}e^{i\mathbf{k}\cdot(\mathbf{r}'-\mathbf{r})}\big[(\omega - \mathbf{k}\cdot\mathbf{v}')\mathbf{E} + \mathbf{k}(\mathbf{v}'\cdot\mathbf{E})\big] \cdot \frac{\partial f_0(\mathbf{v}')}{\partial \mathbf{v}'}dt'. \tag{13.156}$$

Also, to be clear, $f_1(\mathbf{v})$ is the amplitude of the first-order perturbation to f. Furthermore, in order to ensure convergence of the integral, ω is assumed to have a small imaginary part, corresponding to wave growth, so that the wave fields vanish at $t' = -\infty$.

The next step is to define the trajectory that connects $(\mathbf{r}', \mathbf{v}', t')$ to $(\mathbf{r}, \mathbf{v}, t)$. In defining the trajectory, we find that the time history of the trajectory only depends on $t' - t$. We therefore define a new variable $t'' = t - t'$, but immediately drop the double prime. This new variable should not be confused with t in Eq. (13.155), for example, as it is simply appears as an integration variable in Eq. (13.156).

We now assume the conventional orthogonal coordinate system where the magnetic field $\mathbf{B}_0$ defines the z-axis and the perpendicular wave vector $\mathbf{k}_\perp$ defines the x-axis. The past-history vector that connects $\mathbf{v}'$ to $\mathbf{v} = (v_x, v_y, v_z)$ is

$$\mathbf{V}(t) = (v_x \cos\Omega t + v_y \sin\Omega t, -v_x \sin\Omega t + v_y \cos\Omega t, v_z). \tag{13.157}$$

Because of our new integration variable t, the velocity defined by Eq. (13.157) appears to rotate in a left-handed sense for increasing t. But it should be remembered that time goes backward as t increases. A side note may be in order here: $v_z = v_\|$, and, depending on context, we use either to denote this component of the velocity vector. We usually use v_z when the other vector components are also given in cartesian coordinates.

The corresponding vector for the position of the particle is given by

$$\mathbf{R}(t) = \mathbf{r} - \mathbf{r}' = \left(\frac{v_x}{\Omega}\sin\,\Omega t + \frac{v_y}{\Omega}(1 - \cos\,\Omega t), -\frac{v_x}{\Omega}(1 - \cos\,\Omega t) + \frac{v_y}{\Omega}\sin\,\Omega t, v_z t\right). \quad (13.158)$$

The past-history integral (13.156) now becomes

$$f_1(\mathbf{v}) = -\frac{q}{m\omega}\int_0^\infty e^{i\omega t}e^{-i\mathbf{k}\cdot\mathbf{R}(t)}\left[\left(\omega - \mathbf{k}\cdot\mathbf{V}(t)\right)\mathbf{E} + \mathbf{k}\left(\mathbf{V}(t)\cdot\mathbf{E}\right)\right]\cdot\frac{\partial\mathbf{v}}{\partial\mathbf{V}(t)}\cdot\frac{\partial f_0(\mathbf{v})}{\partial\mathbf{v}}dt, \quad (13.159)$$

where we have used Jeans's theorem, which states $f_0(\mathbf{v}') = f_0(\mathbf{v})$, and $\partial\mathbf{v}/\partial\mathbf{V}(t)$ is a tensor:

$$\left.\frac{\partial\mathbf{v}}{\partial\mathbf{V}(t)}\right|_{ij} = \frac{\partial v_j}{\partial V_i(t)} = \begin{pmatrix} \cos\,\Omega t & \sin\,\Omega t & 0 \\ -\sin\,\Omega t & \cos\,\Omega t & 0 \\ 0 & 0 & 1 \end{pmatrix}. \quad (13.160)$$

In order to derive a wave dispersion relation, we use Eq. (13.159) to derive the perturbation charge density per species,

$$\rho_q = \int q f_1(\mathbf{v})d^3v, \quad (13.161)$$

if we assume the waves are electrostatic, or the perturbation current density,

$$\mathbf{j} = \int q\mathbf{v}f_1(\mathbf{v})d^3v, \quad (13.162)$$

for electromagnetic waves.

While the set of equations (13.157)–(13.162) are quite concise, they can be difficult to solve for a general distribution function. Nevertheless, we shall explore the consequences of the past-history integral further in addressing three topics: Landau resonance, gyro-resonance, and Bernstein modes. We include the latter since they are frequently observed in planetary magnetospheres and, from a pedagogical standpoint, they show how to proceed in deriving a kinetic wave dispersion relation.

13.5.2 Landau resonance

In order to explore the Landau resonance we make several assumptions. First, we assume an electrostatic perturbation. In that case the terms in the square brackets in Eq. (13.159) that are dependent on $\mathbf{V}(t)$ can be ignored, since they arose from the wave magnetic field contribution to the Lorentz force. We further assume that the wave is propagating along the magnetic field. The only term that remains in $\mathbf{R}(t)$ is therefore the z-term, and the only term that remains in Eq. (13.160) is the $\partial v_{||}/\partial V_{||}(t)$ term (=1). We can consequently perform the integral in Eq. (13.159), and

$$f_1(\mathbf{v}) = -i\frac{q}{m}E_{||}\frac{\partial f/\partial v_{||}}{\omega - k_{||}v_{||}}. \quad (13.163)$$

In Eq. (13.163) we have replaced $f_0(\mathbf{v})$ with f, and we shall use this notation going forward.

Substituting Eq. (13.163) into Eq. (13.161) yields

$$\rho_{\mathrm{q}} = -i\frac{q^2}{m}E_{||}\int\frac{\partial f/\partial v_{||}}{\omega - k_{||}v_{||}}d^3v. \quad (13.164)$$

Setting the denominator in Eq. (13.164) to zero corresponds to the Landau resonance (13.149). Before specifically discussing the Landau resonance, we want to demonstrate the equivalence between the dispersion relation derived from kinetic theory and from fluid theory. To do this we assume a distribution of the form

$$f = n_0\delta(v_{||} - v_{\mathrm{d}})g(v_\perp), \quad (13.165)$$

where n_0 is the density of the particles represented by f, $\delta(v_{||} - v_{\mathrm{d}})$ is the Dirac delta function, corresponding to a distribution that has no thermal spread in the parallel direction and is drifting with a velocity v_d, and $g(v_\perp)$ is an arbitrary function of $v_\perp$ that satisfies the condition

$$2\pi\int_0^\infty g(v_\perp)v_\perp dv_\perp = 1. \quad (13.166)$$

Equation (13.164) becomes

$$\rho_q = -i\omega_p^2 \varepsilon_0 E_{\|} \int \frac{\partial \delta(v_{\|} - v_d)/\partial v_{\|}}{\omega - k_{\|} v_{\|}} dv_{\|}, \quad (13.167)$$

where ω_p is the species plasma frequency.

We can integrate Eq. (13.167) by parts. We shall further assume an ambient cold stationary electron distribution with density n_e, and a drifting distribution with density n_b and drift velocity v_b, corresponding to the distribution used earlier for the two-stream instability. Poisson's equation further states that

$$\sum_s \rho_{qs} = i\varepsilon_0 k_{\|} E_{\|}, \quad (13.168)$$

where we have summed over species, as indicated by the subscript s. Equation (13.167) then becomes

$$1 = \frac{\omega_{pe}^2}{\omega^2} + \frac{\omega_{pb}^2}{(\omega - \mathbf{k} \cdot \mathbf{v}_b)^2}, \quad (13.169)$$

which is identical to Eq. (13.134).

Clearly, then, the kinetic formalism also gives the fluid results discussed earlier, but has the additional feature that Eq. (13.164) includes the possibility of wave–particle interactions through the resonance given by $\omega = k_{\|} v_{\|}$.

In order to understand why the resonance leads to wave growth, we need to perform some analysis of functions of complex variables. For simplicity, we assume that f corresponds to a single species (nominally electrons, since for high-frequency waves we can ignore the ions), and further that f is the **reduced velocity distribution**, obtained by integrating over perpendicular velocity. Consequently, f only depends on parallel velocity. This is equivalent to assuming that the distribution function is given by $f(v_{\|})g(v_{\perp})$, with $g(v_{\perp})$ satisfying the condition given by Eq. (13.166). Thus, from Eqs. (13.167) and (13.168), the relative permittivity of the plasma is given by

$$\kappa(\omega, k_{\|}) = 1 - \frac{q^2}{\varepsilon_0 m k_{\|}^2} \int_{-\infty}^{\infty} \frac{\partial f/\partial v_{\|}}{v_{\|} - \omega/k_{\|}} dv_{\|}, \quad (13.170)$$

and the dispersion relation corresponds to a solution of

$$\kappa(\omega, k_{\|}) = 0. \quad (13.171)$$

Because we are considering electrostatic waves, $\kappa(\omega, k_{\|})$ is a scalar quantity. For electromagnetic waves, $\boldsymbol{\kappa}(\boldsymbol{\omega}, \mathbf{k}_{\|})$ would be a tensor (cf. Eq. (13.12)).

Cauchy's theorem states that, for a complex variable z,

$$\oint h(z) dz = 0 \quad (13.172)$$

if $h(z)$ is an analytic function of z throughout the area enclosed by the integration contour. We can therefore consider the integration in Eq. (13.170) to be part of a closed contour that closes at infinity. When we performed the past-history integration, we assumed that the frequency had a small positive imaginary part (i.e., growing waves, $\gamma > 0$) to ensure that the integral converged. If we now consider $v_{\|}$ to be a complex variable, the pole corresponding to the resonance lies in the upper half-plane corresponding to $Im(v_{\|}) > 0$. Thus we close the contour in the lower half-plane to satisfy Eq. (13.172).

If, however, $\gamma \leq 0$, the contour now encloses half ($\gamma = 0$) or all ($\gamma < 0$) of the pole. The residual theorem states that

$$\oint \frac{h(z)}{z - a} dz = i2\pi h(a) \quad (13.173)$$

when the contour encloses the pole at $z = a$ in a counter-clockwise sense. Thus the integral in Eq. (13.170) would have discontinuities for $\gamma \leq 0$. From a physical argument, $\kappa(\omega, k_{\|})$ should be continuous. Otherwise the resultant wave dispersion would be very different depending on the sign of γ, or if $\gamma = 0$. The method by which $\kappa(\omega, k_{\|})$ is made to be continuous is through the **Landau prescription.**

The Landau prescription is shown in Figure 13.11. The prescription is to distort the path of integration so that the path can still close in the lower half-plane, but not include the pole at $v_{\|} = \omega/k_{\|}$.

Based on this prescription, the integral in Eq. (13.170) becomes

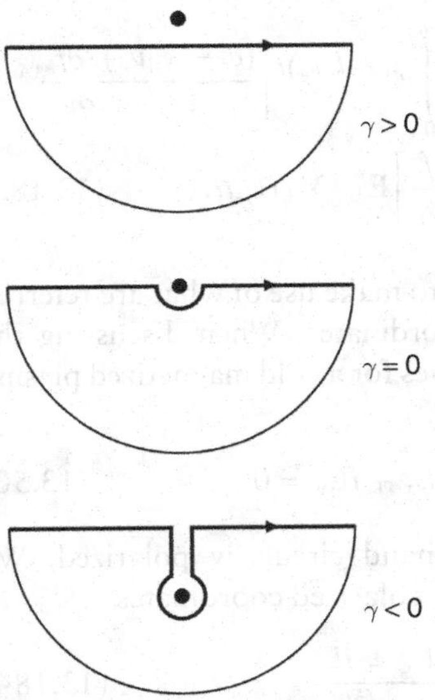

FIGURE 13.11. Distortion of the integration path that ensures that $\kappa(\omega, k_\parallel)$ is a continuous function for $\gamma \leq 0$.

$$\int_{-\infty}^{\infty} \frac{\partial f/\partial v_\parallel}{v_\parallel - \omega/k_\parallel} dv_\parallel = \mathrm{P}\int_{-\infty}^{\infty} \frac{\partial f/\partial v_\parallel}{v_\parallel - \omega/k_\parallel} dv_\parallel$$

$$+ i\pi \frac{\partial f}{\partial v_\parallel}\Big|_{v_\parallel=\omega/k_\parallel}, \tag{13.174}$$

where "P" indicates that the integral is the principal part, evaluated with ω real.

We can now evaluate Eq. (13.170). We assume γ is small and use Taylor expansion to estimate the growth rate; Eq. (13.170) becomes

$$\kappa(\omega, k_\parallel) = \kappa(\omega_\mathrm{r}, k_\parallel) - \gamma \frac{\partial \kappa_\mathrm{i}(\omega_\mathrm{r}, k_\parallel)}{\partial \omega_\mathrm{r}}$$

$$+ i\gamma \frac{\partial \kappa_\mathrm{r}(\omega_\mathrm{r}, k_\parallel)}{\partial \omega_\mathrm{r}} - i\pi \frac{q^2}{\varepsilon_0 m k_\parallel^2} \frac{\partial f}{\partial v_\parallel}\Big|_{v_\parallel=\omega/k_\parallel}, \tag{13.175}$$

where the subscript "r" indicates the real part and subscript "i" indicates the imaginary part.

Both $\kappa(\omega, k_\parallel)$ and $\kappa(\omega_\mathrm{r}, k_\parallel)$ are assumed to satisfy Eq. (13.171), and we shall also assume that the correction to the real part of Eq. (13.175) is small. Then, from the imaginary part,

$$\gamma = \pi \frac{q^2}{\varepsilon_0 m k_\parallel^2} \frac{\partial f/\partial v_\parallel|_{v_\parallel=\omega/k_\parallel}}{\partial \kappa_\mathrm{r}(\omega_\mathrm{r}, k_\parallel)/\partial \omega_\mathrm{r}}. \tag{13.176}$$

The growth rate is positive if both $\partial f/\partial v_\parallel$ and $\partial \kappa_\mathrm{r}(\omega_\mathrm{r}, k_\parallel)/\partial \omega_\mathrm{r}$ are positive. When $\partial[\omega_\mathrm{r}\kappa_\mathrm{r}(\omega_\mathrm{r}, k_\parallel)]/\partial \omega_\mathrm{r} > 0$, the corresponding wave is known as a positive energy density wave. Stix (1962) discusses how this quantity relates to the energy density of electrostatic waves. Since $\kappa_\mathrm{r}(\omega_\mathrm{r}, k_\parallel) = 0$ from Eq. (13.171), that $\partial \kappa_\mathrm{r}(\omega_\mathrm{r}, k_\parallel)/\partial \omega_\mathrm{r} > 0$ is sufficient to determine that an electrostatic wave is a positive energy density wave. Plasma oscillations in a cold plasma are positive energy density waves.

Assuming the waves have positive energy density, Eq. (13.176) shows that a distribution that has $\partial f/\partial v_\parallel > 0$ at the resonance will be unstable. This is known as a "bump on tail" instability and is the resonant counterpart to the two-stream instability discussed earlier. Indeed, as we have seen here, the function defined by Eq. (13.170) can include both a fluid instability if the distribution consists of two low-temperature distributions streaming relative to each other, or the resonance-driven instability, when the streaming distribution is of sufficiently low density not to affect the dispersion, except for the growth rate itself.

To understand why a positive slope in the distribution function leads to wave growth, we note that the effect of wave instabilities is to remove structure in the distribution function, and waves therefore move particles down velocity space gradients, in a process known as **quasi-linear diffusion.** If the slope in the distribution is positive, the diffusion process removes energy from the particles. That energy is taken up by the waves, and the waves grow.

Given the ubiquity of integrals such as that shown in Eq. (13.170), the general form for a maxwellian distribution has been published as the **plasma dispersion function** (Fried and Conte, 1961). In particular, for a one-dimensional maxwellian,

$$f = \frac{n_0}{\pi^{1/2} v_\mathrm{T}} \exp(-v^2/v_\mathrm{T}^2). \tag{13.177}$$

Equation (13.170) becomes

$$\kappa(\omega, k_{\|}) = 1 - \frac{\omega_{\mathrm{p}}^2}{k_{\|}^2 v_{\mathrm{T}}^2} Z'\left(\frac{\omega}{k_{\|} v_{\mathrm{T}}}\right), \qquad (13.178)$$

where $Z'(\zeta) = \partial Z(\zeta)/\partial \zeta$, the first derivative of the plasma dispersion function.

In Section 13.3.1, we noted that electrostatic waves are observed upstream of planetary bow shocks. These waves are generated by reflected particles whose reduced velocity distribution acts as a bump on tail distribution. In the electron foreshock, reflected electrons generate plasma oscillations; in the ion foreshock, reflected ions generate ion acoustic waves.

13.5.3 Gyro-resonance

Having shown how the past-history integral (13.159) leads to Landau resonance for parallel propagating electrostatic waves, we shall now explore gyro-resonance. To do this we again assume parallel propagation ($k_{\perp} = 0$), but now consider transverse waves ($E_{\|} = 0$). In anticipation of our results, we note that, from Eq. (13.160),

$$\mathbf{E}_{\perp} \cdot \frac{\partial \mathbf{v}}{\partial \mathbf{V}(t)} \cdot \frac{\partial f}{\partial \mathbf{v}} = \mathbf{E}_{\perp} \cdot \frac{\partial \mathbf{v}_{\perp}}{\partial \mathbf{V}_{\perp}(t)} \cdot \frac{\partial f}{\partial \mathbf{v}_{\perp}} \qquad (13.179)$$

and

$$\mathbf{k}_{\|} \cdot \frac{\partial \mathbf{v}}{\partial \mathbf{V}(t)} \cdot \frac{\partial f}{\partial \mathbf{v}} = \mathbf{k}_{\|} \cdot \frac{\partial \mathbf{v}_{\|}}{\partial \mathbf{V}_{\|}(t)} \cdot \frac{\partial f}{\partial \mathbf{v}_{\|}} = k_{\|} \frac{\partial f}{\partial v_{\|}}, \qquad (13.180)$$

where $\mathbf{E}_{\perp} = (E_x, E_y, 0)$.

As before, $f = f_0(\mathbf{v}) = f_0(v_{\perp}, v_{\|})$, since f is a function of the constants of the motion. Consequently,

$$\frac{\partial f}{\partial v_x} = \frac{v_x}{v_{\perp}} \frac{\partial f}{\partial v_{\perp}} \text{ and } \frac{\partial f}{\partial v_y} = \frac{v_y}{v_{\perp}} \frac{\partial f}{\partial v_{\perp}}, \qquad (13.181)$$

and Eq. (13.179) becomes

$$\mathbf{E}_{\perp} \cdot \frac{\partial \mathbf{v}}{\partial \mathbf{V}(t)} \cdot \frac{\partial f}{\partial \mathbf{v}} = \frac{\mathbf{E}_{\perp} \cdot \mathbf{V}_{\perp}(t)}{v_{\perp}} \frac{\partial f}{\partial v_{\perp}}. \qquad (13.182)$$

The past-history integral (13.159) now becomes

$$f_1(\mathbf{v}) = -\frac{q}{m\omega} \int_0^{\infty} e^{i(\omega - k_{\|} v_{\|})t} \left[\frac{(\omega - k_{\|} v_{\|})}{v_{\perp}} \frac{\partial f}{\partial v_{\perp}} + k_{\|} \frac{\partial f}{\partial v_{\|}}\right] \mathbf{E}_{\perp} \cdot \mathbf{V}_{\perp}(t) dt. \qquad (13.183)$$

The next step is to make use of what are referred to as polarized coordinates. When discussing the R- and L-mode waves for a cold magnetized plasma we noted that when

$$E_x + iE_y = 0 \qquad (13.50)$$

the wave is right-hand circularly polarized. We therefore define the polarized coordinates

$$E_{\pm} = \frac{E_x \pm iE_y}{\sqrt{2}}, \qquad (13.184)$$

with the plus sign corresponding to the electric field of a left-hand circularly (LHC) polarized wave field and the minus sign corresponding to that of a right-hand circularly (RHC) polarized wave field.

With these polarized coordinates, having assumed transverse wave modes, we now make use of Eq. (13.162) to specify the first-order wave current, and

$$j_{\pm} = -\frac{q^2}{m\omega} \int_{v} \int_{t=0}^{\infty} e^{i(\omega - k_{\|} v_{\|})t} v_{\pm} \left[\frac{(\omega - k_{\|} v_{\|})}{v_{\perp}} \frac{\partial f}{\partial v_{\perp}} + k_{\|} \frac{\partial f}{\partial v_{\|}}\right] \mathbf{E}_{\perp} \cdot \mathbf{V}_{\perp}(t) d^3v dt. \qquad (13.185)$$

From Eqs. (13.157) and (13.181), for the first term in square brackets in Eq. (13.185),

$$\int_{-\infty}^{\infty} \int_{-\infty}^{\infty} \frac{v_{\pm}}{v_{\perp}} \frac{\partial f}{\partial v_{\perp}} \mathbf{E}_{\perp} \cdot \mathbf{V}_{\perp}(t) dv_x dv_y$$
$$= \int_{-\infty}^{\infty} \int_{-\infty}^{\infty} \frac{1}{\sqrt{2}} \left(\frac{\partial f}{\partial v_x} \pm i \frac{\partial f}{\partial v_y}\right) [E_x (v_x \cos \Omega t + v_y \sin \Omega t) + E_y (v_y \cos \Omega t + v_x \sin \Omega t)] dv_x dv_y. \qquad (13.186)$$

The distribution function $f(v)$ must vanish at $v = \pm\infty$, where here v is the integration variable, and, through integration by parts,

$$\int_{-\infty}^{\infty} v_y \frac{\partial f}{\partial v_x} dv_x = 0 = \int_{-\infty}^{\infty} v_x \frac{\partial f}{\partial v_y} dv_y. \qquad (13.187)$$

In addition,

$$\int_{-\infty}^{\infty}\int_{-\infty}^{\infty} v_x \frac{\partial f}{\partial v_x} dv_x\, dv_y = -\int_{-\infty}^{\infty}\int_{-\infty}^{\infty} f dv_x\, dv_y$$

$$= \int_{-\infty}^{\infty}\int_{-\infty}^{\infty} v_y \frac{\partial f}{\partial v_y} dv_x\, dv_y = \pi \int_0^{\infty} v_\perp^2 \frac{\partial f}{\partial v_\perp} dv_\perp, \qquad (13.188)$$

where

$$\int_{-\infty}^{\infty}\int_{-\infty}^{\infty} dv_x dv_y = 2\pi \int_0^{\infty} v_\perp dv_\perp \qquad (13.189)$$

because $f(v_\perp, v_\parallel)$ does not depend on the phase angle φ, where $v_x = v_\perp \cos\varphi$ and $v_y = v_\perp \sin\varphi$.

From Eqs. (13.187) and (13.188), Eq. (13.186) becomes

$$\int_{-\infty}^{\infty}\int_{-\infty}^{\infty} \frac{v_\pm}{v_\perp}\frac{\partial f}{\partial v_\perp} \mathbf{E}_\perp \cdot \mathbf{V}_\perp(t) dv_x dv_y$$

$$= \frac{1}{2}\int_{-\infty}^{\infty}\int_{-\infty}^{\infty} v_\perp \frac{\partial f}{\partial v_\perp} E_\pm e^{\pm i\Omega t} dv_x dv_y. \qquad (13.190)$$

For the second term in square brackets in Eq. (13.185), we note that

$$\int_{-\infty}^{\infty}\int_{-\infty}^{\infty} v_i v_j f\, dv_x dv_y = \int_{-\infty}^{\infty}\int_{-\infty}^{\infty} \frac{v_\perp^2}{2}\delta_{ij} f\, dv_x dv_y \quad (13.191)$$

because the distribution is assumed to have no net streaming velocity in either the x- or y-direction.

On expanding the $\mathbf{E}_\perp \cdot \mathbf{V}_\perp(t)$ term, as was done in Eq. (13.186), we therefore find

$$\int_{-\infty}^{\infty}\int_{-\infty}^{\infty} v_\pm f \mathbf{E}_\perp \cdot \mathbf{V}_\perp(t) dv_x\, dv_y$$

$$= \frac{1}{2}\int_{-\infty}^{\infty}\int_{-\infty}^{\infty} v_\perp^2 f E_\pm e^{\pm i\Omega t} dv_x\, dv_y. \qquad (13.192)$$

From Eqs. (13.190) and (13.192), and after performing the integral over t, Eq. (13.185) becomes

$$j_\pm = -i\frac{q^2}{2m\omega} \int v_\perp \frac{(\omega - k_\parallel v_\parallel)\partial f/\partial v_\perp + k_\parallel v_\perp \partial f/\partial v_\parallel}{\omega - k_\parallel v_\parallel \pm \Omega} E_\pm d^3 v, \qquad (13.193)$$

and the denominator in Eq. (13.193) gives the gyro-resonance (13.150) with $n = \pm 1$. We show further that for parallel propagating transverse modes the sign depends on the polarization of the wave fields.

Equation (13.7) (or Eq. (13.38)) gives the governing equation for an electromagnetic harmonic perturbation. For simplicity, we again only consider electrons; for parallel propagating transverse modes, Eq. (13.7) becomes

$$(\mu^2 - 1)E_\pm = \frac{q^2}{2\varepsilon_0 m\omega^2} \int v_\perp \frac{(\omega - k_\parallel v_\parallel)\partial f/\partial v_\perp + k_\parallel v_\perp \partial f/\partial v_\parallel}{\omega - k_\parallel v_\parallel \pm \Omega} E_\pm d^3 v, \qquad (13.194)$$

where we have made use of the polarized coordinates.

One solution of Eq. (13.194) is $E_+ = 0$ and

$$\mu^2 = 1 + \frac{q^2}{2\varepsilon_0 m\omega^2} \int v_\perp \frac{(\omega - k_\parallel v_\parallel)\partial f/\partial v_\perp + k_\parallel v_\perp \partial f/\partial v_\parallel}{\omega - k_\parallel v_\parallel - \Omega} d^3 v. \qquad (13.195)$$

This solution corresponds to the RHC mode. The solution corresponding to the LHC mode has the opposite sign in the gyro-resonance.

We now make the assumption that

$$f = \frac{n_0}{\pi v_\perp}\delta(v_\perp)\delta(v_\parallel), \qquad (13.196)$$

noting that

$$\int_0^{\infty} \delta(v_\perp) dv_\perp = \frac{1}{2}. \qquad (13.197)$$

Because the second term in the numerator in Eq. (13.195) includes $v_\perp^2$, this term vanishes on integrating over $v_\perp$. The first is evaluated through integration by parts, and we find

$$\mu^2 = 1 - \frac{\omega_{pe}}{\omega(\omega - \Omega)}. \quad (13.198)$$

This is the R-mode dispersion relation, as we would expect.

The assumed distribution function in Eq. (13.196) would not result in any wave growth as there are no resonant particles. But we can use this as a starting point to determine the conditions for a gyro-resonant instability. By analogy with Eq. (13.174), we assume that Eq. (13.195) consists of a principal part and an imaginary part associated with the pole of the integral. For simplicity, we assume the principal part is given by Eq. (13.198), and Eq. (13.195) becomes

$$\mu^2 = 1 - \frac{\omega_{pe}}{\omega(\omega - \Omega)} + i\pi^2 \frac{q^2}{\varepsilon_0 m \omega^2}$$
$$\int_0^\infty v_\perp^2 \left[\left(v_\parallel - \frac{\omega}{k_\parallel} \right) \frac{\partial f}{\partial v_\perp} - v_\perp \frac{\partial f}{\partial v_\parallel} \right] \Bigg|_{v_\parallel = v_{res}} dv_\perp, \quad (13.199)$$

with

$$v_{res} = (\omega - \Omega)/k_\parallel, \quad (13.200)$$

and the term in square brackets in Eq. (13.199) is the pole from the integration over $v_\parallel$.

We can simplify Eq. (13.199) further by defining a pitch angle α relative to the parallel phase velocity of the wave, i.e.,

$$\tan\alpha = \frac{v_\perp}{v_\parallel - \omega/k_\parallel}. \quad (13.201)$$

Equation (13.199) becomes

$$\mu^2 = 1 - \frac{\omega_{pe}}{\omega(\omega - \Omega)} + i\pi^2 \frac{q^2}{\varepsilon_0 m \omega^2} \int_0^\infty v_\perp^2 \frac{\partial f}{\partial \alpha} \Bigg|_{v_\parallel = v_{res}} dv_\perp. \quad (13.202)$$

Thus gyro-resonance driven instabilities are inherently associated with pitch-angle gradients in the phase space density, with the pitch angle defined relative to the parallel phase velocity of the wave.

We shall again make the small growth-rate approximation, cf. Eq. (13.175), and use Taylor expansion with $\omega = \omega_r + i\gamma$. Then, from the imaginary part of Eq. (13.202),

$$\gamma \left(2 + \frac{\omega_{pe}^2 \Omega}{\omega_r(\omega_r - \Omega)^2} \right) = -\pi^2 \frac{q^2}{\varepsilon_0 m \omega_r} \int_0^\infty v_\perp^2 \frac{\partial f}{\partial \alpha} \Bigg|_{v_\parallel = v_{res}} dv_\perp. \quad (13.203)$$

Since the term in parentheses on the left-hand side of Eq. (13.203) is positive, wave growth occurs if

$$\int_0^\infty v_\perp^2 \frac{\partial f}{\partial \alpha} \Bigg|_{v_\parallel = v_{res}} dv_\perp < 0; \quad (13.204)$$

i.e., the gradient in the distribution function with respect to the pitch angle defined in the frame moving at the phase velocity of the wave must be negative for waves to grow.

We came to this result based on the cold plasma R-mode dispersion relation. This mode includes the whistler mode, and gyro-resonance is typically associated with the growth of whistler-mode waves in the Earth's magnetosphere. But we can generalize the result. A more general form of Eq. (13.202) would be

$$D(\omega, k_\parallel) = -i\pi^2 \frac{q^2}{\varepsilon_0 m \omega^2} \int_0^\infty v_\perp^2 \frac{\partial f}{\partial \alpha} \Bigg|_{v_\parallel = v_{res}} dv_\perp, \quad (13.205)$$

where $D(\omega, k_\parallel)$ is the dispersion relation, not to be confused with the displacement D. In the absence of the gyro-resonance term on the right-hand side of Eq. (13.205), the wave dispersion is given by $D(\omega, k_\parallel) = 0$. By convention we have assumed $D(\omega, k_\parallel) = F(\omega, k_\parallel) - \mu^2$, where $F(\omega, k_\parallel)$ is a function that depends on the plasma properties. For the R-mode considered here,

$$F(\omega, k_\parallel) = 1 - \frac{\omega_{pe}^2}{\omega(\omega - \Omega)}. \quad (13.206)$$

Similarly to the Landau resonance discussed earlier, for small growth rates Eq. (13.205) yields

$$\gamma = \frac{-\pi^2 \dfrac{q^2}{\varepsilon_0 m\omega^2} \displaystyle\int_0^\infty v_\perp^2 \left.\frac{\partial f}{\partial \alpha}\right|_{v_\parallel = v_{\text{res}}} dv_\perp}{\partial D_r(\omega_r, k_\parallel)/\partial \omega_r}, \tag{13.207a}$$

or, using the form of the integral in Eq. (13.199),

$$\gamma = \frac{-\pi^2 \dfrac{q^2}{\varepsilon_0 m\omega^2} \displaystyle\int_0^\infty v_\perp^2 \left[\left(v_\parallel - \frac{\omega}{k_\parallel}\right)\frac{\partial f}{\partial v_\perp} - v_\perp \frac{\partial f}{\partial v_\parallel}\right]\bigg|_{v_\parallel = v_{\text{res}}} dv_\perp}{\partial D_r(\omega_r, k_\parallel)/\partial \omega_r}. \tag{13.207b}$$

Having determined the growth rate for gyro-resonance, we shall now make some observations concerning the resonance condition (13.200) itself. First, more than one species can be in gyro-resonance with a wave, and we can extend Eqs. (13.207a) and (13.207b) by summing the numerator over species. As an example, we consider the whistler mode.

For the whistler mode, the wave frequency is less than the electron gyro-frequency. Consequently, gyro-resonance between a parallel propagating whistler-mode wave and an electron requires that the wave and particle move in opposite directions so that the Doppler shift increases the wave frequency as observed by the electron, i.e., $k_\parallel v_\parallel < 0$.

Positively charged ions, on the other hand, gyrate in a left-handed sense with respect to the magnetic field. These ions would interact strongly with a parallel propagating whistler-mode wave if the particle is moving faster than the wave, so that the sense of polarization of the wave changes sign in the particle frame, i.e., $v_\parallel > \omega/k_\parallel$. The difference in the resonance condition affects how temperature anisotropy, for example, results in wave growth (see Problem 13.5).

As a last comment on gyro-resonance, the R-mode also has a superluminous branch, and for that mode the term $(\omega/k_\parallel)(\partial f/\partial v_\perp)$ dominates in Eq. (13.207b). But, on integrating this term by parts, the net effect is to cause damping. To see this we note that even if $\partial f/\partial v_\perp > 0$ for some range of $v_\perp$, the particles with higher perpendicular velocity must have $\partial f/\partial v_\perp < 0$, otherwise the moment integrals will not converge (for convergence $f \to 0$ as $v_\perp \to \infty$). Because the $\partial f/\partial v_\perp$ term is weighted by $v_\perp^2$ in Eq. (13.207b), the particles with higher perpendicular velocity contribute more to the integral. If, however, relativistic effects are included in the gyro-resonance condition, the resonance condition becomes

$$\omega - k_\parallel v_\parallel = \Omega/\Gamma = \Omega(1 - v^2/c^2)^{1/2}, \tag{13.208}$$

where Γ is the Lorentz factor. As noted in Chapter 11, this modification of the gyro-resonance is an essential factor in the generation of auroral kilometric radiation (AKR) as the Lorentz factor limits the maximum value of $v_\perp$ in the integral in Eq. (13.207b), and the damping introduced by the particles at higher perpendicular velocity is reduced. It should be noted that AKR is observed in the R-X mode, and for electrons to be in gyro-resonance with this mode ω_R must be close to Ω_e. Furthermore, if we use the relativistic electron mass in Eq. (13.53), then for sufficiently low densities the relativistic correction gives $\omega_R < \Omega_e$, and in that case Eq. (13.208) can be satisfied with $k_\parallel = 0$. This allows for stronger growth for the waves, and is an essential feature of the AKR source region (see Chapter 11).

13.5.4 Bernstein modes

As the last topic to be discussed in this section on kinetic theory and wave instabilities, we derive the dispersion relation for the Bernstein modes. Although we shall not specifically address how these waves are generated, we shall use the derivation to show the general approach that is used when deriving a wave dispersion relation using kinetic theory.

We again make use of the past-history integral (13.159). At least initially, we make the assumption that $\mathbf{k} = (k_\perp, 0, k_\parallel)$, allowing for obliquely propagating waves. We further assume that the waves are electrostatic, and Eq. (13.159) becomes

$$f_1(\mathbf{v}) = -\frac{q}{m}\int_0^\infty e^{i(\omega - k_\parallel v_\parallel)t} e^{-i\frac{k_\perp v_x}{\Omega}\sin\Omega t} e^{-i\frac{k_\perp v_y}{\Omega}(1-\cos\Omega t)} \left[E_x\left(\frac{\partial f}{\partial v_x}\cos\Omega t + \frac{\partial f}{\partial v_y}\sin\Omega t\right) + E_\parallel \frac{\partial f}{\partial v_\parallel}\right] dt. \tag{13.209}$$

The wave electric field E_x introduces the gyro-resonance $\omega - k_{\|}v_{\|} \pm \Omega = 0$ through the $\cos \Omega t$ and $\sin \Omega t$ terms, and the exponential terms that depend on $k_\perp$ introduce higher-order gyro-resonances. This can be seen on considering the generating function for the Bessel function of the first kind,

$$e^{-iz\sin \Omega t} = \sum_{n=-\infty}^{\infty} e^{-in\Omega t} J_n(z). \qquad (13.210)$$

Indeed, Eq. (13.210) is used when deriving a general kinetic dispersion relation. In that case, however, the x- and y-components of the velocity are usually replaced with $v_x = v_\perp \cos \varphi$ and $v_y = v_\perp \sin \varphi$, where φ is a phase angle, and the velocity space integration includes this angle, i.e.,

$$\int d^3v = \int_{v_\perp=0}^{\infty} \int_{v_\|=-\infty}^{\infty} \int_{\varphi=0}^{2\pi} v_\perp dv_\perp dv_\| d\varphi. \qquad (13.211)$$

Rather than perform the past-history integration and then evaluate the charge density through Eq. (13.161), we instead carry out the velocity space integration first. To do this we assume

$$f = \frac{n_0}{\pi^{3/2} v_{T\perp}^2 v_{T\|}} e^{-v_\perp^2/v_{T\perp}^2} e^{-v_\|^2/v_{T\|}^2}, \qquad (13.212)$$

which is a bimaxwellian.

If the distribution function has the form given in Eq. (13.212), then the integration over $v_\|$ can again be expressed in terms of the plasma dispersion function, but, because of the summation in Eq. (13.210), the terms will be of the form

$$Z_n = Z\left(\frac{\omega - n\Omega}{k_\| v_{T\|}}\right), \qquad (13.213)$$

or first-order derivatives of this equation. At this stage, however, we make the simplifying assumption that $k_\| = 0$.

The resultant charge density is given by

$$\rho_q = -i\frac{n_0 q^2}{m\pi v_{T\perp}^2} \frac{k_\perp E_x}{\Omega} \int_{v_x=-\infty}^{\infty} \int_{v_y=-\infty}^{\infty} \int_{t=0}^{\infty} e^{i\omega t} e^{-i\frac{k_\perp v_x}{\Omega}\sin \Omega t} e^{-i\frac{k_\perp v_y}{\Omega}(1-\cos \Omega t)} e^{-v_\perp^2/v_{T\perp}^2} \sin \Omega t dv_x dv_y dt, \qquad (13.214)$$

where we have integrated by parts for v_x and v_y, and performed the integration over $v_\|$.

The next step is to note that, since $v_\perp^2 = v_x^2 + v_y^2$, the integrations over v_x and v_y correspond to Fourier transforms of a gaussian, and we find

$$\rho_q = -i\frac{n_0 q^2}{m} \frac{k_\perp E_x}{\Omega} \int_{t=0}^{\infty} e^{i\omega t} e^{-\frac{k_\perp^2 v_{T\perp}^2}{2\Omega^2}(1-\cos \Omega t)} \sin \Omega t dt. \qquad (13.215)$$

On defining

$$\lambda = \frac{k_\perp^2 v_{T\perp}^2}{2\Omega^2}, \qquad (13.216)$$

Eq. (13.215) becomes

$$\rho_q = -\frac{n_0 q^2}{m} \frac{k_\perp E_x}{\Omega} e^{-\lambda} \int_{t=0}^{\infty} \left(\frac{e^{i(\omega+\Omega)t} - e^{i(\omega-\Omega)t}}{2}\right) e^{\lambda \cos \Omega t} dt. \qquad (13.217)$$

Making use of the generating function for the modified Bessel function,

$$e^{\lambda \cos \Omega t} = \sum_{n=-\infty}^{\infty} e^{-in\Omega t} I_n(\lambda); \qquad (13.218)$$

then

$$\rho_q = -i\frac{n_0 q^2}{m} \frac{k_\perp E_x}{\Omega} e^{-\lambda} \sum_{n=-\infty}^{\infty} \frac{I_{n+1}(\lambda) - I_{n-1}(\lambda)}{2(\omega - n\Omega)}. \qquad (13.219)$$

The recurrence relation for the modified Bessel function yields

$$I_{n+1}(\lambda) - I_{n-1}(\lambda) = -\frac{2n}{\lambda} I_n(\lambda), \qquad (13.220)$$

and, if we now assume that only electrons contribute to the charge density, from Eq. (13.161) the dispersion relation becomes

$$\frac{\omega_{\rm pe}^2}{|\Omega_{\rm e}|\lambda}e^{-\lambda}\sum_{n=-\infty}^{\infty}\frac{nI_n(\lambda)}{\omega-n|\Omega_{\rm e}|}=1, \qquad (13.221)$$

where we have now adopted the notation used previously to indicate the electron gyro-frequency, $\Omega=|\Omega_{\rm e}|$. Equation (13.221) is the dispersion relation for the Bernstein modes, with solutions shown in Figure 13.12, where we have assumed $\omega_{\rm pe}/|\Omega_{\rm e}|=\sqrt{10}$.

For small λ, $I_n(\lambda)\approx(\lambda/2)^n/n!$, and, except for $n=\pm1$, $I_n(\lambda)$ can become small enough that Eq. (13.221) is satisfied even when $\omega\approx n|\Omega{\rm e}|$. For $n=\pm1$, we instead find the upper hybrid resonance $\omega=\left(\omega_{\rm pe}^2+\Omega_{\rm e}^2\right)^{1/2}\approx3.32|\Omega_{\rm e}|$ for $\omega_{\rm pe}/|\Omega_{\rm e}|=\sqrt{10}$. This mode should be present, as this is the cold plasma limit to Eq. (13.221), where, in taking the limit, $v_{T\perp}\to0$ sufficiently rapidly such that $\lambda\to0$, even though $k_\perp$ can be large. Finally, for large λ, $e^{-\lambda}I_n(\lambda)\approx1/(2\pi\lambda)^{1/2}$, and again Eq. (13.221) can be satisfied when $\omega\approx n|\Omega_{\rm e}|$. Below the upper hybrid resonance, waves are present over the entire band between the gyro-harmonics. Above the upper hybrid resonance, there are gaps where no waves are present.

In closing this section, we again note that we have not discussed how these waves are generated.

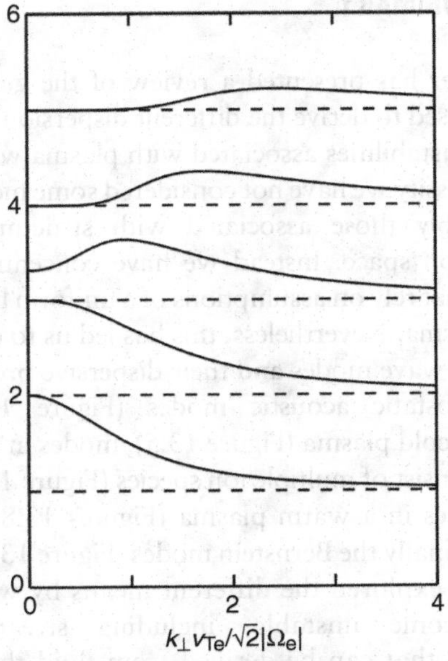

FIGURE 13.12. Dispersion curves for the Bernstein modes. The curves are plotted as a function of $\lambda^{1/2}$.

In order to do so, we would have to retain the resonances dropped by setting $k_\parallel=0$. But the interested reader could use the approach presented here to obtain the complete dispersion relation, including wave growth. There are also several texts, including Clemmow and Dougherty (1969), that present derivations of the full conductivity tensor for both maxwellian and more general distributions.

13.6 SUMMARY

This chapter has presented a review of the general approach used to derive the different dispersion relations and instabilities associated with plasma waves. Out of necessity we have not considered some modes, most notably those associated with structure in configuration space. Instead we have concentrated on waves that rely on assumptions of a uniform background plasma. Nevertheless, this has led us to quite an array of wave modes and their dispersive properties: electrostatic acoustic modes (Figure 13.1); modes in a cold plasma (Figure 13.5), modes in plasmas that consist of multiple ion species (Figure 13.7); MHD modes in a warm plasma (Figures 13.8 and 13.9); and finally the Bernstein modes (Figure 13.12).

We also explored the different means by which waves become unstable, including streaming instabilities that can be derived from fluid theory and the Landau and gyro-resonances, which are inherently kinetic instabilities. The streaming instabilities take energy from the relative motion of the two plasma species, while the resonant instabilities draw energy from gradients present in the velocity space distribution.

The different modes and instabilities appear in a variety of space plasmas. Acoustic modes are observed upstream of planetary bow shocks, where the waves act to scatter and thermalize the reflected particles. High-frequency cold plasma modes, especially the R-X mode, are generated by resonant instabilities in regions where particles are accelerated to form auroras. These waves make magnetized planets radio astronomical objects. Electrostatic waves such as those at the upper hybrid resonance are associated with gradients in the plasma, at the Earth's plasmapause, for example. Bernstein modes are frequently observed near the plasmapause. The lower-frequency whistler-mode wave is a ubiquitous feature of planetary magnetospheres. At the Earth, whistler-mode waves appear as chorus, which has a well-defined structure in frequency, and the more unstructured hiss. Whistler-mode waves can be generated through lightning, and also within the magnetosphere, where the waves act to scatter particles in pitch angle. Some of these particles are lost to the atmosphere. The multi-ion wave modes scatter particles in pitch angle as well. MHD waves are the means by which momentum is transmitted as waves from one plasma region to another.

In closing, we note that the formalism presented in Section 13.5.1, where we present the concept of the past-history integration of the linearized Vlasov equation, is the basis for deriving the kinetic dispersion relation for electromagnetic waves in a variety of plasmas. Readers are encouraged to use the Space Physics Exercises to explore waves in a multicomponent cold plasma, MHD, and shock waves.

Additional reading

BOYD, T. J. M. and J. J. SANDERSON (1969). *Plasma Dynamics*. London: Nelson. This gives a concise introduction to wave propagation in a magnetized plasma, including the derivation of the Appleton–Hartree relation and Clemmow–Mullaly–Allis (CMA) diagrams.

CLEMMOW, P. C. and J. P. DOUGHERTY (1969). *Electrodynamics of Particles and Plasmas*. Reading, MA: Addison-Wesley. This extends the discussion in Boyd and Sanderson (1969), describing the derivation of the plasma dispersion relation including kinetic effects. They discuss concepts such as Landau damping and gyro-resonance, and also warm plasma modes, such as the Bernstein modes, that require a kinetic approach in deriving the dispersion relation. Section 13.5 in this chapter largely follows the approach used by Clemmow and Dougherty.

STIX, T. H. (1962). *The Theory of Plasma Waves*. New York: McGraw-Hill. The classical reference that describes wave energy flow and the relationship between the group velocity and wave energy flux. This includes the case of electrostatic waves, which still carry energy even though they have no magnetic field. In the context of this chapter, that electrostatic waves carry energy is important for Landau resonance.

Problems

13.1 Sketch the group velocity $\partial\omega/\partial k$ and the phase speed ω/k versus ω of the whistler mode, and discuss how this may explain the rate at which the frequency decreases with time for a lightning-generated whistler.

13.2 Show that the Appleton–Hartree dispersion relation (13.101) gives the X-mode dispersion relation (13.62) for perpendicular propagation,

$$\mu^2 = 1 - \frac{\omega_{pe}^2}{\omega^2}\frac{\omega^2 - \omega_{pe}^2}{\omega^2 - \omega_{UHR}^2}.$$

13.3 When an elliptically polarized electromagnetic wave propagates through a magnetized plasma, the orientation of the polarization rotates as it travels. This is known as Faraday rotation. Why does it occur? (Only treat the case of parallel propagation.)

13.4 Using the Plasma Waves option of the Space Physics Exercises,* select the Dispersion Relation portion. Use a magnetic field strength of 5 nT. Four ion species can be specified. For species #1, set Mass = 1, Charge State =1, and Number Density = 5. For species #2 set Mass = 4, Charge State =1, and Number Density = 2. For species #3 set Mass = 16, Charge State = 1, Number Density =1. Set the number density for species #4 to 0. These settings are equivalent to those used in Figure 13.7. Set the Wave Propagation Angle to 0 and press the calculate button. Now use the mouse button to select a frequency near the point where one of the R-mode and L-mode dispersion surfaces crosses in the upper right graph. Note the frequency is reset in the selection. Enter different values of frequency in the box, pressing calculate to reset the polar plot. Find the frequency for which both modes have the same wave number for parallel propagation. This is the crossover frequency. Change the propagation angle to 75° and verify that the modes still change polarity at the crossover frequency.

Now change the plot type to Phase Velocity. Again enter frequencies above and below the crossover frequency. Above the crossover frequency one mode can propagate at 90°. The other mode shows a characteristic "figure 8" phase velocity, where waves only propagate within a restricted range of angles. Which mode shows the figure 8? What happens for waves just below the crossover frequency?
Repeat the analysis for the other crossover frequency, using the Dispersion Relation plot to determine the crossover frequency and the Phase Velocity plot to investigate the changes in the dispersion above and below the crossover.

* Space Physics Exercises are at http://spacephysics.ucla.edu.

13.5 Show through geometric construction why gyro-resonant electrons lose energy (in the plasma frame) on moving to a lower pitch angle (in the wave frame) when in resonance with whistler-mode wave and hence require either a loss cone or $T_\perp/T_\parallel > 1$. Similarly, show why ions require $T_\perp/T_\parallel > 1$.

13.6 Show that the relativistic gyro-resonance results in an ellipse in velocity space, with the ratio between the semi-major and semi-minor axes given by

$$\frac{v_\perp^2}{v_\parallel^2} = \frac{\Omega_e^2 + c^2 k_\parallel^2}{\Omega_e^2}.$$

Also show that the gyro-resonance can occur for $k_\parallel = 0$ provided $\omega < |\Omega_e|$.

Appendix A.1

Notation, vector identities, and differential operators

A.1.1 NOTATION

The coordinate notations used in this text are undoubtedly familiar to most. Boldface is used for vectors; in handwritten material, vectors may be underlined or have arrows placed above them. For spatial coordinates, we may use cartesian, spherical, or cylindrical coordinate systems. The following usages are common (where the first two forms are alternative usages for cartesian coordinates):

$$\mathbf{x} = (x, y, z) \text{ or } \mathbf{x} = (x_1, x_2, x_3); \mathbf{v} = (v_x, v_y, v_z). \tag{A1.1}$$

Here the three components of the vector are the projections of the vectors on each of the three orthogonal axes *x, y, z* or 1, 2, 3. The axes are given in the right-handed sense so that the vector cross product of *x* and *y* is along the positive *z*-direction.

In the two following different representations of a vector:

$$\mathbf{r} = (r, \theta, \varphi) \text{ or } \mathbf{r} = (\rho, \theta, z), \tag{A1.2}$$

the end point is specified by three numbers, but each has a quite different meaning. The left-hand triad shows a scalar length r, a polar angle θ, and an azimuthal angle φ, where θ would be measured from the positive *z*- or the 3-axis and φ is generally measured in the *x–y* plane starting at *x*. In the right-hand representation, ρ is the distance away from the *z*-axis measured in the *x–y* plane and is a positive scalar value, θ is the angle around the *z*-axis measured from the *x*-axis, and *z* is a signed number giving the location along the *z*-axis.

Integrals may be taken over a three-dimensional spatial volume (e.g., $d\mathbf{x} = d^3x = dx\, dy\, dz$), over a three-dimensional volume in velocity space (e.g., $d\mathbf{v} = d^3v = dv_x\, dv_y\, dv_z$), or over a spherical volume $d\mathbf{r} = r^2 dr\, d\theta\, d\varphi$.

Vectors can be combined in various ways. The dot product produces a scalar from a pair of vectors:

$$\mathbf{a} \cdot \mathbf{b} = a_x b_x + a_y b_y + a_z b_z. \tag{A1.3}$$

The dot product is equal to the product of the length of the vectors times the cosine of the angle between them.

The cross product of two vectors produces a new vector:

$$\mathbf{a} \times \mathbf{b} = a_y b_z - a_z b_y,\ \ a_z b_x - a_x b_z,\ \ a_x b_y - a_y b_x. \tag{A1.4}$$

The magnitude of the cross product is the product of the length of the vectors times the sine of the angle between them. The time derivative at constant spatial position is written as $\partial f(x, y, z, t)/\partial t$, and the *x*-derivative at constant *y, z*, and time is written as $\partial f(x, y, z, t)/\partial x$.

Vector derivatives are also very common in our work. These can be created from derivatives in different directions combined with unit vectors that identify the direction in which the derivative is taken. The resultant vector is referred to as a vector operator and is very convenient in writing complicated equations in only a few lines. The operators we shall use are defined as follows:

$$\nabla f(x,y,z,t) = \left(\frac{\partial f(x,y,z,t)}{\partial x}, \frac{\partial f(x,y,z,t)}{\partial y}, \frac{\partial f(x,y,z,t)}{\partial z}\right) \tag{A1.5}$$

which is called the gradient of f;

$$\nabla \cdot \mathbf{A} = \left(\frac{\partial A_x(x,y,z,t)}{\partial x} + \frac{\partial A_y(x,y,z,t)}{\partial y} + \frac{\partial A_z(x,y,z,t)}{\partial z}\right), \tag{A1.6}$$

which is called the divergence of **A**; and

$$\nabla \times \mathbf{A}(x,y,z,t) = \left(\left(\frac{\partial A_z(x,y,z,t)}{\partial y} - \frac{\partial A_y(x,y,z,t)}{\partial z}\right), \left(\frac{\partial A_x(x,y,z,t)}{\partial z} - \frac{\partial A_z(x,y,z,t)}{\partial x}\right), \left(\frac{\partial A_y(x,y,z,t)}{\partial x} - \frac{\partial A_x(x,y,z,t)}{\partial y}\right)\right), \tag{A1.7}$$

which is called the curl of **A**.

Both ∇f and $\nabla \times \mathbf{A}$ are vectors. The vector operator ∇ can be treated almost like any other vector provided that the order of the operators and operands in the expression is retained. One can verify this statement by examining the foregoing forms, recognizing that the divergence is a dot product of two vectors and that the curl is a cross product of two vectors. The divergence and curl are of interest not only as mathematical constructs, but also because they have physical significance. Some remarks on the physical significance of these operators follow Eq. (A1.42).

A.1.2 VECTOR IDENTITIES

The following formulas are taken from Huba (2009).

Notations f and g are scalar functions; **A**, **B**, etc. are vector functions; **T** is a tensor.

$$\mathbf{A}\cdot\mathbf{B}\times\mathbf{C} = \mathbf{B}\cdot\mathbf{C}\times\mathbf{A} = \mathbf{C}\cdot\mathbf{A}\times\mathbf{B}, \tag{A1.8}$$

$$\mathbf{A}\times(\mathbf{B}\times\mathbf{C}) = (\mathbf{C}\times\mathbf{B})\times\mathbf{A} = \mathbf{B}\,(\mathbf{A}\cdot\mathbf{C}) - \mathbf{C}\,(\mathbf{A}\cdot\mathbf{B}), \tag{A1.9}$$

$$\mathbf{A}\times(\mathbf{B}\times\mathbf{C}) + \mathbf{B}\times(\mathbf{C}\times\mathbf{A}) + \mathbf{C}\times(\mathbf{A}\times\mathbf{B}) = 0, \tag{A1.10}$$

$$(\mathbf{A}\times\mathbf{B})\cdot(\mathbf{C}\times\mathbf{D}) = (\mathbf{A}\cdot\mathbf{C})(\mathbf{B}\cdot\mathbf{D}) - (\mathbf{A}\cdot\mathbf{D})(\mathbf{B}\cdot\mathbf{C}), \tag{A1.11}$$

$$(\mathbf{A}\times\mathbf{B})\times(\mathbf{C}\times\mathbf{D}) = (\mathbf{A}\times\mathbf{B}\cdot\mathbf{D})\mathbf{C} - (\mathbf{A}\times\mathbf{B}\cdot\mathbf{C})\mathbf{D}, \tag{A1.12}$$

$$\nabla(fg) = \nabla(gf) = f\nabla g + g\nabla f, \tag{A1.13}$$

$$\nabla\cdot(f\mathbf{A}) = f\nabla\cdot\mathbf{A} + \mathbf{A}\cdot\nabla f, \tag{A1.14}$$

$$\nabla\times(f\mathbf{A}) = f\nabla\times\mathbf{A} + \nabla f\times\mathbf{A}, \tag{A1.15}$$

$$\nabla\cdot(\mathbf{A}\times\mathbf{B}) = \mathbf{B}\cdot\nabla\times\mathbf{A} - \mathbf{A}\cdot\nabla\times\mathbf{B}, \tag{A1.16}$$

$$\nabla\times(\mathbf{A}\times\mathbf{B}) = \mathbf{A}(\nabla\cdot\mathbf{B}) - \mathbf{B}(\nabla\cdot\mathbf{A}) + (\mathbf{B}\cdot\nabla)\mathbf{A} - (\mathbf{A}\cdot\nabla)\mathbf{B}, \tag{A1.17}$$

$$\mathbf{A}\times(\nabla\times\mathbf{B}) = (\nabla\mathbf{B})\cdot\mathbf{A} - (\mathbf{A}\cdot\nabla)\mathbf{B}, \tag{A1.18}$$

$$\nabla(\mathbf{A}\cdot\mathbf{B}) = \mathbf{A}\times(\nabla\times\mathbf{B}) + \mathbf{B}\times(\nabla\times\mathbf{A}) + (\mathbf{A}\cdot\nabla)\mathbf{B} + (\mathbf{B}\cdot\nabla)\mathbf{A}, \tag{A1.19}$$

$$\nabla^2 f = \nabla\cdot\nabla f, \tag{A1.20}$$

$$\nabla^2\mathbf{A} = \nabla(\nabla\cdot\mathbf{A}) - \nabla\times\nabla\times\mathbf{A}, \tag{A1.21}$$

$$\nabla\times\nabla f = 0, \tag{A1.22}$$

$$\nabla\cdot\nabla\times\mathbf{A} = 0. \tag{A1.23}$$

A second-order tensor **T** can be written in a number of different ways. It is often convenient to express such a tensor in terms of two vectors, **A** and **B**, which allows one to write **T** in dyadic form:

$$\mathbf{T} = \mathbf{AB} \text{ or } T_{ij} = A_i B_j. \tag{A1.24}$$

In cartesian coordinates, the divergence of a tensor is a vector with components

$$(\nabla\cdot\mathbf{T})_i = \sum_j (\partial T_{ji}/\partial x_j), \tag{A1.25}$$

$$\nabla\cdot(\mathbf{BA}) = (\mathbf{B}\cdot\nabla)\mathbf{A} + \mathbf{A}(\nabla\cdot\mathbf{B}), \tag{A1.26}$$

$$\nabla\cdot(f\mathbf{T}) = \nabla f\cdot\mathbf{T} + f\nabla\cdot\mathbf{T}. \tag{A1.27}$$

Let $\mathbf{r} = \hat{\mathbf{e}}_x x + \hat{\mathbf{e}}_y y + \hat{\mathbf{e}}_z z$ be the radius vector of magnitude r from the origin to the point x, y, z. Then

$$\nabla\cdot\mathbf{r} = 3, \tag{A1.28}$$

$$\nabla\times\mathbf{r} = 0, \tag{A1.29}$$

$$\nabla r = \mathbf{r}/r, \tag{A1.30}$$

$$\nabla(1/r) = -\mathbf{r}/r^3, \qquad (A1.31)$$

$$\nabla \cdot (\mathbf{r}/r^3) = 4\pi\delta(\mathbf{r}). \qquad (A1.32)$$

If V is a volume enclosed by a surface S and $d\mathbf{S} = \hat{\mathbf{n}}\, dS$, where $\hat{\mathbf{n}}$ is the unit normal outward from V,

$$\int_V dV \nabla f = \int_S d\mathbf{S}\, f, \qquad (A1.33)$$

$$\int_V dV \nabla \cdot \mathbf{A} = \int_S d\mathbf{S} \cdot \mathbf{A}, \qquad (A1.34)$$

$$\int_V dV \nabla \cdot \mathbf{T} = \int_S d\mathbf{S} \cdot \mathbf{T}, \qquad (A1.35)$$

$$\int_V dV \nabla \times \mathbf{A} = \int_S d\mathbf{S} \times \mathbf{A}, \qquad (A1.36)$$

$$\int_V dV (f\nabla^2 g - g\nabla^2 f) = \int_S d\mathbf{S} \cdot (f\nabla g - g\nabla f), \qquad (A1.37)$$

$$\int_V dV (\mathbf{A} \cdot \nabla \times \nabla \times \mathbf{B} - \mathbf{B} \cdot \nabla \times \nabla \times \mathbf{A}) = \int_S d\mathbf{S} \cdot (\mathbf{B} \times \nabla \times \mathbf{A} - \mathbf{A} \times \nabla \times \mathbf{B}). \qquad (A1.38)$$

If S is an open surface bounded by the contour of C, of which the line element is $d\mathbf{l}$,

$$\int_V d\mathbf{S} \times \nabla f = \oint_C d\mathbf{l}\, f, \qquad (A1.39)$$

$$\int_V d\mathbf{S} \cdot \nabla \times \mathbf{A} = \oint_C d\mathbf{l} \cdot \mathbf{A}, \qquad (A1.40)$$

$$\int_V (d\mathbf{S} \times \nabla) \times \mathbf{A} = \oint_C d\mathbf{l} \times \mathbf{A}, \qquad (A1.41)$$

$$\int_V d\mathbf{S} \cdot (\nabla f) \times (\nabla g) = \oint_C f\, dg = -\oint_C g\, df. \qquad (A1.42)$$

The integral relations (A1.34) and (A1.40) are helpful in understanding the physical significance of the divergence and the curl operators. The names "curl" and "divergence" are suggestive and are particularly easy to visualize if the vector $\mathbf{A}$ is set equal to $\mathbf{v}$, the vector velocity in an incompressible fluid; $\mathbf{v}$ is assumed to be a function of position. Consider Eq. (A1.34) applied to the divergence of $\mathbf{v}$. From $\int_v dV \nabla \cdot \mathbf{v} = \oint_s d\mathbf{S} \cdot \mathbf{v}$, we find that the integral of $\nabla \cdot \mathbf{v}$ over a small volume at any location in the fluid is equal to the integral of the normal component of fluid flow across the surface. The divergence of the fluid is non-vanishing at a particular point if the net fluid flow into an infinitesimal volume surrounding the point differs from the net fluid flow out of the volume, but, in an incompressible fluid, that cannot happen. This means that the divergence must vanish, a condition for incompressible flow. The situation is readily visualized by drawing streamlines of the flow and noting that the same number of streamlines enter and exit the volume.

The curl can be understood in an analogous way by setting $\mathbf{A}$ equal to $\mathbf{v}$ in Eq. (A1.40) and taking the surface integral over a surface around a point of interest. Then $\int_s d\mathbf{S} \cdot \nabla \times \mathbf{v} = \oint_c d\mathbf{l} \cdot \mathbf{v}$, and the integral of the curl of $\mathbf{v}$ is seen to be equal to the integral of $\mathbf{v}$ around the curve bounding the surface containing the point. The integral around the closed contour C is called the circulation. If the curl of $\mathbf{v}$ is non-vanishing, there is net circulation around the point. Once again, streamlines are helpful in visualizing the situation. If some of the streamlines encircle the point of interest, the circulation and the curl of the velocity are non-vanishing.

In space physics, one often needs to consider the curl and the divergence of electromagnetic fields. In such situations, the analogs of streamlines are field lines, which are contours in space that are everywhere parallel to either the electric or the magnetic field. From Maxwell's laws (see Chapter 2), we find that $\mathbf{E}$ can have a finite divergence. This means that it is possible to have more field lines that extend outward from a point than inward (or vice versa). However, if there is a net divergence of $\mathbf{E}$ at some point in space, there must a net charge. The situation is different for magnetic fields. Magnetic field lines emerging from a spatial point must always close back on the point, which means that the magnetic field is divergence free.

A.1.3 DIFFERENTIAL OPERATORS IN CURVILINEAR COORDINATES

Analyses of physical problems can often be greatly simplified by taking advantage of symmetry. For example, the form of the Coulomb potential is substantially more complicated in cartesian coordinates than in spherical coordinates. The axial symmetry of a dipole field is more straightforwardly represented in spherical or cylindrical coordinates than in cartesian coordinates. However, operator expressions in non-cartesian (or curvilinear) coordinate systems must be treated with care. This is because one must take into account the fact that distance scales vary with position in a curvilinear coordinate system (e.g., lengths in the azimuthal direction in a spherical coordinate system depend on the colatitude θ and radial distance as $r\,\Delta\varphi \sin\theta$), and unit vectors point in different directions at different points in space (e.g., when $\theta = 90°$, the radial direction is along y). Thus, for spherical and cylindrical coordinates, the expressions for the operators do not have the simple forms given in Eqs. (A1.5)–(A1.7). Next, we give the forms of the principal operator expressions in several important coordinate systems.

A.1.3.1 Cylindrical coordinates

DIVERGENCE:

$$\nabla \cdot \mathbf{A} = \frac{1}{r}\frac{\partial}{\partial r}(rA_r) + \frac{1}{r}\frac{\partial A_\varphi}{\partial \varphi} + \frac{\partial A_z}{\partial z}. \qquad \text{(A1.43)}$$

GRADIENT:

$$(\nabla f)_r = \frac{\partial f}{\partial r}, (\nabla f)_\varphi = \frac{1}{r}\frac{\partial f}{\partial \varphi}, (\nabla f)_z = \frac{\partial f}{\partial z}. \qquad \text{(A1.44)}$$

CURL:

$$\begin{aligned}
(\nabla \times \mathbf{A})_r &= \frac{1}{r}\frac{\partial A_z}{\partial_\varphi} - \frac{\partial A_\varphi}{\partial z},\\
(\nabla \times \mathbf{A})_\varphi &= \frac{\partial A_r}{\partial z} - \frac{\partial A_z}{\partial r},\\
(\nabla \times \mathbf{A})_z &= \frac{1}{r}\frac{\partial}{\partial r}(rA_\varphi) - \frac{1}{r}\frac{\partial A_r}{\partial \varphi}.
\end{aligned} \qquad \text{(A1.45)}$$

LAPLACIAN:

$$\nabla^2 f = \frac{1}{r}\frac{\partial}{\partial r}\left(r\frac{\partial f}{\partial r}\right) + \frac{1}{r^2}\frac{\partial^2 f}{\partial \varphi^2} + \frac{\partial^2 f}{\partial z^2}. \qquad \text{(A1.46)}$$

LAPLACIAN OF A VECTOR:

$$\begin{aligned}
(\nabla^2 \mathbf{A})_r &= \nabla^2 A_r - \frac{2}{r^2}\frac{\partial A_\varphi}{\partial \varphi} - \frac{A_r}{r^2},\\
(\nabla^2 \mathbf{A})_\varphi &= \nabla^2 A_\varphi + \frac{2}{r^2}\frac{\partial A_r}{\partial \varphi} - \frac{A_\varphi}{r^2},\\
(\nabla^2 \mathbf{A})_z &= \nabla^2 A_z.
\end{aligned} \qquad \text{(A1.47)}$$

COMPONENTS OF $(\mathbf{A} \cdot \nabla)\mathbf{B}$:

$$\begin{aligned}
(\mathbf{A} \cdot \nabla \mathbf{B})_r &= A_r\frac{\partial B_r}{\partial r} + \frac{A_\varphi}{r}\frac{\partial B_r}{\partial \varphi} + A_z\frac{\partial B_r}{\partial z} - \frac{A_\varphi B_\varphi}{r},\\
(\mathbf{A} \cdot \nabla \mathbf{B})_\varphi &= A_r\frac{\partial B_\varphi}{\partial r} + \frac{A_\varphi}{r}\frac{\partial B_\varphi}{\partial \varphi} + A_z\frac{\partial B_\varphi}{\partial z} + \frac{A_\varphi B_r}{r},\\
(\mathbf{A} \cdot \nabla \mathbf{B})_z &= A_r\frac{\partial B_z}{\partial r} + \frac{A_\varphi}{r}\frac{\partial B_z}{\partial \varphi} + A_z\frac{\partial B_z}{\partial z}.
\end{aligned} \qquad \text{(A1.48)}$$

DIVERGENCE OF A TENSOR:

$$\begin{aligned}
(\nabla \cdot \mathbf{T})_r &= \frac{1}{r}\frac{\partial}{\partial r}(rT_{rr}) + \frac{1}{r}\frac{\partial}{\partial \varphi}(T_{\varphi r}) + \frac{\partial T_{zr}}{\partial z} - \frac{1}{r}T_{\varphi\varphi},\\
(\nabla \cdot \mathbf{T})_\varphi &= \frac{1}{r}\frac{\partial}{\partial r}(rT_{r\varphi}) + \frac{1}{r}\frac{\partial}{\partial \varphi}(T_{\varphi\varphi}) + \frac{\partial T_{z\varphi}}{\partial z} + \frac{1}{r}T_{\varphi r},\\
(\nabla \cdot \mathbf{T})_z &= \frac{1}{r}\frac{\partial}{\partial r}(rT_{rz}) + \frac{1}{r}\frac{\partial}{\partial \varphi}(T_{\varphi z}) + \frac{\partial T_{zz}}{\partial z}.
\end{aligned} \qquad \text{(A1.49)}$$

A.1.3.2 Spherical coordinates

DIVERGENCE:

$$\nabla \cdot \mathbf{A} = \frac{1}{r^2}\frac{\partial}{\partial r}(r^2 A_r) + \frac{1}{r\sin\theta}\frac{\partial}{\partial \theta}(A_\theta \sin\theta) + \frac{1}{r\sin\theta}\frac{\partial A_\varphi}{\partial \varphi} \tag{A1.50}$$

GRADIENT:

$$(\nabla f)_r = \frac{\partial f}{\partial r}, (\nabla f)_\theta = \frac{1}{r}\frac{\partial f}{\partial \theta}, (\nabla f)_\varphi = \frac{1}{r\sin\theta}\frac{\partial f}{\partial \theta} \tag{A1.51}$$

CURL:

$$\begin{aligned}
(\nabla \times \mathbf{A})_r &= \frac{1}{r\sin\theta}\frac{\partial}{\partial \theta}(A_\varphi \sin\theta) - \frac{1}{r\sin\theta}\frac{\partial A_\theta}{\partial \varphi} \\
(\nabla \times \mathbf{A})_\theta &= \frac{1}{r\sin\theta}\frac{\partial A_r}{\partial \varphi} - \frac{1}{r}\frac{\partial}{\partial r}(rA_\varphi) \\
(\nabla \times \mathbf{A})_r &= \frac{1}{r}\frac{\partial}{\partial r}(rA_\theta) - \frac{1}{r}\frac{\partial A_r}{\partial \theta}
\end{aligned} \tag{A1.52}$$

LAPLACIAN:

$$\nabla^2 f = \frac{1}{r^2}\frac{\partial}{\partial r}\left(r^2 \frac{\partial f}{\partial r}\right) + \frac{1}{r^2 \sin\theta}\frac{\partial}{\partial \theta}\left(\sin\theta \frac{\partial f}{\partial \theta}\right) + \frac{1}{r^2 \sin^2\theta}\frac{\partial^2 f}{\partial \varphi^2} \tag{A1.53}$$

LAPLACIAN OF A VECTOR:

$$\begin{aligned}
(\nabla^2 \mathbf{A})_r &= \nabla^2 A_r - \frac{2A_r}{r^2} - \frac{2}{r^2}\frac{\partial A_\theta}{\partial \theta} - \frac{2A_\theta \cot\theta}{r^2} - \frac{2}{r^2 \sin\theta}\frac{\partial A_\varphi}{\partial \varphi} \\
(\nabla^2 \mathbf{A})_\theta &= \nabla^2 A_\theta + \frac{2}{r^2}\frac{\partial A_r}{\partial \theta} - \frac{A_\theta}{r^2 \sin^2\theta} - \frac{2\cos\theta}{r^2 \sin^2\theta}\frac{\partial A_\varphi}{\partial \varphi} \\
(\nabla^2 \mathbf{A})_\varphi &= \nabla^2 A_\varphi - \frac{A_\varphi}{r^2 \sin^2\theta} + \frac{2}{r^2 \sin\theta}\frac{\partial A_r}{\partial \varphi} + \frac{2\cos\theta}{r^2 \sin^2\theta}\frac{\partial A_\theta}{\partial \varphi}
\end{aligned} \tag{A1.54}$$

COMPONENTS OF $(\mathbf{A} \cdot \nabla)\mathbf{B}$:

$$\begin{aligned}
(\mathbf{A} \cdot \nabla \mathbf{B})_r &= A_r \frac{\partial B_r}{\partial r} + \frac{A_\theta}{r}\frac{\partial B_r}{\partial \theta} + \frac{A_\varphi}{r\sin\theta}\frac{\partial B_r}{\partial \varphi} - \frac{A_\theta B_\theta + A_\varphi B_\varphi}{r} \\
(\mathbf{A} \cdot \nabla \mathbf{B})_\theta &= A_r \frac{\partial B_\theta}{\partial r} + \frac{A_\theta}{r}\frac{\partial B_\theta}{\partial \theta} + \frac{A_\varphi}{r\sin\theta}\frac{\partial B_\theta}{\partial \varphi} + \frac{A_\theta B_r}{r} - \frac{A_\varphi B_\varphi \cot\theta}{r} \\
(\mathbf{A} \cdot \nabla \mathbf{B})_\varphi &= A_r \frac{\partial B_\varphi}{\partial r} + \frac{A_\theta}{r}\frac{\partial B_\varphi}{\partial \theta} + \frac{A_\varphi}{r\sin\theta}\frac{\partial B_\varphi}{\partial \varphi} + \frac{A_\varphi B_r}{r} + \frac{A_\varphi B_\theta \cot\theta}{r}
\end{aligned} \tag{A1.55}$$

DIVERGENCE OF A TENSOR:

$$\begin{aligned}
(\nabla \cdot \mathbf{T})_r &= \frac{1}{r^2}\frac{\partial}{\partial r}(r^2 T_{rr}) + \frac{1}{r\sin\theta}\frac{\partial}{\partial \theta}(T_{\theta r}\sin\theta) + \frac{1}{r\sin\theta}\frac{\partial T_{\varphi r}}{\partial \varphi} - \frac{1}{r}(T_{\theta\theta} + T_{\varphi\varphi}) \\
(\nabla \cdot \mathbf{T})_\theta &= \frac{1}{r^2}\frac{\partial}{\partial r}(r^2 T_{r\theta}) + \frac{1}{r\sin\theta}\frac{\partial}{\partial \theta}(T_{\theta\theta}\sin\theta) + \frac{1}{r\sin\theta}\frac{\partial T_{\varphi\theta}}{\partial \varphi} + \frac{T_{\theta r}}{r} - \frac{\cot\theta}{r}T_{\varphi\varphi} \\
(\nabla \cdot \mathbf{T})_\varphi &= \frac{1}{r^2}\frac{\partial}{\partial r}(r^2 T_{r\varphi}) + \frac{1}{r\sin\theta}\frac{\partial}{\partial \theta}(T_{\theta\varphi}\sin\theta) + \frac{1}{r\sin\theta}\frac{\partial T_{\varphi\varphi}}{\partial \varphi} + \frac{T_{\varphi r}}{r} + \frac{\cot\theta}{r}T_{\varphi\theta}
\end{aligned} \tag{A1.56}$$

A.1.4 SELECTED INTEGRALS

$$\int_{-\infty}^{\infty} dx\, e^{-ax^2} = \left(\frac{\pi}{a}\right)^{1/2}$$

$$\int_{-\infty}^{\infty} dx\, x^{2n} e^{-ax^2} = \frac{2n!}{2^{2n} n!}\left(\frac{\pi}{a^{2n+1}}\right)^{1/2}$$

Appendix A.2

Fundamental constants and plasma parameters of space physics

A.2.1 FUNDAMENTAL CONSTANTS

mass of a proton	1.6726×10^{-27} kg
mass of an electron	9.1095×10^{-31} kg
electron to proton mass ratio	1836.2
speed of light in vacuum	2.9979×10^{8} m s^{-1}
gravitational constant	6.672×10^{-11} N m^{2} kg^{-2}
Stefan–Boltzmann constant	5.6703×10^{-8} J m^{-2} s^{-1} K^{-4}
Boltzmann constant	1.3807×10^{-23} J K^{-1}
electronvolt	1.6022×10^{-19} J
electronic charge	1.6022×10^{-19} C
temperature of 1 eV particle	1.1605×10^{4} K
permittivity of free space, ε_0	8.8542×10^{-12} F m^{-1}
permeability of free space, μ_0	4×10^{-7} H m^{-1}

A.2.2 FUNDAMENTAL PLASMA PARAMETERS IN PRACTICAL UNITS

electron gyro-frequency [Hz]	$28B$ [nT]
proton gyro-frequency [Hz]	$0.01525B$ [nT]
electron plasma frequency [Hz]	$8980\, n_e^{1/2}$ [cm^{-3}]; $8.98 n_e^{1/2}$[m^{-3}]
proton plasma frequency [Hz]	$210\, n_p^{1/2}$ [cm^{-3}]; $0.21\, n_p^{1/2}$ [m^{-3}]
equivalent temperature of electron passing through 1 V of potential drop	11 605 K
electron thermal speed[1] [km s^{-1}]	$5.50\, T_e^{1/2}$ [K]; $593\, T_e^{1/2}$ [eV]
proton thermal speed[1] [km s^{-1}]	$0.129\, T_p^{1/2}$ [K]; $13.90\, T_p^{1/2}$ [eV]
sound speed[2] [km s^{-1}]	$0.117\, T_e^{1/2}$ [K]
electron gyro-radius [km]	$0.0221\, T_e^{1/2}$ [K] $\cdot$ B^{-1} [nT]
proton gyro-radius [km]	$0.947 T_i^{1/2}$ [K] $\cdot$ B^{-1} [nT]
electron inertial length [km]	$5.31\, n_e^{1/2}$ [cm^{-3}]
proton inertial length [km]	$228\, n_i^{1/2}$ [cm^{-3}]
Debye length [cm]	$6.90\, T_e^{1/2}$ [K] $\cdot$ $n^{-1/2}$ [cm^{-3}]
particles in Debye cube	$329\, T_e^{3/2}$ [K] $\cdot$ $n^{-1/2}$ [cm^{-3}]
Alfvén velocity [km s^{-1}]	$21.8\, B$ [nT] $\cdot$ $n_p^{-1/2}$ [$(m_i/m_p)^{-1/2}$cm^{-3}]
magnetic pressure [pPa]	$0.398B^2$ [nT]
proton dynamic pressure [pPa]	$0.00167n$ [cm^{-3}] $\cdot$ v^2 [km s^{-1}]
beta	$3.47 \times 10^{-5} n$ [cm^{-3}] $\cdot$ T [K] $\cdot$ B^{-2}[nT]
neutral charged particle collision frequency[3] [s^{-1}]	$4 \times 10^{-5} n_0$ [cm^{-3}] $\cdot$ $T_q^{1/2}$ [K] $\cdot$ ($m_q^{-1/2}$ $\cdot$ m_p)
Bremsstrahlung from hydrogen-like plasma [W m^{-2}]	$1.46 \times 10^{-30} n_e$ [cm^{-3}] $\cdot$ $T_e^{1/2}$ [K] $\cdot$ $\sum$ [$Z^2\, N(Z)$]
cyclotron radiation [W m^{-2}]	$5.41 \times 10^{-31} B^2$ [nT] $\cdot$ n_e [cm^{-3}] $\cdot$ T_e [K]
cyclotron radiation [W m^{-2}]($T_e = T_i$; $\beta = 1$)	$3.79 \times 10^{-35} n_e^2$ [cm^{-3}] $\cdot$ T_e^2 [K]
magnetic field from an infinite wire, B_θ [nT]	$0.200I$ [A] $\cdot$ r^{-1} [km]

[1] Most probable speed (see Figure 2.37). [2] Assumes $T_e \gg T_i$.
[3] Approximate; mass of charged particle q expressed in proton masses.

Appendix A.3

Geophysical coordinate transformations

A.3.1 INTRODUCTION

Many different coordinate systems are used in experimental and theoretical work on solar-terrestrial relationships. These coordinate systems are used to display satellite trajectories, boundary locations, and vector field measurements. The need for more than one coordinate system arises from the fact that often various physical processes are better understood, experimental data more ordered, or calculations more easily performed in one or another of the various systems. Frequently it is necessary to transform from one to another of these systems. It is possible to derive the transformation from one coordinate system to another in terms of trigonometric relations between angles measured in each system by means of the formulas of spherical trigonometry. However, the use of this technique can be very tricky and can result in rather complex relationships.

Another technique is to find the required Euler rotation angles and construct the associated rotation matrices. Then these rotation matrices can be multiplied to give a single transformation matrix. The vector-matrix formalism is attractive because it permits a shorthand representation of the transformation and because it allows multiple transformations to be performed by matrix multiplication and the inverse transformation to be derived readily.

The matrices required for coordinate transformations need not be derived from Euler rotation angles. This appendix explains another approach and describes the most common coordinate systems in use in the field of solar-terrestrial relationships.

A.3.2 GENERAL REMARKS

In defining a coordinate system, we generally choose two quantities: the direction of one of the axes (two angles) and an orientation for the other two axes in the plane perpendicular to this direction (a third angle). This latter orientation is often specified by requiring one of the two remaining axes to be perpendicular to some direction. A fortunate feature of rotation matrices (the matrix that transforms a vector from one system to another) is that the inverse is simply its transpose. Thus, if the matrix **A** transforms the vector $\mathbf{V}^a$ measured in system a to $\mathbf{V}^b$ measured in system b, then the matrix that transforms $\mathbf{V}^b$ into $\mathbf{V}^a$ is $\mathbf{A}^T$. Thus we can write

$$\mathbf{A}\cdot\mathbf{V}^a = \mathbf{V}^b,$$

$$\mathbf{A}^T\cdot\mathbf{V}^b = \mathbf{V}^a.$$

The simplest way to obtain the transformation matrix A is to find the directions of the three new coordinate axes for system b in the old system (system a). If the direction cosines of the new x-direction expressed in the old system are (x_1, x_2, x_3), those of the new y-direction are (y_1, y_2, y_3), and those of the new z-direction are (z_1, z_2, z_3), then the rotation matrix is formed by these three vectors as rows:

$$\begin{pmatrix} x_1 & x_2 & x_3 \\ y_1 & y_2 & y_3 \\ z_1 & z_2 & z_3 \end{pmatrix}\cdot\begin{pmatrix} V_x^a \\ V_y^a \\ V_z^a \end{pmatrix} = \begin{pmatrix} V_x^b \\ V_y^b \\ V_z^b \end{pmatrix}.$$

Similarly, the transformation from system b to a is

$$\begin{pmatrix} x_1 & y_1 & z_1 \\ x_2 & y_2 & z_2 \\ x_3 & y_3 & z_3 \end{pmatrix} \cdot \begin{pmatrix} V_x^b \\ V_y^b \\ V_z^b \end{pmatrix} = \begin{pmatrix} V_x^a \\ V_y^a \\ V_z^a \end{pmatrix}.$$

The following properties of rotation matrices are useful for error checking. (1) Each row and column is a unit vector. (2) The dot products of any two rows or any two columns is zero. (3) The cross product of any two rows or columns equals the third row or column or its negative (row 1 cross row 2 equals row 3; row 2 cross row 1 equals minus row 3).

A.3.3 COORDINATE SYSTEMS

A.3.3.1 Geocentric equatorial inertial system

A.3.3.1.1 DEFINITION

The geocentric equatorial inertial (GEI) coordinate system has its x-axis pointing from the Earth toward the first point of Aries (the position of the Sun at the vernal equinox). This direction is the intersection of the Earth's equatorial plane and the ecliptic plane, and thus the x-axis lies in both planes. The z-axis is parallel to the rotation axis of the Earth, and y completes the right-handed orthogonal set ($\mathbf{y} = \mathbf{z} \times \mathbf{x}$).

A.3.3.1.2 USES

This is the system commonly used in astronomy and satellite orbit calculations. The angles right ascension and declination are measured in this system. If (V_x, V_y, V_z) is a vector in GEI with magnitude V, then its right ascension, α, is $\tan^{-1}(V_y/V_x)$, $0° \le \alpha \le 180°$ if $V_y \ge 0$, and $180° \le \alpha \le 360°$ if $V_y \le 0$. Its declination, θ, is $\sin^{-1}(V_z/V)$, $-90° \le \theta \le 90°$.

A.3.3.2 Geographic coordinates

A.3.3.2.1 DEFINITION

The geographic coordinate system (GEO) is defined so that its x-axis is in the Earth's equatorial plane, but is fixed with the rotation of the Earth, so that it passes through the Greenwich meridian (0° longitude). Its z-axis is parallel to the rotation axis of the Earth, and its y-axis completes a right-handed orthogonal set ($\mathbf{y} = \mathbf{z} \times \mathbf{x}$).

A.3.3.2.2 USES

This system is used for defining the positions of ground observatories and transmitting and receiving stations. Longitude and latitude in this system are defined in the same way as right ascension and declination in GEI. Longitude is measured positively moving eastward. Universal time (UT) is defined as 12 hours minus the longitude of the Sun converted from degrees to hours by dividing by 15. Local time is the universal time plus the geographic longitude of the observer converted to hours. Universal time is the local time of the Greenwich meridian.

A.3.3.2.3 TRANSFORMATION

Because the GEO and GEI coordinate systems have the z-axis in common, we need only know the position of the first point in Aries (the x-axis of GEI) relative to the Greenwich meridian to determine the required transformation. If we let the angle between the Greenwich meridian and the first point of Aries measured eastward from the first point of Aries in the Earth's equator be θ, then the first point of Aries is at $(\cos\theta, -\sin\theta, 0)$ in the GEO system, and the transformation from GEO to GEI is

$$\begin{pmatrix} \cos\theta & -\sin\theta & 0 \\ \sin\theta & \cos\theta & 0 \\ 0 & 0 & 1 \end{pmatrix} \cdot \begin{pmatrix} V_x \\ V_y \\ V_z \end{pmatrix}_{\text{GEO}} = \begin{pmatrix} V_x \\ V_y \\ V_z \end{pmatrix}_{\text{GEI}},$$

and the inverse transformation is

$$\begin{pmatrix} \cos\theta & \sin\theta & 0 \\ -\sin\theta & \cos\theta & 0 \\ 0 & 0 & 1 \end{pmatrix} \cdot \begin{pmatrix} V_x \\ V_y \\ V_z \end{pmatrix}_{\text{GEI}} = \begin{pmatrix} V_x \\ V_y \\ V_z \end{pmatrix}_{\text{GEO}}.$$

The angle θ is, of course a function of the time of day and the time of year, since the Earth spins 366.25 times per year around its axis in inertial space, rather than 365.25 times. Thus, the duration of a day, relative to inertial space (a sidereal day), is less than 24 h. The angle θ is called Greenwich mean sidereal time and can be calculated by means of the formulas given in Section A.3.5.

A Sun-fixed geographic system can be constructed using the longitude of the Sun in place of the geographic longitude. This could be useful for studying quantities that vary with local time.

A.3.3.3 Geomagnetic coordinates

A.3.3.3.1 DEFINITION

The geomagnetic coordinate system (MAG) is defined so that its z-axis is parallel to the magnetic dipole axis. For January 1, 2016, the geographic coordinates of the dipole axis from the International Geomagnetic Reference Field (IGRF12) are 9.207° colatitude and −72.732° east longitude. Thus the z-axis is (0.04749, −0.15278, 0.98712) in geographic coordinates. The y-axis of this system is perpendicular to the geographic poles, such that if **D** is the dipole position and **P** is the South Pole, $\mathbf{y} = \mathbf{D} \times \mathbf{P}/|\mathbf{D} \times \mathbf{P}|$.

A.3.3.3.2 USES

This system is often used for defining the positions of magnetic observatories. Also, it is a convenient system in which to do field-line tracing when current systems, in addition to the Earth's internal field, are being considered. The magnetic longitude is measured eastward from the x-axis, and magnetic latitude is measured from the equator in magnetic meridians, positive northward and negative southward. Thus, if (V_x, V_y, V_z) is a vector in the MAG system with magnitude V, then its magnetic longitude, λ, is tan (V_y/V_x), $0° \le \lambda \le 180°$ if $V_y \ge 0$, and $180° \le \lambda \le 360°$ if $V_y \le 0$. Its magnetic latitude, θ, is $\sin^{-1} V_z/V$, $-90° \le \theta \le 90°$. Except near the poles, magnetic longitude generally will be about 70° greater than geographic longitude. The magnetic local time is defined in this system as the magnetic longitude of the observer minus the magnetic longitude of the Sun expressed in hours plus 12 h.

A.3.3.3.3 TRANSFORMATION

This system is fixed in the rotating Earth, and thus the transformation from the geographic coordinate system to the geomagnetic system is constant. From the foregoing definitions, we obtain

$$\begin{pmatrix} 0.29301 & -0.94263 & -0.15999 \\ 0.95493 & 0.29683 & 0 \\ 0.04749 & -0.15278 & 0.98712 \end{pmatrix} \cdot \begin{pmatrix} V_x \\ V_y \\ V_z \end{pmatrix}_{\mathrm{GEO}} = \begin{pmatrix} V_x \\ V_y \\ V_z \end{pmatrix}_{\mathrm{MAG}}.$$

A.3.3.4 Geocentric solar ecliptic system

A.3.3.4.1 DEFINITION

The geocentric solar ecliptic (GSE) system has its x-axis pointing from the Earth toward the Sun, and its y-axis is chosen to be in the ecliptic plane pointing toward dusk (thus opposing planetary motion). Its z-axis is parallel to the ecliptic pole. Relative to an inertial system, this system has a yearly rotation.

A.3.3.4.2 USES

This system is used to display satellite trajectories, interplanetary magnetic field observations, and data on solar-wind velocity. The system is useful for the latter display because the aberration of the solar wind caused by the Earth's motion can easily be removed in this system. The velocity of the Earth is approximately 30 km s^{-1} in the $-y$-direction. Because the only important effect of the Earth's orbital motion in solar-terrestrial relationships is to cause the aberration, other choices for the orientation of the y- and z-axes about the x-axis have been used. These are to be discussed later. Longitude, as with the geographic system, is measured in the x–y plane from the x-axis toward the y-axis, and latitude is the angle out of the x–y plane, positive for positive z-components.

A.3.3.4.3 TRANSFORMATION

The most commonly required transformation into the GSE system from those discussed thus far is that from the GEI system. The direction of the ecliptic pole (0, −0.3978, 0.9175) is constant in the GEI system. The x-axis, the direction of the Sun, can be obtained in GEO from the subroutine given in Section A.3.5. If this direction is (S_1, S_2, S_3), then the y-axis in GEI (y_1, y_2, y_3) is

$$(0, \ -0.3978, \ 0.9175) \times (S_1, S_2, S_3),$$

and the transformation is

$$\begin{pmatrix} S_1 & S_2 & S_3 \\ y_1 & y_2 & y_3 \\ 0 & -0.3978 & 0.9175 \end{pmatrix} \cdot \begin{pmatrix} V_x \\ V_y \\ V_z \end{pmatrix}_{\mathrm{GEI}} = \begin{pmatrix} V_x \\ V_y \\ V_z \end{pmatrix}_{\mathrm{GSE}}.$$

A.3.3.5 Geocentric solar equatorial system

A.3.3.5.1 DEFINITION

The geocentric solar equatorial (GSEQ) system, like the GSE system, has its x-axis pointing toward the Sun from the Earth. However, instead of having its y-axis in the ecliptic plane, the GSEQ y-axis is parallel to the Sun's equatorial plane, which is inclined to the ecliptic. We note that, because the x-axis is in the ecliptic plane and therefore is not necessarily in the Sun's equatorial plane, the z-axis of this system will not necessarily be parallel to the Sun's axis of rotation. However, the Sun's axis of rotation must lie in the x–z plane. The z-axis is chosen to be in the same sense as the ecliptic pole (i.e., northward).

A.3.3.5.2 USES

This system has been used extensively to display interplanetary magnetic field data. We note that this system is useful for ordering data controlled by the Sun, and therefore it offers an improvement over the use of the GSE system for studying the interplanetary magnetic field and the solar wind. However, for studying the interaction of the interplanetary magnetic field with the Earth, a third system is more relevant.

A.3.3.5.3 TRANSFORMATION

The rotation axis of the Sun, **R**, has a right ascension of –74.0° and a declination of 63.8°. Thus **R** is (0.1217, –0.424, 0.897) in GEI. To transform from GEI to GSEQ, we must know the position of the Sun (S_1, S_2, S_3) in GEI (see Section A.3.5). Then the y-axis in GEI (y_1, y_2, y_3) is parallel to **R** × **S**. Note that because the cross product of two unit vectors is not a unit vector unless they are perpendicular to each other, this cross product must be normalized. Finally the x-axis in GEI (z_1, z_2, z_3) = **S** × **y**. Then

$$\begin{pmatrix} S_1 & S_2 & S_3 \\ y_1 & y_2 & y_3 \\ z_1 & z_2 & z_3 \end{pmatrix} \cdot \begin{pmatrix} V_x \\ V_y \\ V_z \end{pmatrix}_{\mathrm{GEI}} = \begin{pmatrix} V_x \\ V_y \\ V_z \end{pmatrix}_{\mathrm{GSEQ}}.$$

Because both the GSE and GSEQ coordinate systems have their x-axes directed toward the Sun, they differ only by a rotation about the x-axis. Thus the transformation matrix from GSE to GSEQ must be of the form

$$\begin{pmatrix} 1 & 0 & 0 \\ 0 & \cos\theta & -\sin\theta \\ 0 & \sin\theta & \cos\theta \end{pmatrix} \cdot \begin{pmatrix} V_x \\ V_y \\ V_z \end{pmatrix}_{\mathrm{GSE}} = \begin{pmatrix} V_x \\ V_y \\ V_z \end{pmatrix}_{\mathrm{GSEQ}}.$$

If the transformations from GEI to GSE and GEI to GSEQ are both known, then the angle θ can be determined by examining the angle between the y-axes in the two systems, or the z-axes (i.e., the angle between the vectors formed by the second row of each matrix, or the third row). If these transformation matrices are not available, θ can be calculated from the following formula:

$$\sin\theta = \frac{\mathbf{S} \cdot (-0.032, -0.112, -0.048)}{|(0.1217, -0.424, 0.897) \times \mathbf{S})}$$

where **S**, the direction the Sun in GEI, can be calculated from the formulas in Section A.3.5. Because the Sun's spin axis is inclined 7.25° to the ecliptic, θ ranges from –7.25° (on approximately December 5) to 7.25° (on June 5) each year. The Sun's spin axis is directed most nearly toward the Earth on approximately September 5, at which time the Earth reaches its most northerly heliographic latitude. At this time, θ equals zero.

A.3.3.6 Geocentric solar magnetospheric system

A.3.3.6.1 DEFINITION

The geocentric solar magnetospheric (GSM) system, like both the GSE and GSEQ systems, has its x-axis from the Earth to the Sun. The y-axis is defined to be perpendicular to the Earth's magnetic dipole, so that the x–z plane contains the dipole axis. The positive z-axis is chosen to be in the same sense as the northern magnetic pole. The difference between the GSM system and the GSE and GSEQ systems is simply a rotation about the x-axis.

A.3.3.6.2 USES

This system is useful for displaying magnetopause and shock-boundary positions, magnetosheath and magnetotail magnetic fields, and magnetosheath solar-wind velocities, because the orientation of the magnetic dipole axis alters the otherwise cylindrical symmetry of the solar-wind flow. It is also used in models of magnetopause currents. It reduces the three-dimensional motion of the Earth's dipole in GEI, GSE, and so forth, to motion in a plane (the x–z plane). The angle of the north magnetic pole to the GSM z-axis is called the dipole tilt angle and is positive when the north magnetic pole is tilted toward the Sun. In addition to a yearly period due to the motion of the Earth about the Sun, this coordinate system rocks about the solar direction with a 24 hour period. We note that, because the y-axis is perpendicular to the dipole axis, the y-axis is always in the magnetic equator, and, because it is perpendicular to the Earth–Sun line, it is in the dawn–dusk meridian (pointing toward dusk). GSM longitude is measured in the x–y plane from x toward y, and latitude is the angle northward from the x–y plane. However, another set of spherical polar angles is sometimes used. Here, the angle between the vector and the x-axis, called the solar zenith angle (SZA), is the polar angle, and the angle of the projected vector in the y–z plane is the azimuthal angle. It is measured from the positive y-axis toward the positive z-axis. If the interplanetary magnetic field is being described, these two angles are often called the cone and clock angles, respectively.

A.3.3.6.3 TRANSFORMATION

To transform from GEI to GSM, we need to know both the direction to the Sun in GEI and the orientation of the Earth's dipole axis. The direction of the Sun (S_1, S_2, S_3), can be obtained from Section A.3.5. The direction of the dipole **D** must be obtained by transforming from geographic coordinates (see Section A.3.3.2). In geographic coordinates, the dipole is at 9.207° colatitude and –72.732° east longitude (IGRF epoch 2016.0). Thus, **D** in geographic coordinates is (0.04749, -0.15278, 0.98712). If **D′** is **D** transformed into GEI, the y-axis is

$$\frac{\mathbf{D}' \times \mathbf{S}}{|\mathbf{D}' \times \mathbf{S}|}.$$

We note that the normalizing factor occurs because **D′** and **S** are not necessarily perpendicular. Finally, **z** is **S × y**, and the transformation becomes

$$\begin{pmatrix} S_1 & S_2 & S_3 \\ y_1 & y_2 & y_3 \\ z_1 & z_2 & z_3 \end{pmatrix} \cdot \begin{pmatrix} V_x \\ V_y \\ V_z \end{pmatrix}_{GEI} = \begin{pmatrix} V_x \\ V_y \\ V_z \end{pmatrix}_{GSM}.$$

The transformation matrix between GSM and GSE or GSEQ is of the form

$$\begin{pmatrix} 1 & 0 & 0 \\ 0 & \cos\theta & -\sin\theta \\ 0 & \sin\theta & \cos\theta \end{pmatrix}.$$

Because θ changes with both time of day and time of year, it is not derivable from a simple equation. If the transformation matrix from GEI to GSE, $\mathbf{A}_{GSE}$, and that from GEI to GSM, $\mathbf{A}_{GSM}$, are both known, then the transformation from GSM to GSE is simple: $\mathbf{A}_{GSE} \cdot \mathbf{A}_{GSM}^T$, where $\mathbf{A}_{GSM}^T$ is the transpose of $\mathbf{A}_{GSM}$. An analogous formula holds for the transformation from GSM to GSEQ. We note that the amplitude of the diurnal variation of θ is $\pm 11.0°$, which is added to an annual variation of $\pm 23.5°$.

A.3.3.7 Solar magnetic coordinates

A.3.3.7.1 DEFINITION

In solar magnetic (SM) coordinates, the z-axis is chosen parallel to the north magnetic pole, and the y-axis is perpendicular to the Earth–Sun line toward dusk. The difference between this system and the GSM system is a rotation about the y-axis. The amount of rotation is simply the dipole tilt angle, as defined in the preceding section. We note that in this system the x-axis does not point directly at the Sun. Like the GSM system, the SM system rotates with both a yearly period and a daily period with respect to inertial coordinates.

A.3.3.7.2 USES

The SM system is useful for ordering data controlled more strongly by the Earth's dipole field than by the solar wind. It has been used for magnetopause cross sections and magnetospheric magnetic fields. We note that, because the dipole axis and the z-axis of this system are parallel, the cartesian components of the dipole magnetic field are particularly simple in this system (see Chapter 7).

A.3.3.7.3 TRANSFORMATION

As for GSM, the transformation from GEI to SM requires a knowledge of the Earth–Sun direction **S** and the dipole direction **D** in GEI. Having obtained these as in Section A.3.5, we find **y** = **D** × **S**/|**D** × **S**| and **x** = **y** × **D**. Then the transformation becomes

$$\begin{pmatrix} x_1 & x_2 & x_3 \\ y_1 & y_2 & y_3 \\ D_1 & D_2 & D_3 \end{pmatrix} \cdot \begin{pmatrix} V_x \\ V_y \\ V_z \end{pmatrix}_{\mathrm{GEI}} = \begin{pmatrix} V_x \\ V_y \\ V_z \end{pmatrix}_{\mathrm{SM}}.$$

The transformation from GSM to SM is simply a rotation about the y-axis by the dipole tilt angle. Thus,

$$\begin{pmatrix} \cos\mu & 0 & -\sin\mu \\ 0 & 1 & 0 \\ \sin\mu & 0 & \cos\mu \end{pmatrix} \cdot \begin{pmatrix} V_x \\ V_y \\ V_z \end{pmatrix}_{\mathrm{GSM}} = \begin{pmatrix} V_x \\ V_y \\ V_z \end{pmatrix}_{\mathrm{SM}}.$$

A.3.3.8 Other planets

The coordinate systems we have discussed are all designed for use on or around the Earth. However, they can be generalized for the other planets. Instead of the ecliptic plane in which the Earth orbits, we use the orbital plane of the planet to define the x–y plane. The generalization of the GSE coordinate system thus becomes the Venus solar orbital and the Mars solar orbital at Venus and Mars, respectively. Likewise, at the magnetized planets, the GSM system can be adapted by keeping the y-axis perpendicular to the plane containing the magnetic dipole axis and the solar direction. This is particularly useful at Jupiter, which has a strong dipole field tilted at close to 10° to its rotation axis and a well-developed magnetic tail. Jupiter's rotation axis has a small tilt to its orbital plane. Saturn too has a strong dipole field and well-developed tail. However, its rotational axis and magnetic axes are aligned. Hence, the kronian equivalent of GSM and Sun-fixed geographic coordinates are identical. In studying unmagnetized planets, comets, and phenomena upstream from the bow shocks of even the magnetized planets, it is useful to use a coordinate system whose orientation is ordered by the magnetic field. The most common choice is to orient the coordinate system so that x points to the Sun and the projection of the magnetic field in the y–z plane lies along the y-axis. Because the solar wind flows approximately in the $-x$-direction, the electric field of the solar wind points in the $+z$-direction. Any newly created ions at a comet or near a planet will initially be accelerated in the z-direction. It is suggested that such a coordinate system be called the Venus solar interplanetary (VSI) for Venus, and similarly for other planets.

A.3.4 HELIOSPHERIC COORDINATE SYSTEMS

When a spacecraft is in heliocentric orbit, multiple different coordinate systems become useful. They can be referenced to inertial space, to the location of the Earth, to the Sun's rotation axis, to a Carrington prime median that rotates with the Sun, or be ordered by the location of the measurement in analogy to the coordinate systems useful around the planets.

A.3.4.1 Heliocentric Aries ecliptic

In the heliocentric Aries ecliptic (HAE) system, z is along the north ecliptic pole and the x-axis is from the center of the Sun to the first point of Aries (defined when discussing GEI coordinates). The y-axis completes the xyz right-handed orthogonal system. If, for example, a heliospheric spacecraft encountered a meteor stream, this coordinate system would order the locations of those observations.

A.3.4.2 Heliocentric Earth ecliptic

The heliocentric Earth ecliptic (HEE) system is a rotation about the ecliptic pole so that the location of the Earth defines the longitude in the ecliptic plane. In other words, x goes from the center of the Sun through the Earth. This is useful for missions like STEREO A and B that orbit the Sun in the ecliptic plane away from Earth. However, the rotation axis of the Sun is tilted with respect to the ecliptic pole, and a solar latitude difference might arise in using this system of which the user was not aware. Hence it should be used with caution.

A.3.4.3 Heliocentric Earth equatorial

In the heliocentric Earth equatorial (HEEQ) system, the z-axis is along the solar rotation axis, and the y-axis is in the direction $\mathbf{z} \times \mathbf{R}$, where $\mathbf{R}$ is the vector from the Sun to the Earth. The x-direction completes the right-handed orthogonal coordinate system and lies in the plane of $\mathbf{z}$ and $\mathbf{R}$. A point on the solar equator will vary from positive z and positive latitudes to negative as it passes the Earth at different times of the year. Spacecraft simultaneously measuring at longitudes away from the Earth will sense different latitudes. An active region or coronal hole might be seen at one spacecraft during a solar rotation but not at the other spacecraft.

A.3.4.4 Carrington coordinates

In the CARR (Carrington) system, the z-axis is along the solar rotation axis; y is perpendicular to the Carrington prime meridian; and x lies in the Carrington prime meridian. The prime meridian rotates with the average rotation rate of the Sun. As seen from Earth, the Carrington longitude, which is seen on the Sun, increases in the direction of solar rotation; the usual right-handed sense decreases with time due to the solar rotation. This coordinate system is useful for comparing maps of the solar magnetic field as it evolves over a solar cycle.

A.3.4.5 RTN coordinates

The radial outflow from the Sun dominates the structure of the solar wind and the interplanetary magnetic field. Hence the radial direction is useful for ordering solar-wind measurements. As noted before, the Sun's rotation axis is tilted with respect to the ecliptic plane. Thus the Earth and any spacecraft will move above and below the Sun's equator as it orbits the Sun even if it is located in the ecliptic plane. The solar magnetic field is statistically ordered by the latitude so we define the T-direction in this coordinate system as in the $\mathbf{z} \times \mathbf{R}$ direction, where $\mathbf{z}$ is along the northward pointing solar rotation axis and $\mathbf{R}$ is the vector from the center of the Sun to the point of observation. Thus T is roughly in the direction of the Earth's orbital motion about the Sun. The N-direction completes the RTN orthogonal right-handed triad, and is roughly along the northward ecliptic pole. This coordinate system is used extensively on missions in which the study of the magnetic structure of the heliosphere is of concern.

A.3.5 LOCAL COORDINATE SYSTEMS

As discussed for the RTN coordinate system, it is advantageous to define a coordinate system oriented with reference to the observing site. For example, when one is on the surface of the Earth, a coordinate system oriented with respect to the local vertical is useful for many purposes. In this section we describe several such coordinate systems. Some of these are ordered by the "dominant" terrestrial magnetic field; others are oriented along a normal to a plane that defines a large-scale discontinuity in the plasma being studied. The advantage of the coordinate system is to reduce the problem to variations in only one or two dimensions.

A.3.5.1 Dipole meridian system

A.3.5.1.1 DEFINITION

As with the SM system, the z-axis of the dipole meridian (DM) system is chosen along the north magnetic dipole axis. However, the y-axis is chosen to be perpendicular to a radius vector to the point of observation, rather than the Sun. The positive y-direction is chosen to be eastward, so that the x-axis is directed outward from the dipole. This is a local coordinate system in that it varies with position; however, because the x–z plane contains the dipole magnetic field, it is quite useful.

A.3.5.1.2 USES

The DM system is used to order data controlled by the dipole magnetic field where the influence of the solar-wind interaction with the magnetosphere is weak. It has been used extensively to describe the distortions of the magnetospheric field in terms of the two angles, declination and inclination, that can be easily derived from measurements in this system. The inclination I is simply the angle that the field makes with the radius vector minus 90°. Thus, if $\mathbf{R}$ is the unit vector from the center of the Earth to the point of observation in the DM system (we note that in this system $R_y = 0$), and $\mathbf{b}$ is the direction of the magnetic field in the DM system, then $I = \cos^{-1}(R_x b_s + R_z b_z) - 90°$. The declination, D, is measured about the radius vector, with $D = 0°$ in the x–z plane and positive D angles for positive b_y. Thus, $D = \tan^{-1}[b_y/(R_x b_z - R_z b_x)]$, $0° \le D \le 180°$ for $0° \le b_y \le 1$, and $0° \ge D \ge 180°$ for $0 \ge b_y \ge -1$. As in the SM system, the cartesian components of the dipole field can be expressed very simply in this system. In particular, $B_y = 0$, by definition.

A.3.5.1.3 TRANSFORMATION

To transform from any system to the DM system, we must know the dipole axis $\mathbf{D}$ in this system and the unit position vector of the point of observation relative to the center of the Earth. Because $\mathbf{y}$ is perpendicular to $\mathbf{R}$ and $\mathbf{D}$, $\mathbf{y} = (\mathbf{D} \times \mathbf{R})/(|\mathbf{D} \times \mathbf{R}|)$, and $\mathbf{x} = \mathbf{D} \times \mathbf{y}$. Thus

$$\begin{pmatrix} x_1 & x_2 & x_3 \\ y_1 & y_2 & y_3 \\ D_1 & D_2 & D_3 \end{pmatrix} \cdot \begin{pmatrix} V_x \\ V_y \\ V_z \end{pmatrix} = \begin{pmatrix} V_x \\ V_y \\ V_z \end{pmatrix}_{\mathrm{DM}}.$$

We note that this transformation usually is particularly straightforward from geographic coordinates, because the geographic latitude and longitude of a point of observation are often known, and the dipole is fixed in geographic coordinates. From geomagnetic coordinates, it is a simple rotation about the z-axis by the angle between the projections of the Sun and the local radius vector in the magnetic equator.

A.3.5.2 Surface magnetic measurements

Two local coordinate systems that are used for surface magnetic data differ slightly from the geographic (GEO) and geomagnetic (MAG) systems in the sense that they use the local vertical for their z-direction, but order their x- and y-axes according to geographic and geomagnetic directions, respectively. The first is called simply the XYZ system and is oriented so that the z-direction is downward and the x-direction points to geographic north. The y-direction points to geographic east. The second is the HDZ system, in which the z-direction is vertically downward, the H-direction points to the north magnetic pole, with D roughly eastward orthogonal to H. Caution is urged in the use of this coordinate system, because in some, but not all, instances, researchers express H and D as a magnitude and an angle, rather than as two components.

A.3.5.3 Boundary normal coordinate systems

Across an infinitesimally thin boundary, the normal component of the magnetic field is continuous, or nearly so, because the magnetic field is divergenceless, and thus it is often useful to express solar-terrestrial data in a boundary normal coordinate system. Situations in which this approach is helpful include studies of the bow shock, the magnetopause, and plane waves. The key to using this approach is determining an accurate boundary normal.

A.3.5.3.1 SHOCK NORMAL COORDINATES

The normal direction to the bow shock may be satisfactorily defined geometrically under most circumstances. With interplanetary shocks and the occasional bow shock crossing, one may wish to use the so-called coplanarity theorem to derive the shock normal direction, as discussed in Chapter 6. This technique makes use of the fact that the direction of the change in magnetic field across the shock is perpendicular to the normal to the shock, because the component along the shock normal is constant. Further, the upstream magnetic field (ahead of the shock structure), the shock normal, and the downstream magnetic field (downstream of the shock structure) are in the same plane. Hence, the cross product of the upstream and downstream fields must be perpendicular to the shock normal. Thus, the triple cross product of the upstream field times the downstream field and then times the difference of the two fields, when properly normalized, is along the normal direction. This does not work for parallel or perpendicular shocks when the fields upstream and downstream are parallel or perpendicular. More sophisticated techniques are possible when plasma velocity data are available or when there are observations from multiple spacecraft. For example, timing data from four spacecraft are sufficient to obtain the orientation and velocity of a planar boundary.

A.3.5.3.2 MAGNETOPAUSE NORMAL COORDINATES

Under most circumstances, the normal to the magnetopause, like the normal to the bow shock, can be obtained satisfactorily using a geometric model. The normal can also be obtained from magnetic field observations if the field behaves according to certain assumptions. For example, often the magnetopause appears to be a tangential discontinuity, with no connection of magnetic field across the surface of the magnetopause. Under such conditions the magnetic fields on either side of the boundary are tangential to the boundary, and the direction of the normal can be obtained from the cross product of these two directions. Even when the magnetic field connects across the magnetopause it is possible to determine a normal to the boundary by finding the direction in which the magnetic field remains most nearly constant or has a minimum variance. This direction should be the normal to the magnetopause because the magnetic field is divergenceless and across a thin discontinuity and it should not change in the direction perpendicular to the surface. If the change in magnetic field across the boundary occurs along a single direction, however, this technique becomes indeterminate because there are two orthogonal directions in which the field change is very small. The true normal may be along either direction or anywhere in the plane defined by these two directions.

A.3.5.3.3 PRINCIPAL AXIS COORDINATE SYSTEM

By solving an eigenvalue problem, it is possible to calculate the rotation matrix that rotates a vector time series into a coordinate system in which the direction of the coordinate axes are those of the maximum variance, the minimum variance, and an intermediate variance. The eigenvectors are the rows of the transformation matrix, and the eigenvalues are the variances along each direction. These directions are called the principal axes. These can often be used at the magnetopause, but not at shocks. The change in the magnetic field at shocks, which is mandated by the Rankine–Hugoniot equations, is along a single direction in the plane of the upstream magnetic field and the normal. Variation of the field perpendicular to this plane frequently occurs because of waves that do not necessarily propagate along the shock normal.

Principal axis coordinates are most useful for waves that are circularly or elliptically polarized, such as whistler-mode waves or ion cyclotron waves. At very low frequencies, well below the ion cyclotron frequency, waves tend to be linearly polarized and not amenable to this analysis.

A.3.6 CALCULATIONS OF THE POSITION OF THE SUN

In this section we present a simple subroutine to calculate the position of the Sun in GEI coordinates. It is accurate for the years 1901–2099, to within 0.006°. The inputs are the year, the day of the year, and the seconds of the day in UT. The outputs are Greenwich mean sidereal time in degrees, the ecliptic longitude, and the apparent right ascension and declination of the Sun in degrees. The listing of this program in Fortran follows. We note that the cartesian coordinates of the vector from the Earth to the Sun are

$$X = \cos(SRASN)\cos(SDEC)$$
$$Y = \sin(SRASN)\cos(SDEC)$$
$$Z = \sin(SDEC)$$

```
  SUBROUTINE SUN (IYR, IDAY, SECS, GST,
     SLONG, SRASN, SDEC)
C PROGRAM TO CALCULATE SIDEREAL,
     TIME AND POSITION OF THE SUN
C GOOD FOR YEARS 1901 THROUGH 2099.
     ACCURACY 0.006 DEGREE
3 INPUT IS IYR, IDAY (INTEGERS), AND
     SECS, DEFINING UNIVERSAL TIME
C OUTPUT IS GREENWICH MEAN SIDEREAL
     TIME (GST) IN DEGREES,
C LONGITUDE ALONG ECLIPTIC (SLONG),
     AND APPARENT RIGHT ASCENSION
C AND DECLINATION (SRASN, SDEC) OF
     THE SUN, ALL IN DEGREES.
```

```
DATA RAD /57.29578/
DOUBLE PRECISION DJ, FDAY
IF (IYR.LT.1901.OR.IYR.GT.2099) RETURN
FDAY = SECS/86400
DJ = 365*(IYR – 1900) + (IYR – 1901)/4
   + IDAY + FDAY – 0.5D0
T = DJ/36525
VL = DMOD(279.696678 + 0.9856473354*DJ,
   360.D0)
GST = DMOD(279.690983 + 0.9856473354*DJ
   + 360.*FDAY + 180., 360.D0)
G = DMOD(358.475845 + 0.985600267*DJ,
   360.D0)/RAD
SLONG = VL + (1.91946 – 0.004789*T)*
   SIN(G) = 0.020094* SIN(2.*G)
OBLIQ = (23.45229 – 0.0130125*T)/RAD
SLP = (SLONG – 0.005686)/RAD
SIND = SIN(OBLIQ)*SIN(SLP)
COSD = SQRT(1. – SIND**2)
SDEC = RAD*ATAN(SIND/COSD)
SRASN = 180. – RAD*ATAN2(COTAN
   (OBLIQ)*SIND/COSD, – COS(SLP)/COSD)
RETURN
END
```

Appendix A.4

Time series analysis techniques

A.4.1 INTRODUCTION

The interpretation of time series can be assisted with a few widely used techniques that should be mastered by the space physicist. These include: minimum variance analysis, used to define the normals to boundaries and other discontinuities; power spectral analysis, used to investigate periodicities of fluctuations; and wave analysis techniques, used to determine the detailed behavior of a wave as an aid to mode identification. Here we briefly describe these techniques as they are applied to magnetic field data. However, some of them can be more generally applied.

A.4.2 PRINCIPAL AXIS ANALYSIS OF MAGNETIC FIELD VARIANCE

There are situations in which it is very helpful to examine magnetic field data in the principal axes of their variance matrix. In particular, the direction of minimum variance is often informative because it can often lead to a two-dimensional picture of a process. One useful application can be found at tangential discontinuities, where the magnetic field is steady on one side, as it is in the magnetosphere, while on the other side the direction is quite variable, lying against the plane of the discontinuity in various directions. Here there would be strong variance in a plane and little variance along the normal to the boundary. Another situation occurs in an electromagnetic plane wave propagating in the **k**-direction. Because $\nabla \cdot \mathbf{B} = 0$, there is no variance in the k-direction, i.e. $\mathbf{k} \cdot \delta\mathbf{b} = 0$. Determining the minimum variance direction gives the wave propagation direction. However, in a situation in which the magnetic field changes principally in only one direction, such as at a shock, the technique is not useful.

To find the direction of minimum variance, one first calculates the variance matrix

$$V = \begin{pmatrix} \sum(\delta b_{ix})^2 & \sum(\delta b_{ix})(\delta b_{iy}) & \sum(\delta b_{ix})(\delta b_{iz}) \\ \sum(\delta b_{iy})(\delta b_{ix}) & \sum(\delta b_{iy})^2 & \sum(\delta b_{iy})(\delta b_{iz}) \\ \sum(\delta b_{iz})(\delta b_{ix}) & \sum(\delta b_{iz})(\delta b_{iy}) & \sum(\delta b_{iz})^2 \end{pmatrix} \tag{A4.1}$$

where $\delta b_{ix} = B_{ix} - \langle B_x \rangle$ etc., and i is an index indicating the time or location of the measurement.

At this point, we seek the similarity transformation that diagonalizes the variance matrix **V**:

$$\mathbf{X}^{-1}\mathbf{V}\mathbf{X} = \lambda. \tag{A4.2}$$

This can be determined by solving the equation $(\mathbf{V} - \lambda\mathbf{1})\mathbf{R} = 0$, where λ is the eigenvalue, **1** is the unity matrix, and **R** is the eigenvector. This is solved by using the fact that the above equation can have a solution only when the determinant

$$|\mathbf{V} - \lambda\mathbf{1}| = 0. \tag{A4.3}$$

There are three eigenvectors: a maximum, an intermediate, and a minimum, each associated with a different eigenvector.

Figure A4.1 shows ion cyclotron waves in the saturnian magnetosphere at $5.13R_S$. These waves have been detrended and the average removed from

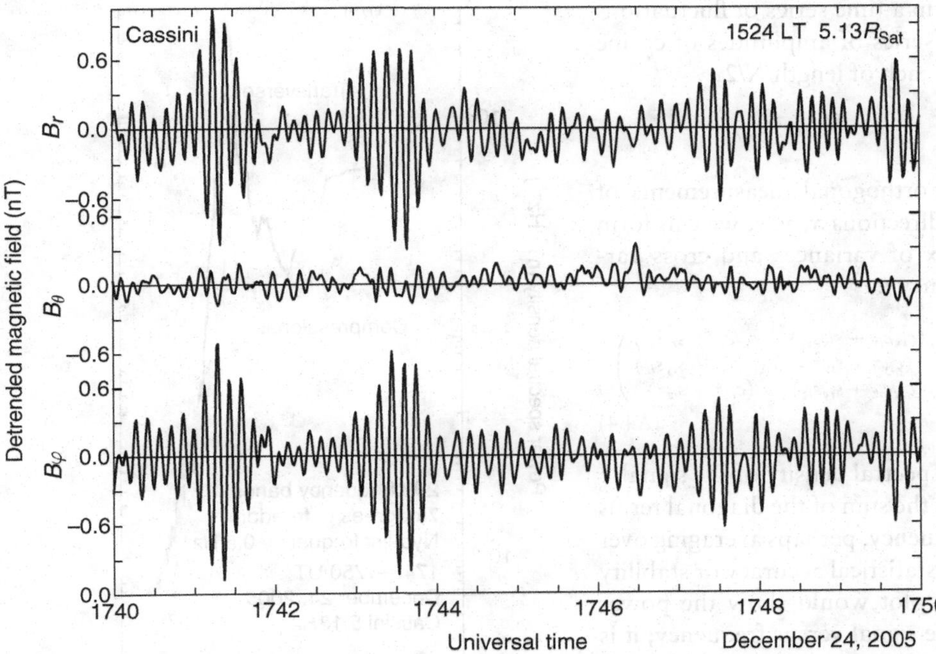

FIGURE A4.1. Detrended magnetic field with 1 s temporal resolution near the saturnian equator at $5.13R_S$, showing the presence of ion cyclotron waves near the H_2O^+ cyclotron frequency. Data are in coordinates r, θ, φ.

the data to avoid the spectrum being contaminated by changes due solely to the motion of the spacecraft. If we calculate the variance matrix for these waves in the frequency band of the waves, we obtain

$$\mathbf{V} = \begin{pmatrix} 0.051 & -0.003 & -0.001 \\ -0.003 & 0.017 & 0.002 \\ -0.001 & 0.002 & 0.057 \end{pmatrix}.$$

The eigenvalues of this real symmetric matrix are 0.057, 0.051, 0.001 nT^2, and the eigenvectors are:

direction of maximum variance	EV1	(0.165	0.039	0.986)
direction of intermediate variance	EV2	(0.0985	–0.049	0.167)
direction of minimum variance	EV3	(0.054	0.998	–0.030)

The directions of maximum and intermediate variance are principally along the φ- and r-directions. This agrees with the observed strength of the signal in these two directions in Figure A4.1. It is, by eye, weakest along the θ-direction, which our solution tells us is the direction of minimum variance. However, the information gained from the variance matrix is much more quantitative than that gained by visual inspection.

A.4.3 POWER SPECTRUM

A time series of measurements may appear random or chaotic until a "power spectrum" is calculated to display its power as a function of frequency or wavelength. The power spectrum can help determine resonant frequencies due to specific plasma instabilities. Peaks or harmonics in the spectrum can be particularly informative, as can the relative phases between quantities such as the magnetic field and plasma bulk velocity or electric field.

A digital signal containing n points sampled every Δt is $n\Delta t$ seconds long. The lowest frequency that can be accurately determined in this finite length time series is $(n\Delta t)^{-1}$ and the highest is $(2\Delta t)^{-1}$ because two data points are required to define a wave. This maximum frequency is called the Nyquist frequency. Fast Fourier transform routines can transform a time series of fluctuations on the x-axis into a series of amplitudes of cosine and sine amplitudes, each of length $N/2$,

$$C_0, C_1 \ldots\ldots C_{n/2-1} \qquad\qquad S_1 \ldots\ldots S_{n/2}.$$

If there are three orthogonal measurements of wave amplitudes in directions x, y, z, we can form the cospectral matrix of variances and cross variances at each frequency i:

$$\mathrm{C}_i = \begin{pmatrix} (c_{ix}^2 + s_{ix}^2) & (c_{ix}c_{iy} + s_{ix}s_{iy}) & (c_{ix}c_{iz} + s_{ix}s_{iz}) \\ (c_{ix}c_{iy} + s_{ix}s_{iy}) & ({c_{iy}}^2 + {s_{iy}}^2) & (c_{iy}c_{iz} + s_{iy}s_{iz}) \\ (c_{ix}c_{iy} + s_{ix}s_{iy}) & (c_{iy}c_{iz} + s_{iy}s_{iz}) & ({c_{iz}}^2 + {s_{iz}}^2) \end{pmatrix} \tag{A4.4}$$

This matrix is the cospectral density of the signal.

If one were to plot the sum of the diagonal terms (the trace) versus frequency, perhaps averaging over frequency to provide statistical accuracy or stability of the estimates, the plot would show the power spectral density of the signal versus frequency; it is usually called the power spectrum. The integral under the power spectrum is the variance of the signal, and its square root is the root mean square of the signal strength. Such a power spectrum for the time series seen in Figure A4.1 is given in Figure A4.2 using seven degrees of freedom with running sums of these successive estimates. Here we show the transverse power and the compressional power separately. We compute the compressional power from the time series of the magnetic field strength. The transverse power is the sum of all power minus the compressional power.

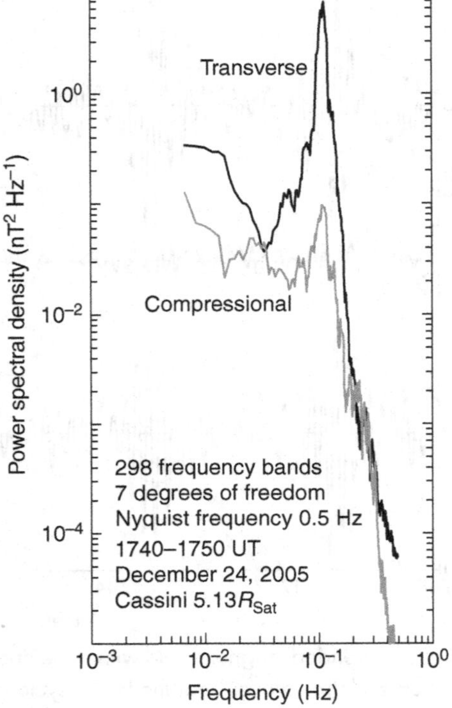

FIGURE A4.2. Power spectrum of signals shown in Figure A4.1. Compressional power is calculated from the time series of field strength. The transverse power is calculated from the sum of the power on the three sensors minus the compressional power.

A.4.4 WAVE ANALYSIS

Although many space physicists stop their analyses when they have calculated the power spectrum, there is much more information available in the variance matrix as a function of frequency. By extending the analysis, we can begin to compare the wave properties with theoretical expectations to determine wave modes and to develop a better understanding of the physical processes at work. We call this "wave analysis." We begin by mining further the cospectral matrix that consists of products of functions with the same phase, sines with sines and cosines with cosines. But it is also possible and informative to examine variances that are products of functions that are 90° out of phase, i.e., sines with cosines. These products inform us about signals in quadrature, circularly polarized signals rather than linearly polarized signals. This matrix is called the quaspectrum. We return first to the cospectrum and recall the analysis in Section A.4.2 in which we determined the normal to the boundary. Here we wish to determine the plane of constant phase as that is perpendicular to the direction of propagation of the wave. This is simply our eigenvalue problem again. We can determine the direction of propagation as a function of frequency by calculating the eigenvectors at each frequency, remembering to average over a

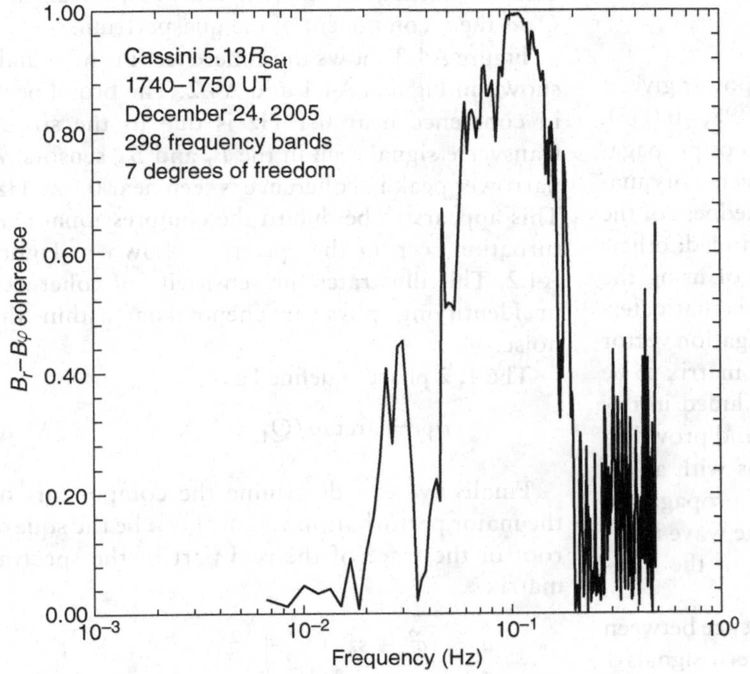

FIGURE A4.3. Coherence between the signals in the B_r and B_φ sensors displayed in Figure A4.1.

frequency range to increase the degrees of freedom in the answer and give a more stable result.

The direction of propagation is not the only parameter that can be deduced. The ratio of the square roots of the eigenvalues in the other two directions measures how elliptical is the signal, and the eigenvectors tell us the direction of the maximum and intermediate variances. If we examine the square root of the eigenvalue ratio, we find an ellipticity of 0.951, which indicates that these waves are very nearly circularly polarized. This ellipticity and its orientation can be diagnostic of the wave type. These properties were exploited by Born and Wolf (1970) to analyze optical signals and by Rankin and Kurtz (1970) to analyze micropulsations. In particular, the latter authors show how the percent polarization of the signals can be calculated.

The cospectrum, however, is only the real part of the spectral matrix. There is also an imaginary part, a quaspectrum based on the quadrature products of the Fourier amplitudes. The quaspectrum can be calculated at each frequency, i, by Eq. (A4.5),

$$Q_i = \begin{pmatrix} 0 & (s_{ix}c_{iy} - s_{iy}c_{ix}) & (s_{ix}c_{iz} - s_{iz}c_{ix}) \\ -(s_{ix}c_{iy} - s_{iy}c_{ix}) & 0 & (s_{iy}c_{iz} - s_{iz}c_{iy}) \\ -(s_{ix}c_{iz} - s_{iz}c_{ix}) & -(s_{iy}c_{iz} - s_{iz}c_{iy}) & 0 \end{pmatrix}. \tag{A4.5}$$

In our example in A4.1, the waves have a quaspectrum over the band of frequencies studied above of

$$\begin{pmatrix} 0.000 & 0.002 & 0.054 \\ -0.002 & 0.000 & -0.003 \\ -0.054 & 0.003 & 0.000 \end{pmatrix}.$$

This is an antisymmetric matrix.

Means (1972) showed that the direction of propagation can be directly obtained from this matrix constructed from the out of phase cross powers

$$\begin{aligned} J_{ixy} &= (s_{ix}c_{iy} - s_{iy}c_{ix}), \\ J_{ixz} &= (s_{ix}c_{iz} - s_{iz}c_{ix}), \\ J_{iyz} &= (s_{iy}c_{iz} - s_{iz}c_{iy}), \end{aligned} \tag{A4.6}$$

and

$$k_{ix} = J_{iyz} / A_i; \; k_{iy} = -J_{ixz} / A_i; \; k_{iz} = J_{ixy} / A_i,$$

where

$$A_i = (J_{ixy}^2 + J_{ixz}^2 + J_{iyz}^2)^{1/2}.$$

Thus we find that the out of phase power gives a direction of propagation of (0.058, 0.998, −0.035), which is only 0.37° from the direction of propagation determined from the in phase power. This analysis immediately determines the handedness of the wave. If $\hat{\mathbf{k}} \cdot \hat{\mathbf{B}} > 0$, the wave is right handed; otherwise it is left handed. An advantage of using the imaginary part of the spectral matrix is that determining the direction of the wave propagation vector and its handedness does not require a matrix to be inverted. Thus it could be readily included in on-board processing of a magnetometer and provide a data compression scheme for missions with a low data rate. Comparing the direction of propagation with the field direction, we find that the wave is left handed and propagating within 5.6° of the magnetic field.

Another useful quantity is the coherence between a pair of signals. The coherence between signals 1 and 2 is defined by

$$coherence_{1,2} = (C_{12}^2 + Q_{12}^2)/C_{11}C_{22}, \qquad \text{(A4.7)}$$

where C_{ij} is the ij-component of the cospectrum and Q_{ij} is the ij-component of the quaspectrum.

Figure A4.3 shows the coherence for the signals shown in Figures A4.1 and A4.2. The broad peak in coherence near 0.1 Hz is due to the strong transverse signal seen in the B_r and B_φ sensors. A narrower peak in coherence is seen near 0.025 Hz. This appears to be due to the compressional perturbation seen in the spectrum shown in Figure A4.2. This illustrates the sensitivity of coherence in identifying physical phenomena within the noise.

The 1, 2 phase is defined as

$$\varphi_{12} = \arctan(Q_{12}/C_{12}). \qquad \text{(A4.8)}$$

Finally we can determine the components of the major perturbation vector. Let a be the square root of the trace of the real part of the spectral matrix:

$$\begin{aligned} a_i &= (c_{ix}^2 + s_{ix}^2 + c_{iy}^2 + s_{iy}^2 + c_{iz}^2 + s_{iz}^2)^{1/2}, \\ L_{ix} &= (c_{ix}^2 + s_{ix}^2)^{1/2}/a_i, \\ L_{iy} &= (c_{ix}c_{iy} + s_{iy}s_{ix})/(a_i^2 L_{ix}), \\ L_{iz} &= (c_{ix}c_{iz} + s_{ix}s_{iz})/(a_i^2 L_{ix}). \end{aligned}$$

Glossary

acoustic modes compressional modes within a plasma that are analogous to sound waves in air.

acoustic wave compression wave in air that carries sound.

active regions localized regions of strong magnetic fields ($\gtrsim$ hundreds of nanotesla) on the Sun, usually composed of a balanced combination of outward and inward magnetic fields that appear as a bipolar pair of adjacent patches or a more complex mixture. These regions do not always produce visible sunspots. Hot loops of coronal material are often seen connecting opposite polarity patches and/or connecting them to other active regions in the vicinity. They are also sites where flares occur and the largest coronal eruptions, known as CMEs, originate.

adiabatic invariants quantities that remain constant as the charged particle moves.

adiabatic relation equation relating pressure and density in situations of conserved energy.

Alfvén (shear) mode low-frequency magnetohydrodynamic wave mode that corresponds to bending (or shearing) of magnetic field lines.

Alfvén speed characteristic speed for shear mode (bending mode) waves in a magnetized plasma, related to the bending of the magnetic field line.

Alfvén wings flux tubes that interact with the ionosphere and atmospheres of bodies in a sub-alfvénic flow. These tubes are bent at an angle governed by the Alfvén speed.

alfvénic auroras name given to auroras associated with low-energy electrons (~100 eV) that are accelerated by parallel electric fields associated with low-frequency shear-mode (Alfvén-mode) waves, as opposed to quasi-static parallel electric fields.

Ampère's law Maxwell equation that relates the curl of the magnetic field to current density.

anomalous cosmic ray cosmic ray ions that have unusual composition, in particular an excess of species such as oxygen, nitrogen, and carbon, and – in contrast to galactic and solar cosmic rays – have a single charge. These are considered to be produced inside the heliosphere from either the neutral interstellar gas that flows in from outside, or from solar system sources such as sputtering or desorption from interplanetary dust.

Appleton–Hartree dispersion relation a form of the cold plasma wave dispersion relation that was originally formulated for wave propagation in the terrestrial ionosphere. This form of the dispersion shows clearly how the refractive index differs from unity.

aurora light emitted by the upper atmosphere upon being excited by collisions of energetic electrons or ions with the atmosphere.

auroral electrojet index difference in the most positive and most negative horizontal field readings at a series of auroral zone stations.

auroral kilometric radiation radio emission from the Earth's auroral zone that has wavelengths of the order 1 km. Planetary analogs include jovian decametric radiation (DAM) and Saturn kilometric radiation (SKR).

auroral substorm sequence of auroral signatures consisting of an initial brightening of an auroral arc and subsequent expansion and break up of the auroral forms.

auroral zone oval region encircling both magnetic poles, defining region of most frequent occurrence of aurora.

Bernoulli's equation relationship connecting specific kinetic energy and specific thermal pressure of a hydrodynamic flow.

Bernstein modes electrostatic waves propagating perpendicular to the ambient magnetic field that are a consequence of including kinetic effects in the wave dispersion relation.

betatron acceleration acceleration of gyrating particles caused by an increasing magnetic field, related to the conservation of the particle magnetic moment.

bi-ion hybrid resonance frequency at which an electromagnetic wave has zero phase velocity in a plasma that has multiple ion species. The bi-ion hybrid frequency occurs between the respective ion gyro-frequencies.

Birkeland current current parallel to magnetic field lines in a planetary magnetosphere.

Boltzmann equation equation that gives the total time derivative of the phase space density in the six-dimensional configuration space and velocity space, including the effect of collisions.

bounce motion motion of a charged particle along the magnetic field that is reflected at mirror points.

bremsstrahlung radiation produced by a decelerated charged particle.

Brunt–Väisälä frequency for upward traveling waves in the atmosphere, the maximum frequency of gravity waves.

Buneman instability a form of two-stream instability where the electrons are streaming relative to the ions.

butterfly diagram plot of sunspot or active region location in solar latitude and time that exhibits a butterfly wing-like pattern with the period of the sunspot cycle. These plots clearly show the progression of sunspots from mid to low latitudes as the cycle progresses.

Carrington rotations the Carrington rotation number is the standard time marker used in studies of the Sun with a period of 27.3 days. Richard Carrington was a British astronomer who characterized and cataloged solar rotations by number, starting in November 1853, even though the Sun's surface exhibits slightly different rotation periods at high and low latitudes.

central meridian line from north to south on the visible solar disk, as seen by an observer through which the line connecting the centers of the Sun and the Earth passes.

centrifugal force fictitious force equal and opposite to the centripetal force that keeps an object moving in a circle, such as the tension on a string on a ball swung in a circle, or solar gravity on the orbiting Earth around the Sun.

Chapman theory mathematical relationship governing the production of an ionosphere and its equilibrium density when the atmosphere is exponential, planar, and horizontally stratified.

charge conservation principle that the total charge in a plasma cannot change; time rate of change of charge density is determined by the divergence of the current density.

chromosphere layer of the Sun's atmosphere, about 2000 km thick, that, along with the transition region above it, comprises the relatively narrow region between the partly neutral gases of the photosphere and the fully ionized, relatively hot, corona. The physical processes that enable this transition are still under investigation. The chromosphere is named for its reddish color seen during total eclipses.

CMA diagram a collection of Friedrichs diagrams showing how wave dispersion properties depend on the plasma parameters, named after Clemmow–Mullaly–Allis.

cold plasma plasma for which the thermal speed associated with the species that constitute the plasma is sufficiently small in comparison to any other characteristic speed that thermal effects can be ignored.

collision frequency rate at which particles collide. Collisions can exchange charge, momentum, and energy.

collisional cross section area of a circle with a radius equal to the sum of the radii of the colliding particles.

collisionless shock thin transition region from supersonic to subsonic flow in a collisionless plasma.

conductivity tensor tensor that gives the current associated with an electric field. This tensor is encountered in two different contexts within space plasmas. The first is within the ionosphere (cf. Hall and Pedersen conductivities). The second is when considering electromagnetic wave dispersion.

continuity equation local conservation equation that relates the time rate of change of a parameter to the divergence of the corresponding flux (e.g., number density to number flux, mass density to mass flux). In the context of space plasmas, the continuity equation refers to number density conservation.

convective zone thick (~0.35 solar radius) layer inside the Sun between the surface (photosphere) and fuel-burning core, where energy generated in the core is partly transformed into mass motions of the material there. These motions may be turbulent but also have underlying, more orderly, circulation cells. This region is thought to be the home of the solar dynamo, where the Sun's main magnetic fields are generated.

corona outermost layer of the Sun's atmosphere, that, at over a million degrees, is fully ionized and expands to great heights (~ several solar radii). The heating of the corona in the absence of collisions is thought to involve the generation and absorption of waves of various kinds, some of which may result from mechanical oscillations at its base.

coronal mass ejection eruption of coronal material into space, sometimes at speeds exceeding 2000 km s^{-1}. These are the cause of the largest plasma and magnetic field disturbances observed in the solar wind, and responsible for most major geomagnetic storms.

Cowling channel region where the Hall current of a secondary current system current opposes the primary Pedersen current from the primary (imposed) electric field. The secondary current system results from a secondary electric field. The secondary electric field is sometimes referred to as a polarization electric field. The primary Hall current and secondary Pedersen current flow in the same direction, resulting in an increase in the current along the primary electric field direction.

Cowling conductivity conductivity that relates the net current in the direction of the primary electric field and the primary (imposed) electric field in a Cowling channel.

critical Mach number Mach number at which the dissipation processes change from being capable of being produced "resistively" to requiring additional processes such as ion reflection and overshoot formation.

crossover frequency frequency for which a parallel propagating wave in a cold plasma switches polarization from right-hand circular to left-hand circular, or vice versa.

curvature drift drift of a particle in a curved magnetic field.

cusps points where magnetic field lines diverge to go in quite different directions.

cyclotron motion gyration of a charged particle around a magnetic field line.

D region lowest layer in the ionosphere.

Debye length characteristic length over which the charge of a particle is shielded from the rest of the plasma.

de Hoffman–Teller frame moving frame of reference that aligns the solar-wind flow to be parallel to the magnetic field.

Dessler–Parker–Sckopke relationship equation giving the energy of the plasma in the magnetosphere from the Dst index.

differential rotation latitude dependence of apparent solar rotation rate based on observations of the motion of features across the solar disk. While the typical mid-latitude period is the Carrington rotation of about 27.3 days, the poles rotate slower by several days and the equator rotates faster by several days, on average. Differential rotation plays a major role in the evolution of the magnetic fields and related features seen on the Sun, and is considered a key factor in models of the solar dynamo.

diffusive shock acceleration scattering in energy of charged particles in the fluctuating field of a shock. A few of the particles encountering the shock find themselves scattered/reflected back and forth between the shock and the waves, or between waves upstream and downstream of the shock. The relative motion of the scattering centers can lead to a statistical acceleration of a particle over time. Quasi-parallel shocks may be particularly good sites of this type of acceleration because of the presence of associated field fluctuations.

dipole magnetic moment strength of the simplest component of a planetary magnetic field. Strength is numerically equal to the radius to the point of observation cubed, times the strength of the magnetic field in the magnetic equator.

dipole tilt angle between the rotation axis of a body and the axis of symmetry of the lowest-order (dipole) field.

dispersion relation equation that gives the relation between the wave frequency and the wave vector.

dissociative recombination process in which an ionized molecule is split upon becoming neutralized by an electron.

distribution function distribution of particles in the six-dimensional phase space, also known as phase space density.

dopplergram image constructed from pictures taken in the red- and blue-shifted wings of a spectral line, indicating the spatial pattern of radial motion away from and toward the observer. Dopplergrams of the Sun make the oscillating surface patterns of granulation cells visible, superposed on the large-scale solar rotation.

drift motion motion of a charged particle across magnetic field lines.

drift-shell splitting separation of drift shells containing particles of differing mirror points in non-rotationally symmetric fields.

Dst index index constructed from the horizontal magnetic field at low latitudes that is proportional to the plasma energy content of the magnetosphere.

E layer middle layer of the ionosphere.

electric field drift particle drift associated with an electric field that is perpendicular to the magnetic field.

electric potential (or electrostatic potential) scalar quantity whose negative gradient gives the irrotational or curl-free component of the electric field.

electromagnetic wave wave that has both electric and magnetic fields.

electron Alfvén layer drift path of electrons that marks the transition from open drift paths to closed drift paths that circle the Earth.

electron and ion momentum equations equations that related the rate of change of momentum in the electron and ion species to the imposed forces, which typically include the electric and magnetic field forces. Plasma pressure can also be included when the momentum equations are written as fluid equations (as opposed to single-particle momentum equations).

electrostatic wave wave that has the wave electric field aligned closely to the wave vector, such that the wave magnetic field can be ignored.

escape velocity speed at which an outward-moving particle or body can escape from the gravitational pull of the parent object of Mass M. The escape velocity from the Sun's surface is 618 km s^{-1}.

Euler's equation part of a set of quasi-linear hyperbolic equations governing adiabatic and inviscid flow.

exobase bottom of the exosphere above which point an atmosphere is collisionless.

exosphere collisionless portion of an atmosphere.

F corona diffuse glow surrounding the Sun that comes from dust scattering sunlight.

F layer highest layer of the ionosphere.

Faraday's law Maxwell equation that relates the curl of the electric field to the time rate of change of the magnetic field, also known as the induction equation.

fast magnetosonic wave magnetohydrodynamic (MHD) wave that compresses both the density and magnetic field. This is the fastest of the three MHD waves.

fast mode fastest of the three Alfvén modes.

Fermi acceleration acceleration of a particle in a magnetic mirror, or other structure in which the particle bounces between two mirrors, caused by the mirror points moving closer together. Related to conservation of the second adiabatic invariant.

field-aligned current current parallel to magnetic field lines in a planetary magnetosphere.

filaments photospheric material that is suspended above the nominal photosphere in sometimes large threads channeled between opposite polarity magnetic field areas on the Sun. These are thought to be lifted by magnetic forces, and are frequently used to indicate the presence of a stressed field configuration in a host active region. Filament formation sometimes precedes flares and/or coronal mass ejections.

first adiabatic invariant magnetic moment of a charged particle defined as the perpendicular kinetic energy divided by the magnetic field strength.

flare impulsive brightening in the low corona associated with active regions. Flares produce enhanced (by orders of magnitude) emissions in energetic photon fluxes from their locale, including extreme ultraviolet and x-rays. They are thought to be a signature of the localized release of magnetic stresses, resulting in low coronal heating. They are sometimes, especially in large flare cases, accompanied by energetic particle emissions and coronal mass ejections. There is a special scale for describing flare magnitudes, based on their emissions (see Table 4.3).

fluid theory theory that treats the constituent species within a plasma as a fluid characterized by bulk parameters such as density, flow velocity, and pressure.

flux rope twisted bundle of magnetic flux.

foreshock region upstream from a planetary standing shock in which energetic particles accelerated at or near the shock are present.

Friedrichs diagram diagram that shows how a wave phase velocity depends on direction of propagation within a magnetized plasma. Can also be used to show the group velocity.

frozen-in theorem statement that the magnetic field can be treated as if it were frozen to the fluid.

Fukushima's theorem theorem that states that the combined effect of a vertical field-aligned current and the corresponding horizontal closure currents in a uniformly conducting ionosphere are such that there is no magnetic field signature below the ionosphere for the combined current system.

galactic cosmic rays charged particle with highly relativistic energy produced in distant regions of the galaxy by events such as supernovas. Usually the flux of such particles changes slowly due to the 22 year variation in the solar magnetic field relative to the local galactic magnetic field.

galilean (non-relativistic) transformation in this frame velocities are assumed to be much lower than the speed of light such that the electric field is frame dependent, while the magnetic field, and hence current densities, are not.

Gauss's law Maxwell equation that relates the divergence of the electric field to charge density.

generalized Ohm's law a more complex form of an Ohm's law, i.e., an equation that relates current density to the electric field which takes into account the presence of a magnetic field and electron inertia and pressure.

geocorona region of the neutral atmosphere dominated by hydrogen.

geomagnetic storm period of disturbed magnetospheric activity as signaled by the Dst index.

gradient drift drift of a particle associated with a gradient of the magnetic field magnitude. The gradient must be perpendicular to the magnetic field.

gravity wave wave in which the restoring force is due to gravity acting on the displaced mass.

guide-field reconnection magnetic reconnection when the magnetic fields on either side of the current sheet are not antiparallel.

guiding center average motion of a particle. The particle motion can be separated into the guiding center motion and circular gyration around the guiding center.

gyration circular motion of a particle in a plane perpendicular to the magnetic field.

gyro-frequency frequency at which the particle gyrates around the magnetic field direction.

gyro-resonance resonance between a particle and a wave such that the Doppler-shifted frequency is a multiple of the particle gyro-frequency.

Hale cycle cycle of solar magnetic activity, taking into account the polarity of the magnetic field at the solar poles as well as the active region pattern of leading (in the sense of solar rotation) polarity matching the local polar field in the rising phase of activity. While the sunspot cycle is about 11 years, the Hale cycle, over which the solar field polarities reverse and are restored, is about 22 years.

Hall conductivity conductivity in a collisional plasma that gives the current density perpendicular to the magnetic field and the electric field.

Hall current current that flows in the collisional ionosphere that is perpendicular to both the ambient magnetic field and the perpendicular electric field.

harmonic perturbation assumption used to derive a wave dispersion relation, where the wave is assumed to be a plane wave and the wave fields are first-order perturbations.

Harris current sheet simple self-consistent model of the plasma sheet.

heat flux third-order moment that gives the energy flux associated with a skew in the distribution function. In solar-wind physics, the flux contained in the narrow field-aligned beam that is observed, in addition to a nearly isotropic thermal component, in the interplanetary plasma electron distributions.

helioseismology technique whereby sound waves detected on the solar surface are used, in a manner similar to solid body waves in a planet, to sound the solar interior.

homopause top of the portion of the atmosphere with uniform composition.

horseshoe distribution characteristic electron phase space distribution that is observed in the auroral acceleration region. The horseshoe shape is a combination of acceleration by a parallel electric field, the magnetic mirror force, and loss of some of the particles into the atmosphere.

hybrid simulation numerical simulation of a magnetized plasma where ion motion is tracked but electrons are treated as a fluid.

idealized Ohm's law simplified version of the generalized Ohm's law where only the electric field and $\mathbf{u} \times \mathbf{B}$ terms remain.

impact ionization process of charging ions by collisions.

induced magnetosphere magnetic structure formed when the solar-wind flow interacts with a body with a conducting shell, such as an ionosphere.

induced magnetotail region downstream from the planet in an induced magnetosphere.

intermediate mode magnetohydrodynamic wave, also known as the shear Alfvén mode, that bends the flow and the magnetic field direction but does not compress the density or magnetic field.

interplanetary field enhancements events in the solar wind in which a magnetic field strength rises and the field becomes twisted in the direction of the upstream electric field.

interplanetary magnetic field magnetic field found in interplanetary space, and throughout the heliosphere. Its most basic geometry, the Parker spiral, is determined by solar rotation acting on the magnetized radial plasma outflow of the solar wind – into which it is "frozen."

interplanetary shocks collisionless shocks found in the solar wind that result from solar-wind fast–slow stream interactions and fast coronal mass ejections. These are usually distinguished from solar coronal shocks, planetary bow shocks, and the heliospheric termination shock, although all of these are examples of collisionless shocks in space plasmas.

invariant latitude latitude at which a magnetic field line intersects the surface of the Earth.

ion cyclotron waves electromagnetic waves oscillating at frequencies lower than the ion cyclotron frequency in the rest frame of a plasma.

ionopause boundary between the magnetized plasma in the magnetosheath and a planetary ionosphere.

ionosphere partially ionized region of the atmosphere that influences electromagnetic wave propagation and supports electrical current flow.

Jeans escape flux flux of atmospheric atoms that has sufficient upward speed to escape the Earth's gravitational field at the exobase.

Jeans's theorem any distribution function that is a function of the constants of the motion automatically satisfies the Vlasov equation.

jovian plasma torus dense plasma torus produced in the interaction of Io's atmosphere with the jovian corotating magnetosphere. Added to by other moons.

Joy's law observed rule of bipolar active region orientation, where the leading (in the sense of right-handed solar rotation) polarity patch is typically the same polarity as the average polar field in the same hemisphere, and tilted toward the equator by several degrees relative to the trailing patch. This order prevails mainly during the rising phase of the solar activity cycle, before the solar polar magnetic field reverses around solar maximum.

K corona solar coronal glow produced by sunlight scattering off its electrons. This is what is generally seen in coronagraph pictures and during total eclipses in the form of streamers and rays.

Kelvin–Helmholtz instability unstable boundaries driven by flow parallel to the boundary.

Kepler's laws three laws that govern planetary motion.

kinetic energy energy of a mass moving at speed v is $\frac{1}{2}mv^2$.

kinetic theory theory that uses the phase space density to characterize the properties of the different species within a plasma.

Knight relation relationship between the accelerating electric potential and the resultant field-aligned current for a precipitating electron distribution.

Landau damping resonant damping of waves where the wave phase speed along the magnetic field matches the parallel velocity of the particles.

Landau prescription method by which an integration contour is distorted to ensure that the function arising from the integration is continuous regardless of the sign of the imaginary part of the variable for which a resonance occurs in the integral.

Landau resonance a particle and a wave having the same parallel velocity with respect to the ambient magnetic field.

Larmor radius radius of circle described by a particle as it gyrates around the magnetic field, also known as the gyro-radius.

law of gravity the attraction of two bodies is proportional to the product of the masses and inversely proportional to the square of the distance between them.

linearization mathematical method where the governing equations only contain zero-order and first-order (small) terms.

Liouville theorem theorem that states that phase space density is constant along a particle trajectory in the six-dimensional phase space.

longitudinal propagation plasma wave that is propagating parallel to the ambient magnetic field.

longitudinal wave plasma wave that has the wave electric field parallel to the wave vector. Synonymous with an electrostatic wave.

Lorentz force law equation that relates the rate of change of a particle's momentum to the electric and magnetic fields.

loss cone hyperbola modification to the boundary in phase space that separates that portion of phase space for which particles are lost into the atmosphere versus those particles that are reflected by the magnetic mirror force. The effect of additional downward acceleration is to change the shape of the boundary into a hyperbola when displayed in two-dimensional velocity space.

lower hybrid resonance frequency at which an electromagnetic wave in a cold plasma has zero phase velocity, characterized by the interaction of both electrons and ions with the wave.

Mach number in a gas, the velocity of a species relative to the speed of sound.

magnetic barrier magnetic field buildup outside a planetary ionosphere that is in balance with the thermal pressure of the ionosphere.

magnetic carpet small-scale magnetic field structure detected on the solar surface, associated with the boundaries of granular and supergranular cells and motions there. Often referred to as "salt and pepper" as the fields of both polarities are well mixed. Also characteristic of "quiet Sun" regions.

magnetic cloud form of coronal ejecta involved in a coronal mass ejection (CME) that locally appears to observers as a large (fractions of astronomical units) magnetic flux rope. About one-third of coronal mass ejection-related disturbances at 1 AU have this appearance, which is thought to be a general feature of most CMEs.

magnetic dipole simplest magnetic field possible from a closed electric current.

magnetic lobes regions in the tail above and below the plasma sheet where the magnetic pressure dominates.

magnetic moment dipole moment of a charged particle as it gyrates around the magnetic field; this is also the first adiabatic invariant.

magnetization current current carried by a plasma because the particles have a magnetic moment, making the plasma diamagnetic.

magnetodisk radially stretched magnetic field lines pulled perpendicular to the planetary rotation axis by a rapidly rotating magnetosphere that is mass loaded by sources such as moon atmospheres.

magnetoyhydrodynamic waves three low-frequency MHD wave modes (fast, intermediate, and slow) that enable the communication of pressure and stress in a plasma.

magnetohydrodynamics the fluid theory that relates hydrodynamic quantities, such as bulk flow and mass density, to the electromagnetic forces.

magnetometer device to measure the strength and direction of a magnetic field.

magnetopause boundary between a planetary magnetic field and the shocked solar-wind plasma.

magnetosheath region of compressed plasma flowing around a magnetosphere between the magnetopause and the standing shock.

magnetosphere volume around a magnetized planet in which magnetic field lines have at least one end rooted in the planet.

magnetotail distorted portion of a magnetosphere in the antisolar direction.

mantle boundary layer on tail where low-density solar-wind plasma is seen.

mass loading the addition of newly formed plasma from a neutral gas to flowing plasma that increases the mass of the flowing plasma. The new plasma may be formed by photoionization or impact ionization. Charge exchange between species of different composition, such as solar wind protons flowing through neutral oxygen, will also result in mass loading.

Maunder minimum period between about 1645 and 1715 when visible sunspots appeared to be absent on the solar disk. This period also saw a period of extreme cold in Europe, often referred to as the "Little Ice Age," resulting in ongoing speculations about a possible physical connection between the two events.

Maxwell's equations set of four equations that govern the behavior of electric and magnetic fields, and how they are related to charge and current densities.

mirror mode plasma wave mode in the solar wind often associated with compressions. Mirror-mode waves are ubiquitous in planetary magnetosheaths, where they exhibit diamagnetic behavior and appear as symmetric field dips or enhancements interleaved in the large-scale draping of the interplanetary field around the front of the solar-wind obstacle.

mirror-mode waves compressional waves that can be produced by pancake distributions of ions.

neutral point point in the tail current sheet where the magnetic field strength approaches zero.

optical depth path length during which the intensity of light decreases by e^{-1}.

ordinary and extraordinary modes waves that propagate perpendicular to the ambient magnetic field in a cold plasma. The wave dispersion for the ordinary (O) mode does not depend on the gyro-frequency. The gyro-frequency affects the dispersion of the extraordinary (X) mode.

overshoot bow shock phenomenon where the magnetic field at a shock front increases above the value predicted by the Rankine–Hugoniot relations.

particle orbit theory theory that investigates the motion of particles in different electric and magnetic field configurations. The theory assumes the particles can be treated as test particles.

Pc 3–4 waves ultralow-frequency waves in the period range 10–150 s, many of which are produced in the solar wind upstream of the shock and convected to the magnetopause.

peak in the ionosphere, a region of local maximum electron density.

peculiar velocity another name for random velocity.

Pedersen conductivity conductivity in a collisional plasma that gives the current density along the perpendicular electric field direction.

Pedersen current current that flows along the perpendicular magnetic field in the collisional ionosphere.

phase space six-dimensional space where three dimensions are related to position (configuration space) and three are given by velocity (velocity space).

phase space density distribution function of particles in the six-dimensional phase space.

photoelectrons electrons produced when photons ionize the neutral atmosphere or the surface of a spacecraft.

photoionization process of producing ions from neutral atoms by the energy of absorbed photons.

photosphere effective visible surface of the Sun, above which the solar gas becomes optically thin. This surface emits at a black body temperature of ~6000 K. There are many features on the photosphere that have been observed and documented over time, including sunspots. Observations of the magnetic fields there can be related to many features observed above, such as coronal loops and streamers.

Pi 2 waves ultralow-frequency waves in the period range 40–150 s that are initiated at substorm onset in the magnetotail.

plage areas of enhanced emission in chromospheric images of the Sun, associated with the stronger magnetic fields of active regions.

plasma collection of ions and electrons with zero net charge and sufficient density that collective forces dominate.

plasma dispersion function function that includes the kinetic response of a maxwellian to a wave perturbation. Used to derive the dispersion relation for waves propagating in a maxwellian plasma.

plasma frequency characteristic frequency at which the plasma oscillates if the electrons and ions are displaced relative to each other.

plasma sheet region between the tail lobes in which the thermal pressure dominates.

plasma skin depth characteristic scale related to the speed of light and the plasma frequency.

plasmapause location at which the near-equator plasma density falls substantially with increased distance from the Earth.

plasmasphere extension of the ionosphere upwards to the equator where, at quiet times, the plasma density builds to a steady-state value.

polar cusp region of low field strength at the magnetopause between magnetic flux in the closed magnetosphere and the tail field lines where the shocked solar-wind plasma gains access to the ionosphere.

potential energy in a gravitational field, the energy an object will gain if it falls in that field.

potential field source surface the usually spherical surface at ~2.5 solar radii that is used in potential field (current free) calculations of the coronal magnetic field structure. This map is based on a photospheric magnetic field map as the inner boundary condition and solutions of the Laplace equation for the field in the volume between. The solutions usually assume the coronal field is radial at the source surface, approximating the effect of the solar-wind outflow on the field.

primary particle colliding particle, electron, ion, or neutral.

prominences filaments seen at the limbs of the solar disk.

proton whistler lightning-generated wave originally propagating along the magnetic field in the ionosphere as a right-handed wave but converted to left-handed at the crossover frequency, then being absorbed at the proton gyro-frequency as it propagates higher.

quasi-linear diffusion diffusion of particles in velocity space that is the average of the motion induced by waves.

quasi-longitudinal approximation approximation that assumes that a wave in a cold plasma is propagating nearly parallel to the magnetic field. Also known as the quasi-parallel approximation.

quasi-parallel approximation an approximation to the Appleton–Hartree dispersion relation that yields a dispersion relation that is used to characterize the whistler mode, especially the dispersion of lightning-generated whistler waves.

R and L modes right- and left-hand circularly polarized waves that propagate parallel to the ambient magnetic field in a cold plasma.

radiation belts regions of intense fluxes of energetic particles trapped in the magnetic field.

radiative transfer propagation of photons through the atmosphere.

radiative zone region within ~0.35 solar radius of the center of the Sun, where the energy generated by fusion reactions is transmitted mainly by electromagnetic waves.

range-energy relation depth of penetration of particles as a function of energy.

rarefaction structure in flow where the flow is expanding with time and distance.

reconnection process in which electrons can no longer follow the magnetic field lines, allowing the field lines to reconnect with different partners. This alters the magnetic stress in a plasma, possibly leading to sudden acceleration.

reduced velocity distribution distribution function that has been integrated over perpendicular velocity so that the distribution is a function of parallel velocity only.

region 1 region of field-aligned current into the auroral oval on its poleward side.

region 2 region of field-aligned current into the auroral oval on its equatorward side.

relative permittivity quantity that specifies the electromagnetic displacement relative to the electric field.

ring current current circling the Earth in the magnetospheric equator due to the pressure of dense energetic plasma.

second adiabatic invariant invariant associated with a particle moving between two magnetic mirror points, also known as the bounce invariant. Equal to integral of the parallel momentum of a particle along the magnetic field over a closed path between mirror points.

secondary electron electron produced in a collision.

sector boundaries interplanetary magnetic field direction reversals observed at a particular site in the solar wind. These are thought to be controlled by the polarity pattern of the solar magnetic field that is carried outward with the solar wind, and have different appearances at different sites. At times of low solar activity, they represent observer crossings of the heliospheric current sheet.

shear Alfvén mode same as the intermediate mode.

shock drift mechanism acceleration process where a charged particle "surfs" along the shock front, thereby experiencing a net electric field acceleration. This shock acceleration process favors the field geometry in quasi-perpendicular shocks.

single-particle motion another name for particle orbit theory.

six-dimensional phase space density combination of configuration space (x, y, z) and velocity space (v_x, v_y, v_z).

slow magnetosonic wave magnetohydrodynamic wave that bends the flow along the magnetic field and compresses the magnetic field while reducing the plasma density.

solar constant total radiated solar flux. It has been established by full-Sun observations across a broad part of the electromagnetic spectrum that the radiative (photon) flux at Earth from the modern Sun changes minimally (by <0.2%) over the solar cycle and from cycle to cycle with a small (~0.1%) solar cycle variation. It is used in some climate models to study the solar variability versus other impacts.

solar cycle approximately 11 year period variation in the number of sunspots.

solar dynamo process by which the solar magnetic field is generated in the solar interior. The convection zone is thought to be the seat of the main solar dynamo. According to theory, dynamo generation of magnetic fields requires turbulent convection and rotation acting on a seed field.

solar energetic particles ions (mainly protons) and electrons in the keV–GeV energy range that originate either in solar flares or at interplanetary shocks.

solar radio emissions that part of the Sun's radiative output that occurs at radio wavelengths. This includes thermal emissions as well as non-thermally generated waves from sources such as flares. The flares and the shocks generated by coronal mass ejections produce distinctive bursts of emission with different time dependences and frequencies.

solar-terrestrial physics discipline studying effects of Sun on Earth from electromagnetic waves outside the visible-light band of wavelengths and from plasma, magnetic fields, and energetic particles from the Sun.

solar wind plasma outflow from the Sun that carries the interplanetary magnetic field and fills the heliosphere.

source surface nominal surface around the Sun where the solar wind starts. Often used to describe the potential field source surface (PFSS) concept.

South Atlantic anomaly weak region in the Earth's magnetic field near the east coast of

South America where charged particles are lost as they drift closer to the atmosphere.

speed of sound characteristic velocity for magnetohydrodynamic waves, related to the plasma pressure. (*See also* Alfvén speed.)

spicules dynamic narrow jets of material seen shooting outward in the chromosphere. These may somehow contribute to the heating of the corona and the acceleration of the solar wind.

standard flare model picture suggesting the basic physical setting of a flare and the associated processes, based on many types and years of flare observations. The model emphasizes magnetic field geometry and reconnection sites.

strahl highly field-aligned, beamed portion of the interplanetary electron heat flux.

stratopause location of the warmest temperature in the stratosphere.

stratosphere region above the troposphere where temperature increases with increasing altitude.

streamer blowout form of coronal mass ejection that has the appearance in coronagraph images of an initially steady coronal streamer erupting outward.

substorm a period of increased plasma flow and particle acceleration associated with a charge in the magnetic stress pattern in a magnetotail.

substorm process sudden release of magnetic flux from the tail lobes resulting in increased closed magnetic flux in the night magnetosphere.

sudden ionospheric disturbances Earth ionospheric disturbances produced by sudden changes in ionization when the solar (extreme-ultraviolet to x-ray wavelength) photon flux increases by orders of magnitude during a flare.

summation convention a convention that facilitates vector and tensor manipulation, where any index that repeats means that the operation applies to all of the components of the constituent tensors or vectors.

sunspot cycle approximately 11 year period over which the number of sunspots visible on the solar disk increases and then decreases in a cyclical manner.

sunspots dark spots on the visible photosphere associated with strong magnetic field locations. While sunspots are actually quite bright, they are cooler than their surroundings and so appear relatively dark. Sunspots are found within active regions.

superluminous a wave with phase speed faster than the speed of light.

synoptic map global maps of the Sun produced in various emissions, and quantities derived from emissions such as magnetic fields. These must be built up over the ~27.3 day solar rotation period to capture the entire global behavior – leading to some limitations in interpreting them as "snapshots."

tachocline region of high-velocity shear in the solar interior, between the differentially rotating convection zone and the rigidly rotating radiative core. The tachocline is often considered a key element in the solar dynamo.

third adiabatic invariant drift shell invariant, where the magnetic flux enclosed by a drift path is constant.

Titan–plasma interaction flowing plasma interacts with the Titan ionosphere and atmosphere when Titan is inside the corotating magnetospheric plasma, when Titan is in the flowing magnetosheath, and when it is in the solar wind.

torsional oscillation azimuthal oscillation associated with latitudinal variation of differential rotation on the surface not being smoothly varying from equator to pole, but contains a band of deviations (faster/slower rotation) around the latitude in each hemisphere where active regions are located.

total solar irradiance similar to the solar constant, but measured in power per unit area at Earth's surface.

transition region narrow, irregular, and physically complex layer of the Sun's atmosphere above the chromosphere, where the final transition to coronal conditions, including its million degree temperature, takes place.

transverse propagation describes a plasma wave that is propagating perpendicular to the ambient magnetic field.

transverse wave plasma wave that has the wave electric field perpendicular to the wave vector. A transverse wave is also electromagnetic, but the reverse need not be the case.

tropopause region where temperatures reach a minimum between the troposphere and the stratosphere.

turbopause homopause above which mixing ceases and the atmosphere becomes inhomogeneous in composition.

two-stream instability instability that occurs in a plasma when two species are streaming relative to each other.

upper hybrid resonance frequency at which an electromagnetic wave in a cold plasma has zero phase velocity. Occurs above the electron plasma frequency and gyro-frequency.

Vlasov equation the Boltzmann equation for which the collisional term is set to zero.

vorticity curl of the flow velocity. A circular flow, or vortex, has vorticity. A flow shear, while not strictly circular, has a curl in the flow velocity and hence a vorticity.

whistler right-handed electromagnetic wave propagating below the electron cyclotron frequency.

Zeeman splitting splitting of a spectral line that takes place when the emitting gas is immersed in a strong magnetic field. This splitting allows the inference and measurement of the magnetic fields on the solar surface.

References

Alfvén, H. (1957). On the theory of comet tails. *Tellus*, 9, 92–96.

Alfvén, H. (1968). Some properties of magnetospheric neutral surfaces. *J. Geophys. Res.*, **73**, 4379–4381.

Akasofu, S.-I. (1968). *Polar and Magnetospheric Substorms*. New York: Springer-Verlag.

Arridge, C. S., C. T. Russell, K. K. Khurana *et al.* (2007). Mass of Saturn's magnetodisk: Cassini observations. *J. Geophys. Res. Lett.*, **34**, L09108, doi:09110.01029/02006GL028921.

Avrett, E. H. and R. Loeser (2008). Models of the solar chromosphere and transition region from SUMER and HRTS observations: formation of the extreme-ultraviolet spectrum of hydrogen, carbon, and oxygen. *Astrophys. J. Suppl. Ser.*, **175**(1), 229–276.

Axford, W. I. (1962). The interaction between the solar wind and the Earth's magnetosphere. *J. Geophys. Res.*, **67**, 3791–3796.

Bame, S. J., J. R. Asbridge, W. C. Feldman, M. D. Montgomery, and P. D. Kearney (1975). Solar wind heavy ion abundances. *Solar Phys.*, **43**, 463–473.

Banks, P. M. and G. Kockarts (1973). *Aeronomy*. New York: Academic Press.

Blanc, M., S. Bolton, J. Bradley *et al.* (2002). Magnetospheric and plasma science with Cassini-Huygens. *Space Sci. Rev.*, **104**, 253–346.

Blanco-Cano, X., N. Omidi, and C. T. Russell (2003). Hybrid simulations of solar wind interaction with magnetized asteroids: comparison with Galileo observations near Gaspra and Ida. *J. Geophys. Res.*, **108**(A5), 1216, doi:10.1029/2002JA009618.

Blanco-Cano, X., N. Omidi, and C. T. Russell (2004). How to make a magnetosphere. *Astron. Geophys.*, **45**, 3.14–3.17.

Born, M. and E. Wolf (1970). *Principles of Optics*, 4th edn. New York: Pergamon, pp. 544–548.

Bothmer, V. and R. Schwenn (1998). The structure and origin of magnetic clouds in the solar wind. *Ann. Geophys.*, **16**, 1–24.

Brekke, A. and A. Egeland (1983). *The Northern Light: From Mythology to Space Research*. Berlin: Springer-Verlag.

Burton, R. K., R. L. McPherron, and C. T. Russell (1975). An empirical relationship between interplanetary conditions and Dst. *J. Geophys. Res.*, **80**(31), 4204–4214.

Cahill, L. J. and V. L. Patel (1967). The boundary of the geomagnetic field, August to November 1961. *Planet. Space Sci.*, **15**, 997–1033.

Carpenter, D. L. (1963). Whistler evidence of a 'knee' in the magnetospheric ionization density profile. *J. Geophys. Res.*, **68**, 1675–1682.

Carr, T. D., M. D. Desch, and J. K. Alexander (1983). Phenomenology of magnetospheric radio emissions, in *Physics of the Jovian Magnetosphere*. Ed. A. J. Dessler. Cambridge: Cambridge University Press, pp. 226–284.

Chapman, S. and J. Bartels (1940). *Geomagnetism*. London: Oxford University Press.

Chapman, S. and V. C. A. Ferraro (1930). A new theory of magnetic storms. *Nature*, **126**, 129.

Chi, P. J. and C. T. Russell (2005). Travel-time magnetoseismology: magnetospheric sounding by timing the tremors in space. *Geophys. Res. Lett.*, **32**, L18108, doi:10.1029/2005GL023441.

Clemmow, P. C. and J. P. Dougherty (1969). *Electrodynamics of Particles and Plasmas*. Reading, MA: Addison-Wesley Publ. Co.

Connors, M., C. T. Russell, and V. Angelopoulos (2011). Magnetic flux transfer in the April 5, 2010 Galaxy 15 substorm: an unprecedented observation. *Ann. Geophys.*, **29**, 619–622.

Coroniti, F. V. (1970). Dissipation discontinuities in hydromagnetic shock waves. *J. Plasma Phys.*, **4**, 265.

COWLING, T. G. (1957). Dynamo theories of cosmic fields. *Vistas Astron.*, **1**, 313–322.

CRAVENS, T. E., H. SHINAGAWA, and A. F. NAGY (1984). The evolution of large-scale magnetic fields in the ionosphere of Venus. *Geophys. Res. Lett.*, **11**, 267.

CROOKER, N. U., G. L. SISCOE, S. SHODHAN, D. F. WEBB, J. T. GOSLING, and E. J. SMITH. (1993). Multiple heliospheric current sheets and coronal streamer belt dynamics. *J. Geophys. Res.*, **98**, 9371–9381.

DECKER, R. B. (1988). Computer modeling of test particle-acceleration at oblique shocks. *Space Sci. Rev.*, **48**, 195–262.

DESSLER, A. J. and E. N. PARKER (1959). Hydromagnetic theory of magnetic storms. *J. Geophys. Res.*, **64**, 2239–2259.

DUNGEY, J. W. (1961). Interplanetary magnetic field and the auroral zones. *Phys. Rev. Lett.*, **6**, 47.

DUNGEY, J. W. (1963). The structure of the exosphere or adventures in velocity space, in *Geophysics: The Earth's Environment*. Eds. C. DEWITT, J. HIEBLOT, and A. LEBEAU. New York: Gordon and Breach, pp. 505–550.

DUVALL, T. L. JR. and A. C. BIRCH (2010). The vertical component of the supergranular motion. *Astrophys. J. Lett.*, **725**, L47–L51.

EGELAND, A. and W. J. BURKE (2013). *Carl Størmer: Auroral Pioneer*. Astrophysics and Space Science Library, vol. 393. Berlin, Heidelberg: Springer, pp. 29–107.

ELPHIC, R. C. and C. T. RUSSELL (1979). ISEE-1 and -2 magnetometer observations of the magnetopause, in *Magnetospheric Boundary Layers*. Ed. B. BATTRICK. Volume ESA SP-148. Paris: European Space Agency, pp. 43–50.

ELPHIC, R. C., C. T. RUSSELL, J. A. SLAVIN, and L. H. BRACE (1980). Observations of the dayside ionopause and ionosphere of Venus. *J. Geophys. Res.*, **85**(A13), 7679–7696, doi: 10.1029/JA085iA13p07679.

ENDEVE, E., T. E. HOLZER, and E. LEER (2003a). 2D MHD models of the large scale solar corona, in *Solar Wind Ten, Proc. Tenth Int. Solar Wind Conf.* Eds. M. VELLI, R. BRUNO, and F. MALARA. AIP Conf. Proc. 679. College Park, MD: American Institute of Physics, p. 331.

ENDEVE, E., E. LEER, and T. E. HOLZER (2003b). Two-dimensional magnetohydrodynamic models of the solar corona: mass loss from the streamer belt. *Astrophys. J.*, **589**, 1040–1053.

FAIRFIELD, D. H. (1971). Average and unusual locations of the Earth's magnetopause and bow shock. *J. Geophys. Res.*, **76**(28), 6700–6716.

FALTHAMMAR, C. G. (1966). On transport of trapped particles in outer magnetosphere. *J. Geophys. Res.*, **71**, 1487.

FARRIS, M. H. and C. T. RUSSELL (1994). Determining the standoff distance of the bow shock: Mach number dependence and use of models. *J. Geophys. Res.*, **99**, 17 681–17 689.

FARRIS, M. H., C. T. RUSSELL, R. J. FITZENREITER, and K. W. OGILVIE (1994). The subcritical, quasi-parallel, switch-on shock. *Geophys. Res. Lett.*, **21**, 837–840.

FELDSTEIN, Y. I. and G. V. STARKOV (1967). Dynamics of auroral belt and polar geomagnetic disturbances. *Planet. Space Sci.*, **15**, 209–229.

FISK, L. A. (1971). Solar modulation of galactic cosmic rays. *J. Geophys. Res.*, **76**, 221.

FRANK, L. A. (1967). On the extraterrestrial ring current during geomagnetic storms. *J. Geophys. Res.*, **72**, 3753–3767.

FRANK, L. A. (1971). Plasma in the Earth's polar magnetosphere. *J. Geophys. Res.*, **76**(22), 5202–5219.

FRIED, B. D. and S. D. CONTE (1961). *The Plasma Dispersion Function: The Hilbert Transform of the Gaussian*. New York: Academic Press.

FUJII, R., O. AMM, A. YOSHIKAWA, A. IEDA, and H. VANHAMÄKI (2011). Reformulation and energy flow of the Cowling channel. *J. Geophys. Res.*, **116**, A02305, doi:10.1029/2010JA015989.

FUKUSHIMA, N. (1969). Equivalence in ground magnetic effect of Chapman-Vestine's and Birkeland-Alfvén's electric current system for polar magnetic storms. *Rep. Ionos. Space Res. Jap.*, **23**, 219–227.

FUKUSHIMA, N. (1976). Generalized theorem for no ground magnetic effect of vertical currents connected with Pedersen currents in the uniform-conductivity ionosphere. *Rep. Ionos. Space Res. Jap.*, **30**, 35–40.

GARY, G. A. (2001). Plasma beta over an active region: rethinking the paradigm. *Solar Phys.*, **203**, 71–86.

GE, Y. S. and C. T. RUSSELL (2006). Polar survey of magnetic field in near tail: reconnection rare inside 9 R_E. *Geophys. Res. Lett.*, **33**, L02101, doi:10.1029/2005GL024574.

GENDRIN, R. (1961). Le guidage des whistlers par le champ magnetique. *Planet. Space Sci.*, **5**, 274–282, doi:10.1016/0032-0633(61)90096-4.

GLOECKLER, G. and L. A. FISK (2006). In *Physics of the Inner Heliosheath: Voyager Observations, Theory, and Future Prospects, 5th IGPP Int. Astrophysics Conf.* Eds. J. HEERIKHUISEN, V. FLORINSKI, G. P. ZANK, and N. P. POGORELOV. AIP Conf. Proc. 858. College Park, MD: Institute of Physics, pp. 153–158.

GOLUB L. and J. M. PASACHOFF (1997). *The Solar Corona*. Cambridge: Cambridge University Press.

Global Oscillation Network Group (2015). GONG Data Archive, gong.nso.edu/data. Accessed September 10, 2015.

GOODRICH, C. C. and J. D. SCUDDER (1984). The adiabatic energy change of plasma electrons and the frame dependence of the cross-shock potential at collisionless magnetosonic shock waves. *J. Geophys. Res.*, **89**, 6654–6662.

GOPALSWAMY, N. and M. L. KAISER (2002). Solar eruptions and long wavelength radio bursts: the 1997 May 12 event. *Adv. Space Res.*, **29**, 307–312.

GOSLING, J. T. and V. J. PIZZO (1999). Formation and evolution of corotating interaction regions and their three dimensional structure. *Space Sci. Rev.*, **89** (1–2), 21–52.

GRINGAUZ, K. I. (1969). Low-energy plasma in the Earth's magnetosphere. *Rev. Geophys.*, **7**, 339–378.

HEIKKILA, W. J. and J. D. WINNINGHAM (1971). Penetration of magnetosheath plasma to low altitudes through the dayside magnetospheric cusps. *J. Geophys. Res.*, **76**(4), 883–891.

HILL, F., P. B. STARK, R. T. STEBBINS *et al.* (1996). The solar acoustic spectrum and eigenmode parameters. *Science*, **272**, 1292–1295.

HOLZER, T. E., M. G. MCLEOD, and E. J. SMITH (1966). Preliminary results from OGO-1 search coil magnetometer: boundary positions and magnetic noise spectrum. *J. Geophys. Res.*, **71**, 1481–1486.

HUBA, J. D. (2009). *Revised NRL Plasma Formulary*. Washington, D.C.: Naval Research Laboratory.

HUDDLESTON, D. E., C. T. RUSSELL, M. G. KIVELSON, K. K. KHURANA, and L. BENNETT (1998). Location and shape of the Jovian magnetopause and bow shock. *J. Geophys. Res.*, **103**, 20 075–20 082.

HUNDHAUSEN A. J., J. T. BURKEPILE, and O. C. ST. CYR (1994). Speeds of coronal mass ejections: SMM observations from 1980 and 1984–1989. *J. Geophys. Res.*, **99**, 6543–6552.

IIJIMA, T. and T. A. POTEMRA (1978). Large-scale characteristics of field-aligned currents associated with substorms. *J. Geophys. Res.*, **83**, 599–615.

IP, W.-H. and W. I. AXFORD (1982). Theories of physical processes in the cometary comae and ion tails, in *Comets*. Ed. L. L. WILKENING. Tucson, AZ: University of Arizona Press, pp. 588–634.

JOHNSON, C. Y. (1969). Ion and neutral composition of the ionosphere, *Ann. IQSY.*, **5**, 197–213.

JOY, S. P., M. G. KIVELSON, R. J. WALKER, K. K. KHURANA, C. T. RUSSELL, and T. OGINO (2002). Probabilistic models of the Jovian magnetopause and bow shock locations. *J. Geophys. Res.*, **107**(A10), 1309, doi:1310.1029/2001JA009146.

KANTROWITZ, A. and H. E. PETSCHEK (1966). MHD characteristics and shock waves, in *Plasma Physics in Theory and Application*. Ed. W. B. KUNKEL. New York: McGraw-Hill, pp. 148–206.

KARIMABADI, H., W. DAUGHTON, and J. D. SCUDDER (2007). Multiscale structure of the electron diffusion region. *Geophys. Res. Lett.*, **34**, L13104, doi:10.1029/2007GL030306.

KELLOGG, P. J. (1962). Flow of plasma around the Earth. *J. Geophys. Res.*, **67**, 3805–3811.

KENNEL, C. F. and H. E. PETSCHEK (1966). Limit on stably trapped particle fluxes. *J. Geophys. Res.*, **71**, 1.

KENNEL, C. F., J. P. EDMISTON, and T. HADA (1985). A quarter century of collisionless shock research, in *Collisionless Shocks in the Heliosphere: A Tutorial Review*. Eds. R. G. Stone and B. T. Tsurutani. Geophysical Monograph Series, vol. 34. Washington, D.C.: American Geophysical Union, p. 1–36.

Kivelson, M. G. and C. T. Russell (1995). *Introduction to Space Physics*. Cambridge: Cambridge University Press.

Knight, S. (1973). Parallel electric fields. *Planet. Space Sci.*, **21**, 741–750.

Knudsen, W. C., K. Spenner, R. C. Whitten, and L. K. Miller (1980). Ion energetics in the Venus nightside ionosphere. *Geophys. Res. Lett.*, **7**, 1045–1048, doi:10.1029/GL007i012p01045.

Knudsen, W. C., K. L. Miller, and K. Spenner (1982). Improved Venus ionopause altitude calculation and comparison with measurement. *J. Geophys. Res.*, **87**(A4), 2246–2254.

Kurth, W. S. and D. A. Gurnett (1991). Plasma waves in planetary magnetospheres. *J. Geophys. Res.*, **96**(18), 977.

Landau L. D. and E. M. Lifshitz (1960). *Mechanics*. Reading, MA: Addison-Wesley.

Lang, K. R. (2010). *NASA's Cosmos*. www.ase.tufts.edu/cosmos. Website accessed September 10, 2015.

Lario, D. (2005). Advances in modeling gradual solar energetic particle events. *Adv. Space Res.*, **36**, 2279–2288.

Le, G. and C. T. Russell (1994). The thickness and structure of the high beta magnetopause current layer. *Geophys. Res. Lett.*, **21**, 2451–2454.

Le, G., X. Blanco-Cano, C. T. Russell *et al.* (2001). Electromagnetic ion cyclotron waves in the high-altitude cusp: polar observations. *J. Geophys. Res.*, **106**, 19 067–19 079.

Le, G., C. T. Russell, and K. Takahashi (2004). Morphology of the ring current derived from magnetic field observations. *Ann. Geophys.*, **22**, 1267–1295.

Lean, J. (1991). Variations in the sun's radiative input. *Rev. Geophys.*, **29**, 505–535.

Lee, C. O., J. G. Luhmann, D. Odstrcil *et al.* (2009). The solar wind at 1 AU during the declining phase of solar cycle 23: comparison of 3D numerical model results with observations. *Solar Phys.*, **254**(1), 155–183.

Leroy, M. M. and A. Mangeney (1984). A theory of energization of solar wind electrons by the Earth's bow shock. *Annales Geophys.*, **2**, 449–456.

Li, Y. and J. G. Luhmann (2006). Coronal magnetic field topology over filament channels: implications for coronal mass ejection initiations. *Astrophys. J.*, **648**, 732–740.

Li, Y., J. G. Luhmann, B. J. Lynch, and E. K. J. Kilpua (2011). Cyclic reversal of magnetic cloud poloidal field. *Solar Phys.*, **270**(1), 331–346.

Lindsay, G. M., J. G. Luhmann, C. T. Russell, and J. T. Gosling (1999). Relationships between coronal mass ejection speeds from coronagraph images and interplanetary characteristics of associated interplanetary coronal mass ejections. *J. Geophys.Res.*, **104**, 12515–12524.

Luhmann, J. G. (1977). Auroral bremsstrahlung spectra in the atmosphere. *J. Atmos. Terr. Phys.*, **39**, 595.

Luhmann, J. G. (1986). The solar wind interaction with Venus. *Space Sci. Rev.*, **44**, 241–306.

Luhmann, J. G. (1990). *The Solar Wind Interaction with Unmagnetized Planets: A Tutorial. Geophysical Monograph Series*, vol. 58. Washington, D.C.: American Geophysical Union.

Luhmann, J. G. (1991). The solar wind interaction with Venus and Mars: cometary analogies and contrasts, in *Cometary Plasma Processes*. Ed. A. Johnstone. Washington, D. C.: American Geophysical Union, doi:10.1029/GM061p0005.

Luhmann, J. G. and L. H. Brace (1991). Near-Mars space. *Rev. Geophys.* **29**, 121.

Luhmann, J. G. and R. C. Elphic (1985). On the dynamo generation of flux ropes in the Venus ionosphere. *J. Geophys. Res.*, **90**, 12 047–12 056.

Luhmann, J. G., R. J. Walker, C. T. Russell, N. U. Crooker, J. R. Spreiter, and S. S. Stahara (1984). Patterns of potential magnetic field merging sites on the dayside magnetopause. *J. Geophys. Res.*, **89**, 1739–1742.

Luhmann, J. G., C. T. Russell, F. L. Scarf, L. H. Brace, and W. C. Knudsen (1987). Characteristics of the Marslike limit of the Venus-solar wind interaction, *J. Geophys. Res.*, **92**(A8), 8545–8557, doi:10.1029/JA092iA08p08545.

Luhmann, J. G., C. T. Russell, and N. A. Tsyganenko (1998). Disturbances in Mercury's magnetosphere: are the Mariner 10 "substorms" simply driven? *J. Geophys. Res.*, **103**, 9113–9119.

Luhmann, J. G., D. W. Curtis, P. Schroeder *et al.* (2008). STEREO IMPACT investigation goals, measurements, and data products overview. *Space Sci. Rev.*, **136**, 117–184, doi 10.1007/s11214-007-9170-x.

Luhmann, J. G., D. Ulusen, S. A. Ledvina *et al.* (2012). Investigating magnetospheric interaction effects on Titan's ionosphere with the Cassini Orbiter Ion Neutral Mass Spectrometer, Langmuir Probe, and Magnetometer observations during targeted flybys. *Icarus*, **219**, 535–555.

McComas, D. J., F. Allegrini, P. Boschler *et al.* (2009). Global observations of the interstellar interaction from the Interstellar Boundary Explorer (IBEX). *Science*, **326**, 959–962.

McDiarmid, I. B., J. R. Burrows, and M. D. Wilson (1979). Large scale magnetic perturbations and particle measurements at 1400 km on the dayside. *J. Geophys. Res.*, **84**, 1431–1441.

McPherron, R. L., C. T. Russell, and M. P. Aubry (1973). Satellite studies of magnetospheric substorms on August 15, 1968. IX. Phenomenological model for substorms, *J. Geophys. Res.*, **78**, 3131–3149.

Manka, R. H. and F. C. Michel (1970). Lunar atmosphere as a source of argon-40 and other lunar surface elements. *Science*, **169**, 278–280.

Marsch, E. (2006). Kinetic physics of the solar corona and solar wind. *Living Rev. Sol. Phys.*, 3, 1.

Marsch, E., R. Schwenn, H. Rosenbauer, K.-H. Muehlhaeuser, W. Pilipp, and F. M. Neubauer (1982). Solar wind protons – three-dimensional velocity distributions and derived plasma parameters measured between 0.3 and 1 AU. *J. Geophys. Res.*, **87**, 52–72.

Mayaud, P. N. (1980). *Derivation, Meaning and Use of Geomagnetic Indices*. Washington, D.C.: American Geophysical Union.

Means, J. D. (1972) Use of the three-dimensional covariance matrix in analyzing the polarization properties of plane waves. *J. Geophys. Res.*, **77**, 28.

Nagy, A. F., T. E. Cravens, J.-H. Yee, and A. I. F. Stewart (1981). Hot oxygen atoms in the upper atmosphere of Venus. *Geophys. Res. Lett.*, 8(6), 629–632.

Newbury, J. A. and C. T. Russell (1996). Observations of a very thin collisionless shock. *Geophys. Res. Lett.*, **23**, 781–784.

Odera, T. J., D. V. Swol, C. T. Russell, and C. A. Green (1991). Pc 3, 4 magnetic pulsations observed simultaneously in the magnetosphere and at multiple ground stations. *Geophys. Res. Lett.*, **18**, 1671–1674.

Omidi, N., X. Blanco-Cano, C. T. Russell, H. Karimabadi, and M. Acuna (2002). Hybrid simulations of solar wind interaction with magnetized asteroids: general characteristics. *J. Geophys. Res.*, **107**, 1487.

Omidi, N., X. Blanco-Cano, and C. T. Russell (2005). Global hybrid simulations of the bow shock, in *The Physics of Collisionless Shocks*. Eds. G. Li, G. P. Zank, and C. T. Russell. AIP Conf. Proc. 781. College Park, MD: American Insitute of Physics, pp. 27–31.

OMNIWeb Plus (2015). omniweb.gsfc.nasa.gov. Accessed September 10, 2015.

Orlowski, D. S., C. T. Russell, and R. P. Lepping (1992). Wave phenomena in the upstream region of Saturn. *J. Geophys. Res.*, **97**, 19 187–19 199.

Palmer, H. P., R. D. Davies, and M. I. Large (1962). *Radio Astronomy Today*. Manchester: Manchester University Press.

Park, C. G. (1974). A morphological study of substorm-associated disturbances in the ionosphere. *J. Geophys. Res.*, **79**, 2821–2827.

Petrinec, S. M. and C. T. Russell (1997). Hydrodynamic and MHD equations across the bow shock and along the surfaces of planetary obstacles. *Space Sci. Rev.*, **79**, 757–791.

Petschek, H. E. (1964). Magnetic field annihilation, in *AAS-NASA Symposium on the Physics of Solar Flares*. Ed. W. N. Hess. Washington, D.C.: NASA, SP-50, pp. 425–439.

Phillips, J. L., J. G. Luhmann, and C. T. Russell (1984). Growth and maintenance of large-scale magnetic fields in the dayside Venus ionosphere. *J. Geophys. Res.*, **89**(A12), 10 676–10 684, doi:10.1029/JA089iA12p10676.

Podgorny, I. M. (1976). Laboratory experiments: intrusion into the magnetic field, in *Physics of Solar Planetary Environment*. Ed. D. J. Williams. Washington, D.C.: American Geophysical Union, pp. 241–254.

Prölss, G. W. (2003). *Physics of the Earth's Space Environment*. Berlin: Springer.

Rankin, D. and R. Kurtz (1970). Statistical study of micropulsation polarizations *J. Geophys. Res.*, **75**, 5444.

Ratcliffe, J. A. (1972). *An Introduction to the Ionosphere and Magnetosphere*. Cambridge: Cambridge University Press.

Rees, M. H. (1989). *Physics and Chemistry of the Upper Atmosphere*. Cambridge: Cambridge University Press.

Rishbeth, H. and O. K. Garriott (1969). *Introduction to Ionospheric Physics. International Geophysics Series*, vol. 14. New York: Academic Press.

Robinson, R. M., R. R. Vondrak, K. Miller, T. Dabbs, and D. Hardy (1987). On calculating ionospheric conductances from the flux and energy of precipitating electrons. *J. Geophys. Res.*, **92**, 2565–2569.

Roederer, J. G. (1967). On the adiabatic motion of energetic particles in a model magnetosphere. *J. Geophys. Res.*, **72**, 981–992.

Roederer, J. G. (1970). *Dynamics of Geomagnetically Trapped Radiation*. Berlin: Springer-Verlag.

Ronnmark, K. G. (1982). *WHAMP Waves in Homogeneous, Anisotropic, Multicomponent Plasmas. Kiruna Geophysical Institute Report*, 179. Kiruna, Sweden: Kiruna Geophysical Institute.

Rossi, B. and S. Olbert (1970). *Introduction to Space Physics*. New York: McGraw-Hill.

Russell, C. T. (1972). The configuration of the magnetosphere, in *Critical Problems of Magnetospheric Physics*. Ed. E. R. Dyer. Washington, D.C.: IUCSTP, National Academy of Sciences, pp. 1–16.

Russell, C. T. (2000). The solar wind interaction with the Earth's magnetosphere: a tutorial. *IEEE Trans. Plasma Sci.*, **28**, 1818–1830.

Russell, C. T. and R. C. Elphic (1979). Observations of the flux ropes in the Venus ionosphere. *Nature*, **279**, 616.

Russell, C. T. and B. K. Fleming (1976). Magnetic pulsations as a probe of the interplanetary magnetic field: a test of the Borok B-index. *J. Geophys. Res.*, **81**, 5882–5886.

Russell, C. T. and E. W. Greenstadt (1979). Initial ISEE magnetometer results: shock observation. *Space Sci. Rev.*, **23**, 3–37.

Russell, C. T. and M. M. Hoppe (1983). Upstream waves and particles. *Space Sci. Rev.*, **34**, 155–172.

Russell, C. T. and B. R. Lichtenstein (1975). On the source of lunar limb compressions. *J. Geophys. Res.*, **80**(34), 4700–4711.

Russell, C. T. and R. L. McPherron (1973). The magnetotail and substorms, *Space Sci. Rev.*, **15**, 205–266.

Russell, C. T. and R. M. Thorne (1970). On the structure of the inner magnetosphere. *Cosmic Electrodyn.*, **1**, 67–89.

Russell, C. T., C. R. Chappell, M. D. Montgomery, M. Neugebauer, and F. L. Scarf (1971). OGO-5 observations of the polar cusp on November 1, 1968. *J. Geophys. Res.*, **76**(28), 6743–6764.

Russell, C. T., M. M. Mellott, E. J. Smith, and J. H. King (1983). Multiple spacecraft observations of interplanetary shocks: four spacecraft determinations of shock normal. *J. Geophys. Res.*, **88**, 4739–4748.

Russell, C. T., D. N. Baker, and J. A. Slavin (1988). The magnetosphere of Mercury, in *Mercury*. Eds. F. Vilas, C. Chapman, and M. Matthews. Tucson: University of Arizona Press, pp. 514–561.

RUSSELL, C. T., P. SONG, and R. P. LEPPING (1989). The Uranian magnetopause: lessons from Earth. *Geophys. Res. Lett.*, **16**, 1485–1488.

RUSSELL, C. T., R. P. LEPPING, and C. W. SMITH (1990). Upstream waves at Uranus. *J. Geophys. Res.*, **95**(A3), 2273–2279.

RUSSELL, C. T., P. J. CHI, V. ANGELOPOULOS *et al.* (1999a). Comparison of three techniques of determining the resonant frequency of geomagnetic pulsations. *J. Atmos. Solar-Terr. Phys.*, **61**, 1289–1297.

RUSSELL, C. T., D. E. HUDDLESTON, K. K. KHURANA, and M. G. KIVELSON (1999b). Observations at the inner edge of the Jovian current sheet: evidence for a dynamic magnetosphere. *Planet. Space Sci.*, **47**, 521–527.

RUSSELL, C. T., M. G. KIVELSON, W. S. KURTH, and D. A. GURNETT (2000). Implications of depleted flux tubes in the jovian magnetosphere. *Geophys. Res. Lett.*, **27**, 3133–3136.

RUSSELL, C. T., Y. L. WANG, and J. RAEDER (2003). Possible dipole tilt dependence of dayside magnetopause reconnection. *Geophys. Res. Lett.*, **30**(18), 1937, doi:10.1029/2003GL017725.

RUSSELL, C. T., J. G. LUHMANN, and L. K. JIAN (2010). How unprecedented a solar minimum? *Rev. Geophys.*, **48**, RG2004, doi:10.1029/2009RG000316.

SAUNDERS, M. A. and C. T. RUSSELL (1986). Average dimension and magnetic structure of the distant Venus magnetotail. *J. Geophys. Res.*, **91**(A5), 5589–5604, doi:10.1029/JA091iA05p05589.

SCHATTEN, K. H., J. M. WILCOX, and N. F. NESS (1969). A model of interplanetary and coronal magnetic fields. *Solar Phys.*, **6**(3), 442–445.

SCHUNK, R. W. and A. F. NAGY (1980). Ionospheres of the terrestrial planets. *Rev. Geophys. Space Phys.*, **18**, 813–852.

SCHUNK, R. W. and A. F. NAGY (2009). *Ionospheres: Physics, Plasma Physics, and Chemistry*. Cambridge: Cambridge University Press.

SCHOLER, M., I. SIDORENKO, C. H. JAROSCHEK, and R. A. TREUMANN (2003). Onset of collisionless magnetic reconnection in thin current sheets: three-dimensional particle simulations. *Phys. Plasmas*, **10**, 3521–3527.

SCHULZ, M. and L. J. LANZEROTTI (1974). *Particle Diffusion in the Radiation Belts*. Berlin: Springer-Verlag.

SCKOPKE, N. (1966). A general relation between the energy of trapped particles and the disturbance field near the Earth. *J. Geophys. Res.*, **71**, 3125–3130.

SCURRY, L. and C. T. RUSSELL (1990). Geomagnetic activity for northward interplanetary magnetic fields: Am index response. *Geophys. Res. Lett.*, **17**, 1065–1068.

SCURRY, L. and C. T. RUSSELL (1991). Proxy studies of energy transfer in the magnetosphere. *J. Geophys. Res.*, **96**, 9541–9548.

SEXL, R., G. MARKS, K. BETHGE, E. STREERUWITZ, and I. RAAB (1980). *Materie in Raum und Zeit*, Einfuhrung in die Physik, 3. Frankfurt am Main: M. Diesterweg, O. Salle, Aarau, Sauerlander.

SHINAGAWA, H. and T. E. CRAVENS (1989). A one-dimensional multi-species magnetohydrodynamic model of the dayside ionosphere of Mars. *J. Geophys. Res.*, **94**, 6506–6517.

SHINAGAWA, H., T. E. CRAVENS, and A. F. NAGY (1987). A one-dimensional time-dependent model of the magnetized ionosphere of Venus. *J. Geophys. Res.*, **92**, 7317–7330.

SHUE, J.-H., P. SONG, C. T. RUSSELL *et al.* (1998). Magnetopause location under extreme solar wind conditions. *J. Geophys. Res.*, **103**, 17 691–17 700.

SOLANKI, S. K., T. WENZLER, and D. SCHMITT (2008). Moments of the latitudinal dependence of the sunspot cycle: a new diagnostic of dynamo models. *Astron. Astrophys.*, **483**, 623–632.

SONG, P. and V. M. VASYLIŪNAS (2011). Heating of the solar atmosphere by strong damping of Alfvén waves. *J. Geophys. Res.*, **116**, A09104.

SONG, P., R. C. ELPHIC, C. T. RUSSELL, J. T. GOSLING, and C. A. CATTELL (1990). Structure and properties of the subsolar magnetopause for northward IMF: ISEE observations. *J. Geophys. Res.*, **95**, 6375–6387.

Song, P., V. M. Vasyliunas, and L. Ma (2005). A three-fluid model of solar wind-magnetosphere-ionosphere-thermosphere coupling, in *Multiscale Coupling of Sun-Earth Processes*. Eds. A. T. Y. Lui, Y. Kamide, and G. Consolini. Amsterdam: Elsevier, pp. 447–456.

Sonnerup, B. U. Ö., G. Paschmann, I. Papamastorakis *et al.* (1981). Evidence for magnetic field reconnection at the Earth's magnetopause. *J. Geophys. Res.*, **86**, 10 049–10 067.

Southwood, D. J., M. G. Kivelson, R. J. Walker, and J. A. Slavin (1980). Io and its plasma environment. *J. Geophys. Res.*, **85**(A11), 5959–5968, doi: 10.1029/JA085iA11p05959.

Speiser, T. W. (1965). Particle trajectories in model current sheets 1. Analytical solutions. *J. Geophys. Res.*, **70**, 4219–4226.

Spjeldvik, W. N. and P. L. Rothwell (1983). The Earth's radiation belts, in *Handbook of Geophysics and the Space Environment*. Ed. A. S. Jursa. Air Force Geophysics Laboratory Technical Report 88–0240. Springfield, VA: Air Force Geophysics Laboratory, Air Force Systems Command.

Spreiter, J. R. and S. S. Stahara (1980). A new predictive model for determining solar wind-terrestrial planet interactions. *J. Geophys. Res.*, 85(A12), 6769–6777.

Spreiter, J. R., A. L. Summers, and A. Y. Alksne (1966). Hydromagnetic flow around the magnetosphere. *Planet. Space Sci.*, **14**, 223–253.

Spreiter, J. R., M. C. Marsh, and A. L. Summers (1970). Hydromagnetic aspects of solar wind flow past the moon. *Cosmic Electrodyn.*, 1(1), 5–50.

Stix, T. H. (1962). *The Theory of Plasma Waves*. New York: McGraw-Hill.

Storey, L. R. O. (1953). An investigation of whistling atmospherics. *Phil. Trans. R. Soc. Lond. A*, **246**, 113–141, doi:10.1098/rsta.1953.0011.

Strangeway, R. J., C. T. Russell, C. W. Carlson *et al.* (2000). Cusp field-aligned currents and ion outflows. *J. Geophys. Res.*, **105**, 21 129–21 142.

Sweet, P. A. (1958).The neutral point theory of solar flares, in *Electromagnetic Phenomena in Cosmical Physics*. Ed. B. Lehnert. Proc. IAU Symp. 6. Cambridge: Cambridge University Press, pp. 123–134.

Taylor, H. A., H. C. Brinton, and I. M. W. Pharo (1968). Contraction of the plasmasphere during geomagneticially disturbed periods. *J. Geophys. Res.*, **73**, 961.

Theis, R. F., L. H. Brace, and H. G. Mayr (1980). Empirical models of the electron temperature and density in the Venus ionosphere. *J. Geophys. Res.*, **85**, 7787–7794.

Troitskaya, V. A., T. A. Plyasova-Bakunina, and A. V. Guglielmi (1971). Relationship between Pc2-4 pulsations and the interplanetary magnetic field. *Dokl. Acad. Nauk.*, **SSSR197**, 1312. (In Russian.)

Tsyganenko, N. A. (1987). Global quantitative models of the geomagnetic field in the cislunar magnetosphere for different disturbance levels. *Planet. Space Sci.*, **35**, 1347–1358.

Tsyganenko, N. A. (1989a). A solution of the Chapman-Ferraro problem for an ellipsoidal magnetopause. *Planet. Space Sci.*, **37**, 1037–1046.

Tsyganenko, N. A. (1989b). A magnetospheric magnetic field model with a warped tail current sheet. *Planet. Space Sci.*, **37**, 5–20.

Tsyganenko, N. A. and A. V. Usmanov (1982). Determination of the magnetospheric current system parameters and development of experimental geomagnetic field models based on data from IMP and HEOS satellite. *Planet Space Sci.*, **30**, 985–998.

Unti, T. and G. Atkinson (1968). Two-dimensional Chapman-Ferraro problem with neutral sheet. 1. The boundary. *J. Geophys. Res.*, **73**, 7319–7327.

Vasyliunas, V. M. (1968). Low energy electrons in the magnetosphere as observed by OGO-1 and OGO-3, in *Physics of the Magnetosphere*. Ed. R. L. Carovillano. Dordrecht: Reidel, pp. 622–650.

Vasyliunas, V. M. (1983). Plasma distribution and flow, in *Physics of the Jovian Magnetosphere*. Ed. A. J. Dessler. London: Cambridge University Press, pp. 395–453.

VERNAZZA, J. E., E. H. AVRETT, and R. LOESER (1973). Structure of the solar chromosphere. 1. Basic computations and summary of the results. *Astrophys. J.*, **184**, 605–631.

WALT, M. (1996). Source and loss processes for radiation belt particles, in *Radiation Belts: Models and Standards*. Eds. J. F. LEMAIRE, D. HEYNDERICKX, and D. N. BAKER. *Geophysical Monograph Series*, vol. 97. Washington, D.C.: American Geophysical Union, p. 1.

WANG, C. and J. D. RICHARDSON (2001). Energy partition between solar wind protons and pickup ions in the distant heliosphere: a three-fluid approach. *J. Geophys. Res.*, **106**(A12), 29 401–29 407.

WANG, Y. L., J. RAEDER, and C. T. RUSSELL (2004). Plasma depletion layer: magnetosheath flow structure and forces. *Ann. Geophys.*, **22**, 1773–1776.

WHIPPLE, E. C., JR. (1978). (U, B, K) coordinates: a natural system for studying magnetospheric convection. *J. Geophys. Res.*, **83**, 4318–4326.

Wilcox Solar Observatory (wso.stanford.edu).

WOLF, R. A. (1983). The quasi-static (slow-flow) region of the magnetosphere, in *Solar Terrestrial Physics*. Eds. R. L. CAROVILLANO and J. M. FORBES. Dordrecht: Reidel, pp. 303–368.

WOODS, L. C. (1969). On the structure of collisionless magnetoplasma shock waves at supercritical Alfvén-Mach numbers. *J. Plasma Phys.*, 3, 435–447.

ZHANG, T. L., W. BAUMJOHANN, J. DU *et al.* (2010). Hemispherical asymmetry of the magnetic field wrapping pattern of the Venusian magnetotail. *Geophys. Res. Lett.*, **37**, L14202.

Index

Printed in the United States
by Baker & Taylor Publisher Services

Printed in the United States
by Baker & Taylor Publisher Services